Functions

Definition: A function f of a real number x is a rule that assigns a single real number $f(x)$ to each x in the domain of the function. This definition can be characterized as a mapping or as a set of ordered pairs.

Constant Function: $f(x) = a$

Linear Function: $f(x) = mx + b$

Standard form:	$Ax + By + C = 0$
Point-slope form:	$y - k = m(x - h)$
Slope-intercept form:	$y = mx + b$
Two-point form:	$y - y_1 = \left(\dfrac{y_2 - y_1}{x_2 - x_1}\right)(x - x_1)$ or $\begin{vmatrix} x & y & 1 \\ x_1 & y_1 & 1 \\ x_2 & y_2 & 1 \end{vmatrix} = 0$
Intercept form:	$\dfrac{x}{a} + \dfrac{y}{b} = 1$
Horizontal line:	$y = k$
Vertical line:	$x = h$

Quadratic Function: $f(x) = ax^2 + bx + c,\ a \neq 0$

Absolute Value Function: $f(x) = |x| = \begin{cases} x \text{ if } x \geq 0 \\ -x \text{ if } x < 0 \end{cases}$

Polynomial Function: $f(x) = a_nx^n + a_{n-1}x^{n-1} + a_{n-2}x^{n-2} + \cdots + a_2x^2 + a_1x + a_0,\ a_n \neq 0$

Rational Function: $f(x) = P(x)/D(x)$, where $P(x)$ and $D(x)$ are polynomial functions and $D(x) \neq 0$.

Asymptotes: Let $f(x) = P(x)/D(x)$, where $P(x)$ and $D(x)$ have no common factors. Then:

1. The *vertical line* given by $x = r$ is an asymptote if $D(r) = 0$.
2. The *horizontal line* given by $y = k$ is an asymptote if $f(x) \to k$ as $|x| \to \infty$.
3. The *slant line* given by $y = mx + b$ is an asymptote if $f(x) = mx + b + \dfrac{R(x)}{D(x)}$; that is, if the degree of P is one more than the degree of D.

Exponential Function: $f(x) = b^x \quad (b > 0,\ b \neq 1)$

Laws of exponents: Let a and b be nonzero real numbers and let m and n be any real numbers, except that the form 0^0 and division by zero are excluded.

FIRST LAW: $b^m \cdot b^n = b^{m+n}$

SECOND LAW: $\dfrac{b^m}{b^n} = b^{m-n}$

THIRD LAW: $(b^n)^m = b^{mn}$

FOURTH LAW: $(ab)^m = a^mb^m$

FIFTH LAW: $\left(\dfrac{a}{b}\right)^m = \dfrac{a^m}{b^m}$

Logarithmic Function: $f(x) = \log_b x \quad (b > 0,\ b \neq 1)$

Laws of logarithms:

FIRST LAW: $\log_b AB = \log_b A + \log_b B$

SECOND LAW: $\log_b \dfrac{A}{B} = \log_b A - \log_b B$

THIRD LAW: $\log_b A^p = p \log_b A$

Log of both sides theorem: $\log_b A = \log_b B \Leftrightarrow A = B$

Change of base theorem: $\log_a x = \dfrac{\log_b x}{\log_b a}$

Trigonometric Functions: Let θ be an angle in standard position with a point $P(x, y)$ on the terminal side a distance of $r = \sqrt{x^2 + y^2}$ from the origin $(r \neq 0)$. Then the trigonometric functions are defined as follows:

$\cos\theta = \dfrac{x}{r}$ $\qquad \sin\theta = \dfrac{y}{r}$ $\qquad \tan\theta = \dfrac{y}{x} \quad (x \neq 0)$

$\sec\theta = \dfrac{r}{x} \quad (x \neq 0)$ $\qquad \csc\theta = \dfrac{r}{y} \quad (y \neq 0)$ $\qquad \cot\theta = \dfrac{x}{y} \quad (y \neq 0)$

Precalculus Mathematics

A FUNCTIONAL APPROACH THIRD EDITION

Precalculus Mathematics

A FUNCTIONAL APPROACH THIRD EDITION

Karl J. Smith, Ph.D.

BROOKS/COLE PUBLISHING COMPANY
MONTEREY, CALIFORNIA

Brooks/Cole Publishing Company
A Division of Wadsworth, Inc.

Printed in the United States of America

10 9 8 7 6 5 4 3

Library of Congress Cataloging-in-Publication Data

Smith, Karl J.
Precalculus mathematics.

Includes index.
1. Functions. I. Title.
QA331.S617 1985 512'.1 85-14678
ISBN 0-534-05232-0

Sponsoring Editor: *Craig Barth*
Editorial Assistant: *Eileen Galligan*
Production Editor: *Ellen Brownstein*
Manuscript Editor: *Pete Shanks*
Permissions Editor: *Carline Haga*
Interior Design: *Vernon T. Boes*
Cover Design: *Jerry Takigawa*
Art Coordinator: *Michèle Judge*
Interior Illustration: *Carl Brown*
Typesetting: *Polyglot Compositors, Singapore*
Cover Printing: *Lehigh Press, Pennsauken, New Jersey*
Printing and Binding: *R. R. Donnelley & Sons Co., Harrisonburg, Virginia*

This Book is Dedicated
with Love to My Daughter,
Melissa Ann

PREFACE

This book is designed to prepare students for the study of calculus. My main goal in writing it has been to make the material clear and interesting for the reader, and at the same time to provide those skills necessary to ensure success in calculus. There is no magic key to success in this course; it will come only by spending time with the material and by working many, many problems. The book, however, makes it easy to know what is important:

- Important terms are presented in boldface type.
- Important ideas are enclosed in a box.
- Common pitfalls and helpful hints and explanations are shown in italic.
- Color is used in a functional way to help you "see" what to do next.
- The most important goals of each section are listed as objectives at the end of each chapter.
- Answers to the odd-numbered problems are provided to let you know if you are on the right track; many solutions and hints are also provided and more than 350 graphs are presented in the answer section alone to provide additional examples.
- Problems are graded and presented in sets so that you can work some of the simpler problems before progressing to the more challenging ones.
- Review problems are provided for each Chapter Objective.

There is no substitute for working problems in mathematics. For that reason I have devoted an extraordinary amount of effort to writing the problems for this text. The sets are divided into A, B, and C problems according to the level of difficulty. A wide variety of applied problems is provided to help the students develop the type of problem-solving ability required in calculus. This edition also offers an innovative idea in problem sets. That is the idea of uniform problem sets throughout the book. This means that the instructor can make "standard assignments" consisting of the same problems being assigned from section to section. For example, I have written the problem sets so that a typical twenty-problem assignment *Problems 3–60, multiples of 3* will take the average student about two to three hours to complete and *in every problem set in the text* this assignment will give the student practice in every essential skill. I have listed below some typical "standard assignments" which could be used throughout the book.

Standard 1–2 hour assignment
Problems 5–50, multiples of 5
Standard 2–3 hour assignment
Problems 3–60, multiples of 3
Spiral 2–3 hour assignment

First day:	5–60 multiples of 5 from Problem Set 1
Second day:	5–60 multiples of 5 from Problem Set 2
	7–28 multiples of 7 from Problem Set 1
Third day:	5–60 multiples of 5 from Problem Set 3
	7–28 multiples of 7 from Problem Set 2
	6–24 multiples of 6 from Problem Set 1
Fourth day:	5–60 multiples of 5 from Problem Set 4
	7–28 multiples of 7 from Problem Set 3
	6–24 multiples of 6 from Problem Set 2
	⋮

These standard assignments do not apply to the review problems at the end of each chapter. There is more specific discussion of these standard assignments in the Instructors Manual. You will notice from these standard assignments that the student is generally required to work at the A and B levels, but each assignment usually includes one C level problem.

The topics covered in this book are those most needed by beginning calculus students. **The main concepts unifying the material are the notions of a function and the graphs of various functions.** Although it is assumed that students have had some high school algebra, the central ideas from algebra—factoring, solving equations, simplifying algebraic expressions, the laws of exponents—are integrated into this text where appropriate since these topics are often the very ones that cause difficulty for the beginning calculus student. It is not necessary for students to have had trigonometry, since an entire trigonometry course is presented in Chapters 5 and 6. If the students are well prepared in trigonometry, these chapters could be skipped over lightly or even omitted.

A great deal of flexibility is possible in the instructor's selection of topics presented in this book. For this reason, more material is provided here than can be used in a single semester or quarter. For ease of planning, however, I have written the book so that each section represents approximately one day of class material. The following table describes the book's contents.

In addition to completely reworked problem sets, this new edition offers several changes from the second edition. Complex numbers are now integrated into the text, rather than isolated in a separate chapter. They can still easily be omitted; any use of complex numbers is well marked so there will be no problems if you choose to do so. Integrating complex numbers makes the discussion of the Quadratic Formula, polynomial functions, Fundamental Theorem of Algebra, and De Moivre's Theorem flow much more smoothly. Other material which has been amplified or added includes new sections on quadratic equations, quadratic inequalities, linear programming, addition laws, double-angle and half-angle identities, and a second chapter on analytic geometry including an expanded treatment of vectors. The locations of other sections in the text have changed: The discussion of partial fractions is now found with that of rational

Chapter	Number of class days (minimum–maximum)	Comments
1. Fundamental Concepts	2–7	This chapter is required for the rest of the material in the course.
2. Functions	5–8	This chapter is required for the rest of the material in the course. Section 2.4 is required only for Section 5.4 and Chapter 10, however, and Section 2.6 is required only for Section 5.5.
3. Polynomial and Rational Functions	6–10	This material is required in Section 7.1.
4. Exponential and Logarithmic Functions	4–6	This material is required in Section 10.6.
5. Trigonometric Functions	4–7	Even if trigonometry has been assumed, it is suggested that Section 5.2 be briefly reviewed and that graphing of the trigonometric functions in Section 5.4 be covered.
6. Analytic Trigonometry	7–12	
7. Systems and Matrices	6–12	Sections 7.1 and 7.8 require Chapter 3. This chapter is not required in the remainder of the book.
8. Additional Topics in Algebra	2–7	Sections 8.1, 8.2, and 8.3–8.5 are independent of each other. This chapter is not required for the remainder of the book.
9. Conic Sections	4–6	Most calculus books contain material in analytic geometry, but a rudimentary knowledge of the conic sections is often assumed. Section 9.9 requires trigonometry.
10. Additional Topics in Analytic Geometry	3–7	Sections 10.1–10.3, 10.4–10.5, and 10.6 are independent of each other. Section 10.6 requires Chapter 4 and Sections 10.5–10.6 require trigonometry.

expressions; vectors are now in the new analytic geometry chapter; the content of the chapter on exponential and logarithmic functions has been rearranged; and mathematical induction and the Binomial Theorem have been given more emphasis. You will find the remainder of the book basically the same as the second edition.

Because of the great availability of inexpensive electronic calculators, the computational aspects of the logarithmic function in Chapter 4 have been minimized. Likewise, the introduction and evaluation of the trigonometric functions in Chapter 5 assumes that calculators are available. The 1981–1983 edition of the Mathematics Examinations of the College Board now advocates the use of calculators and beginning in January 1983 has allowed the use of calculators on AP and CLEP calculus examinations:

> The College Board Mathematics Advisory Committee supports the appropriate use of hand-held calculators in mathematics instruction. Students preparing for college mathematics should be encouraged to purchase scientific calculators and to use them in their school mathematics courses. The use of calculators on examinations, however, is a complicated issue involving fairness to students who have had unequal access to calculators.*

This book advocates the use of calculators but gives alternative solution techniques for those who do not have a calculator. Complete tables are also provided in Appendix B for those who do not have calculators.

Three main types of logic are used on calculators: arithmetic logic, algebraic logic (recognizes the order-of-operations convention from algebra), and RPN logic. It is suggested that you use a calculator with algebraic or RPN logic. RPN calculators are characterized by an [ENTER] or [SAVE] key. If you are using a calculator with arithmetic logic, you must use parentheses or pick out the operations to be performed first. For example, consider the problem $2 + 3 \cdot 4$, which illustrates the differences among the three types of logic:

ALGEBRAIC LOGIC	RPN LOGIC	ARITHMETIC LOGIC
[2]	[2]	[3]
[+]	[ENTER]	[×]
[3]	[3]	[4]
[×]	[ENTER]	[=]
[4]	[4]	[+]
[=]	[×]	[2]
	[+]	[=]

If you input [2] [+] [3] [×] [4] [=] on a calculator with arithmetic logic, you will obtain the incorrect answer 20. For this reason, a calculator with algebraic or RPN logic is recommended.

The correct output is 14.

In this book, we will be using a Texas Instruments TI-30 to illustrate algebraic logic and a Hewlett-Packard HP-35 to illustrate RPN logic. However, you should always consult the owner's manual for the calculator you own.

My thanks go to the reviewers of the first edition of this book: Marjorie S. Freeman of the University of Houston, Downtown College; Thomas Green of Contra Costa College; James Householder of California State University, Humboldt; Raymond McGivney of the University of Hartford; Bill Orr of San Bernardino Valley College; David Tabor of the University of Texas; and Robert Webber of Longwood College.

---

*The Mathematics Examinations of the College Board Booklet for 1981–1983, page 1.

I would like to thank the reviewers of the second edition: James Bailey of the University of Toledo; Roy Bergstrom of the University of the Pacific; Virginia Buchanan of Southwest Texas State University; Lee R. Clancy of Golden West College; John S. Cross of the University of Northern Iowa; Daniel Drucker of Wayne State University; Marjorie S. Freeman of the University of Houston, Downtown College; James Householder of California State University, Humboldt; Laurence P. Maher, Jr. of North Texas State University; Joan Mahmud of Bergen Community College; James McMurdo of Shasta College; Gordon D. Mock of Western Illinois University; Hubert J. Pollick of Killgore College; Richard Redner of the University of Tulsa; Thomas Strommer of Oxford College of Emery University; Donald Taranto of California State University, Sacramento; and Terry Wilson of San Jacinto College.

I would also like to thank the reviewers of the third edition: David Bush, Shasta College, Redding, California; Joel Cunningham, Susquehanna University, Selinsgrove, Pennsylvania; Daniel Drucker, Wayne State University, Detroit; Vivian Fielder, Fisk University, Nashville; Sheldon Gordon, Suffolk Community College, Selden, New York; Steve Hinthorne, Central Washington University at Ellensburg; John Kuisti, Michigan Technological University at South Range; Bill Pelletier, Wayne State University, Detroit; Donald Taranto, California State University at Sacramento; and Terry Wilson, San Jacinto College, Pasadena, Texas.

Special thanks go to my wife, Linda, and our children, Melissa and Shannon, for their love and support while I was working on this book.

Karl J. Smith
Sebastopol, California

CONTENTS

*Optional Section

Precalculus Mathematics

A FUNCTIONAL APPROACH THIRD EDITION

1

Fundamental Concepts

Mathematics is the queen of the sciences and arithmetic the queen of mathematics. She often condescends to render service to astronomy and other natural sciences, but in all relations she is entitled to the first rank.

KARL GAUSS

Karl Gauss (1777–1855)

* Optional Section

Karl Gauss is considered to be one of the three greatest mathematicians of all time, along with Archimedes and Newton. Gauss had a great admiration for Archimedes, but could not understand how Archimedes failed to invent the positional numeration system. Howard Eves quotes Gauss as saying, "To what heights would science now be raised if Archimedes had made that discovery!" Because of the greatness of Gauss, many stories and anecdotes about him have survived. He was a child prodigy, who graduated from college at the age of 15 and proved what was to become the Fundamental Theorem of Algebra for his doctoral thesis at the age of 22. He published only a small portion of the ideas that seemed to storm his mind because he felt that each published result had to be complete, concise, polished, convincing, and with every trace of the analysis by which the results were reached removed from the article. His motto was "Few, but ripe." As you begin your study of this course, it seems appropriate to begin with a quotation by Gauss in which he describes the joy of learning. This quotation is from a letter dated September 2, 1808 to his friend Wolfgang Bolyai:

"It is not knowledge, but the act of learning, not possession but the act of getting there, which grants the greatest enjoyment. When I have clarified and exhausted a subject, then I turn away from it, in order to go into darkness again; the never-satisfied man is so strange—if he has completed a structure, then it is not in order to dwell in it peacefully, but in order to begin another. I imagine the world conqueror must feel thus, who, after one kingdom is scarcely conquered, stretches out his arms for another."

CHAPTER OVERVIEW This chapter reviews many of the preliminary ideas from previous courses and sets the foundation for the remainder of this course. Numbers, linear and quadratic equations and inequalities, as well as a two-dimensional coordinate system are introduced in this chapter. This chapter has 25 objectives which are listed at the end of the chapter on pages 38–40.

1.1 REAL NUMBERS

You are, no doubt, familiar with various sets of numbers, as well as with certain properties of those numbers. Table 1.1 gives a brief summary of some of the more common subsets of the set of **real numbers**.

TABLE 1.1 Sets of numbers

Name	Symbol	Set	Examples
Counting numbers or natural numbers	**N**	$\{1, 2, 3, 4, \ldots\}$	86; 1,986,412; $\sqrt{16}$; $\sqrt{1}$; $\sqrt{100}$;... $\sqrt{a}$ is the nonnegative real number b so that $b^2 = a$. It is called the *principal square root* of a. Some square roots are natural numbers.
Whole numbers	**W**	$\{0, 1, 2, 3, 4, \ldots\}$	0; 86; 49; $\frac{8}{4}$; $\frac{16{,}425}{25}$;... $\frac{a}{b}$ means $a \div b$. It is called the *quotient* of a and b. Some quotients are whole numbers.
Integers	**Z**	$\{\ldots, -2, -1, 0, 1, 2, 3, \ldots\}$	-8; 0; $\frac{-16}{4}$; 43,812; -96; $-\sqrt{25}$;...
Rational numbers	**Q**	Numbers that can be written in the form p/q, where p and q are integers with $q \neq 0$. They are characterized as numbers whose decimal representations either terminate or repeat.	$\frac{2}{3}$; $\frac{4}{17}$; 0.863214; 0.866...; 5; 0; $\frac{-19}{10}$; -16; $\sqrt{\frac{1}{4}}$; 3.1416; $8.\overline{6}$; $0.1\overline{56}$;... An overbar indicates repeating decimals. Some square roots are also rational.

(continued)

TABLE 1.1 (*continued*)

Name	Symbol	Set	Examples
Irrational numbers	**Q′**	Numbers whose decimal representations do not terminate and do not repeat.	4.1234567891011 . . . ; 6.31331333133331 . . . ; $\sqrt{2}$; $\sqrt{3}$; $\sqrt{5}$; π; $\frac{\pi}{2}$; . . . π is the ratio of the circumference of a circle to its diameter. It is sometimes approximated by 3.1416, but this is a rational approximation of an irrational number. We write $\pi \approx 3.1416$ to mean the numbers are *approximately* equal. Square roots of most numbers are irrational.
Real numbers	**R**	Numbers that are either rational or irrational.	All examples listed above are real numbers. Not all numbers are real numbers, however. Some of these will be considered in Section 1.3; they are called *complex numbers*.

The real numbers can most easily be visualized by using a **one-dimensional coordinate system** called a **real number line** (Figure 1.1). A **one-to-one correspondence** is established between all real numbers and all points on a real number line:

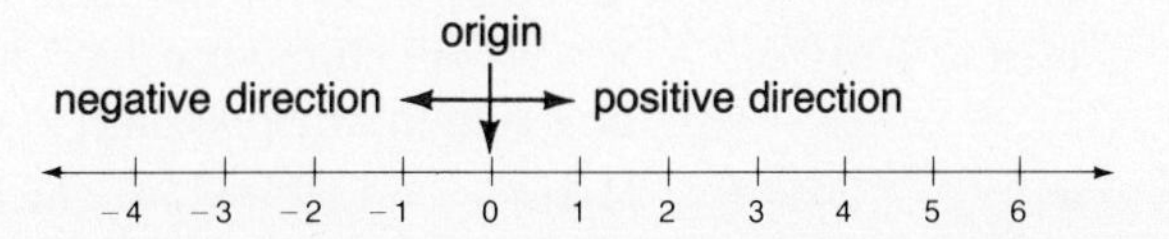

Figure 1.1 A real number line

1. Every point on the line corresponds to precisely one real number.
2. Every real number corresponds to precisely one point.

A point associated with a particular number is called the **graph** of that number. Numbers associated with points to the right of the origin are called **positive real numbers** and those to the left are called **negative real numbers**. Numbers are called **opposites** if they are plotted an equal distance from the origin. The **opposite of a real number a** is denoted by $-a$. Notice that if a is positive, then $-a$ is negative; and if a is negative, then $-a$ is positive. It is a common error to think of $-a$ as a negative number; it might be negative, but it might also be positive. If, say, $a = -5$, then $-a = -(-5) = 5$, which is positive.

EXAMPLE 1 Graph the following numbers on a real number line:

$$5; -2; 2.5; 1.313311333111\ldots; \tfrac{2}{3}; \pi; -\sqrt{2}$$

Solution When graphing, the exact positions of the points are usually approximated.

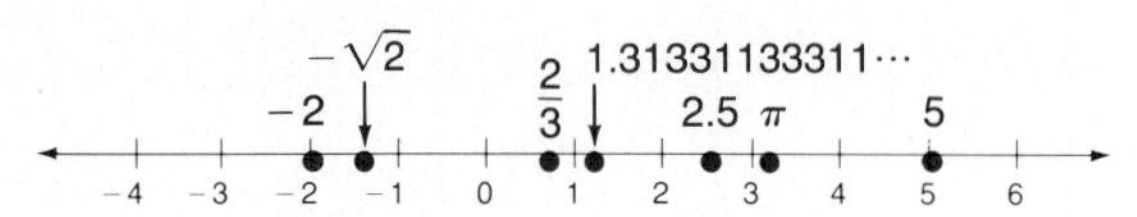

■

There are certain relationships between real numbers with which you should be familiar:

Less Than $a < b$, read "*a is less than b*," means $b - a$ is positive. On a number line, it means the graph of a is to the left of the graph of b.

Greater Than $a > b$, read "*a is greater than b*," means $b < a$ or $a - b$ is positive. On a number line, it means the graph of a is to the right of the graph of b.

Equal to $a = b$, read "*a is equal to b*," means that a and b represent the same real number. On a number line, the graphs of both a and b are the same point.

If a and b represent any real numbers, then they are related by a property called **Trichotomy** or **Property of Comparison**.

Property of Comparison

> Given any two real numbers a and b, exactly one of the following holds:
>
> 1. $a = b$ 2. $a > b$ 3. $a < b$

EXAMPLE 2 Illustrate the Property of Comparison by replacing the □ by =, >, or <.

a. $-3 \square -7$ **b.** $\frac{2}{5}$ 0.4 **c.** $\pi \square 3.1416$ **d.** $\sqrt{2} \square 1.4142$

Solution

a. $-3 > -7$ Think of a number line. The graph of -3 is to the right of the graph of -7; we see that -3 is greater than -7 (so ">" is the answer).

b. $\frac{2}{5} = 0.4$ $.4 = \frac{4}{10} = \frac{2}{5}$, so the numbers are equal.

c. $\pi < 3.1416$ $\pi \approx 3.1415926$, so use "<," less than. (A good approximation of π is found by pressing [π] on a calculator.)

d. $\sqrt{2} > 1.4142$ $(1.4142)^2 = 1.99996164$, which is less than 2. Alternatively, you can press [2] [√] on a calculator for a display of `1.4142136`, so $1.4142136 > 1.4142$. ■

The Property of Comparison establishes the order on a real number line. If you let $b = 0$, for example, then $a = 0$, $a < 0$, or $a > 0$. Using the definition of less than and greater than, it follows that

$a > 0$ if and only if a is positive.

$a < 0$ if and only if a is negative.

In addition to the Property of Comparison, there are four properties of equality that are used in mathematics:

Properties of Equality

Let a, b, and c be real numbers.

REFLEXIVE PROPERTY:	1. $a = a$.
SYMMETRIC PROPERTY:	2. If $a = b$, then $b = a$.
TRANSITIVE PROPERTY:	3. If $a = b$ and $b = c$, then $a = c$.
SUBSTITUTION PROPERTY:	4. If $a = b$, then a may be replaced by b (or b by a) throughout any statement without changing the truth or falsity of the statement.

EXAMPLE 3 Identify the property of equality illustrated by each:

a. $(a + b)(c + d) = (a + b)(c + d)$.
b. If $(a + b)(c + d) = (a + b)c + (a + b)d$ and $(a + b)c + (a + b)d = ac + bc + ad + bd$, then $(a + b)(c + d) = ac + bc + ad + bd$.
c. If $a(b + c) = ab + ac$, then $ab + ac = a(b + c)$.
d. If $a = 3$ and $(a + 2)(a + 5) = 40$, then $(3 + 2)(3 + 5) = 40$.

Solution **a.** Reflexive **b.** Transitive **c.** Symmetric **d.** Substitution ■

Finally, consider some properties of the real numbers. When we are adding real numbers the result is called the **sum** and the numbers added are called **terms**. When multiplying real numbers the result is called the **product** and the numbers multiplied are called the **factors**. The result from subtraction is called the **difference**; the result from division is the **quotient**. The real numbers, together with the relation of equality and the operations of addition and multiplication, satisfy what are called **field properties**.

Field Properties for the Set R of Real Numbers

Let a, b, and c be real numbers.

	Addition Properties	*Multiplication Properties*
CLOSURE:	$a + b$ is a unique real number.	ab is a unique real number.
COMMUTATIVE:	$a + b = b + a$	$ab = ba$
ASSOCIATIVE:	$(a + b) + c = a + (b + c)$	$(ab)c = a(bc)$
IDENTITY:	There exists a unique real number zero, denoted by 0, such that $a + 0 = 0 + a = a$.	There exists a unique real number one, denoted by 1, such that $a \cdot 1 = 1 \cdot a = a$.
INVERSE:	For each real number a, there is a unique real number $-a$ such that $a + (-a) = (-a) + a = 0$.	For each *nonzero* real number a, there is a unique real number $1/a$ such that $a\left(\frac{1}{a}\right) = \left(\frac{1}{a}\right)a = 1$
DISTRIBUTIVE:	$a(b + c) = ab + ac$	

A set is said to be **closed** if it satisfies the closure property, as shown in Example 4.

EXAMPLE 4 The set $\{-1, 0, 1\}$ is closed for multiplication since

$$(-1)(-1) = 1 \qquad (0)(1) = 0$$
$$(-1)(0) = 0 \qquad (0)(0) = 0$$
$$(-1)(1) = -1 \qquad (1)(1) = 1$$

But it is not closed for addition since

$$1 + 1 = 2$$

which is not in the set. ■

EXAMPLE 5 Identify the field property illustrated.

a. $4 \cdot 3$ is a real number. **b.** $\frac{1}{4}(4 \cdot 3) = (\frac{1}{4} \cdot 4)3$
c. $(\frac{1}{4} \cdot 4)3 = 1 \cdot 3$ **d.** $1 \cdot 3 = 3$
e. $4 + (3 + 5) = (3 + 5) + 4$ **f.** $4 + (3 + 5) = (4 + 3) + 5$

Solution
a. Closure property
b. Associative property for multiplication
c. Inverse for multiplication
d. Identity for multiplication
e. Commutative property for addition
f. Associative property for addition ■

PROBLEM SET 1.1

A

Classify each example in Problems 1–6 as a natural number, whole number, integer, rational number, irrational number, real number, or none of these. Notice from Table 1.1 that each number listed may be in more than one of these sets.

1. a. -5 **b.** $\frac{13}{2}$ **c.** $\sqrt{25}$
d. $\sqrt{20}$ **e.** 2π

2. a. 9 **b.** $\sqrt{16}$ **c.** $\frac{0}{5}$
d. $\frac{5}{0}$ **e.** $0.\overline{3}$

3. a. $0.\overline{5}$ **b.** $0.\overline{463}$ **c.** $\sqrt{1000}$
d. 0.281 **e.** $\frac{\pi}{6}$

4. a. $\sqrt{\frac{1}{8}}$ **b.** $\sqrt{\frac{1}{9}}$ **c.** 0.5656
d. $0.5656\ldots$ **e.** $\frac{\pi}{3}$

5. a. $\sqrt{169}$ **b.** $\sqrt{200}$ **c.** π
d. 3.14192 **e.** -12.5

6. a. $-\sqrt{4}$ **b.** $\frac{3}{5}$ **c.** $\frac{4}{9}$
d. $\sqrt{\frac{4}{9}}$ **e.** $\frac{\pi}{4}$

Graph each number given in Problems 7–12 on a real number line.

7. $-3; -\sqrt{3}; \frac{5}{4}; 2; 0.1234567891011\ldots$

8. $-1; \frac{4}{5}; -\sqrt{5}; 2\frac{1}{2}; 1.03469217$

9. $-4; -\frac{-5}{3}; 0; \sqrt{2}; -1.343443444\ldots$

10. $0; \frac{\pi}{6}; \frac{\pi}{-3}; \frac{\pi}{2}; \pi$ **11.** $-\pi; -\frac{\pi}{2}; \frac{\pi}{3}; 0; 0.5$

12. $-\pi; 0.25; 0; \frac{\pi}{4}; 1$

Illustrate the property of comparison in Problems 13–18 by replacing the □ by =, >, or <.

13. a. $-10 \;\square\; -3$ **b.** $6 - 9 \;\square\; 9 - 6$

14. a. $-4 \;\square\; -8$ **b.** $\frac{5}{4} \;\square\; 1.2$

15. a. $-\frac{8}{3} \square -2.66$ **b.** $\frac{35}{21} \square \frac{-30}{-18}$

16. a. $\sqrt{3} \square 1.7$ **b.** $\frac{1}{3} \square 0.33333333$

17. a. $\frac{3}{4} + \frac{4}{5} \square \frac{31}{20}$ **b.** $\pi \square \frac{22}{7}$

18. a. $\frac{2}{3} + \frac{5}{2} \square \frac{19}{6}$ **b.** $\frac{2}{3} \square 0.66666667$

Identify the property of equality or field property illustrated by Problems 19–36.

19. $14x + 8x = 14x + 8x$

20. $10y + 4y = 10y + 4y$

21. $14x + 8x = (14 + 8)x$

22. $10y + 4y = (10 + 4)y$

23. $(14 + 8)x = 22x$

24. $(10 + 4)y = 14y$

25. If $14x + 8x = (14 + 8)x$ and $(14 + 8)x = 22x$, then $14x + 8x = 22x$.

26. If $10y + 4y = (10 + 4)y$ and $(10 + 4)y = 14y$, then $10y + 4y = 14y$.

27. If $14x + 8y = 22$ and $y = b$, then $14x + 8b = 22$.

28. If $22 = 14x + 8y$, then $14x + 8y = 22$.

29. 4π is a real number.

30. $\frac{1}{6}$ is a real number.

31. $\pi \cdot \frac{1}{\pi} = 1$

32. $\frac{1}{\frac{1}{6}}$ is a real number.

33. $5x(a + b) = 5xa + 5xb$

34. $(a + b)5x = 5x(a + b)$

35. $\frac{\sqrt{3} + 2}{\sqrt{5} + 1} = \frac{\sqrt{3} + 2}{\sqrt{5} + 1} \cdot \frac{\sqrt{5} - 1}{\sqrt{5} - 1}$

36. $\frac{4}{5} = \frac{4}{5} \cdot \frac{3}{3}$

B

37. Is the set $\{0, 1\}$ closed for addition?

38. Is the set $\{0, 1\}$ closed for multiplication?

39. Is the set $\{-1, 0, 1\}$ closed for subtraction?

40. Is the set $\{-1, 0, 1\}$ closed for nonzero division?

41. Is the set $\{0, 3, 6, 9, 12, \ldots\}$ closed for addition?

42. Is the set $\{0, 3, 6, 9, 12, \ldots\}$ closed for multiplication?

43. Is the set $\{0, 3, 6, 9, 12, \ldots\}$ closed for nonzero division?

44. Find an example showing that the operation of subtraction on **R** is not commutative.

45. Find an example showing that the operation of division on the set of nonzero real numbers is not commutative.

46. Is the operation of subtraction on **R** associative?

47. Is the operation of division on the set of nonzero real numbers associative?

C

48. What is the additive inverse of the real number $2 + \sqrt{3}$?

49. What is the additive inverse of the real number $\frac{\pi}{3} + 1$?

50. What is the multiplicative inverse of the real number $2 + \sqrt{3}$?

51. What is the multiplicative inverse of the real number $\frac{\pi}{3} + 1$?

52. What is the multiplicative inverse of the real number $\sqrt{5} + 1$?

53. Which of the field properties are satisfied by the set of natural numbers?

54. Which of the field properties are satisfied by the set of whole numbers?

55. Which of the field properties are satisfied by the set of integers?

56. Which of the field properties are satisfied by the set of rational numbers?

57. Which of the field properties are satisfied by the set of irrational numbers?

58. Which of the field properties are satisfied by the set of real numbers?

59. Which of the field properties are satisfied by the set $H = \{1, -1, i, -i\}$, where i is a number such that $i^2 = -1$?

60. Which of the field properties are satisfied by the set of nonnegative multiples of 5, namely $F = \{0, 5, 10, 15, 20, 25, \ldots\}$?

1.2 INTERVALS, INEQUALITIES, AND ABSOLUTE VALUES

A **linear equation in one variable** is an equation that can be written in the form

$$ax + b = 0 \qquad (a \neq 0)$$

where x is a **variable** and a and b are any real numbers. An **open** or **conditional equation** is an equation containing a variable that may be either true or false, depending on the replacement for the variable. A **root** or a **solution** is a replacement for the variable that makes the equation true. We also say that the root **satisfies** the equation. The **solution set** of an open equation is the set of all solutions of the equation. To **solve an equation** means to find its solution set. If there are no values for the variable that satisfy an equation, then the solution set is said to be **empty** and is denoted by $\varnothing$. If every replacement of the variable makes the equation true, then the equation is called an **identity**.

You will need to know how to solve linear inequalities involving variables. There are other inequality relationships besides less than and greater than. For example:

Less Than or Equal to $a \leq b$, read "*a is less than or equal to b*," means that either $a < b$ or $a = b$ (but not both).

Greater Than or Equal to $a \geq b$, read "*a is greater than or equal to b*," means that either $a > b$ or $a = b$ (but not both).

Between $b < a < c$, read "*a is between b and c*," means *both* $b < a$ and $a < c$. Additional between relationships are also used:

$$b \leq a \leq c \quad \text{means} \quad b \leq a \text{ and } a \leq c$$
$$b \leq a < c \quad \text{means} \quad b \leq a \text{ and } a < c$$
$$b < a \leq c \quad \text{means} \quad b < a \text{ and } a \leq c$$

In this book, when we use the word *between* we mean strictly between, namely $b < a < c$.

Graphs of linear inequality statements with a single variable are drawn on a one-dimensional coordinate system. For example, $x < 3$ denotes the interval shown in Figure 1.2a. Notice that $x \neq 3$, and this fact is shown by an open circle as the endpoint of the ray. Compare this with $x \leq 3$ (Figure 1.2b) in which the endpoint $x = 3$ is included. Notice that to sketch the graph you darken (or color) the appropriate portion.

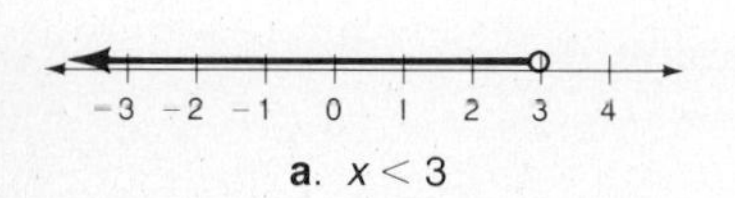

a. $x < 3$

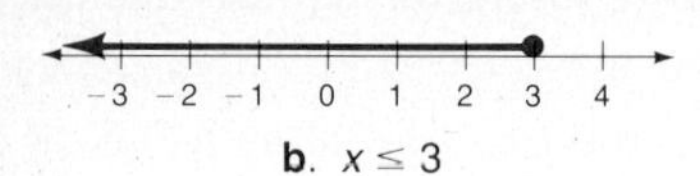

b. $x \leq 3$

-3 -2 -1 0 1 2 3 4

c. $x > 3$

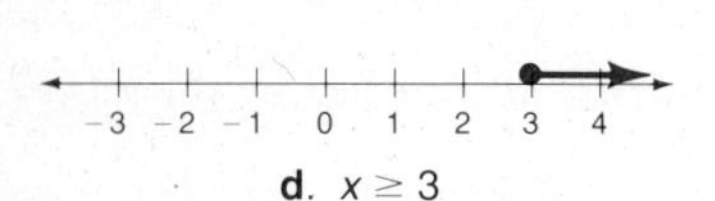

d. $x \geq 3$

Figure 1.2 Graph of inequality statements

The between relationships define **intervals** on a number line. An interval is said to be **closed** if it includes both endpoints; it is **open** if it does not include either endpoint. Study the terminology shown in Figure 1.3. In all cases, a and b are called the **endpoints** of the interval.

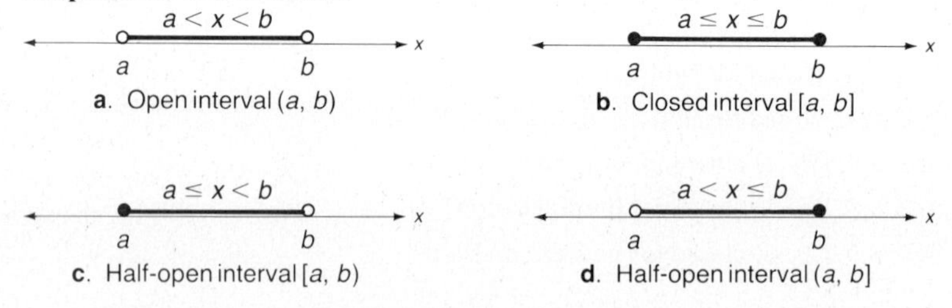

Figure 1.3 Interval notation

Note the correct use of **interval notation**: (a, b), $[a, b]$, $[a, b)$, and $(a, b]$

EXAMPLE 1 Write interval notation for each given inequality statement. Also graph each of the given inequalities.

a. $-6 \leq x < 5$; $[-6, 5)$ -6 5

b. $7 < x < 8$; $(7, 8)$ 7 8

c. $-9 \leq x \leq -2$; $[-9, -2]$

d. $-5 < x \leq 0$; $(-5, 0]$ ■

EXAMPLE 2 Graph each of the given intervals.

a. $[1, 3]$

b. $(2, 5]$

c. $[0, 4)$

d. $(-1, 4)$ ■

Sometimes intervals of unlimited extent in one or both directions are also denoted by using interval notation and the symbol ∞ as shown in Example 3. Do not interpret ∞ (infinity) as a number; as illustrated by Example 3, it is merely a notational device.

EXAMPLE 3 Write interval notation for each given inequality statement. Also graph each of the given inequalities.

a. $x > 2$; $(2, \infty)$

b. $x \geq 2$; $[2, \infty)$

c. $x < 2$; $(-\infty, 2)$

d. $x \leq 2$; $(-\infty, 2]$

e. x is any real number; $(-\infty, \infty)$ ■

A **linear inequality in one variable** is an inequality that can be written in one of the following forms:

$$ax + b < 0 \qquad ax + b \leq 0 \qquad ax + b > 0 \qquad ax + b \geq 0$$

where x is a variable and a is a nonzero real number and b is a real number. Linear inequalities are solved by using the following principles stated for less than, but they are also true for $\leq$, $>$, and $\geq$. These symbols are referred to as the **order** of the inequality.

Properties of Inequality

TRANSITIVITY:
If $a < b$ and $b < c$, then $a < c$ for any real numbers a, b, and c.

ADDITION PRINCIPLE:
If $a < b$, then $a + c < b + c$ for any real numbers a, b, and c.

POSITIVE MULTIPLICATION PRINCIPLE:
If $a < b$, and c is a positive real number then $ac < bc$.

NEGATIVE MULTIPLICATION PRINCIPLE:
If $a < b$, and c is a negative real number then $ac > bc$.

Procedure for Solving Linear Inequalities

The procedure and terminology for solving linear inequalities are identical to the procedure and terminology for solving linear equations except for one fact: If you multiply or divide by a negative number, the order of the inequality is reversed. And remember, from the definition of greater than, that if you reverse the left and right sides of an inequality, the order is also reversed.

EXAMPLE 4 Solve the inequality $5x - 3 \geq 7$ and state the solution set using interval notation.

Solution

$$5x - 3 \geq 7$$
$$5x - 3 + \mathbf{3} \geq 7 + \mathbf{3}$$
$$5x \geq 10$$
$$x \geq 2$$

Solution: $\mathbf{[2, \infty)}$ ■

EXAMPLE 5 Solve the given inequality and write the solution set using interval notation.

Solution

$2(4 - 3t) < 5t - 14$	Eliminate parentheses.
$8 - 6t < 5t - 14$	Isolate the variable on one side.
$-6t < 5t - 22$	
$-11t < -22$	
$t > 2$	Reverse the order of the inequality when you divide by a negative number.

Solution: $\mathbf{(2, \infty)}$ ■

Sometimes two inequality statements are combined, as with

$$-3 \leq 2x + 1 \leq 5$$

This means that $-3 \leq 2x + 1$ *and* $2x + 1 \leq 5$. It also means that $2x + 1$ is between -3 and 5. When solving an inequality of this type, you can frequently work both inequalities simultaneously as shown in Examples 6 and 7.

EXAMPLE 6

$$-3 \leq 2x + 1 \leq 5$$
$$-3 - \mathbf{1} \leq 2x + 1 - \mathbf{1} \leq 5 - \mathbf{1}$$
$$-4 \leq 2x \leq 4$$
$$\frac{-4}{\mathbf{2}} \leq \frac{2x}{\mathbf{2}} \leq \frac{4}{\mathbf{2}}$$
$$-2 \leq x \leq 2$$

Your goal is now to isolate the variable in the middle of the between statement. Each inequality statement is equivalent to the preceding one. Whatever you do to one part, you do to all three parts.

Solution: $\mathbf{[-2, 2]}$ ■

EXAMPLE 7

$$-5 \leq 1 - 3x < 10$$
$$-6 \leq -3x < 9$$
$$\frac{-6}{\mathbf{-3}} \geq \frac{-3x}{\mathbf{-3}} > \frac{9}{\mathbf{-3}}$$

Inequality reverses when you multiply or divide by a negative.

$$2 \geq x > -3$$

However, $2 \geq x > -3$ is not convenient for converting to interval notation, so it is rewritten as

$$-3 < x \leq 2$$

Solution: $\mathbf{(-3, 2]}$ ■

A very important idea in mathematics involves the notion of **absolute value**. If a is a real number, then its graph on a number line is some point, call it A. The distance between A and the origin is the geometric interpretation of the **absolute value of *a***.

Absolute Value

The absolute value of a real number a is denoted by $|a|$ and is defined by

$$|a| = a \qquad \text{if } a \geq 0$$
$$|a| = -a \qquad \text{if } a < 0$$

Thus $|5| = 5$, $|-5| = 5$, $|0| = 0$, and $|-\pi| = \pi$. Notice that $|a|$ is nonnegative for all values of a.

EXAMPLE 8 $|\pi - 3| = \boldsymbol{\pi - 3}$ since $\pi - 3 > 0$; use the first part of the definition of absolute value ■

EXAMPLE 9 $|\pi - 4| = -(\pi - 4)$ since $\pi - 4 < 0$; use the second part of the definition of absolute value
$= \mathbf{4 - \pi}$ ■

EXAMPLE 10 $|w^2 + 1| = \mathbf{w^2 + 1}$ since $w^2 + 1$ is positive for all real values of w ■

EXAMPLE 11 $|-4 - t^2| = -(-4 - t^2)$ since $-4 - t^2$ is negative for all real values of t
$= \mathbf{t^2 + 4}$ ■

Since $|a|$ can be interpreted as the distance between the point A whose coordinate is a and the origin, it is a straightforward derivation to show that the distance between any two points on a number line can be expressed as an absolute value.

Distance on a Number Line

Let x_1 and x_2 be the coordinates of two points P_1 and P_2, respectively, on a number line. The distance d between P_1 and P_2 is

$$d = |x_2 - x_1|$$

Notice that if a is the coordinate of P_1 and 0 is the coordinate of P_2, then $d = |a - 0| = |a|$; and if 0 is the coordinate of P_1 and a is the coordinate of P_2, then $d = |0 - a| = |-a|$. Since d is the same distance for both these calculations,

$$|a| = |-a|$$

It also follows that $d = |x_2 - x_1| = |x_1 - x_2|$ and that $|x^2| = |x|^2$.

EXAMPLE 12 Find the distance between the points whose coordinates are given.

a. (-8) and (2); $d = |2 - (-8)| = |10| = \mathbf{10}$
b. (108) and (-34); $d = |-34 - 108| = |-142| = \mathbf{142}$
c. $(-\pi)$ and $(-\sqrt{2})$; $d = |-\sqrt{2} + \pi| = \boldsymbol{\pi - \sqrt{2}}$
d. $(-\pi)$ and $(-\sqrt{10})$; $d = |-\sqrt{10} + \pi| = -(-\sqrt{10} + \pi) = \boldsymbol{\sqrt{10} - \pi}$ ■

Since $|x - a|$ can be interpreted as the distance between x and a on the number line, an equation of the form

$$|x - a| = b$$

has two values of x that are a given distance from a when represented on a number line. For example, $|x - 5| = 3$ states that x is 3 units from 5 on a number line. Thus x is either 2 or 8.

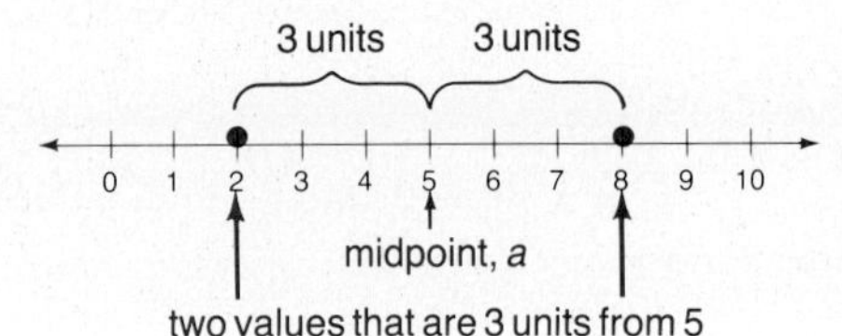

You can verify this conclusion by proving the following property.

Absolute Value Equations

> $|a| = b$ where $b \geq 0$ if and only if $a = b$ or $a = -b$.

Proof There are two parts to a proof using the words "if and only if." **(1)** If $|a| = b$, then $a = b$ or $a = -b$. **(2)** If $a = b$ or $a = -b$, then $|a| = b$. We will prove the first part here and leave the second part as a problem. That is, suppose $|a| = b$; we now wish to show that $a = b$ or $a = -b$. Begin with the property of comparison to compare the real number a with the real number zero:

$$a > 0 \qquad a = 0 \qquad \text{or} \qquad a < 0$$

If $a > 0$ or if $a = 0$, then $|a| = a$ and

$\lvert a\rvert = b$	Given.
$a = b$	Substitution of a for $\lvert a\rvert$.

If $a < 0$, then $|a| = -a$ and

$\lvert a\rvert = b$	Given.
$-a = b$	Substitution of $-a$ for $\lvert a\rvert$.
$a = -b$	Multiplication of both sides by -1.

■

EXAMPLE 13 Solve $|x + 5| = 2$.

Solution

$$x + 5 = 2 \qquad \text{or} \qquad x + 5 = -2$$
$$x = -3 \qquad \text{or} \qquad x = -7$$

Solution: $\{-3, -7\}$ ■

EXAMPLE 14 Solve $|x + 5| = -2$.

Solution The absolute value of every real number is nonnegative, so the solution set is empty. ■

If you are solving an absolute value equation with absolute values on both sides of the equation, the following property may be useful.

Procedure for Solving Absolute Value Equations

> $|a| = |b|$ if and only if $a = b$ or $a = -b$.

You are asked to prove this in Problem 59 of Problem Set 1.2.

EXAMPLE 15 Solve $|x + 5| = |3x - 4|$.

Solution Use the given property of absolute value.

$$\begin{aligned} x + 5 &= 3x - 4 \\ -2x &= -9 \\ x &= \frac{9}{2} \end{aligned} \qquad \text{or} \qquad \begin{aligned} x + 5 &= -(3x - 4) \\ x + 5 &= -3x + 4 \\ 4x &= -1 \\ x &= -\frac{1}{4} \end{aligned}$$

Solution: $\{\frac{9}{2}, -\frac{1}{4}\}$ ■

Since $|x - 5| = 3$ states that x is 3 units from 5, the inequality $|x - 5| < 3$ states that x is any number less than 3 units from 5.

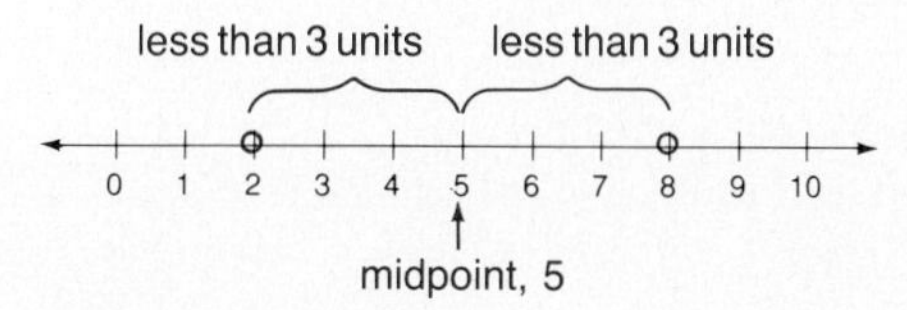

On the other hand, $|x - 5| > 3$ states that x is any number greater than 3 units from 5.

more than 3 units from 5   more than 3 units from 5

0 1 2 3 4 5 6 7 8 9 10

midpoint, 5

These properties are summarized by four absolute-value inequality properties where a and b are real numbers and $b > 0$.

Absolute Value Inequalities

$\lvert a\rvert < b$	if and only if	$-b < a < b$
$\lvert a\rvert \leq b$	if and only if	$-b \leq a \leq b$
$\lvert a\rvert > b$	if and only if	$a > b$ or $a < -b$
$\lvert a\rvert \geq b$	if and only if	$a \geq b$ or $a \leq -b$

Proof *Prove:* If $|a| < b$, then $-b < a < b$. By definition:

$$|a| = a \qquad \text{if } a \geq 0$$
$$|a| = -a \qquad \text{if } a < 0$$

Case i: If $a \geq 0$

$\lvert a\rvert < b$	Given.
$a < b$	Substitute a for $\lvert a\rvert$.

Case ii: If $a < 0$

$\lvert a\rvert < b$	Given.
$-a < b$	Substitute $-a$ for $\lvert a\rvert$.
$0 < a + b$	Add a to both sides.
$-b < a$	Subtract b from both sides.

Therefore $-b < a$ and $a < b$ as seen from cases *i* and *ii*. This is the same as saying $-b < a < b$. The proofs of the other parts are similar and are left for the exercises. □

EXAMPLE 16 Solve $|2x - 3| \leq 4$.

Solution $-4 \leq 2x - 3 \leq 4$. Now solve the *between* relationship:

$$-4 + \mathbf{3} \leq 2x - 3 + \mathbf{3} \leq 4 + \mathbf{3}$$
$$-1 \leq 2x \leq 7$$
$$\frac{-1}{\mathbf{2}} \leq \frac{2x}{\mathbf{2}} \leq \frac{7}{\mathbf{2}}$$
$$-\frac{1}{2} \leq x \leq \frac{7}{2}$$

Solution: $[-\frac{1}{2}, \frac{7}{2}]$ ■

EXAMPLE 17 Solve $|4 - x| < -3$.

Solution Absolute values must be nonnegative, so the solution set is empty. ■

EXAMPLE 18 Solve $|x + 3| > 4$.

Solution

$$x + 3 > 4 \quad \text{or} \quad x + 3 < -4$$
$$x > 1 \quad \text{or} \quad x < -7$$

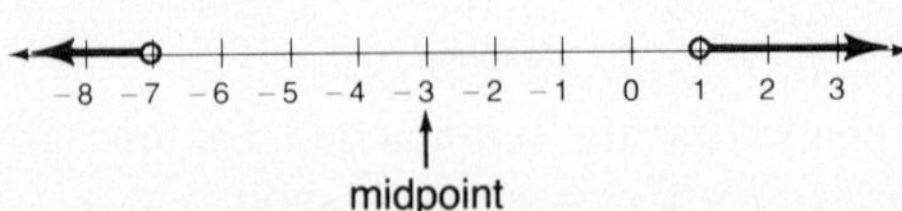

Do not attempt to write this as $1 < x < -7$ since transitivity implies $1 < -7$ (which is not true). Remember: $a < x < c$ means that x satisfies *both* the conditions $a < x$ and $x < c$, and transitivity implies that $a < c$.

Interval notation may be used to express the solution set when more than one part of a number line is included if you use the union symbol $\cup$. The solution is

$$(-\infty, -7) \cup (1, \infty)$$ ■

PROBLEM SET 1.2

A

Write interval notation and graph each of the inequalities in Problems 1–4.

1. a. $3 < x < 7$ b. $-4 < x < -1$ c. $-2 \leq x \leq 6$ d. $-3 < x \leq 0$
2. a. $-6 \leq x \leq 0$ b. $-3 < x \leq 3$ c. $-2 \leq x \leq 2$ d. $-5 \leq x < 1$
3. a. $x \leq -3$ b. $x \geq -2$ c. $x < 0$ d. $2 < x$
4. a. $3 < x \leq 6$ b. $5 < x$ c. $10 > x$ d. $-0.01 \leq x \leq 0.01$

Graph each of the intervals in Problems 5–8.

5. a. $[-3, 2]$ b. $(-2, 2)$ c. $(-\infty, 3]$ d. $[-2, \infty)$
6. a. $(-3, 4]$ b. $[-3, 4)$ c. $[-4, 3)$ d. $(2, \infty)$
7. a. $(-2, 0) \cup (3, 5)$ b. $[-8, 3) \cup [5, \infty)$ c. $(-\infty, -3] \cup (0, 3]$ d. $[-5, -1) \cup (0, 5]$
8. a. $(-\infty, 0) \cup (0, \infty)$ b. $(-\infty, 2) \cup (2, \infty)$ c. $(-\infty, 1) \cup (1, \infty)$ d. $(-\infty, 6) \cup (6, \infty)$

Write each of the intervals in Problems 9–12 as an inequality. Use x as the variable.

9. a. $[-4, 2]$ b. $[-1, 2]$ c. $(0, 8)$ d. $(-5, 3]$
10. a. $(-3, -1)$ b. $[-5, 2)$ c. $[5, 9]$ d. $(-6, -2)$
11. a. $(-\infty, 2)$ b. $(6, \infty)$ c. $(-1, \infty)$ d. $(-\infty, 3]$
12. a. $(-\infty, -6]$ b. $(-\infty, 0)$ c. $[5, \infty)$ d. $(-1, \infty)$

Write each expression in Problems 13–15 without using the symbol for absolute value.

13. a. $|\pi - 2|$ b. $|\pi - 5|$ c. $|2\pi - 6|$ d. $|2\pi - 7|$
14. a. $|x^2|$ b. $|x^2 + 4|$ c. $|-3 - x^2|$ d. $|-2 + \sqrt{5}|$
15. a. $|\sqrt{2} - 1|$ b. $|\sqrt{2} - 2|$ c. $\left|\frac{\pi}{6} - 1\right|$ d. $\left|\frac{2\pi}{3} - 1\right|$

Find the distance between the points whose coordinates are given in Problems 16–18.

16. a. (8) and (15) b. (-103) and (6) c. (-143) and (-120)
17. a. $(-\pi)$ and (-3) b. (23) and (-96) c. $(-\sqrt{5})$ and (-3)
18. a. (π) and (-4) b. $(\sqrt{3})$ and $(-\pi)$ c. $(-\pi)$ and $(-\sqrt{2})$

Solve the equations in Problems 19–30.

19. $|x| = 5$
20. $|x| = 10$
21. $|x| = -1$
22. $|x| = -4$
23. $|x - 3| = 4$
24. $|x + 9| = 3$
25. $|x - 9| = 15$
26. $|2x + 4| = -8$
27. $|2x + 4| = -12$
28. $|3 - 2x| = 7$
29. $|5x + 4| = 6$
30. $|5 - 3x| = 14$

B

Solve the inequalities in Problems 31–54 and write each solution in interval notation.

31. $3x - 9 \geq 12$
32. $-x > -36$
33. $-3x \geq 123$
34. $3(2 - 4x) \leq 0$
35. $5(3 - x) > 3x - 1$
36. $-5 \leq 5x \leq 25$
37. $-8 < 5x < 0$
38. $3 \leq -x < 8$
39. $-3 \leq -x < -1$
40. $-5 < 3x + 2 \leq 5$
41. $-7 \leq 2x + 1 < 5$
42. $-4 < 1 - 5x \leq 11$
43. $-5 \leq 3 - 2x < 18$
44. $|x - 5| \leq 1$
45. $|x - 3| \leq 5$
46. $|x - 8| \leq 0.01$
47. $|x - 3| < 0.001$
48. $|3x + 2| < -3$
49. $|2x + 1| < -1$
50. $|3 - x| < 5$
51. $|4 - 3x| < 3$
52. $|x + 1| > 3$
53. $|2x + 7| > 5$
54. $|3 - 2x| > 5$
55. ***Business*** If Amex Automobile Rental charges \$35 per day and 45¢ per mile, how many miles can you drive and keep the cost under \$125 per day?
56. ***Business*** A saleswoman is paid a salary of \$300 plus a 40% commission on sales. How much does she need to sell in order to have an income of at least \$2000?
57. ***Chemistry*** A certain experiment requires that the temperature be between 20° and 30° C. If Fahrenheit and Celsius degrees are related by the formula $C = \frac{5}{9}(F - 32)$, what are the permissible temperatures in Fahrenheit?
58. ***Economics*** An economist estimates that the consumer price index will grow by 14%, give or take 1%. Let c represent the growth rate of the consumer price index.
    a. Write the condition of this problem as an inequality.
    b. Write the condition of this problem as an absolute value statement.

59. ***Business*** The management of a certain company needs to monitor the activities of salespersons whose sales are not the usual \$15,000 per week. That is, if the sales of a certain employee are less than \$10,000 or greater than \$20,000, the company needs to monitor the amount of time spent on the job. Let s represent the amount of weekly sales.

a. Write the condition of this problem as an inequality.
b. Write the condition of this problem as an absolute value statement.

C

Solve the inequalities in Problems 60–65 and write the solution in interval notation.

60. $\left|\frac{5-x}{2}\right| > 1$

61. $\left|\frac{3-x}{2}\right| > 3$

62. $\frac{5}{|x-2|} \le 1$

63. $\frac{3}{|x+1|} \ge 3$

64. $|2x-1| + |x| - 3 = 0$

65. $|3x+1| + |x| - 5 = 0$

66. Prove that if b is a nonnegative real number and if $a = b$ or $a = -b$, then $|a| = b$.

67. If $|a| \le b$, show that $-b \le a \le b$.

68. If $|a| > b$, show that either $a > b$ or $a < -b$.

69. If $|a| \ge b$, show that either $a \ge b$ or $a \le -b$.

70. Prove that $|a| = |b|$ if and only if $a = b$ or $a = -b$.

1.3 COMPLEX NUMBERS*

To find the roots of certain equations, you must sometimes consider the square roots of negative numbers. Since the set of real numbers does not allow for such a possibility, a number that is *not a real number* is defined. This number is denoted by the symbol i.

The Imaginary Unit

The number i, called the **imaginary unit**, is defined as a number with the following properties:

$$i^2 = -1 \quad \text{and} \quad \sqrt{-a} = i\sqrt{a} \quad (a > 0)$$

With this number you can write the square roots of any negative numbers as the product of a real number and the number i. Thus $\sqrt{-9} = i\sqrt{9} = 3i$. For any positive real number b, $\sqrt{-b} = i\sqrt{b}$. We now consider a new set of numbers.

Complex Numbers

The set of all numbers of the form

$$a + bi$$

with real numbers a and b and i the imaginary unit is called the set of **complex numbers**.

If $b = 0$, then $a + bi = a + 0i = a$, which is a **real number**; thus the real numbers form a subset of the complex numbers. If $a \ne 0$ and $b \ne 0$, then $a + bi$ is called an **imaginary number**; if $a = 0$ and $b \ne 0$, then $a + bi = 0 + bi = bi$ is called a **pure imaginary number**. If $a = 0$ with $b = 1$, then $a + bi = 0 + 1 \cdot i = i$, which is the imaginary unit.

* Complex numbers can be omitted from this text. All parts of the book requiring the use of complex numbers will be marked as optional.

A complex number is **simplified** if it is written in the form $a + bi$, where a and b are simplified real numbers and i is the imaginary unit. In order to work with complex numbers, you will need definitions of equality along with the usual arithmetic operations. Let $a + bi$ and $c + di$ be complex numbers. Values that could cause division by zero are excluded.

Operations with Complex Numbers

EQUALITY: $a + bi = c + di$ if and only if $a = c$ and $b = d$

ADDITION: $(a + bi) + (c + di) = (a + c) + (b + d)i$

SUBTRACTION: $(a + bi) - (c + di) = (a - c) + (b - d)i$

MULTIPLICATION: $(a + bi)(c + di) = (ac - bd) + (ad + bc)i$

DIVISION: $\dfrac{a + bi}{c + di} = \dfrac{(ac + bd) + (bc - ad)i}{c^2 + d^2} = \dfrac{ac + bd}{c^2 + d^2} + \dfrac{bc - ad}{c^2 + d^2}i$

It is not necessary to memorize these definitions, since you can deal with two complex numbers as you would any binomials as long as you remember that $i^2 = -1$. Also notice $i^3 = i \cdot i^2 = -i$ and $i^4 = i^2 \cdot i^2 = 1$.

EXAMPLE 1 Simplify each expression.

a. $(4 + 5i) + (3 + 4i) = \mathbf{7 + 9i}$

b. $(2 - i) - (3 - 5i) = \mathbf{-1 + 4i}$

c. $(5 - 2i) + (3 + 2i) = \mathbf{8}$ or $8 + 0i$

d. $(4 + 3i) - (4 + 2i) = i$ or $0 + i$

e.
$$\begin{aligned}(2 + 3i)(4 + 2i) &= 8 + 16i + 6i^2\\ &= 8 + 16i - 6\\ &= \mathbf{2 + 16i}\end{aligned}$$

f.
$$\begin{aligned}i^{94} &= i^{4\cdot 23 + 2}\\ &= (i^4)^{23}(i^2)\\ &= 1 \cdot i^2\\ &= \mathbf{-1}\end{aligned}$$

g.
$$\begin{aligned}i^{125} &= i^{4\cdot 31 + 1}\\ &= (i^4)^{31}(i^1)\\ &= \mathbf{i}\end{aligned}$$

h.
$$\begin{aligned}i^{1883} &= i^{4\cdot 470 + 3}\\ &= (i^4)^{470}(i^3)\\ &= i^3\\ &= \mathbf{-i}\end{aligned}$$

■

EXAMPLE 2 Verify that multiplying $(a + bi)(c + di)$ in the usual algebraic way gives the same result as that shown in the definition of multiplication of complex numbers.

Solution

$$\begin{aligned}(a + bi)(c + di) &= ac + adi + bci + bdi^2\\ &= ac + (ad + bc)i - bd\\ &= (ac - bd) + (ad + bc)i\end{aligned}$$

■

EXAMPLE 3 Simplify $(4 - 3i)(4 + 3i)$.

Solution

$$\begin{aligned}(4 - 3i)(4 + 3i) &= 16 - 9i^2\\ &= 16 + 9\\ &= \mathbf{25}\end{aligned}$$

■

Example 3 gives a clue for dividing complex numbers. The definition would be difficult to remember, so instead we use the idea of *conjugates*. The complex numbers $a + bi$ and $a - bi$ are called **complex conjugates**, and each is the conjugate of the other:

$$\begin{aligned}(a + bi)(a - bi) &= a^2 - b^2i^2 \\ &= a^2 + b^2\end{aligned}$$

which is a real number. Thus simplify a quotient by using the conjugate of the denominator, as illustrated by Examples 4 to 6.

EXAMPLE 4 Simplify $\frac{15 - 5i}{2 - i}$.

Solution

multiply by 1

$$\frac{15 - 5i}{2 - i} = \frac{15 - 5i}{2 - i} \cdot \frac{\mathbf{2 + i}}{\mathbf{2 + i}}$$

conjugates

$$= \frac{30 + 5i - 5i^2}{4 - i^2}$$

$$= \frac{35 + 5i}{5}$$

$$= \mathbf{7 + i}$$

You can check this by multiplying $(2 - i)(7 + i)$:

$$\begin{aligned}(2 - i)(7 + i) &= 14 - 5i - i^2 \\ &= 15 - 5i\end{aligned}$$

■

EXAMPLE 5

$$\frac{6 + 5i}{2 + 3i} = \frac{6 + 5i}{2 + 3i} \cdot \frac{\mathbf{2 - 3i}}{\mathbf{2 - 3i}}$$

$$= \frac{12 - 8i - 15i^2}{4 - 9i^2}$$

$$= \frac{27}{13} - \frac{8}{13}i$$

■

EXAMPLE 6 Verify that the conjugate method gives the same result as the definition of division of complex numbers.

Solution

$$\frac{a + bi}{c + di} = \frac{a + bi}{c + di} \cdot \frac{\mathbf{c - di}}{\mathbf{c - di}}$$

$$= \frac{ac - adi + bci - bdi^2}{c^2 - d^2i^2}$$

$$= \frac{(ac + bd) + (bc - ad)i}{c^2 + d^2}$$

$$= \frac{ac + bd}{c^2 + d^2} + \frac{bc - ad}{c^2 + d^2}i$$

■

PROBLEM SET 1.3

A

Simplify the expressions in Problems 1–36.

1. $\sqrt{-36}$
2. $\sqrt{-100}$
3. $\sqrt{-49}$
4. $\sqrt{-8}$
5. $\sqrt{-20}$
6. $\sqrt{-24}$
7. $(3 + 3i) + (5 + 4i)$
8. $(6 - 2i) + (5 + 3i)$
9. $(5 - 3i) - (5 + 2i)$
10. $(3 + 4i) - (7 + 4i)$
11. $(4 - 2i) - (3 + 4i)$
12. $5i - (5 + 5i)$
13. $5 - (2 - 3i)$
14. $-2(-4 + 5i)$
15. $6(3 + 2i) + 4(-2 - 3i)$
16. $4(2 - i) - 3(-1 - i)$
17. $i(5 - 2i)$
18. $i(2 + 3i)$
19. $(3 - i)(2 + i)$
20. $(4 - i)(2 + i)$
21. $(5 - 2i)(5 + 2i)$
22. $(8 - 5i)(8 + 5i)$
23. $(3 - 5i)(3 + 5i)$
24. $(7 - 9i)(7 + 9i)$
25. $-i^2$
26. $-i^3$
27. i^3
28. i^4
29. $-i^4$
30. $-i^5$
31. $-i^6$
32. i^6
33. i^{11}
34. i^{236}
35. $-i^{1980}$
36. i^{1982}

B

Simplify the expressions in Problems 37–57.

37. $(6 - 2i)^2$
38. $(3 + 3i)^2$
39. $(4 + 5i)^2$
40. $(1 + i)^3$
41. $(3 - 5i)^3$
42. $(2 - 3i)^3$
43. $\dfrac{-3}{1 + i}$
44. $\dfrac{5}{4 - i}$
45. $\dfrac{2}{1 - i}$
46. $\dfrac{5}{i}$
47. $\dfrac{2}{i}$
48. $\dfrac{3}{-i}$
49. $\dfrac{-2i}{3 + i}$
50. $\dfrac{3i}{5 - 2i}$
51. $\dfrac{-i}{2 - i}$
52. $\dfrac{1 - 6i}{1 + 6i}$
53. $\dfrac{4 - 2i}{3 + i}$
54. $\dfrac{5 + 3i}{4 - i}$
55. $\dfrac{1 + 3i}{1 - 2i}$
56. $\dfrac{3 - 2i}{5 + i}$
57. $\dfrac{2 + 7i}{2 - 7i}$

C

58. Simplify $(1.9319 + 0.5176i)(2.5981 + 1.5i)$.
59. Simplify $\dfrac{-3.2253 + 8.4022i}{3.4985 + 1.9392i}$.

Let $z_1 = (1 + \sqrt{3}) + (2 + \sqrt{3})i$ *and* $z_2 = (2 + \sqrt{12}) + \sqrt{3}i$. *Perform the indicated operations in Problems 60–65.*

60. $z_1 + z_2$
61. $z_1 - z_2$
62. $z_1 z_2$
63. z_1/z_2
64. $(z_1)^2$
65. $(z_2)^2$

For each of Problems 66–68 let $z_1 = a + bi$, $z_2 = c + di$, *and* $z_3 = e + fi$.

66. Prove the commutative laws for complex numbers. That is, prove that:

$$z_1 + z_2 = z_2 + z_1$$
$$z_1 z_2 = z_2 z_1$$

67. Prove the associative laws for complex numbers. That is, prove that:

$$z_1 + (z_2 + z_3) = (z_1 + z_2) + z_3$$
$$z_1(z_2 z_3) = (z_1 z_2) z_3$$

68. Prove the distributive law for complex numbers. That is, prove that:

$$z_1(z_2 + z_3) = z_1 z_2 + z_1 z_3$$

1.4 QUADRATIC EQUATIONS

A **quadratic equation in one variable** is an equation that can be written in the form

$$ax^2 + bx + c = 0 \qquad (a \neq 0)$$

where x is a variable and a, b, and c are real numbers. Quadratic equations can be solved by several methods. The simplest method can be used if the quadratic expression $ax^2 + bx + c$ is factorable over the integers.* The solution depends on the following property of zero.

* In this section we will consider only simple factoring types from beginning algebra. For a comprehensive review of factoring, see Section 3.1 (page 86).

Property of Zero

$AB = 0$ if and only if $A = 0$ or $B = 0$ (or both).

Thus, if the product of two factors is zero, then at least one of the factors is zero. If a quadratic is factorable, this property provides a method of solution.

EXAMPLE 1 Solve $x^2 = 2x + 15$.

Solution

$$x^2 - 2x - 15 = 0$$

$$(x + 3)(x - 5) = 0$$

$$x + 3 = 0 \quad \text{or} \quad x - 5 = 0$$

Since the product is zero, one of the factors must be zero.

$$x = -3 \quad \text{or} \quad x = 5$$

Solution: $\{-3, 5\}$ ■

Solution of Quadratic Equations by Factoring

To solve a quadratic equation that can be expressed as a product of linear factors:

1. rewrite all nonzero terms on one side of the equation;
2. factor the expression;
3. set each of the factors equal to zero;
4. solve each of the linear equations;
5. write the solution set which is the union of the solution sets of the linear equations.

Completing the Square

When the quadratic is not factorable, other methods must be employed. One such method depends on the square-root property.

Square-Root Property

If $P^2 = Q$, then $P = \pm\sqrt{Q}$.

The equation $x^2 = 4$ could be rewritten as $x^2 - 4 = 0$, factored, and solved. However, the square-root property can be used, as illustrated below.

Using the square-root property:

$$x^2 = 4$$

$$x = \pm\sqrt{4}$$

$$x = \pm 2$$

Using factoring:

$$x^2 - 4 = 0$$

$$(x + 2)(x - 2) = 0$$

$$x = -2 \quad \text{or} \quad x = 2$$

The square root property can be derived by using the following property of square roots and an absolute value equation.

> For all real numbers x,
>
> $$\sqrt{x^2} = |x|$$

This means, if $x \geq 0$, then $\sqrt{x^2} = x$ and if $x < 0$, then $\sqrt{x^2} = -x$. For example, $\sqrt{2^2} = |2| = 2$ and $\sqrt{(-2)^2} = |-2| = 2$. The importance of this property, however, is that it can be applied to any quadratic! This is because every quadratic may be expressed in the form $P^2 = Q$ by isolating the variable terms and **completing the square**, as illustrated in the following examples.

EXAMPLE 2 Solve $x^2 = 2x + 15$.

Solution

$x^2 - 2x = 15$ — Isolate the variable.

$x^2 - 2x + 1 = 15 + 1$ — Complete the square by adding the square of half the coefficient of x to both sides.

$(x - 1)^2 = 16$ — Factor.

$x - 1 = \pm 4$ — Use the square-root property.

$x = 1 \pm 4$ — Solve for x.

$x = 1 + 4$ or $x = 1 - 4$

$x = 5$ or $x = -3$

Solution: $\{\mathbf{5, -3}\}$ ■

EXAMPLE 3 Solve $4x^2 - 4x - 7 = 0$.

Solution

$4x^2 - 4x = 7$ — Isolate the variable.

$x^2 - x = \frac{7}{4}$ — Divide by 4 so that the coefficient of the squared term is 1.

$x^2 - x + \left(\frac{1}{2}\right)^2 = \left(\frac{1}{2}\right)^2 + \frac{7}{4}$ — Complete the square by adding the square of half the coefficient of x to both sides.

$\left(x - \frac{1}{2}\right)^2 = 2$ — Factor.

$x - \frac{1}{2} = \pm\sqrt{2}$ — Remember $\pm$.

$x = \frac{1 \pm 2\sqrt{2}}{2}$ — Solve for x. Note $\pm$ means two different solutions.

Solution: $\left\{\frac{1 + 2\sqrt{2}}{2}, \frac{1 - 2\sqrt{2}}{2}\right\}$ ■

Quadratic Formula

The process of completing the square is often cumbersome. However, if *any* quadratic $ax^2 + bx + c = 0$, $a \neq 0$, is considered, completing the square can be used to derive a formula for x in terms of the coefficients a, b, and c. The formula can then be used to solve all quadratics, even those that are nonfactorable.

$$ax^2 + bx + c = 0$$

$$ax^2 + bx = -c \qquad \text{Isolate the variable.}$$

$$x^2 + \frac{b}{a}x = -\frac{c}{a} \qquad \text{Divide by } a.$$

$$x^2 + \frac{b}{a}x + \left(\frac{b}{2a}\right)^2 = \left(\frac{b^2}{4a^2}\right) - \frac{c}{a} \qquad \text{Complete the square; } \tfrac{1}{2} \text{ of } \frac{b}{a} \text{ is } \frac{b}{2a}.$$

$$\left(x + \frac{b}{2a}\right)^2 = \frac{b^2 - 4ac}{4a^2} \qquad \text{Factor.}$$

$$x + \frac{b}{2a} = \pm\sqrt{\frac{b^2 - 4ac}{4a^2}} \qquad \text{Use the square root property.}$$

$$x + \frac{b}{2a} = \pm\frac{\sqrt{b^2 - 4ac}}{2a}$$

$$x = -\frac{b}{2a} \pm \frac{\sqrt{b^2 - 4ac}}{2a} \qquad \text{Solve for } x.$$

$$x = \frac{-b \pm \sqrt{b^2 - 4ac}}{2a}$$

Quadratic Formula

If $ax^2 + bx + c = 0$, $a \neq 0$, then

$$x = \frac{-b \pm \sqrt{b^2 - 4ac}}{2a}$$

EXAMPLE 4 Solve $5x^2 + 2x - 2 = 0$.

Solution

$$x = \frac{-2 \pm \sqrt{4 - 4(5)(-2)}}{2(5)}$$

$$= \frac{-2 \pm 2\sqrt{1 + 10}}{2(5)}$$

$$= \frac{-1 \pm \sqrt{11}}{5}$$

Solution: $\left\{\dfrac{-1 \pm \sqrt{11}}{5}\right\}$ ■

EXAMPLE 5 Solve for x:

$$5x^2 + 2x - (w + 4) = 0$$

Solution

$$x = \frac{-2 \pm \sqrt{4 - 4(5)(-w - 4)}}{2(5)}$$

$$= \frac{-2 \pm 2\sqrt{5w + 21}}{2(5)}$$

There are some steps not shown here. Can you fill in the details?

$$= \mathbf{\frac{-1 \pm \sqrt{5w + 21}}{5}}$$

When solving equations, it is not necessary to recopy the answer using solution set notation. If you leave your answer as shown in this example, it will be understood that the solution set is

$$\left\{\frac{-1 \pm \sqrt{5w + 21}}{5}\right\}$$

■

EXAMPLE 6 Solve $5x^2 + 2x + 2 = 0$.

Solution

$$x = \frac{-2 \pm \sqrt{4 - 4(5)(2)}}{2(5)} = \frac{-2 \pm 2\sqrt{-9}}{10}$$

Since the square root of a negative number is not a real number, the solution set is **empty over the reals**. On the other hand, if you are using the domain consisting of the complex numbers (Section 1.3), then you can solve this equation.

$$x = \frac{-2 \pm 2(3i)}{10} = \frac{2(-1 \pm 3i)}{10} = \frac{-1 \pm 3i}{5} = \mathbf{-\frac{1}{5} \pm \frac{3}{5}i}$$

■

Since the quadratic formula contains a radical, the sign of the radicand will determine whether the roots will be real or nonreal. This radicand is called the **discriminant** of the quadratic, and its properties are summarized in the box.

The Discriminant of the Quadratic

If $ax^2 + bx + c = 0$, $a \neq 0$, then $b^2 - 4ac$ is called the ***discriminant***.
If $b^2 - 4ac < 0$, there are *no real* solutions.
If $b^2 - 4ac = 0$, there is *one real* solution.
If $b^2 - 4ac > 0$, there are *two real* solutions.

Suppose you wish to check your answers for the above examples, say Example 4. One method is to substitute the roots into the original equation. This may involve considerable effort:

$$x = \frac{-1 \pm \sqrt{11}}{5}$$

Substituting,

$$5\left(\frac{-1+\sqrt{11}}{5}\right)^2 + 2\left(\frac{-1+\sqrt{11}}{5}\right) - 2 \stackrel{?}{=} 0 \qquad \text{and}$$

$$5\left(\frac{-1-\sqrt{11}}{5}\right)^2 + 2\left(\frac{-1-\sqrt{11}}{5}\right) - 2 \stackrel{?}{=} 0$$

Instead, suppose we represent the roots by r_1 and r_2. Then:

$$x = r_1 \quad \text{or} \quad x = r_2$$

$$x - r_1 = 0 \quad \text{or} \quad x - r_2 = 0$$

$$(x - r_1)(x - r_2) = 0$$

$$x^2 - r_1x - r_2x + r_1r_2 = 0$$

$$x^2 - (r_1 + r_2)x + r_1r_2 = 0$$

Sum and Product of the Roots of a Quadratic Equation

$$x^2 - \begin{pmatrix}\text{sum of}\\ \text{the roots}\end{pmatrix}x + \begin{pmatrix}\text{product of}\\ \text{the roots}\end{pmatrix} = 0$$

Comparing this result to $ax^2 + bx + c = 0$ or

$$x^2 + \frac{b}{a}x + \frac{c}{a} = 0,$$

we can see that

$$\textbf{sum of roots} = -\frac{\mathbf{b}}{\mathbf{a}} \qquad \text{and} \qquad \textbf{product of roots} = \frac{\mathbf{c}}{\mathbf{a}}$$

For Example 4:

$$5x^2 + 2x - 2 = 0$$

$$x^2 + \frac{2}{5}x - \frac{2}{5} = 0$$

sum of roots:

$$\frac{-1+\sqrt{11}}{5} + \frac{-1-\sqrt{11}}{5} = -\frac{2}{5}$$

product of roots:

$$\left(\frac{-1+\sqrt{11}}{5}\right)\left(\frac{-1-\sqrt{11}}{5}\right) = \frac{1-11}{25} = \frac{-10}{25} = -\frac{2}{5}$$

The answer checks since

$$x^2 + \frac{2}{5}x - \frac{2}{5} = 0$$

opposite of the sum of the roots (points to $\frac{2}{5}$ in $\frac{2}{5}x$)

same as the product of the roots (points to $-\frac{2}{5}$)

EXAMPLE 7 Check the answer for Example 5 using the sum and product method.

Solution

$$5x^2 + 2x - (w + 4) = 0$$

$$x^2 + \frac{2}{5}x + \frac{-w-4}{5} = 0$$

($\frac{2}{5}$: opposite of sum of roots; $\frac{-w-4}{5}$: product of roots)

$$r_1 = \frac{-1 + \sqrt{5w + 21}}{5} \qquad r_2 = \frac{-1 - \sqrt{5w + 21}}{5}$$

$$r_1 + r_2 = \frac{-1 + \sqrt{5w + 21} - 1 - \sqrt{5w + 21}}{5} = -\frac{2}{5} \qquad \text{Checks}$$

$$r_1 r_2 = \left(\frac{-1 + \sqrt{5w + 21}}{5}\right)\left(\frac{-1 - \sqrt{5w + 21}}{5}\right)$$

$$= \frac{1 - (5w + 21)}{25}$$

$$= \frac{-5w - 20}{25}$$

$$= \frac{5(-w - 4)}{25}$$

$$= \frac{-w - 4}{5} \qquad \text{Checks}$$

■

PROBLEM SET 1.4

A

Solve each equation in Problems 1–12 by factoring.

1. $x^2 + 2x - 15 = 0$
2. $x^2 - 8x + 12 = 0$
3. $x^2 + 7x - 18 = 0$
4. $2x^2 + 5x - 12 = 0$
5. $10x^2 - 3x - 4 = 0$
6. $6x^2 + 7x - 10 = 0$
7. $9x^2 - 34x - 8 = 0$
8. $4x^2 + 12x + 9 = 0$
9. $9x^2 - 24x + 16 = 0$
10. $6x^2 = 5x$
11. $6x^2 = 12x$
12. $12x^2 = 48x$

Solve each equation in Problems 13–24 by completing the square.

13. $x^2 + 4x - 5 = 0$
14. $x^2 - x - 6 = 0$
15. $x^2 + 2x - 8 = 0$
16. $x^2 + x - 6 = 0$
17. $x^2 + 7x + 12 = 0$
18. $x^2 + 5x + 6 = 0$
19. $x^2 - 10x - 2 = 0$
20. $x^2 + 2x - 15 = 0$
21. $x^2 - 3x = 1$
22. $x^2 - 4x = 2$
23. $6x^2 = x + 2$
24. $x^2 = 4x + 2$

*Solve each equation in Problems 25–45 over the set of real numbers.**

25. $x^2 + 5x - 6 = 0$
26. $x^2 + 5x + 6 = 0$
27. $x^2 - 10x + 25 = 0$
28. $x^2 + 6x + 9 = 0$
29. $12x^2 + 5x - 2 = 0$
30. $2x^2 - 6x + 5 = 0$
31. $5x^2 - 4x + 1 = 0$
32. $2x^2 + x - 15 = 0$
33. $4x^2 - 5 = 0$
34. $3x^2 - 1 = 0$
35. $3x^2 = 7x$
36. $7x^2 = 3$
37. $3x^2 = 5x + 2$
38. $3x^2 - 2 = 5x$
39. $5x = 3 - 4x^2$
40. $4x^2 = 12x - 9$

* If you have covered Section 1.3, you can solve these over the set of complex numbers.

41. $3x = 1 - 2x^2$ **42.** $5x^2 = 3x - 4$

43. $\sqrt{5} - 4x^2 = 3x$ **44.** $x = \sqrt{2} - 2x^2$

45. $3x^2 - 4x = \sqrt{5}$

B

Solve the equations in Problems 46–55 for x in terms of the other variable.

46. $2x^2 + x - w = 0$ **47.** $2x^2 + wx + 5 = 0$

48. $3x^2 + 2x + (y + 2) = 0$ **49.** $3x^2 + 5x + (4 - y) = 0$

50. $4x^2 - 4x + (1 - t^2) = 0$

51. $4x^2 - (3t + 10)x + (6t + 4) = 0, t > 2$

52. $2x^2 + 3x + 4 - y = 0$ **53.** $y = 2x^2 + x + 6$

54. $(x - 3)^2 + (y - 2)^2 = 4$

55. $x^2 - 6x + y^2 - 4y + 9 = 0$

56. ***Business*** A small manufacturer of citizens' band radios determines that the price of each item is related to the number of items produced per day. The manufacturer knows that (a) the maximum number that can be produced is 10 items; (b) the price should be $400 - 25x$ dollars; (c) the overhead (the cost of producing x items) is $5x^2 + 40x + 600$ dollars; and (d) the daily profit is then found by subtracting the overhead from the revenue:

$$\begin{aligned}\text{profit} &= \text{revenue} - \text{cost}\\ &= (\text{number of items})(\text{price per item}) - \text{cost}\\ &= x(400 - 25x) - (5x^2 + 40x + 600)\\ &= 400x - 25x^2 - 5x^2 - 40x - 600\\ &= -30x^2 + 360x - 600\end{aligned}$$

What is the number of radios produced if the profit is zero?

57. ***Physics*** Suppose you throw a rock at 48 ft/sec from the top of the Sears Tower in Chicago and the height in feet, y, from the ground after x sec is given by

$$y = -16x^2 + 48x + 1454$$

a. What is the height of the Sears Tower?

b. How long will it take (to the nearest tenth of a second) for the rock to hit the ground?

58. ***Physics*** If an object is shot up from the ground with an initial velocity of 256 ft/sec, its distance in feet above the ground at the end of t sec is given by $d = 128t - 16t^2$ (neglecting air resistance). Find the length of time for which $d \geq 240$.

59. ***Physics*** Find the length of time the projectile described in Problem 58 will be in the air.

C

60. ***Engineering*** Many materials, such as brick, steel, aluminum, and concrete, expand due to increases in temperature. This is why fillers are placed between the cement slabs in sidewalks. Suppose you have a 100-ft roof truss securely fastened at both ends, and assume that the buckle is linear. (It is not, but this assumption will serve as a worthwhile approximation.) Let the height of the buckle be x ft. If the percentage of swelling is y, then, for each half of the truss,

$$\begin{aligned}\text{new length} &= \text{old length} + \text{change in length}\\ &= 50 + (\text{percentage})(\text{length})\\ &= 50 + \left(\frac{y}{100}\right)50\\ &= 50 + \frac{y}{2}\end{aligned}$$

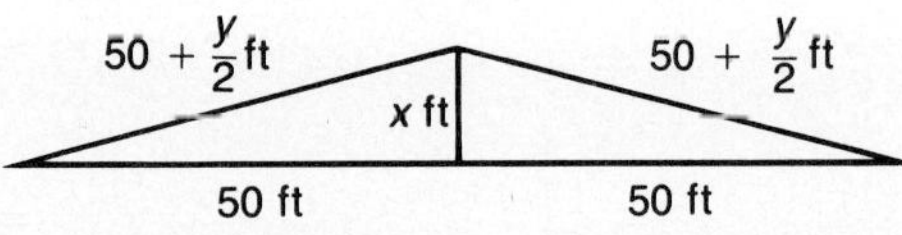

These relationships are shown in the figure. Then, by the Pythagorean theorem,

$$x^2 + 50^2 = \left(50 + \frac{y}{2}\right)^2$$

$$x^2 + 50^2 = \frac{(100 + y)^2}{4}$$

$$4x^2 + 4 \cdot 50^2 = 100^2 + 200y + y^2$$

$$4x^2 - y^2 - 200y = 0$$

Solve this equation for x and then calculate the amount of buckling (to the nearest inch) for the following materials:

a. Brick; $y = 0.03$

b. Steel; $y = 0.06$

c. Aluminum; $y = 0.12$

d. Concrete; $y = 0.05$

61. ***Space Science*** Suppose a model rocket weighs $\frac{1}{4}$ lb. Its engine propels it vertically to a height of 52 ft and a speed of 120 ft/sec at burnout. If its parachute fails to open, determine the approximate time to fall to earth according to the following equation for free fall in a vacuum:

$$h = h_0 + v_0 t - \tfrac{1}{2}gt^2$$

where h is the height (in feet) at time t, h_0 and v_0 are the height (in feet) and velocity (in feet per second) at the time selected at $t = 0$, and g is approximately 32 ft/sec². (For this problem, $h_0 = 52$ ft and $v_0 = 120$ ft/sec.)

62. If $ax^2 + bx + c = 0$, show that the following are roots:

$$r_1 = \frac{2c}{-b + \sqrt{b^2 - 4ac}} \quad \text{and} \quad r_2 = \frac{2c}{-b - \sqrt{b^2 - 4ac}}$$

1.5 QUADRATIC INEQUALITIES

A **quadratic inequality in one variable** is an inequality that can be written in the form

$$ax^2 + bx + c < 0 \qquad (a \neq 0)$$

where x is a variable and a, b, and c are any real numbers. The symbol $<$ can be replaced by $\leq$, $>$, or $\geq$ and it is still called a quadratic inequality in one variable.

The procedure for solving a quadratic inequality is similar to that for solving a quadratic equality. First, use the properties of inequality to obtain a zero on one side of the inequality. The next step is to factor, if possible, the quadratic expression on the left. For example:

$$x^2 - 4 \geq 0$$
$$(x - 2)(x + 2) \geq 0$$

Find the values of x that make the inequality valid. A value for which a factor is zero is called a **critical value** of x. The critical values for this example are 2 and -2. For *every other value of* x the inequality is either positive or negative. Next, plot the critical values on a number line as in Figure 1.4. These critical values divide the number line into three intervals. Choose a sample value from each interval. Evaluate each factor to determine the sign only—it is not necessary to complete the arithmetic to find its sign.

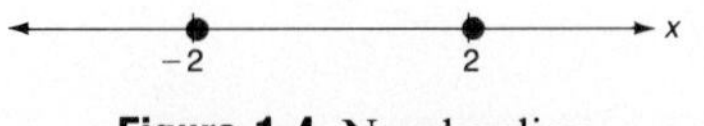

Figure 1.4 Number line with critical values

Sample value	Factor	Sign of factor
This is *your* choice ↓	This is done mentally ↓	
$x = -100$	$x - 2 = -100 - 2 \cdots\cdots\rightarrow -$ $x + 2 = -100 + 2 \cdots\cdots\rightarrow -$	positive product
$x = 0$	$x - 2 = 0 - 2 \cdots\cdots\rightarrow -$ $x + 2 = 0 + 2 \cdots\cdots\rightarrow +$	negative product
$x = 100$	$x - 2 = 100 - 2 \cdots\cdots\rightarrow +$ $x + 2 = 100 + 2 \cdots\cdots\rightarrow +$	positive product

$x-2$: − | − | +
$x+2$: − | + | +
−2 2 x
pos | neg | pos

Figure 1.5 Procedure for solving $x^2 - 4 \geq 0$

This procedure can be summarized as in Figure 1.5. You want

$$(x - 2)(x + 2) \geq 0$$

Since this is positive or zero, you pick out the parts of Figure 1.5 labeled positive and also include those that are zero (namely the critical values). If the given inequality is of the form $\leq$ or $\geq$, the endpoints are included; if it has the form $<$ or $>$, the endpoints are excluded. The solution set is $x \leq -2$ or $x \geq 2$ or, using interval notation, $(-\infty, -2] \cup [2, \infty)$.

EXAMPLE 1 Solve the inequality $2x^2 < 5 - 9x$.

Solution

$$2x^2 + 9x - 5 < 0$$
$$(2x - 1)(x + 5) < 0$$

Solution: $-5 < x < \frac{1}{2}$ or $\mathbf{(-5, \frac{1}{2})}$

$2x - 1$:	$-$	$-$	$+$
$x + 5$:	$-$	$+$	$+$
$(2x - 1)(x + 5)$:	positive	negative	positive

-5 $\quad$ $\frac{1}{2}$ $\quad$ x

■

EXAMPLE 2 Solve the inequality $x^2 + 2x - 4 < 0$.

Solution The term on the left is in simplified form and cannot be easily factored. Therefore proceed by considering $x^2 + 2x - 4$ as a single factor. To find the critical values, find the values for which the factor $x^2 + 2x - 4$ is zero.

$$x^2 + 2x - 4 = 0$$

$$x = \frac{-2 \pm \sqrt{4 - 4(1)(-4)}}{2} = \frac{-2 \pm 2\sqrt{5}}{2} = -1 \pm \sqrt{5}$$

Plot the critical values and check the sign of the factor in each interval:

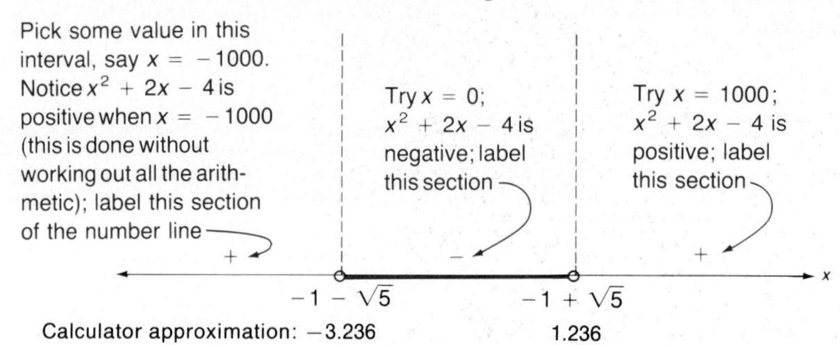

The solution is $-1 - \sqrt{5} < x < -1 + \sqrt{5}$ or, using interval notation,

$$\mathbf{(-1 - \sqrt{5}, -1 + \sqrt{5})}$$

■

The factoring procedure used in this section for solving quadratic inequalities can be used for other inequalities that are in factored form, even though they may not be quadratic.

EXAMPLE 3 Solve the inequality $(x - 5)(2 - x)(2 - 3x) > 0$.

Solution

$x - 5$:	$-$	$-$	$-$	$+$
$2 - x$:	$+$	$+$	$-$	$-$
$2 - 3x$:	$+$	$-$	$-$	$-$
	negative	positive	negative	positive

$\frac{2}{3}$ $\quad$ 2 $\quad$ 5 $\quad$ x

Solution: $\mathbf{(\frac{2}{3}, 2) \cup (5, \infty)}$

■

EXAMPLE 4 Solve

$$\frac{3x(x+1)(x-2)}{(x-1)(x+3)} \geq 0$$

Solution Plot the critical values of 0, -1, 2, 1, and -3. These are included (because it is $\geq$), but *values that cause division by zero* ($x = 1, x = -3$) need to be **excluded**.

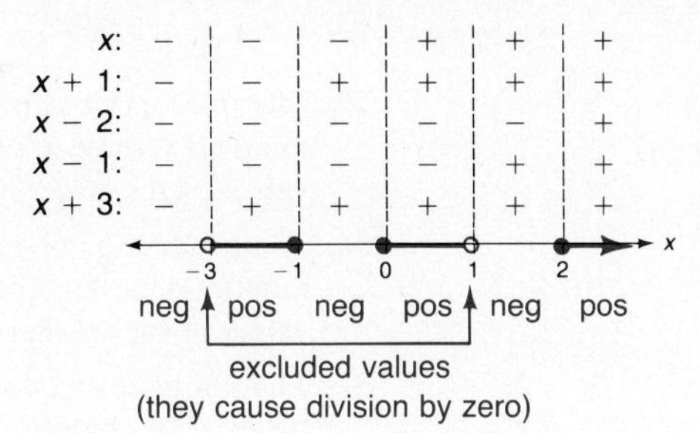

Solution: $(-3, -1] \cup [0, 1) \cup [2, \infty)$ ■

Be careful about following the procedure described above. It is tempting to attempt these solutions without testing intervals on a number line. You must also take care not to divide both sides by a variable. For example, if

$$x^2 > x$$

and you divide both sides by x to obtain $x > 1$, you will have made a very common mistake, which can be shown by a counterexample. Suppose $x = -2$. Then

$$x^2 > x$$
$$(-2)^2 > -2$$
$$4 > -2$$

which is true. But if you divide both sides by $x = -2$ you obtain

$$\frac{4}{-2} > \frac{-2}{-2}$$
$$-2 > 1$$

which is false! Using numbers you can easily see the mistake; the order of inequality was not reversed as it should have been.

PROBLEM SET 1.5

A

Solve the inequalities in Problems 1–54 and write each answer using interval notation.

1. $x(x+3) < 0$
2. $x(x-3) \geq 0$
3. $(x-6)(x-2) \geq 0$
4. $(y+2)(y-8) \leq 0$
5. $(x-8)(x+7) < 0$
6. $(x+1)(x+6) > 0$
7. $(x+2)(2x-1) \leq 0$
8. $(x-2)(2x+1) \leq 0$
9. $(3x+2)(x-3) > 0$
10. $(3x-2)(x+2) > 0$
11. $(x+2)(8-x) \leq 0$
12. $(2-x)(x+8) \geq 0$
13. $(1-3x)(x-4) < 0$
14. $(2x+1)(3-x) > 0$
15. $x(x-3)(x+4) \leq 0$
16. $x(x+3)(x-4) \geq 0$

17. $(x-2)(x+3)(x-4) \geq 0$

18. $(x+1)(x-2)(x+3)<0$

19. $(x+1)(2x+5)(7-3x)>0$

20. $(x-2)(3x+2)(3-2x)<0$

21. $\dfrac{x+2}{x}<0$ 22. $\dfrac{x}{x+3}<0$ 23. $\dfrac{x}{x-8}>0$

24. $\dfrac{x-8}{x}>0$ 25. $\dfrac{x-2}{x+5} \leq 0$ 26. $\dfrac{x+2}{4-x} \geq 0$

B

27. $\dfrac{x(2x-1)}{5-x}>0$ 28. $\dfrac{x}{(2x+3)(x-2)}<0$

29. $\dfrac{1}{x(x-3)(x+2)} \leq 0$ 30. $\dfrac{1}{x(x+1)(5-x)} \geq 0$

31. $x^2 \geq 9$ 32. $x^2 \geq 4$

33. $x^2+9 \geq 0$ 34. $x^2+2x-3<0$

35. $x^2-x-6>0$ 36. $x^2-7x+12>0$

37. $5x-6 \geq x^2$ 38. $4 \geq x^2+3x$

39. $5-4x \geq x^2$ 40. $x^2+2x-1<0$

41. $x^2-2x-2<0$ 42. $x^2-8x+13>0$

43. $2x^2+4x+5 \geq 0$ 44. $x^2-2x-6 \leq 0$

45. $x^2+3x-7 \geq 0$ 46. $\dfrac{(x-3)(x+1)}{(x-2)(x-1)} \leq 0$

47. $\dfrac{x(x+5)(x-3)}{(x+3)(x-4)} \geq 0$ 48. $\dfrac{x(x-6)(2-x)}{(x-4)(3-x)} \leq 0$

C

49. $\dfrac{2}{x-2} \leq \dfrac{3}{x+3}$ 50. $\dfrac{1}{x+1} \geq \dfrac{2}{x-1}$

51. $\dfrac{x-3}{3x-1} \geq \dfrac{x+3}{2x+1}$ 52. $\dfrac{x-7}{x^2-4x-21}<0$

53. $\dfrac{x-4}{x^2-5x+6} \leq 0$ 54. $\dfrac{x^2+4x+3}{x+2} \geq 0$

55. The product of two numbers is at least 340. One of the numbers is three less than the other. What are the possible values of the larger number?

56. The product of two numbers is no larger than 300. One number is five larger than the other. What are the possible values of the smaller number?

57. The quotient of two numbers is positive. The divisor is three larger than the dividend. What are the possibilities for the smaller number?

58. Two numbers have a negative quotient. What are the possibilities for the dividend if it is five larger than the divisor?

59. A rectangular area is to be fenced. If the space is twice as long as it is wide, for what dimensions is the area numerically greater than the perimeter?

60. A rectangular area three times as long as it is wide is to be fenced. For what dimensions is the perimeter numerically greater than the area?

1.6 TWO-DIMENSIONAL COORDINATE SYSTEM AND GRAPHS

There is a one-to-one correspondence between the real numbers and points on a real number line, called a one-dimensional coordinate system. A **two-dimensional coordinate system** can be introduced by considering two perpendicular coordinate lines in a plane. Usually one of the coordinate lines is horizontal with the positive direction to the right; the other is vertical with the positive direction upward. These coordinate lines are called **coordinate axes** and the point of intersection is called the **origin**. Notice from Figure 1.6 that the axes divide the plane into four parts called the **first, second, third,** and **fourth quadrants**. This two-dimensional coordinate system is also called a **Cartesian coordinate system** in honor of René Descartes, who was the first to describe such a coordinate system in mathematical detail.

Points of a plane are denoted by ordered pairs. The term **ordered pair** refers to two real numbers represented by (a, b), where a is called the **first component** and b is the **second component**. The order in which the components are listed is important, since $(a, b) \neq (b, a)$ if $a \neq b$.

Figure 1.6 Cartesian coordinate system

Equality of Ordered Pairs

$(a, b) = (c, d)$ if and only if $a = c$ and $b = d$

The notation (a, b) was used to denote an interval on a number line in the last section. You should know from the discussion whether a one-dimensional or a two-dimensional coordinate system is involved.

The horizontal number line is called the **x axis** (sometimes called the *axis of abscissas*), and x represents the first component of the ordered pair. The vertical number line is called the **y axis** (sometimes called the *axis of ordinates*), and y represents the second component of the ordered pair. The plane determined by the x and y axes is called a **coordinate plane**, **Cartesian plane**, or **xy plane**. When we refer to a point (x, y), we are referring to a point in the coordinate plane whose abscissa is x and whose ordinate is y. To **plot a point** (x, y) means to locate the point with coordinates (x, y) in the plane and represent its location by a dot.

Finding the distance between points in two dimensions requires the Pythagorean Theorem.

Pythagorean Theorem

A triangle with sides a, b, and c is a right triangle if and only if

$$a^2 + b^2 = c^2$$

Let $P_1(x_1, y_1)$ and $P_2(x_2, y_2)$ be any two distinct points in a plane. If $x_1 = x_2$, then P_1P_2 is a *vertical line segment*; if $y_1 = y_2$, then P_1P_2 is a *horizontal line segment* as shown in Figure 1.7.

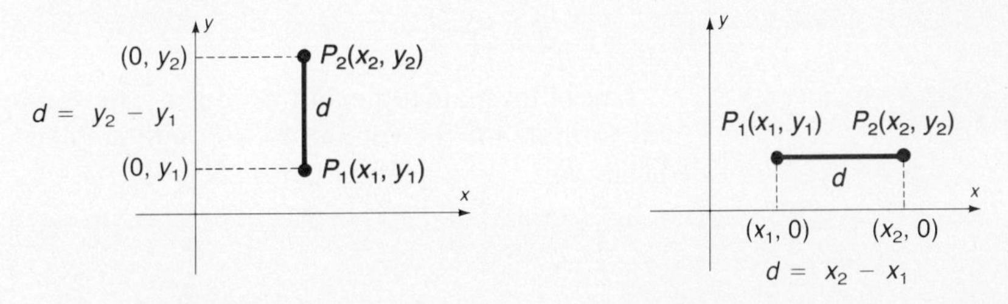

Figure 1.7 Vertical and horizontal line segments

With these special cases, it is easy to find the distance d between P_1 and P_2 because these distances correspond directly to distances on a one-dimensional coordinate system as discussed in the last section. Study Figure 1.7 and see Problems 68 and 69 of Problem Set 1.6 for the details.

Since these special cases are considered in the problem set, we will focus our attention on the general case in which P_1 and P_2 do not lie on the same horizontal or

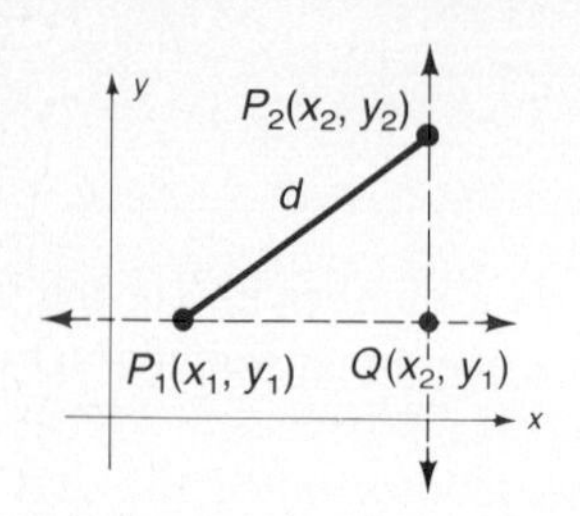

Figure 1.8 Distance between points

vertical line. Draw a line through P_1 parallel to the x axis and another through P_2 parallel to the y axis. These lines intersect at a point Q with coordinates (x_2, y_1) as shown in Figure 1.8. The distance P_1Q is $|x_2 - x_1|$; the distance QP_2 is $|y_2 - y_1|$. By the Pythagorean Theorem,

$$d^2 = |x_2 - x_1|^2 + |y_2 - y_1|^2$$

Thus, since d is nonnegative and $|a|^2 = a^2$ for every real number a, the distance from P_1 to P_2 is given by the following formula.

Distance Formula

If $P_1(x_1, y_1)$ and $P_2(x_2, y_2)$ are any two points, then the distance d from P_1 to P_2 is

$$d = \sqrt{(x_2 - x_1)^2 + (y_2 - y_1)^2}$$

EXAMPLE 1 Find the distance between $(-3, 2)$ and $(-1, -6)$.

Solution

$$\begin{aligned} d &= \sqrt{[(-1) - (-3)]^2 + (-6 - 2)^2} \\ &= \sqrt{4 + 64} \\ &= 2\sqrt{17} \end{aligned}$$

■

To find the **midpoint** M of a segment $P_1(x_1, y_1)$ and $P_2(x_2, y_2)$, you simply average the coordinates of the two endpoints.

Midpoint Formula

The midpoint M between point $P_1(x_1, y_1)$ and $P_2(x_2, y_2)$ is

$$M = \left(\frac{x_1 + x_2}{2}, \frac{y_1 + y_2}{2}\right)$$

EXAMPLE 2 Find the midpoint of the segment connecting $(-3, 2)$ and $(-1, -6)$.

Solution

$$\begin{aligned} M &= \left(\frac{(-3) + (-1)}{2}, \frac{(2) + (-6)}{2}\right) \\ &= (-2, -2) \end{aligned}$$

■

One of the main topics of this course is the graphing of certain relations. Consider the formula for the volume V of a right circular cone whose radius is one-half its height h:

$$V = \frac{\pi h^3}{12}$$

A table of values showing the volumes for different heights is given here:

Height	1	2	3	4	5	6
Volume	$\frac{\pi}{12}$	$\frac{2\pi}{3}$	$\frac{9\pi}{4}$	$\frac{16\pi}{3}$	$\frac{125\pi}{12}$	18π

Do you see how these values were obtained? If $h = 1$, then

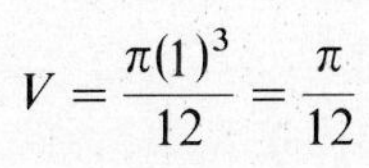

$$V = \frac{\pi(1)^3}{12} = \frac{\pi}{12}$$

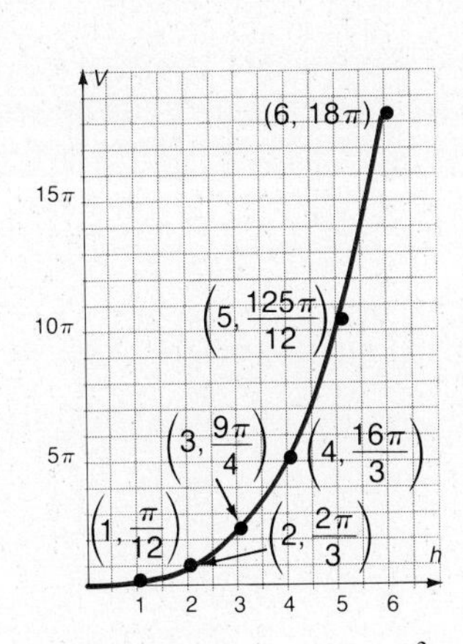

Figure 1.9 Graph of $V = \dfrac{\pi h^3}{12}$ for $0 \le h \le 6$

If $h = 2$, then

$$V = \frac{\pi(2)^3}{12} = \frac{8\pi}{12} = \frac{2\pi}{3}$$

And so on. This table can be written as a set of ordered pairs: $(1, \pi/12)$, $(2, 2\pi/3)$, $(3, 9\pi/4), \ldots$. A set of ordered pairs is called a **relation**. In this example, the ordered pairs are (h, V) so that $V = \pi h^3/12$. If an ordered pair has components that, when substituted for their corresponding variables, yield a true equation, we say that the ordered pair **satisfies** the equation. For example, $(h, V) = (1, \pi/12)$ satisfies the equation $V = \pi h^3/12$. By plotting the points whose coordinates are shown in the table and then connecting them with a smooth curve, you arrive at the **graph** of this relation, a portion of which is shown in Figure 1.9.

In general, when we speak of a **graph of a relation** we mean there is a one-to-one correspondence between the set of all ordered pairs (x, y) in the relation and the set of all points with coordinates (x, y) that lie on the curve.

EXAMPLE 3 Graph $2x + 3y + 6 = 0$ by plotting points.

Solution One of the most efficient methods for graphing a curve by plotting points is to solve the equation for y and then substitute values for x, arranging the ordered pairs in table form.

$$2x + 3y + 6 = 0$$

$$3y = -2x - 6$$

$$y = -\tfrac{2}{3}x - 2$$

x	y	
0	-2	You choose the x values.
3	-4	Do you see why we chose 3 and not 1 or 2?
-3	0	
-6	2	This table entry represents the point $(-6, 2)$.

Plot the points $(0, -2)$, $(3, -4), \ldots$ and draw a smooth curve connecting them:

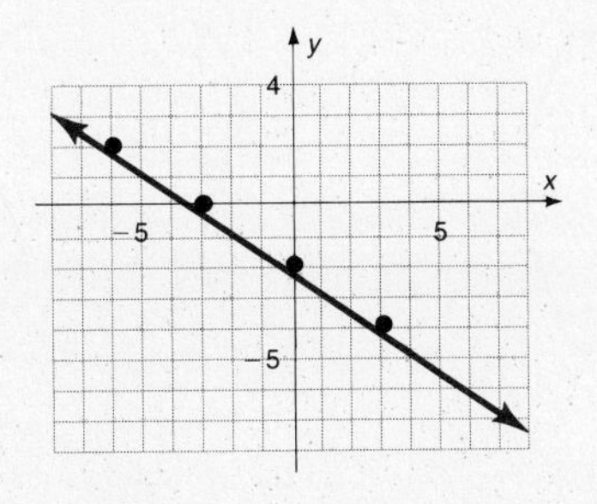

■

EXAMPLE 4 Graph $y = x^2$ by plotting points.

Solution

x	y
0	0
1	1
2	4
-1	1
-2	4

■

EXAMPLE 5 Graph $y = |x|$ by plotting points.

Solution

x	y
0	0
1	1
-1	1
2	2
-2	2

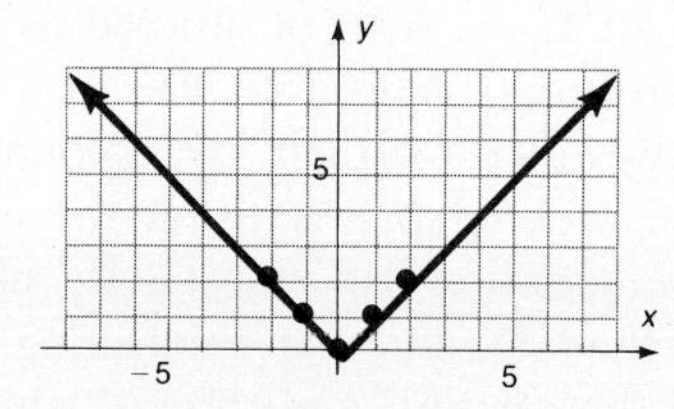

■

The method of drawing the graph of a relation as shown here—that of plotting points—is very primitive, and one of the primary purposes of this book is to develop more efficient methods of graphing curves than simply plotting points.

The first property of a graph we will consider is called **symmetry**. The *idea* of symmetry is the *idea* of mirror images. A graph or curve is **symmetric with respect to a line**, for example, if the graph is the same on both sides of that line as shown in Figures 1.10 and 1.11.

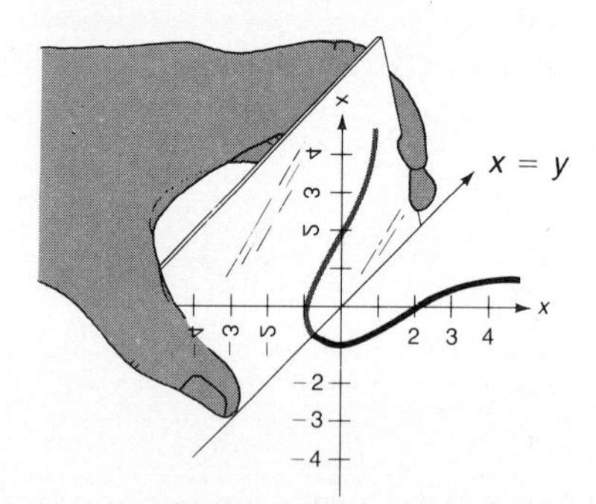

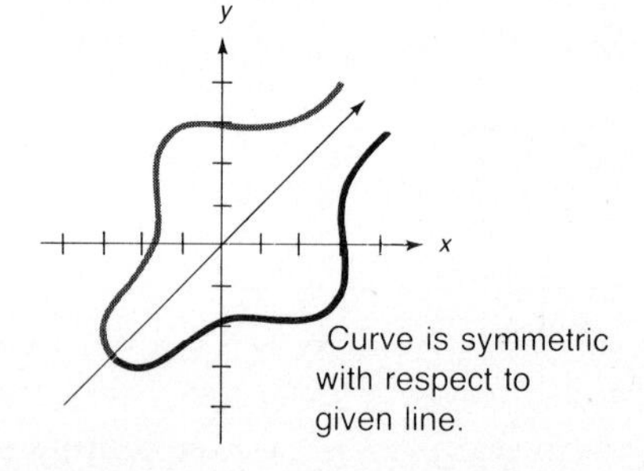

Figure 1.10 Symmetry with respect to a given line

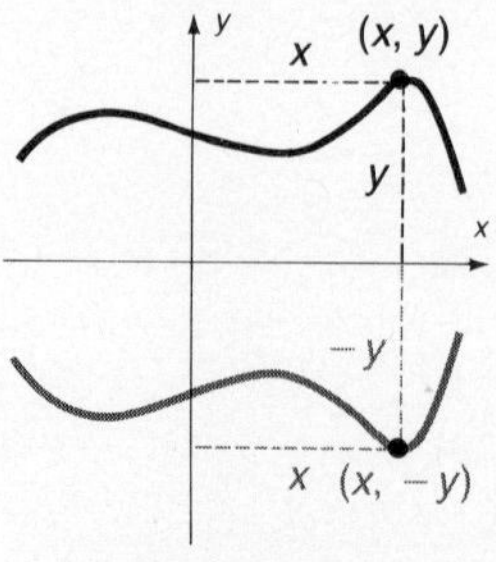

a. Symmetry with respect to the x axis

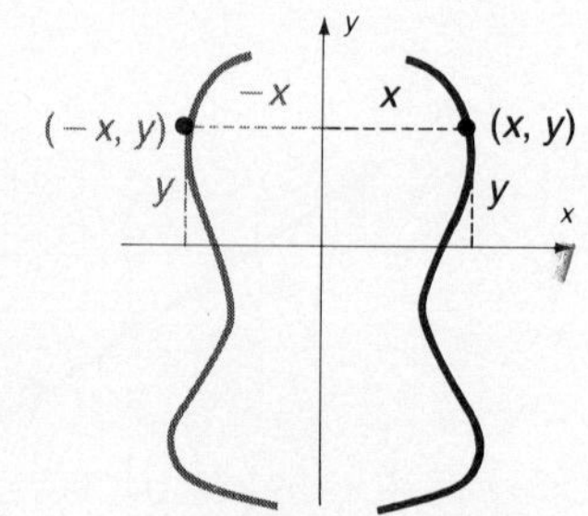

b. Symmetry with respect to the y axis

Figure 1.11 Symmetry with respect to the x axis and the y axis

Symmetry with Respect to a Coordinate Axis

The graph of a relation is **symmetric with respect to the x axis** if substitution of $-y$ for y does not change the set of coordinates satisfying the relation. The graph of a relation is **symmetric with respect to the y axis** if substitution of $-x$ for x does not change the set of coordinates satisfying the relation.

EXAMPLE 6 Draw the reflection of the given curve as it would appear in the mirror.

Solution The answer is shown as a dashed curve. Your paper would look like the graph on the right.

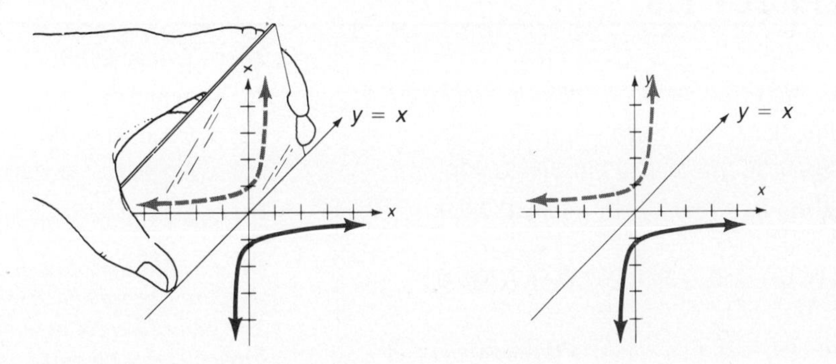

■

EXAMPLE 7 Draw the given curve so that it is: (**a**) symmetric with respect to the x axis, and (**b**) symmetric with respect to the y axis.

Solution

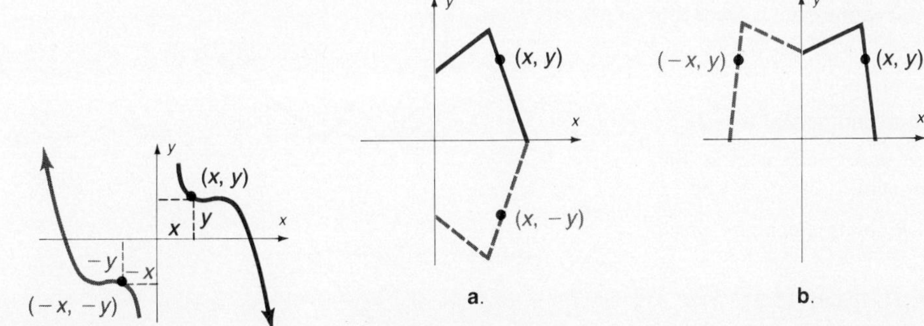

■

Figure 1.12 Symmetry with respect to the origin

Some graphs possess another type of symmetry called symmetry with respect to the origin (Figure 1.12).

Symmetry with Respect to the Origin

The graph of a relation is **symmetric with respect to the origin** if the simultaneous substitution of $-x$ for x and $-y$ for y does not change the set of coordinates satisfying the relation.

EXAMPLE 8 Draw the given curve so that it is symmetric with respect to the origin.

Solution

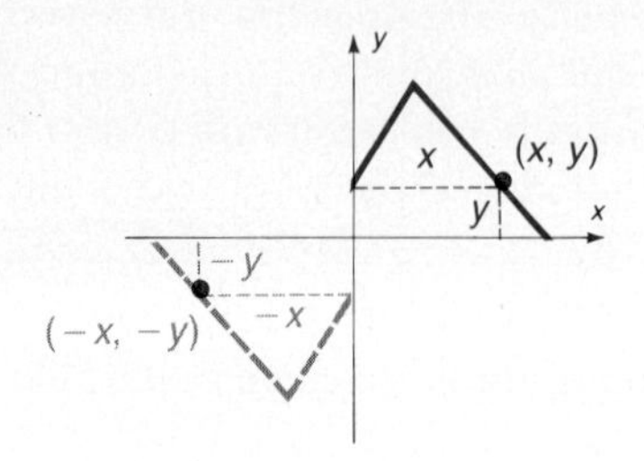

■

PROBLEM SET 1.6

A

Plot the points whose coordinates are given in Problems 1–6.

1. $A(3,4)$; $B(6,2)$; $C(-3,-5)$; $D(-4,6)$; $E(-3,0)$
2. $F(4,-3)$; $G(7,2)$; $H(0,5)$; $I(-3,0)$; $J(-5,-2)$
3. $A(1,175)$; $B(-3,-125)$; $C(-2,-150)$; $D(2,25)$; $E(0,100)$
4. $F(250,-1)$; $G(-350,-2)$; $H(-150,3)$; $I(300,4)$; $J(-50,-5)$
5. $A(\frac{\pi}{6},1)$; $B(\frac{-2\pi}{3},-2)$; $C(-\pi,-1)$; $D(\frac{\pi}{4},3)$; $E(-\frac{\pi}{4},-2)$
6. $F(-\frac{\pi}{3},\frac{1}{2})$; $G(\frac{\pi}{4},\frac{3}{4})$; $H(\frac{3\pi}{2},-\frac{1}{2})$; $I(\pi,1)$; $J(-\pi,-1)$
7. Plot 5 points (x,y) whose second component is the opposite of the first component.
8. Plot five points (x,y) whose first and second components are equal.
9. Plot five points (x,y) whose second component is twice the first component.
10. Plot five points (x,y) whose first component is 3.
11. Plot five points (x,y) whose second component is -2.
12. Plot five points (x,y) whose second component is the opposite of twice the first component.

Find the distance between the points whose coordinates are given in Problems 13–21.

13. $(5,1)$ and $(8,5)$
14. $(1,4)$ and $(13,9)$
15. $(-2,4)$ and $(0,0)$
16. $(0,0)$ and $(5,-2)$
17. $(4,5)$ and $(3,-1)$
18. $(-2,1)$ and $(-1,-5)$
19. $(7x,5x)$ and $(3x,2x)$, $x<0$
20. $(x,5x)$ and $(-3x,2x)$, $x<0$
21. $(x,5x)$ and $(-3x,2x)$, $x>0$

Find the midpoint of the segment connecting the points whose coordinates are given in Problems 22–30.

22. $(5,1)$ and $(8,5)$
23. $(1,4)$ and $(13,9)$
24. $(-2,4)$ and $(0,0)$
25. $(0,0)$ and $(5,-2)$
26. $(4,5)$ and $(3,-1)$
27. $(-2,1)$ and $(-1,-5)$
28. $(x,5x)$ and $(3x,2x)$, $x<0$
29. $(x,5x)$ and $(-3x,2x)$, $x<0$
30. $(x,5x)$ and $(-3x,2x)$, $x>0$

Draw a reflection of each curve given in Problems 31–34. To do this, draw coordinate axes on your paper. Next draw the line $x=y$ and the curve as shown. Finally imagine a mirror as shown here and draw the curve as it would look in this mirror.

31.

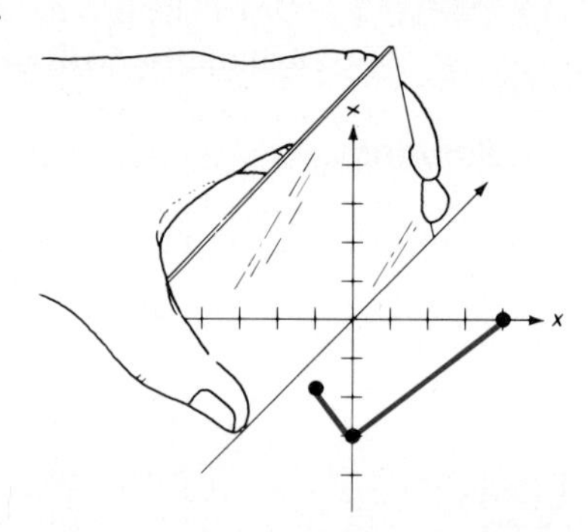

32.

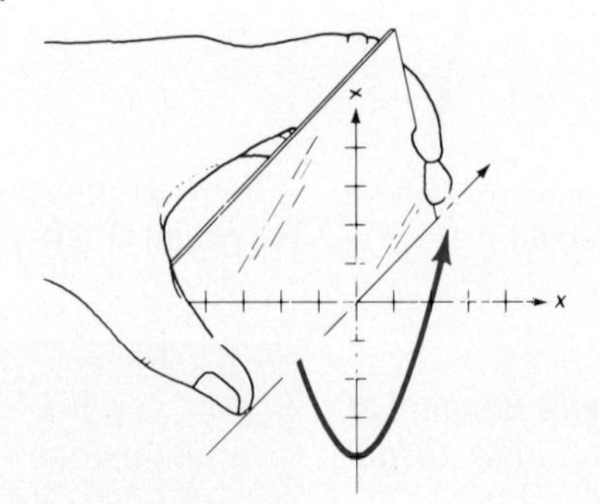

33.

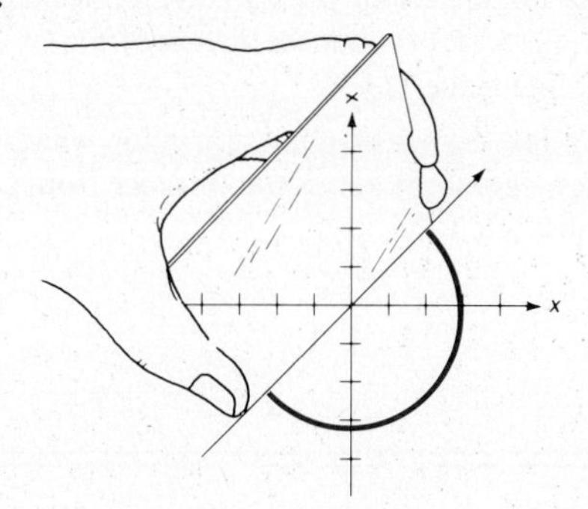

34.

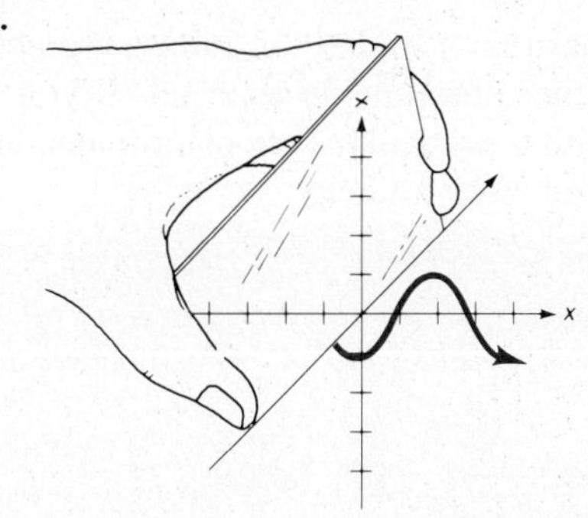

49.

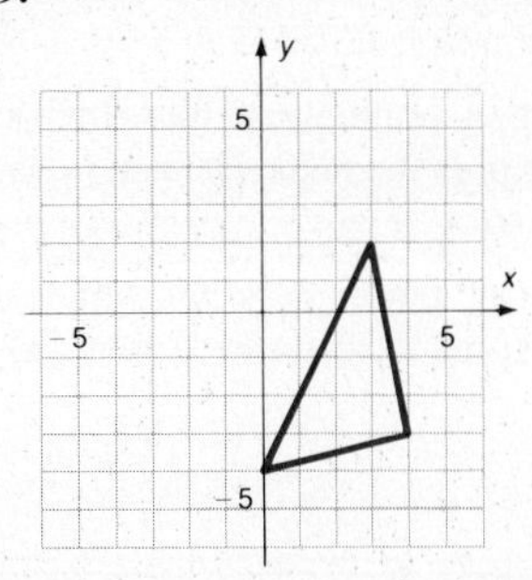

50.

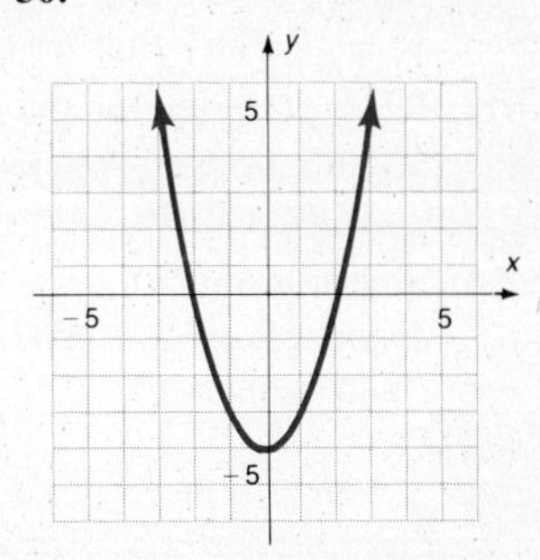

51.

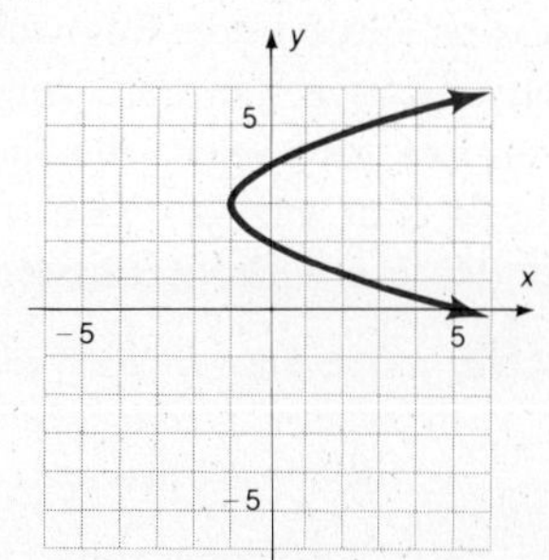

52.

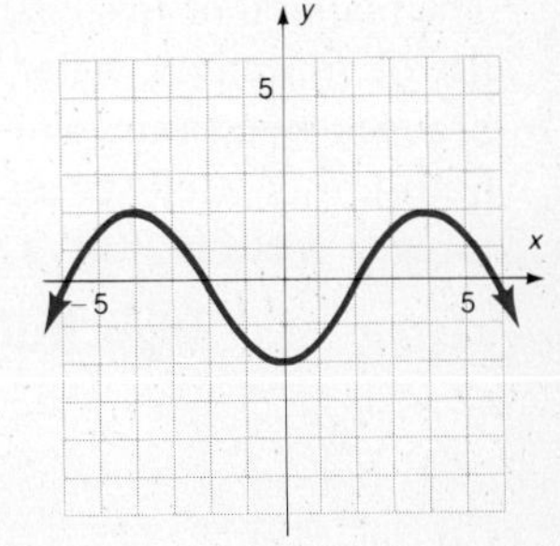

Graph each relation in Problems 35–46 by plotting points.

35. $x + y + 3 = 0$
36. $3x - y + 1 = 0$
37. $3x + 2y + 6 = 0$
38. $4x - 5y + 2 = 0$
39. $y = -x^2$
40. $2x^2 + y = 0$
41. $x^2 + 3y = 0$
42. $x^2 - 2y = 0$
43. $y = -|x|$
44. $y = |x + 2|$
45. $y = 2|x|$
46. $y = |x| + 2$

B

In Problems 47–52, draw a curve so that it is symmetric to the given curve with respect to: (a) the x axis; (b) the y axis; and (c) the origin.

47.

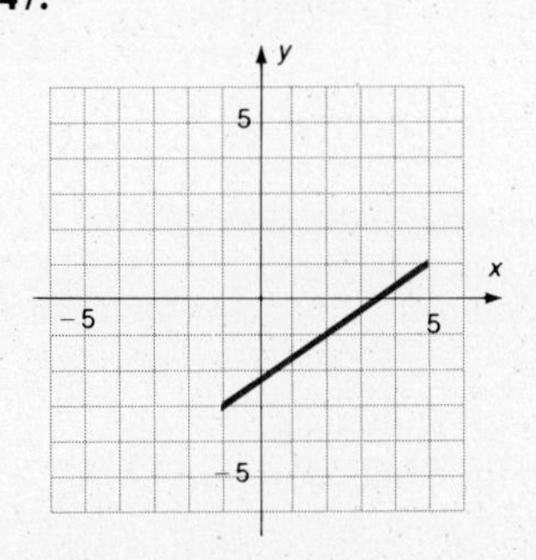

48.

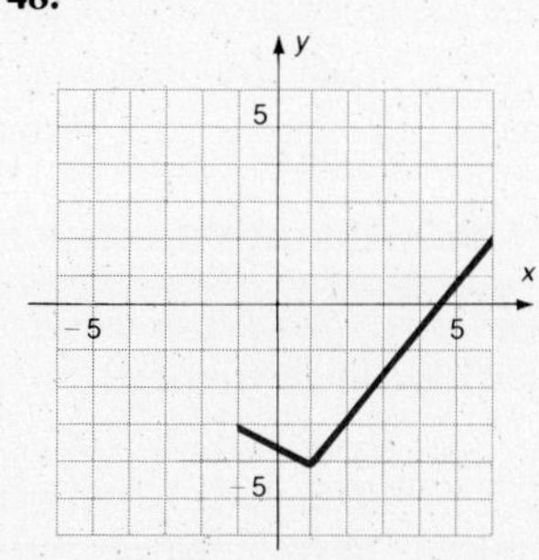

The sides of a triangle satisfy $a^2 + b^2 = c^2$ if and only if it is a right triangle. Which of the triangles whose vertices are given in Problems 53–58 are right triangles?

53. $(1, 3), (7, 1), (7, 10)$
54. $(0, 0), (6, 4), (2, 10)$
55. $(0, 0), (4, 3), (-3, 8)$
56. $(-6, 0), (-5, -6), (0, -5)$
57. $(3, 2), (7, -4), (4, -6)$
58. $(8, -1), (14, 2), (6, 3)$
59. Find a formula for the set of points (x, y) for which the distance from (x, y) to $(2, 3)$ is 7.
60. Find a formula for the set of points (x, y) for which the distance from (x, y) to $(-3, -5)$ is 5.
61. Find a formula for the set of points (x, y) for which the distance from (x, y) to $(-4, 1)$ is 3.
62. Find all points on the x axis that are 8 units from the point $(2, 4)$.
63. Find all points on the y axis that are 8 units from the point $(2, 4)$.
64. Find all points on the y axis that are 5 units from the point $(-3, -1)$.
65. Draw the graph of the relation $A = \pi r^2$.
66. Draw the graph of the relation $V = \frac{2}{3}\pi R^2$.
67. Draw the graph of the relation $h = \frac{a}{2}\sqrt{3}$.
68. Let $P_1(x_1, y_1)$ and $P_2(x_2, y_2)$ be two points such that $y_1 = y_2$. Show that the distance from P_1 to P_2 is $|x_2 - x_1|$.

69. Let $P_1(x_1, y_1)$ and $P_2(x_2, y_2)$ be two points such that $x_1 = x_2$. Show that the distance from P_1 to P_2 is $|y_2 - y_1|$.

70. Find a formula for the set of points (x, y) for which the distance from (x, y) to $(3, 0)$ plus the distance from (x, y) to $(-3, 0)$ equals 10.

71. Find a formula for the set of points (x, y) for which the distance from (x, y) to $(4, 0)$ plus the distance from (x, y) to $(-4, 0)$ equals 10.

72. Find a formula for the set of points (x, y) for which the distance from (x, y) to $(5, 0)$ minus the distance from (x, y) to $(-5, 0)$, in absolute value, is 6.

73. Find a formula for the set of points (x, y) for which the distance from (x, y) to $(0, 5)$ minus the distance from (x, y) to $(0, -5)$, in absolute value, is 8.

CHAPTER 1 SUMMARY

The material of this chapter is reviewed in the following list of objectives. After each objective there are some practice questions. For a sample test, select the first question of each set and check your answers with the answer section. For a sample test without answers, use the second question of each set. Additional practice is given by the other questions in each set. If you are having trouble with a particular type of problem, look back to that section for extra help.

1.1 REAL NUMBERS

Objective 1 Be familiar with the counting numbers, whole numbers, integers, rational numbers, irrational numbers and real numbers. Classify each of the following numbers using the above listed sets.

1. $\frac{14}{7}, \sqrt{144}, 6.\overline{2}, \pi$
2. $\sqrt{2.25}, \sqrt{30}, 6.545454\ldots, \frac{\pi}{6}$
3. $3.\overline{1}, \frac{5\pi}{6}, \frac{22}{7}, \sqrt{10}$
4. $3.1416, 4.513, \sqrt{1.69}, \sqrt{12}$

Objective 2 Graph numbers on a number line. Graph each of the following sets of numbers on the same real number line.

5. $\frac{14}{7}, \sqrt{144}, 6.\overline{2}, \pi$
6. $\sqrt{2.25}, \sqrt{30}, 6.545454\ldots, \frac{\pi}{6}$
7. $3.\overline{1}, \frac{5\pi}{6}, \frac{22}{7}, \sqrt{10}$
8. $3.1416, 4.513, \sqrt{1.69}, \sqrt{12}$

Objective 3 Use <, >, and = relationships. Illustrate the property of comparison by replacing □ by =, <, or >.

9. $\frac{5}{8}$ □ 0.625
10. $\frac{5}{9}$ □ 0.555
11. $\frac{5}{7}$ □ $\frac{8}{11}$
12. $\sqrt{2}$ □ 1.414

Objective 4 Know the reflexive, symmetric, transitive, and substitution properties. Complete the given statement so that the requested property of equality is demonstrated.

13. $a(b + c) =$ ______________ (reflexive property)
14. If $a(b + c) = ab + ac$, then ______________ (symmetric property)
15. If $a(b + c) = ab + ac$ and $ab + ac = 5$, then ______________ (transitive property)
16. If $a(b + c) = 5$ and $a = 3$, then ______________ (substitution property)

Objective 5 Be familiar with the real number properties. Complete the given statement so that the requested field property is demonstrated.

17. $a(b + c) =$ ______________ (commutative property for multiplication)
18. $a(b + c) =$ ______________ (commutative property for addition)
19. $a(b + c) =$ ______________ (distributive property)
20. $a(b + c) =$ ______________ (identity property for multiplication)

1.2 INTERVALS, INEQUALITIES, AND ABSOLUTE VALUES

Objective 6 Write interval notation and graph inequalities.

21. $-4 < x < 2$ **22.** $-12 \leq x < -8$ **23.** $x > -3$ **24.** $4 \geq x$

Objective 7 Graph inequalities given interval notation.

25. $[-5, -2]$ **26.** $[-3, 0)$ **27.** $[-1, \infty)$ **28.** $(-\infty, 4) \cup (4, \infty)$

Objective 8 Write interval notation using inequality notation. Write each of the intervals as an inequality. Use x as the variable.

29. $[-8, -5)$ **30.** $[-2, 0)$ **31.** $(-\infty, 3)$ **32.** $[3, \infty)$

Objective 9 Solve linear inequalities. Leave your answer in interval notation.

33. $3x + 2 \leq 14$ **34.** $-9 \leq -x$
35. $5 \leq 1 - x < 9$ **36.** $-0.001 \leq x + 2 \leq 0.001$

Objective 10 Know the definition of absolute value. Write expressions without using absolute value notation.

37. $|-\sqrt{11}|$ **38.** $|4 - \sqrt{11}|$ **39.** $|3 - \sqrt{11}|$ **40.** $|2\pi - 9|$

Objective 11 Find the distance between points on a number line.

41. (3) and (-5) **42.** (-5) and (-1) **43.** $(-\pi)$ and (2) **44.** (4) and $(-\sqrt{5})$

Objective 12 Solve absolute value equations.

45. $|x| = 8$ **46.** $|x| = -5$ **47.** $|2x + 3| = 8$ **48.** $|2 - 3x| = 11$

Objective 13 Solve absolute value inequalities.

49. $|x - 4| < 5$ **50.** $|3x + 1| \leq 5$ **51.** $|5 - 2x| \leq 25$ **52.** $|x - 5| \geq 3$

*1.3 COMPLEX NUMBERS

Objective 14 Define a complex number. Simplify expressions involving complex numbers.

53. $-i^7$ **54.** $(2 - 3i) - (5 + 6i)$ **55.** $(2 + 5i)(2 - 5i)$ **56.** $\dfrac{1 - 8i}{3 + 2i}$

1.4 QUADRATIC EQUATIONS

Objective 15 Solve quadratic equations by factoring.

57. $x^2 - x - 12 = 0$ **58.** $x^2 - 10x + 24 = 0$ **59.** $(3 - x)(5 + 2x) = 0$
60. $x^2 - 100 = 0$

Objective 16 Solve quadratic equations by completing the square.

61. $x^2 - 2x - 15 = 0$ **62.** $x^2 + 6x + 8 = 0$ **63.** $x^2 + 9x + 20 = 0$
64. $x^2 - 3x + 1 = 0$

*Objective 17 Know the quadratic formula. Solve quadratic equations over the set of real numbers.***

65. $x^2 - 5x + 3 = 0$ **66.** $2x^2 - 5x - 3 = 0$ **67.** $x^2 + 2x - 5 = 0$
68. $3x^2 + 2x + 1 = 0$

1.5 QUADRATIC INEQUALITIES

Objective 18 Solve quadratic inequalities. Write your answer using interval notation.

69. $3x^2 - 2x - 1 < 0$ **70.** $3 + 5x \geq 2x^2$ **71.** $x^2 + 2x + 1 \geq 0$ **72.** $x^2 - x - 1 \leq 0$

Objective 19 Solve inequalities in factored form. Write your answer using interval notation.

73. $x(3 - x)(x + 1) < 0$ **74.** $\dfrac{x + 5}{x - 9} \leq 0$

---

* Optional section.

** Find the solution over the set of complex numbers if you covered Section 1.3.

75. $x(x-2)^2(x+1) > 0$

76. $\dfrac{(2x-1)(x-2)}{(x+1)(x+2)} \geq 0$

1.6 TWO-DIMENSIONAL COORDINATE SYSTEMS AND GRAPHS

Objective 20 *Plot points on a Cartesian coordinate system.*

77. $\left(\dfrac{\pi}{2}, 0\right)$ **78.** $\left(\dfrac{2\pi}{3}, -\dfrac{1}{2}\right)$ **79.** $\left(\dfrac{\pi}{4}, \dfrac{\sqrt{2}}{2}\right)$ **80.** $\left(\dfrac{5\pi}{6}, \dfrac{-\sqrt{3}}{2}\right)$

Objective 21 *Know the Pythagorean Theorem and the distance formula. Find the distance between points in a plane.* Find the distance between each pair of points.

81. (α, β) and (γ, δ)
82. (x, x) and $(5x, 4x)$, where $x > 0$
83. (x, x) and $(5x, 4x)$, where $x < 0$
84. $(-3, -2)$ and $(1, -4)$

Objective 22 *Find the midpoint of a segment.* Find the midpoint of the segment connecting the given points.

85. (α, β) and (γ, δ)
86. (x, x) and $(5x, 4x)$, where $x > 0$
87. (x, x) and $(5x, 4x)$, where $x < 0$
88. $(-3, -2)$ and $(1, -4)$

Objective 23 *Know the definition of a relation and the terminology of graphing in two dimensions.* Fill in the blanks.

89. A relation is ______________________________.
90. By a graph of a relation we mean ______________________________.
91. If an ordered pair has components that, when substituted for their corresponding variables, yield a true equation, we say that the ordered pair ______________ the equation.
92. Draw a Cartesian coordinate system, label the axes, origin, and quadrants by number.

Objective 24 *Graph relations specified by an equation by plotting points.*

93. $2x - y + 3 = 0$ **94.** $y = -\frac{2}{3}x$ **95.** $y = -\frac{2}{3}x^2$ **96.** $y = -\frac{2}{3}|x|$

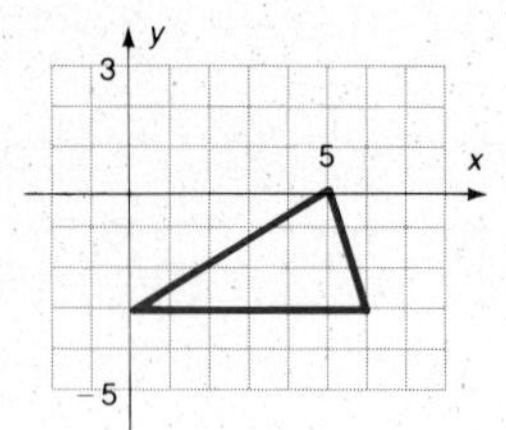

Objective 25 *Draw curves showing symmetry with respect to a line; the x axis; the y axis; and the origin.* Given the curve below, draw a curve so that it has the indicated symmetry.

97. x axis **98.** y axis **99.** origin **100.** the line $x = y$

2

Functions

Leonhard Euler (1707–1783)

Nature herself exhibits to us measurable and observable quantities in definite mathematical dependence; the conception of a function is suggested by all the processes of nature where we observe natural phenomena varying according to distance or to time. Nearly all the "known" functions have presented themselves in the attempt to solve geometrical, mechanical, or physical problems.

J. T. MERTZ
History of European Thought in the Nineteenth Century

Euler calculated without any apparent effort, just as men breathe and as eagles sustain themselves in the air.

F. Arago, quoted *in In Mathematical Circles* by Howard Eves

The word *function* was used as early as 1694 by the universal genius of the seventeenth century, Gottfried von Leibniz (1646–1716), to denote any quantity connected with a curve. The notion was generalized and modified by Johann Bernoulli (1667–1748) and by Leonhard Euler, the most prolific mathematical writer in history. Euler's father was a pastor whose avocation was mathematics, and he had a profound influence on his son, Leonhard, not only in the boy's early religious training, but also in his direction toward mathematics. Throughout his life Euler was both a frequent contributor to research journals and a superb textbook writer who was widely known for his clarity, detail, and completeness. His early religious training gave him a simple, unquestioning faith that enabled him to accept his blindness with courage. Euler's work with functions was later expanded by the mathematician P. G. Lejeune-Dirichlet (1805–1859). Around 1815, functions were being considered that were not "nice" and it was Dirichlet who, in 1837, suggested a very broad definition of a function, the one, in fact, that leads to the definition used in this chapter. Dirichlet proposed a very "badly behaved" function: when x is rational, let $y = c$, and when x is irrational, let $y = d \neq c$. This function, often known as Dirichlet's function, is so pathological that it is often used to test hypotheses about functions.

CHAPTER OVERVIEW The central idea for this course is the notion of a function, and you are introduced to this concept in this chapter. Sections 2.3 and 2.5 give you specific examples of functions; the other sections present essential properties of functions in general which will be used throughout the course. This chapter has 24 objectives which are listed on pages 80–82.

2.1 FUNCTIONS

The material presented in this book is designed to prepare you for the study of calculus. As the title suggests, the main thread that will lead you through this book is the concept of a *function*. You have, no doubt, been introduced to this idea before, probably in algebra. The notion of a correspondence between sets is a common idea. The price of a stock, for example, can be determined by looking at the daily quote in the newspaper; the height of a bridge can be determined by dropping a rock and measuring the time it takes to hit the bottom; and the surface area of a balloon can be determined if you know its radius. All of these are everyday examples of functions.

In the last chapter a relation was defined as a set of ordered pairs (x, y). For a function, consider a correspondence, or a mapping, between two sets X and Y so that x is a member of X and y is a member of Y.

Function as a Mapping

A **function** f is a mapping that assigns to each element x of X a unique element y of Y. The element y is called the **image** of x under f and is denoted by $f(x)$. The set X is called the **domain** of the function. The set of all images of elements of X is called the **range** of the function.

If a function f maps a set X *into* a set Y, it can be illustrated as shown in Figure 2.1.

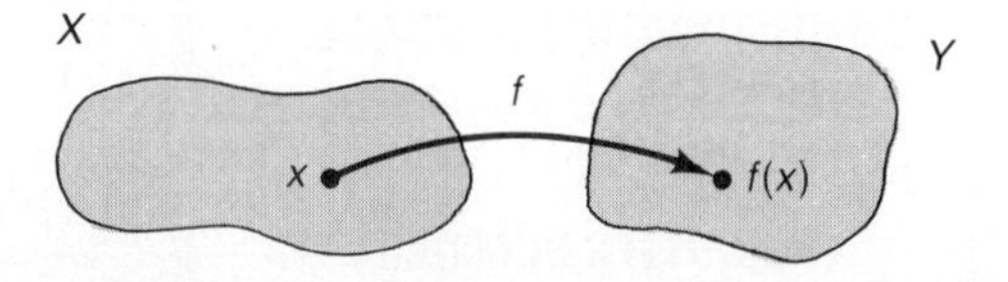

Figure 2.1 A function as a mapping

The sets X and Y are represented by points within regions in a plane. We have shown them as different sets, but they may have elements in common or may even be equal. Notice that the range of f is a subset of Y and does not have to include all of Y. If the range and Y are equal, however, then X maps *onto* Y. The symbol $f(x)$ is read "f of x" and does not mean multiplication; it represents that unique element of Y which corresponds to the element x in X.

EXAMPLE 1

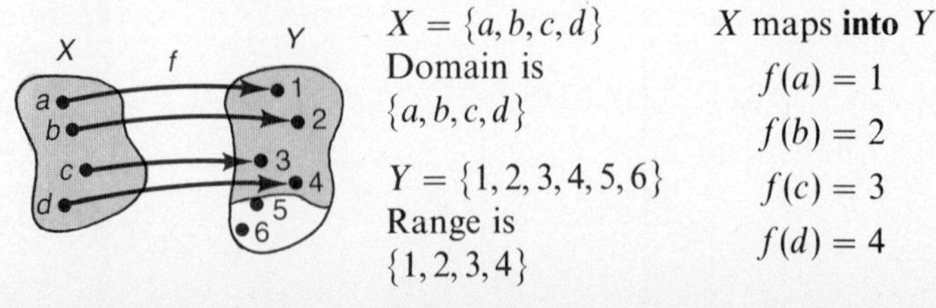

$X = \{a, b, c, d\}$
Domain is $\{a, b, c, d\}$

$Y = \{1, 2, 3, 4, 5, 6\}$
Range is $\{1, 2, 3, 4\}$

X maps **into** Y

$f(a) = 1$
$f(b) = 2$
$f(c) = 3$
$f(d) = 4$

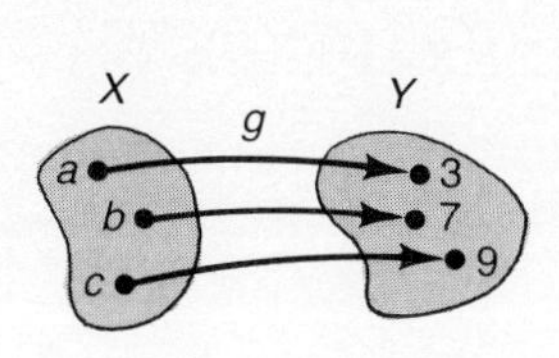

$X = \{a, b, c\}$ $Y = \{3, 7, 9\}$
Domain is $\{a, b, c\}$ Range is $\{3, 7, 9\}$

X maps **onto** Y

$g(a) = 3$

$g(b) = 7$

$g(c) = 9$

■

EXAMPLE 2 It is possible that different elements in the domain have the same image as shown by this example.

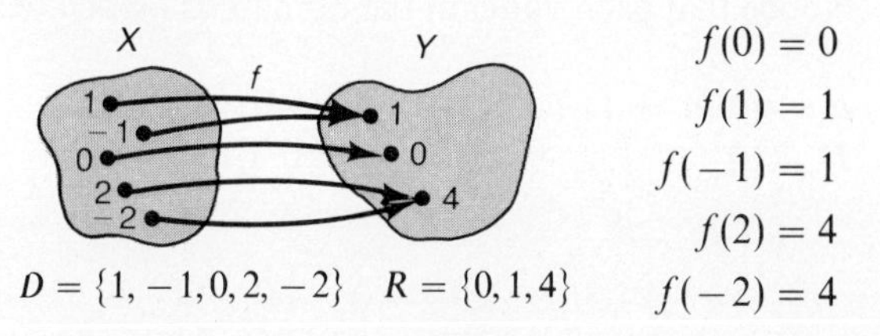

$D = \{1, -1, 0, 2, -2\}$ $R = \{0, 1, 4\}$

$f(0) = 0$

$f(1) = 1$

$f(-1) = 1$

$f(2) = 4$

$f(-2) = 4$ ■

If the images are always different (as they were in Example 1), then the function is called *one-to-one*.

One-to-One Function

If f maps X into Y so that for any distinct elements x_1 and x_2 of X, $f(x_1) \neq f(x_2)$, then f is a **one-to-one function** of X into Y.

Example 3 illustrates a mapping that is not a function.

EXAMPLE 3

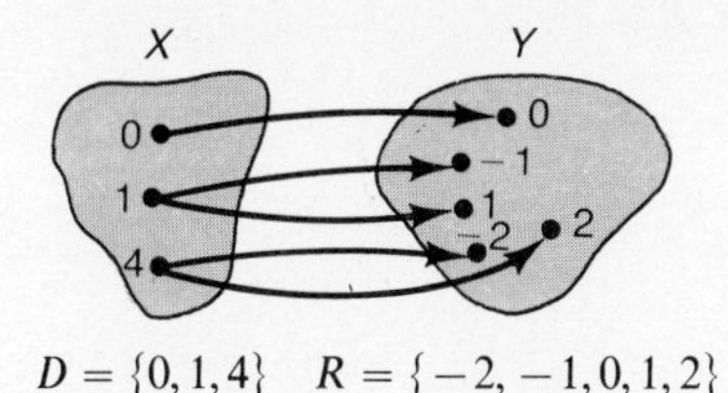

$D = \{0, 1, 4\}$ $R = \{-2, -1, 0, 1, 2\}$

This is not a function because 1 is associated with more than one image (so is 4). ■

It is perhaps more common to describe a function as a set of ordered pairs than it is to describe it as a mapping. It must, however, possess a certain property as described in the following alternative definition of a function.

Function as a Set of Ordered Pairs

A **function** is a set of ordered pairs for which each member x of the domain is associated with exactly one member $f(x)$ of the range.

EXAMPLE 4 $f = \{(0,2), (1,5), (2,8), (3,11)\}$
$D = \{0, 1, 2, 3\}$ $\quad R = \{2, 5, 8, 11\}$

D	R	
$0 \to$	2	$f(0) = 2$
$1 \to$	5	$f(1) = 5$
$2 \to$	8	$f(2) = 8$
$3 \to$	11	$f(3) = 11$

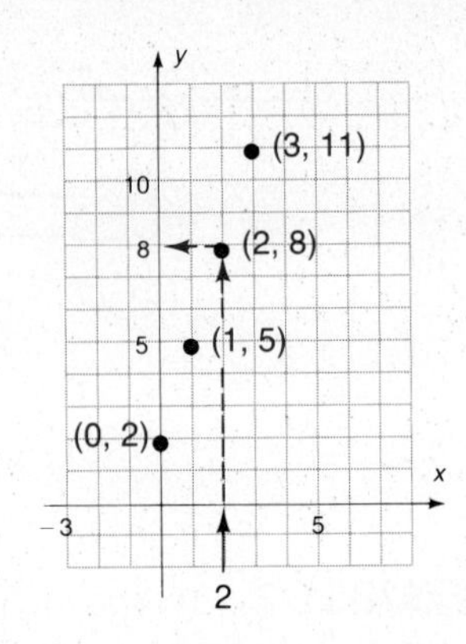

Notice that each value in the domain is associated with a single value in the range. ■

EXAMPLE 5 $g = \{(0,0), (1,1), (-1,1), (2,4), (-2,4)\}$
$D = \{0, 1, -1, 2, -2\}$ $\quad R = \{0, 1, 4\}$

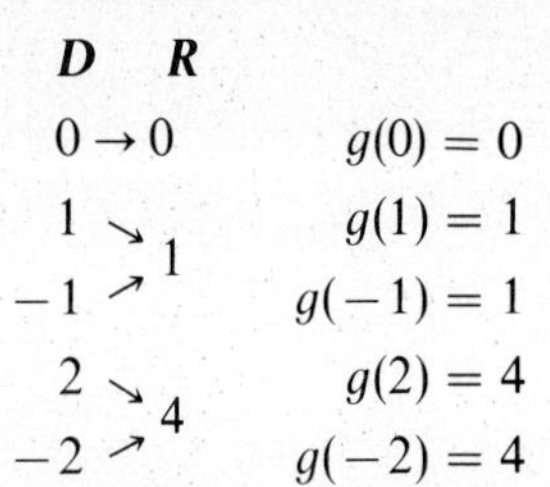

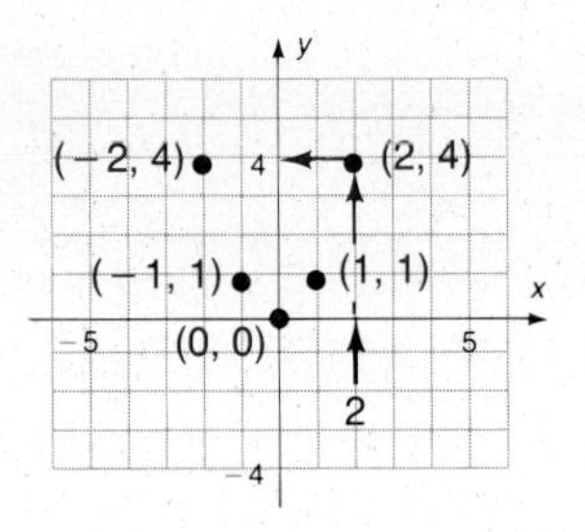

Notice that in Example 4 f is one-to-one while in this Example g is not. Not all sets of ordered pairs are functions, as illustrated in Example 6. ■

EXAMPLE 6 $\{(0,0), (1,1), (1,-1), (4,2), (4,-2)\}$
$D = \{0, 1, 4\}$ $\quad R = \{0, 1, -1, 2, -2\}$

D	R
$0 \to$	0
$1 \nearrow$	1
$1 \searrow$	-1
$4 \nearrow$	2
$4 \searrow$	-2

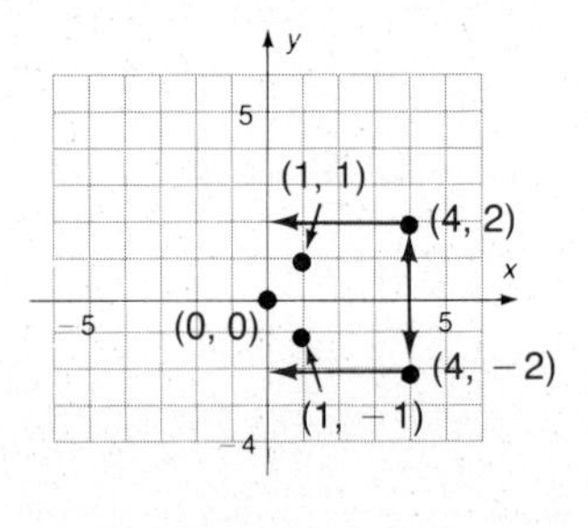

Do not use $f(x)$ notation unless the set of ordered pairs is a function. ■

The **graph of a function** f is the set of all points $(x, f(x))$ in a coordinate plane, where x is in the domain of f. That is, the graph of f can be described as the set of all points $P(x, y)$ such that $y = f(x)$. Since a function is a special type of relation, the graphing of functions is a special case of the graphing of relations (discussed in Chapter 1). The graph of a typical function f is shown in Figure 2.2. Notice that the graph of a function is such that for each a in the domain there is only *one point* $(a, f(a))$ on the graph. This means that every vertical line passes through the graph of a function in at most one point. This is the so-called *vertical line test* for the graphs of functions as illustrated in Example 7.

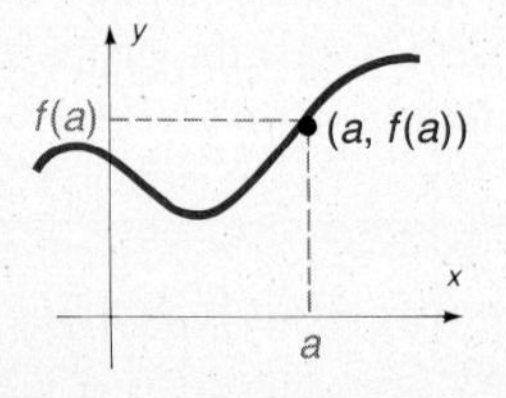

Figure 2.2 Graph of a function

EXAMPLE 7 Which of the following graphs are functions?

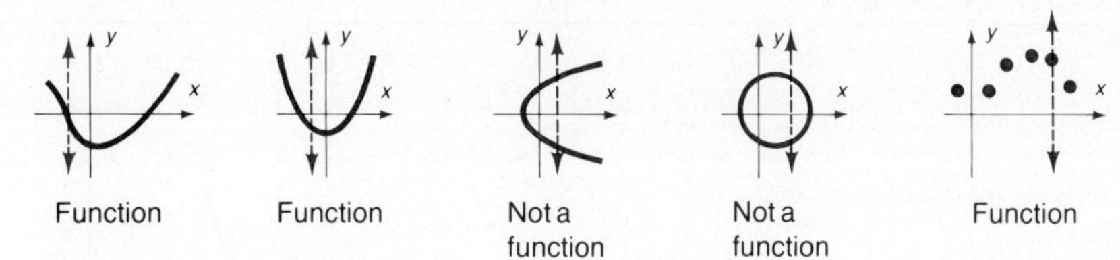

Solution Function Function Not a function Not a function Function

Imagine a vertical line sweeping from left to right across the plane—if it passes through more than one point of the graph at one time, then the graph does not represent a function. ■

There are several points of the graph of a function that will be of particular importance to us. The first are the intercepts. These points are called intercepts because they are the places where a curve intercepts the coordinate axes.

Intercepts

If the number zero is in the domain of f, then $f(0)$ is called the **y intercept** of the graph of f and is the point $(0, f(0))$. If a is a real number in the domain such that $f(a) = 0$, then a is called an **x intercept** and is the point $(a, 0)$. Any number x such that $f(x) = 0$ is called a **zero of the function.**

EXAMPLE 8 Find the domain, range, and intercepts for g defined by the graph.

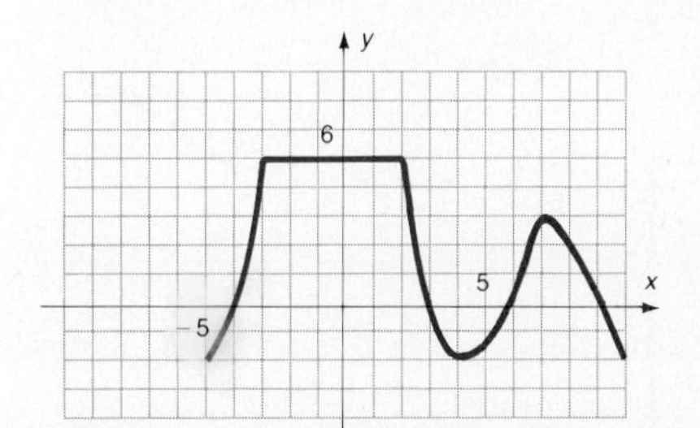

Domain: $-5 \le x \le 10$
Range: $-2 \le y \le 5$
***y* intercept**: $(0, 5)$; we usually say simply that the y intercept is 5.
***x* intercepts**: $(-4, 0)$, $(3, 0)$, $(6, 0)$, and $(9, 0)$; the zeros of the function g are -4, 3, 6, and 9. ■

Notice that a function might have several x intercepts, or several zeros, but a function can have at most only one y intercept. As you progress through this book, you will encounter a variety of special types of functions. The first types we will mention are functions that are increasing, decreasing, or constant.

Increasing, Decreasing, and Constant Functions

Let S be a subset of the domain of a function f. Then:

f is **increasing** on S if $f(x_1) < f(x_2)$ whenever $x_1 < x_2$ in S.
f is **decreasing** on S if $f(x_1) > f(x_2)$ whenever $x_1 < x_2$ in S.
f is **constant** on S if $f(x_1) = f(x_2)$ for every x_1 and x_2 in S.

EXAMPLE 9 Consider the function g graphed in Example 8. Note that g is increasing on $[-5, -3]$; g is constant on $[-3, 2]$; g is decreasing on $[2, 4]$; g is increasing on $[4, 7]$; and g is decreasing on $[7, 10]$. ■

PROBLEM SET 2.1

A

Let $X = \{1, 2, 3\}$ and $Y = \{1, 2, 3\}$. Classify the sets in Problems 1–6 as onto, one-to-one, function, or not a function. More than one, or none, of these terms may apply.

1. $\{(1, 1), (2, 2)\ (3, 3)\}$

2. $\{(1, 1), (2, 1), (3, 1)\}$

3. $\{(1, 1), (1, 2), (1, 3)\}$

4. $\{(2, 3), (2, 1), (2, 2)\}$

5. $\{(2, 3), (1, 2), (3, 3)\}$

6. $\{(1, 3), (2, 1), (3, 2)\}$

Let $X = \{a, b, c, d, e\}$ and $Y = \{1, 2, 3, 4, 5\}$. Classify the sets of Problems 7–12 as onto, one-to-one, function, or not a function. More than one, or none, of these terms may apply.

7. $f = \{(a, 3), (b, 5), (c, 2), (d, 1), (e, 3)\}$

8. $F = \{(a, 1), (b, 2), (c, 3), (d, 4), (e, 5)\}$

9. $g = \{(a, 4), (b, 4), (d, 4), (e, 4)\}$

10. $h = \{(a, 3), (b, 2), (c, 1), (d, 4), (a, 4)\}$

11. $G = \{(a, 5), (a, 4), (a, 3), (a, 2), (a, 1)\}$

12. $H = \{(a, 5), (b, 4), (c, 3), (d, 4), (e, 5)\}$

State whether each set in Problems 13–29 is or is not a function.

13. $\{(8, 2), (7, 1), (6, 3), (5, 1)\}$

14. $\{(5, 2), (7, 3), (1,6), (7, 4)\}$

15. $\{1, 2, 3, 4\}$

16. $\{6, 9, 12, 15\}$

17. $\{(x, y) \mid y = 4x + 3\}$*

18. $\{(x, y) \mid y \leq 4x + 3\}$

19.

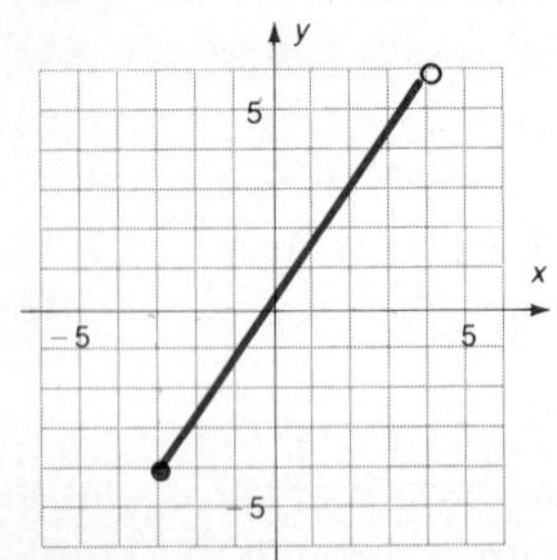

20.

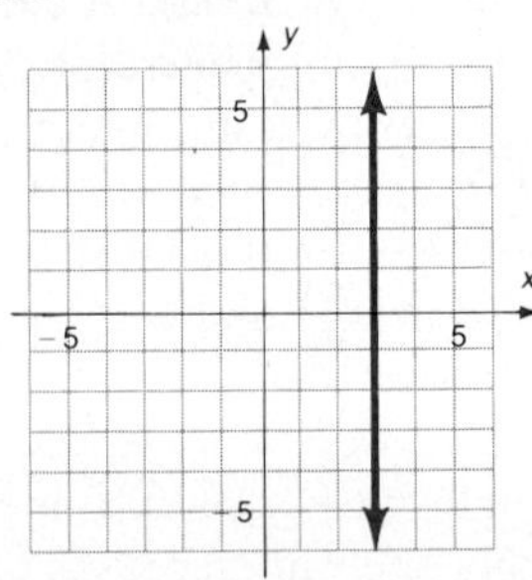

21.

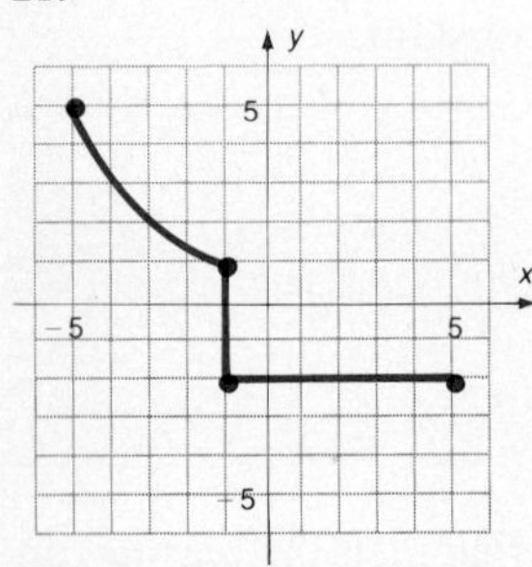

22.

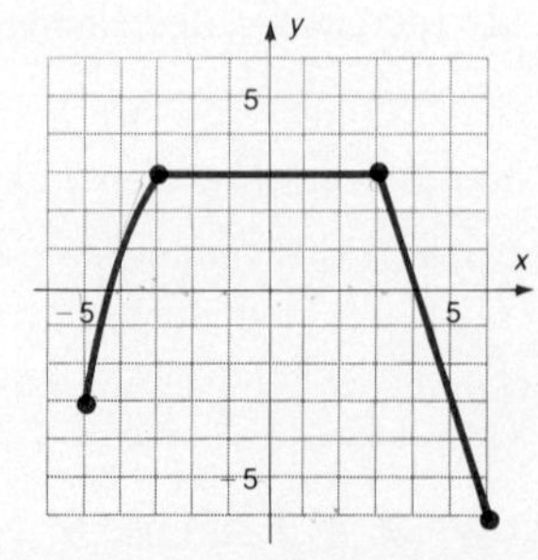

23.

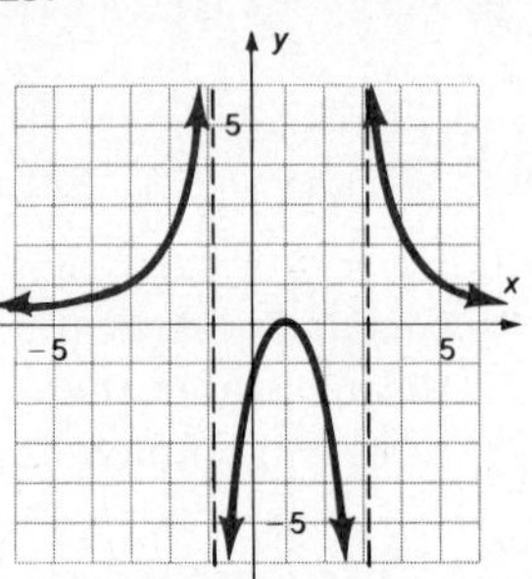

24.

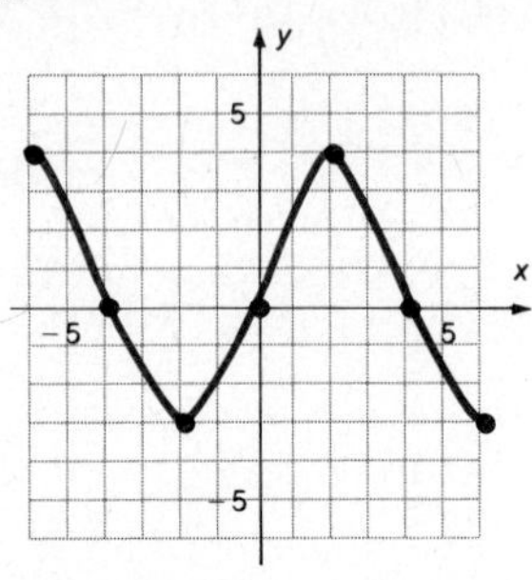

25. Let $y = -1$ if x is a rational number, and let $y = 1$ if x is an irrational number.

26. Let $y = 1$ if x is positive and $y = -1$ if x is negative.

27. Suppose that x is the closing price of Xerox stock on July 1 of year y.

28. Suppose that y is the closing price of IBM stock on January 2 of year x.

29. Let A be the area of a cross-sectional slice of an orange whose circumference is C.

30. See Figure 2.3a. If point A has coordinates $(2, f(2))$, what are the coordinates of P and Q?

(a) Graph of f

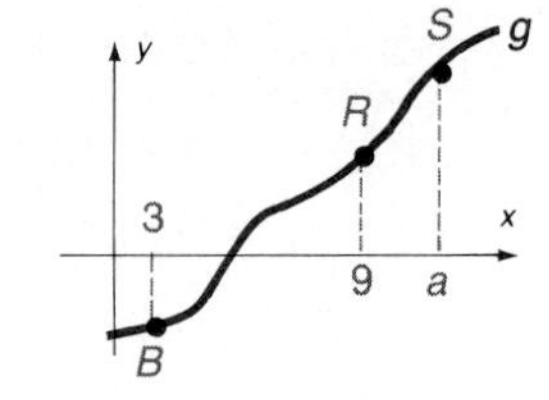

(b) Graph of g

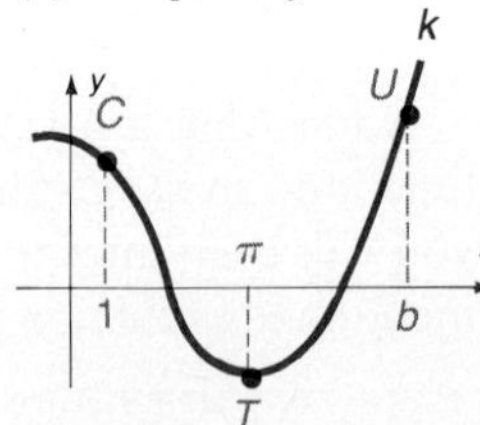

(c) Graph of k

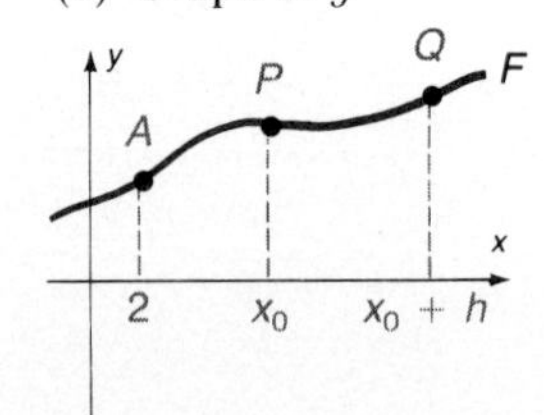

(d) Graph of F

Figure 2.3

31. See Figure 2.3b. If point B has coordinates $(3, g(3))$, what are the coordinates of R and S?

32. See Figure 2.3c. If point C has coordinates $(1, k(1))$, what are the coordinates of T and U?

33. See Figure 2.3d. If point A has coordinates $(2, F(2))$, what are the coordinates of P and Q?

* This is called set builder notation and is read as "the set of all ordered pairs (x, y) such that $y = 4x + 3$."

34. See Figure 2.3e. If point B has coordinates $(3, G(3))$, what are the coordinates of R and S?

35. See Figure 2.3f. If point C has coordinates $(1, K(1))$, what are the coordinates of T and U?

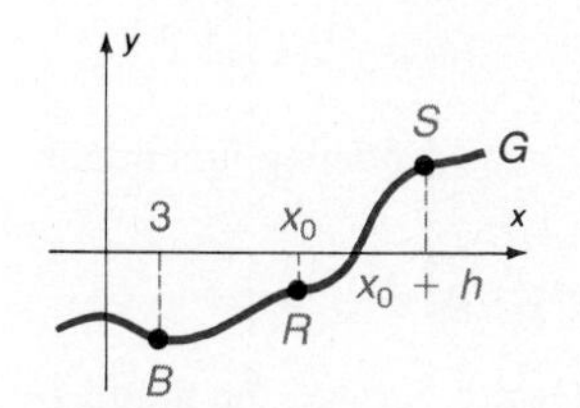

(e) Graph of G

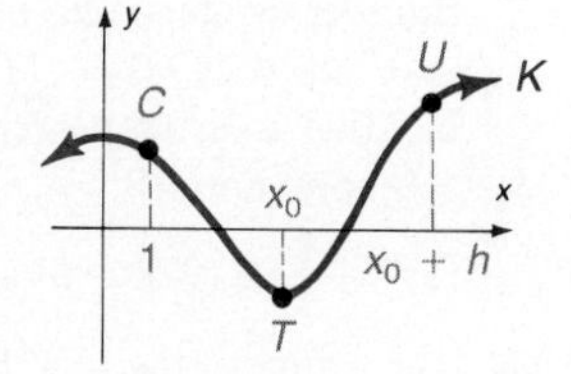

(f) Graph of K

Figure 2.3

B

Find the domain, range, and intercepts of the functions defined by the graphs indicated in Problems 36–44. Also tell where the function is increasing, decreasing, and constant.

36. Graph of Problem 19

37. Graph of Problem 22

38. Graph of Problem 24

39.

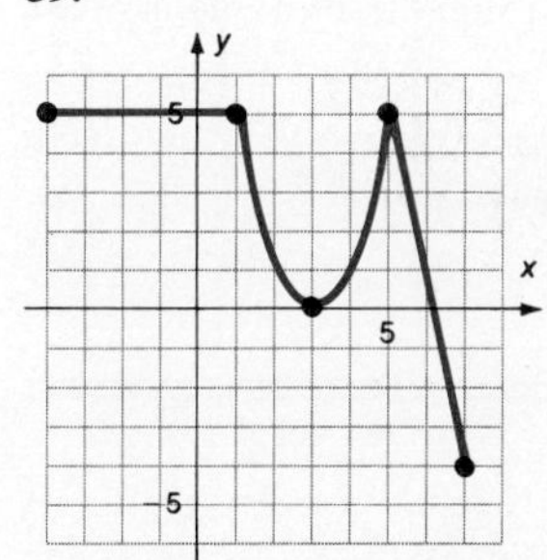

40.

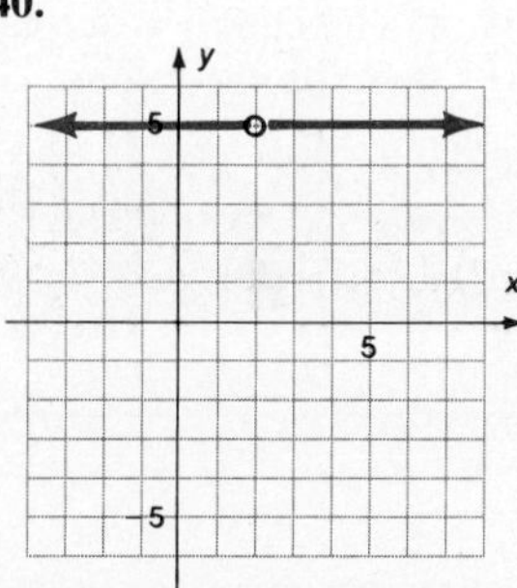

41.

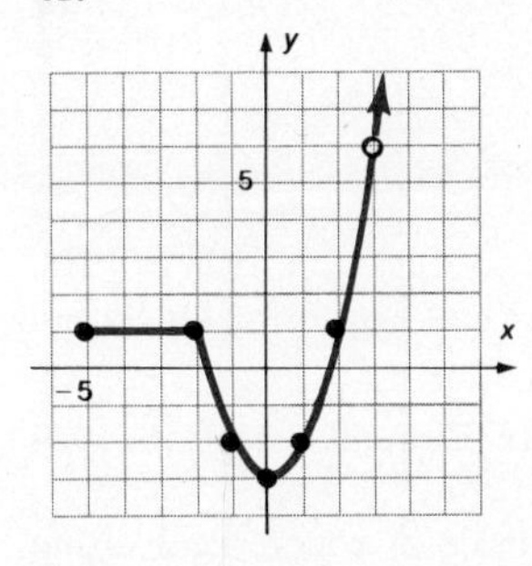

42.

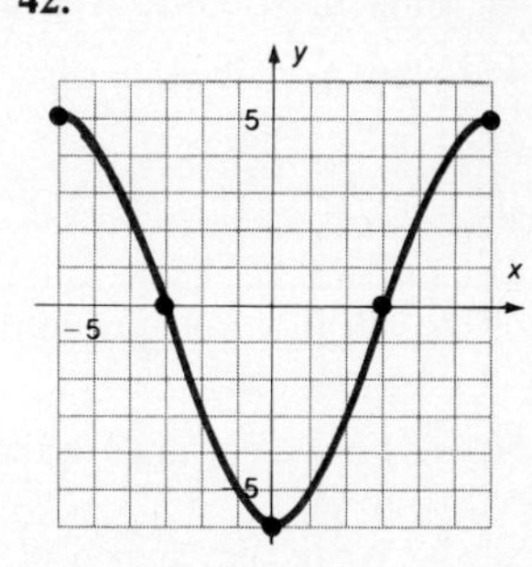

43.

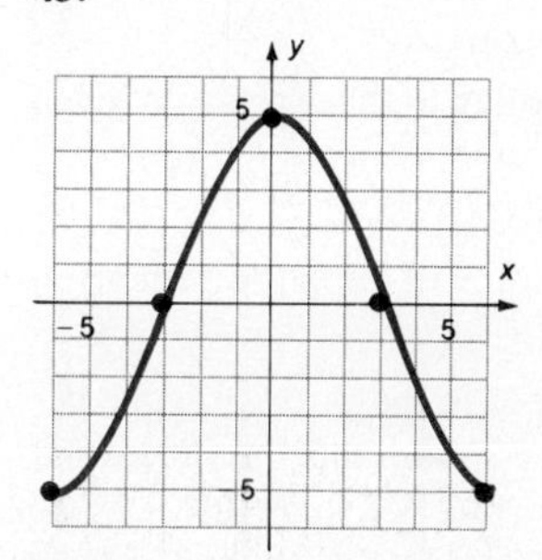

44.

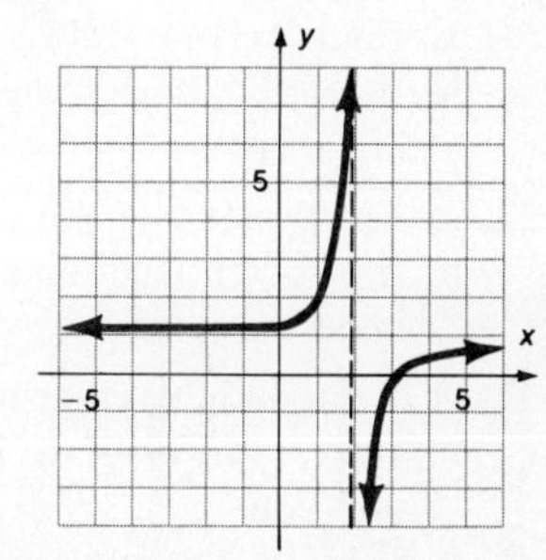

For Problems 45–54, use the following table, which reflects the purchasing power of the dollar from October 1944 to October 1984 (Source: Bureau of Labor Statistics, Consumer Division). Let x represent the year, let the domain be the set {1944, 1954, 1964, 1974, 1984}, and let

$r(x)$ = *price of 1 lb of round steak*

$s(x)$ = *price of a 5-lb bag of sugar*

$b(x)$ = *price of a loaf of bread*

$c(x)$ = *price of 1 lb of coffee*

$e(x)$ = *price of a dozen eggs*

$m(x)$ = *price of $\frac{1}{2}$ gal of milk*

$g(x)$ = *price of 1 gal of gasoline*

Year	Round steak (1 lb)	Sugar (5-lb bag)	Bread (loaf)	Coffee (1 lb)	Eggs (1 dozen)	Milk ($\frac{1}{2}$ gal)	Gasoline (1 gal)
1944	$0.45	$0.34	$0.09	$0.30	$0.64	$0.29	$0.21
1954	0.92	0.52	0.17	1.10	0.60	0.45	0.29
1964	1.07	0.59	0.21	0.82	0.57	0.48	0.30
1974	1.78	2.08	0.36	1.31	0.84	0.78	0.53
1984	2.15	1.49	1.29	2.69	1.15	1.08	1.10

45. Find: **a.** $r(1954)$ **b.** $m(1954)$

46. Find: **a.** $g(1944)$ **b.** $c(1984)$

47. Find $s(1984) - s(1944)$.

48. Find $b(1984) - b(1944)$.

49. **a.** Find the change in the price of eggs from 1944 to 1984.
    b. Write the change in the price of eggs using functional notation.

50. **a.** Find the change in the price of round steak from 1944 to 1984.
    b. Write the change in the price of round steak using functional notation.

51. **a.** Find $[g(1944 + 40) - g(1944)]/40$.
    b. In words, attach some meaning to the expression given in part **a.**

52. **a.** Find $[m(1944 + 40) - m(1944)]/40$.
    b. In words, attach some meaning to the expression given in part **a.**

53. **a.** What is the average increase in the price of sugar per year from 1944 to 1954? Write this in functional notation.
    b. What is the average increase in the price of sugar per year from 1944 to 1964? Write this in functional notation.
    c. What is the average increase in the price of sugar per year from 1944 to 1974? Write this in functional notation.
    d. What is the average increase in the price of sugar per year from 1944 to 1984? Write this in functional notation.
    e. What is the average increase in the price of sugar per year from 1944 to $1944 + h$, where h is an unspecified number of years? Write this in functional notation.

54. Repeat Problem 53 for coffee instead of sugar.

55. According to the U.S. Public Health Service, the number of marriages in the United States was about 2,176,000 in 1977 and about 2,495,000 in 1982. Let $M(x)$ represent the number of marriages in year x.
    a. Find $[M(1982) - M(1977)]/5$.
    b. Give a verbal description for the following functional expression:

$$[M(1977 + h) - M(1977)]/h$$

56. According to the U.S. Public Health Service, the number of divorces in the United States was about 1,097,000 in 1977 and about 1,180,000 in 1982. Let $D(x)$ represent the number of divorces in year x.
    a. Find $[D(1982) - D(1977)]/5$.
    b. Give a verbal description for the following functional expression:

$$[D(1977 + h) - D(1977)]/h$$

C

57. If F is a one-to-one function mapping X onto Y, and the domain of F contains exactly five elements, what can you conclude about the set Y?

58. If a function f is increasing throughout its domain, prove that f is one-to-one.

59. If a function f is decreasing throughout its domain, prove that f is one-to-one.

60. If g maps X into Y, and the range of g is a set S (where S is a subset of Y), explain why g maps X onto S.

2.2 FUNCTIONAL NOTATION

The notation for functions introduced in the last section is fundamental for this course and for calculus. This section provides additional examples and practice in using functional notation. Remember:

The function is denoted by f

x is a member of the domain

$$f(x)$$

$f(x)$ is a member of the range

The *function* is denoted by f; $f(x)$ is the *number* associated with x. Sometimes functions are defined by expressions such as

$$f(x) = 3x + 2 \qquad \text{or} \qquad g(x) = x^2 + 4x + 3$$

To emphasize the difference between f and $f(x)$, some books use $f: x \to 3x + 2$ to define functions, but this book simply uses the notation $f(x) = 3x + 2$ to mean the set of all ordered pairs (x, y) such that $y = 3x + 2$.

EXAMPLE 1 Given f and g defined by $f(x) = 3x + 2$ and $g(x) = x^2 + 4x + 3$, find the indicated values: **(a)** $f(1)$; **(b)** $g(2)$; **(c)** $g(-3)$; **(d)** $f(-3)$.

Solution **a.** The symbol $f(1)$ represents the second component of the ordered pair of the function f with first component 1. Replace x by 1 in the expression

$$\begin{aligned} f(x) &= 3x + 2 \\ &\updownarrow \quad \updownarrow \\ f(1) &= 3(1) + 2 \\ &= 5 \end{aligned}$$

b. $g(2)$:

$$\begin{aligned} g(x) &= x^2 + 4x + 3 \\ &\updownarrow \quad \updownarrow \quad \updownarrow \\ g(2) &= (2)^2 + 4(2) + 3 \\ &= 4 + 8 + 3 \\ &= 15 \end{aligned}$$

c.

$$\begin{aligned} g(-3) &= (-3)^2 + 4(-3) + 3 \\ &= 9 - 12 + 3 \\ &= 0 \end{aligned}$$

d.

$$\begin{aligned} f(-3) &= 3(-3) + 2 \\ &= -9 + 2 \\ &= -7 \end{aligned}$$

■

The members of the domain of a function may also be represented by variables, as shown in Example 2.

EXAMPLE 2 Let F and G be defined by $F(x) = x^2 + 1$ and $G(x) = (x + 1)^2$.

a. $F(w) = w^2 + 1$

b.

$$\begin{aligned} G(t) &= (t + 1)^2 \\ &= t^2 + 2t + 1 \end{aligned}$$

c.

$$\begin{aligned} F(w + 3) &= (w + 3)^2 + 1 \\ &= w^2 + 6w + 9 + 1 \\ &= w^2 + 6w + 10 \end{aligned}$$

d.

$$\begin{aligned} G(x - 2) &= [(x - 2) + 1]^2 \\ &= (x - 1)^2 \\ &= x^2 - 2x + 1 \end{aligned}$$

e.

$$\begin{aligned} F(w + h) &= (w + h)^2 + 1 \\ &= w^2 + 2wh + h^2 + 1 \end{aligned}$$

f. $G(x + h) = [(x + h) + 1]^2$
$$= x^2 + xh + x + xh + h^2 + h + x + h + 1$$
$$= x^2 + 2xh + h^2 + 2x + 2h + 1$$ ■

In calculus, functional notation is used to carry out manipulations such as those shown in Example 3.

EXAMPLE 3 Find $\dfrac{f(x + h) - f(x)}{h}$ for each function.

a. $f(x) = x^2$, where $x = 5$

$$\frac{f(5 + h) - f(5)}{h} = \frac{(5 + h)^2 - 5^2}{h}$$

$$= \frac{25 + 10h + h^2 - 25}{h}$$

$$= \frac{(10 + h)h}{h}$$

$$= 10 + h$$

b. $f(x) = 2x^2 + 1$, where $x = 1$

$$\frac{f(1 + h) - f(1)}{h} = \frac{[2(1 + h)^2 + 1] - [2(1)^2 + 1]}{h}$$

$$= \frac{[2(1 + 2h + h^2) + 1] - (2 + 1)}{h}$$

$$= \frac{2h^2 + 4h + 3 - 3}{h}$$

$$= 2h + 4$$

c. $f(x) = x^2 + 3x - 2$

$$\frac{f(x + h) - f(x)}{h} = \frac{[(x + h)^2 + 3(x + h) - 2] - [x^2 + 3x - 2]}{h}$$

$$= \frac{x^2 + 2xh + h^2 + 3x + 3h - 2 - x^2 - 3x + 2}{h}$$

$$= \frac{2xh + h^2 + 3h}{h}$$

$$= 2x + 3 + h$$ ■

Functional notation can be used to work a wide variety of applied problems, as shown by Example 4 and again in the problem set.

EXAMPLE 4 If an object is dropped from a certain height, it is known that it will fall a distance of s ft in t sec according to the formula

$$s = 16t^2$$

This formula can be represented by $f(t) = 16t^2$.

a. How far will the object fall in the first second?

$$\begin{aligned} f(1) &= 16 \cdot 1^2 \\ &= 16 \quad \text{or} \quad \mathbf{16\ ft} \end{aligned}$$

b. How far will it fall in the *next* 2 sec?

$$\begin{aligned} f(1 + 2) &= 16 \cdot 3^2 \\ &= 144 \quad \text{or} \quad 144 \text{ ft in } 3 \text{ sec} \end{aligned}$$

So the answer to the question is

$$\begin{aligned} f(3) - f(1) &= 144 - 16 \\ &= 128 \quad \text{or} \quad \mathbf{128\ ft} \end{aligned}$$

c. How far will it fall during the time $t = 1$ sec to $t = 1 + h$ sec?

$$\begin{aligned} f(1 + h) - f(1) &= 16(1 + h)^2 - 16 \\ &= 16 + 32h + 16h^2 - 16 \\ &= \mathbf{(32h + 16h^2)\ ft} \end{aligned}$$

d. What is the average rate of change of distance (in feet per second, fps) during the time $t = 1$ sec to $t = 3$ sec?

$$\frac{f(3) - f(1)}{3 - 1} = \frac{128}{2} = \mathbf{64\ fps}$$

e. What is the average rate of change of distance during the time $t = 1$ sec to $t = 1 + h$ sec?

$$\begin{aligned} \frac{f(1 + h) - f(1)}{h} &= \frac{32h + 16h^2}{h} \\ &= \mathbf{(32 + 16h)\ fps} \end{aligned}$$

f. What is the average rate of change of distance during the time $t = x$ sec to $t = x + h$ sec?

$$\begin{aligned} \frac{f(x + h) - f(x)}{(x + h) - x} &= \frac{f(x + h) - f(x)}{h} \qquad \text{Does this look familiar?} \\ &= \frac{16(x + h)^2 - 16x^2}{h} \\ &= \frac{16x^2 + 32xh + 16h^2 - 16x^2}{h} \\ &= \mathbf{(32x + 16h)\ fps} \end{aligned}$$

■

The variable t in Example 4 represents an arbitrary number from the domain of f and is often called the **independent variable**. The variable s, which represents a number from the range of f, is called a **dependent variable** since its value depends on the value assigned to h.

It is also important to make note of the domain of a function. In this book, unless otherwise specified, the domain is the set of real numbers for which the given function is meaningful. For Example 4, the domain is the set of all nonnegative real numbers, since the formula is meaningless for negative time. If a function f is **undefined at x**, it means that x is not in the domain of f.

EXAMPLE 5 Find the domain for the given functions:

a. $f(x) = 2x - 1$ **b.** $g(x) = \dfrac{(2x - 1)(x + 3)}{x + 3}$

c. $h(x) = \sqrt{x + 1}$ **d.** $F(x) = \sqrt{2 - 3x - 2x^2}$

e. $G(x) = 2x - 1, x \neq -3$ **f.** $r(x) = 2 - \dfrac{x}{x}$

Solution **a.** All real numbers; $(-\infty, \infty)$

b. All real numbers except $x = -3$ because if $x = -3$, then the expression is meaningless. That is, set the denominator $(x + 3)$ equal to zero $(x + 3 = 0)$, and solve $(x = -3)$. The domain is $(-\infty, -3) \cup (-3, \infty)$. This domain is usually denoted simply by $x \neq -3$.

c. Here h has meaning if $x + 1$ is nonnegative. That is,

$$x + 1 \geq 0$$
$$x \geq -1$$

This domain can be described by writing $[-1, \infty)$.

d. F has meaning if $2 - 3x - 2x^2$ is nonnegative.

$$2 - 3x - 2x^2 \geq 0$$
$$(1 - 2x)(2 + x) \geq 0$$

+ − neg | + + pos | − + neg

−2, $\frac{1}{2}$

Domain: $[-2, \frac{1}{2}]$

e. The domain has $x = -3$ explicitly eliminated. This means that the domain is $(-\infty, -3) \cup (-3, \infty)$ or simply $x \neq -3$.

f. $r(0)$ is meaningless, but $r(x) = 1$ for $x \neq 0$, so the domain is all real numbers except 0. ■

Equality of Functions

Two functions f and g are **equal** if and only if

1. f and g have the same domain.
2. $f(x) = g(x)$ for all x in the domain.

Compare Examples 5*a* and 5*b* where $f(x) = 2x - 1$ and $g(x) = \dfrac{(2x-1)(x+3)}{x+3}$.

In algebra you wrote

$$\frac{(2x-1)(x+3)}{x+3} = 2x - 1$$

but it is *not* true that $f = g$, since their domains are not the same. However, $g = G$ (from Example 5*e*, $G(x) = 2x - 1$, $x \neq -3$) because both conditions of the definition are met.

EXAMPLE 6 $f(x) = \dfrac{(x-3)(x+5)}{x+5}$; $\quad g(x) = x - 3$

$f \neq g$ since the domain of f is all reals except -5 and the domain of g is all real numbers. ■

EXAMPLE 7 $f(x) = \dfrac{(2x-5)(x+1)}{x+1}$; $\quad g(x) = 2x - 5, x \neq -1$

$f = g$ since the domain of both f and g is all reals except $x = -1$ and $f(x) = g(x)$ for all x in the domain. ■

Functions are sometimes classified as *even* or *odd*.

Even and Odd Functions

A function f is called

even if $f(-x) = f(x)$ and

odd if $f(-x) = -f(x)$.

Just as not every real number is even or odd (2 is even, 3 is odd, but 2.5 is neither), not every function is even or odd.

EXAMPLE 8 Classify the given functions as even, odd, or neither.

a. $f(x) = x^2$ is **even** since
$$\begin{aligned} f(-x) &= (-x)^2 \\ &= x^2 \\ &= f(x) \end{aligned}$$

b. $g(x) = x^3$ is **odd** since
$$\begin{aligned} g(-x) &= (-x)^3 \\ &= -x^3 \\ &= -[x^3] \\ &= -g(x) \end{aligned}$$

c. $h(x) = x^2 + x$ is **neither** since
$$\begin{aligned} h(-x) &= (-x)^2 + (-x) \\ &= \underbrace{x^2 - x} \end{aligned}$$

This is neither $h(x)$ nor $-h(x)$. ■

PROBLEM SET 2.2

A

In Problems 1–12, let $f(x) = 2x + 1$ and $g(x) = 2x^2 - 1$. Find the requested values.

1. a. $f(0)$ b. $f(2)$ c. $f(-3)$ d. $f(\sqrt{5})$ e. $f(\pi)$
2. a. $f(1)$ b. $g(1)$ c. $f(\sqrt{3})$ d. $g(\sqrt{3})$ e. $g(\pi)$
3. a. $f(w)$ b. $g(w)$ c. $g(t)$ d. $g(v)$ e. $f(m)$
4. a. $f(t)$ b. $f(p)$ c. $f(t + 1)$ d. $g(t + 1)$ e. $f(t^2)$
5. a. $f(1 + \sqrt{2})$ b. $g(1 + \sqrt{2})$ c. $g(t + 3)$ d. $f(t^2 + 2t + 1)$ e. $g(m - 1)$
6. a. $f(x + 2)$ b. $g(x + 2)$ c. $f(t + h)$ d. $g(t + h)$ e. $f(x + h)$
7. $\dfrac{f(t + 3) - f(t)}{3}$
8. $\dfrac{f(t + h) - f(t)}{h}$
9. $\dfrac{f(x + h) - f(x)}{h}$
10. $\dfrac{g(t + 2) - g(t)}{2}$
11. $\dfrac{g(t + h) - g(t)}{h}$
12. $\dfrac{g(x + h) - g(x)}{h}$

In Problems 13–19, compute the given value where $f(x) = x^2 - 1$ and $g(x) = 2x + 5$.

13. a. $f(w)$ b. $f(h)$ c. $f(w + h)$ d. $f(w) + f(h)$
14. a. $g(s)$ b. $g(t)$ c. $g(s + t)$ d. $g(s) + g(t)$
15. a. $f(x^2)$ b. $f(\sqrt{x})$ c. $f(x + h)$ d. $f(-x)$
16. a. $g(x^2)$ b. $g(\pi)$ c. $g(x + \pi)$ d. $g(-x)$
17. $\dfrac{g(x + h) - g(x)}{h}$
18. $\dfrac{f(t + h) - f(t)}{h}$
19. $\dfrac{f(x + h) - f(x)}{h}$

B

In Problems 20–28, find $\dfrac{f(x + h) - f(x)}{h}$ for the given function f.

20. $f(x) = 9x + 3$
21. $f(x) = 5 - 2x$
22. $f(x) = |x|$
23. $f(x) = |2x + 1|$
24. $f(x) = 5x^2$
25. $f(x) = 3x^2 + 2x$
26. $f(x) = 2x^2 + 3x - 4$
27. $f(x) = \dfrac{1}{x}$
28. $f(x) = \dfrac{x + 1}{x - 1}$

Find the domain for the functions defined by the equations in Problems 29–38 and leave your answer in interval notation.

29. $f(x) = 3x + 1$
30. $g(x) = 3x + 1, x \neq 2$
31. $h(x) = \dfrac{(3x + 1)(x + 2)}{x + 2}$
32. $F(x) = \dfrac{(2x + 1)(x - 1)}{x^2 + 1}$
33. $G(x) = \sqrt{2x + 1}$
34. $H(x) = \sqrt{1 - 3x}$
35. $f(x) = \sqrt{2 - x - x^2}$
36. $g(x) = \sqrt{2 + x - x^2}$
37. $h(x) = \dfrac{3x^2 - 4x - 4}{x^2 - 4}$
38. $h(x) = \dfrac{x^2 - 3x + 2}{x^2 + 2x - 3}$

State whether the functions f and g are equal in Problems 39–44.

39. $f(x) = \dfrac{2x^2 + x}{x}$; $g(x) = 2x + 1$
40. $f(x) = \dfrac{2x^2 + x}{x}$; $g(x) = 2x + 1, x \neq 0$
41. $f(x) = \dfrac{2x^2 + x - 6}{x - 2}$; $g(x) = 2x + 3, x \neq 2$
42. $f(x) = \dfrac{3x^2 - 7x - 6}{x - 3}$; $g(x) = 3x + 2, x \neq 3$
43. $f(x) = \dfrac{3x^2 - 5x - 2}{x - 2}$; $g(x) = 3x + 1$
44. $f(x) = \dfrac{(3x + 1)(x - 2)}{x - 2}, x \neq 6$;
    $g(x) = \dfrac{(3x + 1)(x - 6)}{x - 6}, x \neq 2$

Classify the functions defined in Problems 45–53 as even, odd, or neither.

45. $f_1(x) = x^2 + 1$
46. $f_2(x) = \sqrt{x^2}$
47. $f_3(x) = \dfrac{1}{3x^3 - 4}$
48. $f_4(x) = x^3 + x$
49. $f_5(x) = \dfrac{1}{(x^3 + 3)^2}$
50. $f_6(x) = \dfrac{1}{(x^3 + x)^2}$
51. $f_7(x) = |x|$
52. $f_8(x) = |x| + 3$
53. $f_9(x) = 5$
54. ***Business*** A firm determines that the total cost C (in dollars) of producing x units of a certain product is given by

$$C(x) = -0.02x^2 + 4x + 500 \qquad (0 \leq x \leq 150)$$

Find $C(50)$ and $C(100)$.

55. ***Business*** What is the average cost per unit in Problem 54 if 50 units are produced? Repeat for 100 units. What is the

per-unit increase in cost for the increase from 50 to 100 units?

56. ***Business*** What is the per-unit increase in cost in Problem 54 for an increase from 50 units to 51 units? Compare this answer with the answer to Problem 55.

57. ***Business*** What is the per-unit increase in cost in Problem 54 for an h-unit increase in cost above a production level of x units?

58. ***Physics*** Let d be a function that represents the distance an object falls (neglecting air resistance) in t sec. It can be shown that $d(t) = 16t^2$. Find the average rate that the object falls for the intervals of time given:
a. From $t = 2$ to $t = 6$ **b.** From $t = 2$ to $t = 4$
c. From $t = 2$ to $t = 3$ **d.** From $t = 2$ to $t = 2 + h$
e. From $t = x$ to $t = x + h$

59. ***Physics*** In Problem 58, give a physical interpretation for

$$\frac{d(x + h) - d(x)}{h}$$

C

60. If $f(x) = x^2$, then

$$f\left(\frac{1}{x}\right) = \left(\frac{1}{x}\right)^2 = \frac{1}{x^2} = \frac{1}{f(x)}$$

Give an example of a function for which

$$f\left(\frac{1}{x}\right) \neq \frac{1}{f(x)}$$

61. If $f(x) = x$, then $f(x^2) = [f(x)]^2$. Give an example of a function for which $f(x^2) \neq [f(x)]^2$.

62. If f is an even function, show that the graph of f is symmetric with respect to the y axis.

63. If f is an odd function, show that the graph of f is symmetric with respect to the origin.

2.3 LINEAR FUNCTIONS

The first type of function we will consider in this book is one with which you have had some experience in beginning algebra.

Linear Function

> A function f is a **linear function** if
>
> $$f(x) = mx + b$$
>
> where m and b are real numbers.

Notice that if $m = 0$, then $f(x) = b$, which we called a constant function in Section 2.1. If the domain of a constant function is the set of real numbers, then the graph of $f(x) = b$ is a **horizontal line** as shown in Figure 2.4.

Let $P_1(x_1, y_1)$ and $P_2(x_2, y_2)$ be any points on a line, and suppose that $x_1 = x_2$. Then the line is parallel to the y axis and is called a **vertical line** (Figure 2.5). Notice that vertical lines are not functions.

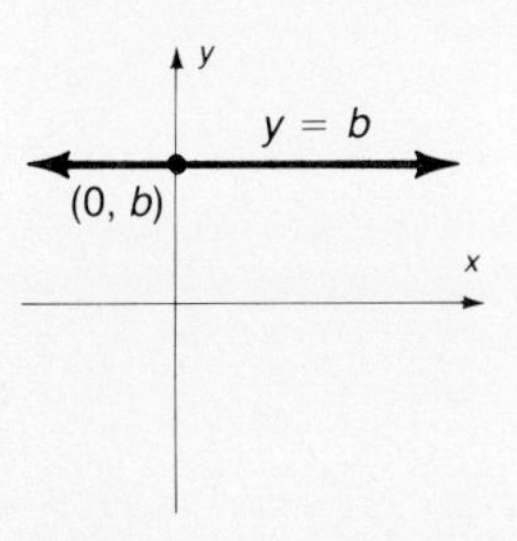

Figure 2.4 A horizontal line

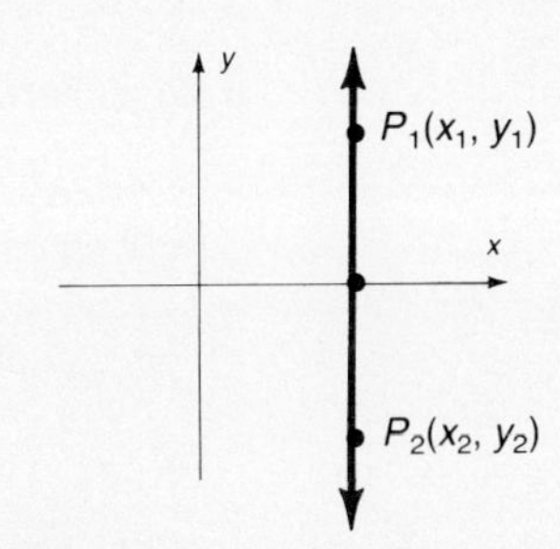

Figure 2.5 A vertical line is not a function

The steepness of a line is specified by using the idea of slope. If $x_1 \neq x_2$, then the slope is defined as follows.

Slope of a Line

Let $P_1(x_1, y_1)$ and $P_2(x_2, y_2)$ be distinct points on a line such that $x_1 \neq x_2$. Then

$$\textbf{Slope} = \frac{\text{vertical change}}{\text{horizontal change}} = \frac{y_2 - y_1}{x_2 - x_1}$$

The numerator $y_2 - y_1$ is often called the **rise** and the denominator $x_2 - x_1$ the **run** from P_1 to P_2. If you use functional notation for the points $P_1(x_1, f(x_1))$ and $P_2(x_2, f(x_2))$, then the slope is found by

$$\text{Slope} = \frac{f(x_2) - f(x_1)}{x_2 - x_1}$$

EXAMPLE 1 Sketch the line passing through the points whose coordinates are given. Then find the slope of each line.

a. $(2, -3)$ and $(-1, 2)$ **b.** $(-4, -1)$ and $(1, 3)$
c. $(-3, 4)$ and $(5, 4)$ **d.** $(-3, 2)$ and $(-3, 4)$
e. $(3, f(3))$ and $(3 + h, f(3 + h))$ **f.** $(a, f(a))$ and $(a + h, f(a + h))$

Solution **a.** $m = \dfrac{2 - (-3)}{-1 - 2} = \dfrac{5}{-3} = -\dfrac{5}{3}$

negative slope

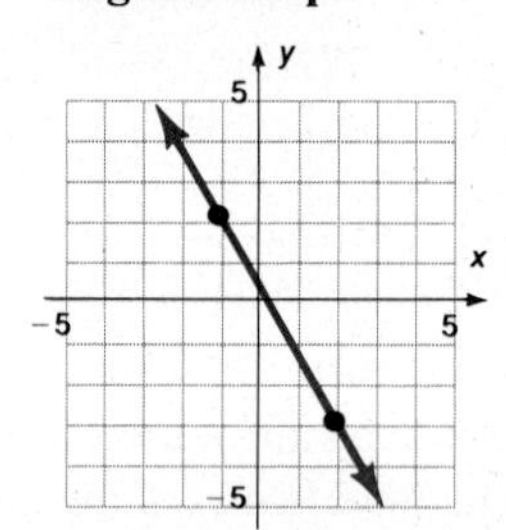

b. $m = \dfrac{3 - (-1)}{1 - (-4)} = \dfrac{4}{5}$

positive slope

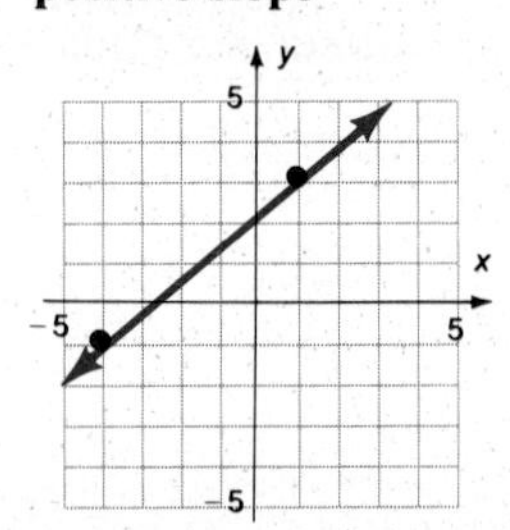

c. $m = \dfrac{4 - 4}{5 + 3} = 0$

0 slope; horizontal line

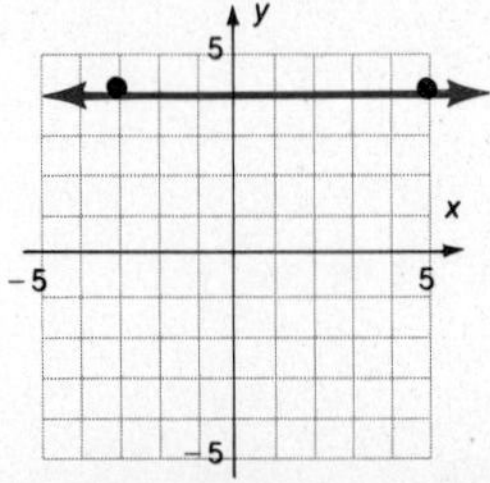

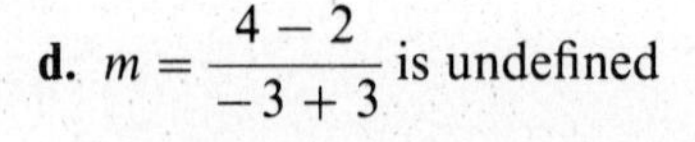

d. $m = \dfrac{4 - 2}{-3 + 3}$ is undefined

undefined slope; vertical line

e. $m = \dfrac{f(3+h) - f(3)}{3 + h - 3} = \dfrac{f(3+h) - f(3)}{h}$

arbitrary slope

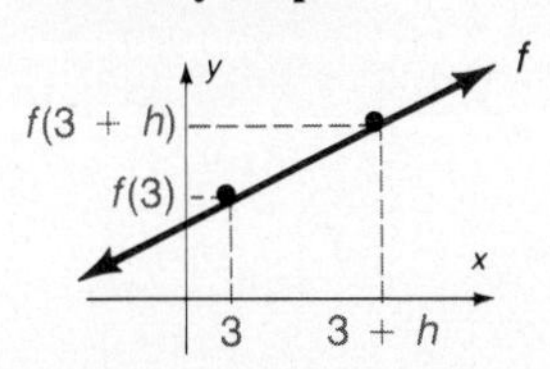

f. $m = \dfrac{f(a+h) - f(a)}{h}$

arbitrary slope

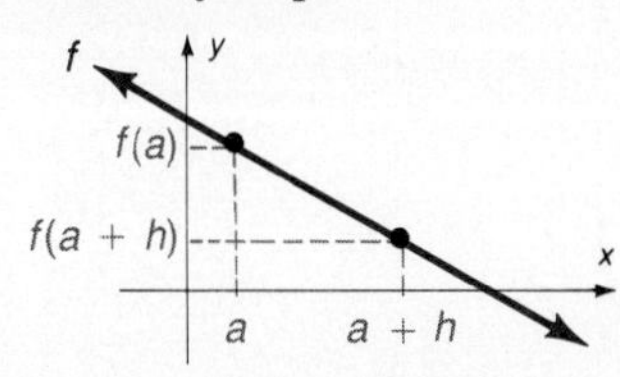

■

Two properties relating lines and slopes can be proved geometrically.

Parallel and Perpendicular Lines

Let L_1 and L_2 be two nonvertical lines with slopes m_1 and m_2, respectively. Then:

1. L_1 and L_2 are parallel if and only if $m_1 = m_2$.
2. L_1 and L_2 are perpendicular if and only if $m_1 m_2 = -1$.

EXAMPLE 2 Show that the points $Q(-3, 7)$, $U(8, 2)$, $A(4, -3)$, and $D(-7, 2)$ are the corners of a parallelogram $QUAD$.

Solution

$$m_{QU} = \frac{2 - 7}{8 + 3} = \frac{-5}{11}$$

$$m_{DA} = \frac{-3 - 2}{4 + 7} = \frac{-5}{11}$$

Thus $\overline{QU}$ is parallel to $\overline{DA}$.

$$m_{UA} = \frac{-3 - 2}{4 - 8} = \frac{-5}{-4} = \frac{5}{4}$$

$$m_{QD} = \frac{2 - 7}{-7 + 3} = \frac{-5}{-4} = \frac{5}{4}$$

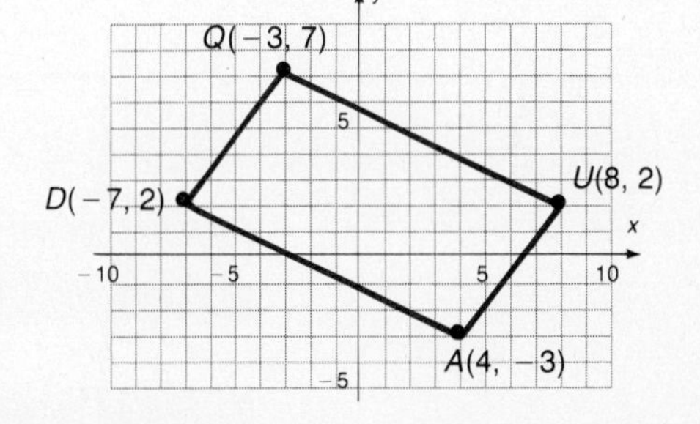

Thus $\overline{UA}$ is parallel to $\overline{QD}$. Since opposite sides of quadrilateral $QUAD$ are parallel, it is a parallelogram. Notice that even though it is helpful to draw the graph, you cannot use the figure to prove your argument. ■

EXAMPLE 3 Are the diagonals of $QUAD$ in Example 2 perpendicular?

Solution

$$m_{QA} = \frac{-3 - 7}{4 + 3} = \frac{-10}{7}$$

$$m_{DU} = \frac{2 - 2}{8 + 7} = 0$$

Since $m_{QA} m_{DU} \neq -1$, the segments $\overline{QA}$ and $\overline{DU}$ are not perpendicular. ■

Consider the linear function $f(x) = mx + b$. Then

$$\text{Slope} = \frac{\text{rise}}{\text{run}} = \frac{f(x_2) - f(x_1)}{x_2 - x_1}$$

$$= \frac{(mx_2 + b) - (mx_1 + b)}{x_2 - x_1}$$

$$= \frac{m(x_2 - x_1)}{x_2 - x_1} = m$$

Thus the slope of the graph of a linear function is m.

The y intercept for the linear function is found when $x = 0$:

$$f(0) = m(0) + b = b$$

Thus the y intercept of a linear function is b.

Slope-Intercept Form of the Equation of a Line

The graph of the equation $y = mx + b$ is a line having slope m and having y intercept b.

This form of the equation of a line can be used for graphing certain lines for which it is not convenient to plot points. This procedure is summarized in Figure 2.6.

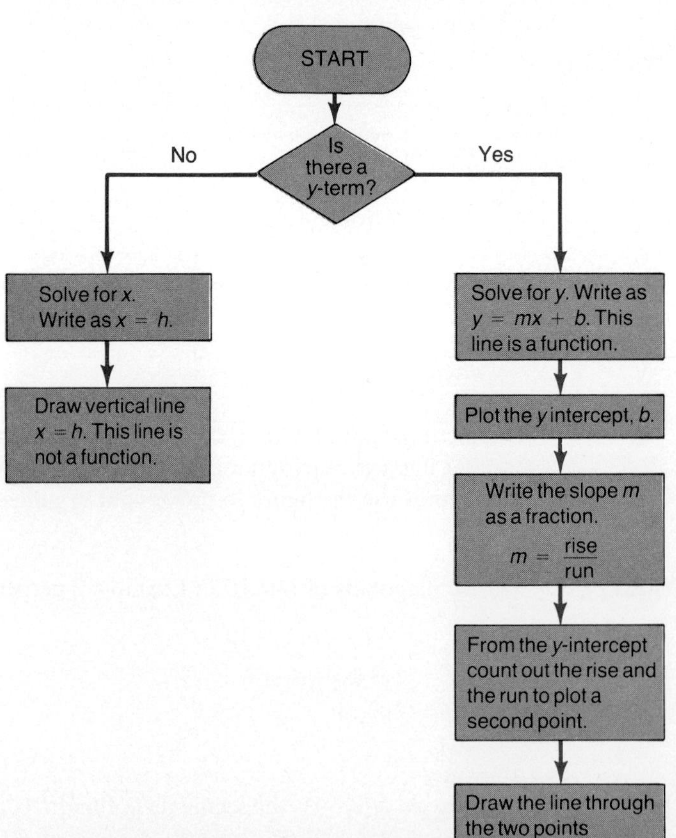

Figure 2.6 Procedure for graphing a line by using the slope-intercept form.

EXAMPLE 4 Graph $y = \frac{1}{2}x + 3$.

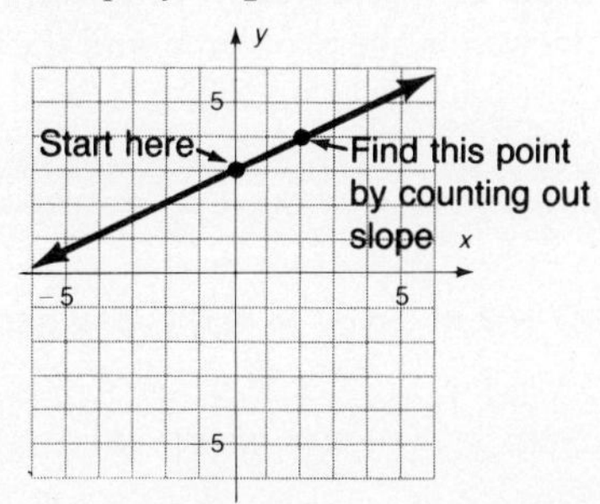

Solution By inspection, the y intercept is 3 and the slope is $\frac{1}{2}$; the line is graphed by first plotting the y intercept $(0, 3)$, then finding a second point by counting out the slope: up 1 and over 2. ■

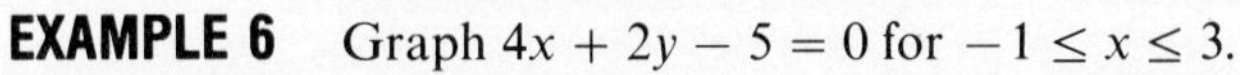

EXAMPLE 5 Graph $2x + 3y - 6 = 0$.

Solution Solve for y:

$$3y = -2x + 6$$

$$y = -\frac{2}{3}x + 2$$

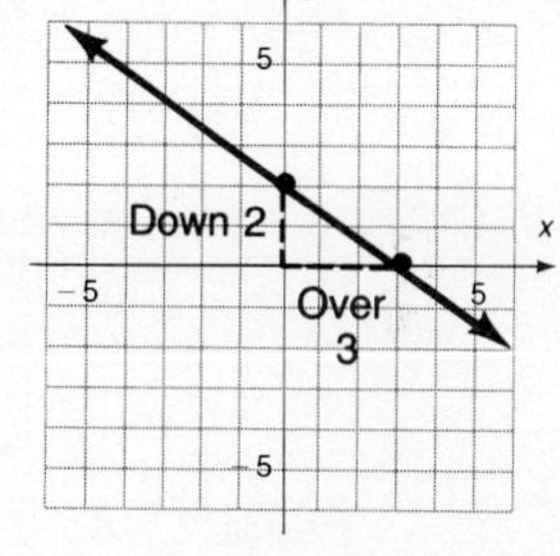

The y intercept is 2 and the slope is $-\frac{2}{3}$ the line is graphed as shown ■

EXAMPLE 6 Graph $4x + 2y - 5 = 0$ for $-1 \leq x \leq 3$.

Solution Solve for y:

$$2y = -4x + 5$$

$$y = -2x + \frac{5}{2}$$

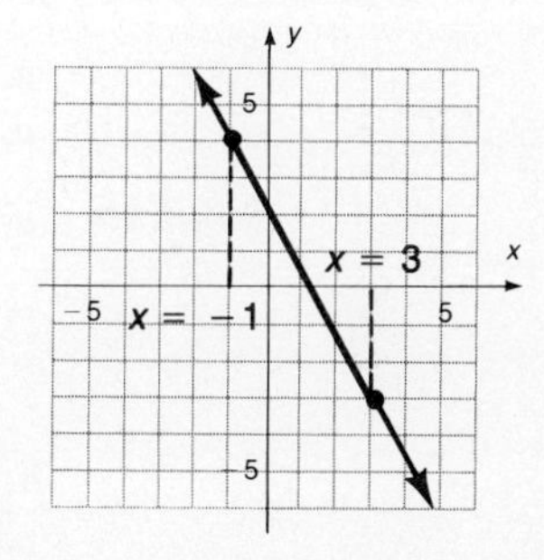

The y intercept is $\frac{5}{2}$ and the slope is -2; this line is shown as a dotted line. Because of the restriction on the domain you draw that part of the line with x values between -1 and 3 (inclusive) as shown by the colored line segment. ■

A variation of graphing linear functions is seen when we graph absolute value functions.

EXAMPLE 7 Graph $y = |x|$

Solution First apply the definition of absolute value:

$$y = x \quad \text{if} \quad x \geq 0$$

$$y = -x \quad \text{if} \quad x < 0$$

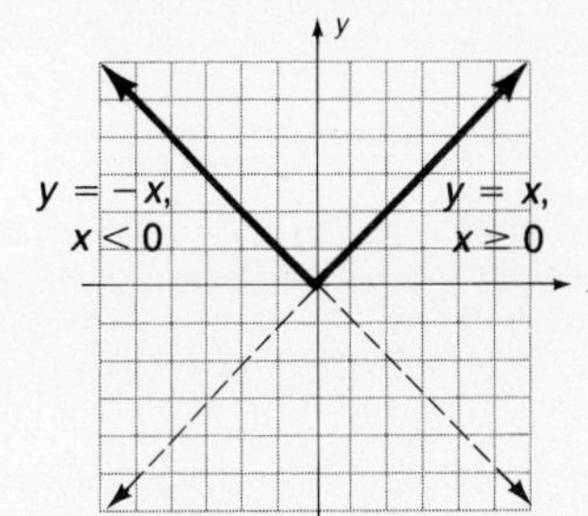

Now graph these two equations with restrictions as you did in Example 6. The graph of $y = |x|$ is the part shown in color. ■

In algebra you studied several forms of the equation of a line. The derivation of some of these is reviewed in the problems, and the forms are stated below for review.

Forms of a Linear Equation

STANDARD FORM: $\mathbf{Ax + By + C = 0}$, where (x, y) is any point on the line and A, B, and C are constants, A and B not both zero.

SLOPE-INTERCEPT FORM: $\mathbf{y = mx + b}$, where m is the slope and b is the y

POINT-SLOPE FORM: $\mathbf{y - k = m(x - h)}$, where m is the slope and (h, k) is a point on the line.

TWO-POINT FORM: $\mathbf{y - y_1 = \left(\dfrac{y_2 - y_1}{x_2 - x_1}\right)(x - x_1)}$, where (x_1, y_1) and (x_2, y_2) are points on the line.

INTERCEPT FORM: $\mathbf{\dfrac{x}{a} + \dfrac{y}{b} = 1}$, where $(a, 0)$ and $(0, b)$ are the x and y intercepts, respectively.

HORIZONTAL LINE passing through the point (h, k): $\mathbf{y = k}$.

VERTICAL LINE passing through the point (h, k): $\mathbf{x = h}$.

EXAMPLE 8 Find the equation of the line using the given information. Leave your answer in standard form.

a. y intercept 4; slope 6
b. Slope 4; passing through $(-3, 2)$
c. Passing through $(2, 3)$ and $(5, 7)$
d. No slope; passing through $(7, -3)$

Solution **a.** Since you are given the slope and the intercept, use the slope-intercept form: $y = mx + b$, where $b = 4$ and $m = 6$:

In standard form, $\mathbf{6x - y + 4 = 0}$.

b. Use the point-slope form, where $h = -3$, $k = 2$, and $m = 4$:

$$y - k = m(x - h)$$
$$y - 2 = 4(x + 3)$$
$$y - 2 = 4x + 12$$

In standard form, $\mathbf{4x - y + 14 = 0}$.

c. Use the two-point form:

$$y - 3 = \left(\frac{7 - 3}{5 - 2}\right)(x - 2)$$

This is the point (2, 3), but you could also use (5,7) to obtain the same result.

$$y - 3 = \frac{4}{3}(x - 2)$$

$$3y - 9 = 4x - 8$$

In standard form, $\mathbf{4x - 3y + 1 = 0}$.

d. Do not confuse no slope (vertical line) with zero slope (horizontal line). This is a vertical line, so the equation has the form $x = h$ when it passes through (h, k). Thus

$$x = 7$$

In standard form, $\mathbf{x - 7 = 0}$. ■

PROBLEM SET 2.3

A

Sketch the line passing through the points whose coordinates are given in Problems 1–9. Also find the slope of each line.

1. $(2, 3)$ and $(5, 6)$
2. $(0, 7)$ and $(3, 0)$
3. $(-1, -2)$ and $(4, 11)$
4. $(4, -2)$ and $(7, -3)$
5. $(-6, -4)$ and $(-9, 3)$
6. $(6, 0)$ and $(-3, 0)$
7. $(0, 0)$ and $(0, 3)$
8. $(4, -3)$ and $(4, 1)$
9. $(-1, 2)$ and $(3, 2)$

Graph the lines whose equations are given in Problems 10–21 by finding the slope and y intercept.

10. $y = 3x + 3$
11. $y = -4x - 1$
12. $y = \frac{2}{3}x + \frac{4}{3}$
13. $y = \frac{1}{5}x - \frac{6}{5}$
14. $y = 40x$
15. $y = 300x$
16. $x - 4 = 0$
17. $y + 2 = 0$
18. $5x - 4y - 8 = 0$
19. $x - 3y + 2 = 0$
20. $100x - 250y + 500 = 0$
21. $2x - 5y - 1200 = 0$

Graph the line segments or the absolute value functions in Problems 22–27.

22. $3x + y - 2 = 0, -7 \le x \le 1$
23. $2x - 2y - 6 = 0, 5 \le x \le 9$
24. $5x - 3y - 9 = 0, -3 \le x \le 1$
25. $y = 2|x|$
26. $y = -3|x|$
27. $y = |4x|$

Use slopes to decide whether the coordinates given in Problems 28–30 are vertices of a right triangle.

28. $T(-6, -4), R(6, 12), I(-4, 7)$
29. $A(1, -1), N(4, 1), G(0, 7)$
30. $L(-4, 6), E(-10, 2), S(-3, -1)$

Use slopes to decide whether the coordinates given in Problems 31–34 are vertices of a parallelogram.

31. $R(3, 0), E(6, 7), C(2, 9), T(-3, 3)$
32. $A(-1, 5), N(2, 3), G(-3, -4), E(-6, -2)$
33. $P(1, 10), A(-4, 5), R(-3, -2), L(2, 3)$
34. $E(-3, 6), L(9, 11), G(14, 1), R(2, -4)$
35. Are the diagonals of the quadrilateral in Problem 33 perpendicular?
36. Are the diagonals of the quadrilateral in Problem 34 perpendicular?

B

Find the equation of the line satisfying the given conditions in Problems 37–52. Give your answer in standard form.

37. y intercept 6; slope 5
38. y intercept -3; slope -2
39. y intercept 0; slope 0
40. y intercept 5; slope 0
41. Slope 3; passing through $(2, 3)$
42. Slope -1; passing through $(-4, 5)$
43. Slope $\frac{1}{2}$; passing through $(3, 3)$
44. Slope $\frac{2}{5}$; passing through $(5, -2)$
45. Passing through $(-4, -1)$ and $(4, 3)$
46. Passing through $(4, -2)$ and $(4, 5)$
47. Passing through $(5, 6)$ and $(1, -2)$
48. Passing through $(5, 6)$ and $(7, 6)$
49. Passing through $(2, 4)$ parallel to $2x + 3y - 6 = 0$
50. Passing through $(-1, -2)$ parallel to $x - 2y + 4 = 0$
51. Passing through $(-1, -2)$ perpendicular to $x - 2y + 4 = 0$
52. Passing through $(2, 4)$ perpendicular to $2x + 3y - 6 = 0$
53. Consider Figure 2.7a.
  a. What are the coordinates of A and B?
  b. What is the slope of the line passing through A and B?

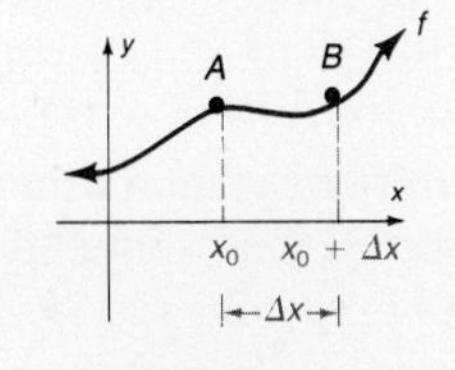

(a) Graph of f

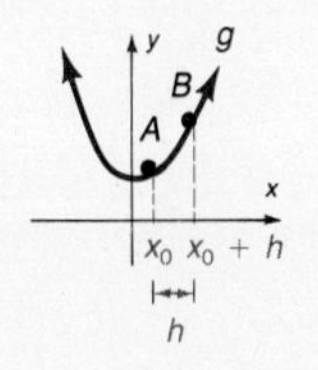

(b) Graph of g

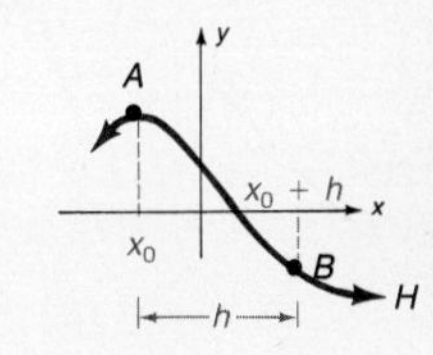

(c) Graph of H

Figure 2.7

54. Consider Figure 2.7b.
  a. What are the coordinates of A and B?
  b. What is the slope of the line passing through A and B?

55. Consider Figure 2.7c.
  a. What are the coordinates of A and B?
  b. What is the slope of the line passing through A and B?

Problems 56–62 provide some real-world examples of line graphs. One way of finding the equation of the line is to write two data points from the given information, and then use those two points to write the equation. Use the given information to write a standard-form equation of the line described by the problem.

56. The demand for a certain product is related to the price of the item. Suppose a new line of stationery is tested at two stores. It is found that 25 boxes are sold within a month if they are priced at \$1 and 15 boxes priced at \$2 are sold in the same time. Let x be the price and y be the number of boxes sold.

57. An important factor that is related to the demand for a product is the supply. The amount of the stationery in Problem 56 that can be supplied is also related to the price. At \$1 each 10 boxes can be supplied and at \$2 each 20 boxes can be supplied. Let x be the price and y be the number of boxes sold.

58. The population of Florida was roughly 6.8 million in 1970, and 9.7 million in 1980. Let x be the year (let 1950 be the base year; that is $x = 0$ represents 1950 and $x = 10$ represents 1960) and y be the population. Use this equation to predict the population in 1990.

59. The population of Texas was roughly 11.2 million in 1970, and 14.2 million in 1980. Let x be the year (let 1950 be the base year; that is, $x = 0$ represents 1950 and $x = 10$ represents 1960) and y be the population. Use this equation to predict the population in 1990.

60. It costs \$90 to rent a car if you drive 100 miles and \$140 if you drive 200 miles. Let x be the number of miles driven and y the total cost of the rental. Use this equation to find how much it would cost if you drove 394 miles.

61. It costs \$60 to rent a car if you drive 50 miles and \$60 if you drive 260 miles. Let x be the number of miles and y be the total cost of the rental. Use this equation to find how much it would cost if you drove 394 miles.

62. Suppose it costs \$100 for maintenance and repairs to drive a three-year-old car 1000 miles and \$650 for maintenance and repairs to drive it 6500 miles. Let x be the number of miles and y be the cost for repairs and maintenance.

63. Begin with the slope-intercept form and derive the point-slope form:

$$y - k = m(x - h)$$

64. Begin with the point-slope form and derive the two-point form:

$$y - y_1 = \left(\frac{y_2 - y_1}{x_2 - x_1}\right)(x - x_1)$$

65. Begin with the two-point form and derive the intercept form:

$$\frac{x}{a} + \frac{y}{b} = 1$$

66. Prove that if $m > 0$, then the linear function is an increasing function throughout its domain.

67. Prove that if $m < 0$, then the linear function is a decreasing function throughout its domain.

2.4 TRANSLATION OF FUNCTIONS

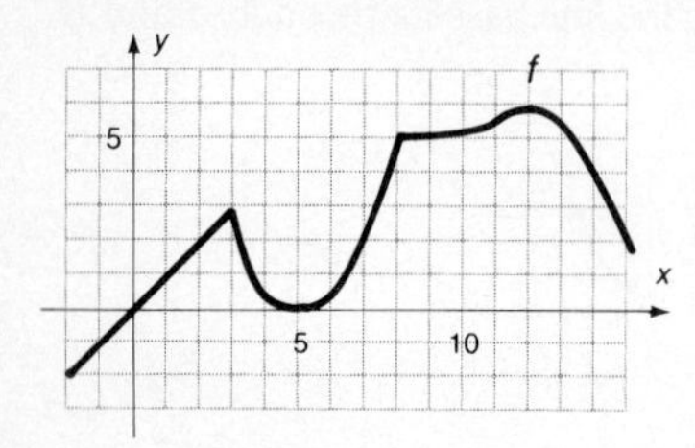

Figure 2.8 Graph of f

Consider the function f defined by the graph in Figure 2.8.
It is possible to shift the entire curve up, down, right, or left, as shown in Figure 2.9.

Instead of considering the curve shifting relative to fixed axes, consider the effect of shifting the axes. If the coordinate axes are shifted up k units, the origin of this new coordinate system would correspond to the point $(0, k)$ in the old coordinate system. If the axes are shifted to the right h units, the origin would correspond to the point $(h, 0)$ in the old system. A horizontal shift of h units followed by a vertical shift of k units would shift the new coordinate axes so that the origin corresponds to a point (h, k) on the old axes. Suppose a *new* coordinate system with origin at (h, k) is drawn and the new axes are labeled x' and y', as shown in Figure 2.10.

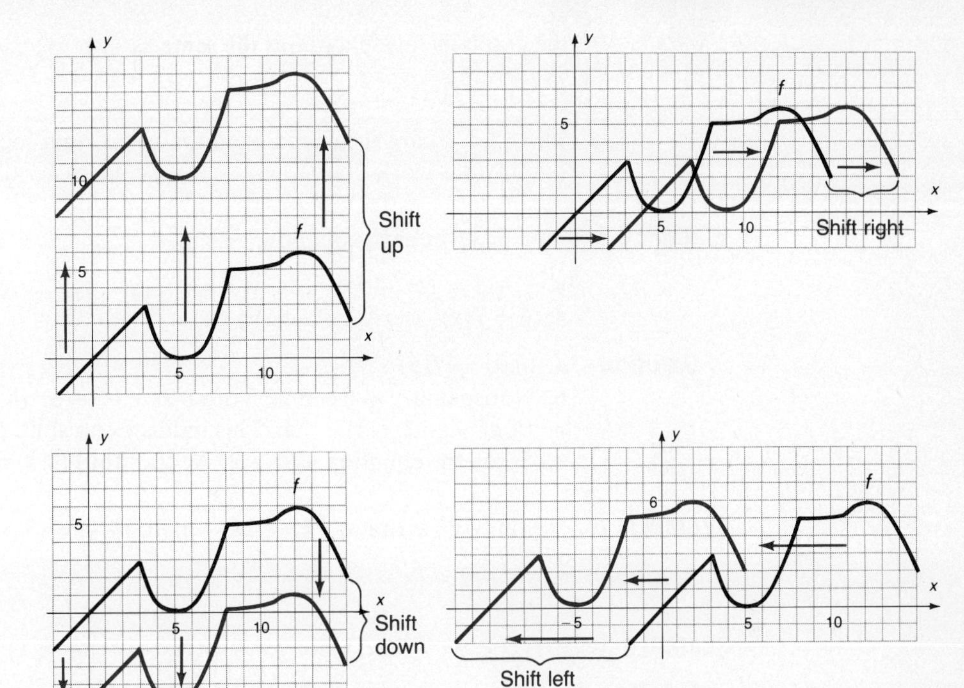

Figure 2.9 Shifting the graph of f

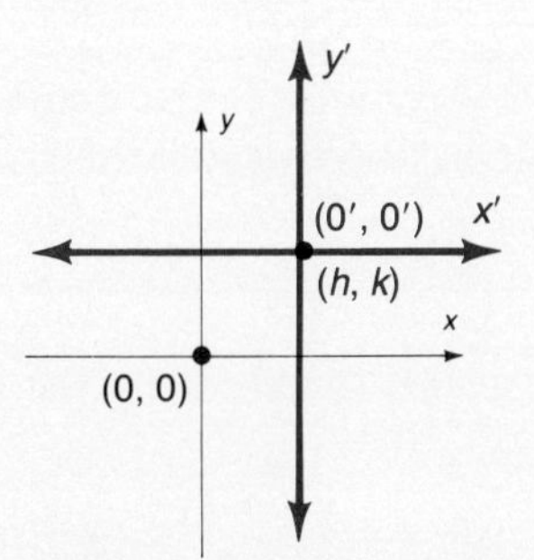

Figure 2.10 Shifting the axes to (h, k)

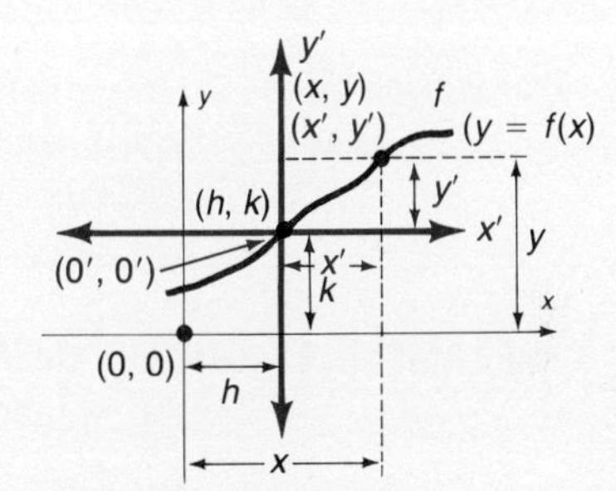

Figure 2.11 Comparison of coordinate axes

Every point on a given curve can now be denoted in two ways (see Figure 2.11):

1. As (x, y) measuring from the old origin
2. As (x', y') measuring from the new origin (color)

To find the relationship between (x, y) and (x', y'), consider the graph shown in Figure 2.11:

$$\begin{aligned} x &= x' + h \\ y &= y' + k \end{aligned} \qquad \text{or} \qquad \begin{aligned} x' &= x - h \\ y' &= y - k \end{aligned}$$

This says that if you are given any function

$$y - k = f(x - h)$$

the graph of this function is the same as

$$y' = f(x')$$

where (x', y') are measured from the new origin located at (h, k). This can greatly simplify our work since $y' = f(x')$ is usually easier to graph than $y - k = f(x - h)$.

EXAMPLE 1 Find (h, k) for each equation.

a. $y - 5 = f(x - 7)$ **b.** $y + 6 = f(x - 1)$ **c.** $y + 1 = f(x + 3)$
d. $y = f(x)$ **e.** $y - 6 = f(x) + 15$

Solution **a.** $(h, k) = (7, 5)$
b. Notice that $y + 6$ can be written as $y - (-6)$; $(h, k) = (1, -6)$.
c. $(h, k) = (-3, -1)$ **d.** This indicates no shift; $(h, k) = (0, 0)$.
e. Write the equation as $y - 21 = f(x)$; thus $(h, k) = (0, 21)$. ■

EXAMPLE 2 If $f(x) = x^2$, write $y - k = f(x - h)$ for $(h, k) = (3, -2)$.

Solution

$$y + 2 = (x - 3)^2$$ ■

EXAMPLE 3 If $f(x) = 3x^2 + 5x$, write $y - k = f(x - h)$ for $(h, k) = (-\sqrt{2}, \pi)$.

Solution

$$y - \pi = 3(x + \sqrt{2})^2 + 5(x + \sqrt{2})$$ ■

Translation of Axes

> The equation $y - k = f(x - h)$ is the graph of $y = f(x)$ on a system of coordinate axes that has been **translated** h units horizontally and k units vertically.

EXAMPLE 4 Given f defined by $y = f(x)$ as shown by the graph in Figure 2.12, graph the following functions.

a. $y = f(x - 3)$ **b.** $y + 2 = f(x)$ **c.** $y - 4 = f(x + 5)$

Solution **a.** Since $(h, k) = (3, 0)$, the shift is 3 units to the right.
b. Since $(h, k) = (0, -2)$, the shift is 2 units down.
c. Since $(h, k) = (-5, 4)$, the shift is 5 units to the left and 4 units up.

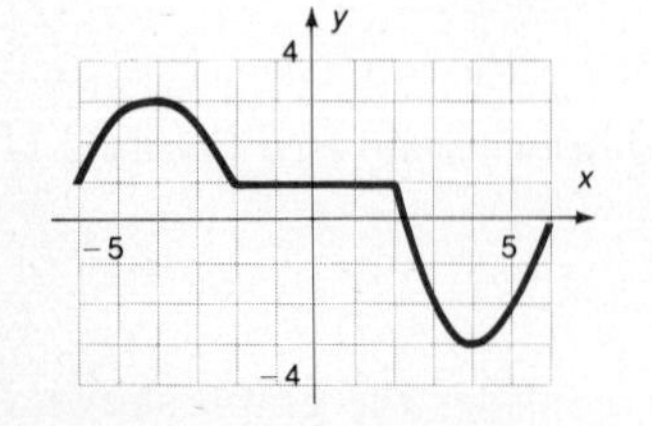

Figure 2.12 Graph of f

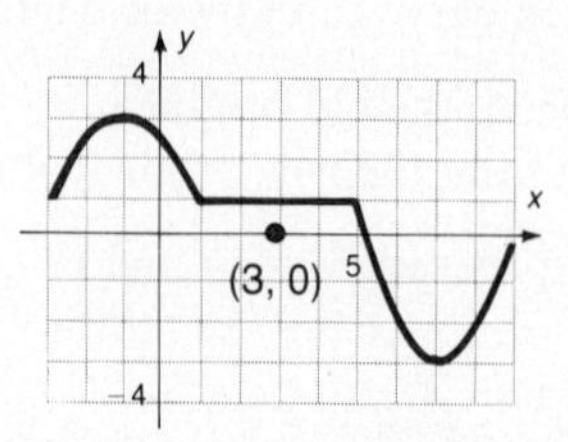

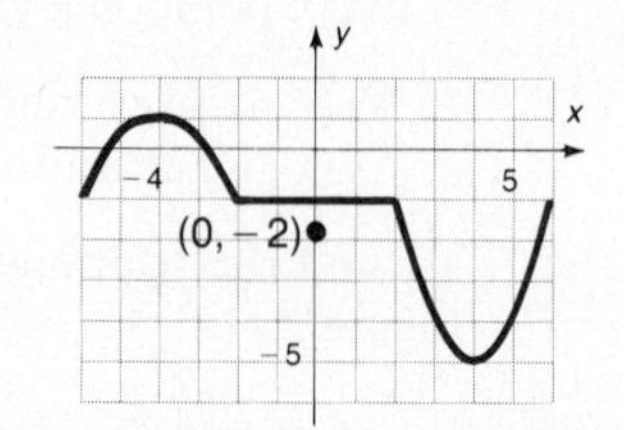

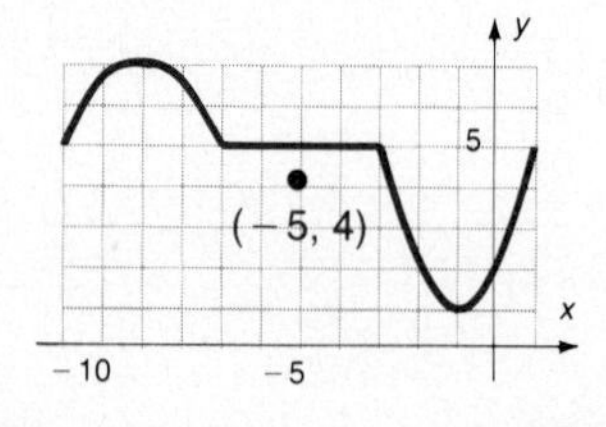

■

EXAMPLE 5 Substitute $x' = x + 3$ and $y' = y - 6$ into the given equations.

a. $y - 6 = (x + 3)^2$ **b.** $y = 5(x + 3)^2 + 6$
c. $y = 6 - 2(x + 3)^2$ **d.** $y - 6 = 7(x + 3)^2 + 2(x + 3)$

Solution
a. $y' = x'^2$
b. Rewrite as $y - 6 = 5(x + 3)^2$; thus $y' = 5x'^2$.
c. Rewrite as $y - 6 = -2(x + 3)^2$; thus $y' = -2x'^2$.
d. $y' = 7x'^2 + 2x'$ ■

Translations are useful because they sometimes allow us to take a computationally difficult equation and rewrite it in terms of a simpler equation. For example, recall the two previously considered curves shown in Figure 2.13.

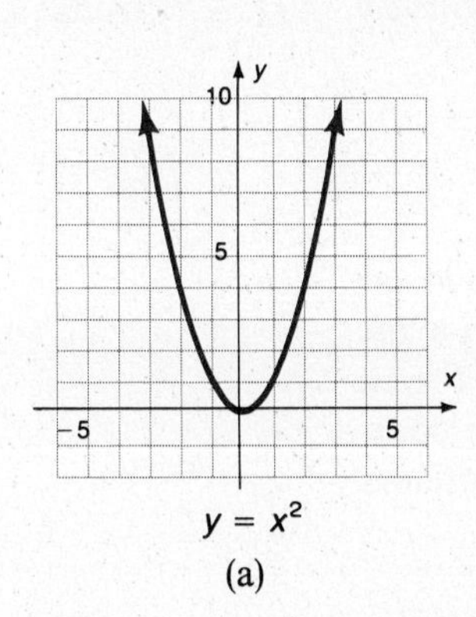

$y = x^2$
(a)

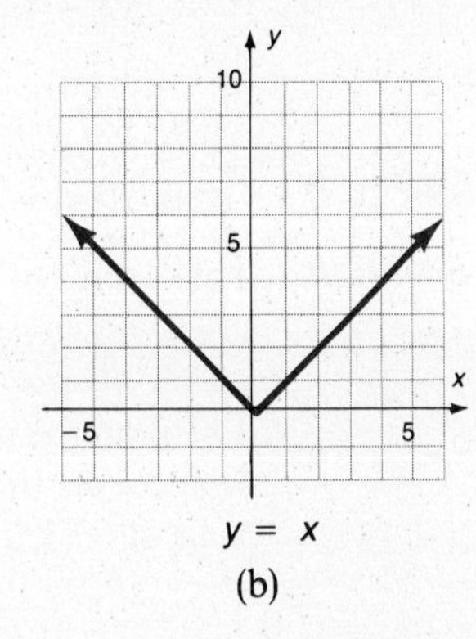

$y = x$
(b)

Figure 2.13 Standard parabola and absolute value curves

We can now draw the graphs of some rather complicated equations by using these curves and the idea of translation.

EXAMPLE 6 Graph $y - 3 = (x - 2)^2$ **(a)** by plotting points and **(b)** by using a translation.

Solution a. By plotting points

x	y
0	7
1	4
2	3
3	4
4	7
5	12
6	19

You will not appreciate the other method unless *you* actually *do* the arithmetic of this problem by this method

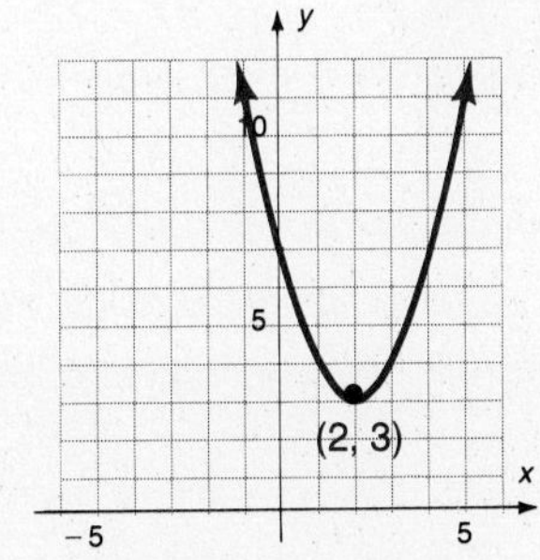

Graph of $y - 3 = (x - 2)^2$

b. By shifting axes

$(h, k) = (2, 3)$

Plot this point. Now, plot points in $y' = x'^2$ or draw the curve in Figure 2.13a using the point (2, 3) as the starting point.

x'	y'
1	1
−1	1
2	4
−2	4
3	9
−3	9

This table of values can be done mentally. Use (2, 3) as the origin when counting out these values. ■

EXAMPLE 7 Graph $y - \frac{1}{2} = (x - \frac{3}{2})^2$.

Solution First, plot $(\frac{3}{2}, \frac{1}{2})$. Next, *count* out the same table of values generated for part b of the previous example. Remember, you should be able to generate these values mentally since you are using the equation

$$y' = x'^2$$

■

Graph of $y - \frac{1}{2} = (x - \frac{3}{2})^2$

As you can see, the graph of $y - k = f(x - h)$ is done in two steps:

1. Plot (h, k).
2. Graph the simpler curve $y' = f(x')$ by using (h, k) as the new origin.

EXAMPLE 8 Graph $f(x) = |x - 3| + 2$.

Solution This can be rewritten as $y - 2 = |x - 3|$, which is the graph of $y = |x|$ (see Figure 2.13b) on the system of coordinate axes that is translated to $(h, k) = (3, 2)$.

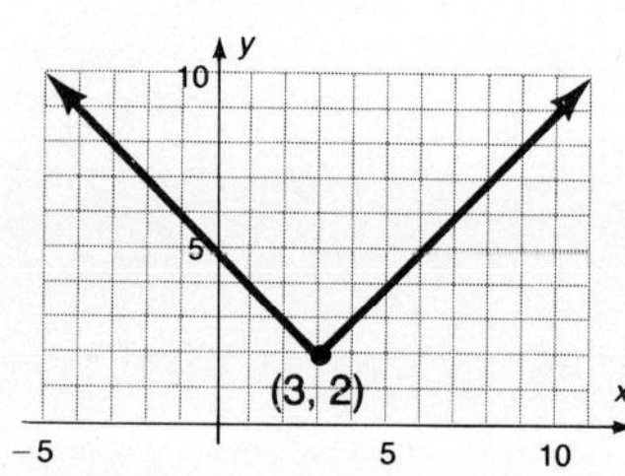

Graph of $f(x) = |x - 3| + 2$

■

PROBLEM SET 2.4

A

Find (h, k) for each of the functions given in Problems 1–12.

1. $y - 3 = f(x - 6)$
2. $y - 5 = f(x + 3)$
3. $y + 1 = f(x - 6)$
4. $y = f(x - 4)$
5. $y - \sqrt{2} = f(x)$
6. $y = f(x) + 6$
7. $y = f(x - 3) - 4$
8. $y = f(x + 1) + 2$
9. $y = 4f(x)$
10. $y = 2f(x + \sqrt{2})$
11. $6y = f(x + 3)$
12. $y = 2f\left(x + \frac{\pi}{2}\right) - \frac{\pi}{3}$

If $f(x) = x^2$, write $y - k = f(x - h)$ for the (h, k) given in Problems 13–17.

13. $(2, 3)$
14. $(-3, 0)$
15. $(0, -1)$
16. $\left(-\pi, \frac{\pi}{4}\right)$
17. $(-\sqrt{3}, -2)$

If $f(x) = |x|$, write $y - k = f(x - h)$ for the (h, k) given in Problems 18–22.

18. $(6, 1)$
19. $(0, -6)$
20. $(-\pi, 0)$
21. $(\sqrt{2}, -\sqrt{3})$
22. $\left(-\frac{\pi}{2}, -\frac{\sqrt{2}}{2}\right)$

Substitute $x' = x - 5$ and $y' = y + 1$ into the equations given in Problems 23–31.

23. $y + 1 = (x - 5)^2$
24. $y + 1 = \frac{2}{3}(x - 5)^2$
25. $y + 1 = -5(x - 5)^2$
26. $y = (x - 5)^2 - 1$
27. $y = -2(x - 5)^2 - 1$
28. $y = -1 + 3(x - 5)^2$
29. $y + 1 = |x - 5|$
30. $y + 1 = \frac{2}{3}|x - 5|$
31. $y = -2|x - 5| - 1$

B

Let f, g, and h be the functions whose graphs are shown in Figure 2.14. Graph the functions indicated by the equations in Problems 32–49.

32. $y + 3 = f(x)$
33. $y + 2 = g(x)$
34. $y + \frac{1}{2} = h(x)$
35. $y - 5 = g(x)$
36. $y - 1 = h(x)$
37. $y + \pi = f(x)$
38. $y = h\left(x - \frac{\pi}{2}\right)$
39. $y = g(x + 3)$
40. $y = f(x - \sqrt{5})$
41. $y + 4 = f(x - 3)$
42. $y + 1 = g(x - 4)$
43. $y - 2 = h(x + \pi)$

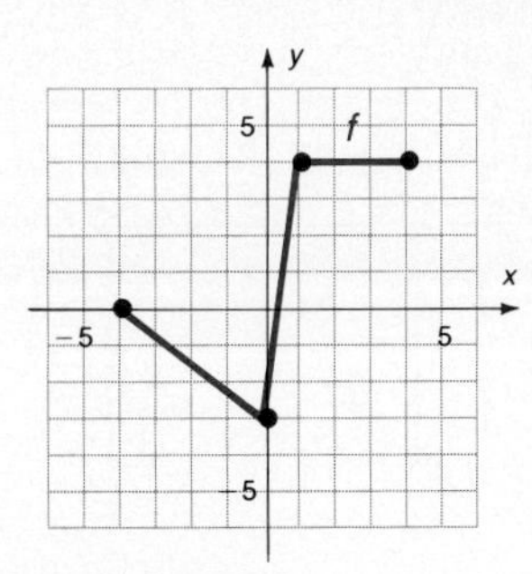

Figure 2.14 (a) Graph of f (b) Graph of g (c) Graph of h

44. $y - 3 = g(x + 5)$

45. $y - 2 = h(x + 3)$

46. $y + 2 = f(x - 4)$

47. $y - \frac{21}{5} = g\left(x - \frac{15}{2}\right)$

48. $y - \sqrt{3} = f(x + \sqrt{2})$

49. $y + \pi = h\left(x - \frac{3\pi}{2}\right)$

Graph the functions given in Problems 50–66.

50. $y - 3 = (x - 2)^2$

51. $y = (x + 3)^2$

52. $y = x^2 + 1$

53. $y - 1 = |x - 7|$

54. $y = |x + \pi|$

55. $y = |x| + 6$

56. $y + \sqrt{3} = |x - \sqrt{2}|$

57. $y - \frac{\pi}{4} = (x + \pi)^2$

58. $y + 2 = (x + \sqrt{3})^2$

59. $y - \sqrt{2} = (x + \sqrt{5})^2$

60. $y - \sqrt{2} = (x - \sqrt{3})^2$

61. $y - \pi = \left|x + \frac{\pi}{6}\right|$

C

62. $y + 3 = (x + 8)^2$, such that $-14 \le x \le -8$

63. $y - 2 = (x + 3)^2$, such that $-7 \le x \le -2$

64. $y - 12 = \left(x + \frac{25}{2}\right)^2$, such that $y > -10$

65. $y + 3 = (x + 3)^2$, such that $y < 6$

66. $y + 12 = (x - 8)^2$, such that $y < 4$

2.5 QUADRATIC FUNCTIONS

The second type of function to be considered in this chapter is the quadratic function.

Quadratic Function

> A function f is a **quadratic function** if
>
> $$f(x) = ax^2 + bx + c$$
>
> where a, b, and c are real numbers and $a \neq 0$.

If $b = c = 0$, however, the quadratic function has the form

$$y = ax^2$$

and has a graph called a **standard-position parabola**. You graphed functions of this type by plotting points in Section 1.6. We begin by sketching two additional standard-position parabolas in order to make some generalizations.

EXAMPLE 1 Sketch the graph of $y = 3x^2$.

Solution Find some ordered pairs satisfying the equation:

x	0	1	2	3	...
y	0	3	12	27	...

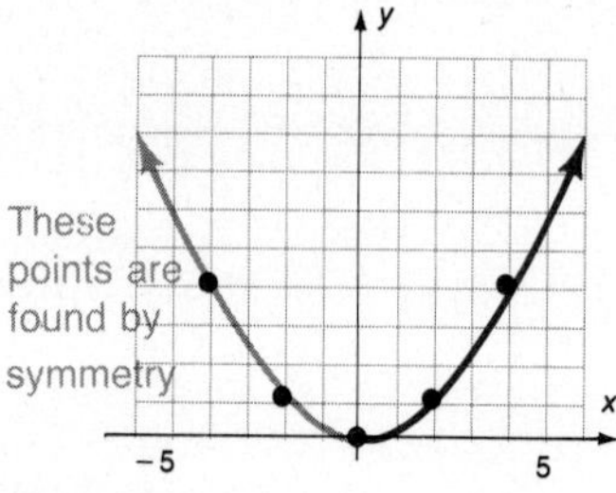

Figure 2.15 Graph of $y = 3x^2$

Also, if $y = f(x)$ then

$$f(-x) = 3(-x)^2 = 3x^2$$

so the graph is symmetric with respect to the y axis. Plot the points represented by the ordered pairs, use symmetry, and draw a smooth graph as shown in Figure 2.15. ∎

EXAMPLE 2 Sketch the graph of $y = -\frac{1}{2}x^2$.

x	0	1	2	3	4
y	0	$-\frac{1}{2}$	-2	$-\frac{9}{2}$	-8

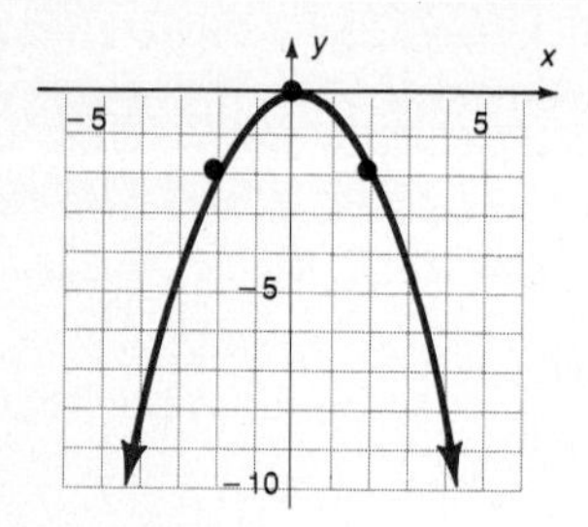

Figure 2.16 Graph of $y = -\frac{1}{2}x^2$

Plot these points and use symmetry as shown in Figure 2.16. ∎

We will now make some general observations based on the special case $y = ax^2$:

1. The graph has a shape of a curve which is called a **parabola**.
2. The point $(0, 0)$ is the lowest point if the parabola opens up $(a > 0)$; $(0, 0)$ is the highest point if the parabola opens down $(a < 0)$. This highest or lowest point is called the **vertex**.
3. The parabola is **symmetric** with respect to the vertical line passing through the vertex.
4. Relative to a fixed scale, the magnitude of a determines the "wideness" of the parabola: Small values of $|a|$ yield "wide" parabolas; large values of $|a|$ yield "narrow" parabolas.

For graphs of parabolas of the form

$$y - k = a(x - h)^2$$

you simply translate the axes to the point (h, k) and then graph the parabola $y' = ax'^2$, as shown in Example 3. The point (h, k) is the vertex. Using functional notation, this can be written

$$f(x) = a(x - h)^2 + k$$

EXAMPLE 3 Sketch the graph of $y + 5 = 3(x + 2)^2$.

Solution By inspection, $(h, k) = (-2, -5)$ and the standard-position parabola is $y = 3x^2$. The table of values (or points to plot) is the same as shown for Example 1, which is repeated below; the only difference here is that you count out these points from $(-2, -5)$ instead of from the origin.

Table from Example 1

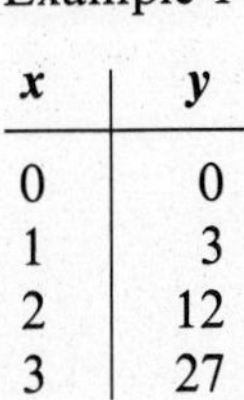

x	y
0	0
1	3
2	12
3	27

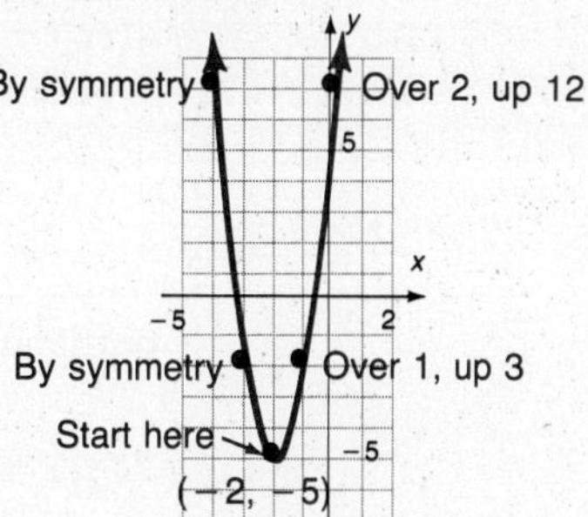

Graph of $y + 5 = 3(x + 2)^2$ ■

Now you are ready to consider the general quadratic function

$$y = ax^2 + bx + c \qquad (a \neq 0)$$

Consider Example 3 above: $y + 5 = 3(x + 2)^2$. This was graphed by translating the axes to $(-2, -5)$ and then considering $y' = 3x'^2$. Suppose you rewrite the given equation as

$$y + 5 = 3(x + 2)^2$$
$$y + 5 = 3(x^2 + 4x + 4)$$
$$y + 5 = 3x^2 + 12x + 12$$
$$y = 3x^2 + 12x + 7$$

The last form is the general quadratic form, where $a = 3, b = 12$, and $c = 7$. Suppose you are given this form and told to graph the curve. Then, to reverse the process, you must **complete the square** (review Section 1.4).

To complete the square, follow these steps:

Step 1 Subtract c (the constant term) from both sides:

General Form	*Example*
$y = ax^2 + bx + c$	$y = 2x^2 + 12x + 7$
$y - c = ax^2 + bx$	$y - 7 = 2x^2 + 12x$

Step 2 Factor the a term from the expression on the right (remember $a \neq 0$):

$$y - c = a\left(x^2 + \frac{b}{a}x\right) \qquad y - 7 = 2(x^2 + 6x)$$

Step 3 To complete the square on the number inside parentheses:

$$x^2 + \frac{b}{a}x + ? = (x + ?)^2 \qquad x^2 + 6x + ? = (x + ?)^2$$

you square one-half the coefficient of the x term:

$$\left(\frac{b}{2a}\right)^2 = \frac{b^2}{4a^2} \qquad \left(\frac{6}{2}\right)^2 = 9$$

Then add a times this number to both sides of the original equation:

$$y - c + \frac{b^2}{4a} = a\left(x^2 + \frac{b}{a}x + \frac{b^2}{4a^2}\right) \qquad y - 7 + 18 = 2(x^2 + 6x + 9)$$

$\frac{1}{2}\cdot\frac{b}{a}$ squared — $\frac{1}{2}\cdot 6$ squared

distributive property — distributive property

$a\cdot\frac{b^2}{4a^2} = \frac{b^2}{4a}$ — $2\cdot 9 = 18$

Add $\frac{b^2}{4a}$ to both sides — Add 18 to both sides

Step 4 Factor the right-hand side as a perfect square and simplify the left-hand side:

$$y + \frac{-4ac}{4a} + \frac{b^2}{4a} = a\left(x + \frac{b}{2a}\right)^2 \qquad y - 7 + 18 = 2(x + 3)^2$$

common denominator if fractions are involved

$$\left(y + \frac{b^2 - 4ac}{4a}\right) = a\left(x + \frac{b}{2a}\right)^2 \qquad y + 11 = 2(x + 3)^2$$

This is of the form

$$y - k = a(x - h)^2$$

and is called the **general form of a parabola**. This equation can be graphed by doing a translation as shown in Example 4.

EXAMPLE 4 Sketch the graph of $y = x^2 + 6x + 10$.

Solution Complete the square:

$$y - 10 = x^2 + 6x$$

$$y - 10 + \mathbf{9} = x^2 + 6x + \mathbf{9}$$

Since $\frac{1}{2}\cdot 6 = 3$ and $3^2 = 9$, you add 9 to both sides.

$$y - 1 = (x + 3)^2$$

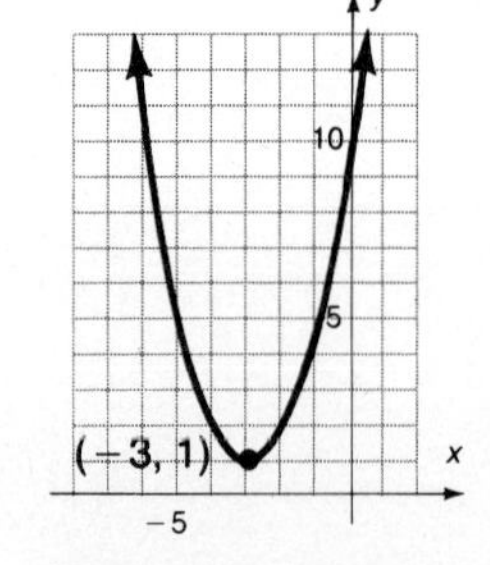

Figure 2.17 Graph of $y = x^2 + 6x + 10$

The vertex is at $(-3, 1)$, the parabola opens up, and the graph is shown in Figure 2.17. ■

Notice from Example 4 above that $c = 10$ (compare the given equation with the form $y = ax^2 + bx + c$) and that the y intercept is $(0, 10)$. In general, if $x = 0$ then

$$y = a(0)^2 + b(0) + c$$
$$= c$$

which shows that the **y intercept for a quadratic function is (0, *c*)**. This often serves as a checkpoint for graphs such as the one shown in Example 4.

EXAMPLE 5 Sketch the graph of $y = 1 - 5x - 2x^2$

Solution

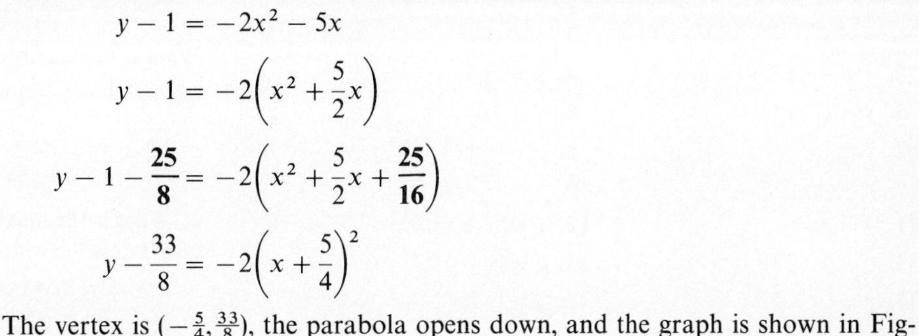

$$y - 1 = -2x^2 - 5x$$

$$y - 1 = -2\left(x^2 + \frac{5}{2}x\right)$$

$$y - 1 - \mathbf{\frac{25}{8}} = -2\left(x^2 + \frac{5}{2}x + \mathbf{\frac{25}{16}}\right)$$

$$y - \frac{33}{8} = -2\left(x + \frac{5}{4}\right)^2$$

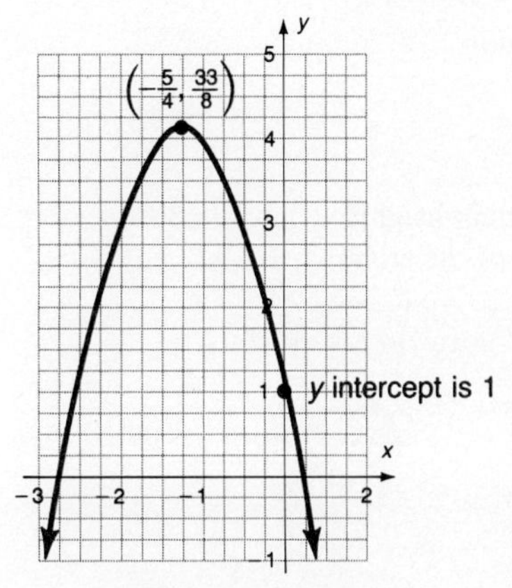

Figure 2.18 Graph of $y = 1 - 5x - 2x^2$

The vertex is $(-\frac{5}{4}, \frac{33}{8})$, the parabola opens down, and the graph is shown in Figure 2.18. For fractions, you can sometimes choose a scale that is more convenient than one square per unit. Notice in Figure 2.18 that there are four squares per unit. Also, remember that once you have found the vertex, you simply have to graph $y = -2x^2$ translated to the point $(-\frac{5}{4}, \frac{33}{8})$. ■

Many applications are concerned with finding the maximum or minimum value of a function f. If f is a quadratic function defined by

$$y - k = a(x - h)^2$$

then the **maximum or minimum value is at the vertex**. That is, if $x = h$ then the maximum value of $f(x) = a(x - h)^2 + k$ is $y = k$. You can see this is true because $(x - h)^2$ is necessarily nonnegative for all x and zero if and only if $x = h$. For Example 5, the maximum value of $y = \frac{33}{8}$, which occurs for $x = -\frac{5}{4}$.

EXAMPLE 6 A small manufacturer of CB radios determines that the price of each item is related to the number of items produced. Suppose that x items are produced per day where the maximum number that can be produced is 10 items, and that the profit, in dollars, is determined to be

$$P(x) = -30(x - 6)^2 + 480$$

How many radios should be produced in order to maximize the profit?

Solution The profit function has a graph which opens down. The maximum profit is found at the vertex of this parabola as shown in Figure 2.19.

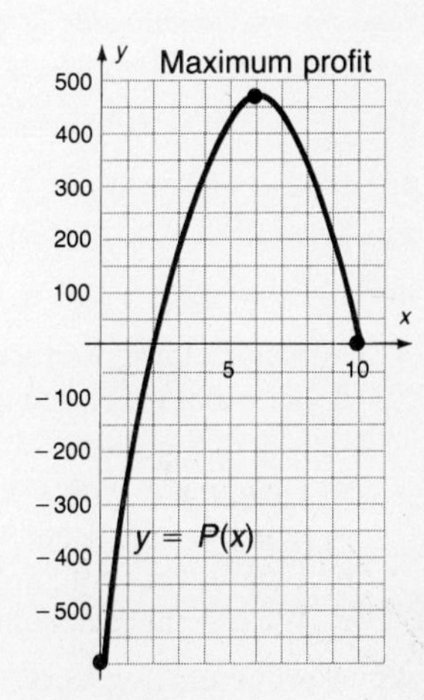

Figure 2.19 Graph of $P(x) = -30(x - 6)^2 + 480$

The vertex of the function

$$P(x) - 480 = -30(x - 6)^2$$

is (6, 480) so the maximum profit of \$480 is obtained when 6 radios are manufactured. Notice also from the graph that if there were a strike and no radios could be produced, the daily profit would be -600 (this is a \$600 per day *loss*). ■

PROBLEM SET 2.5

A

Sketch the graph of each equation given in Problems 1–24.

1. $y = x^2$
2. $y = -x^2$
3. $y = -2x^2$
4. $y = 2x^2$
5. $y = -5x^2$
6. $y = 5x^2$
7. $y = \frac{1}{3}x^2$
8. $y = -\frac{1}{3}x^2$
9. $y = \frac{1}{10}x^2$
10. $y = -\frac{1}{10}x^2$
11. $y = \pi x^2$
12. $y = -2\pi x^2$
13. $y = (x - 3)^2$
14. $y = (x - 1)^2$
15. $y = (x + 2)^2$
16. $y = -2(x - 1)^2$
17. $y = \frac{1}{4}(x - 1)^2$
18. $y = \frac{1}{2}(x + 1)^2$
19. $y = \frac{1}{3}(x + 2)^2$
20. $y - 2 = (x - 1)^2$
21. $y - 2 = 3(x + 2)^2$
22. $y - 2 = -\frac{3}{5}(x - 1)^2$
23. $y + 3 = \frac{2}{3}(x + 2)^2$
24. $y - 1 = \frac{1}{3}(x - 4)^2$

B

Sketch the graph of each equation given in Problems 25–40.

25. $y = x^2 + 4x + 4$
26. $y = x^2 + 6x + 9$
27. $y = x^2 + 2x - 3$
28. $y = 2x^2 - 4x + 5$
29. $y = 2x^2 - 4x + 4$
30. $y = 2x^2 + 8x + 5$
31. $y = 2x^2 + 8x + 5$
32. $y = 3x^2 - 30x + 76$
33. $y = \frac{1}{2}x^2 + 2x - 1$
34. $y = \frac{1}{2}x^2 + 4x + 10$
35. $y = \frac{1}{2}x^2 - x + \frac{5}{2}$
36. $y = \frac{1}{2}x^2 - 3x + \frac{3}{2}$
37. $x^2 - 6x - 2y - 1 = 0$
38. $x^2 + 2x + 2y - 3 = 0$
39. $2x^2 - 12x - 3y + 6 = 0$
40. $2x^2 - 4x - 3y + 11 = 0$

Find the maximum value of y for the functions defined by the equations in Problems 41–46.

41. $y = -4x^2 - 8x - 1$
42. $y = -3x^2 + 6x - 5$
43. $10x^2 - 160x + y + 655 = 0$
44. $6x^2 + 84x + y + 302 = 0$
45. $9x^2 + 6x + 81y - 53 = 0$
46. $100x^2 - 120x + 25y + 41 = 0$
47. A manufacturer produces high-quality boats at a profit, in dollars, that is determined to be
$$P(x) = -10(x - 375)^2 + 1{,}156{,}250$$
   a. How many boats should be produced in order to maximize the profit?
   b. What is the profit (or loss) if no boats are produced?
   c. What is the maximum profit?
48. A profit function, P, is
$$P(x) = -10(x - 75)^2 + 3750$$
   Find the maximum profit.
49. The profit function for a ratchet flange is
$$P(x) = -2(x - 25)^2 + 650$$
   What is the maximum profit?
50. An arch has an equation
$$y - 18 = -\frac{2}{81}x^2$$
   a. What is the maximum height?
   b. What is the width of the arch?

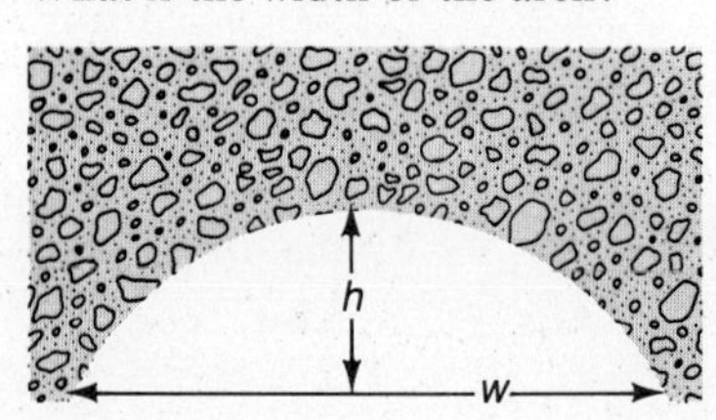

51. What is the height of the arch in Problem 50 at the following places?
   a. Nine feet from the center
   b. Eighteen feet from the center
52. One side of a storage yard is against a building. The other three sides of the rectangular yard are to be fenced with 36 ft of fencing. How long should the sides be to produce the greatest area with the given length of fence? What is the area obtained?
53. The sum of the length and the width of a rectangular area is 50 m. Find the greatest area possible, and find the dimensions of the figure.
54. ***Physics*** The highest bridge in the world is the bridge over the Royal Gorge of the Arkansas River in Colorado. It is 1053 ft above the water. If a rock is projected vertically upward from this bridge with an initial velocity of 64 fps, the height h of the object above the river at time t is described by the function
$$h(t) = -16t^2 + 64t + 1053$$
   What is the maximum height possible for a rock projected vertically upward from the bridge with an initial velocity of 64 fps? After how many seconds does it reach that height?
55. ***Physics*** In 1974, Evel Knievel attempted a skycycle ride across the Snake River. Suppose the path of the skycycle is given by the equation
$$d(x) = -0.0005x^2 + 2.39x + 600$$
   where $d(x)$ is the height in feet above the canyon floor for a

horizontal distance of x units from the launching ramp. What was Knievel's maximum height?

C

Sketch the graph of each equation given in Problems 56–61.

56. $y = x^2 - 5x + 2$

57. $2x^2 - x - y + 3 = 0$

58. $2x^2 - 8x + 3y + 20 = 0$

59. $4x^2 - 20x - 16y + 33 = 0$

60. $4x^2 + 24x - 27y - 17 = 0$

61. $25x^2 - 30x - 5y + 2 = 0$

62. ***Business*** A small manufacturer of CB radios determines that the price of each item is related to the number of items produced. If x items are produced per day, and the maximum number that can be produced is 10 items, then the price should be

$$400 - 25x \text{ dollars}$$

It is also determined that the overhead (the cost of producing x items) is

$$5x^2 + 40x + 600 \text{ dollars}$$

The daily profit is then found by subtracting the overhead from the revenue:

$$\begin{aligned} \text{profit} &= \text{revenue} - \text{cost} \\ &= (\text{number of items})(\text{price per item}) - \text{cost} \\ &= x(400 - 25x) - (5x^2 + 40x + 600) \\ &= 400x - 25x^2 - 5x^2 - 40x - 600 \\ &= -30x^2 + 360x - 600 \end{aligned}$$

A negative profit is called a loss.

a. What is the domain for f?

b. For what values of the domain does the manufacturer show a positive profit?

c. Does the manufacturer ever show a loss? For what values?

d. If there were a strike and production were brought to a halt, what would be the value of x? What would be the profit (or loss) for this situation?

e. How many items should the manufacturer produce per day, and what should be the expected daily profit in order to maximize the profit?

63. ***Physics (General Interest)***

a. In most states, drivers are required to know approximately how long it takes to stop their cars at various speeds. Suppose you estimate three car lengths for 30 mph and six car lengths for 60 mph. One car length per 10 mph assumes a linear relationship between speed and distance covered. Write a linear equation where x is the speed of the car and y is the distance traveled by the car in feet. (Assume that one car length = 20 ft.)

b. The scheme for the stopping distance of a car given in part *a* is convenient but not accurate. The stopping distance is more accurately approximated by the quadratic equation

$$y = 0.071x^2$$

This says that your car requires four times as many feet to stop at 60 mph as at 30 mph. Doubling your speed quadruples the braking distance. Graph this quadratic equation and the linear equation you found in part *a* on the same coordinate axes.

c. Comment on the results from part *b*. At about what speed are the two measures the same?

2.6 COMPOSITE AND INVERSE FUNCTIONS

Suppose a farmer sells eggs to Safeway. If the farmer's price for a dozen eggs is x dollars, then there is a function f that can be used to describe $f(x)$, the total cost of those eggs to Safeway. Note that x and $f(x)$ are not the same because Safeway must pay for ordering, shipping, and distributing the eggs. Moreover, in order to determine the price to the consumer, Safeway must add an appropriate markup. Suppose this markup function is called g. Then the price of the eggs to the consumer is

$$g[f(x)] \qquad \text{and not} \qquad g(x)$$

since the markup must be on Safeway's *total* cost of the eggs and not just on the price the farmer charges for the eggs. This process of evaluating a function of a function illustrates the idea of composition of functions.

Consider two functions f and g such that f is a function from X to Y and g is a function from Y to Z as illustrated by Figure 2.20.

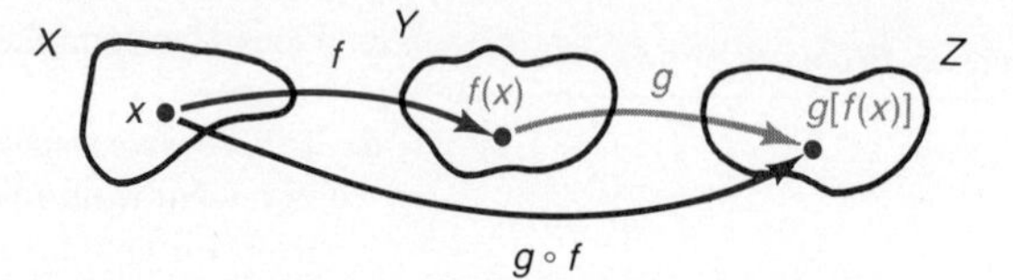

Figure 2.20 Composition of f and g

The image of x in the set Z is the number $g[f(x)]$ and defines a function from X to Z called the **composition of functions f and g.**

Composite Function

> Let X, Y, and Z be sets of real numbers. Let f be a function from X to Y and g be a function from Y to Z. Then the **composite function $g \circ f$** is the function from X to Z defined by
>
> $$(g \circ f)(x) = g[f(x)]$$

EXAMPLE 1 If $f(x) = x^2$ and $g(x) = x + 4$, find:

a. $(g \circ f)(x)$ **b.** $(f \circ g)(x)$ **c.** $(g \circ f)(-1)$ **d.** $(f \circ g)(5)$

Solution

a. $(g \circ f)(x) = g[f(x)]$
$= g[x^2]$
$= x^2 + 4$

b. $(f \circ g)(x) = f[g(x)]$
$= f[x + 4]$
$= (x + 4)^2$
$= x^2 + 8x + 16$

c. $(g \circ f)(-1) = (-1)^2 + 4$
$= 5$

d. $(f \circ g)(5) = 5^2 + 8(5) + 16$
$= 81$

■

Notice from Examples 1a and 1b that $g \circ f \neq f \circ g$. Note also that when we are applying the definition of $g \circ f$, it is not necessary that the domain of g be all of Y—it must simply contain the range of f. For this reason you may sometimes want to restrict x to some subset of X so that $f(x)$ is in the domain of g as illustrated by Figure 2.21.

Domain of f Range of f

f

$(g \circ f)(x_1)$ is not defined since $f(x)$ is not in the domain of g

x_1

x_2

Part of f maps into this region

$f(x_1)$

$f(x_2)$

Part of f maps into this region

$(g \circ f)(x_2)$

f

g

$g \circ f$ is defined for this region since it is the domain of g

Domain of g (shaded)

Range of g

Figure 2.21 Composition $g \circ f$ is defined for the subset of the domain of f for which $g \circ f$ is defined (shaded portion of the domain of f)

EXAMPLE 2 $f = \{(0,0), (-1,1), (-2,4), (-3,9), (5,25)\}$

$g = \{(0,-5), (-1,0), (2,-3), (4,-2), (5,-1)\}$

a. Find $g \circ f$. **b.** Find $f \circ g$.

Solution **a.**

f	g
$0 \to 0 \cdots \to 0 \to -5$	
$-1 \to 1 \cdots \to$ Not defined, so go back to domain of f and exclude -1	
$-2 \to 4 \cdots \to 4 \to -2$	
$-3 \to 9 \cdots \to$ Not defined; exclude -3 from the domain of $g \circ f$	
$5 \to 25 \cdots \to$ Not defined; exclude 5 from the domain of $g \circ f$	

$$g \circ f = \{(0,-5), (-2,-2)\}$$

Notice that -1, -3, and 5 are excluded from the domain of $g \circ f$ even though they are in the domain of f.

b.

g	f
$0 \to -5 \cdots \to$ Not defined	
$-1 \to 0 \cdots \to 0 \to 0$	
$2 \to -3 \cdots \to -3 \to 9$	
$4 \to -2 \cdots \to -2 \to 4$	
$5 \to -1 \cdots \to -1 \to 1$	

$$f \circ g = \{(-1,0), (2,9), (4,4), (5,1)\}$$

■

Inverse Functions

If f is a one-to-one function from X onto Y, then a function g from Y to X is called the **inverse function of f** whenever

$$(g \circ f)(x) = x \qquad \text{for every } x \text{ in } X$$

and

$$(f \circ g)(x) = x \qquad \text{for every } x \text{ in } Y$$

EXAMPLE 3 Show that f and g defined by

$$f(x) = 5x + 4 \qquad \text{and} \qquad g(x) = \frac{x-4}{5}$$

are inverse functions.

Solution You must show that f and g are inverse functions in two parts:

$$(g \circ f)(x) = g(5x + 4) = \frac{(5x + 4) - 4}{5} = \frac{5x}{5} = x$$

and

$$(f \circ g)(x) = f\left(\frac{x - 4}{5}\right) = 5\left(\frac{x - 4}{5}\right) + 4 = (x - 4) + 4 = x$$

Thus $(g \circ f)(x) = (f \circ g)(x) = x$, so f and g are inverse functions. ■

Once you are certain that a function g is the inverse of a function f, you can denote it by f^{-1} so that

Notation for Inverse Functions

$$(f^{-1} \circ f)(x) = (f \circ f^{-1})(x) = x$$

Be careful about this notation. The symbol f^{-1} means the *inverse* of f and does not mean 1 *divided* by f.

Now consider a function defined by a set of ordered pairs $y = f(x)$. The image of x is y, as shown in Figure 2.22.

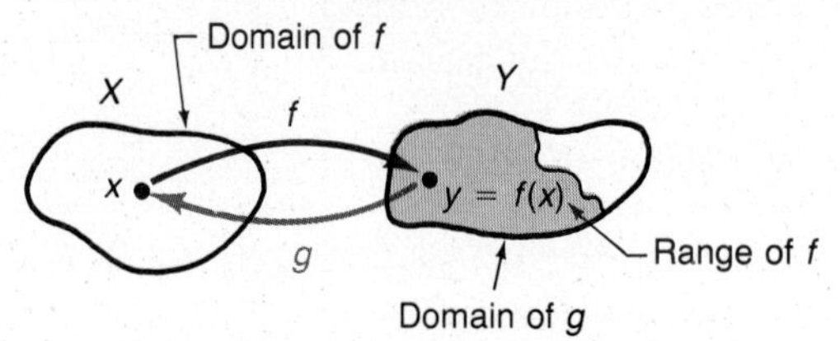

Figure 2.22 Inverse functions

If y is a member of the domain of the function g, then $g(y) = x$, as illustrated in Figure 2.22. This means that if you think of a function as a set of ordered pairs (x, y), the inverse of f is the set of ordered pairs with the components (y, x).

EXAMPLE 4 Let $f = \{(0, 3), (1, 5), (3, 9), (5, 13)\}$. Find the inverse of f.

Solution The inverse simply reverses the ordered pairs: $f^{-1} = \{(3, 0), (5, 1), (9, 3), (13, 5)\}$ ■

Example 5 tells us how to find the inverse if the function is defined by a graph.

EXAMPLE 5 Consider the function f defined by the graph in Figure 2.23.

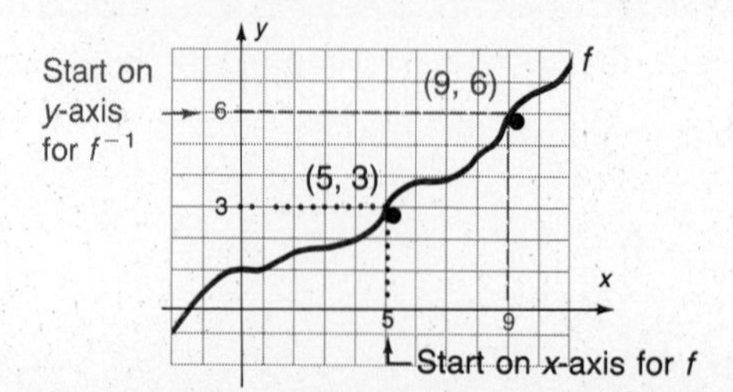

Figure 2.23 Graph of f

a. Find $f(5)$. **b.** Find $f^{-1}(6)$.

Solution **a.** Use the graph as shown in Figure 2.23:

This is a member of the domain of f—locate on x axis.

$$f(5) = 3$$

This is found by following dotted arrows in Figure 2.23.

This is a member of the domain of f^{-1}—locate on y axis.

b. $$f^{-1}(6) = 9$$

This is found by following dashed arrows in Figure 2.23. ∎

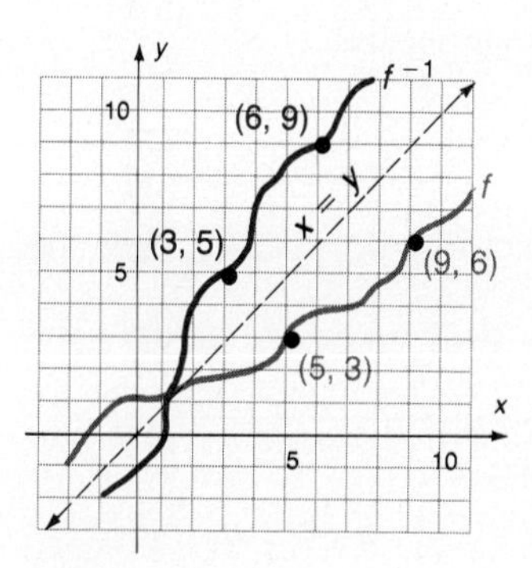

Figure 2.24 Graphs of f and f^{-1}

Although Example 5 illustrates a convenient method for finding ordered pairs in f or f^{-1}, it is not the usual way for representing f^{-1} since the first component of a function is usually plotted on the axis of abscissas. That is, to draw the graph of f^{-1}, *each* ordered pair of f—(5, 3) for example—must be rewritten as (3, 5) and plotted in the usual fashion (see Figure 2.24).

The graph of f and its inverse f^{-1} are symmetric with respect to the line $x = y$. If a function f is defined by an equation such as

$$f(x) = 4x - 3$$

you can find the inverse by first writing this equation using (x, y) notation:

$$y = 4x - 3$$

Next find the inverse by interchanging the x and y components:

$$x = 4y - 3 \qquad \text{Interchange } x \text{ and } y.$$

Finally, solve for y:

$$x + 3 = 4y \qquad \text{Solve for } y.$$

$$y = \frac{x + 3}{4}$$

The result is an equation for the inverse of f:

$$f^{-1}(x) = \frac{x + 3}{4}$$

EXAMPLE 6 Let f be defined by $f(x) = x^2 - 1$ on the interval $[0, \infty)$. Find f^{-1}.

Solution Do you see why some restriction on the domain is necessary? The interval given in this example forces f to be increasing and therefore one-to-one. Without this restriction you have

$$f(3) = 9 - 1 \qquad \text{and} \qquad f(-3) = 9 - 1$$

which means that (3, 8) and (−3, 8) both belong to f—which shows that even though f is a function, it is not one-to-one. If x is a member of the domain $[0, \infty)$, however, then f is one-to-one and the inverse function is defined. Now let $y = f(x)$ so that

	Domain	*Range*
$y = x^2 - 1$	$x \geq 0$	$y \geq -1$

Interchange x and y for the inverse:

$x = y^2 - 1$	$y \geq 0$	$x \geq -1$
	This one is now called the range because it shows the restriction on y.	This one is now called the domain because it shows the restriction on x.

Solve for y:

$$y^2 = x + 1$$

$$y = \pm\sqrt{x + 1} \qquad \text{Reject } y = -\sqrt{x+1} \text{ since } y \text{ is nonnegative.}$$

$$y = \sqrt{x + 1} \qquad x \geq -1$$

Thus

$$f^{-1}(x) = \sqrt{x + 1} \text{ on } [-1, \infty)$$

■

PROBLEM SET 2.6

A

In Problems 1–12, find $(f \circ g)(x)$ and $(g \circ f)(x)$.

1. $f(x) = 2x - 3;\quad g(x) = x + 6$

2. $f(x) = 2x - 3;\quad g(x) = x^2 + 1$

3. $f(x) = 5x - 1;\quad g(x) = \dfrac{x + 1}{5}$

4. $f(x) = x^2;\quad g(x) = x^2 - x$

5. $f(x) = \dfrac{x - 2}{x + 1}$ and $g(x) = x^2 - x$

6. $f(x) = 4x + 1$ and $g(x) = x^3 + 3$

7. $f(x) = \dfrac{1}{x - 1}$ and $g(x) = x^2 - 1$

8. $f(x) = |x|$ and $g(x) = x^2$

9. $f(x) = 3$ and $g(x) = 5x^2$

10. $f = \{(0, 1), (1, 4), (2, 7), (3, 10)\}$ and $g = \{(0, -3), (1, -1), (2, 1), (3, 3)\}$

11. $f = \{(5, 3), (6, 2), (7, 9), (8, 12)\}$ and $g = \{(5, 8), (6, 5), (7, 4), (8, 3)\}$

12. $f = \{(5, 9), (10, 29), (15, 39), (20, 49)\}$ and $g = \{(5, 4), (10, 5), (15, 6), (20, 9)\}$

Determine which pairs of functions defined by the equations in Problems 13–18 are inverses.

13. $f(x) = 5x + 3;\ g(x) = \dfrac{x - 3}{5}$

14. $f(x) = \frac{2}{3}x + 2;\ g(x) = \frac{3}{2}x + 3$

15. $f(x) = \frac{4}{5}x + 4;\ g(x) = \frac{5}{4}x + 3$

16. $f(x) = \dfrac{1}{x},\ x \neq 0;\ g(x) = \dfrac{1}{x},\ x \neq 0$

17. $f(x) = x^2,\ x < 0;\ g(x) = \sqrt{x},\ x > 0$

18. $f(x) = x^2,\ x \geq 0;\ g(x) = \sqrt{x},\ x \geq 0$

Find the inverse function, if it exists, of each function given in Problems 19–34.

19. $f = \{(4, 5), (6, 3), (7, 1), (2, 4)\}$

20. $f = \{(1, 4), (6, 1), (4, 5), (3, 4)\}$

21. $g = \{(1, 3), (2, 5), (4, 6), (5, 9)\}$

22. $g = \{(1, 4), (2, 16), (3, 36), (4, 64)\}$

23. $f(x) = x + 3$

24. $f(x) = 2x + 3$

25. $g(x) = 5x$

26. $g(x) = \frac{1}{5}x$

27. $h(x) = x^2 - 5$

28. $h(x) = \sqrt{x} + 5$

29. $f(x) = x$

30. $f(x) = \dfrac{1}{x}$

31. $f(x) = 6$

32. $g(x) = -2$

33. $f(x) = \dfrac{1}{x - 2}$

34. $f(x) = \dfrac{2x + 1}{x}$

B

In Problems 35–42 find **a.** $f \circ g$ **b.** $g \circ h$ **c.** $(f \circ g) \circ h$ **d.** $f \circ (g \circ h)$

35. $f(x) = x^2,\ g(x) = 2x - 1,\ h(x) = 3x + 2$

36. $f(x) = x^2, g(x) = 3x - 2, h(x) = x^2 + 1$

37. $f(x) = 2x + 4, g(x) = \frac{1}{2}x - 2, h(x) = x^2 + 1$

38. $f(x) = \sqrt{x}, g(x) = x^2, h(x) = x + 2; x > 0$ for f, g, and h

39. $f(x) = 3x + 2, g(x) = 2x - 5, h(x) = x + 1$

40. $f(x) = g(x) = h(x) = x$

41. $f(x) = g(x) = h(x) = 2x$

42. $f(x) = x, g(x) = x^2, h(x) = x^3$

If f is defined by the graph in Figure 2.25, find the values requested in Problems 43–48.

43. a. $f(3)$ b. $f(4)$ c. $f(7)$ d. $f(11)$ e. $f(14)$

44. a. $f(0)$ b. $f^{-1}(4)$ c. $f^{-1}(5)$ d. $f(15)$ e. $f^{-1}(9)$

45. a. $f^{-1}(0)$ b. $f^{-1}(1)$ c. $f^{-1}(6)$ d. $f^{-1}(-2)$ e. $f^{-1}(-6)$

46. a. $f(1)$ b. $f^{-1}(-1)$ c. $f^{-1}(-4)$ d. $f(2)$ e. $f^{-1}(2)$

47. a. $f^{-1}(-5)$ b. $f(5)$ c. $f^{-1}(3)$ d. $f^{-1}(8)$ e. $f(13)$

Figure 2.25 Graph of f

48. a. $f^{-1}(4)$ b. $f(8)$ c. $f^{-1}(-3)$ d. $f(9)$ e. $f^{-1}(7)$

49. a. What is the domain and range of the function f defined by the graph in Figure 2.25?
    b. What is the domain and range of the function f^{-1}?

50. Graph f^{-1} for the function f defined by the graph in Figure 2.25.

Determine which pairs of functions defined by the equations in Problems 51–54 are inverses.

51. $f(x) = 2x^2 + 1, x \geq 0;$ $g(x) = \frac{1}{2}\sqrt{2x - 2}, x \geq 1$

52. $f(x) = 2x^2 + 1, x \leq 0;$ $g(x) = -\frac{1}{2}\sqrt{2x - 2}, x \geq 1$

53. $f(x) = (x + 1)^2, x \geq 1;$ $g(x) = -1 - \sqrt{x}, x \geq 0$

54. $f(x) = (x + 1)^2, x \geq -1;$ $g(x) = -1 + \sqrt{x}, x \geq 0$

55. ***Business*** Suppose that a store sells calculators by marking up the price 20%. That is, the price of an item costing c dollars is

$$p(c) = c + 0.20c$$

Also suppose that the cost c of manufacturing n calculators is $50n + 200$ (dollars). This means that the cost of each calculator depends on the number manufactured according to the function

$$c(n) = \frac{50n + 200}{n} \qquad (n \geq 1)$$

a. Find the price for one calculator if only one calculator is manufactured.
b. Find the price for one calculator if 10 calculators are manufactured.
c. Express the price as a function of the number of calculators produced by finding $p \circ c$.

56. ***Physics*** Suppose that the volume of a certain cone is expressed as a function of its height so that it is defined by

$$V(h) = \frac{\pi h^3}{12}$$

Suppose the height is expressed as a function of time by $h(t) = 2t$.
a. Find the volume for $t = 2$.
b. Express the volume as a function of time by finding $V \circ h$.
c. If the domain of V is $\{h \mid 0 < h \leq 6\}$, find the domain of h. That is, what are the permissible values for t?

57. ***Physics*** The surface area of a spherical balloon is given by

$$S(r) = 4\pi r^2$$

Suppose the radius is expressed as a function of time by $r(t) = 3t$.
a. Find the surface area for $t = 2$.
b. Express the surface area as a function of time by finding $S \circ r$.
c. If the domain of S is $\{r \mid 0 < r < 8\}$, find the domain of r. That is, what are the permissible values for t?

C

Find the inverse of each function given in Problems 58–67.

58. $f(x) = x^2$ on $[0, \infty)$

59. $f(x) = x^2$ on $(-\infty, 0]$

60. $f(x) = x^2 + 1$ on $(-\infty, 0]$

61. $f(x) = x^2 + 1$ on $[0, \infty)$

62. $f(x) = 2x^2$ on $[2, 10]$

63. $f(x) = 2x^2$ on $[-10, -1]$

64. $f(x) = \dfrac{1}{3x + 1}$ on $(-\frac{1}{3}, \infty)$

65. $f(x) = \dfrac{1}{2x - 1}$ on $(\frac{1}{2}, \infty)$

66. $f(x) = x$ on $[0, \infty)$

67. $f(x) = x + 1$ on $[-1, \infty)$

68. If $f(x) = 1 + \dfrac{1}{x}$, find:

a. $(f \circ f)(x)$ b. $(f \circ f \circ f)(x)$ c. $(f \circ f \circ f \circ f)(x)$

69. a. Let $f(x) = \sqrt{x}$. Choose *any* positive x. Find a numerical value for $(f \circ f)(x)$, $(f \circ f \circ f)(x)$, and $(f \circ f \circ f \circ f)(x)$. Suppose this procedure is repeated a large number of times:

$$(f \circ f \circ f \circ \ldots \circ f)(x)$$

Can you predict the outcome for *any* x?

b. Let $f(x) = 2\sqrt{x}$. Answer the questions of part **a**.

c. Let $f(x) = 3\sqrt{x}$. Answer the questions of part **a**.

d. Let $f(x) = k\sqrt{x}$, where k is a positive integer. Answer the questions of part **a**.

CHAPTER 2 SUMMARY

The material of this chapter is reviewed in the following list of objectives. After each objective there are some practice questions. For a sample test, select the first question of each set and check your answers with the answer section. For a sample test without answers, use the second question of each set. Additional practice is given by the other questions in each set. If you are having trouble with a particular type of problem, look back to that section for extra help.

2.1 FUNCTIONS

Objective 1 *Be able to classify examples as onto, one-to-one, functions, or not functions.* Let X and Y be sets of real numbers such that $x \in X$ and $y \in Y$. (The symbol $\in$ means 'is a member of.')

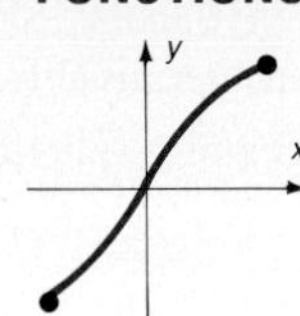

1. $y = 3x + 3$ **2.** $y = 2x^2 + 3$ **3.** $y \leq 2x + 3$

4. The set of ordered pairs defined by the graph at the left.

Objective 2 *Name subsets of the domain of a function in which the function is increasing, decreasing, or constant. Find the intercepts.* Consider the graph shown in the following figure.

5. What is the domain and range of f?

6. Name the subsets of the domain for which f is increasing, f is decreasing, and f is constant.

7. What are the intercepts?

8. What are the coordinates of A and B?

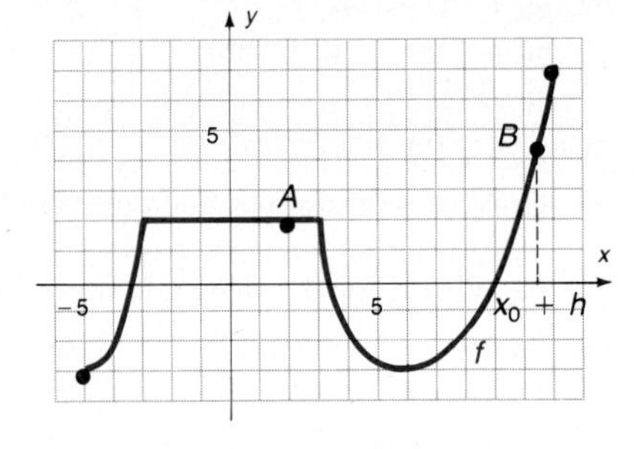

2.2 FUNCTIONAL NOTATION

Objective 3 *Distinguish between the f and $f(x)$ notation. Use functional notation.* Let $f(x) = 3x - 1$ and $g(x) = 5 - x^2$. Find the requested values.

9. a. $f(4)$ **b.** $g(4)$ **10. a.** $f(-6)$ **b.** $g(-6)$

11. a. $f(w)$ **b.** $g(w + h)$ **12. a.** $3f(t)$ **b.** $g(\sqrt{t})$

Objective 4 *Find* $\dfrac{f(x + h) - f(x)}{h}$ *for a given function f.*

13. $f(x) = 5 - x^2$ **14.** $f(x) = 5x + 3$ **15.** $f(x) = 5$ **16.** $f(x) = 3x^2 - 1$

Objective 5 *Find the domain of a given function over the set of real numbers.*

17. $f(x) = \sqrt{x^2 - 7x + 6}$ **18.** $f(x) = \dfrac{(3x + 1)(2x + 5)}{2x + 5}$

19. $f(x) = 3x^2 - 5x + 1$ **20.** $f(x) = \dfrac{6x^2 + 17x + 15}{3x + 1}, x \neq 0$

Objective 6 *Determine when functions are equal.*

21. $f(x) = \dfrac{x^2 + x}{x + 1}$; $g(x) = x$

22. $f(x) = \dfrac{(x + 3)(x - 2)}{x - 2}, x \neq 2$; $g(x) = \dfrac{(x + 3)(x - 2)}{x + 3}, x \neq -3$

23. $f(x) = \sqrt{x^2}$; $g(x) = x$

24. $f(x) = \dfrac{(2x + 1)(3x - 2)}{3x - 2}$; $g(x) = 2x + 1, x \neq 0$

Objective 7 *Classify functions as even, odd, or neither.*

25. $f(x) = 3x^2 + 2x - 5$

26. $f(x) = 2x^3 + 5x$

27. $f(x) = \dfrac{1}{(x + 4)^2}$

28. $f(x) = 7$

2.3 LINEAR FUNCTIONS

Objective 8 *Define a linear function. Graph linear functions.*

29. $y = 5x - 3$

30. $2x + 3y + 6 = 0$

31. $5x - 2y + 8 = 0$

32. $y + 3 = \frac{3}{5}(x - 4)$

Objective 9 *Find the slope of a line.*

33. The line passing through $(2, 2)$ and $(6, -3)$
34. The line passing through $(a, f(a))$ and $(b, f(b))$
35. The line $7x - 5y + 3 = 0$
36. A horizontal line

Objective 10 *Graph line segments.*

37. $4x - 3y + 9 = 0$ for $-3 < x \leq 4$

38. $y = \frac{3}{5}x - 3$ for $-5 \leq x \leq 5$

39. $2x = y - 5$ for $-4 \leq x < 2$

40. $y = 5$ for $-4 < x < 4$

Objective 11 *Graph absolute value relations.*

41. $y = |-x|$

42. $y = -2|x|$

43. $y = \frac{2}{3}|x|$

44. $y = |4x|$

Objective 12 *Determine when given lines or line segments are parallel or perpendicular.*

45. Use slopes to decide whether the triangle with vertices $A(12, -7)$, $B(4, 3)$, $C(-3, -2)$ is a right triangle.
46. What is the slope of a line parallel to $3x - 4y + 7 = 0$?
47. What is the slope of a line perpendicular to $6x - 4y + 3 = 0$?
48. Are the diagonals of the quadrilateral $ABCD$ with vertices $A(3, 4)$, $B(-2, 5)$, $C(-4, -2)$, $D(3, -5)$ perpendicular?

Objective 13 *Know and use the various forms of a linear equation.*

49. State the standard form equation and define the variables.
50. State the point–slope form equation and define the variables.
51. State the slope–intercept form equation and define the variables.
52. State the equation of a vertical line and define the variables.

Objective 14 *Given the graph, or information about the graph, of a line, write the standard-form equation of the line.*

53. y-intercept -9; slope, $\frac{2}{3}$

54. Slope -3; passing through $(3, -2)$

55. Passing through $(-3, -1)$ and $(5, -6)$

56. Passing through $(4, 5)$ and perpendicular to the line $2x - 3y + 8 = 0$

2.4 TRANSLATION OF FUNCTIONS

Objective 15 Find the shift (h, k) *when given an equation* $y - k = f(x - h)$.

57. $y - 6 = f(x + \pi)$
58. $y - 1 = 3(x + 3)^2$
59. $y = 5(x - 4)^2$
60. $y = \frac{2}{3}|x - \sqrt{2}| + 5$

Objective 16 Given $y = f(x)$ *and* (h, k), *write* $y - k = f(x - h)$.

61. $f(x) = 9x^2; (h, k) = (-\sqrt{2}, 3)$
62. $f(x) = 2|x|; (h, k) = (4, -3)$
63. $f(x) = -2x^2; (h, k) = (\pi, 0)$
64. $f(x) = 3x^2 + 2x + 5; (h, k) = (1, 2)$

Objective 17 Substitute $x' = x - h$ *and* $y' = y - k$ *into an equation* $y - k = f(x - h)$. Substitute $x' = x + 3$ and $y' = y - 1$ into the following equations.

65. $y - 1 = 3(x + 3)^2$
66. $y - 1 = 2|x + 3|$
67. $y = 5(x + 3)^2 + 1$
68. $y - 1 = (x + 3)^2 + (x + 3) - 5$

Objective 18 Given a function defined by $y = f(x)$ *and shown by a graph, draw the graph of* $y - k = f(x - h)$.

69. $y - 2 = f(x + 1)$
70. $y + 1 = f(x - 3)$
71. $y = f(x - \pi)$
72. $y + \sqrt{2} = f(x - \sqrt{3})$

2.5 QUADRATIC FUNCTIONS

Objective 19 Define and graph quadratic functions of the form $y - k = a(x - h)^2$.

73. $y - 1 = \frac{3}{4}(x + 3)^2$
74. $y + 1 = -2(x + 2)^2$
75. $y = 10x^2$
76. $y - 1 = 3(x + 3)^2$

Objective 20 Graph quadratic functions by completing the square.

77. $x^2 + 2x - y - 1 = 0$
78. $x^2 + 4x - 2y - 2 = 0$
79. $y = x^2 - 6x + 10$
80. $x^2 - 4x + 5y + 9 = 0$

Objective 21 Find the maximum or minimum value for a quadratic function. Be sure to state whether your answer is a maximum or a minimum value.

81. $2x^2 + 24x + y = 178$
82. $2x^2 + 20x + y + 190 = 0$
83. $x^2 - 10x - y - 825 = 0$
84. $x^2 - 8x - 3y + 3616 = 0$

2.6 COMPOSITE AND INVERSE FUNCTIONS

Objective 22 Find the composition of functions. Let $f(x) = 3x - 1$, $g(x) = 5 - x^2$, and $h(x) = 6$. Find the requested values.

85. $(f \circ g)(x)$
86. $(g \circ f)(x)$
87. $(f \circ h)(x)$
88. $(h \circ f)(x)$

Objective 23 Given two functions, decide whether or not they are inverses. Decide if f and g are inverses.

89. $f(x) = 3x + 6; g(x) = x - 2$
90. $f(x) = 2x + 1; g(x) = x - 2$
91. $f(x) = 3x - 2; g(x) = \dfrac{x - 2}{3}$
92. $f(x) = 2x - 3; g(x) = \dfrac{x + 2}{3}$

Objective 24 Given a one-to-one function, find its inverse.

93. $f(x) = 3x - 1$
94. $g(x) = 5 - x^2$ on $[0, \infty)$
95. $f(x) = \frac{1}{2}x + 5$
96. $g(x) = \sqrt{2x}$ on $[0, \infty)$

3

Polynomial and Rational Functions

The business of concrete mathematics is to discover the equations which express the mathematical laws of the phenomenon under consideration; and these equations are the starting point of the calculus, which must obtain from them certain quantities by means of others.

AUGUSTE COMTE
Positive Philosophy

Amalie Emmy Noether
(1882–1935)

*** Optional Section**

The greatest woman mathematician of all time is Emmy Noether, who is famous for her work in physics and algebra. She obtained her degrees at a time when it was unusual for a woman to attend college. In fact, a leading historian of the day wrote that talk of "surrendering our universities to the invasion of women . . . is a shameful display of moral weakness." Nevertheless, Noether did receive her degrees and significantly influenced the development of abstract algebra. In 1935, Albert Einstein wrote the following tribute:

In the judgement of the most competent living mathematicians, Fraulein Noether was the most significant creative mathematical genius thus far produced since the higher education of women began. In the realm of algebra in which the most gifted of mathematicians have been busy for centuries, she discovered methods which have proved of enormous importance in the development of the present day younger generation of mathematicians.

As quoted in *In Mathematical Circles Revisited* by Howard Eves

CHAPTER OVERVIEW Equation solving is one of the most important topics of elementary algebra. In high school you learned about first-degree equations, and in Chapter 1 we reviewed second-degree equations. In this chapter we consider the solution of polynomial equations of higher degree than two. There are 19 objectives in this chapter, which are listed on pages 135–137.

3.1 POLYNOMIAL FUNCTIONS

The key to working with polynomials is the proper handling of exponents. You have used the laws of exponents extensively in algebra, and they must also be fully understood for the study of calculus.

Definition of Exponents

If b is any real number and n is any natural number, then

$$b^n = \underbrace{b \cdot b \cdot b \cdots b}_{n \text{ factors}}$$

And if $b \neq 0$, then

$$b^0 = 1 \qquad b^{-n} = \frac{1}{b^n}$$

The number b is called the **base**, n is called the **exponent**, and b^n is called the **nth power** of b. The five laws of exponents can now be summarized.

Laws of Exponents

Let a and b be real numbers and let m and n be any integers. Then the five rules listed below govern the use of exponents except that the form 0^0 and division by zero are excluded.

First law: $b^m \cdot b^n = b^{m+n}$ Fourth law: $(ab)^m = a^m b^m$

Second law: $\dfrac{b^m}{b^n} = b^{m-n}$ Fifth law: $\left(\dfrac{a}{b}\right)^m = \dfrac{a^m}{b^m}$

Third law: $(b^n)^m = b^{mn}$

The linear and quadratic functions defined in Chapter 2 are special cases of a more general function called a polynomial function.

Polynomial Function

A function P is a **polynomial function** in x if

$$P(x) = a_n x^n + a_{n-1} x^{n-1} + a_{n-2} x^{n-2} + \cdots + a_1 x + a_0$$

where n is an integer greater than or equal to zero and the coefficients $a_0, a_1, a_2, \ldots, a_{n-1}, a_n$ are real numbers.

In an expression of the form $a_n x^n$, a_n is called the **coefficient** of x^n. And if a_n is the coefficient of the highest power of x, it is called the **leading coefficient** of the polynomial. If a_n and x are nonzero, the polynomial function is said to have **degree** n. Notice that if $n = 2$ (degree 2), then $P(x)$ is a *quadratic function*; if $n = 1$ (degree 1), then $P(x)$ is a *linear function*; and if $n = 0$ (degree 0), then $P(x)$ is a *constant function.* If all the coefficients of a polynomial are 0, it is called the **zero polynomial** and is denoted by 0. The zero polynomial is not assigned a degree.

Each $a_k x^k$ of a polynomial function is called a **term** of the polynomial, and it is customary to arrange the terms of a polynomial in order of decreasing powers of the variable. **Similar terms** are terms with the same variable and the same degree. Two polynomials are **equal** if and only if coefficients of similar terms are the same. If some of the numerical coefficients are negative, they are usually written as subtractions of terms. Thus

$$6 + (-3)x + x^3 + (-2)x^2$$

would customarily be written as

$$x^3 - 2x^2 - 3x + 6$$

Examples 1–4 provide some practice with polynomial notation and operations.

EXAMPLE 1 Let $P(x) = 4x - 5$, $Q(x) = 5x^2 + 2x + 1$, and $R(x) = 3x^3 - 4x^2 + 3x - 2$. Find $P(2)$, $Q(2)$, and $R(2)$.

Solution

$$\begin{aligned} P(2) &= 4(2) - 5 \\ &= 8 - 5 \\ &= \mathbf{3} \end{aligned} \qquad \begin{aligned} Q(2) &= 5(2)^2 + 2(2) + 1 \\ &= 20 + 4 + 1 \\ &= \mathbf{25} \end{aligned}$$

$$\begin{aligned} R(2) &= 3(2)^3 - 4(2)^2 + 3(2) - 2 \\ &= 24 - 16 + 6 - 2 \\ &= \mathbf{12} \end{aligned}$$

■

EXAMPLE 2 Find $Q(x) + R(x)$.

Solution Use the distributive law to combine similar terms:

$$\begin{aligned} Q(x) + R(x) &= (5x^2 + 2x + 1) + (3x^3 - 4x^2 + 3x - 2) \\ &= 3x^3 + \underbrace{5x^2 + (-4)x^2}_{\text{similar terms}} + \underbrace{2x + 3x}_{\text{similar terms}} + \underbrace{1 + (-2)}_{\text{similar terms}} \\ &= \mathbf{3x^3 + x^2 + 5x - 1} \end{aligned}$$

This step is usually done mentally.

■

EXAMPLE 3 Find $Q(x) - R(x)$.

Solution Be careful when subtracting polynomials; to subtract $R(x)$ you must subtract *each* term of $R(x)$. That is, remember that $Q(x) - R(x) = Q(x) + (-1)R(x)$. Thus

$$\begin{aligned} Q(x) - R(x) &= (5x^2 + 2x + 1) - (3x^3 - 4x^2 + 3x - 2) \\ &= 5x^2 + 2x + 1 + (-3x^3) + 4x^2 + (-3x) + 2 \\ &= \mathbf{-3x^3 + 9x^2 - x + 3} \end{aligned}$$

■

EXAMPLE 4 Find $P(x)R(x)$.

Solution Use the distributive property:

$$\begin{aligned} P(x)R(x) &= (4x - 5)(3x^3 - 4x^2 + 3x - 2) \\ &= (4x - 5)(3x^3) + (4x - 5)(-4x^2) + (4x - 5)(3x) + (4x - 5)(-2) \\ &= 12x^4 - 15x^3 - 16x^3 + 20x^2 + 12x^2 - 15x - 8x + 10 \\ &= \mathbf{12x^4 - 31x^3 + 32x^2 - 23x + 10} \end{aligned}$$

■

If a polynomial is written as the product of other polynomials, each polynomial in the product is called a **factor** of the original polynomial. **Factoring** a polynomial means to break up the polynomial into the product of individual factors. The types of factoring problems usually encountered are summarized in Table 3.1. For example, the expression $6x^2 + 3x - 18$ is properly factored by *first* finding the common factor (type 1) and then noting that types 2–6 do not apply. Finally, use type 7 (FOIL) to write

$$\begin{aligned} 6x^2 + 3x - 18 &= 3(2x^2 + x - 6) \\ &= 3(2x - 3)(x + 2) \end{aligned}$$

We usually factor **over the set of integers**, which means that all the numerical coefficients are integers. If the original polynomial has fractional coefficients, you should factor out the fractional part first, as shown in Example 5.

EXAMPLE 5

$$\begin{aligned} \frac{1}{36}x^2 - 9y^4 &= \frac{1}{36}(x^2 - 6^2 \cdot 3^2 y^4) \\ &= \frac{1}{36}[x^2 - (18y^2)^2] \\ &= \mathbf{\frac{1}{36}(x - 18y^2)(x + 18y^2)} \end{aligned}$$

■

Factoring over the set of integers rules out factoring

$$x^2 - 3 = (x - \sqrt{3})(x + \sqrt{3})$$

since the factors do not have integer coefficients. In this book, a polynomial is **completely factored** if all fractions are eliminated by common factoring (as shown in Example 5) and if no further factoring is possible *over the set of integers.* If, after fractional common factors have been removed, a polynomial cannot be factored further, then it is said to be **irreducible**. Examples 6–16 review various factoring techniques.

EXAMPLE 6 $5x + 7xy = x(5 + 7y)$ Common factor x ■

EXAMPLE 7 $7 - 35x = 7(1 - 5x)$ Common factor 7 (note the 1) ■

EXAMPLE 8 $5x(3a - 2b) + y(3a - 2b) = (5x + y)(3a - 2b)$ Common factor $(3a - 2b)$ ■

EXAMPLE 9

$$\begin{aligned} 3x^2 - 75 &= 3(x^2 - 25) && \text{Common factor first} \\ &= 3(x - 5)(x + 5) && \text{Difference of squares} \end{aligned}$$

■

TABLE 3.1
Factoring types

Type	Form	Comments
1. Common factors	$ax + ay + az = a(x + y + z)$	This simply uses the distributive property. It can be applied with any number of terms.
2. Difference of squares	$x^2 - y^2 = (x - y)(x + y)$	The *sum* of two squares cannot be factored in the set of real numbers.
3. Difference of cubes	$x^3 - y^3 = (x - y)(x^2 + xy + y^2)$	This is similar to the difference of squares and can be proved by multiplying the factors.
4. Sum of cubes	$x^3 + y^3 = (x + y)(x^2 - xy + y^2)$	Unlike the sum of squares, the sum of cubes can be factored.
5. Perfect square	$x^2 + 2xy + y^2 = (x + y)^2$ $x^2 - 2xy + y^2 = (x - y)^2$	The middle term is twice the product of xy.
6. Perfect cube	$x^3 + 3x^2y + 3xy^2 + y^3 = (x + y)^3$ $x^3 - 3x^2y + 3xy^2 - y^3 = (x - y)^3$	The numerical coefficients of the terms are: 1 3 3 1 or 1 −3 3 −1
7. Binomial product, FOIL	$acx^2 + (ad + bc)xy + bdy^2$ $= (ax + by)(cx + dy)$ See Examples 12–14 in the text.	This trial-and-error procedure is used with a trinomial. It should be used after types 1–6 have been checked.
8. Grouping	See Example 16 in the text.	After types 1–7 have been checked, you can factor some expressions by grouping.
9. Irreducible (cannot be factored over the set of integers)	Examples arise in every factoring situation. Expressions such as $x^2 + 4$, $x^2 + xy + y^2$, and $x^2 + y^2$ cannot be factored over the set of integers.	When factoring, you are not through until all the factors are irreducible.

EXAMPLE 10

$$\frac{9a^2}{b^2} - (a + 3b)^2 = \frac{1}{b^2}[9a^2 - b^2(a + 3b)^2]$$ Common factor to eliminate fractions

$$= \frac{1}{b^2}\{(3a)^2 - [b(a + 3b)]^2\}$$ This step can be done mentally.

$$= \frac{1}{b^2}[3a - b(a + 3b)][3a + b(a + 3b)]$$ Difference of squares

$$= \frac{1}{b^2}(3a - ab - 3b^2)(3a + ab + 3b^2)$$ Simplify ■

EXAMPLE 11 $(x + 3y)^3 + 8 = [(x + 3y) + 2][(x + 3y)^2 - (x + 3y)(2) + (2)^2]$ Sum of cubes
$= (x + 3y + 2)(x^2 + 6xy + 9y^2 - 2x - 6y + 4)$ ■

EXAMPLE 12 $x^2 - 8x + 15 = (x - 5)(x - 3)$ ■

EXAMPLE 13 $6x^2 + x - 12 = (2x + 3)(3x - 4)$ ■

Binomial Product, sometimes called FOIL—you may have to try several possibilities before you find the correct one. Remember to factor completely; several types may be combined in one problem.

EXAMPLE 14
$$6x^2 - 9x - 15 = 3(2x^2 - 3x - 5)$$
$$= 3(2x - 5)(x + 1)$$ ■

EXAMPLE 15
$$4x^4 - 13x^2y^2 + 9y^4 = (x^2 - y^2)(4x^2 - 9y^2)$$
$$= (x - y)(x + y)(2x - 3y)(2x + 3y)$$ ■

EXAMPLE 16
$$9x^3 + 18x^2 - x - 2 = (9x^3 + 18x^2) - (x + 2)$$
$$= 9x^2(x + 2) - (x + 2)$$
$$= (9x^2 - 1)(x + 2)$$
$$= (3x - 1)(3x + 1)(x + 2)$$

None of the types 1–7 from Table 3.1 seem to apply. Thus try grouping the terms. Some groupings may lead to a factorable form; others may not. ■

An important application of factoring in algebra, and in calculus, is the **Zero Factor Theorem**. It is simply the principle of zero products of Section 1.3 stated for polynomials.

Zero Factor Theorem

If $P(x)$ and $Q(x)$ are polynomials such that

$$P(x)Q(x) = 0$$

then either

$$P(x) = 0 \quad \text{or} \quad Q(x) = 0 \quad \text{(perhaps both)}$$

You can apply this theorem to graph equations expressible in factored form. Suppose you wish to graph the curve whose equation is

$$x^2 - y^2 = 0$$

Since this equation can be factored as

$$(x - y)(x + y) = 0$$

you can use the Zero Factor Theorem to write

$$x - y = 0 \quad \text{or} \quad x + y = 0$$

Thus the graph of $x^2 - y^2 = 0$ is found by graphing both $x - y = 0$ and $x + y = 0$, as shown in Figure 3.1.

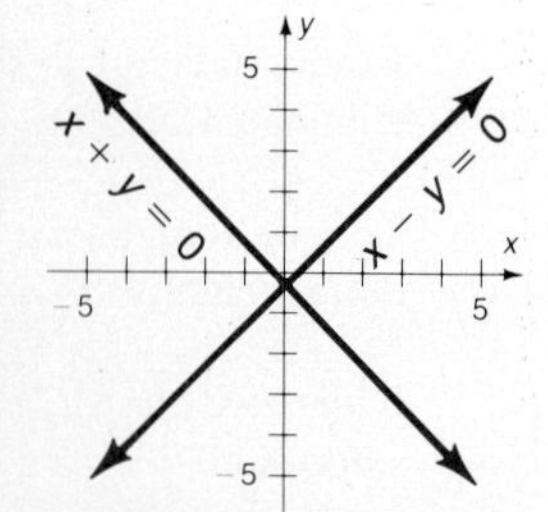

Figure 3.1 Graph of $x^2 - y^2 = 0$

EXAMPLE 17 Graph $3x^3 - 3xy - 2x^2y + 2y^2 = 0$.

Solution First factor by grouping:

$$3x(x^2 - y) - 2y(x^2 - y) = 0$$
$$(x^2 - y)(3x - 2y) = 0$$

Next use the Zero Factor Theorem:

$$x^2 - y = 0$$

or

$$3x - 2y = 0$$

The graph is shown in Figure 3.2.

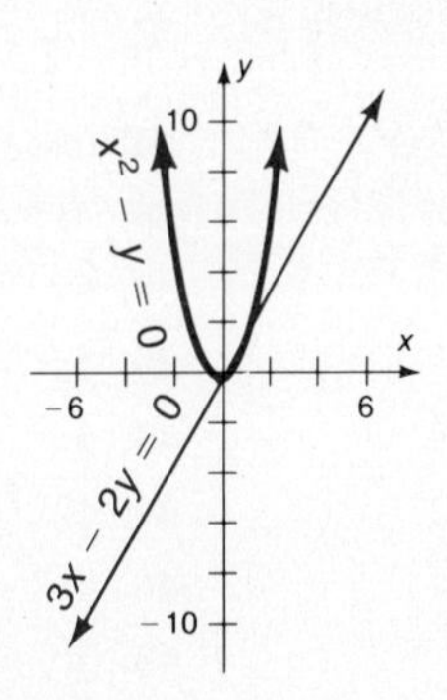

Figure 3.2 Graph of $3x^3 - 3xy - 2x^2y + 2y^2 = 0$ ■

PROBLEM SET 3.1

A

Let $P(x) = 5x + 1$, $Q(x) = 3x^2 - 5x + 2$, *and* $R(x) = x^3 - 4x^2 + x - 4$ *in Problems 1–6. Find the required result.*

1. a. $P(2)$ **b.** $Q(2)$ **c.** $R(2)$

2. a. $P(3)$ **b.** $Q(0)$ **c.** $R(1)$

3. a. $P(5)$ **b.** $Q(1)$ **c.** $R(0)$

4. a. $P(x) + R(x)$ **b.** $Q(x) - P(x)$

5. a. $R(x) - Q(x)$ **b.** $P(x)Q(x)$

6. a. $P(x)Q(x) - R(x)$ **b.** $3Q(x) - 4R(x)$

Write the expressions in Problems 7–12 in simplified polynomial form.

7. a. $(x + 2)(x + 1)$ **b.** $(y - 2)(y + 3)$
c. $(x + 1)(x - 2)$ **d.** $(y - 3)(y + 2)$

8. a. $(a - 5)(a - 3)$ **b.** $(b + 3)(b - 4)$
c. $(c + 1)(c - 7)$ **d.** $(z - 3)(z + 5)$

9. a. $(2x + 1)(x - 1)$ **b.** $(2x - 3)(x - 1)$
c. $(x + 1)(3x + 1)$ **d.** $(x + 1)(3x + 2)$

10. a. $(x + y)(x - y)$ **b.** $(a + b)(a - b)$
c. $(5x - 4)(5x + 4)$ **d.** $(3y - 2)(3y + 2)$

11. a. $(a + 2)^2$ **b.** $(b - 2)^2$
c. $(x + 4)^2$ **d.** $(y - 3)^2$

12. a. $(u + v)^3$ **b.** $(s - 2t)^3$
c. $(1 - 3n)^3$ **d.** $(2x - y)^3$

Factor completely, if possible, the expressions in Problems 13–18.

13. a. $me + mi + my$ **b.** $a^2 - b^2$
c. $a^2 + b^2$ **d.** $a^3 - b^3$

14. a. $a^3 + b^3$ **b.** $s^2 + 2st + t^2$
c. $m^2 - 2mn + n^2$ **d.** $u^2 + 2uv + v^2$

15. a. $a^3 + 3a^2b + 3ab^2 + b^3$ **b.** $p^3 - 3p^2q + 3pq^2 - q^3$
c. $-c^3 + 3c^2d - 3cd^2 + d^3$ **d.** $x^2y + xy^2$

16. a. $(a + b)x + (a + b)y$ **b.** $(4x - 1)x + (4x - 1)3$
c. $x^2 - 2x - 35$ **d.** $2x^2 + 7x - 15$

17. a. $3x^2 - 5x - 2$ **b.** $6y^2 - 7y + 2$
c. $8a^2b + 10ab - 3b$ **d.** $2s^2 - 10s - 48$

18. a. $4y^3 + y^2 - 21y$ **b.** $12m^2 - 7m - 12$
c. $12p^4 + 11p^3 - 15p^2$ **d.** $9x^2y + 15xy - 14y$

B

Write the expressions in Problems 19–24 in simplified polynomial form.

19. a. $(3x - 1)(x^2 + 3x - 2)$ **b.** $(2x + 1)(x^2 + 2x - 5)$

20. a. $(5x + 1)(x^3 - 2x^2 + 3x - 5)$
b. $(4x - 1)(x^3 + 3x^2 - 2x - 4)$

21. a. $(x + 1)(x - 3)(2x + 1)$ **b.** $(2x - 1)(x + 3)(3x + 1)$

22. a. $(x - 2)(x + 3)(x - 4)$ **b.** $(x + 1)(2x - 3)(2x + 3)$

23. a. $(x - 2)^2(x + 1)$ **b.** $(x - 2)(x + 1)^2$

24. a. $(x - 5)^2(x + 2)$ **b.** $(x - 3)^2(3x + 2)$

Factor completely, if possible, the expressions in Problems 25–51.

25. $(x - y)^2 - 1$ **26.** $(2x + 3)^2 - 1$

27. $(5a - 2)^2 - 9$ **28.** $(3p - 2)^2 - 16$

29. $\frac{4}{25}x^2 - (x + 2)^2$ **30.** $\dfrac{4x^2}{9} - (x + y)^2$

31. $\frac{x^6}{y^8} - 169$

32. $\frac{9x^{10}}{4} - 144$

33. $(a + b)^2 - (x + y)^2$

34. $(m - 2)^2 - (m + 1)^2$

35. $2x^2 + x - 6$

36. $3x^2 - 11x - 4$

37. $6x^2 + 47x - 8$

38. $6x^2 - 47x - 8$

39. $6x^2 + 49x + 8$

40. $6x^2 - 49x + 8$

41. $4x^2 + 13x - 12$

42. $9x^2 - 43x - 10$

43. $9x^2 - 56x + 12$

44. $12x^2 + 12x - 25$

45. $4x^4 - 17x^2 + 4$

46. $4x^4 - 45x^2 + 81$

47. $x^6 + 9x^3 + 8$

48. $x^6 - 6x^3 - 16$

49. $(x^2 - \frac{1}{4})(x^2 - \frac{1}{9})$

50. $(x^2 - \frac{1}{9})(x^2 - \frac{1}{16})$

51. $(x^3 + \frac{1}{8})(x^2 - \frac{1}{4})$

Sketch the curves whose equations are given in Problems 52–61.

52. $\frac{x^2}{9} - \frac{y^2}{4} = 0$

53. $\frac{x^2}{16} - \frac{y^2}{25} = 0$

54. $\frac{y^2}{49} = \frac{x^2}{36}$

55. $\frac{y^2}{25} - \frac{x^2}{36} = 0$

56. $2x^3 - 2xy + x^2y - y^2 = 0$

57. $4x^3 - 8xy + 3x^2y - 6y^2 = 0$

58. $(x - y - 1)(3x + 2y - 4) = 0$

59. $(5x - 2y - 6)(x + y - 2) = 0$

60. $(x - y)(x + 2y)(x^2 - y) = 0$

61. $(x^2 - y^2)(2x + y - 3) = 0$

C

Write the expressions in Problems 62–67 in simplified polynomial form.

62. $(2x - 1)^2(3x^4 - 2x^3 + 3x^2 - 5x + 12)$

63. $(x - 3)^3(2x^3 - 5x^2 + 4x - 7)$

64. $(2x + 1)^2(5x^4 - 6x^3 - 3x^2 + 4x - 5)$

65. $(3x^2 + 4x - 3)(2x^2 - 3x + 4)$

66. $(2x^3 + 3x^2 - 2x + 4)^3$

67. $(x^3 - 2x^2 + x - 5)^3$

Factor completely, if possible, the expressions in Problems 68–83.

68. $x^{2n} - y^{2n}$

69. $x^{3n} - y^{3n}$

70. $x^{3n} + y^{3n}$

71. $x^{2n} - 2x^ny^n + y^{2n}$

72. $(x - 2)^2 - (x - 2) - 6$

73. $(x + 3)^2 - (x + 3) - 6$

74. $z^5 - 8z^2 - 4z^3 + 32$

75. $x^5 + 8x^2 - x^3 - 8$

76. $x^2 - 2xy + y^2 - a^2 - 2ab - b^2$

77. $x^2 + 2xy + y^2 - a^2 - 2ab - b^2$

78. $x^2 + y^2 - a^2 - b^2 - 2xy + 2ab$

79. $x^3 + 3x^2y + 3xy^2 + y^3 + a^3 + 3a^2b + 3ab^2 + b^3$

80. $(x + y + 2z)^2 - (x - y + 2z)^2$

81. $(x^2 - 3x - 6)^2 - 4$

82. $2(x + y)^2 - 5(x + y)(a + b) - 3(a + b)^2$

83. $2(s + t)^2 + 3(s + t)(s + 2t) - 2(s + 2t)^2$

3.2 SYNTHETIC DIVISION

In the previous section we considered sums, differences, and products of polynomials. In this section we will consider quotients of polynomials and see how this division can be quickly and easily accomplished using a process called *synthetic division.* This will, in turn, lead us in the next sections to some methods for finding the roots of certain polynomials as well as to some techniques for graphing polynomial functions.

Consider positive integers P and D. If the result of P divided by D is an integer Q, then D is a factor of P. For example, if $P = 15$ and $D = 3$, then $P/D = 5$, and 5 is called the *quotient*. If D is not a factor of P, we will obtain a quotient Q and a remainder R that is less than D so that

$$\frac{P}{D} = Q + \frac{R}{D}$$

For example:

$$\text{If } \frac{15}{3} = 5, \quad \text{then} \quad 15 = 5 \cdot 3$$

$$\text{If } \frac{17}{3} = 5 + \frac{2}{3}, \quad \text{then} \quad 17 = 5 \cdot 3 + 2$$

Notice how division may be checked by multiplying:

$$\text{If } \frac{P}{D} = Q, \quad \text{then} \quad P = QD$$

$$\text{If } \frac{P}{D} = Q + \frac{R}{D}, \quad \text{then} \quad P = QD + R \quad (R < D)$$

Division of polynomials is similar and leads to a result called the **Division Algorithm**.

Division Algorithm

If $P(x)$ and $D(x)$ are polynomials $[D(x) \neq 0]$, then there exist unique polynomials $Q(x)$ and $R(x)$ such that

$$\frac{P(x)}{D(x)} = Q(x) + \frac{R(x)}{D(x)}$$

where $Q(x)$ is a unique polynomial and $R(x)$ is a polynomial such that the degree of $R(x)$ is less than the degree of $D(x)$. The polynomial $Q(x)$ is called the **quotient** and $R(x)$ the **remainder**.

1. If $R(x) = 0$, then $D(x)$ is a factor of $P(x)$.
2. If the degree of $D(x)$ is greater than the degree of $P(x)$, then $Q(x) = 0$.
3. If both sides are multiplied by $D(x)$, the Division Algorithm is then stated in product form:

$$P(x) = Q(x)D(x) + R(x)$$

The question to be considered is how to *find* $Q(x)$ and $R(x)$. The first method is by long division. Let $P(x) = 3x^4 + 7x^3 + 2x^2 + 3x + 5$ and $D(x) = x + 1$. Then:

Multiply $3x^3(x + 1) = 3x^4 + 3x^3$ and write the answer so that similar terms are aligned.

$$\begin{array}{r} \mathbf{3x^3} \\ x + 1 \overline{)3x^4 + 7x^3 + 2x^2 + 3x + 5} \\ \mathbf{3x^4 + 3x^3} \end{array}$$

$3x^3$ was picked so that the terms shown by the arrow are identical.

$$\begin{array}{r} 3x^3 + \mathbf{4x^2} \\ x + 1 \overline{)3x^4 + 7x^3 + 2x^2 + 3x + 5} \\ 3x^4 + 3x^3 \\ \hline 4x^3 + 2x^2 + 3x + 5 \\ \mathbf{4x^3 + 4x^2} \end{array}$$

Multiply $4x^2(x + 1)$ and align similar terms.

Next subtract (or add the opposite).

This must be zero since the terms were identical.

$4x^2$ was picked so that these terms are identical.

Now subtract and repeat the procedure:

$$\begin{array}{r|l} & 3x^3 + 4x^2 - 2x + 5 \\ \hline x+1 & 3x^4 + 7x^3 + 2x^2 + 3x + 5 \\ & \underline{3x^4 + 3x^3} \\ & \quad 4x^3 + 2x^2 + 3x + 5 \\ & \quad \underline{4x^3 + 4x^2} \\ & \qquad -2x^2 + 3x + 5 \\ & \qquad \underline{-2x^2 - 2x} \\ & \qquad\qquad 5x + 5 \\ & \qquad\qquad \underline{5x + 5} \\ & \qquad\qquad\quad 0 \end{array}$$

Do not forget to subtract: $2x^2 - 4x^2 = -2x^2$

Subtract: $3x - (-2x) = 5x$

$0 \leftarrow 5 - 5 = 0$

The remainder is zero, so $x + 1$ is a factor of $3x^4 + 7x^3 + 2x^2 + 3x + 5$. You can check this by verifying that $P(x) = Q(x)D(x)$:

$$\begin{aligned} Q(x)D(x) &= (3x^3 + 4x^2 - 2x + 5)(x + 1) \\ &= 3x^4 + 4x^3 - 2x^2 + 5x + 3x^3 + 4x^2 - 2x + 5 \\ &= 3x^4 + 7x^3 + 2x^2 + 3x + 5 \end{aligned}$$

Since this is $P(x)$, the result checks.

EXAMPLE 1 Let $P(x) = 4x^4 - 6x^2 - 10x + 3$ and $D(x) = x^2 + x - 2$. Find $P(x)/D(x)$.

Solution

$$\begin{array}{r|l} & 4x^2 - 4x + 6 \\ \hline x^2 + x - 2 & 4x^4 + 0x^3 - 6x^2 - 10x + 3 \\ & \underline{4x^4 + 4x^3 - 8x^2} \\ & \quad -4x^3 + 2x^2 - 10x + 3 \\ & \quad \underline{-4x^3 - 4x^2 + 8x} \\ & \qquad 6x^2 - 18x + 3 \\ & \qquad \underline{6x^2 + 6x - 12} \\ & \qquad\quad -24x + 15 \end{array}$$

Notice that there is no x^3 term. Leave a space for this "missing" term, or write $0x^3$.

The remainder is $-24x + 15$. Thus

$$\frac{P(x)}{D(x)} = \mathbf{4x^2 - 4x + 6 + \frac{-24x + 15}{x^2 + x - 2}}$$

■

The process of long division is indeed *long* because of the duplication of symbols when carrying out this process. Consider the following example for dividing by a polynomial of the form $x - r$:

$$\begin{array}{r|l} & 2x^3 - 4x^2 - x + 3 \\ \hline x - 1 & 2x^4 - 6x^3 + 3x^2 + 4x - 3 \\ & \underline{\mathbf{2x^4} - 2x^3} \\ & \quad -4x^3 + \mathbf{3x^2 + 4x - 3} \\ & \quad \underline{\mathbf{-4x^3} + 4x^2} \\ & \qquad -x^2 + \mathbf{4x - 3} \\ & \qquad \underline{\mathbf{-x^2} + x} \\ & \qquad\quad 3x - \mathbf{3} \\ & \qquad\quad \underline{\mathbf{3x} - 3} \\ & \qquad\qquad 0 \end{array}$$

This is first degree, so the quotient is one degree less than the given polynomial.

Notice that each term in color is a repetition of the term directly above, so we could eliminate writing it down.

Also, the position of the term indicates the degree, so it is not even necessary to write down the variable:

The degree of the first term is 1 less than the degree of the given polynomial.

$$
\begin{array}{r|l}
 & 2 - 4 - 1 \quad 3 \\
\hline
-1 & 2 - 6 + 3 + 4 - 3 \\
 & \underline{-2} \\
 & -4 \\
 & \quad \underline{+4} \\
 & \quad -1 \\
 & \qquad \underline{+1} \\
 & \qquad 3 \\
 & \qquad \underline{-3} \\
 & \qquad 0
\end{array}
$$

Interpret this answer by recognizing that it is of degree 1 less than the degree of the dividend. This means the answer is $2x^3 - 4x^2 - x + 3$. That is, these numbers are the coefficients of the quotient polynomial.

There is no reason to spread out the array; it can be compressed into a more efficient form:

$$
\begin{array}{r|rrrrr}
 & \textcircled{2} & \mathbf{-4} & \mathbf{-1} & \mathbf{3} & \\
\hline
-1 & 2 & -6 & +3 & +4 & -3 \\
 & & -2 & +4 & +1 & -3 \\
\hline
 & \uparrow & \mathbf{-4} & \mathbf{-1} & \mathbf{3} & 0
\end{array}
$$

← Delete these terms.
← This line is no longer needed.
This is the remainder.

The top line is the same as the bottom line if you bring down the leading coefficient

$$
\begin{array}{r|rrrrr}
-1 & 2 & -6 & +3 & +4 & -3 \\
 & & -2 & +4 & +1 & -3 \\
\hline
 & 2 & -4 & -1 & 3 & 0
\end{array}
$$

These are the coefficients of the quotient (which begins with a degree 1 less than that of the given polynomial).

This is the remainder.

The process is now fairly compact, but the most common error in this procedure is in the subtraction. Remember: It is usually easier to add the opposite than to subtract, so change the sign of the divisor (-1 to 1 in this example) and add:

$$
\begin{array}{r|rrrrr}
1 & 2 & -6 & 3 & 4 & -3 \\
 & & 2 & -4 & -1 & 3 \\
\hline
 & 2 & -4 & -1 & 3 & 0
\end{array}
$$

Change the sign of this number, and **add**.

The condensed form of the division of a polynomial by $x - r$ is called **synthetic division**. **Notice synthetic division is used only when the divisor is of the form $x - r$** (*r positive or negative*). **If the divisor is not linear, then long division must be used.**

EXAMPLE 2 Divide $x^4 + 3x^3 - 12x^2 + 5x - 2$ by $x - 2$.

Solution

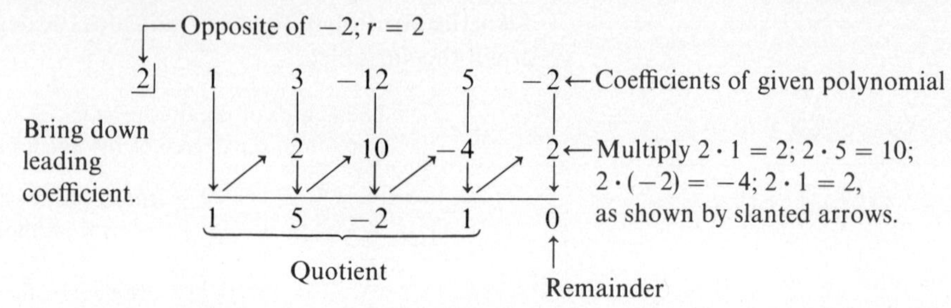

$x^3 + 5x^2 - 2x + 1$ ← Degree one less than the given polynomial

$$\frac{x^4 + 3x^3 - 12x^2 + 5x - 2}{x - 2} = \mathbf{x^3 + 5x^2 - 2x + 1}$$ ■

EXAMPLE 3 Divide $x^4 - 2x^3 - 10x^2 + 26x + 3$ by $x + 3$.

Solution Opposite of $+3$

-3	1	-2	-10	26	3 ← Coefficients
		-3	15	-15	-33
	1	-5	5	11	-30 ↑ Remainder

The answer is $\mathbf{x^3 - 5x^2 + 5x + 11 + \dfrac{-30}{x + 3}}$. ■

EXAMPLE 4 Divide $x^5 - 3x^2 + 1$ by $x - 2$.

Solution

2	1	0	0	-3	0	1	Be sure to include zero coefficients for missing terms.
		2	4	8	10	20	
	1	2	4	5	10	21	← (R)

The answer is $\mathbf{x^4 + 2x^3 + 4x^2 + 5x + 10 + \dfrac{21}{x - 2}}$ ■

PROBLEM SET 3.2

A

Fill in the blanks for the synthetic divisions in Problems 1–10.

1. $\dfrac{3x^3 - 9x^2 + 11x - 10}{x - 2}$

2	3	**b**	11	-10
		6	**c**	10
	a	-3	5	0

The quotient is **d**.

2. $\dfrac{4x^3 - 6x^2 - 8x - 5}{x - 3}$

a	4	-6	**c**	**e**
		b	**d**	**f**
	4	6	10	25

The quotient is **g**.

3. $\dfrac{2x^3 - 4x^2 - 10x + 12}{x + 2}$

a	2	−4	c	e
		b	d	f
	2	−8	6	0

The quotient is g.

4. $\dfrac{5x^3 + 12x^2 + 2x - 4}{x + 3}$

a	5	12	d	f
		−15	9	−33
	b	c	e	g

The quotient is h.

5. $\dfrac{x^3 + 7x^2 + 8x - 20}{x + 4}$

a	1	7	d	f
		−4	−12	16
	b	c	e	g

The quotient is h.

6. $\dfrac{x^4 - 3x^2 + 2x - 7}{x - 1}$

a	b	c	d	e	g
		1	1	−2	0
	1	1	−2	f	h

The quotient is i.

7. $\dfrac{x^4 + 2x^3 - 5x + 2}{x - 1}$

a	b	c	d	e	g
		1	3	3	−2
	1	3	3	f	h

The quotient is i.

8. $\dfrac{x^5 - 1}{x - 1}$

a	b	c	d	e	f	h
		1	1	1	1	1
	1	1	1	1	g	i

The quotient is j.

9. $\dfrac{x^5 - 32}{x + 2}$

a	b	c	d	e	g	j
		−2	4	−8	h	k
	1	−2	4	f	i	l

The quotient is m.

10. $\dfrac{x^4 + 3x^3 - 2x^2}{x + 3}$

a	b	c	e	h	k
		−3	f	i	l
	1	d	g	j	m

The quotient is n.

Use synthetic division to find the quotients in Problems 11–30.

11. $\dfrac{3x^3 - 2x^2 + 4x - 75}{x - 3}$

12. $\dfrac{3x^3 + 2x^2 - 4x + 8}{x + 2}$

13. $\dfrac{x^4 - 6x^3 + x^2 - 8}{x + 1}$

14. $\dfrac{x^4 - 3x^3 + x + 6}{x - 2}$

15. $\dfrac{2x^4 - 15x^2 + 8x - 3}{x + 3}$

16. $\dfrac{2x^4 - 15x - 2}{x - 2}$

17. $\dfrac{4x^5 - 3x^4 - 5x^3 + 4}{x - 1}$

18. $\dfrac{4x^3 - 3x^2 - 5x + 2}{x + 1}$

19. $\frac{3x^3 - 2x^2 + 4x - 24}{x - 2}$

20. $\frac{3x^3 + 2x^2 + 4x + 24}{x + 2}$

21. $\frac{x^3 + 5x^2 - 2x - 24}{x + 4}$

22. $\frac{x^3 + 3x^2 - 6x - 8}{x - 2}$

23. $\frac{x^3 - 4x^2 - 17x + 60}{x - 5}$

24. $\frac{x^5 - 3x^4 + x - 3}{x - 3}$

25. $\frac{4x^4 + 4x^3 - 15x^2 + 7}{x - 1}$

26. $\frac{5x^3 - 21x^2 - 13x - 35}{x - 5}$

27. $\frac{x^4 - x^3 + x^2 - x - 4}{x + 1}$

28. $\frac{2x^4 + 6x^3 - 4x^2 - 11x + 3}{x + 3}$

29. $\frac{3x^4 + 10x^3 - 8x^2 - 5x - 20}{x + 4}$

30. $\frac{x^4 - 9x^3 + 20x^2 - 15x + 18}{x - 6}$

49. $\frac{x^5 + 3x^3 - 3x^4 - 16x^2 + 21x - 6}{x - 3}$

50. $\frac{5x^4 - x + x^5 - 1}{x + 5}$

51. $\frac{x^4 - 12x^2 + 4x + 15}{(x + 1)(x - 3)}$

52. $\frac{x^4 - x^3 - 12x^2 + 28x - 16}{(x - 2)(x + 4)}$

53. $\frac{2x^3 - 3x^2 - 11x + 6}{(x - 3)(x + 2)}$

54. $\frac{2x^3 - 3x^2 - 8x - 3}{(2x - 3)(x + 1)}$

55. $\frac{2x^4 + 5x^3 - 16x^2 - 423x + 1116}{(x - 3)(x + 4)}$

56. $\frac{2x^4 + 5x^3 - 16x^2 - 49x - 30}{(x + 1)(x + 3)}$

B

In Problems 31–36, use long division to find $Q(x)$ and $R(x)$ if $P(x)$ is divided by $D(x)$.

31. $P(x) = 4x^4 - 14x^3 + 10x^2 - 6x + 2$; $D(x) = 2x - 1$
32. $P(x) = x^5 - x^3 + x^2 + x - 1$; $D(x) = x^2 + x$
33. $P(x) = x^5 - x^3 + x^2 + 1$; $D(x) = x^2 + x$
34. $P(x) = x^2 + 2x + 1$; $D(x) = x^3$
35. $P(x) = 6x^4 - 11x^3 + 6x^2 - 2x + 5$; $D(x) = 2x^2 + x + 1$
36. $P(x) = 2x^4 + 5x^3 - 16x^2 - 45x - 18$; $D(x) = x^2 + x - 2$

In Problems 37–56, use synthetic division to find $Q(x)$.

37. $\frac{3x^3 - 7x^2 - 5x + 2}{x - 3}$

38. $\frac{2x^3 - 3x^2 + 4x - 10}{x - 2}$

39. $\frac{x^4 - 3x^3 - 4x^2 + 2x - 5}{x - 4}$

40. $\frac{x^4 - 5x^3 + 2x^2 - x + 3}{x - 2}$

41. $\frac{5x^4 + 10x^3 - 20x^2 - 12x - 2}{x + 3}$

42. $\frac{2x^4 + 5x^3 + 2x^2 + 5x + 2}{x + 2}$

43. $\frac{x^5 - 3x^4 + 2x^2 - 5}{x + 2}$

44. $\frac{x^4 - 20x^2 - 10x - 50}{x - 5}$

45. $\frac{4x^3 - x^2 + 2x - 1}{x + 1}$

46. $\frac{6x^4 - x^2 + 1}{x - 3}$

47. $\frac{5x^5 - 2x + 1}{x + 2}$

48. $\frac{5x^4 + 7x^3 - 27x^2 + 14x - 12}{x - 1}$

C

57. Devise a procedure for using synthetic division on

$$\frac{x^4 + 7x^3 + 5x^2 - 23x + 10}{x^2 + 4x - 5}$$

58. Devise a procedure for using synthetic division on

$$\frac{2x^4 - 7x^3 - 4x^2 + 27x - 18}{x^2 - x - 6}$$

59. Find K so that $x^4 + Kx^3 + 7x^2 - 2x + 8$ has no remainder when divided by $x + 2$.

60. Find h and k so that $x^4 + hx^3 - kx + 15$ has no remainder when divided by $x - 1$ and $x + 3$.

61. Let $P(x) = 2x^3 - x^2 + 5x + 3$.
    a. Divide $P(x)$ by $2x + 1$ using long division.
    b. Divide $P(x)$ by $x + \frac{1}{2}$ using synthetic division.
    c. Compare your answers to parts **a** and **b**.
    d. Notice that, in part **b**, the divisor of part **a** was divided by 2. If a divisor in any division problem is divided by 2, how will the quotient be affected?
    e. Write out a general procedure for dividing $P(x)$ by $ax + b$ synthetically.

62. Check the procedure you outlined in Problem 61**e** by applying it to the following problems (divide synthetically):
    a. $2x^3 + 7x^2 + x - 1$ divided by $2x + 1$
    b. $6x^3 - 13x^2 + 14x - 12$ divided by $2x - 3$
    c. $6x^3 + x^2 + 3x + 1$ divided by $2x + 1$
    d. $3x^3 - 10x^2 - 19x + 5$ divided by $3x + 5$

3.3 GRAPHING POLYNOMIAL FUNCTIONS

Graphs of linear and quadratic functions were discussed in Chapter 2, so we begin this section by considering a process for graphing a cubic function. Suppose you wish to graph the cubic function

$$f(x) = 2x^3 - 3x^2 - 12x + 17$$

Since there is a one-to-one correspondence between points on the graph and ordered pairs satisfying this equation, you can begin by plotting points:

$$\begin{aligned} f(0) &= 2 \cdot 0^3 - 3 \cdot 0^2 - 12 \cdot 0 + 17 \\ &= 17 && (0, 17) \text{ is on the graph.} \\ f(1) &= 2 \cdot 1^3 - 3 \cdot 1^2 - 12 \cdot 1 + 17 \\ &= 2 - 3 - 12 + 17 \\ &= 4 && (1, 4) \text{ is on the graph.} \\ f(-1) &= 2(-1)^3 - 3(-1)^2 - 12(-1) + 17 \\ &= -2 - 3 + 12 + 17 \\ &= 24 && (-1, 24) \text{ is on the graph.} \end{aligned}$$

But you can see that this procedure could be very tedious by the time you find enough points to determine the shape of the curve. Instead, consider the **Remainder Theorem**.

Remainder Theorem

When a polynomial $f(x)$ is divided by $x - r$, the remainder is equal to $f(r)$.

To verify this theorem, recall the Division Algorithm:

$$\frac{P(x)}{D(x)} = Q(x) + \frac{R(x)}{D(x)} \qquad \text{or} \qquad P(x) = Q(x)D(x) + R(x)$$

In this context, $P(x) = f(x)$, $D(x) = x - r$, and $R(x)$ is a constant since the degree of $R(x)$ must be less than the degree of $D(x)$, which is 1. The Division Algorithm can be restated as

$$f(x) = Q(x)(x - r) + R$$

where R represents the remainder. Now

$$\begin{aligned} f(r) &= Q(r)(r - r) + R \\ &= R \end{aligned}$$

since $Q(r)(r - r) = Q(r) \cdot 0 = 0$. Points on the curve can therefore be found by synthetic division.

EXAMPLE 1 Find several points of the curve defined by the function $f(x) = 2x^3 - 3x^2 - 12x + 17$ by using synthetic division.

Solution Since the same polynomial is to be evaluated repeatedly, it is not necessary to recopy it each time. The work can be arranged as shown below:

	2	−3	−12	17	point
1	2	−1	−13	4	(1, 4) ← Try to do the work mentally: $1 \cdot 2 + (-3) = -1$; $1 \cdot (-1) + (-12) = -13$; and so on.
−1	2	−5	−7	24	(−1, 24)
2	2	1	−10	−3	(2, −3)
−2	2	−7	2	13	(−2, 13)
0				17	(0, 17) ← Why is $f(0)$ always equal to the constant term?
3	2	3	−3	8	(3, 8)
−3	2	−9	15	−28	(−3, −28)
4	2	5	8	49	(4, 49)
−4	2	−11	32	−111	(−4, −111)

■

After you have found enough points, which is a much less tedious procedure when you use synthetic division, you can connect these points to draw a smooth curve. This property, called **continuity**, is studied extensively in calculus, but for our purposes we can state the following useful theorem even though its proof requires calculus.

Intermediate-Value Theorem for Polynomial Functions

> If f is a polynomial function on $[a, b]$ such that $f(a) \neq f(b)$, then $f(x)$ takes on every value between $f(a)$ and $f(b)$ over the interval $[a, b]$.

This theorem allows you to connect the points found by synthetic division in order to draw a smooth curve.

EXAMPLE 2 Use the information in Example 1 as well as the Intermediate-Value Theorem to sketch the graph of the polynomial function

$$f(x) = 2x^3 - 3x^2 - 12x + 17$$

Solution Plot the points from Example 1, paying attention to the scales on the axes so that they accommodate most of the values obtained. Connect the points to draw a smooth curve as shown in Figure 3.3.

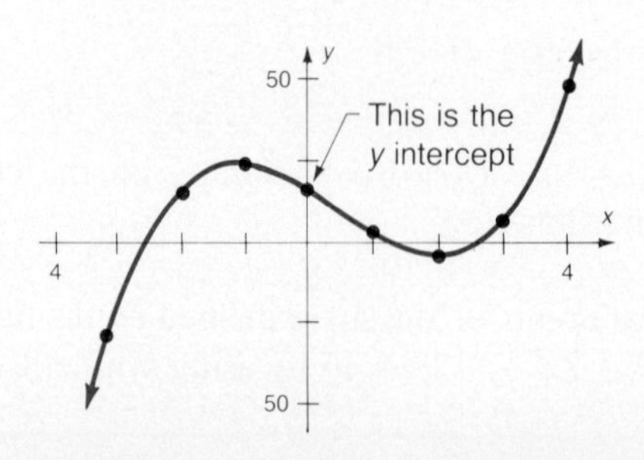

Figure 3.3 Graph of $f(x) = 2x^3 - 3x^2 - 12x + 17$

■

EXAMPLE 3 Sketch the graph of $f(x) = 3x^4 - 8x^3 - 30x^2 + 72x + 47$.

Solution *You* select the integers that are convenient.

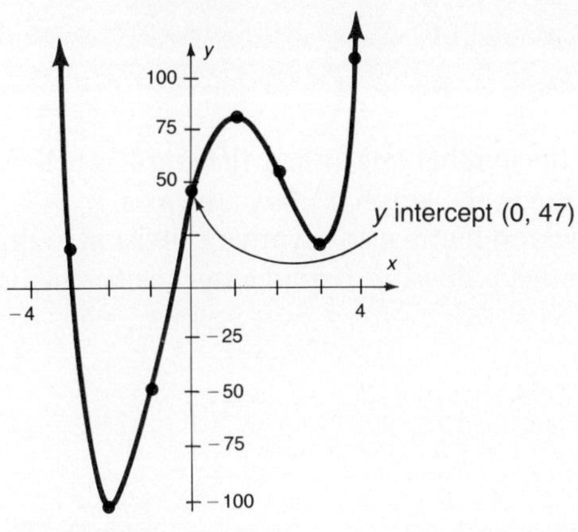

Figure 3.4 Graph of $f(x) = 3x^4 - 8x^3 - 30x^2 + 72x + 47$

	3	**−8**	**−30**	**72**	**47**	
−4	3	−20	50	−128	559	(−4, 559)
−3	3	−17	21	9	20	(−3, 20)
−2	3	−14	−2	76	−105	(−2, −105)
−1	3	−11	−19	91	−44	(−1, −44)
0					47	(0, 47)
1	3	−5	−35	37	84	(1, 84)
2	3	−2	−34	4	55	(2, 55)
3	3	1	−27	−9	20	(3, 20)
4	3	4	−14	16	111	(4, 111)

Plot the points and draw a smooth curve as in Figure 3.4. ■

The graphs of polynomial functions have *turning points* where the functions change from increasing to decreasing or from decreasing to increasing. A polynomial of degree n has at most $n - 1$ turning points. In calculus you will find how to locate the exact position of these turning points. For now, however, you will need to rely on synthetic division and plotting points to sketch graphs. If a polynomial function can be factored into linear factors, the number of plotted points can be reduced as illustrated by Example 4.

EXAMPLE 4 Sketch the graph of $y = (x + 1)(x - 2)(3x - 2)$.

Solution Begin by locating the *critical values* for the polynomial function. Recall that these are the x values for which $y = 0$ (the x-intercepts):

$$(x + 1)(x - 2)(3x - 2) = 0 \quad \leftarrow y = 0$$

$$x = -1, 2, \frac{2}{3}$$

These critical values divide the x-axis into four regions:

$-1 \qquad \frac{2}{3} \qquad 2$

$x < -1 \qquad -1 < x < \frac{2}{3} \qquad \frac{2}{3} < x < 2 \qquad x > 2$

Since $y = 0$ for $x = -1$, $x = \frac{2}{3}$, and $x = 2$, it follows that y must be either positive or negative in each of the four regions. That is, you need to determine whether y is positive or negative in *each* of the four regions. To do this, select an x value in *each* region and substitute that value into the function, as shown below:

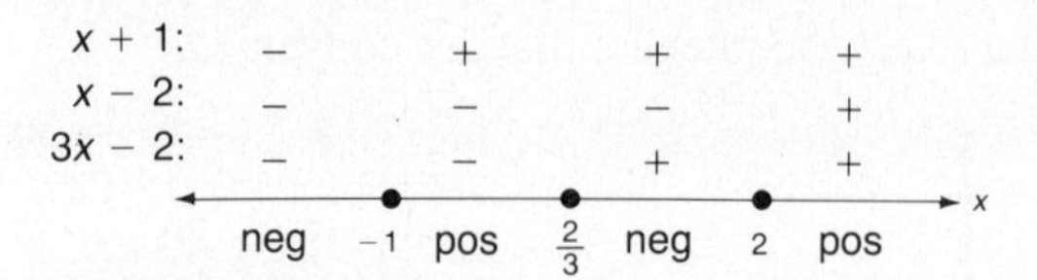

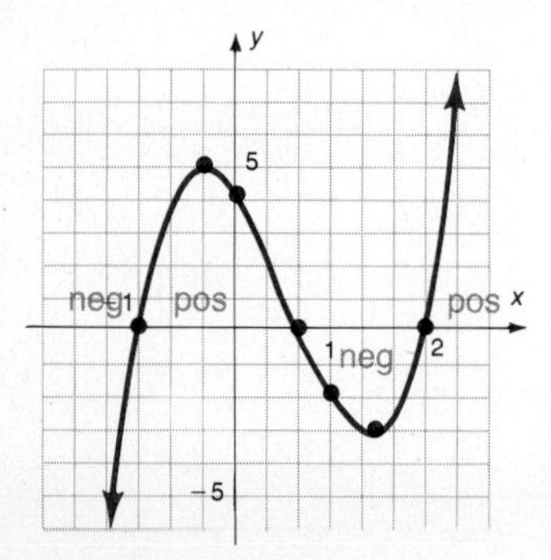

Figure 3.5 Graph of $y = (x+1)(x-2)(3x-2)$ or $y = 3x^3 - 5x^2 - 4x + 4$

Wherever you have labeled the number line as *neg*, the graph is below the axis; and wherever you have labeled it *pos*, the graph is above the axis.

Plot some points (to the nearest tenth) using synthetic division to draw the graph, as shown in Figure 3.5. Synthetic division requires the coefficients in polynomial (not factored) form. Thus,

$$\begin{aligned} y &= (x+1)(x-2)(3x-2) \\ &= 3x^3 - 5x^2 - 4x + 4 \end{aligned}$$

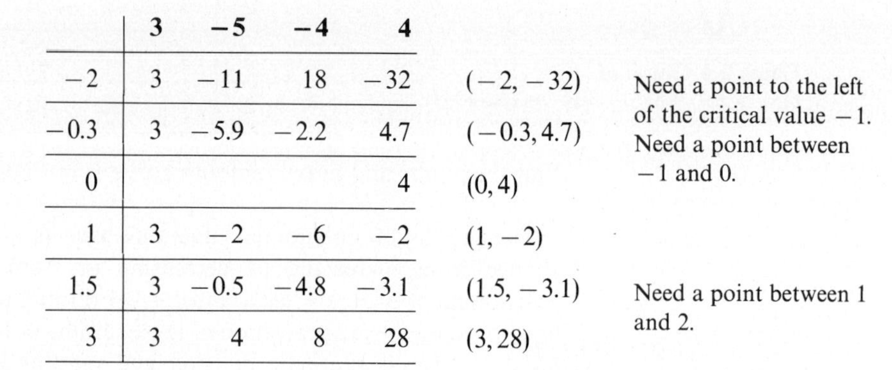

	3	−5	−4	4		
−2	3	−11	18	−32	(−2, −32)	Need a point to the left of the critical value −1.
−0.3	3	−5.9	−2.2	4.7	(−0.3, 4.7)	Need a point between −1 and 0.
0				4	(0, 4)	
1	3	−2	−6	−2	(1, −2)	
1.5	3	−0.5	−4.8	−3.1	(1.5, −3.1)	Need a point between 1 and 2.
3	3	4	8	28	(3, 28)	

■

PROBLEM SET 3.3

A

Use synthetic division to find the values specified for the functions given in Problems 1–12.

1. $f(x) = 5x^3 - 7x^2 + 3x - 4$
   a. $f(1)$ b. $f(-1)$ c. $f(0)$
   d. $f(6)$ e. $f(-4)$
2. $g(x) = 4x^3 + 10x^2 - 120x - 350$
   a. $g(1)$ b. $g(-1)$ c. $g(0)$
   d. $g(5)$ e. $g(-5)$
3. $P(x) = x^4 - 10x^3 + 20x^2 - 23x - 812$
   a. $P(1)$ b. $P(-1)$ c. $P(0)$
   d. $P(3)$ e. $P(7)$
4. $f(t) = 3t^4 + 5t^3 - 8t^2 - 3t$
   a. $f(0)$ b. $f(1)$ c. $f(2)$
   d. $f(-2)$ e. $f(-4)$
5. $g(t) = 4t^4 - 3t^3 + 5t - 10$
   a. $g(0)$ b. $g(-1)$ c. $g(-2)$
   d. $g(5)$ e. $g(-3)$
6. $P(t) = t^4 - t^3 - 39t - 70$
   a. $P(0)$ b. $P(1)$ c. $P(-1)$
   d. $P(-5)$ e. $P(7)$
7. $f(h) = 8h^4 - 6h^3 + 5h^2 + 4h - 3$
   a. $f(0)$ b. $f(1)$ c. $f(\frac{1}{2})$
   d. $f(-\frac{1}{2})$ e. $f(-3)$
8. $g(h) = 16h^4 + 64h^3 + 19h^2 - 81h + 18$
   a. $g(0)$ b. $g(-2)$ c. $g(-3)$
   d. $g(\frac{1}{4})$ e. $g(\frac{3}{4})$
9. $P(x) = 4x^4 - 8x^3 - 43x^2 + 29x + 60$
   a. $P(-1)$ b. $P(1)$ c. $P(4)$
   d. $P(\frac{3}{2})$ e. $P(-\frac{5}{2})$

10. $f(x) = (x-2)(x+3)(2x-5)$
  a. $f(2)$ b. $f(-3)$ c. $f(\frac{5}{2})$
  d. $f(1)$ e. $f(-2)$

11. $g(x) = (x-1)(x-4)(2x+1)$
  a. $g(1)$ b. $g(4)$ c. $g(-\frac{1}{2})$
  d. $g(-2)$ e. $g(-1)$

12. $h(x) = (x+1)(x+2)(3x-1)$
  a. $h(-1)$ b. $h(1)$ c. $h(-2)$
  d. $h(2)$ e. $h(\frac{1}{3})$

Sketch the graph of each polynomial function in Problems 13–20.

13. $f(x) = x^3 - 3x^2 + 10$
14. $f(x) = x^3 + 3x^2 + 11$
15. $f(x) = 2x^3 - 3x^2 - 12x + 3$
16. $f(x) = 2x^3 + 3x^2 - 12x + 48$
17. $f(x) = x^3 - 2x^2 + x - 5$
18. $f(x) = x^3 + 4x^2 - 3x + 2$
19. $f(x) = x^3 - 3x^2 + 10$
20. $f(x) = x^3 + 3x^2 + 11$

B

Sketch the graph of each polynomial function in Problems 21–48.

21. $f(x) = x^3 - 6x^2 + 9x - 9$
22. $f(x) = 2x^3 - 3x^2 - 36x + 78$
23. $f(x) = 4x^4 - 8x^3 - 43x^2 + 29x + 60$
24. $f(x) = 16x^4 + 64x^3 + 19x^2 - 81x + 18$
25. $f(x) = x^4 - 7x^2 - 2x + 2$
26. $f(x) = x^4 - 14x^3 + 58x^2 - 46x - 9$
27. $f(x) = x^6 - 4x^4 - 4x^2 + 4$
28. $f(x) = x^5 + 2x^4 - 5x^3 - 10x^2 + 4x + 8$
29. $y = 3x^4 - x^3 - 14x^2 + 4x + 8$
30. $y = 5x^4 + 3x^3 - 22x^2 - 12x + 8$
31. $y = x^4 - x^3 - 3x^2 + 2x + 4$
32. $y = x^4 - 2x^2 - 4x + 3$
33. $y = x^4 - 7x^2 - 2x + 2$
34. $y = x^4 - 14x^3 + 58x^2 - 46x - 9$
35. $y = (x-1)(x+1)(x+3)$
36. $y = (x-1)(x-4)(x+3)$
37. $y = (x+1)(x+3)(2x-5)$
38. $y = (x-1)(x-4)(2x+1)$
39. $y = (x+1)(x+2)(3x-1)$
40. $y = x(x-3)(x+3)$
41. $y = 3x^2(x-3)(x+1)$
42. $y = 5x^2(x-4)(x+2)$
43. $y = x^2(x^2-1)$
44. $y = x^2(x^2-4)$
45. $f(x) = 3x^4 - 7x^3 + 5x^2 + x - 10$
46. $f(x) = 8x^4 + 12x^3 - 3x^2 + 4x + 20$
47. $f(x) = x^5 - 3x^4 + 2x^3 - 7x + 15$
48. $f(x) = x^5 - 5x^4 + 3x^3 + x^2$
49. Graph $y = x^3$ and $y = -x^3$ on the same axes.
50. Graph $y = x^3$, $y = \frac{1}{2}x^3$, and $y = \frac{1}{100}x^3$ on the same axes.
51. Graph $y = x^3$ and $y - 2 = (x-1)^3$ on the same axes.
52. Graph $y = x^4$ and $y = -x^4$ on the same axes.
53. Graph $y = x^4$, $y = \frac{1}{10}x^4$, and $y = \frac{1}{100}x^4$ on the same axes.
54. Graph $y = x^4$ and $y + 4 = (x-2)^4$ on the same axes.

C

55. Using Problems 49–51, discuss the graph of $y - k = a(x-h)^3$ by comparing it with the graph of $y = x^3$.
56. Using Problems 52–54, discuss the graph of $y - k = a(x-h)^4$ by comparing it with the graph of $y = x^4$.
57. On the same axes, graph $y = x^n$ for $-1 \le x \le 1$, where n is equal to:
  a. 0 b. 2 c. 4 d. 6
58. Repeat Problem 57 for $-4 \le x \le 4$ and $-500 \le y \le 500$.
59. On the same axes, graph $y = x^n$ for $-1 \le x \le 1$, where n is equal to:
  a. 1 b. 3 c. 5 d. 7
60. Repeat Problem 59 for $-4 \le x \le 4$ and $-2000 \le y \le 2000$.
61. Using the results from Problems 57–60, make some conjectures concerning the graph of $y = x^n$.
62. Using the results from Problems 57–61, make a conjecture about the graph of $y - k = (x-h)^n$.

3.4 REAL ROOTS OF POLYNOMIAL EQUATIONS

If

$$P(x) = a_n x^n + a_{n-1}x^{n-1} + \cdots + a_2x^2 + a_1x + a_0 \qquad (a_n \neq 0)$$

then the *roots* or *solutions* of $P(x) = 0$ are values of x that satisfy this equation. Such an equation is called a **polynomial equation**. Recall from Chapter 1 that a is called a *zero* of a function P if $P(a) = 0$. Thus we speak of the **roots** of a polynomial equation and the **zeros** of a polynomial function. If a zero is a real number, then it is an **x intercept** of the graph of the polynomial function.

Consider the polynomial equation

$$4x^4 - 8x^3 + 43x^2 + 29x + 60 = 0$$

To solve this equation, find the values of x that make it true. As a first step, write the equation in factored form if possible. From the Division Algorithm, we know that if $R = 0$ and

$$P(x) = Q(x)(x - r) + R$$

then $P(x) = Q(x)(x - r)$. This says that $x - r$ is a factor of $P(x)$. But since you are looking for values of x such that $P(x) = 0$,

$$0 = Q(x)(x - r)$$

Notice that this equation is satisfied by $x = r$ and leads to a result called the **Factor Theorem**. The polynomial equation $Q(x) = 0$ is called a **depressed equation** of the polynomial equation $P(x) = 0$.

Factor Theorem

> If r is a root of the polynomial equation $P(x) = 0$, then $x - r$ is a factor of $P(x)$. Moreover, if $x - r$ is a factor of $P(x)$, then r is a root of the polynomial equation $P(x) = 0$.

The Factor Theorem is a generalization of the method of solving quadratic equations by factoring. Remember: If $P \cdot Q = 0$, then $P = 0$ or $Q = 0$ (perhaps both). Suppose you wish to solve

$$x^2 - x - 2 = 0$$

You can do so by factoring:

$$(x - 2)(x + 1) = 0$$

Then each factor is set equal to zero to find

$$x = 2, -1$$

Suppose further that the same factor appears more than once in the factoring process, as in

$$(x - 2)(x - 2) = 0$$

Then there is a single root (which occurs twice) to give

$$x = 2$$

In this chapter it will be useful to attach some terminology to a repeated root. If a factor $x - r$ occurs exactly k times in the factorization of $P(x)$, then r is called a **zero of multiplicity k**. If a factor $x - r$ occurs exactly k times in the factorization of $P(x)$ in the equation $P(x) = 0$, then r is called a **root of multiplicity k**. In the preceding illustration, 2 is a root of multiplicity 2. If

$$f(x) = (x - 1)(x - 3)(x - 4)(x - 1)(x - 3)(x - 1)$$

the **zeros** are 1 (multiplicity 3), 3 (multiplicity 2), and 4. Notice that this is an *expression*, so we use the word *zero*. On the other hand,

$$(x-1)(x-3)(x-4)(x-1)(x-3)(x-1)=0$$

has **roots** 1 (multiplicity 3), 3 (multiplicity 2), and 4. This is an equation, so we use the word *root* (or *solution*).

The relationship between roots of the polynomial equation $f(x)=0$ and factors of $f(x)$, as described by the Factor Theorem, leads to a statement concerning the number of roots to expect for a polynomial equation of degree n. Suppose $P(x)=0$ is a polynomial equation in which $P(x)$ is a third-degree polynomial. It is impossible for $P(x)=0$ to have more than three roots. If it had more, say four roots (r_1, r_2, r_3, r_4), then the Factor Theorem provides

$$P(x)=a_4(x-r_1)(x-r_2)(x-r_3)(x-r_4)$$

so that $P(x)$ is a *fourth-degree* polynomial. Thus a third-degree polynomial equation cannot have more than three roots.

Root Limitation Theorem

A polynomial function f of degree n has, at most, n distinct zeros.

The related question—Does it necessarily have three zeros?—is answered by the Fundamental Theorem of Algebra, which is considered in the next section.

As you saw in the previous section, there is a close relationship between the zeros of a polynomial function and its graph. Suppose, by synthetic division, you find for two values a and b that $P(a)$ and $P(b)$ are opposite in sign. Then the Intermediate-Value Theorem for polynomial functions tells you that there is a value r such that $a<r<b$ and $P(r)=0$.

Location Theorem

If f is a polynomial function such that $f(a)$ and $f(b)$ are opposite in sign, then there is at least one real zero on the interval between a and b.

Next you need some reasonable method for finding the zeros, since you cannot simply find them by trial and error, even with the location theorem and synthetic division. The best that mathematicians can offer is a theorem that provides a list of *possible* rational roots of the polynomial equation $f(x)=0$. Not every number on the list will be a root, but every rational root of the polynomial equation will appear someplace on the list. Given this *finite* list (which admittedly might be large), you can check values from this list using synthetic division and the location theorem.

Rational Root Theorem

If $P(x)=a_nx^n+a_{n-1}x^{n-1}+\cdots+a_1x+a_0$ has integer coefficients and p/q (where p/q is reduced) is a rational zero, then p is a factor of a_0 and q is a positive factor of a_n.

To use this theorem, make a list of *all* factors of a_0 and divide these integers by the factors of a_n. Notice also, if $a_n=1$, then all rational zeros are integers which divide

a_0. The procedure for finding all possible rational roots is not very difficult if you work systematically.

EXAMPLE 1 List all possible rational roots of $x^3 - x^2 - 4x + 4 = 0$.

Solution

p: $a_0 = 4$, with factors $1, -1, 2, -2, 4, -4$

q: $a_n = 1$, with factor 1

Shorten these lists by using the $\pm$ sign: p: $\pm 1, \pm 2, \pm 4$

Form all possible fractions:

$$\frac{p}{q}: \frac{1}{1}, \frac{-1}{1}, \frac{2}{1}, \frac{-2}{1}, \frac{4}{1}, \frac{-4}{1}$$

Simplifying and not bothering to rewrite those that are repeated, we find that the possible rational roots are ± 1, ± 2, ± 4. ■

EXAMPLE 2 List all possible rational zeros of $P(x) = 4x^4 - 8x^3 - 43x^2 + 29x + 60$.

Solution

$p(a_0 = 60)$: ± 1, ± 2, ± 3, ± 4, ± 5, ± 6, ± 10, ± 12, ± 15, ± 20, ± 30, ± 60

$q(a_n = 4)$: 1, 2, 4

$$\frac{p}{q}: \pm 1, \pm\frac{1}{2}, \pm\frac{1}{4}, \pm 2, \pm 3, \pm\frac{3}{2}, \pm\frac{3}{4}, \pm 4, \pm 5, \pm\frac{5}{2}, \pm\frac{5}{4},$$

$$\pm 6, \pm 10, \pm 12, \pm 15, \pm\frac{15}{2}, \pm\frac{15}{4}, \pm 20, \pm 30, \pm 60$$

Note: If a factor is already listed, it is not repeated. For example, $\pm\frac{2}{4}$ is not listed separately from $\pm\frac{1}{2}$. ■

You can see from the examples that the list of possible rational zeros may be quite large, but it is a *finite* list, so with enough time and effort the entire list could be checked. Usually this is not necessary if you first pick the values that are easiest to check. (That is, do not start with the fractions or large numbers.) There is also another theorem, called the **Upper and Lower Bound Theorem**, that helps to rule out many of the values listed with the Rational Root Theorem. If f is a polynomial function, then a real number b is called an **upper bound** for the polynomial equation $f(x) = 0$ if there is no root, or solution, larger than b. A real number a is a **lower bound** if there is no solution less than a.

Upper and Lower Bound Theorem

If $a > 0$ and, in the synthetic division of $P(x)$ by $x - a$, all the numbers in the last row have the *same sign*, then a is an *upper bound* for the roots of $P(x) = 0$.

If $b < 0$ and, in the synthetic division of $P(x)$ by $x - b$, the numbers in the last row *alternate in sign*, then b is a *lower bound* for the roots of $P(x) = 0$.

EXAMPLE 3 Solve $2x^4 - 5x^3 - 8x^2 + 25x - 10 = 0$.

Solution

$p(a_0 = -10)$: ± 1, ± 2, ± 5, ± 10

$q(a_n = 2)$: 1, 2

$$\frac{p}{q}: \pm 1, \pm\frac{1}{2}, \pm 2, \pm 5, \pm\frac{5}{2}, \pm 10$$

	2	**−5**	**−8**	**25**	**−10**
1	2	−3	−11	14	4
−1	2	−7	−1	26	−36
2	2	−1	−10	5	0

Begin with the values that are easiest to check. In Examples 3–5 we will use this shaded portion to tell you the value checked is not a root.
$x - 2$ is a factor.

Since $x - 2$ is a factor, you can write the polynomial equation as

$$(x-2)(2x^3 - x^2 - 10x + 5) = 0$$

Now focus your attention on the depressed equation

$$2x^3 - x^2 - 10x + 5 = 0$$

by using these coefficients in the synthetic division process. It is not necessary to recopy these coefficients in your work.

	2	−1	−10	5	
−2	2	−5	0	5	
5	2	9	35	180	← All sums are positive, so 5 is an upper bound. No larger values need to be checked.
−5	2	−11	45	−220	← Sums have alternating signs, so −5 is a lower bound. No smaller values need be checked.
$\frac{1}{2}$	2	0	−10	0	← $x - \frac{1}{2}$ is a factor. The resulting depressed equation is now quadratic, so you can stop the synthetic division.

$$(x-2)(x-\tfrac{1}{2})(2x^2 - 10) = 0$$
$$2(x-2)(x-\tfrac{1}{2})(x^2 - 5) = 0$$

Note: The resulting quadratic may not have rational roots. In fact, you may need the Quadratic Formula for its solution.

The roots are **$2, \frac{1}{2}, \sqrt{5}, -\sqrt{5}$**. ■

EXAMPLE 4 Solve $4x^4 - 8x^3 - 43x^2 + 29x + 60 = 0$.

Solution The list of possible rational roots is shown in Example 2.

	4	**−8**	**−43**	**29**	**60**	
1	4	−4	−47	−18	42	
−1	4	−12	−31	60	0	← $x + 1$ is a factor, so −1 is a root. Do not recopy, but use these *coefficients* as you continue synthetic division.
2	4	−4	−39	−18		
−2	4	−20	9	42		
−3	4	−24	41	−63		← −3 is a lower bound; since the Location Theorem says there is a root between −2 and −3, look on the list of possible rational roots for the next number to try.
$-\frac{5}{2}$	4	−22	24	0		

$x + \frac{5}{2}$ is a factor. The resulting depressed equation is quadratic, so now you can write the given polynomial equation in factored form and complete the factoring process directly.

$$4x^4 - 8x^3 - 43x^2 + 29x + 60 = 0$$

$$(x+1)\left(x+\frac{5}{2}\right)(4x^2 - 22x + 24) = 0$$

Do you see where these factors come from in the synthetic division process?

$$(x+1)\left(x+\frac{5}{2}\right)(2)(2x^2 - 11x + 12) = 0$$

$$(x+1)\left(x+\frac{5}{2}\right)(2)(2x - 3)(x - 4) = 0$$

Now use the Factor Theorem.

The roots are **-1, $-\frac{5}{2}$, $\frac{3}{2}$, and 4.** ■

EXAMPLE 5 Solve $8x^5 - 44x^4 + 86x^3 - 73x^2 + 28x - 4 = 0$.

Solution

$p(a_0 = -4)$: $\pm 1, \pm 2, \pm 4$

$q(a_n = 8)$: 1, 2, 4, 8

$\frac{p}{q}$: $\pm 1, \pm\frac{1}{2}, \pm\frac{1}{4}, \pm\frac{1}{8}, \pm 2, \pm 4$

	8	**−44**	**86**	**−73**	**28**	**−4**	
1	8	−36	50	−23	5	1	
−1	8	−52	138	−211	239	−243	← Lower bound
2	8	−28	30	−13	2	0	← $x - 2$ is a factor.
4	8	4	46	171	686		← Upper bound
2	8	−12	6	−1	0		← $x - 2$ is a factor.
2	8	4	14	27			← Upper bound; this is a new upper bound because we are working with the depressed equation.
$\frac{1}{2}$	8	−8	2	0			← $x - \frac{1}{2}$ is a factor.

Note: Do not forget possible multiple roots. Just because $x - 2$ was a factor once does not mean it cannot be again.

$$(x-2)(x-2)\left(x-\frac{1}{2}\right)(8x^2 - 8x + 2) = 0$$

$$2(x-2)(x-2)\left(x-\frac{1}{2}\right)(4x^2 - 4x + 1) = 0$$

$$2(x-2)(x-2)\left(x-\frac{1}{2}\right)(2x - 1)(2x - 1) = 0$$

The roots are 2 and $\frac{1}{2}$; 2 is a root of multiplicity 2 and $\frac{1}{2}$ is a root of multiplicity 3. ■

In Examples 3 and 4 the degree was 4 and in both cases you found four roots. The Root Limitation Theorem tells you that there cannot be more roots. Even when you find multiple roots as illustrated by Example 5, you know you have found all the real roots. Suppose, however, you attempt to solve a polynomial equation that has fewer *real* roots than the degree. In such a situation you will not know whether you cannot find the real roots because although they *are not* rational they are still real (that is, are x intercepts) or because they are simply not real numbers. The following theorem will help you to answer this dilemma by telling you when you have found all the positive or negative real roots. When applying this theorem, remember that a root of multiplicity m is counted as m roots.

Descartes' Rule of Signs

Let $P(x)$ be a polynomial with real coefficients written in descending powers of x. Count the number of sign changes in the signs of the coefficients.

1. The number of positive real zeros is equal to the number of sign changes or is equal to that number decreased by an even integer.
2. The number of negative real zeros is equal to the number of sign changes in $P(-x)$ or is equal to that number decreased by an even integer.

EXAMPLE 6

$$f(x) = 2x^4 - 5x^3 - 8x^2 + 25x - 10 \qquad \text{From Example 3}$$

(sign changes marked 1, 2, 3)

There are three sign changes, so there are three or one positive real zeros. Next calculate $f(-x)$:

$$\begin{aligned} f(-x) &= 2(-x)^4 - 5(-x)^3 - 8(-x)^2 + 25(-x) - 10 \\ &= 2x^4 + 5x^3 - 8x^2 - 25x - 10 \end{aligned}$$

(sign change marked 1)

There is one sign change on the coefficients of $f(-x)$, so there is one negative real zero. Compare this result with the answer you found in Example 3:

$\underbrace{2, \tfrac{1}{2}, \sqrt{5}}_{\text{3 positive zeros}}, \; -\sqrt{5} \leftarrow$ 1 negative zero

■

EXAMPLE 7

$$f(x) = 8x^5 - 44x^4 + 86x^3 - 73x^2 + 28x - 4 = 0 \qquad \text{From Example 5}$$

(sign changes marked 1, 2, 3, 4, 5)

There are five sign changes here, so there are five, three, or one positive real roots.

$$f(-x) = -8x^5 - 44x^4 - 86x^3 - 73x^2 - 28x - 4 = 0$$

Here there are no sign changes, so there are no negative real roots. Compare this result with the answer you found in Example 5: $2, \frac{1}{2}$. There seems to be a discrepancy, but remember that 2 has multiplicity two and $\frac{1}{2}$ has multiplicity three, so *think*:

$\underbrace{2, 2, \frac{1}{2}, \frac{1}{2}, \frac{1}{2}}$

5 positive real roots
and no negative real roots

■

The following example ties together some of the ideas of this section and the preceding section.

EXAMPLE 8 Solve $x^4 - 3x^2 - 6x - 2 = 0$.

Solution $p(a_0 = -2)$: $\pm1, \pm2$ $\quad \frac{p}{q}$: $\pm1, \pm2$

$q(a_n = 1)$: 1

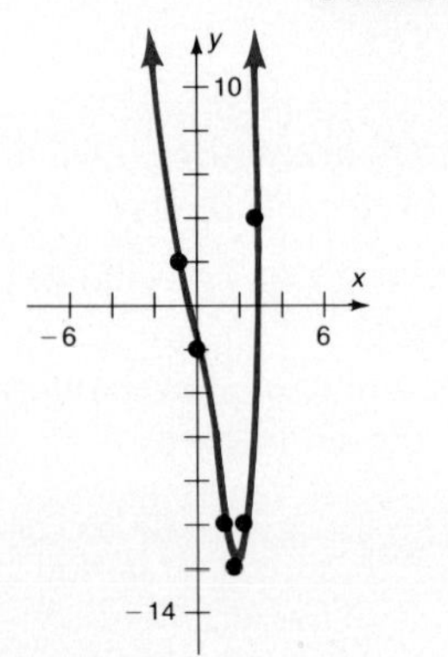

Figure 3.6 Graph of $y = x^4 - 3x^2 - 6x - 2$

	1	**0**	**−3**	**−6**	**−2**
1	1	1	−2	−8	−10
2	1	2	1	−4	−10
−1	1	−1	−2	−4	2
−2	1	−2	1	−8	14 ← Lower bound

There are no rational roots (we have tried all the numbers on our list). Next verify the type of roots by using Descartes' Rule of Signs:

$f(x) = x^4 - 3x^2 - 6x - 2$

1

One positive root

$f(-x) = x^4 - 3x^2 + 6x - 2$

1 2 3

Three or one negative root(s)

Using synthetic division to find some additional points: $(0, -2)$ for the y intercept; $(-3, 70)$; and $(3, 34)$ for the upper bound. Plot the known points as shown in Figure 3.6. You would expect roots between -1 and 0 as well as between 2 and 3. Moreover, because of the upper and lower bounds, as well as Descartes' Rule of Signs, you would expect these to be the only real roots. Since these roots are not rational you will not be able to find them exactly. You can, however, approximate them to any desired degree of accuracy by using synthetic division. For the root between -1 and 0:

A calculator would be very helpful

	1	**0**	**−3**	**−6**	**−2**	
−0.5	1	−0.5	−2.75	−4.625	0.3125	Root between −0.5 and −0.4 since one is negative and the other is positive
−0.4	1	−0.4	−2.84	−4.864	−0.0544	
−0.41	1	−0.41	−2.8319	−4.8389	−0.0161	Root between −0.41 and −0.42
−0.42	1	−0.42	−2.8236	−4.8141	0.0219	

Continue in this fashion to approximate the root to any degree of accuracy desired. (This would be a good problem for a computer.) Repeat the procedure for the root between 2 and 3 (it is about 2.41). Thus the real roots (to the nearest tenth) are **−0.4, 2.4.** ■

PROBLEM SET 3.4

A

In Problems 1–12, find the zeros of the polynomial and state the multiplicity of each zero.

1. $f(x) = (x - 2)(x + 3)^2$
2. $f(x) = (x + 1)^2(x - 3)^2$
3. $f(x) = x^3(2x - 3)^2$
4. $f(x) = x^2(3x + 1)^3$
5. $f(x) = (x^2 - 1)^2(x + 2)$
6. $f(x) = (x - 5)(x^2 - 4)^3$
7. $f(x) = (x^2 + 2x - 15)^2$
8. $f(x) = (6x^2 + 7x - 3)^2$
9. $f(x) = x^4 - 8x^3 + 16x^2$
10. $f(x) = x^4 + 6x^3 + 9x^2$
11. $f(x) = (x^3 - 9x)^2$
12. $f(x) = (x^3 - 25x)^2$

In Problems 13–24, use Descartes' Rule of Signs to state the number of possible positive and negative real roots.

13. $x^4 - 3x^3 + 7x^2 - 19x + 15 = 0$
14. $3x^3 - 7x^2 + 5x + 7 = 0$
15. $2x^5 + 6x^4 - 3x + 12 = 0$
16. $x^3 + 2x^2 - 5x - 6 = 0$
17. $x^3 + 3x^2 - 4x - 12 = 0$
18. $2x^3 + x^2 - 13x + 6 = 0$
19. $2x^3 - 3x^2 - 32x - 15 = 0$
20. $x^4 - 12x^3 + 54x^2 - 108x + 81 = 0$
21. $x^4 + 3x^3 - 20x^2 - 3x + 18 = 0$
22. $x^4 - 13x^2 + 36 = 0$
23. $2x^2 + 6x - 3x^3 - 4 = 0$
24. $5x^3 - 2x^4 + x^2 - 7 = 0$

Find the possible rational roots for the polynomial equations in Problems 25–36.

25. $x^4 - 3x^3 + 7x^2 - 19x + 15 = 0$
26. $3x^3 - 7x^2 + 5x + 7 = 0$
27. $2x^5 + 6x^4 - 3x + 12 = 0$
28. $x^3 + 2x^2 - 5x - 6 = 0$
29. $x^3 + 3x^2 - 4x - 12 = 0$
30. $2x^3 + x^2 - 13x + 6 = 0$
31. $2x^3 - 3x^2 - 32x - 15 = 0$
32. $x^4 - 12x^3 + 54x^2 - 108x + 81 = 0$
33. $x^4 + 3x^3 - 20x^2 - 3x + 18 = 0$
34. $x^4 - 13x^2 + 36 = 0$
35. $5x^2 + 6x^3 - 2x - 1 = 0$
36. $10x^2 - 8x^3 + 17x - 10 = 0$

B

Solve the polynomial equations in Problems 37–56.

37. $x^3 - x^2 - 4x + 4 = 0$
38. $2x^3 - x^2 - 18x + 9 = 0$
39. $x^3 - 2x^2 - 9x + 18 = 0$
40. $x^3 + 2x^2 - 5x - 6 = 0$
41. $x^3 + 3x^2 - 4x - 12 = 0$
42. $2x^3 + x^2 - 13x + 6 = 0$
43. $2x^3 - 3x^2 - 32x - 15 = 0$
44. $x^4 - 12x^3 + 54x^2 - 108x + 81 = 0$
45. $x^4 + 3x^3 - 19x^2 - 3x + 18 = 0$
46. $x^4 - 13x^2 + 36 = 0$
47. $x^3 + 15x^2 + 71x + 105 = 0$
48. $x^3 - 15x^2 + 74x - 120 = 0$
49. $8x^3 - 12x^2 - 66x + 35 = 0$
50. $12x^3 + 16x^2 - 7x - 6 = 0$
51. $x^5 + 8x^4 + 10x^3 - 60x^2 - 171x - 108 = 0$
52. $x^5 + 6x^4 + x^3 - 48x^2 - 92x - 48 = 0$
53. $x^7 + 2x^6 - 4x^5 - 2x^4 + 3x^3 = 0$
54. $x^6 - 3x^4 + 3x^2 - 1 = 0$
55. $x^6 - 12x^4 + 48x^2 - 64 = 0$
56. $x^7 + 3x^6 - 4x^5 - 16x^4 - 13x^3 - 3x^2 = 0$
57. Does there exist a real number that exceeds its cube by 1?
58. ***Consumer*** The dimensions of a rectangular box are consecutive integers, and its volume is 720 cm^3. What are the dimensions of the box?
59. ***Engineering*** A 2-cm-thick slice is cut from a cube, leaving a volume of 384 cm^3. What is the length of a side of the original cube?
60. ***Engineering*** A rectangular sheet of tin with dimensions 3×5 m has equal squares cut from its four corners. The resulting sheet is folded up on the sides to form a topless box. Find all possible dimensions of the cutout square to the nearest 0.1 m such that the box has a volume of 1 m^3.

C

Italian mathematicians discovered the algebraic solution of cubic and quartic equations in the sixteenth century. At that time they would challenge one another to solve certain equations. Problems 61–64 were such challenge problems. Find the real roots in each problem to the nearest tenth.

61. In 1515, Scipione del Ferro solved the cubic equation $x^3 + mx + n = 0$ and revealed his secret to his pupil Antonio Fior. At about the same time, Tartaglia solved the equation $x^3 + px^2 = n$. Fior thought Tartaglia was bluffing and challenged him to a public contest of solving cubic equations. According to the historian Howard Eves, Tartaglia triumphed completely. Solve the cubic $x^3 + px^2 = n$, where $p = 5$ and $n = 21$.
62. Girolamo Cardano stole the solution of the cubic equation from Tartaglia and published it in his *Ars Magna.*

Tartaglia protested, but Cardano's pupil, Ludovico Ferrari (who solved the biquadratic equation), claimed that both Cardano and Tartaglia stole it from del Ferro. According to the historian Howard Eves, there was a dispute from which Tartaglia was lucky to have escaped alive. One of the problems in *Ars Magna* was $x^3 - 63x = 162$. Solve this cubic.

63. Cardano solved the quartic $13x^2 = x^4 + 2x^3 + 2x + 1$. Find the real roots for this equation.

64. In 1540, Cardano was given the problem "Divide 10 into three parts such that they shall be in continued proportion and that the product of the first two shall be 6." Let x, y, and z be the three parts. Then $x + y + z = 10$. Also,

$$\frac{x}{y} = \frac{y}{z} \quad \text{and} \quad xy = 6$$

Eliminating x and y, you obtain $z^4 + 6z^2 + 36 = 60z$. Find the real roots for this equation.

3.5 THE FUNDAMENTAL THEOREM OF ALGEBRA*

In 1799, a 22-year-old graduate student named Karl Gauss proved in his doctoral thesis that every polynomial equation has at least one solution in the complex numbers. This, of course, is an assumption that you have made throughout your study of algebra—from the time you solved first-degree equations in beginning algebra until now. It is an idea so basic to algebra that it is called the Fundamental Theorem of Algebra. This theorem strengthens the Root Limitation Theorem of the last section. However, in order to prove this theorem it is necessary to allow the domain to be the set of complex numbers. Remember that the set of complex numbers has the set of real numbers as a subset so that when we speak of complex coefficients of a polynomial equation, we are including all those polynomial equations previously considered in this chapter.

Fundamental Theorem of Algebra

If $P(x)$ is a polynomial of degree $n \geq 1$ with complex coefficients, then $P(x) = 0$ has at least one complex root.

If an equation has one solution, a depressed equation of one degree less may be obtained. That new equation, according to the Fundamental Theorem, has a root. This root may now be used to obtain an equation of lower degree. The result of this process suggests the following theorem.

Number of Roots Theorem

If $P(x)$ is a polynomial of degree $n \geq 1$ with complex coefficients, then $P(x) = 0$ has exactly n roots (if roots are counted according to their multiplicity).

Of course, the roots need not be distinct or real. Consider the following example.

EXAMPLE 1 Show that $x^6 - 2x^3 + x^2 - 2x + 2 = 0$ has at least two nonreal complex roots.

---

* This section is optional and requires complex numbers.

Solution Check Descartes' Rule of Signs:

$$P(x) = x^6 - 2x^3 + x^2 - 2x + 2$$

1 2 3 4

4, 2, or 0 positive real roots

$$P(-x) = x^6 + 2x^3 + x^2 + 2x + 2$$

0 negative real roots

The polynomial equation has six roots, and at most four of these are real (positive). Thus, at least two roots are complex and nonreal. ■

If synthetic division is used on the equation in Example 1, two positive real roots are quickly found:

	1	0	0	−2	1	−2	2	
1	1	1	1	−1	0	−2	0	
1	1	2	3	2	2	0		All positive, upper bound

Notice that $x = 1$ is a root of *multiplicity 2*. When counting the number of roots using Descartes' Rule, multiple roots are *not* counted as a single root. A root of multiplicity 2 counts as two roots and roots of multiplicity n count as n roots.

Note further that the depressed equation produced by the synthetic division has all positive coefficients; thus, all the positive roots have been found. Hence, the polynomial equation $x^6 - 2x^3 + x^2 - 2x + 2 = 0$ has one real root (with multiplicity 2) and four nonreal complex roots.

We can now distinguish between roots and x intercepts of polynomial equations. There are n roots of an nth-degree polynomial equation $P(x) = 0$. The **real roots** correspond to the x intercepts of the graph of $y = P(x)$ whereas the **imaginary roots** do not.

EXAMPLE 2 Solve $x^4 - 2x^3 + x^2 - 8x - 12 = 0$, and draw the graph of $y = x^4 - 2x^3 + x^2 - 8x - 12$.

Solution Check Descartes' Rule of Signs:

$$P(x) = x^4 - 2x^3 + x^2 - 8x - 12$$

1 2 3

3 or 1 positive real roots

$$P(-x) = x^4 + 2x^3 + x^2 + 8x - 12$$

1

1 negative real root

p: $\pm 1, \pm 2, \pm 3, \pm 4, \pm 6, \pm 12$ $\quad \frac{p}{q}$: $\pm 1, \pm 2, \pm 3, \pm 4, \pm 6, \pm 12$

q: 1

	1	−2	1	−8	−12	
1	1	−1	0	−8	−20	
−1	1	−3	4	−12	0	Use this depressed equation.
−1	1	−4	8	−20		Since −1 could be a multiple root, try it again.
3	1	0	4	0		

The depressed equation is

$$x^2 + 4 = 0$$

$$x = \pm 2i$$

The roots are −1, 3, $2i$, $-2i$; this is consistent with the results from Descartes' Rule of Signs (one positive and one negative real root). It is also consistent with the Number of Roots Theorem (degree 4, four roots). Now you would also expect the graph to have two x intercepts corresponding to the real roots. The graph is shown in Figure 3.7.

	1	−2	1	−8	−12
0					−12
1	1	−1	0	−8	−20
−1	1	−3	4	−12	0
2	1	0	1	−6	−24
−2	1	−4	9	−26	40
3	1	1	4	4	0
4	1	2	9	28	100

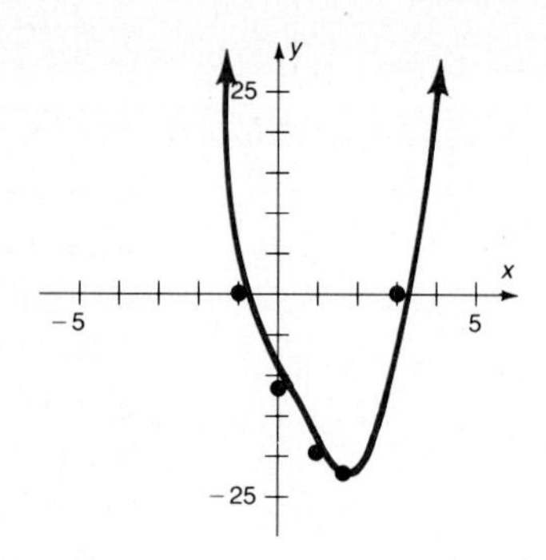

Figure 3.7 Graph of $y = x^4 - 2x^3 + x^2 - 8x - 12$ ■

Another result that is sometimes helpful when finding roots is the Conjugate Pair Theorem.

Conjugate Pair Theorem

1. If $P(x) = 0$ is a polynomial equation with real coefficients, then when $a + bi$ is a root, $a - bi$ is also a root (a and b are real numbers).
2. If $P(x) = 0$ is a polynomial equation with rational coefficients, then when $m + \sqrt{n}$ is a root, $m - \sqrt{n}$ is also a root (m and n are rational numbers and $\sqrt{n}$ is irrational).

EXAMPLE 3 Solve $x^4 - 4x - 1 = 4x^3$ given that $2 + \sqrt{5}$ is a root.

Solution

$$x^4 - 4x^3 - 4x - 1 = 0$$

$2+\sqrt{5}$	1	−4	0	−4	−1
		$2+\sqrt{5}$	1	$2+\sqrt{5}$	1
	1	$-2+\sqrt{5}$	1	$-2+\sqrt{5}$	0

and $2 - \sqrt{5}$ must be a root also:

$$\begin{array}{r|rrrr} 2-\sqrt{5} & 1 & -2+\sqrt{5} & 1 & -2+\sqrt{5} \\ & & 2-\sqrt{5} & 0 & 2-\sqrt{5} \\ \hline & 1 & 0 & 1 & 0 \end{array}$$

The depressed equation is

$$x^2 + 1 = 0$$
$$x = \pm i$$

The roots are $\mathbf{2 \pm \sqrt{5}, \pm i}$. ■

Alternate Solution The synthetic division solution shown above is rather cumbersome, so we offer an alternate procedure. Since $2 + \sqrt{5}$ is a root, then $2 - \sqrt{5}$ must also be a root, which (by the Factor Theorem) says that the original polynomial has factors $x - (2 + \sqrt{5})$ and $x - (2 - \sqrt{5})$. Multiply these factors:

$$\begin{aligned}[x - (2 + \sqrt{5})][x - (2 - \sqrt{5})] &= x^2 - (2 - \sqrt{5})x - (2 + \sqrt{5})x \\ &\quad + (2 - \sqrt{5})(2 + \sqrt{5}) \\ &= x^2 - 4x - 1\end{aligned}$$

You could also use the Sum and Product of the Roots Theorem (Sec. 1.4) to find

$$(2 + \sqrt{5}) + (2 - \sqrt{5}) = 4$$
$$(2 + \sqrt{5})(2 - \sqrt{5}) = 4 - 5 = -1$$

Thus, the quadratic is

$$x^2 \underset{\substack{\uparrow \\ \text{opposite of the} \\ \text{sum of the roots}}}{- 4x} \underset{\substack{\uparrow \\ \text{product of the roots}}}{- 1}$$

Now divide (using long division) this factor into the original polynomial:

$$\begin{array}{r} x^2 + 1 \\ x^2 - 4x - 1 \overline{\smash{\big)} x^4 - 4x^3 + 0x^2 - 4x - 1} \\ \underline{x^4 - 4x^3 - x^2} \\ x^2 \\ \underline{x^2 - 4x - 1} \\ 0 \end{array}$$

Thus, $x^4 - 4x^4 - 4x - 1 = \underbrace{(x^2 - 4x - 1)}\underbrace{(x^2 + 1)}$

$$\begin{array}{ll} x^2 - 4x - 1 = 0 & x^2 + 1 = 0 \\ x = 2 \pm \sqrt{5} & x = \pm i \end{array}$$

Now set each factor equal to zero and solve.

The roots are $\mathbf{2 \pm \sqrt{5}, \pm i}$. ■

EXAMPLE 4 Solve $x^4 - 3x^2 - 6x - 2 = 0$ given that $-1 + i$ is a root.

Solution You can divide by $-1 + i$ synthetically, but instead we will multiply together the known factors:

$$[x-(-1-i)][x-(-1+i)] = x^2-(-1+i)x-(-1-i)x + (-1-i)(-1+i)$$
$$= x^2 + x - ix + x + ix + (1-i^2)$$
$$= x^2 + 2x + 2$$

opposite of sum (2) ↑ ↑ product of roots (2)

Find the other factor(s) by long division:

$$\begin{array}{r} x^2 - 2x \quad - 1 \\ x^2+2x+2 \overline{)\, x^4 + 0x^3 - 3x^2 - 6x - 2} \\ \underline{x^4 + 2x^3 + 2x^2} \qquad\qquad \\ -2x^3 - 5x^2 \qquad\qquad \\ \underline{-2x^3 - 4x^2 - 4x} \qquad \\ -x^2 - 2x \qquad \\ \underline{-x^2 - 2x - 2} \\ 0 \end{array}$$

Thus, $x^4 - 3x^2 - 6x - 2 = (x^2+2x+2)(x^2-2x-1)$.

$$x^2 + 2x + 2 = 0 \qquad\qquad x^2 - 2x - 1 = 0$$

$$x = \frac{-2 \pm \sqrt{4-8}}{2} \qquad\qquad x = \frac{2 \pm \sqrt{4+4}}{2}$$

$$x = -1 \pm i \qquad\qquad x = 1 \pm \sqrt{2}$$

The roots are $\mathbf{-1 \pm i, 1 \pm \sqrt{2}}$.

You might wish to try using synthetic division to check this answer. ■

PROBLEM SET 3.5

A

In Problems 1–6 let $f(x) = x^4 - 6x^3 + 15x^2 - 2x - 10$ to find the requested value.

1. $f(i)$ **2.** $f(-i)$ **3.** $f(\sqrt{2})$

4. $f(-\sqrt{3})$ **5.** $f(2+i)$ **6.** $f(1-\sqrt{3})$

In Problems 7–12 let $f(x) = x^4 - 8x^3 + 21x^2 - 14x - 10$ to find the requested value.

7. $f(-i)$ **8.** $f(i)$ **9.** $f(\sqrt{3})$

10. $f(-\sqrt{2})$ **11.** $f(3-i)$ **12.** $f(1+\sqrt{2})$

In Problems 13–18 let $f(x) = x^4 - 10x^3 + 36x^2 - 58x + 35$ to find the requested value.

13. $f(i)$ **14.** $f(-i)$ **15.** $f(2-i)$

16. $f(2+i)$ **17.** $f(3+\sqrt{2})$ **18.** $f(3-\sqrt{2})$

In Problems 19–24 let $f(x) = x^4 - 6x^3 + 18x^2 - 30x + 25$ to find the requested value.

19. $f(2)$ **20.** $f(3)$ **21.** $f(2+i)$

22. $f(2-i)$ **23.** $f(1+2i)$ **24.** $f(1-2i)$

In Problems 25–30 let $f(x) = 2x^4 - x^3 - 13x^2 + 5x + 15$ to find the requested value.

25. $f(2)$ **26.** $f(-2)$ **27.** $f(\sqrt{5})$

28. $f(-\sqrt{5})$ **29.** $f(i)$ **30.** $f(-i)$

31. Is $1+\sqrt{2}$ a root of $x^3 - 2x^2 + 1 = 0$? If it is, name another root.

32. Is $1+i$ a root of $x^3 - 4x^2 + 6x - 4 = 0$? If it is, name another root.

33. Is $1-2i$ a root of $x^3 - x^2 + 3x + 5 = 0$? If it is, name another root.

34. Is $1 - 2i$ a root of $x^4 - 2x^3 + 4x^2 + 2x - 5$? If it is, name another root.

35. Is $1 + 2i$ a root of $x^4 - 7x^3 + 14x^2 + 2x - 20$? If it is, name another root.

36. Is $2 + \sqrt{5}$ a root of $x^4 - 4x^3 - 5x^2 + 16x + 4$? If it is, name another root.

B

Solve the polynomial equations $P(x) = 0$ and graph the curve $y = P(x)$ in Problems 37–55.

37. $P(x) = x^3 - 8$

38. $P(x) = x^3 - 64$

39. $P(x) = x^3 - 125$

40. $P(x) = x^4 - 64$

41. $P(x) = x^4 - 81$

42. $P(x) = x^4 - 625$

43. $P(x) = x^4 + 9x^2 + 20$

44. $P(x) = x^4 + 10x^2 + 9$

45. $P(x) = x^4 + 13x^2 + 36 = 0$

46. $P(x) = (x^2 - 4x - 1)(x^2 - 6x + 7)$

47. $P(x) = (x^2 - 6x + 10)(x^2 - 8x + 17)$

48. $P(x) = (x^2 + 2x + 5)(x^2 - 3x + 5)$

49. $P(x) = (x^2 - 4x - 1)(x^2 - 3x + 5)$

50. $P(x) = 2x^4 - x^3 + 2x - 1$

51. $P(x) = 2x^3 - 3x^2 + 4x + 3 = 0$

52. $P(x) = x^4 - 7x^3 + 14x^2 + 2x - 20$

53. $P(x) = x^4 - 2x^3 + 4x^2 + 2x - 5$

54. $P(x) = 4x^4 - 10x^3 + 10x^2 - 5x + 1$

55. $P(x) = 27x^4 - 180x^3 + 213x^2 + 62x - 10$

C

Assuming the given value is a root, solve the equations in Problems 56–65.

56. $x^3 - 2x^2 + 4x - 8 = 0;\ 2i$

57. $x^4 + 13x^2 + 36 = 0;\ -3i$

58. $x^4 - 6x^2 + 25 = 0;\ 2 + i$

59. $x^4 - 4x^3 + 3x^2 + 8x - 10 = 0;\ 2 + i$

60. $2x^4 - 5x^3 + 9x^2 - 15x + 9 = 0;\ i\sqrt{3}$

61. $2x^4 - x^3 - 13x^2 + 5x + 15 = 0;\ -\sqrt{5}$

62. $3x^5 + 10x^4 - 8x^3 + 12x^2 - 11x + 2 = 0;\ -2 + \sqrt{5}$

63. $2x^5 + 9x^4 - 3x^2 - 8x - 42 = 0;\ i\sqrt{2}$

64. $x^5 - 11x^4 + 24x^3 + 16x^2 - 17x + 3 = 0;\ 2 + \sqrt{3}$

65. $x^6 + x^5 - 3x^4 - 4x^3 + 4x + 4 = 0;\ -\sqrt{2}$ is a multiple root.

3.6 RATIONAL FUNCTIONS

To evaluate and graph polynomial functions, we used synthetic division and the division algorithm. That is, we considered $P(x)/D(x)$ for polynomials $P(x)$ and $D(x)$, where $D(x) \neq 0$. Now consider this quotient from another viewpoint. An expression of the form $P(x)/D(x)$ is called a *rational function.*

Rational Function

A **rational function** f is the quotient of polynomial functions $P(x)$ and $D(x)$; that is,

$$f(x) = \frac{P(x)}{D(x)} \qquad \text{where } D(x) \neq 0$$

The domain of a rational function is the set of all real x such that $D(x) \neq 0$. When writing a rational function, we will assume this domain. That is, if

$$f(x) = \frac{x + 3}{x - 2}$$

it is not necessary to write $x \neq 2$ since this is implied in the definition.

As an aid to sketching certain rational functions, consider the notion of an asymptote. An **asymptote** is a line having the property that the distance from a point P on the curve to the line approaches zero as the distance from P to the origin

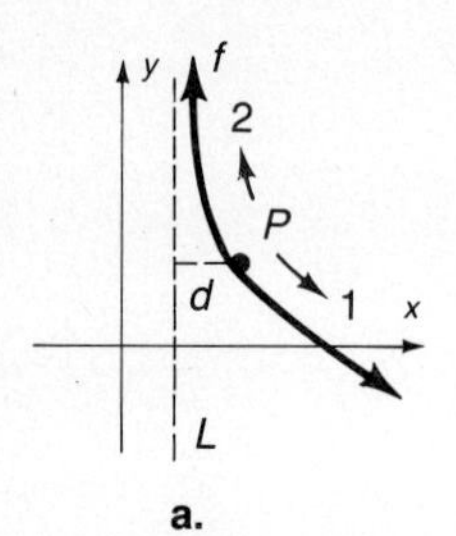

a.

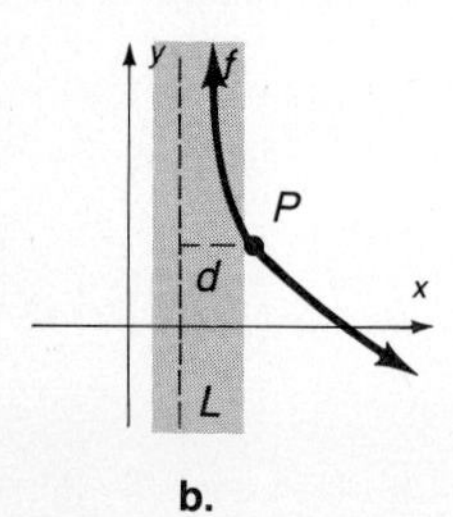

b.

Figure 3.8 An asymptote to a curve

increases without bound and P is *on a suitable part of the curve*. This last phrase (in italics) is best illustrated by considering Figure 3.8, where L is an asymptote for the function f. Consider P and d, the distance from P to the line L, as shown in Figure 3.8a. Now the distance from P to the origin can increase in two ways depending on whether P moves along the curve in direction 1 or direction 2. In direction 1 the distance d increases without bound, but in direction 2 the distance d approaches zero. Thus if you consider the portion of the curve in the shaded region of Figure 3.8b, you see that the conditions of the definition of an asymptote apply. Even though some rational functions may not have asymptotes, there are three types of asymptotes that occur frequently enough to merit consideration. These are *vertical, horizontal,* and *slant asymptotes*. An example of each is shown in Figure 3.9.

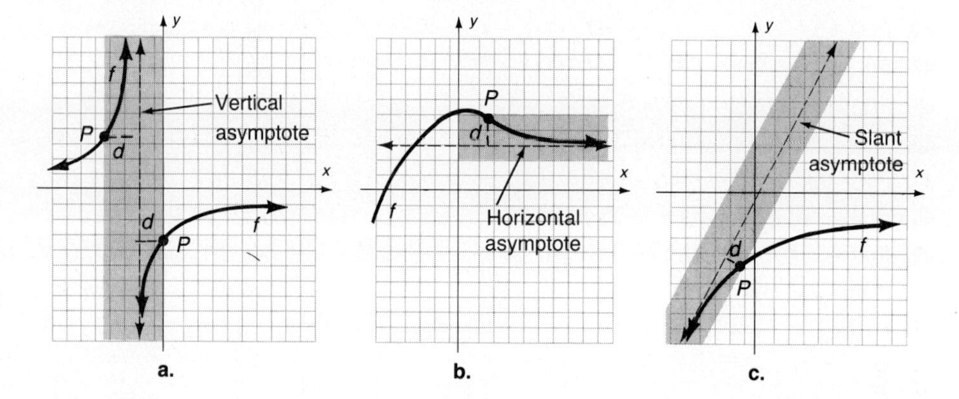

Figure 3.9 Asymptotes. Notice in figure b that it is possible for a curve to pass through one of its asymptotes.

If rational functions have asymptotes, the easiest to find are the vertical asymptotes. If $f(x) = P(x)/D(x)$, where $P(x)$ and $D(x)$ have no common factors, then $|f(x)|$ must get large as x gets close to any value for which $D(x) = 0$ and $P(x) \neq 0$. This means that **if r is a root of $D(x) = 0$, then $x = r$ is the equation of a vertical asymptote.**

EXAMPLE 1 Let

$$f(x) = \frac{1}{x - 3}$$

Notice that $x - 3 = 0$ if $x = 3$, which is the equation of a vertical line. This line is a vertical asymptote. ■

EXAMPLE 2 Let

$$f(x) = \frac{2x^2 - 3x + 5}{x^2 - x - 2}$$

There are no common factors in the numerator and denominator of f, so solve

$$x^2 - x - 2 = 0$$
$$(x - 2)(x + 1) = 0$$
$$x = 2 \text{ or } -1$$

The vertical asymptotes are given by $\mathbf{x = 2}$ and $\mathbf{x = -1}$ since neither of these values also causes the numerator to be zero. ■

EXAMPLE 3 Let

$$f(x) = \frac{x^2 + 2x - 8}{x^2 - 4}$$

Be careful with functions defined by equations like this one. If you do not notice the common factor, you might be led to the incorrect conclusion that $x = 2$ and $x = -2$ are the equations of the vertical asymptotes. Instead, factor *both* numerator and denominator as shown:

$$\begin{aligned} f(x) &= \frac{x^2 + 2x - 8}{x^2 - 4} \\ &= \frac{(x - 2)(x + 4)}{(x - 2)(x + 2)} \\ &= \frac{x + 4}{x + 2} \qquad (x \neq 2) \end{aligned}$$

Do you see why we wrote $x \neq 2$? You might think it should be $x \neq -2$. Remember: $x \neq -2$ is implied by the statement

$$\frac{x + 4}{x + 2}$$

because it would cause division by zero. The condition $x \neq 2$ is not implied, but it is necessary because of the *previous* step (before reducing). You might wish to review the idea of equality of functions in Chapter 2.

From the simplified form $\frac{x + 4}{x + 2}$ you can see that $\mathbf{x = -2}$ is the equation of the vertical asymptote. ■

If the equation can be solved easily for x so that

$$x = \frac{p(y)}{d(y)}$$

where $p(y)$ and $d(y)$ have no common factors, and if r is a root of $d(y) = 0$, then the line with the equation $y = r$ is a horizontal asymptote. Unfortunately, it is not always easy to solve for x, so we need to set some preliminary groundwork to find horizontal asymptotes.

Consider the function f defined by

$$f(x) = \frac{x^3 + 4x^2 + 7x + 6}{x + 2}$$

Notice that -2 is not in the domain of f. But what about values of x "close to -2"? In mathematics this question is phrased another way by asking what happens to

$f(x)$ as x *approaches* -2. Consider the table of values given here (found by using a calculator):

x	$f(x)$	x	$f(x)$
-1	2	-3	6
-1.5	2.25	-2.5	4.25
-1.9	2.81	-2.1	3.21
-1.99	2.9801	-2.01	3.0201
-1.999	2.9981	-2.001	3.002001
$x \to -2^+$	$f(x) \to 3$	$x \to -2^-$	$f(x) \to 3$

We use an arrow $\to$ to mean *approaches.*

$x \to -2^+$ means x is approaching -2 from the right on the number line.

$x \to -2^-$ means x is approaching -2 from the left on the number line.

Notice that f is *not defined* for $x = -2$, but as the value of x gets closer to -2, the value of $f(x)$ approaches 3. We denote this relationship by

$$\lim_{x \to -2} f(x)$$

which is read "**the *limit* of f of x as x approaches -2.**" The concept of a limit of a function is rather sophisticated, and we are considering it only from an intuitive standpoint. This notion will be precisely defined for you in calculus.

EXAMPLE 4 Find $\lim_{x \to 4} x^2$.

Solution Construct a table of values (you can use a calculator):

x	$f(x)$	x	$f(x)$
3	9	5	25
3.5	12.25	4.5	20.25
3.9	15.21	4.1	16.81
3.99	15.9201	4.01	16.0801
3.999	15.992001	4.001	16.008001
$x \to 4^-$	$f(x) \to 16$	$x \to 4^+$	$f(x) \to 16$

$f(x) = x^2$

Thus $x^2 \to 16$ as $x \to 4$. ■

Notice that the limit in Example 4 is the same as $f(4)$. But we are not concerned with the value *at* $x = 4$; we care only about the values of f *close to* $x = 4$. Compare this with the next example, which is not defined for $f(4)$.

EXAMPLE 5 Find $\lim_{x \to 4} \dfrac{x^2 - 16}{x - 4}$.

Solution

x	$f(x)$	x	$f(x)$
3	7	5	9
3.5	7.5	4.5	8.5
3.9	7.9	4.1	8.1
3.99	7.99	4.01	8.01
3.999	7.999	4.001	8.001
$x \to 4^-$	$f(x) \to 8$	$x \to 4^+$	$f(x) \to 8$

$f(x) = \dfrac{x^2 - 16}{x - 4}$

A table of values can be tedious to construct, even with a calculator, especially if the function is very complicated. Instead of a table of values, consider the following argument. Notice that f is not defined at $x = 4$, so you cannot compute $f(4)$. However, since you are concerned only with values *near* 4 (since x is *approaching* 4), you can assume that x is not equal to 4. But if $x \neq 4$, then

$$\frac{x^2 - 16}{x - 4} = \frac{(x - 4)(x + 4)}{x - 4} = x + 4$$

In this form it is easy to see that $x + 4$ is near 8 for values near 4. Thus

$$\frac{x^2 - 16}{x - 4} \to 8 \qquad \text{as } x \to 4$$

■

EXAMPLE 6 Find $\lim_{x \to 1} \dfrac{x + 1}{x^2 - 1}$.

Solution If $x \neq 1$, then

$$\frac{x + 1}{x^2 - 1} = \frac{x + 1}{(x - 1)(x + 1)} = \frac{1}{x - 1}$$

Consider a table of values for $1/(x - 1)$ as x approaches 1:

x	$f(x)$	x	$f(x)$
2	1	0.5	-2
1.5	2	0.9	-10
1.1	10	0.99	-100
1.01	100	0.999	$-1,000$
1.001	1,000	0.9999	$-10,000$
1.0001	10,000	0.99999	$-100,000$
$x \to 1^+$	$f(x) \to \infty$	$x \to 1^-$	$f(x) \to -\infty$

$f(x) = \dfrac{x + 1}{x^2 - 1}$

It appears from the table that $f(x)$ increases without bound as x approaches 1 from the right or decreases without bound as x approaches 1 from the left. In such cases we say that the limit does not exist. Sometimes the symbol ∞ (read "infinity") is used, and "$f(x) \to \infty$" is read as "$f(x)$ becomes infinite." The symbol ∞ does not represent a real number but is used to indicate certain types of functional behavior. ■

EXAMPLE 7 Find the value of $1/x$ as x increases without bound. This is written as

$$\lim_{x \to \infty} \frac{1}{x}$$

Solution

x	$\frac{1}{x}$
1	1
2	0.5
10	0.1
100	0.01
1,000	0.001
10,000	0.0001
$x \to \infty$	$\frac{1}{x} \to 0$

We say that $(1/x) \to 0$ as x increases without bound and symbolize this by

$$\frac{1}{x} \to 0 \qquad \text{as } x \to \infty$$

■

EXAMPLE 8 Find the value of $1/x$ as x decreases without bound. This is written as

$$\lim_{x \to -\infty} \frac{1}{x}$$

Solution

x	$\frac{1}{x}$
-1	-1
-2	-0.5
-10	-0.1
-100	-0.01
$-1,000$	-0.001
$-10,000$	-0.0001
$x \to -\infty$	$\frac{1}{x} \to 0$

We say that $1/x \to 0$ as x decreases without bound and symbolize this by

$$\frac{1}{x} \to 0 \qquad \text{as } x \to -\infty$$

■

Since $1/x \to 0$ for both $x \to \infty$ and $x \to -\infty$ in Examples 7 and 8, we say that $1/x \to 0$ as x increases or decreases without bound and symbolize this by

$$\frac{1}{x} \to 0 \qquad \text{as } |x| \to \infty$$

Examples 7 and 8 also generalize to a useful theorem about limits.

Limit Theorem

Let x be any real number, k any constant, and n a natural number. Then as $|x| \to \infty$:

1. $\dfrac{1}{x} \to 0$ 2. $\dfrac{1}{x^n} \to 0$ 3. $\dfrac{k}{x^n} \to 0$

You can now exploit this theorem to find some limits of rational functions.

EXAMPLE 9 Find $\lim\limits_{x\to\infty} \dfrac{x}{2x+1}$.

Solution You could consider a table of values. Instead multiply the rational expression by 1, written as $\dfrac{\mathbf{1/x}}{\mathbf{1/x}}$, and use the limit theorem:

$$\frac{x}{2x+1}\cdot\frac{\mathbf{1/x}}{\mathbf{1/x}} = \frac{1}{2+(1/x)}$$

Since $1/x \to 0$ as $x \to \infty$, you can see that

$$\frac{1}{2+(1/x)} \to \frac{1}{2+0} \qquad \text{as } x \to \infty$$

Thus

$$\frac{x}{2x+1} \to \frac{1}{2} \qquad \text{as } x \to \infty$$

■

EXAMPLE 10 Let $f(x) = \dfrac{3x^2 - 7x + 2}{7x^2 + 2x + 5}$. Find $\lim\limits_{x\to\infty} f(x)$.

Solution Notice that the largest power of x in the expression is x^2, so multiply the numerator and denominator by $1/x^2$:

$$f(x) = \frac{3x^2 - 7x + 2}{7x^2 + 2x + 5} = \frac{3x^2 - 7x + 2}{7x^2 + 2x + 5}\cdot\frac{\mathbf{1/x^2}}{\mathbf{1/x^2}} = \frac{3 - \dfrac{7}{x} + \dfrac{2}{x^2}}{7 + \dfrac{2}{x} + \dfrac{5}{x^2}}$$

Since k/x and k/x^2 both approach zero as x increases without bound,

$$\frac{3 - (7/x) + (2/x^2)}{7 + (2/x) + (5/x^2)} \to \frac{3 - 0 + 0}{7 + 0 + 0} = \frac{3}{7} \qquad \text{as } x \to \infty$$

Thus $f(x) \to \frac{3}{7}$ as $x \to \infty$. ■

Now you are able to find horizontal asymptotes by using limits. Consider

$$f(x) = \frac{P(x)}{D(x)}$$

which is the usual form for a rational function, where $P(x)$ and $D(x)$ have no common factors. **If $f(x)$ approaches some number k as $|x|$ becomes large without bound, then the line with the equation $y = k$ is a horizontal asymptote.** Remember that the distance between a point on the graph of $y = f(x)$ and the horizontal $y = k$ must approach zero as $|x|$ gets large. Thus

$$f(x) \to k \qquad \text{as } |x| \to \infty$$

EXAMPLE 11 Find the horizontal asymptote for $f(x) = \dfrac{1}{x-3}$.

Solution Find $\lim\limits_{|x|\to\infty} f(x)$:

$$\frac{1}{x-3}\cdot\frac{\mathbf{1/x}}{\mathbf{1/x}} = \frac{\frac{1}{x}}{1-\frac{3}{x}} \to \frac{0}{1-0} = 0 \qquad \text{as } |x| \to \infty$$

This limit is zero and the horizontal asymptote is $y = 0$. ■

EXAMPLE 12 Find the horizontal asymptote for $f(x) = \dfrac{2x^2 - 3x + 5}{x^2 - x - 2}$.

Solution

$$\frac{2x^2 - 3x + 5}{x^2 - x - 2}\cdot\frac{\frac{\mathbf{1}}{\mathbf{x^2}}}{\frac{\mathbf{1}}{\mathbf{x^2}}} = \frac{2 - \frac{3}{x} + \frac{5}{x^2}}{1 - \frac{1}{x} - \frac{2}{x^2}} \to 2 \qquad \text{as } |x| \to \infty$$

Thus the line given by $y = 2$ is a horizontal asymptote. ■

EXAMPLE 13 Find the horizontal asymptote for $f(x) = \dfrac{x^2 + 2x - 8}{x^2 - 4}$.

Solution

$$\frac{(x-2)(x+4)}{(x-2)(x+2)} = \frac{x+4}{x+2}\cdot\frac{\mathbf{1/x}}{\mathbf{1/x}} \qquad \text{Be sure to eliminate common factors.}$$

$$= \frac{1+\frac{4}{x}}{1-\frac{2}{x}} \to 1 \qquad \text{as } |x| \to \infty$$

Thus the line given by $y = 1$ is a horizontal asymptote. ■

Linear asymptotes that are neither horizontal nor vertical are called **slant asymptotes**. Consider $f(x) = P(x)/D(x)$, where $P(x)$ and $D(x)$ have no common factors; there are three possibilities:

1. The degree of $P(x)$ is less than or equal to the degree of $D(x)$.
2. The degree of $P(x)$ is one more than the degree of $D(x)$.
3. The degree of $P(x)$ exceeds the degree of $D(x)$ by more than one.

Consider $\lim\limits_{|x|\to\infty} f(x)$ for these three possibilities.

1. If the degree of $P(x)$ is less than the degree of $D(x)$, then

$$\lim_{|x| \to \infty} \frac{P(x)}{D(x)} = 0.$$

This gives the horizontal asymptote represented by $y = 0$. If the degree of $P(x)$ is the same as the degree of $D(x)$, we have the situation discussed previously for other horizontal asymptotes.

2. If the degree of $P(x)$ is one more than the degree of $D(x)$, then

$$f(x) = \frac{P(x)}{D(x)} = mx + b + \frac{R(x)}{D(x)},$$

where the degree of $R(x)$ is less than the degree of $D(x)$. Then

$$\lim_{|x| \to \infty} \frac{R(x)}{D(x)} = 0,$$

which means that for large values of $|x|$, $f(x)$ is near the line given by $y = mx + b$. This says that the line with the equation $y = mx + b$ is a slant asymptote for the curve given by $y = f(x)$.

3. If the degree of $P(x)$ exceeds the degree of $D(x)$ by more than one, then the quotient is no longer linear; and since asymptotes are lines, we see that there will be no slant asymptotes.

EXAMPLE 14 Find the slant asymptote for $f(x) = \dfrac{2x^2 - 5x + 1}{x - 3}$.

Solution Divide synthetically to find

$$f(x) = 2x + 1 + \frac{4}{x - 3}.$$

Thus, the slant asymptote is given by $y = 2x + 1$. ■

EXAMPLE 15 Find the slant asymptote for $f(x) = \dfrac{3x^3 - 2x^2 + x - 5}{x^2 + 3}$.

Solution Divide:

$$\begin{array}{r} 3x \quad - 2 \\ x^2 + 3 \overline{) 3x^3 - 2x^2 + x - 5} \\ \underline{3x^3 + 9x } \\ - 2x^2 - 8x - 5 \\ \underline{- 2x^2 - 6} \\ - 8x + 1 \end{array}$$

Thus,

$$f(x) = 3x - 2 + \frac{-8x + 1}{x^2 + 3},$$

and the slant asymptote is given by $y = 3x - 2$. ■

Asymptotes for a Rational Function f(x)

> Let $f(x) = \dfrac{P(x)}{D(x)}$, where $P(x)$ and $D(x)$ have no common factors.
>
> 1. The *vertical line* given by $x = r$ is an asymptote if $D(r) = 0$.
> 2. The *horizontal line* given by $y = k$ is an asymptote if
>
> $$\lim_{|x| \to \infty} f(x) = k$$
>
> 3. The *slant line* given by $y = mx + b$ is an asymptote if the degree of $P(x)$ is one more than the degree of $D(x)$ and
>
> $$\frac{P(x)}{D(x)} = mx + b + \frac{R(x)}{D(x)}$$

PROBLEM SET 3.6

A

Find the limits in Problems 1–30.

1. $\lim_{x \to 0} x^3$ **2.** $\lim_{x \to 2} (x^2 - 4)$ **3.** $\lim_{x \to -3} \dfrac{1}{x - 3}$

4. $\lim_{x \to 3} \dfrac{1}{x - 3}$ **5.** $\lim_{x \to 1} \dfrac{1}{x^2 + 1}$ **6.** $\lim_{x \to -1} \dfrac{1}{x^2 + 1}$

7. $\lim_{x \to \infty} 2x$ **8.** $\lim_{x \to \infty} (3x - 4)$ **9.** $\lim_{x \to 2} \dfrac{x^3 - 8}{x - 2}$

10. $\lim_{x \to 3} \dfrac{x^2 + 3x - 10}{x - 2}$ **11.** $\lim_{x \to 3} \dfrac{x^2 - 8x + 15}{x - 3}$

12. $\lim_{x \to -5} \dfrac{x^2 + 3x - 10}{x + 5}$ **13.** $\lim_{x \to 2} \dfrac{x^2 - 1}{x - 2}$

14. $\lim_{x \to 4} \dfrac{2x^2 - 5x - 12}{x - 4}$ **15.** $\lim_{x \to 2} \dfrac{x^3 - 8}{x^2 + 2x + 4}$

16. $\lim_{x \to 2} \dfrac{x^2 + 2x + 4}{x^3 - 8}$ **17.** $\lim_{x \to 2} \dfrac{x + 2}{x^3 + 8}$

18. $\lim_{x \to 2} \dfrac{6 - x}{2x - 15}$ **19.** $\lim_{|x| \to \infty} \dfrac{2x^2 - 5x - 3}{x^2 - 9}$

20. $\lim_{|x| \to \infty} \dfrac{3x - 1}{2x + 3}$ **21.** $\lim_{|x| \to \infty} \dfrac{x^2 + 6x + 9}{x + 3}$

22. $\lim_{|x| \to \infty} \dfrac{6x^2 - 5x + 2}{2x^2 + 5x + 1}$ **23.** $\lim_{x \to \infty} \dfrac{5x + 10{,}000}{x - 1}$

24. $\lim_{x \to -\infty} \dfrac{4x + 10^5}{x + 1}$ **25.** $\lim_{x \to \infty} \left(x + 2 + \dfrac{3}{x - 1}\right)$

26. $\lim_{x \to -\infty} \left(2x - 3 + \dfrac{4}{x + 2}\right)$

27. $\lim_{x \to -\infty} \dfrac{4x^4 - 3x^3 + 2x + 1}{3x^4 - 9}$ **28.** $\lim_{x \to \infty} \dfrac{x^4 + 1}{x^2 - 1}$

29. $\lim_{x \to \infty} \dfrac{3x^3 - 2x^2 + 1}{5x^3 + 3x - 100}$ **30.** $\lim_{x \to 1} \dfrac{x^2 + x + 1}{x^3 - 1}$

B

Find the horizontal, vertical, and slant asymptotes, if any exist, for the functions given in Problems 31–60.

31. $y = \dfrac{1}{x}$ **32.** $y = \dfrac{1}{x} + 2$ **33.** $y = -\dfrac{1}{x} + 1$

34. $y = \dfrac{4}{x^2}$ **35.** $y = \dfrac{2x^2 + 2}{x^2}$ **36.** $y = \dfrac{1}{x - 4}$

37. $y = \dfrac{-1}{x + 3}$ **38.** $y = \dfrac{4x}{x^2 - 2}$ **39.** $y = \dfrac{x^2}{x - 4}$

40. $y = \dfrac{x^3}{(x - 1)^2}$ **41.** $y = \dfrac{-x^2}{x - 1}$ **42.** $y = \dfrac{x^2 + x - 2}{x - 1}$

43. $y = \dfrac{x}{x^2 + x - 6}$ **44.** $y = \dfrac{-x}{x^2 + x - 6}$

45. $y = \dfrac{x^2}{x^3 - x^2 - 20x}$ **46.** $y = \dfrac{x^2}{20x - x^2 - x^3}$

47. $y = \dfrac{x^2 + x - 6}{x + 3}$ **48.** $y = \dfrac{x^2 + 3x - 2}{x^2 + 2x - 8}$

49. $y = \dfrac{x^3 - 2x^2 + x - 2}{(x - 2)(x^2 + 1)}$ **50.** $y = \dfrac{(x - 3)(x^2 + 1)}{x^3 + 3x^2 + x + 3}$

51. $y = \dfrac{(15x^2 + 13x - 6)(x - 1)}{3x^2 - 4x + 1}$

C

52. $y = \dfrac{2x^3 - 3x^2 - 32x - 15}{x^2 - 2x - 15}$

53. $y = \dfrac{3x^3 + 5x^2 - 26x + 8}{x^2 + 2x - 8}$

54. $y = \dfrac{x^3 + 6x^2 + 10x + 4}{x + 2}$

55. $y = \dfrac{x^3 + 9x^2 + 15x - 9}{x + 3}$

56. $y = \dfrac{x^3 + 12x^2 + 40x + 40}{x + 2}$

57. $y = \dfrac{x^3 + 5x^2 + 6x}{x + 3}$

58. $y = \dfrac{x^2}{x^3 - x^2 - 20x}$

59. $y = \dfrac{x^2 + 3x - 2}{x^2 + 2x - 8}$

60. $y = \dfrac{x^2}{20x - x^2 - x^3}$

3.7 GRAPHING RATIONAL FUNCTIONS

One of the primary topics in this book is the graphing of various types of functions. In this section we will consider graphing rational functions. The simplest types are those that reduce to polynomial functions, or polynomial functions with deleted points. Suppose, for example, that a function f is defined by

$$
\begin{aligned}
f(x) &= \frac{2x^2 + 5x - 3}{x + 3} \\
&= \frac{(2x - 1)(x + 3)}{x + 3} \\
&= 2x - 1 \qquad (x \neq -3)
\end{aligned}
$$

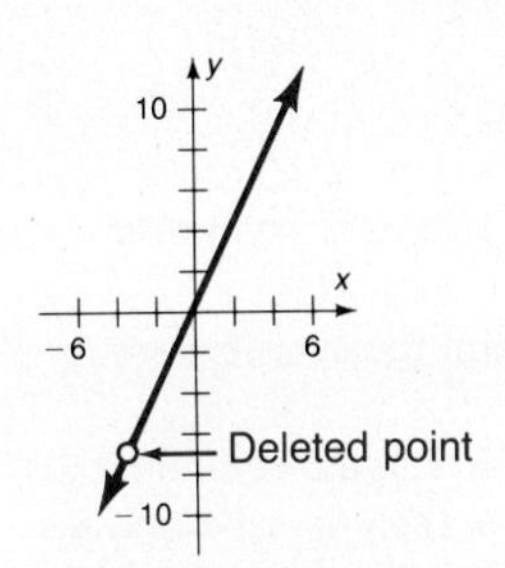

Figure 3.10 Graph of $f(x) = \dfrac{2x^2 + 5x - 3}{x + 3}$

Then the graph of f is the same as for the linear function $y = 2x - 1$ with the point at $x = -3$ deleted from the domain as in Figure 3.10.

EXAMPLE 1 Graph $f(x) = \dfrac{x^3 + 4x^2 + 7x + 6}{x + 2}$.

Solution Simplify if possible:

$$
\begin{aligned}
f(x) &= \frac{(x + 2)(x^2 + 2x + 3)}{x + 2} \\
&= x^2 + 2x + 3 \qquad (x \neq -2)
\end{aligned}
$$

These factors were obtained by synthetic division:

$$
\begin{array}{r|rrrr}
-2 & 1 & 4 & 7 & 6 \\
 & & -2 & -4 & -6 \\
\hline
 & 1 & 2 & 3 & 0
\end{array}
$$

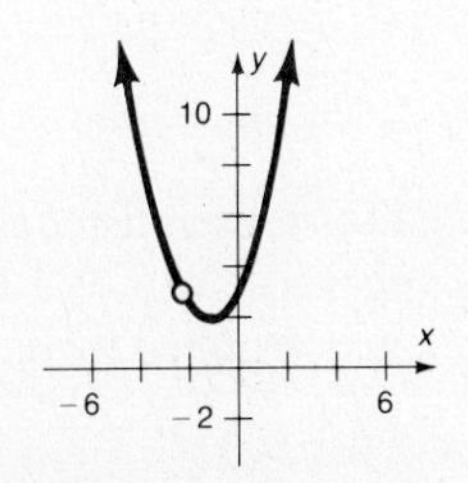

Figure 3.11 Graph of $f(x) = \dfrac{x^3 + 4x^2 + 7x + 6}{x + 2}$

The graph of f is the same as for the quadratic function $y = x^2 + 2x + 3$ with the point at $x = -2$ deleted from the domain. To sketch this parabola, complete the square:

$$
\begin{aligned}
y - 3 &= x^2 + 2x \\
y - 3 + 1 &= x^2 + 2x + 1 \\
y - 2 &= (x + 1)^2
\end{aligned}
$$

The vertex is at $(-1, 2)$ and the parabola opens upward. It is drawn with the point where $x = -2$ deleted as in Figure 3.11. ■

For rational functions that do not reduce to polynomials with deleted points, follow the procedure in Table 3.2.

TABLE 3.2 Procedure for graphing rational functions

Let $f(x) = \dfrac{P(x)}{D(x)}$.

Step	Procedure
1. Reduce.	If $P(x)$ and $D(x)$ have common factors, reduce the rational expression and note deleted points. For the rest of the steps assume that $P(x)$ and $D(x)$ have no common factors.
2. Find and graph the asymptotes, if any.	*vertical asymptote:* $x = r$ if r is a value for which $D(r) = 0$. *horizontal asymptote:* $y = k$ if $f(x) \to k$ as $\lvert x \rvert \to \infty$. *slant asymptote:* $y = mx + b$ if the degree of P is one more than the degree of D.
3. Find the intercepts.	*x intercepts:* set $y = 0$ and solve for x. *y intercept:* set $x = 0$ and solve for y.
4. Find the points where the graph passes through an asymptote.	*vertical asymptote.* Graph will not pass through a vertical asymptote. *horizontal asymptote* $y = k$. Substitute k for y in the equation $y = f(x)$ and solve for x. If $x = h$ is a value for which $y = k$, then the curve passes through the horizontal asymptote at (h, k). *slant asymptote* $y = mx + b$. Substitute $mx + b$ for y in the equation $y = f(x)$ and solve for x. If $x = h$ and $k = mh + b$, then the curve passes through the slant asymptote at (h, k).
5. Check symmetry.	*x axis.* Substitute $-y$ for y; if the equation remains unchanged, then it is symmetric with respect to the x axis. *y axis.* Substitute $-x$ for x; if the equation remains unchanged, then it is symmetric with respect to the y axis. *origin.* Substitute $-x$ for x and $-y$ for y simultaneously; if the equation remains unchanged, then it is symmetric with respect to the origin.
6. Plot selected points.	The plane is now divided into one or more regions with asymptotes as boundaries. The curve will only pass from one region to another at points of intersection of the curve and an asymptote (step 4 above). Plot selected points within each region to determine the shape of the graph within that region. Remember to make use of the fact that the boundaries of these regions are asymptotes and that the distance between the curve and the asymptote must decrease as you move out from the origin. Also remember that you are graphing functions and that no x value can have more than one corresponding y value.

For simple rational functions it is not necessary to use all five steps as listed above, but the more complicated the example, the more steps are necessary. We begin, however, with the simplest rational function $y = 1/x$ which was graphed by plotting points in Chapter 1. We review it, along with several variations, in Figure 3.12.

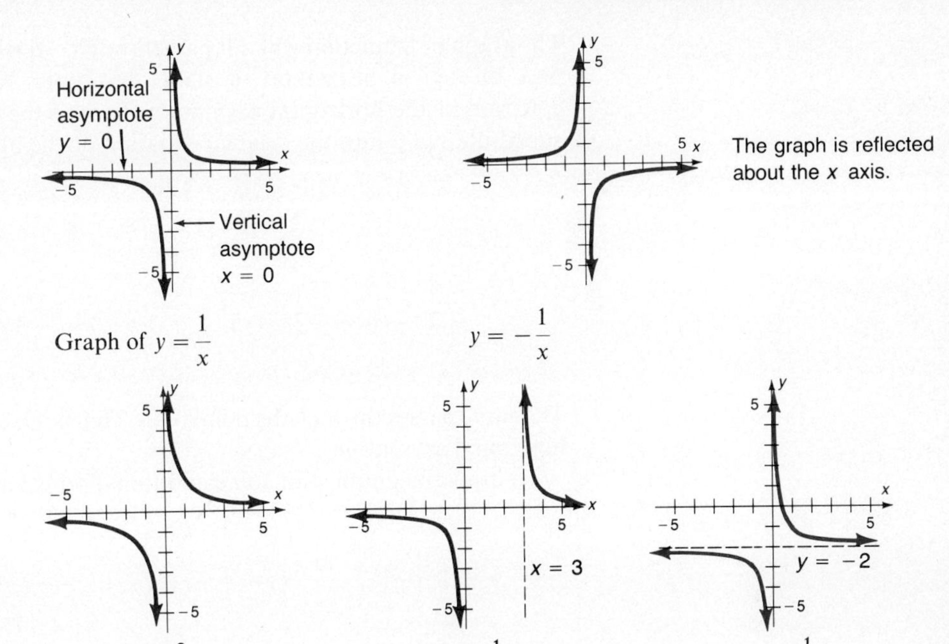

Graph of $y = \frac{1}{x}$ $\quad$ $y = -\frac{1}{x}$

$y = \frac{2}{x}$ Each point is twice as far from the x axis.

$y = \frac{1}{x-3}$ Vertical asymptote is translated 3 units to the right.

$y + 2 = \frac{1}{x}$ Horizontal asymptote is translated 2 units down.

Figure 3.12 Graphs of $y = 1/x$ and four variations.

EXAMPLE 2 Sketch: $y = \dfrac{2x^2 - 3x + 5}{x^2 - x - 2}$

Solution **Reduce**, if possible; then **find the asymptotes**:

$$x^2 - x - 2 = (x - 2)(x + 1) \qquad \text{So the vertical asymptotes are } x = 2 \text{ and } x = -1.$$

The degrees of the numerator and denominator are the same, so the horizontal asymptote is

$$y = \frac{2}{1} = 2 \qquad \text{Horizontal asymptote}$$

Draw the asymptotes and plot some points (see Figure 3.13a). Some convenient points to plot are the x-intercepts (set $y = 0$ and solve for x); the y-intercepts (set $x = 0$ and solve for y); and the points where the graph passes through an asymptote. For $y = 0$, we have

$$0 = \frac{2x^2 - 3x + 5}{x^2 - x - 5}. \qquad \text{Multiply both sides by } x^2 - x - 5.$$

$$0 = 2x^2 - 3x + 5 \qquad \text{No real roots since } b^2 - 4ac = 9 - 4 \cdot 2 \cdot 5 < 0$$

For $x = 0$, we have

$$y = \frac{2 \cdot 0^2 - 3 \cdot 0 + 5}{0^2 - 0 - 2} = -\frac{5}{2} \qquad \text{The } y\text{-intercept is } (0, -\tfrac{5}{2}).$$

The graph of a function will not pass through a vertical asymptote. It can, however, pass through a horizontal or slant asymptote. For this example, $y = 2$ is the equation of the horizontal asymptote. To find the point(s) of intersection, if any, substitute $y = 2$ into the original equation of the function. That is,

$$2 = \frac{2x^2 - 3x + 5}{x^2 - x - 2}$$

$$2x^2 - 2x - 4 = 2x^2 - 3x + 5 \qquad x \neq 2, -1$$

$$-2x - 4 = -3x + 5$$

$$x = 9$$

The curve passes through the point (9, 2). That is, (9, 2) is on the curve and also on the horizontal asymptote.

To draw the graph, plot some additional points as shown in Figure 3.13b.

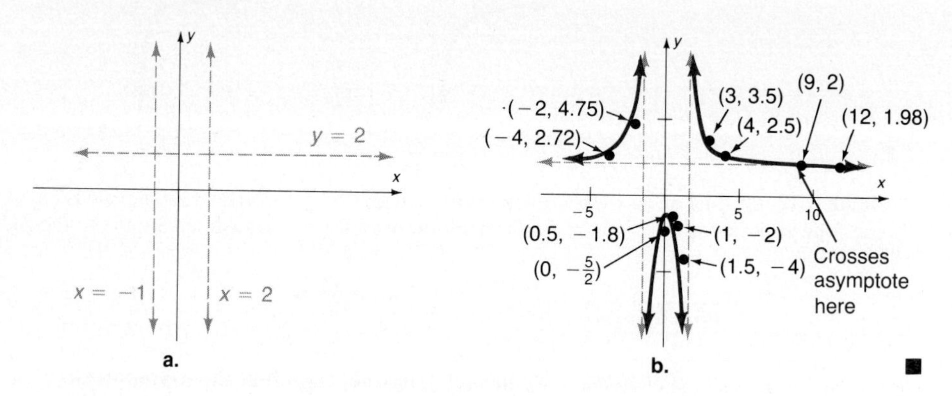

Figure 3.13 Graph of $y = \dfrac{2x^2 - 3x + 5}{x^2 - x - 2}$

The next example illustrates a slant asymptote.

EXAMPLE 3 Let f be a function defined by the equation

$$y = \frac{2x^2 - 5x + 1}{x - 3}$$

Sketch the graph.

Solution There are no horizontal asymptotes; the vertical asymptote is $x = 3$ (by inspection). There is a slant asymptote since the degree of the numerator is 1 more than the degree of the denominator:

	2	−5	1
3	2	1	4
	↑	↑	↑
	m	b	Remainder;

$$f(x) = 2x + 1 + \frac{4}{x - 3}$$

The line $y = mx + b$ is a slant asymptote, so for this example, $y = 2x + 1$ is the slant asymptote.

The y intercept is $(0, -\frac{1}{3})$ and is found by

$$y = \frac{2 \cdot 0^2 - 5 \cdot 0 + 1}{0 - 3} = -\frac{1}{3}$$

The x intercepts are $(2.3, 0)$ and $(.2, 0)$ and are found by

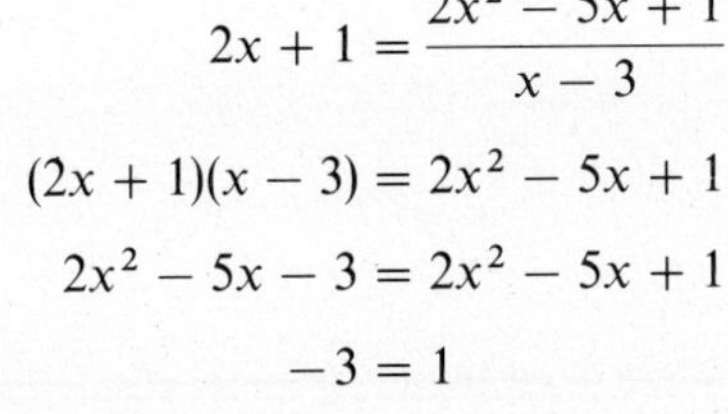

$$0 = \frac{2x^2 - 5x + 1}{x - 3} \qquad \text{Multiply both sides by } x - 3;\ x \neq 3.$$

$$2x^2 - 5x + 1 = 0$$

$$x = \frac{5 \pm \sqrt{17}}{4}$$

The intercepts for the slant asymptote are found by substituting $y = 2x + 1$ into $y = f(x)$. If we do this, however, we find that

$$2x + 1 = \frac{2x^2 - 5x + 1}{x - 3}$$

$$(2x + 1)(x - 3) = 2x^2 - 5x + 1$$

$$2x^2 - 5x - 3 = 2x^2 - 5x + 1$$

$$-3 = 1$$

This is a false equation, so there is no solution; the graph does not pass through the slant asymptote.

Plot some additional points to find the graph as shown in Figure 3.14. ■

Figure 3.14 Graph of $y = \frac{2x^2 - 5x + 1}{x - 3}$

PROBLEM SET 3.7

A

Graph the function defined by each equation given in Problems 1–60.

1. $y = \frac{x^2 - x - 12}{x + 3}$
2. $y = \frac{x^2 + x - 2}{x - 1}$
3. $y = \frac{x^2 - x - 6}{x + 2}$
4. $y = \frac{2x^2 - 13x + 15}{x - 5}$
5. $y = \frac{6x^2 - 5x - 4}{2x + 1}$
6. $y = \frac{15x^2 + 13x - 6}{3x - 1}$
7. $y = \frac{3}{x}$
8. $y = -\frac{3}{x}$
9. $y = \frac{-2}{x}$
10. $y = \frac{1}{x} + 2$
11. $y = \frac{1}{x} + 1$
12. $y = \frac{1}{x} - 3$
13. $y = -\frac{1}{x} + 2$
14. $y = -\frac{1}{x} + 1$
15. $y = -\frac{1}{x} - 3$
16. $y = \frac{1}{x + 2}$
17. $y = \frac{1}{x - 3}$
18. $y = \frac{1}{x - 4}$
19. $y = \frac{-1}{x + 3}$
20. $y = \frac{-1}{x - 2}$
21. $y = \frac{-2}{x + 1}$
22. $y - 3 = \frac{-1}{x + 1}$
23. $y + 1 = \frac{-1}{x - 2}$
24. $y = \frac{4}{x + 2} - 3$
25. $y = \frac{4}{x^2}$
26. $y = \frac{-2}{x^2}$
27. $y = \frac{-3}{x^2}$
28. $y = \frac{2}{(x - 1)^2}$
29. $y = \frac{-1}{(x - 1)^2} + 3$
30. $y = \frac{1}{(x + 1)^2} - 2$

B

31. $y = \dfrac{2x^3 - 3x^2 - 32x - 15}{x^2 - 2x - 15}$ **32.** $y = \dfrac{3x^3 + 5x^2 - 26x + 8}{x^2 + 2x - 8}$

33. $y = \dfrac{x^3 + 6x^2 + 10x + 4}{x + 2}$ **34.** $y = \dfrac{x^3 + 9x^2 + 15x - 9}{x + 3}$

35. $y = \dfrac{x^3 + 12x^2 + 40x + 40}{x + 2}$ **36.** $y = \dfrac{x^3 + 5x^2 + 6x}{x + 3}$

37. $y = \dfrac{2x^2 + 2}{x^2}$ **38.** $y = \dfrac{2x^2 - 1}{x^2}$ **39.** $y = \dfrac{x^2}{x - 4}$

40. $y = \dfrac{4x}{x^2 - 2}$ **41.** $y = \dfrac{-x^2}{x - 1}$ **42.** $y = \dfrac{x^2}{x - 1}$

43. $y = \dfrac{x^2}{x - 2}$ **44.** $y = \dfrac{x^2}{2 - x}$ **45.** $y = \dfrac{2x + 1}{3x - 2}$

46. $y = \dfrac{4x + 3}{3x - 1}$ **47.** $y = \dfrac{3x - 1}{2x + 3}$

48. $y = \dfrac{2x^2 + 3x - 1}{x - 1}$ **49.** $y = \dfrac{3x^2 - 2x + 1}{x - 2}$

50. $y = \dfrac{x^2 + 3x - 2}{x + 1}$ **51.** $y = \dfrac{40{,}000x^2}{110x - x^2}$

52. $y = \dfrac{18{,}000x^2}{100x - x^2}$ **53.** $y = \dfrac{20{,}000x^3}{120x^2 - x^3}$

C

54. $y = \dfrac{x^2}{x^3 - x^2 - 20x}$ **55.** $y = \dfrac{x^2}{20x - x^2 - x^3}$

56. $y = \dfrac{x^2 + 3x - 2}{x^2 + 2x - 8}$ **57.** $y = \dfrac{x}{x^2 - 1}$

58. $y = \dfrac{x^2}{x^2 - 1}$ **59.** $y = \dfrac{x^3}{x^2 - 1}$ **60.** $y = \dfrac{x^4}{x^2 - 1}$

61. Can you see and describe a pattern in Problems 57–60?

3.8 PARTIAL FRACTIONS*

In algebra, rational expressions are added by finding common denominators; for example:

$$\frac{5}{x-2} + \frac{3}{x+1} = \frac{5(x+1) + 3(x-2)}{(x-2)(x+1)}$$

$$= \frac{8x - 1}{(x-2)(x+1)}$$

In calculus, however, it is sometimes necessary to break apart the expression

$$\frac{8x - 1}{(x-2)(x+1)}$$

into two fractions with denominators that are linear. The technique for doing this is called the **method of partial fractions.**

The rational expression

$$f(x) = \frac{P(x)}{D(x)}$$

can be **decomposed** into partial fractions if there are no common factors and if the degree of $P(x)$ is less than the degree of $D(x)$. If the degree of $P(x)$ is greater than or

---

* Optional

equal to the degree of $D(x)$, then use either long division or synthetic division to obtain a polynomial plus a proper fraction. For example,

$$\frac{x^4 + 2x^3 - 4x^2 + x - 3}{x^2 - x - 2} = x^2 + 3x + 1 + \underbrace{\frac{8x - 1}{x^2 - x - 2}}$$

This was found by long division.

Proper fraction: This is the part that is decomposed into partial fractions.

Now look at the proper fraction. There is a theorem that says this proper fraction can be written as a sum,

$$F_1 + F_2 + \cdots + F_j$$

where *each* F_i is of the form

$$\frac{A}{(x - r)^n} \quad \text{or} \quad \frac{Ax + B}{(x^2 + sx + t)^n}$$

We begin by focusing on the first form.

Partial Fraction Decomposition—Linear Factors

Let $f(x) = P(x)/D(x)$, where $P(x)$ and $D(x)$ have no common factors and the degree of $P(x)$ is less than the degree of $D(x)$. Also suppose that $D(x) = (x - r)^n$. Then $f(x)$ can be decomposed into partial fractions:

$$\frac{A_1}{x - r} + \frac{A_2}{(x - r)^2} + \cdots + \frac{A_n}{(x - r)^n}$$

EXAMPLE 1 Decompose $\dfrac{8x - 1}{x^2 - x - 2}$ into partial fractions.

Solution

$$\frac{8x - 1}{x^2 - x - 2} = \frac{8x - 1}{(x - 2)(x + 1)}$$

First, factor the denominator, if possible, and make sure there are no common factors.

$$= F_1 + F_2$$

Break up the fraction into parts, each with a linear factor.

$$= \frac{A}{x - 2} + \frac{B}{x + 1}$$

The task is to find A and B.

$$= \frac{A(x + 1) + B(x - 2)}{(x - 2)(x + 1)}$$

Obtain a common denominator on the right.

Now, multiply both sides of this equation by the least common denominator, which is $(x - 2)(x + 1)$ for this example:

$$8x - 1 = A(x + 1) + B(x - 2)$$

Substitute, one at a time, the values that cause each of the factors in the least common denominator to be zero.

Let $x = -1$:

$$8x - 1 = A(x + 1) + B(x - 2)$$
$$8(-1) - 1 = A(-1 + 1) + B(-1 - 2)$$
$$-9 = 0 + B(-3)$$
$$-9 = -3B$$
$$\mathbf{3 = B}$$

Let $x = 2$:

$$8x - 1 = A(x + 1) + B(x - 2)$$
$$8(2) - 1 = A(2 + 1) + B(2 - 2)$$
$$15 = 3A$$
$$\mathbf{5 = A}$$

If $A = 5$ and $B = 3$, then

$$\frac{8x-1}{(x-2)(x+1)} = \frac{5}{x-2} + \frac{3}{x+1}$$ ■

Example 2 illustrates the process if there is a repeated linear factor.

EXAMPLE 2 Decompose $\dfrac{x^2 - 6x + 3}{(x-2)^3}$ by using the method of partial fractions.

Solution

$$\frac{x^2 - 6x + 3}{(x-2)^3} = \frac{A}{x-2} + \frac{B}{(x-2)^2} + \frac{C}{(x-2)^3}$$

Multiply both sides by $(x - 2)^3$:

$$x^2 - 6x + 3 = A(x - 2)^2 + B(x - 2) + C$$

Let $x = 2$.

$$(2)^2 - 6(2) + 3 = A(2 - 2)^2 + B(2 - 2) + C$$
$$4 - 12 + 3 = 0 + 0 + C$$
$$\mathbf{-5 = C}$$

Notice that with repeated factors we cannot find all the denominators as we did in Example 1. Now substitute $C = -5$ into the original equation and simplify by combining terms on the right side:

$$\begin{aligned} x^2 - 6x + 3 &= A(x - 2)^2 + B(x - 2) - 5 \\ &= A(x^2 - 4x + 4) + B(x - 2) - 5 \\ &= Ax^2 - 4Ax + Bx + 4A - 2B - 5 \\ &= Ax^2 + (-4A + B)x + (4A - 2B - 5) \end{aligned}$$

If the polynomials on the left and right sides of the equality are equal, then the coefficients of the like terms must be equal. That is,

$$x^2 - 6x + 3 = Ax^2 + (-4A + B)x + (4A - 2B - 5)$$

$$A = 1$$

$$-4A + B = -6$$

$$4A - 2B - 5 = 3$$

If $A = 1$, then

$$\begin{aligned} -4A + B &= -6 \\ -4(1) + B &= -6 \\ \mathbf{B} &= \mathbf{-2} \end{aligned}$$

Check: If $A = 1$ and $B = -2$, then

$$\begin{aligned} 4A - 2B - 5 &= 4(1) - 2(-2) - 5 \\ &= 4 + 4 - 5 \\ &= 3 \end{aligned}$$

Thus,

$$\frac{x^2 - 6x + 3}{(x-2)^3} = \frac{1}{x-2} + \frac{-2}{(x-2)^2} + \frac{-5}{(x-2)^3}$$ ■

We will now consider quadratic factors.

Partial Fraction Decomposition—Quadratic Factors

Let $f(x) = P(x)/D(x)$, where $P(x)$ and $D(x)$ have no common factors and the degree of $P(x)$ is less than the degree of $D(x)$. If $D(x) = (x^2 + sx + t)^m$, then $f(x)$ can be decomposed into partial fractions:

$$\frac{A_1x + B_1}{x^2 + sx + t} + \frac{A_2x + B_2}{(x^2 + sx + t)^2} + \cdots + \frac{A_mx + B_m}{(x^2 + sx + t)^m}$$

EXAMPLE 3 Decompose $f(x) = \dfrac{2x^3 + 3x^2 + 3x + 2}{(x^2 + 1)^2}$

Solution

$$\frac{2x^3 + 3x^2 + 3x + 2}{(x^2 + 1)^2} = \frac{Ax + B}{x^2 + 1} + \frac{Cx + D}{(x^2 + 1)^2}$$

Multiply by $(x^2 + 1)^2$:

$$2x^3 + 3x^2 + 3x + 2 = (Ax + B)(x^2 + 1) + Cx + D$$

This time, $x^2 + 1 \neq 0$ in the set of real numbers, so multiply out the right side:

$$\begin{aligned} 2x^3 + 3x^2 + 3x + 2 &= Ax^3 + Bx^2 + Ax + B + Cx + D \\ &= Ax^3 + Bx^2 + (A + C)x + (B + D) \end{aligned}$$

Equate the coefficients of the similar terms on the left and right:

$$A = 2$$
$$B = 3$$
$$A + C = 3 \qquad \text{If } A = 2, \text{ then } 2 + C = 3 \text{ and } \mathbf{C = 1}.$$
$$B + D = 2 \qquad \text{If } B = 3, \text{ then } 3 + D = 2 \text{ and } \mathbf{D = -1}.$$

Thus,

$$\frac{2x^3 + 3x^2 + 3x + 2}{(x^2 + 1)^2} = \frac{2x + 3}{x^2 + 1} + \frac{x - 1}{(x^2 + 1)^2}$$ ■

PROBLEM SET 3.8

A

List the factors of each denominator in the decomposition of the rational expressions in Problems 1–24. For Example 1 these factors are $(x - 2)$ *and* $(x + 1)$*; for Example 2 they are* $(x - 2)$, $(x - 2)^2$, *and* $(x - 2)^3$

1. $\frac{2x + 10}{x^2 + 7x + 12}$
2. $\frac{2x - 14}{x^2 + x - 6}$
3. $\frac{7x - 7}{2x^2 - 5x - 3}$
4. $\frac{4(x - 1)}{x^2 - 4}$
5. $\frac{34 - 5x}{48 - 14x + x^2}$
6. $\frac{x - 7}{20 - 9x + x^2}$
7. $\frac{5x^2 - 5x - 4}{x^3 - x}$
8. $\frac{4x^2 - 7x - 3}{x^3 - x}$
9. $\frac{2x^2 - 18x - 12}{x^3 - 4x}$
10. $\frac{2x - 1}{(x - 2)^2}$
11. $\frac{4x - 22}{(x - 5)^2}$
12. $\frac{x^2 + 5x + 1}{x(x + 1)^2}$
13. $\frac{5x^2 - 2x + 2}{x(x - 1)^2}$
14. $\frac{2x^2 + 7x + 2}{(x + 1)^3}$
15. $\frac{7x - 3x^2}{(x - 2)^3}$
16. $\frac{x}{x^2 + 4x - 5}$
17. $\frac{x}{x^2 - 2x - 3}$
18. $\frac{7x - 1}{x^2 - x - 2}$
19. $\frac{10x^2 - 11x - 6}{x^3 - x^2 - 2x}$
20. $\frac{-17x - 6}{x^3 + x^2 - 6x}$
21. $\frac{12 + 9x - 6x^2}{x^3 - 5x^2 + 4x}$
22. $\frac{x^3}{(x + 1)^2(x - 2)}$
23. $\frac{2x^3 - 3x^2 + 6x - 1}{1 - x^4}$
24. $\frac{2x^3 - 7x^2 + 8x - 7}{x^2 - 4x + 4}$

B

Decompose each fraction in Problems 25–60 by using the method of partial fractions.

25. $\frac{x^2 + 2x + 5}{x^3}$
26. $\frac{3x^2 - 2x + 1}{x^3}$
27. $\frac{2x^2 - 5x + 4}{x^3}$
28. $\frac{1}{(x + 2)(x + 3)}$
29. $\frac{1}{(x + 4)(x + 5)}$
30. $\frac{1}{(x + 3)(x + 2)}$
31. $\frac{7x - 10}{(x - 2)(x - 1)}$
32. $\frac{11x - 1}{(x - 1)(x + 1)}$
33. $\frac{7x + 2}{(x + 2)(x - 4)}$
34. $\frac{2x + 10}{x^2 + 7x + 12}$
35. $\frac{2x - 14}{x^2 + x - 6}$
36. $\frac{7x - 7}{2x^2 - 5x - 3}$
37. $\frac{4(x - 1)}{x^2 - 4}$
38. $\frac{34 - 5x}{48 - 14x + x^2}$
39. $\frac{x - 7}{20 - 9x + x^2}$
40. $\frac{5x^2 - 5x - 4}{x^3 - x}$
41. $\frac{4x^2 - 7x - 3}{x^3 - x}$
42. $\frac{2x^2 - 18x - 12}{x^3 - 4x}$
43. $\frac{2x - 1}{(x - 2)^2}$
44. $\frac{4x - 22}{(x - 5)^2}$
45. $\frac{x^2 + 5x + 1}{x(x + 1)^2}$
46. $\frac{5x^2 - 2x + 2}{x(x - 1)^2}$
47. $\frac{2x^2 + 8x + 3}{(x + 1)^3}$
48. $\frac{7x - 3x^2}{(x - 2)^3}$
49. $\frac{x}{x^2 + 4x - 5}$
50. $\frac{x}{x^2 - 2x - 3}$
51. $\frac{7x - 1}{x^2 - x - 2}$
52. $\frac{10x^2 - 11x - 6}{x^3 - x^2 - 2x}$
53. $\frac{-17x - 6}{x^3 + x^2 - 6x}$
54. $\frac{12 + 9x - 6x^2}{x^3 - 5x^2 + 4x}$

C

55. $\frac{5x^2 - 6x + 7}{(x - 1)(x^2 + 1)}$
56. $\frac{x^2}{(x + 1)(x^2 + 1)}$
57. $\frac{x^3}{(x - 1)^2}$
58. $\frac{x^3}{(x + 1)^2}$
59. $\frac{2x^3 - 3x^2 + 6x - 1}{1 - x^4}$
60. $\frac{2x^3 - 7x^2 + 8x - 7}{x^2 - 4x + 4}$

CHAPTER 3 SUMMARY

The material of this chapter is reviewed in the following list of objectives. After each objective there are some practice questions. For a sample test, select the first question of each set and check your answers with the answer section. For a sample test without answers, use the second question of each set. Additional practice is given by the other questions in each set. If you are having trouble with a particular type of problem, look back to that section for extra help.

3.1 POLYNOMIAL FUNCTIONS

Objective 1 Be familiar with the terminology of polynomials, including the definition and laws of exponents. Fill in the blanks.

1. A function P is a polynomial function in x if ______
2. If b is any real number and n is ______, then $b^n =$ ______
3. $b^m \cdot b^n =$ ______
4. $(ab)^m =$ ______

Objective 2 Add, subtract, and multiply polynomials. Let $P(x) = 3x^3 + 4x^2 - 35x - 12$ and $D(x) = 3x + 1$. Find the requested expressions.

5. $P(x)D(x)$
6. $P(t) + D(t)$
7. $D(w) - P(w)$
8. $[D(s)]^3$

Objective 3 Factor polynomials.

9. $\dfrac{4x^2}{y^2} - (2x + y)^2$
10. $x^4 - 26x^2 + 25$
11. $(x^3 - \frac{1}{8})(8x^3 + 8)$
12. $4x^3 + 8x^2 - x - 2$

Objective 4 Graph polynomial forms that can be factored.

13. $\dfrac{x^2}{4} - \dfrac{y^2}{9} = 0$
14. $(x^2 - y)(x^2 - y^2) = 0$
15. $(2y + x^2)(y - x) = 0$
16. $x^3 + xy + x^2y + y^2 = 0$

3.2 SYNTHETIC DIVISION

Objective 5 Be familiar with the Division Algorithm and do long division of polynomials with and without remainders.

17. $\dfrac{2x^3 + 9x^2 + 5x - 6}{x^2 + 3x - 2}$
18. $\dfrac{12x^4 + 22x^3 + 7x + 4}{3x + 1}$
19. $\dfrac{6x^4 - x^3 + 4x^2 + 3x - 2}{2x + 1}$
20. $\dfrac{3x^3 + 2x^2 - 12x - 8}{x^2 - 4}$

Objective 6 Use synthetic division.

21. $\dfrac{x^4 + 2x^2 - x - 26}{x + 2}$
22. $\dfrac{6x^4 - x^3 + 4x^2 + 3x - 2}{x + \frac{1}{2}}$
23. $\dfrac{3x^3 + 2x^2 - 12x - 8}{x - 2}$
24. $\dfrac{4x^5 - 3x^3 + 2x - 5}{x + 1}$

3.3 GRAPHING POLYNOMIAL FUNCTIONS

Objective 7 Use synthetic division and the Remainder Theorem to find points on the graph of a polynomial function. Let $P(x) = 3x^3 + 4x^2 - 35x - 12$. Find the specified values.

25. $P(2)$
26. $P(-2)$
27. $P(3)$
28. $P(-3)$

Objective 8 Use the Intermediate-Value Theorem for polynomial functions to sketch the graph of a polynomial function.

29. $P(x) = 3x^3 + 4x^2 - 35x - 12$

30. $f(x) = 3x^4 - 8x^3 - 48x^2 + 492$

31. $Q(x) = 3x^3 + 2x^2 - 12x - 8$

32. $g(x) = 6x^4 - x^3 + 3x^2 + 3x - 2$

3.4 REAL ROOTS OF POLYNOMIAL EQUATIONS

Objective 9 Find the zeros of polynomial functions by using the Factor Theorem, and state the multiplicity of each zero.

33. $y = (x - 1)^3(x + 2)^2$

34. $y = (x^2 - 1)^2(x + 1)^3$

35. $y = (x^3 - 16x)^2$

36. $y = (x^2 - 4)^3$

Objective 10 List the possible rational roots of a polynomial equation.

37. $3x^3 + 4x^2 - 35x - 12 = 0$

38. $6x^4 - x^3 + 3x^2 + 3x - 2 = 0$

39. $3x^3 + 2x^2 - 12x - 8 = 0$

40. $x^4 + 2x^2 - x - 26 = 0$

Objective 11 Use Descartes' Rule of Signs to determine the number of positive and negative real roots.

41. $3x^3 + 4x^2 - 35x - 12 = 0$

42. $6x^4 - 13x^3 + 3x^2 + 9x - 5 = 0$

43. $3x^3 + 2x^2 - 12x - 8 = 0$

44. $x^4 + 11x^3 - 25x^2 + 11x - 26 = 0$

Objective 12 Solve polynomial equations over the set of real numbers.

45. $3x^3 + 4x^2 - 35x - 12 = 0$

46. $6x^4 - 13x^3 + 3x^2 + 9x - 5 = 0$

47. $3x^3 + 2x^2 - 12x - 8 = 0$

48. $x^4 + 11x^3 - 25x^2 + 11x - 26 = 0$

*3.5 THE FUNDAMENTAL THEOREM OF ALGEBRA

Objective 13 Evaluate polynomial functions over the set of complex numbers. Let $f(x) = x^5 + x^4 - 2x^3 - 2x^2 - 3x - 3$. Find the requested values.

49. $f(i)$

50. $f(-i)$

51. $f(\sqrt{3})$

52. $f(-\sqrt{3})$

Objective 14 Solve polynomial equations over the set of complex numbers and graph the corresponding polynomial functions.

53. Solve $x^4 - 3x^3 - 9x^2 + 25x - 6 = 0$

54. Graph $f(x) = x^4 - 3x^3 - 9x^2 + 25x - 6$

55. Solve $x^4 - 14x^2 + x^3 - 14x = 0$

56. Graph $g(x) = x^4 - 14x^2 + x^3 - 14x$

Objective 15 Use the Conjugate Pair Theorem to solve polynomial equations. Solve the equations, assuming the given value is a root.

57. $x^4 - 2x^3 + 5x^2 - 6x + 6 = 0;\ 1 + i$

58. $x^4 + 2x^2 - 63 = 0;\ -3i$

59. $x^4 - 2x^3 - x^2 + 10x - 20 = 0;\ 1 + \sqrt{3}i$

60. $x^4 - 6x^3 - 8x^2 + 62x + 15 = 0;\ 2 + \sqrt{5}$

3.6 RATIONAL FUNCTIONS

Objective 16 Find limits of polynomial and rational functions.

61. $\lim_{x \to 2} (x^2 - 3x + 4)$

62. $\lim_{x \to 2} \dfrac{x^3 - 5x^2 + 10x - 8}{x - 2}$

63. $\lim_{x \to 0} \dfrac{x + 2}{x^2 - 4}$

64. $\lim_{|x| \to \infty} \dfrac{3x^2 - 4x - 7}{2x^2 - 9x + 4}$

---

* Optional section

Objective 17 Find the vertical, horizontal, and slant asymptotes for a given rational function.

65. $f(x) = \frac{1}{x-2} + 2$

66. $f(x) = \frac{3x^2 + 2x - 5}{x - 1}$

67. $f(x) = \frac{2x^2 - 3x - 1}{x - 2}$

68. $f(x) = \frac{x^3 - x - 6}{x - 2}$

3.7 GRAPHING RATIONAL FUNCTIONS

Objective 18 Graph rational functions.

69. $f(x) = \frac{1}{x-2} + 2$

70. $f(x) = \frac{3x^2 + 2x - 5}{x - 1}$

71. $f(x) = \frac{2x^2 - 3x - 1}{x^2 - x - 2}$

72. $f(x) = \frac{x^3 - x - 6}{x - 2}$

*3.8 PARTIAL FRACTIONS

Objective 19 Decompose rational expressions by using the method of partial fractions.

73. $\frac{5x^2 - 19x + 17}{(x-1)(x-2)^2}$

74. $\frac{7x - 7}{2x^2 - 9x + 4}$

75. $\frac{2x^2 + 13x - 9}{x^2 + 2x - 15}$

76. $\frac{2x^3 + 2x + 3}{(x^2 + 1)^2}$

---

* Optional section

4

Exponential and Logarithmic Functions

The invention of logarithms and the calculation of the earlier tables form a very striking episode in the history of exact science, and with the exception of the "Principia" *of Newton, there is no mathematical work published in the country which has produced such important consequences, or to which so much interest attaches as to Napier's* "Descriptio."

J.W.L. GLAISHER
Encyclopedia Britannica

John Napier (1550–1617)

John Napier was the Isaac Asimov of his day, having envisioned the tank, the machine gun, and the submarine. He also predicted that the end of the world would occur between 1688 and 1700. He is best known today as the inventor of logarithms. The word *logarithm* means "ratio number" and was adopted by Napier after he had first chosen the term *artificial number.* As you will see in this chapter, today we define a logarithm as an exponent, but historically logarithms were discovered before exponents were in use. Common logarithms, or logs to the base 10, were introduced by a professor at Oxford University named Henry Briggs (1561–1631). Until the advent of the low-cost calculator, logarithms were used extensively in complicated mathematical calculations. The mathematician and historian F. Cajori said in his 1897 *History of Mathematics,* "The miraculous powers of modern calculation are due to three inventions: the Arabic Notation, Decimal Fractions, and Logarithms." Today we would have to add another to those three inventions: the hand-held calculator. This, however, does not minimize the importance of the logarithm. It simply changes the emphasis from calculation to application. This chapter reflects the recent change in emphasis brought about by the calculator. Napier, however, believed that his reputation would rest ultimately on his predictions about the end of the world. He considered logarithms merely an interesting recreational diversion.

CHAPTER OVERVIEW The important ideas of this chapter are solving exponential and logarithmic equations. In order to do this, two new and important functions are defined and discussed. The 16 objectives of this chapter are listed on pages 170–171. Following these objectives is the first Extended Application in the book. This is introduced with a newspaper article; a mathematical discussion follows, in which the techniques of this chapter are used to answer questions related to the article. This application concerns population growth.

4.1 EXPONENTIAL FUNCTIONS

The linear, quadratic, polynomial, and rational functions considered in the first part of this book are all called **algebraic functions**. An algebraic function is a function that can be expressed in terms of algebraic operations alone. If a function is not algebraic, it is called a **transcendental function**. In this chapter, two examples of transcendental functions, *exponential* and *logarithmic* functions, are considered. In the next chapter, other types of transcendental functions will be considered.

Exponential Function

> The function f is an **exponential function** if
>
> $$f(x) = b^x$$
>
> where b is a positive constant other than 1 and x is any real number. The number x is called the **exponent** and b is called the **base**.

Recall that if n is a natural number, then

$$b^n = \underbrace{b \cdot b \cdot b \cdots b}_{n\ factors}, \quad b^0 = 1, \quad b^{-n} = \frac{1}{b^n}$$

This definition of b^n is used in conjunction with the five laws of exponents stated in Section 3.1 on page 84.

The next step is to extend the definition of an exponent to include rational exponents. To do this, you need to recall the definition of roots. If n is a natural number, then an ***n*th root** of a number b is a only if $a^n = b$. If $n = 2$, the root is called a **square root**; if $n = 3$, it is called a **cube root**. The number $\sqrt[n]{a}$ is called the **principal *n*th root** of a. If $n = 2$, it is customary to write $\sqrt{a}$ instead of $\sqrt[2]{a}$. The symbol $\sqrt[n]{a}$ is also called a **radical**; the number a is the **radicand**, and the **index** is the number n. To relate this discussion to exponents, consider $\sqrt{2}$. Suppose we wish to find an x so that

$$\sqrt{2} = 2^x$$

From the definition of square root,

$$2 = (2^x)^2$$

If the properties of exponents are to hold, then

$$2^1 = 2^{2x}$$

We would now like to conclude that the exponents are equal. In this case, it is indeed true. But under what conditions does $x = y$ when

$$b^x = b^y$$

If $b = 1$, you cannot conclude that $x = y$ since

$$1^5 = 1^4$$

but $5 \neq 4$. If $b \neq 1$, however, it can be proved true for all positive real numbers b and is called the **exponential property of equality**.

Exponential Property of Equality

For positive real b ($b \neq 1$):

If $b^x = b^y$, then $x = y$.

We now use this property to conclude that if

$$2^1 = 2^{2x}$$

then $1 = 2x$ and $x = \frac{1}{2}$. This shows that $2^{1/2} = \sqrt{2}$. Another way of showing this same fact is to notice that

$$2^{1/2} \cdot 2^{1/2} = 2^1 = 2$$

and

$$\sqrt{2}\sqrt{2} = 2$$

so $2^{1/2} = \sqrt{2}$.

The use of rational numbers as exponents preserves the laws of exponents and, more important, gives an alternative choice of notation for roots as shown by the following definition.

Rational Exponents

For $b > 0$ and m, n positive integers where m/n is reduced,

$$b^{1/n} = \sqrt[n]{b} \quad \text{and} \quad b^{m/n} = (b^{1/n})^m = \sqrt[n]{b^m} = (\sqrt[n]{b})^m$$

EXAMPLE 1 Simplify the given expressions.

a. $16^{1/2} = (4^2)^{1/2}$
$= 4^1$
$= 4$

b. $-16^{1/2} = -(4^2)^{1/2}$
$= -4$

c. $(-16)^{1/2} = \sqrt{-16}$ is not defined (b must be greater than zero by definition).

d. $(343)^{2/3} = (7^3)^{2/3}$
$= 7^2$
$= 49$

e. $25^{-3/2} = \dfrac{1}{25^{3/2}}$ First use the definition $b^{-p} = \dfrac{1}{b^p}$.

$$= \frac{1}{(5^2)^{3/2}}$$

$$= \frac{1}{5^3}$$

$$= \frac{1}{125}$$ ■

EXAMPLE 2 Use the ordinary rules of algebra to simplify the given expressions.

a. $x(x^{2/3} + x^{1/2}) = x^1x^{2/3} + x^1x^{1/2}$

$= x^{1+(2/3)} + x^{1+(1/2)}$ Recall that $x^px^q = x^{p+q}$, so $x^1x^{2/3} = x^{1+(2/3)}$.

$= x^{5/3} + x^{3/2}$

b. $(x^{1/2} + y^{1/2})(x^{1/2} - y^{1/2}) = x^{1/2}x^{1/2} - x^{1/2}y^{1/2} + x^{1/2}y^{1/2} - y^{1/2}y^{1/2}$

$= x - y$ ■

The next step in enlarging the domain for x in $f(x) = b^x$ requires the following property:

Squeeze Theorem for Exponents

> Suppose b is a real number greater than 1. Then for any real number x there is a unique real number b^x. Moreover, if h and k are any two rational numbers such that $h < x < k$, then
>
> $b^h < b^x < b^k$

This squeeze theorem will give meaning to expressions such as $2^{\sqrt{3}}$. Consider the graph of the function $f(x) = 2^x$ by plotting the points shown in the table as in Figure 4.1.

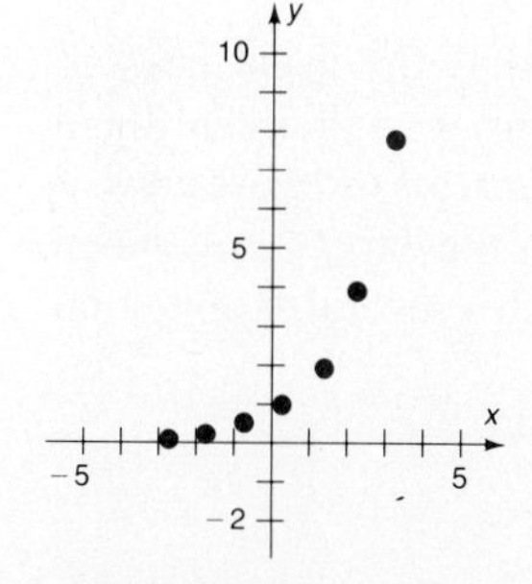

Figure 4.1 Selected points that satisfy $f(x) = 2^x$

x	$y = f(x) = 2^x$
−3	$2^{-3} = \frac{1}{8}$
−2	$2^{-2} = \frac{1}{4}$
−1	$2^{-1} = \frac{1}{2}$
0	$2^0 = 1$
1	$2^1 = 2$
2	$2^2 = 4$
3	$2^3 = 8$

If these points are connected with a smooth curve, as shown in Figure 4.2, you can see that $2^{\sqrt{3}}$ is defined and is between 2^1 and 2^2.

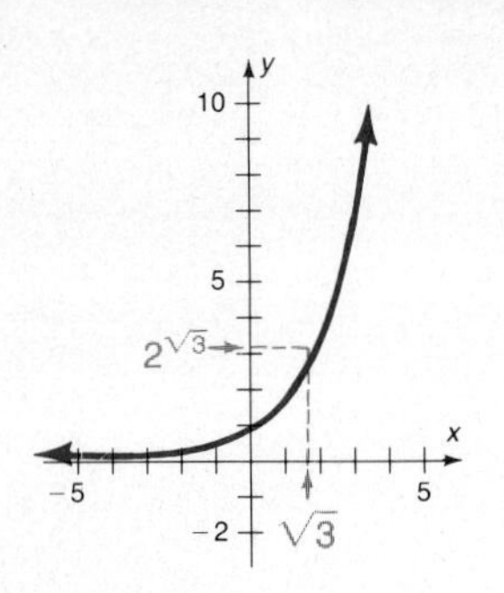

Figure 4.2 Graph of $f(x) = 2^x$

The number $2^{\sqrt{3}}$ can be approximated to any desired degree of accuracy:

$$
\begin{array}{lcl}
 & 1 < \sqrt{3} < 2 & & 2^1 < 2^{\sqrt{3}} < 2^2 \\
 & 1.7 < \sqrt{3} < 1.8 & & 2^{1.7} < 2^{\sqrt{3}} < 2^{1.8} \\
\text{Since} & 1.73 < \sqrt{3} < 1.74 & \text{we have} & 2^{1.73} < 2^{\sqrt{3}} < 2^{1.74} \\
 & 1.732 < \sqrt{3} < 1.733 & & 2^{1.732} < 2^{\sqrt{3}} < 2^{1.733} \\
 & \vdots & & \vdots
\end{array}
$$

Base with irrational exponent is squeezed between same base with rational exponents.

Even using a calculator

[2] [y^x] [3] [√] [=]

gives a *rational* approximation of $\sqrt{3} \approx 1.732050808$ and does not find $2^{\sqrt{3}}$ but rather $2.^{1.732050808}$. This process can be visualized by looking at the portions of Figure 4.2 shown in Figure 4.3.

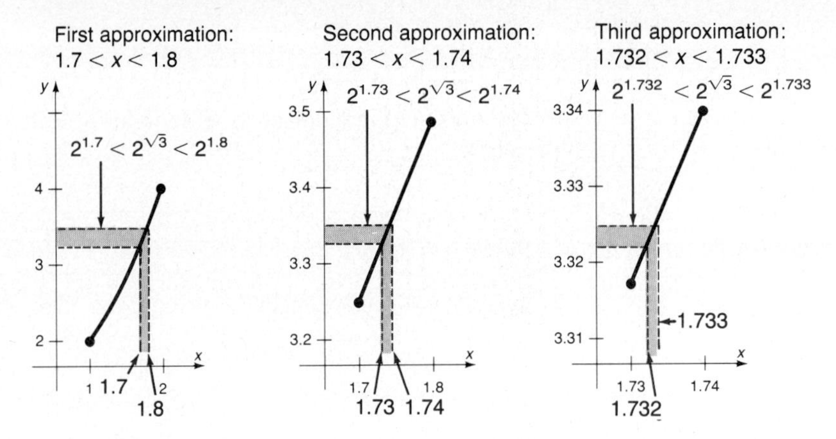

Figure 4.3 Successive approximations for locating $2^{\sqrt{3}}$

Because it is beyond the scope of this book to prove that the usual laws of exponents hold for all real exponents in exponential functions, we will accept them as axioms. Because of the restrictions on b, however, the hypotheses for these laws of exponents with any real exponents must be changed so that they apply only when the bases are positive numbers not equal to 1. Suppose we sketch several exponential functions and observe their behavior.

EXAMPLE 3 Sketch $f(x) = (\frac{1}{2})^x$.

Solution Notice that

$$
\begin{aligned}
y &= \left(\frac{1}{2}\right)^x \\
&= (2^{-1})^x
\end{aligned}
$$

The points are plotted in Figure 4.4 and connected by a smooth curve.

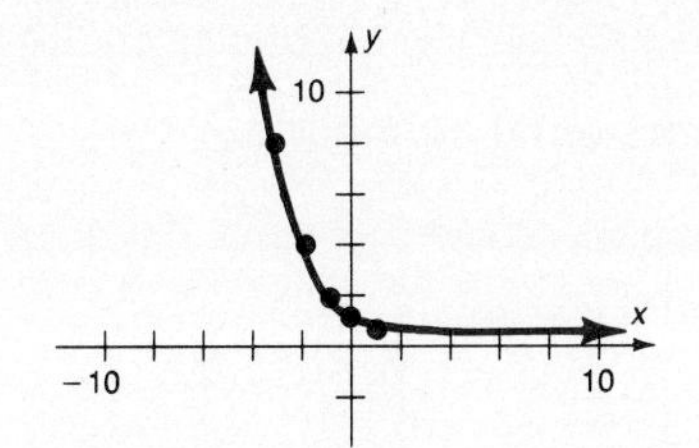

Figure 4.4 Graph of $f(x) = (\frac{1}{2})^x$

x	$f(x) = (\frac{1}{2})^x$
−3	$(2^{-1})^{-3} = 8$
−2	$(2^{-1})^{-2} = 4$
−1	$(2^{-1})^{-1} = 2$
0	$(2^{-1})^{0} = 1$
1	$(\frac{1}{2})^1 = \frac{1}{2}$
2	$(\frac{1}{2})^2 = \frac{1}{4}$
3	$(\frac{1}{2})^3 = \frac{1}{8}$

f is a decreasing function with a horizontal asymptote at $y = 0$. ■

EXAMPLE 4 Sketch $f(x) = 10^x$.

Solution

x	$f(x) = 10^x$
−3	$10^{-3} = \frac{1}{1000}$
−2	$10^{-2} = \frac{1}{100}$
−1	$10^{-1} = \frac{1}{10}$
0	$10^0 = 1$
1	$10^1 = 10$
2	$10^2 = 100$
3	$10^3 = 1000$

f is an increasing function with a horizontal asymptote at $y = 0$.

Figure 4.5 Graph of $f(x) = 10^x$

Notice that it is often necessary to alter the scale for exponential functions. ■

By looking at Figures 4.2, 4.4, and 4.5, we can make some observations regarding the graph of $f(x) = b^x$:

1. It passes through the point (0, 1).
2. $f(x) > 0$ for all x.
3. If $b > 1$, f is an increasing function, and since

$$b^x \to 0 \qquad \text{as } x \to -\infty$$

then $y = 0$ is the equation of a horizontal asymptote.

4. If $b < 1$, f is a decreasing function, and since

$$b^x \to 0 \qquad \text{as } x \to \infty$$

then $y = 0$ is the equation of a horizontal asymptote.

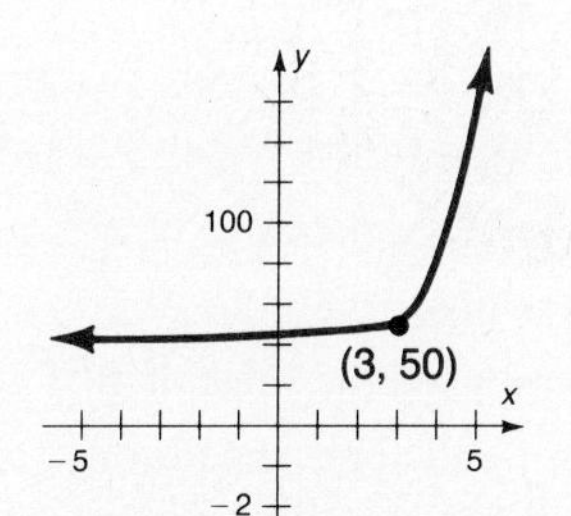

Figure 4.6 Graph of $f(x) = 10^{x-3} + 50$

EXAMPLE 5 Sketch $f(x) = 10^{x-3} + 50$

Solution Write this as

$$y = 10^{x-3} + 50$$
$$y - 50 = 10^{x-3}$$

or

$$y' = 10^{x'} \qquad (x' = x - 3;\ y' = y - 50)$$

Thus, this is the graph shown in Figure 4.6 translated to $(h, k) = (3, 50)$. ■

EXAMPLE 6 Sketch $f(x) = 2^{-x^2}$.

Solution $2^{-x^2} \to 0$ as $|x| \to \infty$, so $y = 0$ is the equation of a horizontal asymptote.

x	$f(x) = 2^{-x^2}$
-3	$2^{-9} = \frac{1}{512}$
-2	$2^{-4} = \frac{1}{16}$
-1	$2^{-1} = \frac{1}{2}$
0	$2^0 = 1$
1	$2^{-1} = \frac{1}{2}$
2	$2^{-4} = \frac{1}{16}$
3	$2^{-9} = \frac{1}{512}$

A calculator could be used to estimate additional points.

x	$f(x) = 2^{-x^2}$
-1.5	0.21
-0.5	0.84
0.5	0.84
1.5	0.21

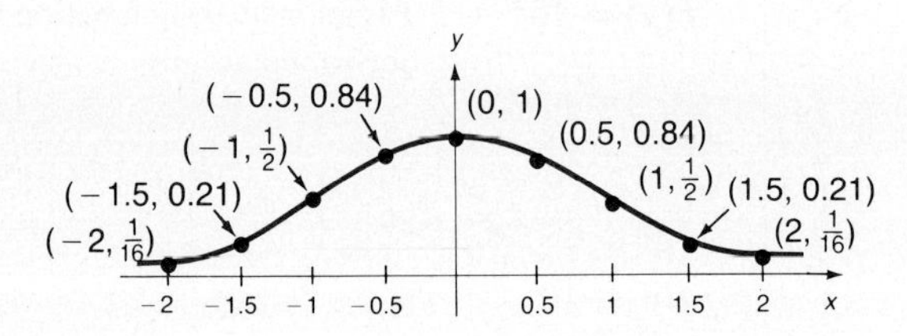

Figure 4.7 Graph of $f(x) = 2^{-x^2}$ ∎

PROBLEM SET 4.1

A

Simplify the expressions in Problems 1–20. Eliminate negative exponents from your answers.

1. $25^{1/2}$ **2.** $-25^{1/2}$ **3.** $-27^{1/3}$ **4.** $-216^{1/3}$
5. $216^{1/3}$ **6.** $8^{2/3}$ **7.** $64^{2/3}$ **8.** $64^{3/2}$
9. $-64^{3/2}$ **10.** $-64^{2/3}$ **11.** $(-64)^{3/2}$ **12.** $7^{1/3} \cdot 7^{2/3}$
13. $8^{4/3} \cdot 8^{-1/3}$ **14.** $10^{4/3}/10^{1/3}$
15. $(2^{1/3} \cdot 3^{1/2})^6$ **16.** $(2^6 \cdot 3^{12})^{1/6}$
17. $1000^{-2/3}$ **18.** $0.001^{-2/3}$
19. $100^{-3/2}$ **20.** $0.01^{-3/2}$

B

Simplify the expressions in Problems 21–30.

21. $x^{1/2}(x^{1/2} + x^{-1/2})$ **22.** $x(x^{1/2} + x^{-1/2})$
23. $x^{2/3}(x^{-2/3} + x^{1/3})$ **24.** $x^{1/4}(x^{3/4} + x^{-1/4})$
25. $(x^{2/3}y^{-1/3})^3$ **26.** $(x^{2^2+1}x^5)^{1/10}$
27. $(x^{1/2} + y^{1/2})^2$ **28.** $(x^{1/2} - y^{1/2})^2$
29. $(x^{1/3} + y^{1/3})(x^{2/3} - x^{1/3}y^{1/3} + y^{2/3})$
30. $(x^{1/3} - y^{1/3})(x^{2/3} + x^{1/3}y^{1/3} + y^{2/3})$

$x^{\frac{1}{3}} = \sqrt[3]{x}$

Sketch the graph of each function given in Problems 31–54.

31. $y = 3^x$ **32.** $y = 4^x$ **33.** $y = 5^x$ **34.** $y = (\frac{1}{3})^x$
35. $y = 4^{-x}$ **36.** $y = 5^{-x}$ **37.** $y = 2^{x-2}$ **38.** $y = 2^{x-1}$
39. $y = 2^{x+3}$ **40.** $y - 2 = 2^x$
41. $y - 3 = 2^x$ **42.** $y + 5 = 2^x$
43. $y - 5 = 2^{x+4}$ **44.** $y - 10 = 2^{x+3}$
45. $y = 2^{x-3} - 10$ **46.** $y = 2^{|x|}$
47. $y = 3^{|x|}$ **48.** $y = 2^{-|x|}$ **49.** $y = 3^{x^2}$ **50.** $y = 2^{x^2}$
51. $y = 10^{x^2}$ **52.** $y = 3^{-x^2}$ **53.** $y = 5^{-x^2}$ **54.** $y = 10^{-x^2}$

55. Use graphical methods to estimate the value of $2^{\sqrt{2}}$.

56. Use graphical methods to estimate the value of $3^{\sqrt{2}}$.

57. Use graphical methods to estimate the value of $10^{\sqrt{2}}$.

58. Graph $y = 10^x$, $-1 \le x \le 1$ and approximate x if:
a. $10^x = 5$ **b.** $10^x = 0.5$ **c.** $10^x = 2$
d. $10^x = 8.4$ **e.** $10^x = -1$

59. In the definition of the exponential function $f(x) = b^x$, we require that b is a positive constant.
a. What happens if $b = 1$? Draw the graph of $f(x) = b^x$, where $b = 1$. Is this an algebraic or a transcendental function?
b. What happens if $b = 0$? Draw the graph of $f(x) = b^x$, where $b = 0$. Is this an algebraic or a transcendental function?

60. In the definition of the exponential function $f(x) = b^x$, we require that b is a positive constant. What happens if $b < 0$, say $b = -2$? For what values of x is f defined? Describe the graph of $f(x)$ in this case.

C

61. ***Physics*** Radioactive argon-39 has a half-life of 4 min. This means that the time required for half of the argon-39 to decompose is 4 min. If we start with 100 milligrams (mg) of argon-39, the amount (A) left after t min is given by

$$A = 100\left(\frac{1}{2}\right)^{t/4}$$

Graph this function.

62. ***Earth Science*** Carbon-14, used for archaeological dating, has a half-life of 5700 years. This means that the time required for half of the carbon-14 to decompose is 5700 years. If we start with 100 mg of carbon-14, the amount (A) left after t years is given by

$$A = 100\left(\frac{1}{2}\right)^{t/5700}$$

Graph this function for $t \geq 0$.

63. ***Social Science*** In 1982 the world population was about 4.56 billion. If we assume a growth rate of 2%, the formula expressing the world population for t years after 1982 is given by

$$P = 4.56(1 + 0.02)^t$$
$$= 4.56(1.02)^t$$

where P is the population in billions. Graph this function for 1982–1992.

64. Use graphical methods to determine which is larger:
   a. $(\sqrt{3})^\pi$ or $\pi^{\sqrt{3}}$ **b.** $(\sqrt{5})^\pi$ or $\pi^{\sqrt{5}}$ **c.** $(\sqrt{6})^\pi$ or $\pi^{\sqrt{6}}$
   d. Consider $(\sqrt{N})^\pi = \pi^{\sqrt{N}}$, where N is a positive real number. From parts **a**–**c**, notice that $(\sqrt{N})^\pi$ is larger for some values of N and $\pi^{\sqrt{N}}$ is larger for others. For $N = \pi^2$,

$$(\sqrt{N})^\pi = \pi^{\sqrt{N}}$$

is obviously true. Using a graphic method, find another value (approximately) for which the given statement is true.

4.2 INTRODUCTION TO LOGARITHMS

Many natural phenomena follow patterns of exponential growth or decay. Human population growth exhibits exponential growth, for example, and is considered at length in a special Extended Application following this chapter. Compound interest provides another application of exponential growth that is very important in business. If a sum of money, called the **principal**, is denoted by P and invested at an annual interest rate of r for t years, then the amount of money present is denoted by A and is found by

$$A = P + I$$

where I denotes the interest. **Interest** is an amount of money paid for the use of another's money. **Simple interest** is found by multiplication:

$$I = Prt$$

For example, \$1000 invested for 3 years at 15% simple interest would generate an interest of \$1000(0.15)(3) = \$450, so the amount present after three years is \$1450.

Most businesses, however, pay interest on the interest, as well as the principal, after a certain period of time. When this is done, it is called **compound interest**. For example, \$1000 invested at 15% annual interest compounded annually for 3 years can be found as follows.

$$\begin{aligned}
\text{First year: } A &= P + I \\
&= P + Pr && I = Prt \text{ and } t = 1 \\
&= P(1 + r) \\
&= \$1000(1 + 0.15) \\
&= \$1150
\end{aligned}$$

Second year: The amount from the first year becomes the principal for the second year:

$$A = \$1150(1 + 0.15)$$
$$= \$1322.50$$

Third year: $A = \$1322.50(1 + 0.15)$

$= \$1520.88$ Rounded to the nearest cent

Notice that the amount with simple interest is \$1450, whereas with compound interest it is \$1520.88. The calculation for compound interest, shown above, can become very tedious (especially for large t), so it is desirable to derive a general formula:

First year: $A = P + I$

$= P + Pr$ $\quad I = Prt$ and $t = 1$

$= P(1 + r)$

Second year: $A = P(1 + r) + I$

$= P(1 + r) + P(1 + r)r$

$= P(1 + r)(1 + r)$ Common factor $P(1 + r)$

$= P(1 + r)^2$

Third year: $A = P(1 + r)^2 + P(1 + r)^2 r$

$= P(1 + r)^2(1 + r)$ Common factor $P(1 + r)^2$

$= P(1 + r)^3$

$\vdots$

This pattern leads to the compound interest formula. The proof requires mathematical induction, which is discussed in Chapter 8.

Compound Interest

> If a principal (or present value) of P dollars is invested at an interest rate of r per period for a total of t periods, then the amount present (or future value) after t periods, A, is given by the formula
>
> $$\mathbf{A = P(1 + r)^t}$$

EXAMPLE 1 If \$12,000 is invested for 5 years at 18% compounded annually, what is the amount present at the end of 5 years?

Solution $P = \$12{,}000$; $r = 0.18$; and $t = 5$. Then

$A = \$12{,}000(1 + 0.18)^5$

$\approx \$12{,}000(2.2877578)$ Use a calculator.

$\approx \$27{,}453.09$

■

EXAMPLE 2 If the interest in Example 1 is compounded monthly, find the amount present.

Solution

$P = \$12{,}000$

$r = \dfrac{0.18}{12}$ — r is the rate per period; in this case the period is monthly, so divide by 12.

$= 0.015$

$t = 5(12)$ — t is the number of periods.

$= 60$

$A = \$12{,}000(1 + 0.015)^{60}$

$\approx \$12{,}000(2.4432198)$ — Use a calculator.

$\approx \$29{,}318.64$ ■

Instead of finding the amount present in Example 2, suppose we want to know how long it will take for \$12,000 to grow to some specified amount. This question gives rise to an equation for which the variable or unknown value is an exponent. Such equations are called **exponential equations**. Consider

$$A = b^x$$

where $b > 1$. How can you solve this equation for x? Notice that

x is the exponent of b that yields A

This can be rewritten

x = exponent of b to get A

It appears that the equation is now solved for x, but this is simply a notational change. The expression "exponent of b to get A" is called, for historical reasons, "the log of A to the base b." That is,

x = log A to the base b

But this phrase is shortened to the notation

$$x = \log_b A$$

The term **log** is an abbreviation for **logarithm**, but even the introduction of this notation does not give us the numerical answer we are looking for. It does solve for x algebraically, however, which is a first step in solving exponential equations. It is important to recognize this as a notational change only:

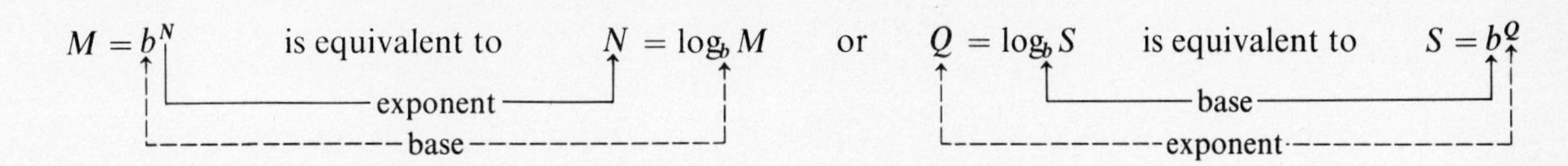

EXAMPLE 3 Change from exponential form to logarithmic form.

a. Remember: The log form solves for the exponent.

$5^2 = 25 \Leftrightarrow \log_5 25 = 2$ Use the symbol $\Leftrightarrow$ to mean "is equivalent to."

Remember: This is the base.

b. $3^2 = 9 \Leftrightarrow \log_3 9 = 2$

base exponent

c. $\frac{1}{8} = 2^{-3} \Leftrightarrow \log_2 \frac{1}{8} = -3$

d. $\sqrt{16} = 4 \Leftrightarrow \log_{16} 4 = \frac{1}{2}$ Remember: $\sqrt{16} = 16^{1/2}$. ■

EXAMPLE 4 Change from logarithmic form to exponential form.

a. $\log_{10} 100 = 2 \Leftrightarrow 10^2 = 100$

base exponent

b. $\log_{10} \frac{1}{1000} = -3 \Leftrightarrow 10^{-3} = \frac{1}{1000}$

c. $\log_3 1 = 0 \Leftrightarrow 3^0 = 1$ ■

To **evaluate** a logarithm means to find a numerical value for the given logarithm. The first ones you are asked to find use the definition of logarithm and the exponential property of equality.

EXAMPLE 5 Evaluate the given logarithms.

a. $\log_2 64$. Since it is usually necessary to supply a variable to convert to exponential form, we will use N in these examples.

$$\log_2 64 = N \quad \text{or} \quad 2^N = 64$$
$$2^N = 2^6$$
$$N = 6$$

Thus $\log_2 64 = 6$.

b. $\log_3 \frac{1}{9}$

$$3^N = \frac{1}{9}$$
$$3^N = 3^{-2}$$
$$N = -2$$

Thus $\log_3 \frac{1}{9} = -2$.

c. $\log_9 27$

$$9^N = 27$$
$$3^{2N} = 3^3$$
$$2N = 3$$
$$N = \frac{3}{2}$$

Thus $\log_9 27 = \frac{3}{2}$.

d. $\log_{10} 1 = 0$ Can you do this mentally?
e. $\log_{10} 10 = 1$
f. $\log_{10} 100 = 2$
g. $\log_{10} 0.1 = -1$ ■

Consider Example **5*d*–*g***; suppose you want to find $\log_{10} 5.03$.

$$\log_{10} 5.03 = x \Leftrightarrow 10^x = 5.03$$

Since 5.03 is between 1 and 10 and

$$10^0 = 1$$

$$10^x = 5.03 \qquad \text{You want to find this } x.$$

$$10^1 = 10$$

the number x should be between 0 and 1 by the Squeeze Theorem for Exponents. Although you *still* do not have the value of x, all is not lost because tables showing approximations for these exponents have been prepared. Base 10 is fairly common, and if a logarithm is to the base 10 it is called a **common logarithm** and written without the subscript 10. Table B.2 in Appendix B at the back of the book is a table of common logarithms. To find log 5.03, locate 5.0 (the first two digits) in the column headed N and then read over to the column headed 3. The result is

$$\log 5.03 \approx 0.7016 \qquad \text{Table values are approximate.}$$

This means that $10^{0.7016} \approx 5.03$.

Calculators have, to a large extent, eliminated the need for extensive log tables. To find x on a calculator that has logarithmic keys, you must push the keys in the indicated order:

[5.03] [log]
↑ ↑
Number first, then log key

Notice that the key for $\log_{10}$ on a calculator is simply labeled *log*. You can always assume that a logarithm is to the base 10 unless it is otherwise specified.

EXAMPLE 6 Evaluate log 7.68 correct to four significant digits using Table B.2 or a calculator.*

Solution **a.** From Table B.2, $\log 7.68 \approx 0.8854$.
b. By calculator,

[7.68] [log] DISPLAY: .88536122

To four significant digits, $x = 0.8854$. ■

Notice that Table B.2 is limited to logarithms of numbers between 1.00 and 9.99. Other logarithms can be found by using essential properties of logarithms. Let A and B be positive real numbers and let b be a positive real number other than 1.

---

* Significant digits are discussed in Appendix A.

Then:

First Law of Logarithms	$\log_b AB = \log_b A + \log_b B$	The log of the product of two numbers is the sum of the logs of those numbers.
Second Law of Logarithms	$\log_b \frac{A}{B} = \log_b A - \log_b B$	The log of the quotient of two numbers is the log of the numerator minus the log of the denominator.
Third Law of Logarithms	$\log_b A^p = p \log_b A$	The log of the pth power of a number is p times the log of that number.

The proofs of these laws of logarithms are easy if you remember that logarithmic equations are equivalent to exponential equations and that the properties of exponents can be applied. The first law of logarithms is a restatement of the first law of exponents:

$$b^x b^y = b^{x+y}$$

Let $A = b^x$ and $B = b^y$. Then $\log_b A = x$ and $\log_b B = y$. The first law concerns the product of A and B, so

$$\begin{aligned} AB &= b^x b^y \\ &= b^{x+y} \end{aligned}$$

Therefore by changing to logarithmic form

$$\begin{aligned} \log_b AB &= x + y \\ &= \log_b A + \log_b B \qquad \text{By substitution} \end{aligned}$$

The second law concerns the quotient A/B $(B \neq 0)$, so

$$\frac{A}{B} = \frac{b^x}{b^y}$$

$$\frac{A}{B} = b^{x-y} \qquad \text{Second law of exponents}$$

Thus

$$\begin{aligned} \log_b \frac{A}{B} &= x - y \\ &= \log_b A - \log_b B \qquad \text{By substitution} \end{aligned}$$

The proof of the third law of logarithms follows from the third law of exponents and is left as an exercise.

EXAMPLE 7 Evaluate log 852 correct to three significant digits by using Table B.2 or a calculator.

Solution **a.** From Table B.2: 852 is not found in the table, so rewrite it as an integral power of 10 times a number between 1 and 10. This form of a number is called **scientific notation** of the number:

$$852 = \underbrace{8.52}_{\uparrow} \times \overset{}{10^2}$$

This number is found in Table B.2. — (8.52)

This part is handled by using the definition of log and the laws of exponents. — (10^2)

$\log 852 = \log(8.52 \times 10^2)$	These steps are usually done mentally, but this first example will show you why it works.
$= \log 8.52 + \log 10^2$	
$= \log 8.52 + 2 \log 10$	
$\approx 0.9304 + 2(1)$	log 8.52 from Table B.2; log 10 from the definition (found by inspection)
$= 2.9304$	
≈ 2.93	Rounded to three significant digits

b. By calculator:

To three significant digits, $\log 852 = 2.93$. ■

If a number n is written in scientific notation as

$$n = M \cdot 10^C \qquad (1 \leq M < 10)$$

then

$$\begin{aligned} \log n &= \log(M \cdot 10^C) \\ &= \log M + C \log 10 \\ &= C + \log M \end{aligned}$$

Since $1 \leq M < 10$, $\log M$ can be found in Table B.2. Log M is called the **mantissa** and C is called the **characteristic**; C can be found by inspection.

EXAMPLE 8 Find x, where $x = \log 2420$.

Solution **a.** By Table B.2: The mantissa is $\log 2.420 = 0.3838$ and the characteristic is 3, so

$$x = \log 2420 \approx 3.3838$$

b. By calculator:

$x =$ [2420] [log] DISPLAY: 3.383815366

To five significant digits, $x = 3.3838$. ■

EXAMPLE 9 Find x, where $\log 2426 = x$.

Solution **a.** By table: The mantissa is $\log 2.426$ and the characteristic is 3. But 2.426 is not listed in the table so proceed as follows:

$$6\left[\begin{array}{l} 2.420 \\ 2.426 \end{array}\right. \quad \begin{array}{c} 0.3838 \\ ? \\ 0.3856 \end{array}\left.\vphantom{\begin{array}{c}0\\0\\0\end{array}}\right] \text{Difference is } 0.0018$$
$$\quad 2.430$$

Take a proportion of the difference between the two logs we know and add it to the smaller one. In this case, we add $\frac{6}{10}$ of 0.0018 or 0.00108. Thus $\log 2426 \approx 3.3849$. This procedure is called **linear interpolation** and is explained more fully in the Extended Application at the end of Chapter 6. In practice, linear interpolation for logarithms is no longer necessary. In this book we will normally only require the accuracy found in a four-place table, such as Table B.2; since 2.426 is closer to 2.430 than to 2.420, the answer $x = \log 2426 \approx 3.3856$ is acceptable. Five- or six-place tables could be used for greater accuracy, but calculators have made them effectively obsolete.

b. By calculator:

$x =$ [2426] [log] DISPLAY: 3.3848907965

To four significant digits, $x = 3.385$. ■

EXAMPLE 10 Find x, where $\log 0.00728 = x$.

Solution **a.** By table: The mantissa is $\log 7.28 = 0.8621$ and the characteristic is -3, so

$$x = \log 0.00728 \approx 0.8621 - 3$$
$$= -2.1379$$

Since mantissas are always positive but characteristics are negative for any number less than 1, the answer is sometimes left in the form

$$0.8621 - 3$$

or

$$0.8621 - 3 = (0.8621 + 10) - 10 - 3$$
$$= 10.8621 - 13$$

b. By calculator:

[0.00728] [log] DISPLAY: −2.137868621

To three significant digits, $x = -2.14$. ■

Let us now return to the compound interest formula used in Examples 1 and 2. If \$12,000 is deposited at 18% and is compounded annually for 5 years the amount present is \$27,453.09 (Example 1), and if it is compounded monthly the amount present is \$29,318.64 (Example 2). A reasonable extension is to ask what happens if the interest is compounded even more frequently than monthly. Can we compound daily, hourly, every minute, or every split second? We certainly can; in fact money can be compounded **continuously**, which means that every instant the newly accumulated interest is used as part of the principal for the next instant. In order to understand these concepts consider the following contrived example. Suppose \$1 is invested at 100% interest for 1 year compounded at different intervals. The compound interest formula for this example is

$$A = \left(1 + \frac{1}{n}\right)^n$$

where n is the number of times of compounding in one year. The calculations of this formula for different values of n are shown in Table 4.1.

TABLE 4.1
Effect of compounding on $1 investment

Number of periods	Formula	Amount
Annually, $n = 1$	$\left(1+\frac{1}{1}\right)^1$	$2.00
Semiannually, $n = 2$	$\left(1+\frac{1}{2}\right)^2$	2.25
Quarterly, $n = 4$	$\left(1+\frac{1}{4}\right)^4$	2.44
Monthly, $n = 12$	$\left(1+\frac{1}{12}\right)^{12}$	2.61
Daily, $n = 360$	$\left(1+\frac{1}{360}\right)^{360}$	2.715
Hourly, $n = 8640$	$\left(1+\frac{1}{8640}\right)^{8640}$	2.7181

If you continue these calculations for even larger n, you will obtain the following results:

$n = 10{,}000$	the formula yields	2.718145926
$n = 100{,}000$		2.718268237
$n = 1{,}000{,}000$		2.718280469
$n = 10{,}000{,}000$		2.718281828
$n = 100{,}000{,}000$		2.718281828

The calculator can no longer distinguish the values of $(1 + 1/n)^n$ for larger n. These values are approaching a particular number. This number, it turns out, is an irrational number so it does not have a convenient decimal representation. (That is, its decimal representation does not terminate and does not repeat.) Mathematicians, therefore, have agreed to denote this number by the symbol e, which is defined as a limit.

The Number e

$$\left(1+\frac{1}{n}\right)^n \to e \quad \text{as } n \to \infty$$

For interest compounded continuously, the following formula is used:

Continuous Interest

$$A = Pe^{rt}$$

You can find e, as well as powers of e, by using Table B.1 in Appendix B or by using a calculator.

EXAMPLE 11 Find e, e^2, and e^{-3}.

Solution **a.** From Table B.1,

$$e \approx 2.718 \qquad e = e^1$$
$$e^2 \approx 7.389$$
$$e^{-3} \approx 0.050$$

b. On a calculator, locate a key labeled e^x. First enter the value for x, then press $\boxed{e^x}$:

$e \approx 2.718$ $\boxed{1}\boxed{e^x}$ DISPLAY: 2.718281828

$e^2 \approx 7.389$ $\boxed{2}\boxed{e^x}$ DISPLAY: 7.389056099

$e^{-3} \approx 0.050$ $\boxed{3}\boxed{+/-}\boxed{e^x}$ DISPLAY: .0497870684 ■

EXAMPLE 12 If the interest in Example 1 is compounded continuously, find the amount present.

Solution Since $P = \$12{,}000$, $r = 0.18$, and $t = 5$,

$$A = \$12{,}000e^{0.18(5)}$$
$$\approx \$12{,}000(2.4596031) \qquad \text{Look up } e^{0.9} \text{ or press: } \boxed{.18}\boxed{\times}\boxed{5}\boxed{=}\boxed{e^x}$$
$$\approx \$29{,}515.24$$

■

You should memorize at least the first six digits of e.

$$e \approx 2.71828$$

Logarithms to the base e are called **natural logarithms** and are denoted by

$$\log_e x = \ln x$$

The expression $\ln x$ is often pronounced "lon x." Tables for natural logarithms are usually given as powers of e and are used as shown in Example 12.

EXAMPLE 13 Find $\ln 3.49$.

Solution **a.** By table: *Think* $\ln 3.49 = \log_e 3.49 = x$ and write this as an exponential equation:

$$e^x = 3.49$$

Now look at Table B.1 (found in Appendix B at the back of the book). Look down the e^x columns until you find an entry equal to (or close to) 3.49; you will see $x = 1.25$ in the column headed x. This means

$$e^{1.25} \approx 3.49$$

Thus $\ln 3.49 \approx 1.25$.

b. By calculator: If your calculator has a $\boxed{\log}$ key, chances are it also has an $\boxed{\ln}$ key:

$\boxed{3.49}\boxed{\ln}$ DISPLAY: 1.249901736

Calculator answers are more accurate than table answers. However, it is important to realize that any answer (whether from the table or a calculator) is only as accurate as the input number (3.49 in this problem). ■

EXAMPLE 14 Find $\ln 0.403$.

Solution **a.** By table: Find 0.403 in Table B.1; it is found for $x = 0.91$ in the column headed e^{-x}. Thus

$$e^{-0.91} \approx 0.403$$

and $\ln 0.403 \approx -0.91$.

b. By calculator:

[.403] [ln] DISPLAY: −.908818717 ■

EXAMPLE 15 Find $\ln 15$.

Solution **a.** By table: Since there is no entry in Table B.1, you can use linear interpolation or find the entry closest to 15. It is 2.71 in the column headed e^x. Thus $\ln 15 \approx 2.71$.

b. By calculator:

[15] [ln] DISPLAY: 2.708050201 ■

Tables for natural logs (or for e^x) are more restrictive than tables for common logs because it is not convenient to use the idea of characteristic and mantissa when working with natural logs. Notice that the limitations on Table B.1 are for

$$0.00 \le x \le 20.086$$

Other values of $\ln x$ are considered in the next section.

PROBLEM SET 4.2

A

Write the equations in Problems 1–12 in logarithmic form.

1. $64 = 2^6$ **2.** $100 = 10^2$ **3.** $1000 = 10^3$ **4.** $64 = 8^2$

5. $125 = 5^3$ **6.** $a = b^c$ **7.** $m = n^p$ **8.** $1 = e^0$

9. $9 = (\frac{1}{3})^{-2}$ **10.** $8 = (\frac{1}{2})^{-3}$ **11.** $\frac{1}{3} = 9^{-1/2}$ **12.** $\frac{1}{2} = 4^{-1/2}$

Write the equations in Problems 13–28 in exponential form.

13. $\log_{10} 10{,}000 = 4$ **14.** $\log 0.01 = -2$

15. $\log 1 = 0$ **16.** $\log x = 2$

17. $\log_e e^2 = 2$ **18.** $\ln e^3 = 3$

19. $\ln x = 5$ **20.** $\ln x = 0.03$

21. $\log_2 \frac{1}{8} = -3$ **22.** $\log_2 32 = 5$

23. $\log_4 2 = \frac{1}{2}$ **24.** $\log_{1/2} 16 = -4$

25. $\log_m n = p$ **26.** $\log_a b = c$

27. $\log_x 8 = 3$ **28.** $\log_x 52 = 2$

Use the definition of logarithm, Tables B.1 and B.2, or a calculator to evaluate the expressions in Problems 29–52.

29. $\log_b b^2$ **30.** $\log_t t^3$ **31.** $\log_e e^4$ **32.** $\log_\pi \sqrt{\pi}$

33. $\log_\pi(1/\pi)$ **34.** $\log_2 8$ **35.** $\log_3 9$ **36.** $\log_{19} 1$

37. $\log 4.27$ **38.** $\log 1.08$ **39.** $\log 8.43$ **40.** $\log 9760$

41. $\log 71{,}600$ **42.** $\log 0.042$ **43.** $\log 0.321$ **44.** $\log 0.0532$

45. $\ln 2.27$ **46.** $\ln 16.77$ **47.** $\ln 2$ **48.** $\ln \frac{1}{8}$

49. $\ln 13$ **50.** $\ln 0.15$ **51.** $\ln 7.3$ **52.** $\ln 10.57$

B

53. Prove the third law of logarithms, $\log_b A^p = p \log_b A$.

54. If \$1000 is invested at 7% compounded annually, how much money will there be in 25 years?

55. If \$1000 is invested at 12% compounded semiannually, how much money will there be in 10 years?

56. If \$1000 is invested at 16% interest compounded continuously, how much money will there be in 25 years?

57. If \$1000 is invested at 14% interest compounded continuously, how much money will there be in 10 years?

58. If \$8500 is invested at 18% interest compounded monthly, how much money will there be in 3 years?

59. If \$3600 is invested at 15% interest compounded daily, how much money will there be in 7 years? (Use a 365-day year; this is called *exact* interest.)

60. If \$10,000 is invested at 14% interest compounded daily, how much money will there be in 6 months? (Use a 360-day year; this is called *ordinary* interest.)

61. ***Business*** An advertising agency conducted a survey and found that the number of units sold, N, is related to the amount a spent on advertising (in dollars) by the following formula:

$$N = 1500 + 300 \ln a \qquad (a \geq 1)$$

a. How many units are sold after spending \$1000?
b. How many units are sold after spending \$50,000?
c. If the company wants to sell 5000 units, how much money will it have to spend?

62. ***Chemistry*** The pH of a substance measures its acidity or alkalinity. It is found by the formula

$$\text{pH} = -\log[\text{H}^+]$$

where $[\text{H}^+]$ is the concentration of hydrogen ions in an aqueous solution given in moles per liter.

a. What is the pH (to the nearest tenth) of a lemon for which $[\text{H}^+] = 2.86 \times 10^{-4}$?
b. What is the pH (to the nearest tenth) of rainwater for which $[\text{H}^+] = 6.31 \times 10^{-7}$?
c. If a shampoo advertises that it has a pH of 7, what is $[\text{H}^+]$?

63. ***Earth Science*** The Richter scale for measuring earthquakes was developed by Gutenberg and Richter. It relates the energy E (in ergs) to the magnitude of the earthquake, M, by the formula

$$M = \frac{\log E - 11.8}{1.5}$$

a. A small earthquake is one that releases 15^{15} ergs of energy. What is the magnitude of such an earthquake on the Richter scale?
b. A large earthquake is one that releases 10^{25} ergs of energy. What is the magnitude of such an earthquake on the Richter scale?

C

64. ***Physics*** The intensity of sound is measured in decibels D and is given by

$$D = \log\left(\frac{I}{I_0}\right)^{10}$$

where I is the power of the sound in watts per cubic centimeter (W/cm^3) and $I_0 = 10^{-16}$ W/cm^3 (the power of sound just below the threshold of hearing). Find the number of decibels of the given sound:

a. A whisper, 10^{-13} W/cm^3
b. Normal conversation, $3.16 \cdot 10^{-10}$ W/cm^3
c. The world's loudest shout by Skipper Kenny Leader, 10^{-5} W/cm^3
d. A rock concert, $5.23 \cdot 10^{-6}$ W/cm^3

65. ***Space Science*** A test of a rocket engine for a certain spacecraft on a launch pad shows the noise level to be 100 decibels outside the spacecraft and 45 decibels inside. How many times greater is the noise intensity outside the spacecraft than inside? See Problem 64 for the appropriate formula.

4.3 LOGARITHMIC FUNCTIONS AND EQUATIONS

The notion of a logarithm, which was introduced in the previous section, can also be considered as a special type of function.

Logarithmic Function

The function f defined by

$$f(x) = \log_b x$$

where $b > 0$, $b \neq 1$, is called the **logarithmic function with base b.**

By relating logarithmic and exponential functions, you will notice that you already know a great deal about the logarithmic function. For example, the graph of $y = \log_2 x$ is the same as the graph of $2^y = x$, as shown in Figure 4.8.

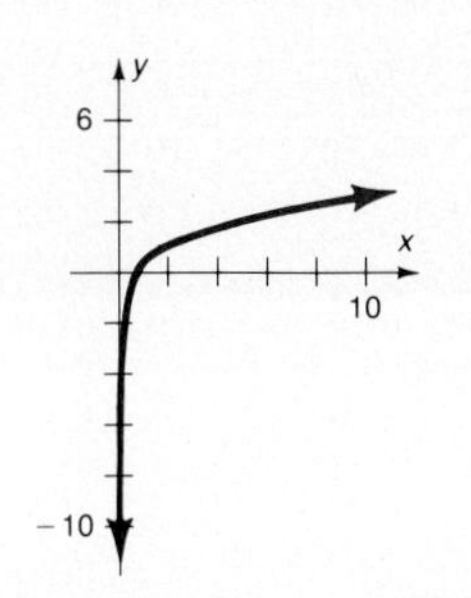

Figure 4.8 Graph of $y = \log_2 x$

y	x
-3	$2^{-3} = \frac{1}{8}$
-2	$2^{-2} = \frac{1}{4}$
-1	$2^{-1} = \frac{1}{2}$
0	$2^0 = 1$
1	$2^1 = 2$
2	$2^2 = 4$
3	$2^3 = 8$

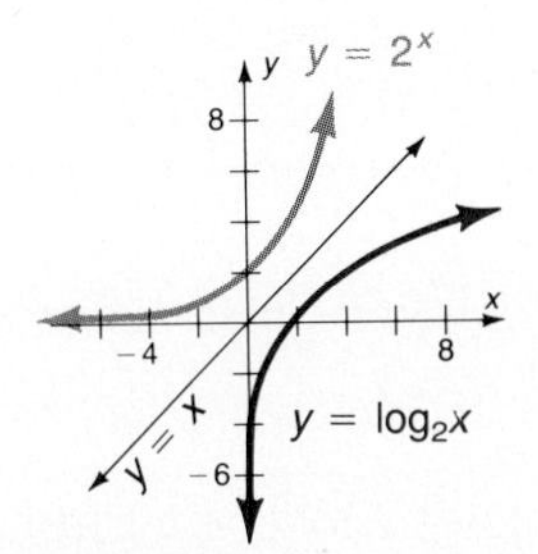

Figure 4.9 Graph of $y = 2^x$, $y = \log_2 x$, and $y = x$

Compare Figure 4.8 with Figure 4.2. The graphs in these figures are shown graphed with $y = x$ in Figure 4.9.

The functions $f(x) = 2^x$ and $g(x) = \log_2 x$ appear to be inverse functions. To prove this, we will need to consider two properties of logarithms that follow from the definition of logarithm:

1. $\log_b b^x = x$ 2. $b^{\log_b x} = x$

Both of these properties follow directly from the definition of logarithms:

$$b^M = N \Leftrightarrow \log_b N = M$$

exponent

Definition: $b^M = N \Leftrightarrow \log_b N = M$

Property 1: $b^x = b^x \Leftrightarrow \log_b b^x = x$

exponent

Property 1 is an exact statement of the definition, only with $M = x$ and $N = b^x$. Since $b^x = b^x$ is true, then $\log_b b^x = x$ is true by the definition of logarithm.

exponent

Definition: $b^M = N \Leftrightarrow \log_b N = M$

Property 2: $b^{\log_b x} = x \Leftrightarrow \log_b x = \log_b x$

exponent

Property 2 is an exact statement of the definition. Let $M = \log_b x$ and $N = x$. Since $\log_b x = \log_b x$, then $b^{\log_b x} = x$ is true by the definition of logarithm.

To show that $f(x) = \log_b x$ and $g(x) = b^x$ are inverse functions, show $(f \circ g)(x) = (g \circ f)(x) = x$:

$$\begin{aligned}(f \circ g)(x) &= f[g(x)] \\ &= f(b^x) \\ &= \log_b b^x \\ &= x \quad \text{By Property 1}\end{aligned}$$

and

$$\begin{aligned}(g \circ f)(x) &= g[f(x)] \\ &= g(\log_b x) \\ &= b^{\log_b x} \\ &= x \quad \text{By Property 2}\end{aligned}$$

This proves the following theorem.

Exponential and Logarithmic Functions Are Inverses

> The exponential and logarithmic functions with base b are inverse functions of one another.

This relationship of inverse functions is needed to find e on several brands of calculators. If a calculator has

[ln x] and [INV]

keys, but no e^x key, how can you find e or e^x? Since

$$y = \ln x \qquad \text{and} \qquad y = e^x$$

are inverse functions, to find e (or e^1) you can press

[1] [INV] [ln x] DISPLAY: 2.718281828

This gives the inverse of the ln x function; that is, it is e^x where the x value is input just prior to pressing these buttons.

If you need $e^{5.2}$ on such a calculator, press

[5.2] [INV] [ln x] DISPLAY: 181.2722419

Properties 1 and 2 also lead to another theorem that allows us to solve logarithmic equations.

Log of Both Sides Theorem

> If A, B, and b are positive real numbers with $b \neq 1$, then
>
> $$\log_b A = \log_b B \Leftrightarrow A = B$$

Proof If $A = B$, then

$$\begin{aligned} \log_b A &= \log_b A \\ &= \log_b B \qquad \text{By substitution} \end{aligned}$$

If $\log_b A = \log_b B$, then

$$\begin{aligned} b^{\log_b B} &= A \qquad \text{By definition of logarithm} \\ b^{\log_b B} &= B \qquad \text{By Property 2} \\ B &= A \qquad \text{By substitution} \end{aligned}$$

□

We now use this theorem and the definition of logarithm to solve exponential equations.

EXAMPLE 1 Solve $\log_2 \sqrt{2} = x$ for x.

Solution For this logarithmic equation, the exponent is the unknown. Apply the definition of logarithm:

$$\begin{aligned} 2^x &= \sqrt{2} \\ 2^x &= 2^{1/2} \\ x &= \tfrac{1}{2} \qquad \text{Exponential property of equality} \end{aligned}$$

■

EXAMPLE 2 Solve $\log_x 25 = 2$.

Solution For this logarithmic equation, the base is the unknown. Apply the definition of logarithm:

$$x^2 = 25$$
$$x = \pm 5$$

Be sure the values you obtain are permissible values for the definition of a logarithm. In this case, $x = -5$ is not a permissible value since a logarithm with a negative base is not defined. Therefore, the solution is $\boldsymbol{x = 5}$. ■

EXAMPLE 3 Solve $\ln x = 5$.

Solution The power itself is the unknown in this logarithmic equation. Apply the definition of logarithm (to the base e):

$$e^5 = x$$

$x \approx 148.41$ PRESS: [5] [e^x] DISPLAY: 148.4131591 ■

EXAMPLE 4 Solve $\log_5 x = \log_5 72$.

Solution Use the log of both sides theorem: $x = 72$. ■

Basically, all logarithmic equations fall into one of the four types illustrated by Examples 1–4. For problems like those in Examples 1–3, apply the definition of logarithm and for those like Example 4, apply the log of both sides theorem. More difficult examples require algebraic simplification using the laws of logarithms (see page 150) in order to write a single logarithmic function on either one or both sides, as illustrated by Examples 5–7.

EXAMPLE 5

$\log_8 3 + \frac{1}{2}\log_8 25 = \log_8 x$	
$\log_8 3 + \log_8 25^{1/2} = \log_8 x$	Third law of logarithms
$\log_8 3 + \log_8 5 = \log_8 x$	$25^{1/2} = \sqrt{25} = 5$
$\log_8(3 \cdot 5) = \log_8 x$	First law of logarithms
$15 = x$	Log of both sides theorem

The solution is 15. ■

EXAMPLE 6

$$5\log_x 2 - \tfrac{1}{2}\log_x 8 = 2 - \tfrac{1}{2}\log_x 2$$

$$\log_x 2^5 - \log_x \sqrt{8} = 2 - \log_x \sqrt{2}$$
$$\log_x 32 - \log_x 2\sqrt{2} + \log_x \sqrt{2} = 2$$
$$\log_x\left(\frac{32}{2\sqrt{2}} \cdot \sqrt{2}\right) = 2$$
$$\log_x 16 = 2$$
$$x^2 = 16 \qquad \text{Definition of logarithm}$$
$$x = \pm 4$$

By the definition of a logarithm, x must be positive, so the solution is 4. ■

EXAMPLE 7

$$\ln x - \tfrac{1}{2}\ln 2 = \tfrac{1}{2}\ln(x + 4)$$
$$\ln x - \ln\sqrt{2} = \ln\sqrt{x + 4}$$
$$\ln\left(\frac{x}{\sqrt{2}}\right) = \ln\sqrt{x + 4}$$
$$\frac{x}{\sqrt{2}} = \sqrt{x + 4}$$
$$\frac{x^2}{2} = x + 4$$
$$x^2 - 2x - 8 = 0$$
$$(x - 4)(x + 2) = 0$$
$$x = 4, -2$$

Notice that $\ln(-2)$ is not defined, so $x = -2$ is an extraneous root. Therefore the solution is 4. ■

For many years logarithms were taught in elementary mathematics primarily as an aid to computation. As you have seen in this chapter, logarithms are useful as functions in mathematics, but with the widespread use of calculators, they are no longer necessary for complicated calculations. In the remaining part of this section we will consider some problems and give their logarithmic and calculator solutions so you can compare the different methods. If you do not have a calculator available, you can work the problems using the four-place log tables in the back of the book.

EXAMPLE 8 Find $25{,}000(1.15)^{30}$ to two significant digits.

Solution By calculator with algebraic logic:

[1.15] [y^x] [30] [×] [25000] [=] DISPLAY: 1655294.299

By calculator with RPN logic:

[30] [ENTER] [1.15] [x^y] [25000] [×] DISPLAY: 1655294.299

By logarithms:

$$A = 25{,}000(1.15)^{30}$$
$$\log A = \log[25{,}000(1.15)^{30}]$$
$$= \log 25{,}000 + 30 \log 1.15$$
$$\approx 4.3979 + 30(0.0607)$$
$$= 6.2189$$

The problem now is to use Table B.2 in reverse by finding the mantissa 0.2189 in the table; it is closest to 1.66. (You can use linear interpolation for a more accurate answer.) Thus if $\log A = 6.2189$, then

$$A = 1.66 \times 10^6$$
$$\approx 1{,}660{,}000$$

To two significant digits the answer is 1,700,000. ■

EXAMPLE 9 Find $13{,}250e^{1.35}$ to three significant digits.

Solution By calculator with algebraic logic:

[1.35] [e^x] [×] [13250] [=] DISPLAY: 51110.88828

By calculator with RPN logic:

[1.35] [e^x] [ENTER] [13250] [×] DISPLAY: 51110.88828

By logarithms:

$$\begin{aligned} A &= 13{,}250e^{1.35} \\ &= (13{,}250)(3.857) && \text{By Table B.1} \\ \log A &= \log 13{,}250 + \log 3.857 \\ &\approx 4.1222 + 0.5862 \\ &= 4.7085 \\ A &\approx 5.11 \times 10^4 \end{aligned}$$

To three significant digits, the answer is 51,100. ■

PROBLEM SET 4.3

A

Solve the equations in Problems 1–21.

1. $\log_5 25 = x$

2. $\log_2 128 = x$

3. $\log \frac{1}{10} = x$

4. $\log_x 84 = 2$

5. $\log_x 28 = 2$

6. $\log_x 50 = 2$

7. $\ln x = 2$

8. $\ln x = 3$

9. $\ln x = 4$

10. $\ln x = \ln 14$

11. $\ln 9.3 = \ln x$

12. $\ln 109 = \ln x$

13. $\log_3 x^2 = \log_3 125$

14. $\ln x^2 = \ln 12$

15. $\log x^2 = \log 70$

16. $\log_2 8\sqrt{2} = x$

17. $\log_3 27\sqrt{3} = x$

18. $\log_x 1 = 0$

19. $\log_x 10 = 0$

20. $\log_2 x = 5$

21. $\log_{10} x = 5$

Use a calculator or logarithms to evaluate the expressions in Problems 22–31. Be sure to round off your answers to the appropriate number of significant digits. (See Appendix A for a discussion of significant digits.)

22. $(14)(351)$

23. $(218)(263)$

24. $(2.00)^4(1245)(277)$

25. $(3.00)^3(182)$

26. $\dfrac{(1979)(1356)}{452}$

27. $\dfrac{(515)(20{,}600)}{200}$

28. $(990)(1117)(342) - 89$

29. $[0.14 + (197)(25.08)](19)$

30. $\dfrac{1.00}{0.005 + 0.020}$

31. $\dfrac{241^2 + 568^2 - 351^2}{2(241)(568)}$

B

Graph the functions given in Problems 32–43.

32. $f(x) = e^x$

33. $f(x) = e^{-x}$

34. $f(x) = -e^{-x}$

35. $y = \log_3 x$

36. $y = \log_{1/2} x$

37. $y = \log_{1/3} x$

38. $y = \log_{10} x$

39. $y = \log_e x$

40. $y = \log_\pi x$

41. $y = \log(x - 4)$

42. $y - 3 = \log x$

43. $y - 1 = \log(x + 2)$

Solve the equations in Problems 44–51.

44. $\log_8 5 + \frac{1}{2}\log_8 9 = \log_8 x$

45. $\log_7 x - \frac{1}{2}\log_7 4 = \frac{1}{2}\log_7(2x - 3)$

46. $\ln 10 - \frac{1}{2}\ln 25 = \ln x$

47. $\frac{1}{2}\ln x = 3\ln 5 - \ln x$

48. $\ln x - \frac{1}{2}\ln 3 = \frac{1}{2}\ln(x + 6)$

49. $2\ln x - \frac{1}{2}\ln 9 = \ln 3(x - 2)$

50. $3\ln\dfrac{e}{\sqrt[3]{5}} = 3 - \ln x$

51. $5\ln\dfrac{e}{\sqrt[5]{2}} = 1 - \ln x$

Business If P dollars are borrowed for n months at a monthly interest rate of r, then the monthly payment is found by the formula

$$M = \frac{Pr}{1 - (1 + r)^{-n}}$$

To use this formula to find M with a calculator with algebraic logic, press:

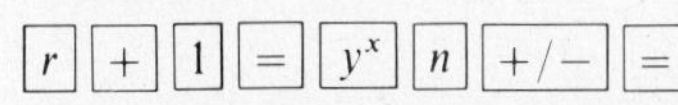

[+/−] [+] [1] [=] [1/x] [×] [P] [×] [r] [=]

Use this information in Problems 52–55.

52. What is the monthly car payment for a new car costing \$12,487 with a down payment of \$2487? The car is financed for 4 years at 12%. (*Hint:* $P = \$10{,}000$ and $r = 0.01$.)

53. A home loan is made for \$110,000 at 12% interest for 30 years. What is the monthly payment? (*Hint:* $P = \$110{,}000$ and $r = 0.01$.)

54. A purchase of \$2430 is financed at 23% for 3 years. What is the monthly payment?

55. A home is financed at $14\frac{1}{2}\%$ for 30 years. If the amount financed is \$125,000, what is the monthly payment?

56. ***Chemistry*** The pH (hydrogen potential) of a solution is given by

$$\text{pH} = \log \frac{1}{[\text{H}^+]}$$

where $[\text{H}^+]$ is the concentration of hydrogen ions in a water solution given in moles per liter. Find the pH (to the nearest tenth) for the substances with the $[\text{H}^+]$ given.

a. Lemon juice, $5.01 \cdot 10^{-3}$ **b.** Milk, $3.98 \cdot 10^{-7}$
c. Vinegar, $7.94 \cdot 10^{-4}$ **d.** Rainwater, $5.01 \cdot 10^{-7}$
e. Seawater, $4.35 \cdot 10^{-9}$

57. ***Psychology*** The "learning curve" describes the rate at which a person learns certain tasks. If a person sets a goal of typing N words per minute (wpm), the length of time t days to achieve this goal is given by

$$t = -62.5 \ln\left(1 - \frac{N}{80}\right)$$

a. How long would it take to learn to type 30 wpm?
b. If we accept this formula, is it possible to learn to type 80 wpm?
c. Solve for N.

58. ***Psychology*** In Problem 57 an equation for learning was given. Psychologists are also concerned with forgetting. In a certain experiment, students were asked to remember a set of nonsense syllables, such as "htm." The students then had to recall the syllables after t sec. The equation that was found to describe forgetting was

$$R = 80 - 27 \ln t \qquad (t \geq 1)$$

where R is the percentage of students who remember the syllables after t sec.

a. What percentage of the students remembered the syllables after 3 sec?
b. In how many seconds would only 10% of the students remember?
c. Solve for t.

59. The equation for the Richter scale relating energy E (in ergs) to the magnitude of the earthquake, M, is given by the formula

$$M = \frac{\log E - 11.8}{1.5}$$

a. Solve for E.
b. What was the energy of the 1974 Alaska earthquake, which measured 8.5 on the Richter scale?

C

60. Consider the following argument:

$4 > 3$	Obviously true
$4 \log_{10} \frac{1}{3} > 3 \log_{10} \frac{1}{3}$	Multiply both sides by $\log_{10} \frac{1}{3}$
$\log_{10}(\frac{1}{3})^4 > \log_{10}(\frac{1}{3})^3$	Property of exponents
$(\frac{1}{3})^4 > (\frac{1}{3})^3$	Theorem of logarithms
$\frac{1}{81} > \frac{1}{27}$	Obviously false

Can you find the error?

61. Graph $y = (1 + \frac{1}{x})^x$ for $x > 0$.

62. Graph: **a.** $c(x) = \dfrac{e^x + e^{-x}}{2}$ **b.** $s(x) = \dfrac{e^x - e^{-x}}{2}$

63. ***Engineering*** The functions c and s of Problem 62 are called the *hyperbolic cosine* and *hyperbolic sine* functions. These are defined by

$$\cosh x = \frac{e^x + e^{-x}}{2} \quad \text{and} \quad \sinh x = \frac{e^x - e^{-x}}{2}$$

a. Show that $\cosh^2 x - \sinh^2 x = 1$.
b. Show that the hyperbolic cosine is an even function and that the hyperbolic sine is an odd function.
c. Show that $\sinh 2x = 2 \sinh x \cosh x$.
d. Graph $y = \cosh x + \sinh x$.
e. Define

$$\tanh x = \frac{\sinh x}{\cosh}$$

Graph $\tanh x$.

4.4 EXPONENTIAL EQUATIONS

In this section, we return to the question that prompted our discussion of logarithms in the first place—How do you solve an exponential equation? The most straightforward method for solving exponential equations applies the exponential property of equality stated in Section 4.1. Recall that if equal bases are raised to some power and the results are equal, then the exponents must also be equal.

EXAMPLE 1 Solve $7^x = 343$ for x.

Solution Write both sides with the same base if possible:

$7^x = 7^3$

$x = 3$ Exponential property of equality ■

EXAMPLE 2 Solve $25^x = 125$.

Solution

$(5^2)^x = 5^3$

$5^{2x} = 5^3$

$2x = 3$ Exponential property of equality

$x = \frac{3}{2}$ ■

EXAMPLE 3 Solve $36^x = \frac{1}{6}$.

Solution

$(6^2)^x = 6^{-1}$

$6^{2x} = 6^{-1}$

$2x = -1$

$x = -\frac{1}{2}$ ■

Not all exponential equations are as easy to solve as those in Examples 1–3. If the bases cannot be forced to be the same, then the exponential equations will fall into one of three types:

Base:	10 (common log)	e (natural log)	b (arbitrary base)
Example:	$10^x = 5$	$e^t = 3.456$	$7^x = 3$

We will work two examples of each of these types. First, consider base 10.

EXAMPLE 4 Solve $10^x = 5$.

Solution Use the definition of logarithm.

$x = \log 5$ Exact answer

$x \approx 0.699$ Approximate answer, found by using tables or a calculator ■

EXAMPLE 5 Solve $10^{5x+3} = 195$.

Solution Rewrite in logarithmic form: $\log 195 = 5x + 3$. Solve for x:

$$\log 195 - 3 = 5x$$

$$x = \frac{1}{5}(\log 195 - 3) \qquad \text{Exact answer}$$

By table: $\log 195 \approx 2.2900$, so

$$x \approx \frac{1}{5}(2.2900 - 3)$$

$$= \frac{1}{5}(-0.71)$$

$$= -0.142$$

By calculator: You can multiply by $\frac{1}{5}$, but it is more common to think of this as division by 5. *Think:*

$$x = \frac{\log 195 - 3}{5}$$

[195] [log] [−] [3] [=] [÷] [5] [=]

↑ If you do not press this equals key, your calculator may assume you want 3 ÷ 5 and not the whole quantity divided by 5.

DISPLAY: −.1419930777

RPN logic: [195] [ln] [ENTER] [3] [−] [5] [÷] ■

The next two examples involve base e.

EXAMPLE 6 Solve $e^{0.06t} = 3.456$ for t.

Solution **a.** By table: Find $e^x = 3.456$; $x = 1.24$. Thus

$$e^{0.06t} \approx e^{1.24}$$

By the exponential property of equality,

$$0.06t = 1.24$$

$$t \approx 20.67$$

b. By calculator: Write in logarithmic form:

$$e^{0.06t} = 3.456$$

$$0.06t = \ln 3.456$$

$$t = \frac{\ln 3.456}{0.06}$$

Algebraic logic:

[3.456] [ln] [÷] [.06] [=] DISPLAY: 20.66853085

RPN logic:

[3.456] [ln] [ENTER] [.06] [÷] DISPLAY: 20.66853085 ■

EXAMPLE 7 Solve $\frac{1}{2} = e^{-0.000425t}$.

Solution **a.** By table: Find $e^x = \frac{1}{2} = 0.5$ in Table B.1. Notice that this is found in the column headed e^{-x} for $x \approx 0.69$. (You can interpolate if better accuracy is needed.) Thus, since $e^{-0.69} \approx \frac{1}{2}$,

$$e^{-0.69} \approx e^{-0.000425t}$$

$$-0.69 \approx -0.000425t$$

$$t \approx \frac{0.69}{0.000425}$$

$$\approx 1624$$

b. By calculator: $\frac{1}{2} = e^{-0.000425t}$ in logarithmic form is

$$\ln 0.5 = -0.000425t$$

$$t = \frac{\ln 0.5}{-0.000425}$$

Algebraic logic:

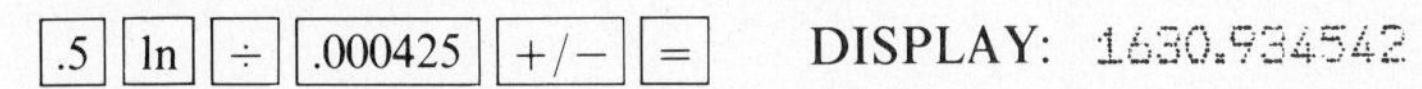

RPN logic:

■

Notice that there is a considerable discrepancy between the answers in Example 7 depending on whether you use Table B.1 or a calculator. Remember, however, that only two significant digits were used in finding the result from Table B.1, so to two significant digits both of these answers are 1600.

Since you do not have tables or calculator keys for bases other than 10 and e, you must proceed differently for an arbitrary base b. You should use the log of both sides theorem and the procedure for solving logarithmic equations. You generally can elect to work with base 10 or base e.

EXAMPLE 8 Solve $7^x = 3$.

Solution

Using base e:

$$\ln 7^x = \ln 3$$

$$x \ln 7 = \ln 3$$

$$x = \frac{\ln 3}{\ln 7}$$

$$\approx \frac{1.099}{1.946} \quad \text{By Table B.1 or by calculator}$$

$$\approx 0.5646$$

Using base 10:

$$\log 7^x = \log 3$$

$$x \log 7 = \log 3$$

$$x = \frac{\log 3}{\log 7}$$

$$\approx \frac{0.4771}{0.8451} \quad \text{By Table B.2 or by calculator}$$

$$\approx 0.5646$$

Same answer (you use only one of these) ■

There is another method for solving the exponential equation of Example 8. This one directly applies the definition of logarithm:

$$7^x = 3 \Leftrightarrow \log_7 3 = x$$

If you had a $\log_7$ key on your calculator or a $\log_7$ table, you would have the answer. Since you do not, you need one final logarithm theorem that changes logarithms from one base to another.

Change of Base Theorem

$$\log_a x = \frac{\log_b x}{\log_b a}$$

Notice that to change from base a to another (possibly more familiar) base b, you simply change the base on the given logarithm from a to b and then divide by the logarithm to the base b of the old base a.

Proof Let $y = \log_a x$

$a^y = x$	Definition of logarithm
$\log_b a^y = \log_b x$	Log of both sides theorem
$y \log_b a = \log_b x$	Third law of logarithms
$y = \dfrac{\log_b x}{\log_b a}$	Divide both sides by $\log_b a (\log_b a \neq 0)$ □

EXAMPLE 9 Change $\log_7 3$ to logarithms with base 10 and evaluate.

Solution

$$\log_7 3 = \frac{\log 3}{\log 7}$$

$$\approx \frac{0.4771}{0.8451}$$

$$\approx 0.5646$$

Compare this result with the one from Example 8. ■

EXAMPLE 10 Change $\log_3 3.84$ to logarithms with base e and evaluate.

Solution

$$\log_3 3.84 = \frac{\ln 3.84}{\ln 3}$$

$$\approx \frac{1.345}{1.099}$$

$$\approx 1.225$$ ■

We are now ready to consider the second of our examples of exponential equations with an arbitrary base.

EXAMPLE 11 Solve $6^{3x+2} = 200$.

Solution Method 1:

$$\log_6 200 = 3x + 2$$
$$\log_6 200 - 2 = 3x$$
$$x = \frac{\log_6 200 - 2}{3}$$
$$= \frac{\frac{\ln 200}{\ln 6} - 2}{3}$$

Algebraic calculator:

[200] [ln] [÷] [6] [ln] [=] [−] [2] [=] [÷] [3] [=] DISPLAY: .3190157417

RPN calculator:

[200] [ln] [ENTER] [6] [ln] [÷] [2] [−] [3] [÷] DISPLAY: .3190157417

Method 2:

$$\log 6^{3x+2} = \log 200 \qquad \text{Log of both sides theorem}$$
$$(3x + 2)\log 6 = \log 200$$
$$3x \log 6 + 2 \log 6 = \log 200$$
$$(3 \log 6)x = \log 200 - 2 \log 6$$
$$x = \frac{\log 200 - 2 \log 6}{3 \log 6}$$

Algebraic calculator:

[200] [log] [−] [2] [×] [6] [log] [=] [÷] [(] [3] [×] [6] [log] [)] [=]

DISPLAY: .3190157417

RPN calculator:

[200] [log] [ENTER] [2] [ENTER] [6] [log] [×] [−] [3] [ENTER] [6] [log] [×] [÷]

DISPLAY: .3190157417

Answer: $x \approx 0.32$ ■

In Section 4.2 we considered continuous compounding of interest. This is an example of continuous *growth*. A similiar application involves continuous *decay*.

Decay Formula $\boldsymbol{A = A_0 e^{-kt}}$

where an initial quantity A_0 decays to an amount A after a time t. The positive constant k depends on the substance.

EXAMPLE 12 If 100 mg of Neptunium 239 (^{239}Np) decays to 73.36 mg after 24 hr, find the value of k in the decay formula for t expressed as days.

Solution Since $A = 73.36$, $A_0 = 100$, and $t = 1$, we have

$$\begin{aligned} 73.36 &= 100e^{-k(1)} \\ 0.7336 &= e^{-k} \\ k &= -\ln 0.7336 \\ &\approx 0.30979136 \end{aligned}$$

■

Radioactive decay is usually specified in terms of its **half-life**, which means the amount of time necessary for one-half of its substance to disintegrate into another substance. The decay formula for half-life is

Half-Life Decay

$$\frac{1}{2} = e^{-kt}$$

EXAMPLE 13 What is the half-life of ^{239}Np? Use the value of k found in Example 4 ($k = 0.3098$).

Solution

$$\frac{1}{2} = e^{-0.3098t}$$

$$\begin{aligned} \ln 0.5 &= -0.3098t \\ t &= \frac{\ln 0.5}{-0.3098} \\ &\approx 2.2374021 \end{aligned}$$

The half-life is about 2.24 days. ■

A very important application of both growth and decay is the prediction of the size of various populations. Population growth is considered at length in a special Extended Application following the chapter review.

PROBLEM SET 4.4

A

Solve the exponential equations in Problems 1–16.

1. $2^x = 128$ **2.** $3^x = 243$ **3.** $8^x = 32$ **4.** $9^x = 27$

5. $125^x = 25$ **6.** $4^x = \frac{1}{16}$ **7.** $27^x = \frac{1}{81}$ **8.** $(\frac{1}{2})^x = 8$

9. $(\frac{1}{2})^x = \frac{1}{8}$ **10.** $(\frac{2}{3})^x = \frac{9}{4}$ **11.** $(\frac{3}{4})^x = \frac{16}{9}$ **12.** $2^{3x+1} = \frac{1}{2}$

13. $3^{4x-3} = \frac{1}{9}$ **14.** $27^{2x+1} = 3$

15. $8^{5x+2} = 16$ **16.** $125^{2x+1} = 25$

Evaluate the logarithms in Problems 17–22 by using the change of base theorem.

17. $\log_5 30$ **18.** $\log_2 15$ **19.** $\log_6 0.1$ **20.** $\log_4 0.05$

21. $\log_\pi 10$ **22.** $\log_\pi 25$

B

Solve the exponential equations in Problems 23–46.

23. $10^x = 42$ **24.** $10^x = 126$

25. $10^x = 0.00325$ **26.** $10^x = 0.0234$

27. $10^{x+3} = 214$ **28.** $10^{x-5} = 0.036$

29. $10^{x-1} = 0.613$ **30.** $10^{4x+1} = 719$

31. $10^{5-3x} = 0.041$ **32.** $10^{2x-1} = 515$

33. $e^{2x} = 10$ **34.** $e^{5x} = \frac{1}{4}$

35. $e^{4x} = \frac{1}{10}$ **36.** $e^{2x+1} = 5.474$

37. $e^{1-2x} = 3$ **38.** $e^{1-5x} = 15$

39. $8^x = 300$ **40.** $2^x = 1000$

41. $5^x = 10$ **42.** $9^x = 0.045$

43. $2^{-x} = 5$ **44.** $4^x = 0.82$

45. $5^{-x} = 8$ **46.** $7^{-x} = 125$

47. If \$1000 is invested at 12% compounded semiannually, how long will it take (to the nearest year) for the money to double?

48. If \$1000 is invested at 16% interest compounded annually, how long will it take (to the nearest year) for the money to quadruple?

49. If \$1000 is invested at 12% interest compounded quarterly, how long will it take (to the nearest quarter) for the money to reach \$2500?

50. If \$1000 is invested at 15% interest compounded continuously, how long will it take (to the nearest year) for the money to triple?

51. If the half-life of cesium-137 is 30 years, find the constant k for which $A = A_0e^{-kt}$ where t is expressed in years.

52. Find the half-life of strontium-90 if $k = 0.0246$, where $A = A_0e^{-kt}$ and t is expressed in years.

53. Find the half-life of krypton if $k = 0.0641$, where $A = A_0e^{-kt}$ and t is expressed in years.

54. The formula used for carbon-14 dating in archaeology is

$$A = A_0\left(\frac{1}{2}\right)^{t/5700} \quad \text{or} \quad P = \left(\frac{1}{2}\right)^{t/5700}$$

where P is the percentage of carbon-14 present after t years. Solve for t.

55. Some bone artifacts were found at the Lindenmeier site in northeastern Colorado and tested for their carbon-14 content. If 25% of the original carbon-14 was still present, what is the probable age of the artifacts? Use the formula given in Problem 54.

56. An artifact was discovered at the Debert site in Nova Scotia. Tests showed that 28% of the original carbon-14 was still present. What is the probable age of the artifact? Use the formula given in Problem 54.

57. An artifact was found and tested for its carbon-14 content. If 12% of the original carbon-14 was still present, what is its probable age? Use the formula given in Problem 54.

58. An artifact was found and tested for its carbon-14 content. If 85% of the original carbon-14 was still present, what is its probable age? Use the formula given in Problem 54.

59. ***Physics*** The atmospheric pressure P in pounds per square inch (psi) is given by

$$P = 14.7e^{-0.21a}$$

where a is the altitude above sea level (in miles). If a city has an atmospheric pressure of 13.23 psi, what is its altitude?

60. ***Earth Science*** In 1975 ($t = 0$), the world use of petroleum, P_0, was 19,473 million barrels of oil. If the world reserves are 584,600 million barrels and the growth rate for the use of oil is k, then the total amount A used during a time interval $t > 0$ is given by

$$A = \frac{P_0}{k}(e^{kt} - 1)$$

How long will it be before the world reserves are depleted if:

a. $k = 0.08$ (8%); **b.** $k = 0.052$ (5.2%)?

61. ***Space Science*** A satellite has an initial radioisotope power supply of 50 watts (W). The power output in watts is given by the equation

$$P = 50e^{-t/250}$$

where t is the time in days. Solve for t.

C

62. ***Physics*** Newton's law of cooling states that an object at temperature B surrounded by air temperature A will cool to a temperature T after t min according to the equation

$$T = A + (B - A)e^{-kt}$$

where k is a constant depending on the item being cooled.

a. If you draw a tub of 120°F water for a bath and let it stand in a 75°F room, what is the temperature of the water after 30 min if $k = 0.01$?

b. What is k for an apple pie taken from a 375°F oven and cooled to 75°F after it is left in a 72°F room for 1 hr?

c. Solve the given equation for t.

63. In calculus it is shown that

$$e^x = 1 + x + \frac{x^2}{2} + \frac{x^3}{2 \cdot 3} + \frac{x^4}{2 \cdot 3 \cdot 4} + \cdots$$

a. What are the next two terms?

b. What is the rth term?

c. Calculate e correct to the nearest thousandth using this equation.

d. Calculate $\sqrt{e} = e^{0.5}$ correct to the nearest thousandth using this equation.

CHAPTER 4 SUMMARY

The material of this chapter is reviewed in the following list of objectives. After each objective there are some practice questions. For a sample test, select the first question of each set and check your answers with the answer section. For a sample test without answers, use the second question of each set. Additional practice is given by the other questions in each set. If you are having trouble with a particular type of problem, look back to that section for extra help.

4.1 EXPONENTIAL FUNCTIONS

Objective 1 Define an exponential function and a rational exponent. Apply the five laws of exponents to rational exponents. Simplify the given expressions (assume that the variables are positive).

1. $125^{2/3}$
2. $(2^{1/2} \cdot 3^{1/3})^6$
3. $27^{2/3}/27^{1/2}$
4. $(x^{1/2} - y^{1/2})(x^{1/2} + y^{1/2})$

Objective 2 Sketch exponential functions.

5. $y = (\frac{1}{2})^x$
6. $y = -2^x$
7. $y = 2^{-x}$
8. $y = e^{-x/2}$

4.2 INTRODUCTION TO LOGARITHMS

Objective 3 Write an exponential equation in logarithmic form.

9. $10^{0.5} = \sqrt{10}$
10. $e^0 = 1$
11. $9^3 = 729$
12. $(\sqrt{2})^3 = 2\sqrt{2}$

Objective 4 Write a logarithmic equation in exponential form.

13. $\log 1 = 0$
14. $\ln \frac{1}{e} = -1$
15. $\log_2 64 = 6$
16. $\log_\pi \pi = 1$

Objective 5 Evaluate common logarithms.

17. $\log 3$
18. $\log 451$
19. $\log .0021$
20. $\log 3^4$

Objective 6 Evaluate natural logarithms.

21. $\ln 3$
22. $\ln 451$
23. $\ln 0.013$
24. $\ln 3^4$

Objective 7 Evaluate logarithms using the definition of logarithm.

25. $\log_3 27$
26. $\log_2 0.125$
27. $\ln e^5$
28. $\log 0.01$

Objective 8 State and prove the laws of exponents. Fill in the blanks.

29. $\log_b AB =$ _______________.
30. $\log_b A - \log_b B =$ _______________.
31. $\log_b A^p =$ _______________.
32. If $A = b^x$, then $x =$ _______________.

4.3 LOGARITHMIC FUNCTIONS AND EQUATIONS

Objective 9 Graph logarithmic functions.

33. $y = \log x$
34. $y = \ln x$
35. $y = \log_3 x$
36. $y + 3 = \log(x - 2)$

Objective 10 Solve logarithmic equations. Solve for x.

37. $\log_5 25 = x$
38. $\log_x (x + 6) = 2$
39. $3 \log 3 - \frac{1}{2} \log 3 = \log \sqrt{x}$
40. $2 \ln \frac{e}{\sqrt{7}} = 2 - \ln x$

Objective 11 Use logarithms or a calculator to carry out complicated calculations. Simplify the given expressions.

41. $(3450)(241)$
42. $\frac{689}{14}$
43. 19^6
44. $\left(1 + \frac{012}{360}\right)^{720}$

4.4 EXPONENTIAL EQUATIONS

Objective 12 Solve exponential equations with base 10.

45. $10^{x+2} = 125$
46. $10^{2x-3} = 0.5$
47. $10^{-x^2} = 0.45$
48. $10^{4-3x} = 15$

Objective 13 Solve exponential equations with base e.

49. $e^{3x} = 50$
50. $e^{x-5} = 0.49$
51. $e^{1-2x} = 690$
52. $e^{x^2} = 9$

Objective 14 *Solve exponential equations with arbitrary bases.*

53. $5^x = 125$ **54.** $5^{2x+1} = 0.2$ **55.** $3^x = 7$ **56.** $7^{1-3x} = 0.048$

Objective 15 *Change logarithms from one base to another. Evaluate the given expressions.*

57. $\log_2 818$ **58.** $\log_5 4.51$ **59.** $\log_3 100$ **60.** $\log_4 \sqrt[3]{4}$

Objective 16 *Solve applied problems involving exponential equations.*

61. The half-life of arsenic-76 is 26.5 hr. Find the constant k for which $A = A_0 e^{-kt}$ where t is expressed in hours.

62. If \$1500 is placed in a $2\frac{1}{2}$-year time certificate paying 13.5% compounded monthly, what is the amount in the account when the certificate matures in $2\frac{1}{2}$ years?

63. If a person's present salary is \$20,000 per year, use the formula $A = P(1 + r)^n$ to determine the salary necessary to equal this salary in 15 years if you assume the 1980 U.S. inflation rate of 13.4%.

64. Solve the formula given in Problem 63 for n.

65. Repeat Problem 63 for the 1984 rate of 4%.

Population Growth

World Population 4 Billion Tonight

BY EDWARD K. DeLONG

WASHINGTON (UPI)—By midnight tonight, the Earth's population will reach the 4 billion mark, twice the number of people living on the planet just 46 years ago, the Population Reference Bureau said Saturday.

The bureau expressed no joy at the new milestone.

It said global birth rates are too high, placing serious pressures on all aspects of future life and causing "major concern" in the world scientific community, and more than one-third of the present population has yet to reach child-bearing age.

The PRB found cause for optimism, however, in that some governments are stressing birth control to blunt the impact of "explosive growth" and the population growth rate dropped slightly in the past year.

"In 1976, each new dawn brings a formidable increase of approximately 195,000 newborn infants to share the resources of our finite world," it said.

One expert warned that a lack of jobs, rather than too little food, may be the "ultimate threat" facing society as the planet becomes more and more crowded.

It took between two and three million years for the human race to hit the one billion mark in 1850, the PRB said. By 1930, 80 years later, the population stood at 2 billion. A mere 31 years after that, in 1961, it was 3 billion. The growth from 3 to the present 4 billion took just 16 years.

The world could find it has 5 billion people by 1989—just 13 years from now—if population growth continues at the present rate of 1.8 per cent a year, said Dr. Leon F. Bouvier, vice president of the private, nonprofit PRB.

Bouvier said the newly calculated growth rate is a little lower than the 1.9 per cent estimated last year. Thanks to that slowdown, the passing of the 4 billion milestone came a year later than some demographers had predicted.

"I really think the rate of growth is going to start declining ever so slightly because of declining fertility," Bouvier said. "I think there is some evidence of progress—ever so slow, much too slow."

The new PRB figures show there were 3,982,815,000 people on Earth on Jan. 1. By March 1 the number had grown to 3,994,812,000, the organization said, and by April 1 the total will be 4,000,824,000.

The bureau said its calculations are based on estimates of 328,000 live births per day minus 133,000 deaths.

A growing number of governments are taking steps to slow growth rates, the PRB said.

Singapore appears likely to meet the goal of the two-child family "well before the target date of 1980," it said, and several states in India, which yearly adds the equivalent of the population of Australia, are considering financial incentives to birth control and mandatory sterilization after the birth of two children.

Dr. Paul Ehrlich of Stanford University, one of several population experts contacted by PRB, said he was sad to realize at the age of 44 he had lived through a doubling of Earth's population. He expressed fear the next 44 years could see population growth halted "by a horrifying increase in death rates."

"At this point, hunger does not seem the greatest issue presented by the ever growing number of people," said Dr. Louis M. Hellman, chief of population staff at the Health, Education and Welfare Department.

"Rather, the threat appears to lie in the increasing numbers who can find no work. As these masses of unemployed migrate toward the cities, they create a growing impetus toward political unrest and instability."

SOURCE: Courtesy of *The Press Democrat*, Santa Rosa, California, March 28, 1976. Reprinted by permission.

Human populations grow according to the exponential equation

$$P = P_0 e^{rt}$$

where P_0 is the size of the initial population, r is the growth rate, t is the length of time, and P is the size of the population after time t.

EXAMPLE 1 On March 28, 1976, the world population reached 4 billion. If the annual growth rate is 2%, what was the expected population on March 28, 1985?

Solution The initial population P_0 is 4 (billion), $r = 0.02$, and $t = 9$. Then

$$\begin{aligned} P &= 4e^{0.02(9)} \\ &= 4e^{0.18} \\ &\approx 4(1.197) \qquad \text{From Table B.1 or a calculator} \\ &= 4.789 \end{aligned}$$

The expected population was 4.8 billion. ■

EXAMPLE 2 When was the world population expected to reach 6 billion, given a growth rate of 2%?

Solution $r = 0.02$, $P_0 = 4$, $P = 6$

$$\begin{aligned} 6 &= 4e^{0.02t} \\ e^{0.02t} &= 1.5 \\ .02t &= \ln 1.5 \\ t &= 50 \ln 1.5 \qquad \text{Divide both sides by } .02 (\tfrac{1}{.02} = 5.) \\ &\approx 20.2733 \qquad \text{By Table B.1 or a calculator} \end{aligned}$$

This is 20 years, 100 days after March 28, 1976, or July 6, 1996. ■

If you can find the present world population (consult a recent almanac or call your library), you can compare the *predicted* population from Example 1 (using 1976 figures) with the actual figures. Suppose they differ; what conclusion can you make?

If you know the actual population of a city, state, country, or even the world for two dates, you can use the population formula to find the **growth rate** over that period. If factors influencing population growth do not change significantly, this growth rate can be used to predict future population growth. The question asked at the end of the preceding paragraph would be answered by saying that the *actual* growth rate was not the 2% assumed in Example 1.

EXAMPLE 3 The 1970 population of San Antonio, Texas, was 654,153 and the 1980 population was 783,296. What was the growth rate of San Antonio for this period?

Solution Since $P_0 = 654{,}153$, $P = 783{,}296$, and $t = 10$, we have

$$\begin{aligned} P &= P_0 e^{rt} \\ \frac{P}{P_0} &= e^{rt} \\ rt &= \ln \frac{P}{P_0} \\ r &= \frac{1}{t} \ln \frac{P}{P_0} \end{aligned}$$

For this example,

$$r = \frac{1}{10} \ln \frac{783{,}296}{654{,}153}$$

[783296] [÷] [654153] [=] [ln x] [÷] [10] [=] DISPLAY: .0180169389

RPN logic:

[783296] [ENTER] [654153] [÷] [ln] [10] [÷]

The growth rate is about 1.8%. ■

Extended Application Problems—Population Growth

1. The world population was 1 billion in 1850 and it took 80 years to reach the 2 billion mark. What was the annual growth rate for this period?
2. The world population was 2 billion in 1930 and it took 31 years to reach 3 billion. What was the annual growth rate for this period?
3. The world population was 3 billion in 1961 and it took 16 years to reach 4 billion. What was the annual growth rate for this period?
4. On March 28, 1976, the world population reached 4 billion. If the annual growth rate is 1.8%, what was the expected population on March 28, 1986?
5. On March 28, 1976, the world population reached 4 billion. If the annual growth rate is 1.8%, when would you expect the world population to reach 5 billion?
6. How long will it take the population to double from 4 to 8 billion if the growth rate is 1.8%?
7. According to the news article, the current growth rate is 1.8%. Is the article correct when it predicts 5 billion people in 13 years?
8. **a.** If San Jose, California, grew from 459,913 in 1970 to 625,763 in 1980, what was the growth rate for this period?
   b. If you use the growth rate found in part *a*, what was the expected population in 1990?
9. **a.** If the population of Boston declined from 641,071 in 1970 to 562,118 in 1980, what was the growth rate for this period? (*Note:* A negative growth rate means the population declined.)
   b. If you use the growth rate found in part *a*, what was the expected population in 1990?
10. **a.** If the population of Denver declined from 514,678 in 1970 to 488,765 in 1980, what was the growth rate for this period? (*Note:* A negative growth rate means the population declined.)
    b. If you use the growth rate found in part *a*, what was the expected population in 1990?
11. ***Research Problem*** From your local Chamber of Commerce, obtain the population figures for your city for 1960, 1970, and 1980.
    a. Find the rate of growth for each period.
    b. Forecast the population of your city for the year 2000. Which of the rates you obtained in part *a* is the most accurate for this purpose?

World Consumption of Petroleum Products

Year	Millions of barrels
1915	43
1920	68
1925	110
1930	150
1935	170
1940	220
1945	260
1950	340
1955	480
1960	780
1965	1130
1970	1670
1975	1890
1980	2400

12. ***Research Problem*** Use the table of oil consumption given here for this problem.

a. Establish that the use of petroleum has grown exponentially.

b. Estimate the rate of growth of the worldwide use of petroleum.

c. Forecast figures for 1985, 1990, 1995, and 2000.

d. Verify forecast figures for years for which data are available.

e. How do these figures compare to estimated world petroleum reserves? Graph use and remaining reserves on one grid. What are your conclusions?

13. List some factors, such as new zoning laws, that could change the growth rate of your city.

14. List some factors, such as a change in the tax laws of your state, that could change the growth rate of your state.

15. List some factors, such as war, that could change the growth rate of a country or the world.

5

Trigonometric Functions

Hipparchus
(about 180–125 B.C.)

Trigonometry contains the science of continually undulating magnitude: meaning magnitude which becomes alternately greater or less, without any termination to succession of increase or decrease ....

AUGUSTUS DEMORGAN
Trigonometry and Double Algebra

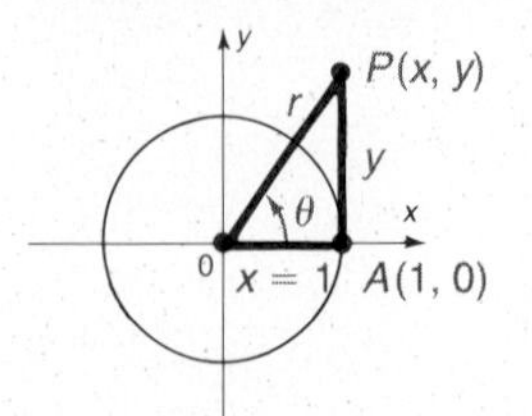

During the second half of the second century B.C., the astronomer Hipparchus of Nicaea compiled the first trigonometric tables in 12 books. It was he who used the 360° of the Babylonians and thus introduced trigonometry with an angle measure that we still use today. Hipparchus' work formed the foundation for Ptolemy's *Mathematical Syntaxis*, the most significant early work in trigonometry. Ptolemy acknowledged the earlier contributions of Hipparchus, whom he described as "a labor-loving and truth-loving man." It should be pointed out that these early works do not use trigonometric ratios or unit circles; rather they use trigonometric lines that take the form of chords of a circle. For a discussion of this method, see Carl Boyer's *A History of Mathematics*.

The Greek discoveries were later lost in Europe. Fortunately, they had been translated into Arabic, and the Arabs had taken them as far as India, which accounts for the interesting origin of the term *sine*. The Hindu mathematician Aryabhata (about 476–550) called it *jyā-adhā* (chord half) and abbreviated it *jyā*, which the Arabs wrote as *jiba* but shortened to *jb*. Later writers erroneously interpreted *jb* to stand for *jaib*, which means cove or bay. When Europeans rediscovered trigonometry through the Arab tradition, they translated the texts into Latin. The Latin equivalent for *jaib* is *sinus*, from which our present word *sine* is derived. Several of our mathematical terms, including *algebra*, are based on misunderstandings of Arabic words—but in this case the Arabs themselves were confused.

CHAPTER OVERVIEW This chapter introduces you to six very important functions in mathematics, called the trigonometric functions. You will need to be very familiar with these functions in order to continue with your studies in mathematics. The 25 objectives of this chapter are listed on pages 215–217.

5.1 ANGLES AND THE UNIT CIRCLE

The circle is of primary importance in the study of trigonometry. Although you are no doubt familiar with a circle, it is worthwhile to review circles briefly at this time. A **circle** is the set of points a given distance, called the **radius**, from a given point called the **center**. Since a circle is not the graph of a function, we will delay discussion of the general equation and properties of a circle until Chapter 10. We do, however, have need for the equation of a circle with radius r and center at the origin. The equation of this circle is derived by using the Distance Formula. Let (x, y) be any point on a circle of radius r. Then

$$(x_2 - x_1)^2 + (y_2 - y_1)^2 = d^2$$ This is the Distance Formula with both sides squared.

$$(x - 0)^2 + (y - 0)^2 = r^2$$ The center is $(0, 0)$, so let $(x_1, y_1) = (0, 0)$; let $(x_2, y_2) = (x, y)$ and $d = r$.

$$x^2 + y^2 = r^2$$

If $r = 1$, this is called the *unit circle.*

Unit Circle

The **unit circle** is the circle with radius 1 and center at the origin. The equation of the unit circle is

$$x^2 + y^2 = 1$$

In mathematics it is often useful to consider functions of angles; however, the definition of an angle depends on the context in which it is being used. In geometry, an angle is usually defined as the union of two rays with a common endpoint. In advanced mathematics courses, a more general definition is used.

Angle

An **angle** is formed by rotating a ray about its endpoint (called the **vertex**) from some initial position (called the **initial side**) to some terminal position (called the **terminal side**). The measure of an angle is the amount of rotation. An angle is also formed if a line segment is rotated about one of its endpoints.

If the rotation of the ray is in a counterclockwise direction, the measure of the angle is called **positive**. If the rotation is in a clockwise direction, the measure is called **negative**. The notation $\angle ABC$ means the measure of an angle with vertex B and points A and C (different from B) on the sides; $\angle B$ denotes the measure of an angle with vertex at B, and a curved arrow is used to denote the direction and amount of rotation, as shown in Figure 5.1. If no arrow is shown, the measure of the

Commonly used Greek letters

Symbol	Name
α	alpha
β	beta
γ	gamma
δ	delta
θ	theta
λ	lambda
ϕ or φ	phi
ω	omega

π (pi) is a lowercase Greek letter that will not be used to represent an angle. It denotes an irrational number approximately equal to 3.141592654.

angle is considered to be the smallest positive rotation. Lowercase Greek letters are also used to denote the angles as well as the measure of angles. For example, θ may represent the angle or the measure of the angle called θ; you will know which is meant by the context in which it is used. Some examples are shown in Figure 5.1.

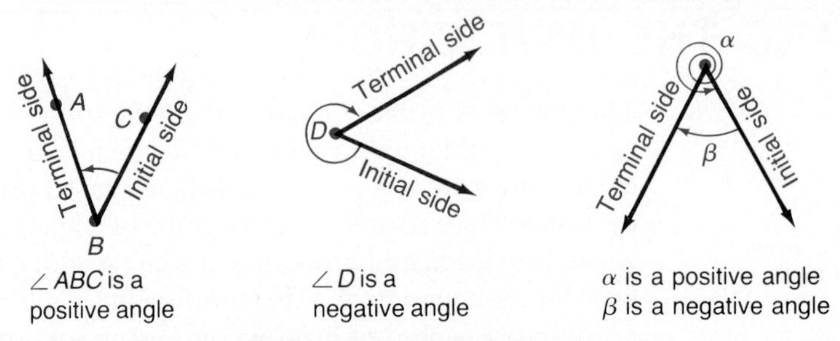

Figure 5.1 Examples of angles

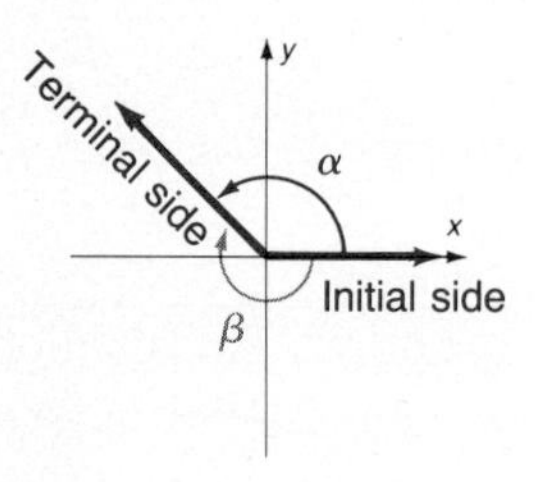

Figure 5.2 Standard-position angles α and β; α is a positive angle and β is a negative angle; α and β are coterminal angles.

A Cartesian coordinate system may be superimposed on an angle so that the vertex is at the origin and the initial side is along the positive x axis. In this case the angle is in **standard position**. Angles in standard position having the same terminal sides are **coterminal angles**. Given any angle α, there is an unlimited number of coterminal angles (some positive and some negative). In Figure 5.2, β is coterminal with α. Can you find other angles coterminal with α?

Several units of measurement are used for measuring angles. Let α be an angle in standard position with a point P not the vertex but on the terminal side. As this side is rotated through one revolution, the trace of the point P forms a circle. The measure of the angle is one revolution, but since much of our work will be with amounts less than one revolution, we need to define measures of smaller angles. Historically the most common scheme divides one revolution into 360 equal parts with each part called a **degree**. Sometimes even finer divisions are necessary, so a degree is divided into 60 equal parts, each called a **minute** ($1^\circ = 60'$). Furthermore, a minute is divided into 60 equal parts, each called a **second** ($1' = 60''$). For most applications, we will write decimal parts of degrees instead of minutes and seconds. That is, 32.5° is preferred over $32^\circ\,30'$.

In calculus and scientific work, another measure for angles is defined. This method uses real numbers to measure angles. Draw a circle with any nonzero radius r. Next measure out an arc with length r. Figure 5.3a shows the case in which $r = 1$

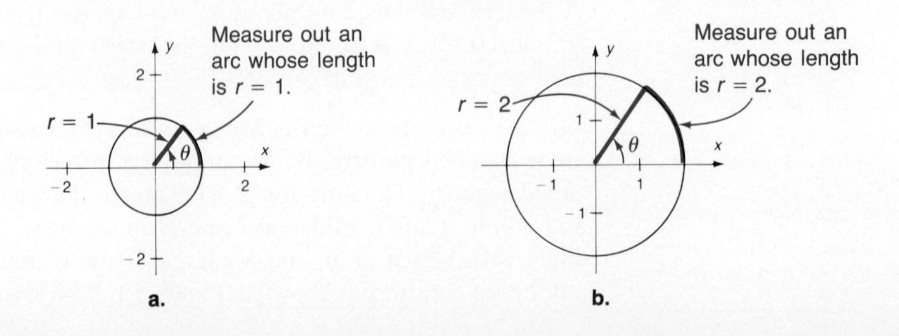

Figure 5.3 Radian measure

and Figure 5.3b shows $r = 2$. Regardless of your choice for r, the angle determined by this arc of length r is the same. (It is labeled θ in Figure 5.3.)

This angle is used as a basic unit of measurement and is called a **radian**. Notice that the circumference C generates an angle of one revolution. Since $C = 2\pi r$, and since the basic unit of measurement on the circle is r,

$$\begin{aligned}\text{One revolution} &= \frac{C}{r}\\ &= \frac{2\pi r}{r}\\ &= 2\pi\end{aligned}$$

Thus $\frac{1}{2}$ revolution is $\frac{1}{2}(2\pi) = \pi$ radians; $\frac{1}{4}$ revolution is $\frac{1}{4}(2\pi) = \frac{\pi}{2}$ radians.

Notice that when measuring angles in radians, you are using *real numbers*. Because radian measure is used so frequently, we agree that **radian measure is understood** when **no units of measure** for an angle are indicated. Figure 5.4 shows a protractor for measuring angles using radian measure.

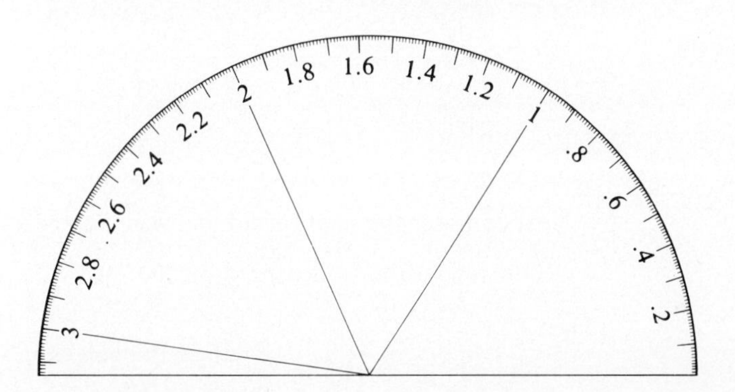

Figure 5.4 Protractor for radian measure

Example 1 asks you to draw several angles using radian measure. Do not try to change these angles to degree measure. Work them directly as shown in the examples. You will have to work at thinking in terms of radian measure. You should memorize the approximate size of an angle of measure 1 radian in much the same way you have memorized the approximate size of an angle of measure 45°.

EXAMPLE 1 Let $r = 1$ and draw the angles with the given measures.

a. $\theta = 2$
(2 radians is understood)

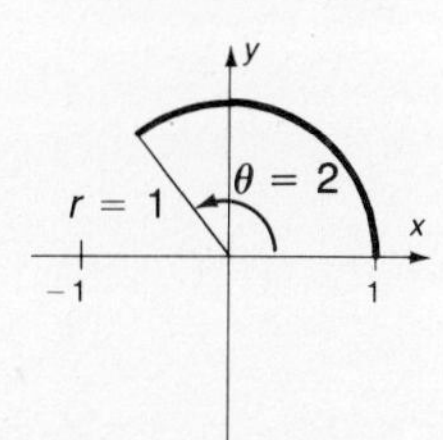

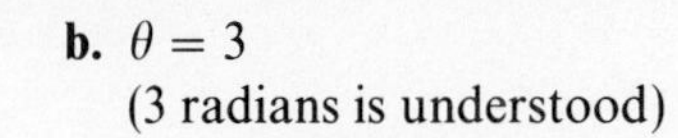

b. $\theta = 3$
(3 radians is understood)

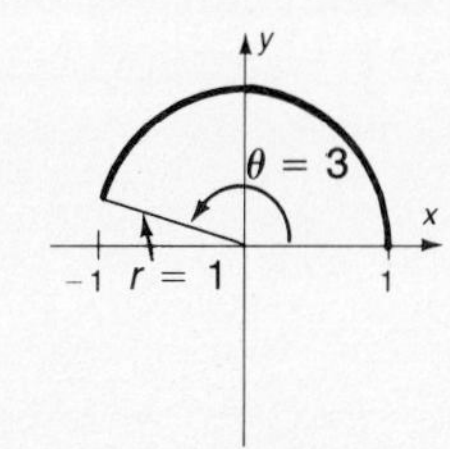

c. $\theta = \sqrt{19}$
$\sqrt{19} \approx 4.36$; since a straight angle is π, use the radian protractor for $\sqrt{19} - \pi \approx 1.22$.

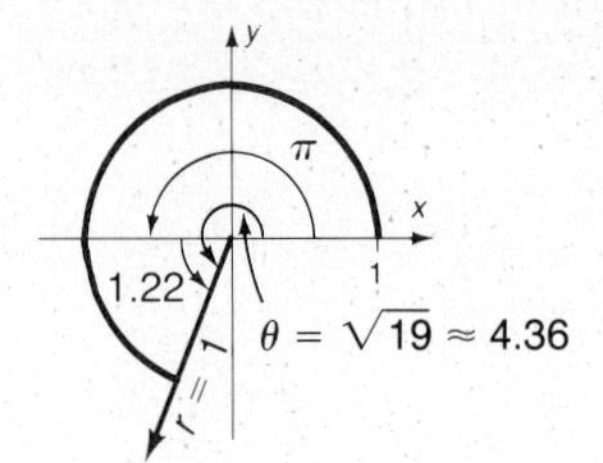

d. $\theta = \sqrt{50}$
$\sqrt{50} \approx 7.07$; since one revolution is 2π, use the radian protractor for $\sqrt{50} - 2\pi \approx 0.79$.

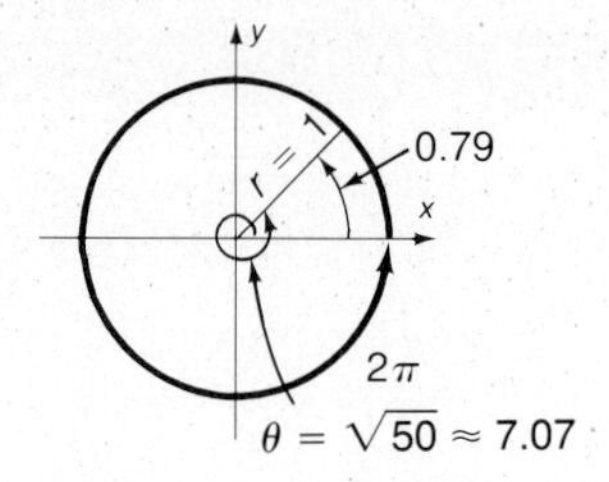

e. $\theta = 0.5$

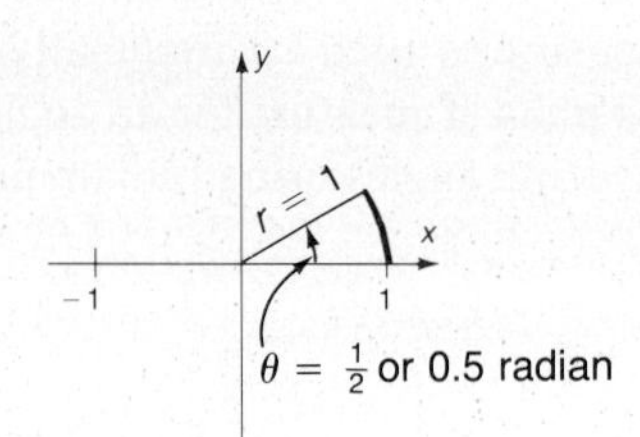

f. $\theta = -\dfrac{\pi}{4}$

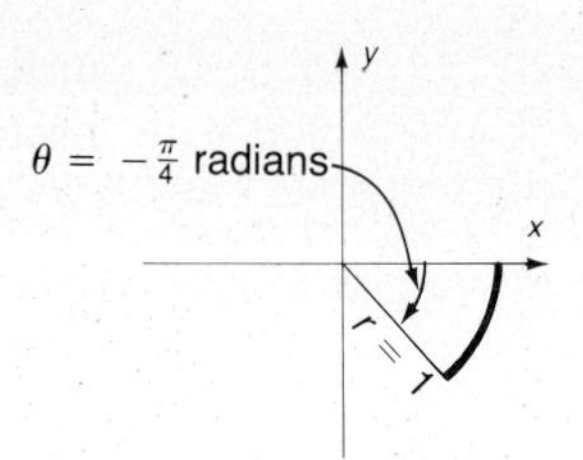

■

Next consider the relationship between degree and radian measure:

One revolution is measured by 360° or by 2π radians

Then

$$\text{Number of revolutions} = \frac{\text{angle in degrees}}{360}$$

$$\text{Number of revolutions} = \frac{\text{angle in radians}}{2\pi}$$

Therefore:

Relationship Between Degree and Radian Measure

$$\frac{\text{angle in degrees}}{360} = \frac{\text{angle in radians}}{2\pi}$$

EXAMPLE 2 **a.** Change 45° to radians.

$$\frac{45}{360} = \frac{\theta}{2\pi}$$

$$\frac{90\pi}{360} = \theta$$

$$\frac{\pi}{4} = \theta$$

An alternative method is to remember that π radians is 180°. That is, since 45° is $\frac{1}{4}$ of 180°, you know that the radian measure is $\frac{\pi}{4}$. As a decimal, θ can be approximated by calculator:

$$\boldsymbol{\theta \approx 0.78539816}$$

b. Change 2.30 to degrees.

$$\begin{aligned}\theta &= \frac{180}{\pi}(2.30)\\ &\approx (57.296)(2.3)\\ &= 131.7808\end{aligned}$$

To the nearest hundredth, the angle is 131.78°. If you have a calculator, you can obtain a much more accurate answer by using a better approximation for $\frac{180}{\pi}$. For example, if 2.3 is exact, then by calculator:

Algebraic logic: [180] [÷] [π] [×] [2.3] [=]

RPN logic: [180] [ENTER] [π] [÷] [2.3] [×]

The result is approximately **131.7802929°**.

c. Change 1° to radians.

$$\frac{1}{360} = \frac{\theta}{2\pi}$$

$$\frac{\pi}{180} = \theta$$

Even though this solution is in the desired form, you might be interested in performing the division on a calculator:

1° ≈ 0.0174532925 radian

d. Change 1 to degrees.

$$\frac{\theta}{360} = \frac{1}{2\pi}$$

$$\theta = \frac{180}{\pi}$$

By calculator:

$\theta \approx$ 57.29577951° or 57°17′45″

Decimal degrees is the preferred form. ■

For the more common measures of angles, it is a good idea to memorize the equivalent degree and radian measures. If you keep in mind that 180° in radian measure is π, the rest of the values will be easy to remember.

Commonly Used Degree and Radian Measures

Degrees	*Radians*
$0°$	0
$30°$	$\frac{\pi}{6}$
$45°$	$\frac{\pi}{4}$
$60°$	$\frac{\pi}{3}$
$90°$	$\frac{\pi}{2}$
$180°$	π
$270°$	$\frac{3\pi}{2}$
$360°$	2π

We now relate the radian measure of an angle to a circle to find the arc length. An **arc** is part of a circle; thus **arc length** is the distance around part of a circle. The arc length corresponding to one revolution is the **circumference** of the circle. Let s be the length of an arc and let θ be the angle measured in radians. Then

$$\text{Angle in revolutions} = \frac{s}{2\pi r}$$

since one revolution has an arc length (circumference of the circle) of $2\pi r$. Also,

$$\text{Angle in radians} = (\text{angle in revolutions})(2\pi)$$

Substituting,

$$\theta = \frac{s}{2\pi r}(2\pi)$$

$$= \frac{s}{r}$$

From this result, we derive the following formula.

Arc Length Formula

The **arc length** cut by a central angle θ (measured in radians) from a circle of radius r is denoted by s and is found by

$$s = r\theta$$

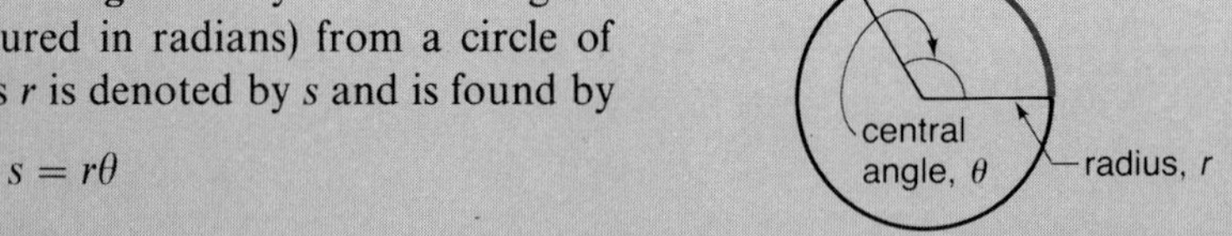

EXAMPLE 3 The length of the arc subtended (cut off) by a central angle of 36° in a circle with a radius of 20 centimeters (cm) is found as follows. First change 36° to radians so that you can use the formula given above.

$$\frac{36}{360} = \frac{\theta}{2\pi}$$

Solving for θ,

$$\frac{\pi}{5} = \theta$$

Thus

$$s = 20\left(\frac{\pi}{5}\right)$$
$$= 4\pi$$

The length of the arc is **4π cm**. This is about 12.6 cm. ■

If the terminal side of an angle coincides with a coordinate axis, the angle is called a **quadrantal angle**. If the angle θ is not a quadrantal angle then we refer to its *reference angle*, which is denoted by θ' throughout this book.

Reference Angle

> Given an angle θ in standard position the **reference angle** θ' is defined as the acute angle the terminal side makes with the x axis.

The procedure for finding the reference angle depends on the quadrant of θ. One example for each quadrant is shown in Figure 5.5.

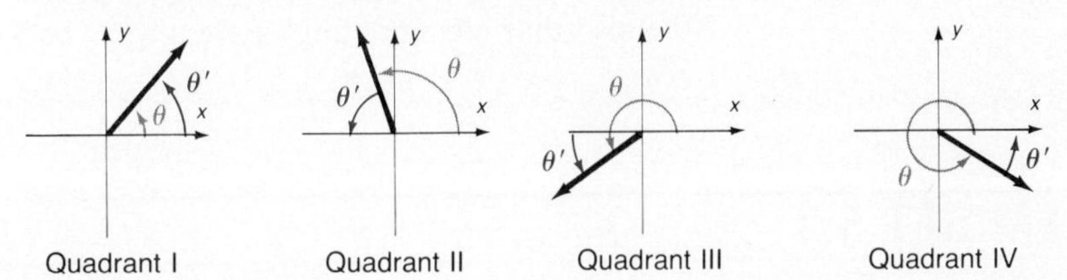

Figure 5.5 Reference angles

EXAMPLE 4 Find the reference angle and draw both the given angle and the reference angle.

a. 210°

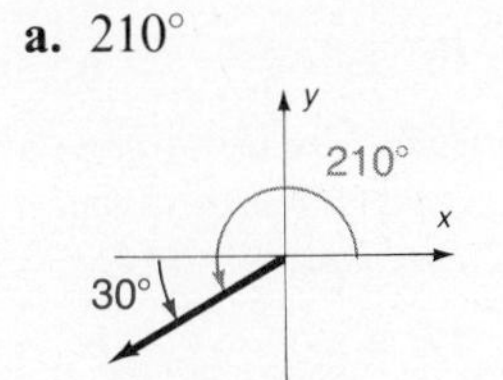

$210° - 180° = 30°$
Reference angle is **30°**.

b. 150°

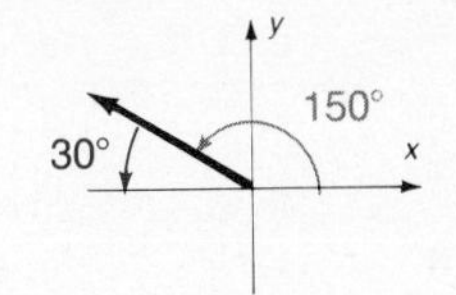

$180° - 150° = 30°$
Reference angle is **30°**.

c. $-\dfrac{5\pi}{3}$

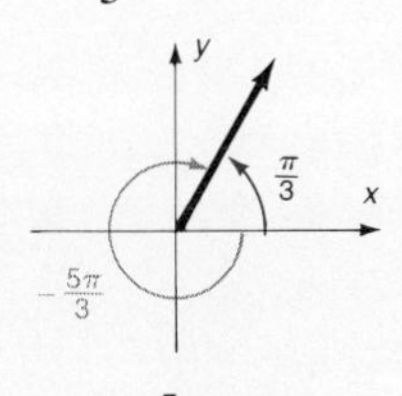

$$2\pi - \frac{5\pi}{3} = \frac{\pi}{3}$$
Reference angle is $\frac{\pi}{3}$.

d. 2.5

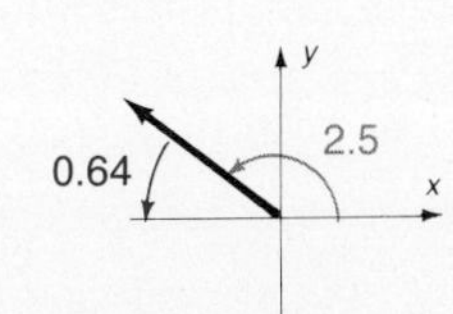

$\pi - 2.5 \approx 0.64$
Reference angle is approximately **0.64**.

e. 812°. If the angle is more than one revolution, first find a nonnegative coterminal angle that is less than one revolution. 812° is coterminal with 92°.

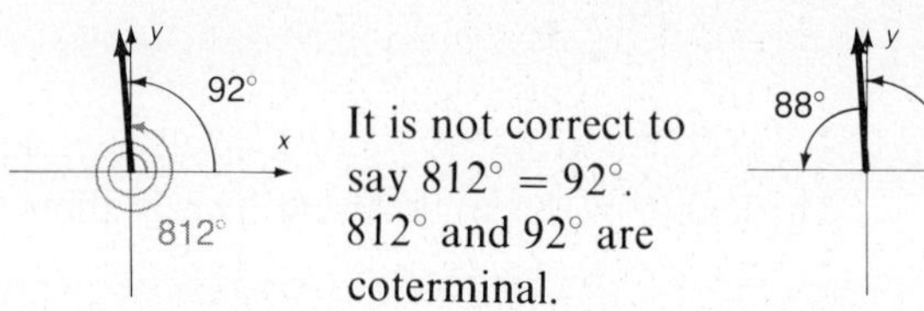

It is not correct to say $812° = 92°$. 812° and 92° are coterminal.

$180° - 92° = 88°$
Reference angle is **88°**.

f. The angle whose measure is 30. 30 is coterminal with **4.867** ($30 - 8\pi \approx 4.867$).

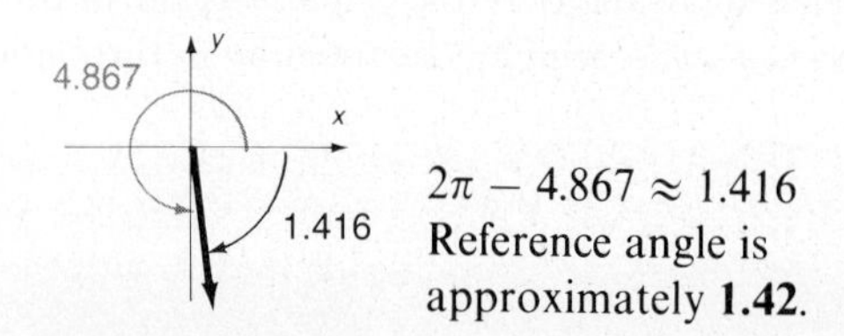

$2\pi - 4.867 \approx 1.416$
Reference angle is approximately **1.42**.

The purpose of this example is to remind you that $30 \neq 30°$. ■

Keep in mind the differences between coterminal angles and reference angles. Note also that reference angles are always between 0 and 90° or 0 and $\frac{\pi}{2}$ (about 1.57).

PROBLEM SET 5.1

A

From memory, give the radian measure for each of the angles whose degree measure is stated, and the degree measure for each of the angles whose radian measure is stated, in Problems 1–4.

1. a. 30° **b.** 90° **c.** 270° **d.** 45°

2. a. 360° **b.** 60° **c.** 180° **d.** 0°

3. a. π **b.** $\frac{\pi}{4}$ **c.** $\frac{\pi}{3}$ **d.** 2π

4. a. $\frac{\pi}{2}$ **b.** 0 **c.** $\frac{3\pi}{2}$ **d.** $\frac{\pi}{6}$

Use the radian protractor in Figure 5.4 to help you sketch each of the angles in Problems 5–10.

5. a. $\frac{\pi}{2}$ **b.** $\frac{\pi}{6}$ **c.** -2.5 **d.** $\sqrt{17}$

6. a. $\frac{2\pi}{3}$ **b.** $\frac{3\pi}{4}$ **c.** -1 **d.** $\sqrt{95}$

7. a. $-\frac{\pi}{4}$ **b.** $\frac{7\pi}{6}$ **c.** -2.76 **d.** $\sqrt{115}$

8. a. $-\frac{3\pi}{2}$ **b.** $\frac{13\pi}{3}$ **c.** -1.2365 **d.** $\sqrt{23}$

9. a. $\frac{\pi}{15}$ **b.** $-\frac{9\pi}{7}$ **c.** -4 **d.** $\sqrt{10}$

10. a. $-\frac{5\pi}{6}$ **b.** $\frac{5\pi}{4}$ **c.** -3 **d.** $\sqrt{19}$

Find the exact value of a positive angle less than one revolution that is coterminal with each of the angles in Problems 11–16.

11. a. 400° **b.** 540° **c.** 750° **d.** 1050°

12. a. $-30°$ **b.** $-200°$ **c.** $-55°$ **d.** $-320°$

13. a. $-120°$ **b.** 500° **c.** $-180°$ **d.** 1000°

14. a. 3π **b.** $\frac{13\pi}{6}$ **c.** $-\pi$ **d.** 7

15. a. $-\frac{\pi}{4}$ **b.** $\frac{17\pi}{4}$ **c.** $\frac{11\pi}{3}$ **d.** -2

16. a. $-\frac{\pi}{6}$ **b.** $-\frac{5\pi}{4}$ **c.** $\frac{15\pi}{6}$ **d.** 8

Find a positive angle less than one revolution correct to four decimal places so that it is coterminal with each of the angles in Problems 17–19.

17. a. 9 **b.** -5 **c.** $\sqrt{50}$ **d.** -6

18. a. 6.2832 **b.** -3.1416 **c.** 30 **d.** $3\sqrt{5}$

19. a. 6.8068 **b.** -0.7854 **c.** 150 **d.** 9.4247

Find the reference angle for the angles given in Problems 20–27. Use the unit of measurement (degrees or radians) given in the problem.

20. a. 150° **b.** 210° **c.** 240° **d.** 330°

21. a. 60° **b.** 120° **c.** 300° **d.** 135°

22. a. $\frac{5\pi}{3}$ **b.** $\frac{7\pi}{6}$ **c.** $\frac{4\pi}{3}$ **d.** $\frac{5\pi}{4}$

23. a. $\frac{11\pi}{12}$ **b.** $\frac{2\pi}{3}$ **c.** $\frac{11\pi}{6}$ **d.** $\frac{\pi}{4}$

24. a. $-30°$ **b.** $-200°$ **c.** $-55°$ **d.** $-320°$

25. a. $-\frac{\pi}{4}$ **b.** $-\pi$ **c.** $-\frac{13\pi}{6}$ **d.** $-\frac{5\pi}{3}$

26. a. 7 **b.** 9 **c.** -5 **d.** -6

27. a. $\sqrt{50}$ **b.** $3\sqrt{5}$ **c.** 6.8068 **d.** -0.7854

B

Change the angles in Problems 28–39 to decimal degrees correct to the nearest hundredth degree.

28. $\frac{2\pi}{9}$ **29.** $\frac{\pi}{10}$ **30.** $\frac{\pi}{30}$ **31.** $\frac{5\pi}{3}$

32. $-\frac{11\pi}{12}$ **33.** $\frac{3\pi}{18}$ **34.** 2 **35.** -3

36. -0.25 **37.** -2.5 **38.** 0.4 **39.** 0.51

Change the angles in Problems 40–45 to radians using exact values.

40. $40°$ **41.** $20°$ **42.** $-64°$ **43.** $-220°$

44. $254°$ **45.** $85°$

Change the angles in Problems 46–51 to radians correct to the nearest hundredth.

46. $112°$ **47.** $314°$ **48.** $-62.8°$ **49.** $350°$

50. $-480°$ **51.** $985°$

In Problems 52–59, find the intercepted arc to the nearest hundredth if the central angle and radius are as given.

52. Angle 1, radius 1 m **53.** Angle 2.34, radius 6 cm

54. Angle 3.14, radius 10 m **55.** Angle $\frac{\pi}{3}$, radius 4 m

56. Angle $\frac{3\pi}{2}$, radius 15 cm **57.** Angle $40°$, radius 7 ft

58. Angle $72°$, radius 10 ft **59.** Angle $112°$, radius 7.2 cm

60. How far does the tip of an hour hand on a clock move in 3 hr if the hour hand is 2.00 cm long?

61. A 50-cm pendulum on a clock swings through an angle of $100°$. How far does the tip travel in one arc?

C

62. ***Surveying*** In about 230 B.C., a mathematician named Eratosthenes estimated the radius of the earth using the following information: Syene and Alexandria in Egypt are on the same line of longitude. They are also 800 km apart. At noon on the longest day of the year, when the sun was directly overhead in Syene, Eratosthenes measured the sun to be $7.2°$ from the vertical in Alexandria. Because of the distance of the earth from the sun, he assumed that the rays were parallel. Thus he concluded that the arc from Syene to Alexandria is subtended by a central angle of $7.2°$ measured at the center of the earth. Using this information, find the approximate radius of the earth.

63. ***Geography*** Omaha, Nebraska, is located at approximately $97°$ west longitude, $41°$ north latitude; Wichita, Kansas, is located at approximately $97°$ west longitude, $37°$ north latitude. Notice that these two cities have about the same longitude. If we know that the radius of the earth is about 6370 kilometers (km), what is the distance between these cities to the nearest 10 km?

64. ***Geography*** Entebbe, Uganda, is located at approximately $33°$ east longitude, and Stanley Falls in Zaire is located at $25°$ east longitude. Both these cities lie approximately on the equator. If we know that the radius of the earth is about 6370 km, what is the distance between the cities to the nearest 10 km?

65. ***Astronomy*** Suppose it is known that the moon subtends an angle of $45.75'$ at the center of the earth (see Figure 5.6). It is also known that the center of the moon is 384,417 km from the surface of the earth. What is the diameter of the moon to the nearest 10 km?

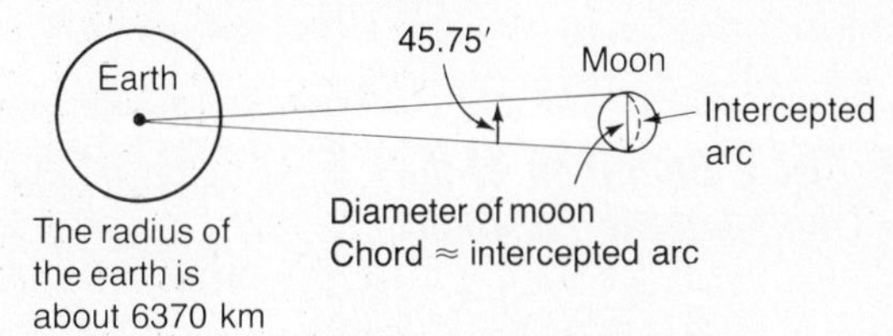

For small central angles with large radii, the intercepted arc is approximately equal to its chord.

Figure 5.6

66. One side of a triangle is 20 cm longer than another, and the angle between them is $60°$. If two circles are drawn with these sides as diameters, one of the points of intersection of the two circles is the common vertex. How far from the third side is the other point of intersection?

5.2 TRIGONOMETRIC FUNCTIONS

To introduce you to the trigonometric functions, we will consider a relationship between angles and circles. Draw a unit circle with an angle θ in standard position, as in Figure 5.7.

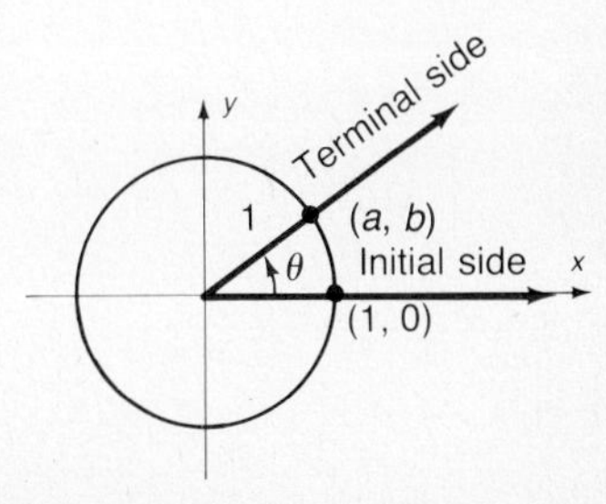

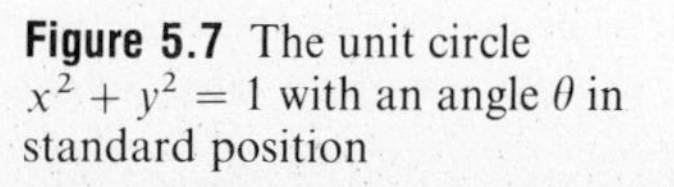

Figure 5.7 The unit circle $x^2 + y^2 = 1$ with an angle θ in standard position

The initial side of θ intersects the unit circle at $(1, 0)$, and the terminal side intersects the unit circle at (a, b). Define functions of θ as follows:

$$c(\theta) = a \quad \text{and} \quad s(\theta) = b$$

EXAMPLE 1 Find **a.** $s(90°)$ **b.** $c(90°)$ **c.** $s(-270°)$ **d.** $s(-\frac{\pi}{2})$ **e.** $c(-3.1416)$

Solution **a.** $s(90°)$. This is the second component of the ordered pair (a, b), where (a, b) is the point of intersection of the terminal side of a 90° standard-position angle and the unit circle. By inspection, it is 1. Thus, $s(90°) = 1$.

b. $c(90°) = 0$.

c. $s(-270°) = 1$. This is the same as part *a* because the angles 90° and −270° are coterminal.

d. $s(-\frac{\pi}{2}) = -1$.

e. $c(-3.1416) \approx c(-\pi) = -1$. ■

The function $c(\theta)$ is called the **cosine function**, and the function $s(\theta)$ is called the **sine function**. These functions, along with four others, make up the **trigonometric functions** or **trigonometric ratios**, which are further examples of *transcendental functions*. One way to define the trigonometric functions is as follows.

Unit Circle Definition of the Trigonometric Functions

Let θ be an angle in standard position with the point (a, b) the intersection of the terminal side of θ and the unit circle. Then the six trigonometric functions, with their standard abbreviations, are defined as follows:

cosine:	$\cos\theta = a$	**secant:**	$\sec\theta = \frac{1}{a} \quad (a \neq 0)$
sine:	$\sin\theta = b$	**cosecant:**	$\csc\theta = \frac{1}{b} \quad (b \neq 0)$
tangent:	$\tan\theta = \frac{b}{a} \quad (a \neq 0)$	**cotangent:**	$\cot\theta = \frac{a}{b} \quad (b \neq 0)$

Notice the condition on the tangent, secant, cosecant, and cotangent functions. These exclude division by 0; for example, $a \neq 0$ means that θ cannot be 90°, 270°, or any angle coterminal to these angles. We summarize this condition by saying that the *tangent and secant are not defined for 90° or 270°*. If $b \neq 0$, then $\theta \neq 0°$, 180°, or any angle coterminal to these angles; thus, *cosecant and cotangent are not defined for 0° or 180°*.

In many applications, you will know the angle measure and will want to find one or more of its trigonometric functions. To do this you will carry out a process called **evaluation of the trigonometric functions**. In order to help you see the relationship between the angle and the function, consider Figure 5.8 and Example 2.

EXAMPLE 2 Evaluate the trigonometric functions of $\theta = 110°$ by drawing the unit circle and approximating the point (a, b).

Solution Draw a unit circle as shown in Figure 5.8.

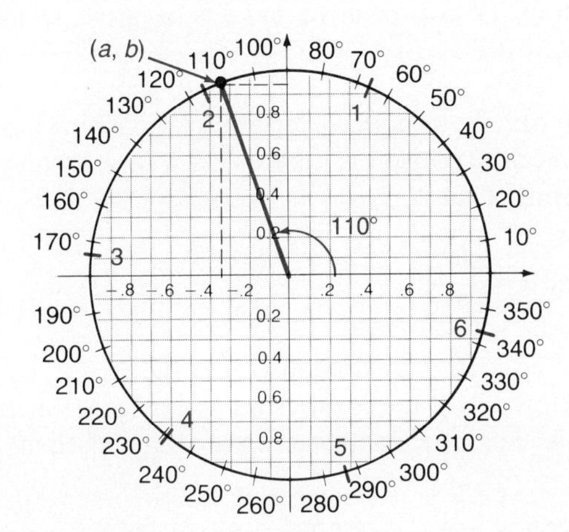

Figure 5.8 Approximate values of circular functions

Next draw the terminal side of the angle 110°. Estimate $a \approx -0.35$ and $b \approx 0.95$. Thus

$$\cos 110° \approx -0.35 \qquad \sec 110° \approx \frac{1}{-0.35} \approx -2.9$$

$$\sin 110° \approx 0.95 \qquad \csc 110° \approx \frac{1}{0.95} \approx 1.1$$

$$\tan 110° \approx \frac{0.95}{-0.35} \approx -2.7 \qquad \cot 110° \approx \frac{-0.35}{0.95} \approx -0.37$$

■

Notice from Example 2 that some of the trigonometric functions of 110° are positive and others are negative. We can make certain predictions about the signs of these functions, as summarized in Table 5.1.

TABLE 5.1
Signs of trigonometric functions

Table 5.1 can be summarized by remembering the following:

Sine positive	All positive
Tangent positive	Cosine positive

	Quadrant I *a* pos *b* pos	*Quadrant II* *a* neg *b* pos	*Quadrant III* *a* neg *b* neg	*Quadrant IV* *a* pos *b* neg
$\cos\theta = a$	pos	neg	neg	pos
$\sin\theta = b$	pos	pos	neg	neg
$\tan\theta = b/a$	pos	neg	pos	neg
Summary	**All pos**	**Sine pos**	**Tangent cos**	**Cosine pos**

Proof In Quadrant I, a and b are both positive, so all six trigonometric functions must be positive. In Quadrant II, a is negative and b is positive, so from the definition of trigonometric functions, all are negative except the sine and cosecant. In Quadrant III, a and b are both negative, so all the functions are negative except tangent and cotangent because those are ratios of two negatives, which are positive. In Quadrant IV, a is positive and b is negative, so all are negative except cosine and secant. □

Easy-to-remember form:

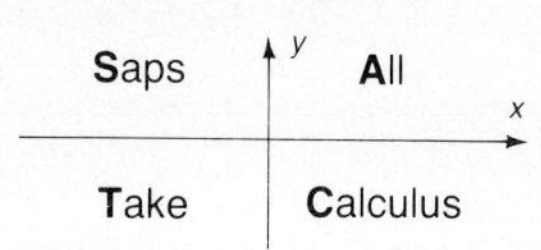

You may have noticed certain relationships among these functions. The first is that since the tangent is the ratio of b to a, it can be found by dividing the sine by the cosine. That is,

$$\tan\theta = \frac{\sin\theta}{\cos\theta}$$

whenever $\cos\theta \neq 0$. Note also that the secant, cosecant, and cotangent functions are reciprocals of the cosine, sine, and tangent functions. This means that if values of θ which cause division by zero are excluded, then

$$\sec\theta = \frac{1}{\cos\theta} \qquad \csc\theta = \frac{1}{\sin\theta} \qquad \cot\theta = \frac{1}{\tan\theta}$$

These are called the **reciprocal** relationships, or **identities**. The term *identity* is defined and discussed in the next chapter, so for now we will just remember which pairs of functions are called reciprocals. This can be confusing because the functions are also paired according to their names: sine and cosine, secant and cosecant, and tangent and cotangent. These pairs are called **cofunctions**. Study Table 5.2 until this terminology is clear to you.

TABLE 5.2
Reciprocal and cofunction relationships

Reciprocals	Cofunctions
cosine and secant	cosine and sine
sine and cosecant	cosecant and secant
tangent and cotangent	cotangent and tangent

Since the method shown in Example 2 for evaluating the trigonometric functions is not very practical, many additional procedures for finding them have been discovered. The most common method today, however, is to use a calculator. The procedure is straightforward, but you must note several details:

1. Note the unit of measure used: degree or radian. Calculators have a variety of ways of changing from radian to degree format—so many ways, in fact, that you will need to consult your owner's manual to find out. Most, however, simply have a switch (similar to an on/off switch) that sets the calculator in either degree or radian mode. From now on, we will assume that you are working in the appropriate radian/degree mode. Remember: If later in the course you suddenly

start obtaining strange answers and have no idea what you are doing wrong, double-check to make sure you are using the proper mode.

2. With most calculators you enter the angle first and then press the button corresponding to the trigonometric function.
3. You must remember which functions are reciprocals since a normal scientific calculator does not have sec θ, csc θ, or cot θ keys. It does, however, have a reciprocal key, which is labeled

   [1/x]

EXAMPLE 3 Find the trigonometric functions of 110° by calculator.

Solution

$\cos 110° \approx -0.34202014$	PRESS: [110] [cos]
$\sin 110° \approx 0.93969262$	PRESS: [110] [sin]
$\tan 110° \approx -2.74747742$	PRESS: [110] [tan]
$\sec 110° \approx -2.9238044$	PRESS: [110] [cos] [1/x]
$\csc 110° \approx 1.06417777$	PRESS: [110] [sin] [1/x]
$\cot 110° \approx -0.36397023$	PRESS: [110] [tan] [1/x]

■

EXAMPLE 4 Find $\csc \pi/12$ by calculator.

Solution

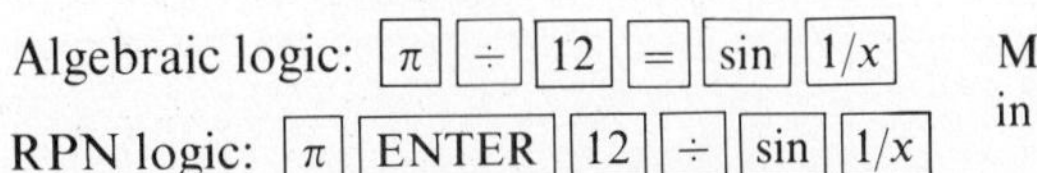

Make sure you are in the radian mode.

The answer now displayed is 3.8637033. ■

EXAMPLE 5 Find tan 70°23′40″.

Solution Switch key to degrees; this measure must first be changed to a decimal degree measure:

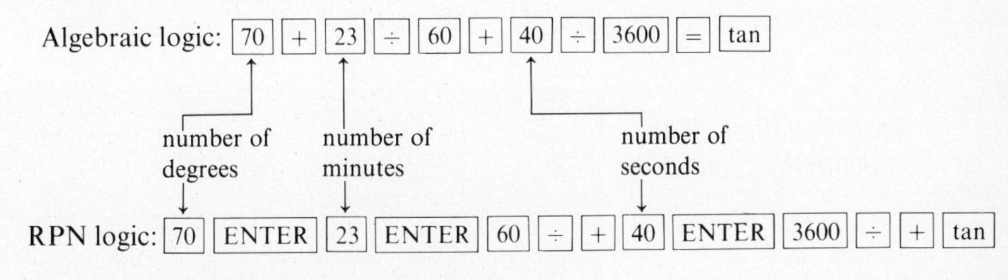

The answer 2.807464819 is displayed. ■

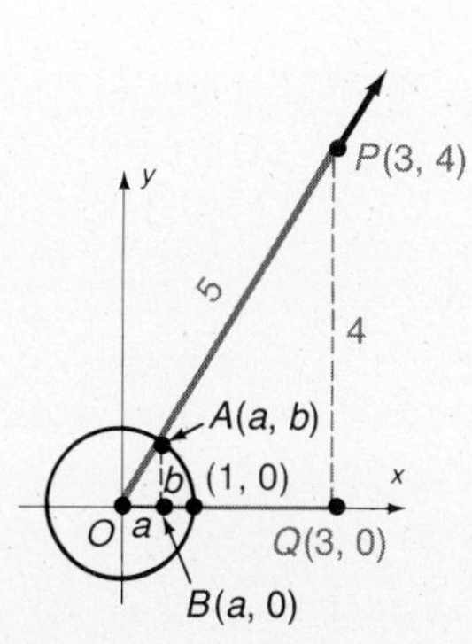

Figure 5.9 Finding (a, b) when a point on the terminal side is known

Suppose you want to find the trigonometric functions of an angle whose terminal side passes through some known point, say (3, 4). To apply the definition, you need to find the point (a, b), as shown in Figure 5.9.

Let (3, 4) be denoted by P and (a, b) by A. Let B be the point $(a, 0)$ and Q be the point (3, 0). ($OA = 1$, $OP = \sqrt{3^2 + 4^2} = 5$). Now consider $\triangle AOB$ and $\triangle POQ$.

Recall from geometry that two triangles are **similar** if two angles of one are congruent to two angles of the other. For these triangles, $\angle OBA$ is congruent to $\angle OQP$ since they are both right angles; also $\angle O$ is congruent to $\angle O$ since equal angles are congruent. Thus these triangles are similar, which is denoted by

$$\triangle AOB \sim \triangle POQ$$

The important property of similar triangles is that corresponding parts of similar triangles are proportional. Thus

$$b = \frac{b}{1} = \frac{4}{5} \qquad \text{and} \qquad a = \frac{a}{1} = \frac{3}{5}$$

Thus $\cos\theta = \frac{3}{5}$, $\sin\theta = \frac{4}{5}$, and $\tan\theta = \frac{4/5}{3/5} = \frac{4}{3}$. The reciprocals are $\sec\theta = \frac{5}{3}$, $\csc\theta = \frac{5}{4}$, and $\cot\theta = \frac{3}{4}$.

If you carry out these steps for a point $P(x, y)$ instead of $(3, 4)$, as shown in Figure 5.10, it is still true that

$$\triangle AOB \sim \triangle POQ$$

Let r be the distance from O to P. That is, let

$$r = \sqrt{x^2 + y^2}$$

Then

$$a = \frac{a}{1} = \frac{x}{r} \qquad \frac{1}{a} = \frac{r}{x}$$

$$b = \frac{b}{1} = \frac{y}{r} \qquad \frac{1}{b} = \frac{r}{y}$$

$$\frac{a}{b} = \frac{x/r}{y/r} = \frac{x}{y} \qquad \frac{b}{a} = \frac{y}{x}$$

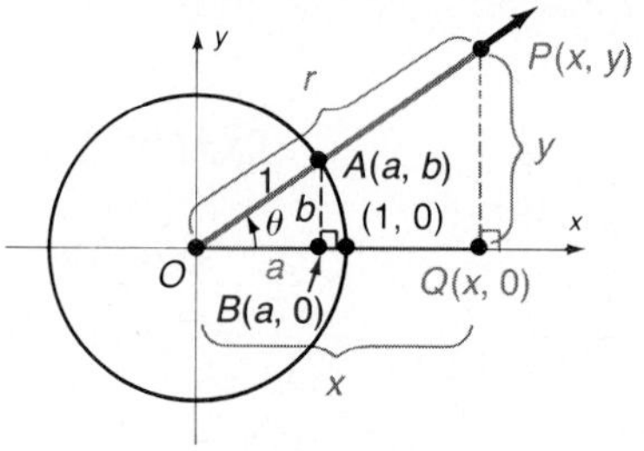

Figure 5.10

These ratios lead to an alternative definition of the trigonometric functions that allows you to choose *any* point (x, y). In practice, it is this definition that is most frequently used.

Ratio Definition of the Trigonometric Functions

Let θ be an angle in standard position with any point $P(x, y)$ on the terminal side a distance r from the origin ($r \neq 0$). Then the six trigonometric functions are defined as follows:

$$\cos\theta = \frac{x}{r} \qquad \sec\theta = \frac{r}{x} \quad (x \neq 0)$$

$$\sin\theta = \frac{y}{r} \qquad \csc\theta = \frac{r}{y} \quad (y \neq 0)$$

$$\tan\theta = \frac{y}{x} \quad (x \neq 0) \qquad \cot\theta = \frac{x}{y} \quad (y \neq 0)$$

EXAMPLE 6 Find the values of the six trigonometric functions for an angle θ in standard position with the terminal side passing through $(-5, -2)$.

Solution $x = -5$, $y = -2$, and $r = \sqrt{25 + 4} = \sqrt{29}$. Thus

$$\cos\theta = \frac{-5}{\sqrt{29}} = \frac{-5}{29}\sqrt{29} \qquad \sec\theta = \frac{-\sqrt{29}}{5}$$

$$\sin\theta = \frac{-2}{\sqrt{29}} = \frac{-2}{29}\sqrt{29} \qquad \csc\theta = \frac{-\sqrt{29}}{2}$$

$$\tan\theta = \frac{-2/\sqrt{29}}{-5/\sqrt{29}} = \frac{2}{5} \qquad \cot\theta = \frac{5}{2}$$

■

Both the unit circle and ratio definitions of the trigonometric functions were given using angle domains. We can extend the definition to include real number domains since radian measure is in terms of real numbers. That is, for

$$\cos\frac{\pi}{2} \qquad \text{or} \qquad \cos 2$$

it does not matter whether $\frac{\pi}{2}$ and 2 are considered as radian measures of angles or simply as real numbers—the functional values are the same.

Trigonometric Functions of Real Numbers

For any real number x, let $x = \theta$, where θ is a standard-position angle measured in radians. Then

$$\cos x = \cos\theta \qquad \sin x = \sin\theta \qquad \tan x = \tan\theta$$

$$\sec x = \sec\theta \qquad \csc x = \csc\theta \qquad \cot x = \cot\theta$$

PROBLEM SET 5.2

A

From the unit circle definition of the trigonometric functions, estimate to one decimal place the numbers in Problems 1–9.

1. $\cos 50°$ **2.** $\sin 20°$ **3.** $\sin 320°$ **4.** $\tan 80°$
5. $\tan(-20°)$ **6.** $\cos(-340°)$ **7.** $\sec 70°$ **8.** $\csc 190°$
9. $\cot 250°$

Use a calculator to evaluate the functions given in Problems 10–27.

10. $\cos 50°$ **11.** $\sin 20°$ **12.** $\tan 80°$ **13.** $\sec 70°$
14. $\csc 150°$ **15.** $\cot 250°$ **16.** $\cos(-34°)$ **17.** $\sin(-95°)$
18. $\tan 56.2°$ **19.** $\cot 78.4°$ **20.** $\sin 1°$ **21.** $\sin 1$
22. $\tan(-1.5)$ **23.** $\cos(-0.48)$ **24.** $\sin(-21.3°)$
25. $\cos(65.21°)$ **26.** $\tan 129°9'12''$ **27.** $\cos 240°8''$

Tell whether each of the functions in Problems 28–39 are positive or negative. You should be able to do this without tables or a calculator.

28. sine, Quadrant I **29.** cosine, Quadrant I
30. tangent, Quadrant II **31.** secant, Quadrant II
32. cosecant, Quadrant III **33.** cotangent, Quadrant IV
34. $\sin 1$ **35.** $\cos 2$ **36.** $\tan 3$
37. $\sec 4$ **38.** $\sin(-1)$ **39.** $\cos(-2)$

Tell in which quadrant(s) a standard-position angle θ could lie if the conditions in Problems 40–45 are true.

40. $\sin\theta > 0$ **41.** $\cos\theta < 0$ **42.** $\tan\theta < 0$
43. $\sin\theta < 0$ and $\tan\theta > 0$ **44.** $\sin\theta > 0$ and $\tan\theta < 0$ **45.** $\cos\theta < 0$ and $\sin\theta < 0$

B

Find the values of the six trigonometric functions for an angle θ in standard position with terminal side passing through the points given in Problems 46–54. Draw a picture showing θ and the reference angle θ'.

46. $(3, 4)$ **47.** $(-3, 4)$
48. $(-3, -4)$ **49.** $(5, 12)$
50. $(-5, -12)$ **51.** $(5, -12)$
52. $(2, -5)$ **53.** $(-6, 1)$
54. $(-4, -5)$

In the next section you will need to simplify some radical expressions. Recall that $\sqrt{x^2} = |x|$. This means that $\sqrt{x^2} = x$ if x is positive and $-x$ if x is negative. Simplify the expressions in Problems 55–63.

55. $\sqrt{2x^2}$ **56.** $\sqrt{9x^2}$
57. $\sqrt{2x^2}$ if x is negative **58.** $\sqrt{2x^2}$ if x is positive
59. $\sqrt{9x^2}$ if x is positive **60.** $\sqrt{9x^2}$ if x is negative
61. $\dfrac{x}{\sqrt{4x^2}}$ if x is positive **62.** $\dfrac{x}{\sqrt{4x^2}}$ if x is negative
63. $\dfrac{\sqrt{16x^2}}{x}$ if x is negative

C

64. a. Let $P(x, y)$ be any point in the plane. Show that $P(r\cos\theta, r\sin\theta)$ is a representation for P, where θ is the standard-position angle formed by drawing ray $\overrightarrow{OP}$.
b. Let $A(\cos\alpha, \sin\alpha)$ and $B(\cos\beta, \sin\beta)$ be any two points on a unit circle. Use the distance formula to show that

$$|AB| = \sqrt{2 - 2(\cos\alpha\cos\beta + \sin\alpha\sin\beta)}$$

65. You will learn in calculus that

$$\sin x = x - \frac{x^3}{3!} + \frac{x^5}{5!} - \frac{x^7}{7!} + \cdots$$

where $n! = n(n-1)(n-2)\cdots 3 \cdot 2 \cdot 1$. Find sin 1 correct to four decimal places by using this equation. (Remember that the 1 in sin 1 refers to radian measure.)

66. You will learn in calculus that

$$\cos x = 1 - \frac{x^2}{2!} + \frac{x^4}{4!} - \frac{x^6}{6!} + \cdots$$

Use this equation to find cos 1 correct to four decimal places.

5.3 VALUES OF THE TRIGONOMETRIC FUNCTIONS

If an angle has a terminal side that coincides with one of the coordinate axes, it is easy to evaluate the trigonometric functions by using the unit circle definition.

EXAMPLE 1 Evaluate the trigonometric functions for $-\frac{5\pi}{2}$.

Solution Since $-\frac{5\pi}{2}$ has a terminal side coinciding with the negative y axis, the intersection of this terminal side and the unit circle is $(0, -1)$. This means that $a = 0$ and $b = -1$. Hence

$$\cos\frac{-5\pi}{2} = a = 0 \qquad \sin\frac{-5\pi}{2} = b = -1$$

$$\csc\frac{-5\pi}{2} = \frac{1}{b} = -1, \qquad \cot\frac{-5\pi}{2} = \frac{a}{b} = 0$$

$\tan\dfrac{-5\pi}{2} = \dfrac{1}{a}$ and $\sec\dfrac{-5\pi}{2} = \dfrac{1}{a}$ are undefined since $a = 0$. ■

There are times when you will not be able to rely on calculator approximations for the trigonometric functions but instead will need to find **exact values**.

EXAMPLE 2 Evaluate the trigonometric functions for $\frac{\pi}{4}$.

Solution

$$\cos\frac{\pi}{4} = \frac{x}{r} \qquad \text{From the ratio definition (this is true for any angle)}$$

$$= \frac{x}{\sqrt{x^2 + y^2}} \qquad r = \sqrt{x^2 + y^2} \text{ for any angle}$$

If $\theta = \frac{\pi}{4}$, then $x = y$ since $\frac{\pi}{4}$ bisects Quadrant I. By substitution,

$$\cos\frac{\pi}{4} = \frac{x}{\sqrt{x^2 + x^2}}$$

$$= \frac{x}{\sqrt{2x^2}}$$

$$= \frac{x}{x\sqrt{2}} \qquad \sqrt{x^2} = |x| = x \text{ since } x \text{ is positive in Quadrant I}$$

$$= \frac{1}{\sqrt{2}}$$

$$= \frac{1}{2}\sqrt{2} \qquad \frac{1}{\sqrt{2}} = \frac{1}{\sqrt{2}}\cdot\frac{\sqrt{2}}{\sqrt{2}} = \frac{\sqrt{2}}{2} = \frac{1}{2}\sqrt{2}$$

Similarly, $\sin\frac{\pi}{4} = \frac{\sqrt{2}}{2}$, $\tan\frac{\pi}{4} = 1$, $\sec\frac{\pi}{4} = \sqrt{2}$, $\csc\frac{\pi}{4} = \sqrt{2}$, and $\cot\frac{\pi}{4} = 1$. ■

EXAMPLE 3 Find the exact values for the trigonometric functions of 30°.

Solution Consider not only the standard-position angle 30° but also the standard-position angle −30°. Choose $P_1(x, y)$ and $P_2(x, -y)$ respectively on the terminal sides. (See Figure 5.11.)

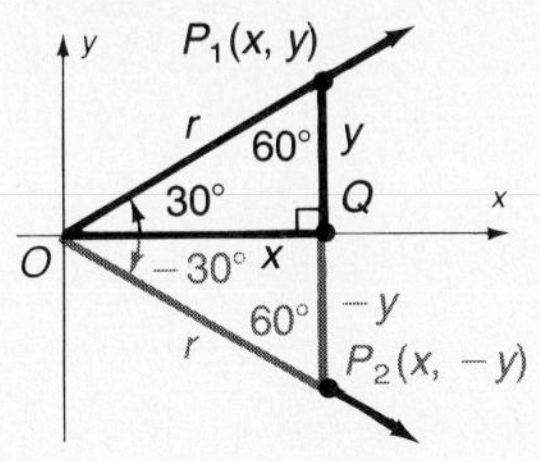

Figure 5.11

Angles OQP_1 and OQP_2 are right angles, so $\angle OP_1Q = 60°$ and $\angle OP_2Q = 60°$. Thus $\triangle OP_1P_2$ is an equiangular triangle. (All angles measure 60°). From geometry we know that an equiangular triangle has sides the same length. Thus $2y = r$. Notice the following relationship between x and y:

$$r^2 = x^2 + y^2$$

$$(2y)^2 = x^2 + y^2 \qquad 2y = r$$

$$3y^2 = x^2 \qquad \sqrt{3}|y| = |x|$$

For 30°, x and y are both positive, so $x = \sqrt{3}y$.

$$\cos 30° = \frac{x}{r} = \frac{\sqrt{3}y}{2y} = \frac{\sqrt{3}}{2} \qquad \sec 30° = \frac{2}{\sqrt{3}} = \frac{2}{3}\sqrt{3}$$

$$\sin 30° = \frac{y}{r} = \frac{y}{2y} = \frac{1}{2} \qquad \csc 30° = 2$$

$$\tan 30° = \frac{y}{x} = \frac{y}{\sqrt{3}y} = \frac{1}{\sqrt{3}} = \frac{\sqrt{3}}{3} \qquad \cot 30° = \sqrt{3}$$

■

The derivation in Example 3 leads to the following result from plane geometry.

30°–60°–90° Triangle Theorem

> In a 30°–60°–90° triangle, the leg opposite the 30° angle equals one-half the hypotenuse and the leg opposite the 60° angle equals one-half the hypotenuse times the square root of 3.

EXAMPLE 4 Evaluate the trigonometric functions for 60°.

Solution Using the 30°–60°–90° triangle theorem, the hypotenuse r is twice the length of the shorter leg, as shown in Figure 5.12.

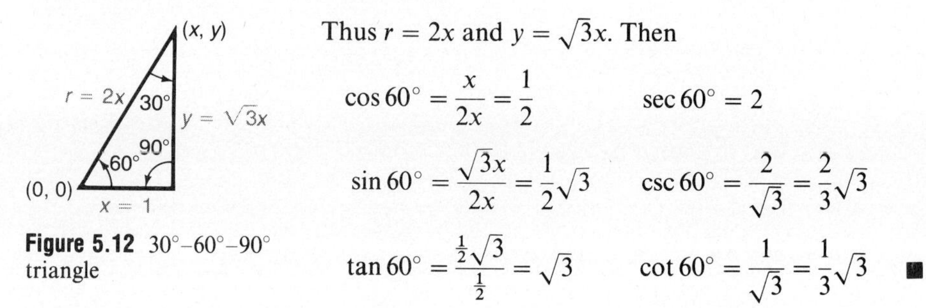

Figure 5.12 30°–60°–90° triangle

Thus $r = 2x$ and $y = \sqrt{3}x$. Then

$$\cos 60° = \frac{x}{2x} = \frac{1}{2} \qquad \sec 60° = 2$$

$$\sin 60° = \frac{\sqrt{3}x}{2x} = \frac{1}{2}\sqrt{3} \qquad \csc 60° = \frac{2}{\sqrt{3}} = \frac{2}{3}\sqrt{3}$$

$$\tan 60° = \frac{\frac{1}{2}\sqrt{3}}{\frac{1}{2}} = \sqrt{3} \qquad \cot 60° = \frac{1}{\sqrt{3}} = \frac{1}{3}\sqrt{3} \quad ■$$

In a manner similar to Examples 1 to 4, a **table of exact values** is constructed (Table 5.3). Since this table of values is used extensively, you should memorize at least the values for $\cos\theta$, $\sin\theta$, and $\tan\theta$ as you did multiplication tables in elementary school.

TABLE 5.3 Exact values

Function \ Angle θ	$0 = 0°$	$\frac{\pi}{6} = 30°$	$\frac{\pi}{4} = 45°$	$\frac{\pi}{3} = 60°$	$\frac{\pi}{2} = 90°$	$\pi = 180°$	$\frac{3\pi}{2} = 270°$
$\cos\theta$	1	$\frac{\sqrt{3}}{2}$	$\frac{\sqrt{2}}{2}$	$\frac{1}{2}$	0	-1	0
$\sin\theta$	0	$\frac{1}{2}$	$\frac{\sqrt{2}}{2}$	$\frac{\sqrt{3}}{2}$	1	0	-1
$\tan\theta$	0	$\frac{\sqrt{3}}{3}$	1	$\sqrt{3}$	undef.	0	undef.
$\sec\theta$	1	$\frac{2}{\sqrt{3}} = \frac{2}{3}\sqrt{3}$	$\frac{2}{\sqrt{2}} = \sqrt{2}$	$\frac{2}{1} = 2$	undef.	$\frac{1}{-1} = -1$	undef.
$\csc\theta$	undef.	$\frac{2}{1} = 2$	$\frac{2}{\sqrt{2}} = \sqrt{2}$	$\frac{2}{\sqrt{3}} = \frac{2}{3}\sqrt{3}$	1	undef.	-1
$\cot\theta$	undef.	$\frac{3}{\sqrt{3}} = \sqrt{3}$	1	$\frac{1}{\sqrt{3}} = \frac{\sqrt{3}}{3}$	0	undef.	0

These are the reciprocals (which is why the exact values are given in reciprocal form as well as in simplified form). In a problem you would use the rationalized form. The reciprocal form makes it easy to remember them.

The values for $\sec\theta$, $\csc\theta$, and $\cot\theta$ do not need to be memorized as separate entries because they are simply the reciprocals of $\cos\theta$, $\sin\theta$, and $\tan\theta$.

You can find exact values of the trigonometric functions that are multiples of those in Table 5.3 by using the idea of a reference angle and the ***reduction principle***:

Reduction Principle

> If t represents any of the six trigonometric functions, then
>
> $$t(\theta) = \pm t(\theta')$$
>
> where θ' is the reference angle of θ and the sign plus or minus depends on the quadrant of the terminal side of the angle θ.

EXAMPLE 5 $\tan 210° = + \tan 30° = \dfrac{\sqrt{3}}{3}$ ■

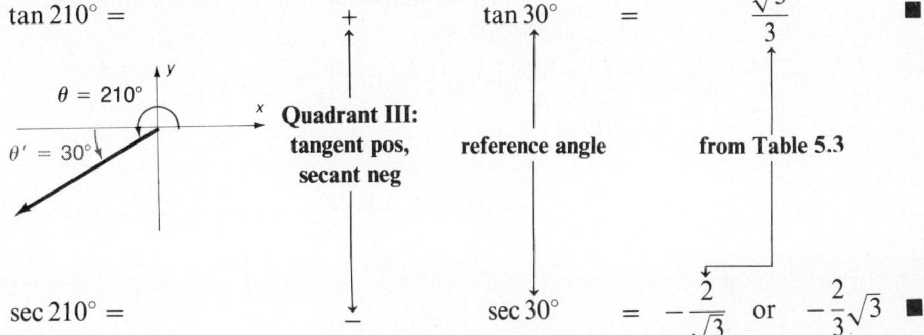

EXAMPLE 6 $\sec 210° = - \sec 30° = -\dfrac{2}{\sqrt{3}}$ or $-\dfrac{2}{3}\sqrt{3}$ ■

EXAMPLE 7 $\csc\dfrac{3\pi}{2} = -1$

If it is a quadrantal angle, then it comes directly from the memorized table. ■

EXAMPLE 8 $\cos 405° = \cos 45° = \frac{1}{2}\sqrt{2}$

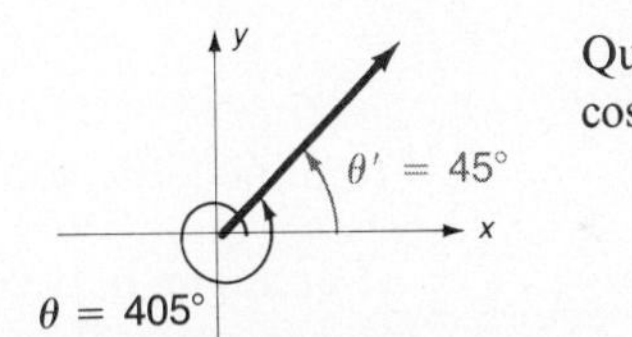

Quadrant I:
cosine positive

Sketch the angle if necessary to find the quadrant and the reference angle. ■

EXAMPLE 9 $\cot\left(-\dfrac{7\pi}{6}\right) = -\cot\dfrac{\pi}{6} = -\sqrt{3}$

$\theta' = \frac{\pi}{6}$, $\theta = -\frac{7\pi}{6}$

Quadrant II:
cotangent negative ■

EXAMPLE 10 $\sec\dfrac{5\pi}{3} = +\sec\dfrac{\pi}{3} = 2$

$\theta = \frac{5\pi}{3}$, $\theta' = \frac{\pi}{3}$

Quadrant IV:
secant positive ■

If you do not have access to a calculator but need to evaluate a trigonometric function of an angle that cannot be found as an exact value, you can use tables of approximate values, such as Table B.3 (in Appendix B at the back of the book). Table B.3 is presented in a concise form that can be used to approximate any trigonometric function for any angle given in degrees or radians, by using reference angles and the reciprocal relationships.

Values of $\sin\theta$ and $\cos\theta$ from 0° to 90° (or 0 to 1.57 radians) can be read directly from Table B.3; their reciprocals give the values of $\sec\theta$ and $\csc\theta$. Values of $\tan\theta$ from 0° to 45° (0 to 0.79) and of $\cot\theta$ from 45° to 90° (0.79 to 1.57) can be read directly; their reciprocals give the remaining values of $\tan\theta$ and $\cot\theta$ in Quadrant I. With a little practice, following the procedures shown in Examples 11–13, you will find the table much less complicated to use than to describe.

For angles between 0° and 45°, find the angle to the nearest tenth in the *left-hand* column headed "Deg." Then read across that row to find the value of the desired function named at the *top*. (If the function is not named at the top, look for its reciprocal.)

EXAMPLE 11 Find tan 22.1° using Table B.3.

Solution A portion of Table B.3 is reproduced here:

Rad	Deg	cos	sin	tan		
.35	20.0	.9397	.3420	.3640	70.0	1.22
		.9391	.3437	.3659	69.9	1.22
			.3453	.3679	69.8	1.22
	22.1 →	Read across to find tan 22.1. →		.4061		

Read down this column until the desired angle is found.

From the table, $\tan 22.1° \approx 0.4061$. ■

For angles between 45° and 90° (0.79 and 1.57), find the desired angle in the *right-hand* column. Then read across that row to find the desired function (or reciprocal) named at the *bottom*.

EXAMPLE 12 Find cot 57.6° using Table B.3.

Solution Another portion of Table B.3 is reproduced here. Disregard the headings at the top when seeking values for angles between 45° and 90°.

		Read across to find cot 57.6.		.6346 ←	57.6	
.61	34.8	.8211	.5707	.6950		
.61	34.9	.8202	.5721	.6976		
.61	35.0	.8192	.5736	.7002	55.0	96
		sin	cos	cot	Deg	Rad

Read up this column until the desired angle is found.

From the table, $\cot 57.6° \approx 0.6346$. ■

EXAMPLE 13 Find tan 5 using Table B.3.

Solution The reduction principle requires that you first find the reference angle:

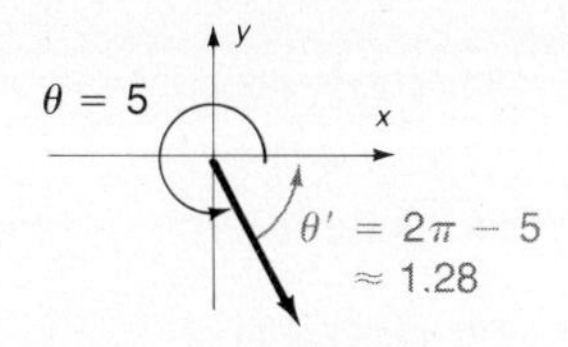

Quadrant IV:
tangent negative

$\tan 5 \approx -\tan 1.28$

The value for tan 1.28 is not listed in Table B.3. However, cot 1.28 is, so we use the reciprocal relationship:

$$-\tan 1.28 = \frac{-1}{\cot 1.28}$$

Look up the column labeled "Rad" at the bottom and then over to the value in the column labeled "cot." Substitute that value in the equation:

$$-\tan 1.28 \approx \frac{-1}{0.3000}$$

$$\approx -3.333$$

Therefore $\tan 5 \approx -3.333$. (Note: On a calculator this -3.380514999) ■

PROBLEM SET 5.3

A

In Problems 1–10, give the exact values in simplified form.

1. **a.** $\tan\frac{\pi}{4}$ **b.** $\cos 0$ **c.** $\sin 60°$ **d.** $\cos 30°$
2. **a.** $\cos 270°$ **b.** $\tan\frac{\pi}{6}$ **c.** $\tan 180°$ **d.** $\sin 45°$
3. **a.** $\sin \pi$ **b.** $\sin\frac{\pi}{2}$ **c.** $\tan 0$ **d.** $\cos\frac{\pi}{4}$
4. **a.** $\sec\frac{\pi}{6}$ **b.** $\csc 0$ **c.** $\sec\frac{\pi}{4}$ **d.** $\sec 0°$
5. **a.** $\csc\frac{\pi}{4}$ **b.** $\cot \pi$ **c.** $\sec\frac{\pi}{3}$ **d.** $\cot\frac{\pi}{6}$
6. **a.** $\tan 90°$ **b.** $\tan 60°$ **c.** $\cos\frac{\pi}{3}$ **d.** $\sec\frac{\pi}{3}$
7. **a.** $\cot 45°$ **b.** $\cos \pi$ **c.** $\sin\frac{3\pi}{2}$ **d.** $\sin 0°$
8. **a.** $\sec \pi$ **b.** $\tan 270°$ **c.** $\sin\frac{\pi}{6}$ **d.** $\csc\frac{3\pi}{2}$
9. **a.** $\cos(-300°)$ **b.** $\sin 390°$ **c.** $\sin\frac{17\pi}{4}$ **d.** $\cos(-6\pi)$
10. **a.** $\cos\frac{9\pi}{2}$ **b.** $\sin(-765°)$ **c.** $\tan(-765°)$ **d.** $\cos 495°$

Use Table B.3 to evaluate the functions in Problems 11–26.

11. **a.** $\sin 34.4°$ **b.** $\cos 54.2°$
12. **a.** $\tan 70.2°$ **b.** $\cot 46.7°$
13. **a.** $\cos 50°$ **b.** $\cot 80°$
14. **a.** $\sin 70°$ **b.** $\tan 20°$
15. **a.** $\tan(-20°)$ **b.** $\sin 190°$
16. **a.** $\cos(-340°)$ **b.** $\cot(-213°)$
17. **a.** $\sin 132.8°$ **b.** $\tan(-25.6°)$
18. **a.** $\cot(-125.6°)$ **b.** $\cos 163.4°$
19. **a.** $\sin 1.20$ **b.** $\cos 0.65$
20. **a.** $\tan 0.51$ **b.** $\cot 1.85$
21. **a.** $\tan 1$ **b.** $\cot 1.5$
22. **a.** $\sin 0.8$ **b.** $\cos 0.5$
23. **a.** $\tan 2.5$ **b.** $\sin 3$
24. **a.** $\cos 4.5$ **b.** $\cot 6$
25. **a.** $\cos(-0.45)$ **b.** $\tan(-2.8)$
26. **a.** $\sin(-3.9)$ **b.** $\cot 10$

B

27. Verify the entries in Table 5.3 for the angle $\frac{\pi}{6}$.
28. Find $\cos\frac{3\pi}{4}$ by using the procedure illustrated in Example 2.

29. Find $\cos \frac{5\pi}{4}$ by using the procedure illustrated in Example 2.

30. Find $\cos 135°$ by choosing an arbitrary point (x, y) on the terminal side of $135°$ and applying the ratio definition of the trigonometric functions.

31. Find $\sin(-\frac{\pi}{4})$ by choosing an arbitrary point (x, y) on the terminal side of $-\frac{\pi}{4}$ and applying the ratio definition of the trigonometric functions.

32. Find $\sin 210°$ by choosing an arbitrary point (x, y) on the terminal side of $210°$ and applying the ratio definition of the trigonometric functions.

33. Find $\cos 210°$ by choosing an arbitrary point (x, y) on the terminal side of $210°$ and applying the ratio definition of the trigonometric functions.

Substitute the exact values for the trigonometric functions in the expressions in Problems 34–59 and simplify. When a trigonometric function is raised to a power, such as $(\sin x)^2$, *it is written as* $\sin^2 x$.

34. $\sin 30° + \cos 0°$

35. $\sin \frac{\pi}{2} + 3\cos \frac{\pi}{2}$

36. $2\cos \frac{\pi}{2}$

37. $\cos \frac{2\pi}{2}$

38. $\sin \frac{2\pi}{4}$

39. $2\sin \frac{\pi}{4}$

40. $\sin^2 60°$

41. $\cos^2 \frac{\pi}{4}$

42. $\sin^2 \frac{\pi}{6} + \cos^2 \frac{\pi}{2}$

43. $\sin^2 \frac{\pi}{2} + \cos^2 \frac{\pi}{2}$

44. $\sin^2 \frac{\pi}{3} + \cos^2 \frac{\pi}{3}$

45. $\sin^2 \frac{\pi}{6} + \cos^2 \frac{\pi}{3}$

46. $\sin \frac{\pi}{6} \csc \frac{\pi}{6}$

47. $\csc \frac{\pi}{2} \sin \frac{\pi}{2}$

48. $\cos(\frac{\pi}{4} - \frac{\pi}{2})$

49. $\cos \frac{\pi}{4} - \cos \frac{\pi}{2}$

50. $\tan(2 \cdot 30°)$

51. $2\tan 30°$

52. $\csc(\frac{1}{2} \cdot 60°)$

53. $\dfrac{\csc 60°}{2}$

54. $\cos(\frac{1}{2} \cdot 60°)$

55. $\sqrt{\dfrac{1 + \cos 60°}{2}}$

56. $\tan(2 \cdot 60°)$

57. $\dfrac{2\tan 60°}{1 - \tan^2 60°}$

58. $\cos(\frac{\pi}{2} - \frac{\pi}{6})$

59. $\cos \frac{\pi}{2} \cos \frac{\pi}{6} + \sin \frac{\pi}{2} \sin \frac{\pi}{6}$

C

60. What is the smaller angle between the hands of a clock at 12:25 P.M.?

61. **a.** If θ is in Quadrant I, then $\theta + \pi$ is in Quadrant III with a reference angle θ. Use this fact and the reduction principle to show that $\sin(\theta + \pi) = -\sin \theta$ if θ is in Quadrant I.
b. Show that $\sin(\theta + \pi) = -\sin \theta$ if θ is in Quadrant II.
c. Show that $\sin(\theta + \pi) = -\sin \theta$ if θ is in Quadrant III.
d. Show that $\sin(\theta + \pi) = -\sin \theta$ if θ is in Quadrant IV.
e. By considering parts *a–d*, show that $\sin(\theta + \pi) = -\sin \theta$ for any angle θ.

62. Show that $\cos(\theta + \pi) = -\cos \theta$ for any angle θ. (*Hint:* See Problem 61.)

63. ***Computer*** If you have access to a computer, write a program that will output a table of trigonometric values for the sine, cosine, and tangent for every degree from 0° to 45°.

5.4 GRAPHS OF THE TRIGONOMETRIC FUNCTIONS

As with the polynomial and rational functions, we are interested in the graphs of the trigonometric functions. We will first determine the general shape of the trigonometric functions by plotting points and then generalize so we can graph the functions without too many calculations concerning points.

To graph $y = \sin x$, begin by plotting familiar values for the sine:

x = real number	0	$\frac{\pi}{6}$	$\frac{\pi}{4}$	$\frac{\pi}{3}$	$\frac{\pi}{2}$	π	$\frac{3\pi}{2}$
$y = \sin x$	0	$\frac{1}{2}$	$\frac{\sqrt{2}}{2}$	$\frac{\sqrt{3}}{2}$	1	0	-1
y (approximate)	0	0.5	0.71	0.87	1	0	-1

We are using exact values here, but you could also use Table B.3 or a calculator to generate these values. (See Problems 3 and 4 in the problem set.) The hardest part of graphing the sine function is deciding on the scales to use for the x and y axes. You may find it convenient to choose 12 squares on the x axis for π units and 10 squares on the y axis for 1 unit. You can then plot additional values by using the reduction principle. Continue the table to include x in Quadrants II, III and IV and plot the points (x, y) as shown in Figure 5.13. The smooth curve that connects these points is called the **sine curve**.

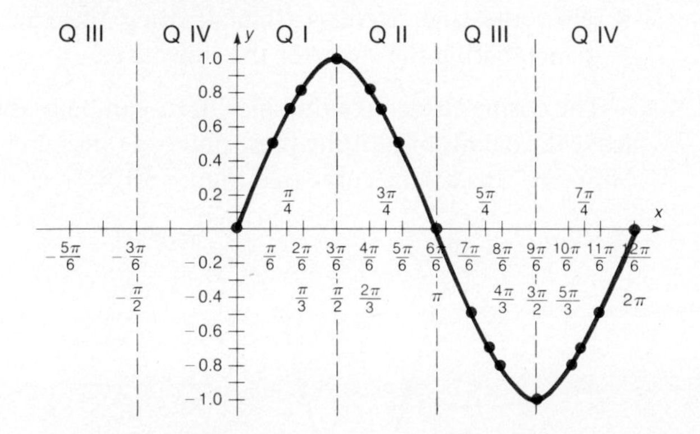

Figure 5.13 Graph of $y = \sin x$ for $0 \le x \le 2\pi$

Notice that when x is in Quadrant I, then $0 < x < \frac{\pi}{2}$, which does *not* correspond to the first quadrant of the graph $y = \sin x$. Figure 5.13 shows the intervals corresponding to the quadrants of the angle x.

If you plot values for $y = \sin x$ outside the interval $[0, 2\pi]$, you will see that this curve repeats the curve already plotted (Figure 5.14). Since $\sin(\theta + 2\pi) = \sin \theta$, we say the **period of the sine is 2π.**

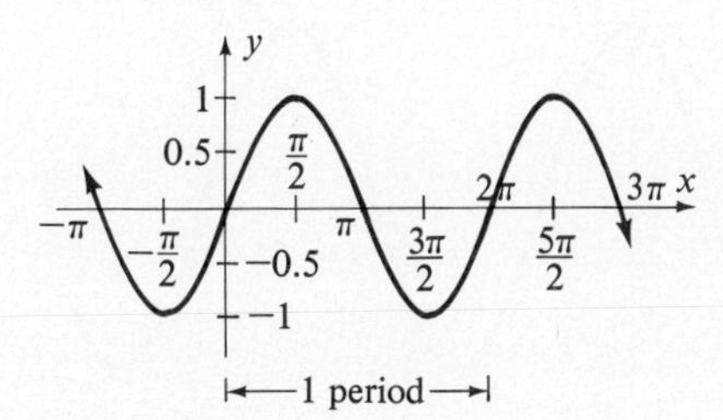

Figure 5.14 Graph of $y = \sin x$

Notice that for the base period shown the sine curve starts at $(0, 0)$, goes *up* to $(\frac{\pi}{2}, 1)$, then *down* to $(\frac{3\pi}{2}, -1)$ passing through $(\pi, 0)$, and then back *up* to $(2\pi, 0)$, which completes one period. The procedure for sketching the sine curve is shown in Figure 5.15.

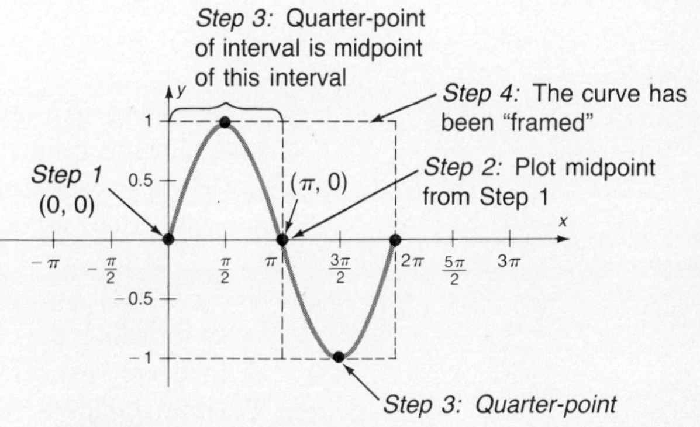

Figure 5.15 Procedure for framing the sine curve

Procedure for framing a sine curve:

1. Plot the endpoints of the base period—namely $(0, 0)$ and $(2\pi, 0)$.
2. Plot the midpoint $(\pi, 0)$.
3. Halfway between these points, plot the highest (up) point $(\frac{\pi}{2}, 1)$, and the lowest (down) point $(\frac{3\pi}{2}, -1)$; these are called the quarter-points. This is easy to remember—the sine curve "goes up and down."
4. Now the sine curve is framed; draw the curve through the plotted points, remembering the shape of the sine curve.

The cosine curve, like the sine curve, can be graphed by plotting points. We will leave the details of plotting these points as an exercise and summarize the results by "framing" the cosine curve (see Figure 5.16).

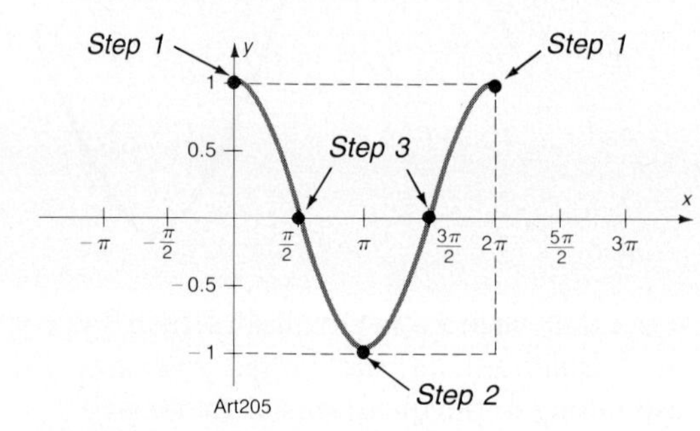

Figure 5.16 Procedure for framing the cosine curve

1. Plot the endpoints of the base period—namely $(0, 1)$ and $(2\pi, 1)$.
2. Plot the midpoint $(\pi, -1)$.
3. Halfway between these points, plot the points $(\frac{\pi}{2}, 0)$ and $(\frac{3\pi}{2}, 0)$.
4. Now the cosine curve is framed; draw the curve through the plotted points.

Since values for x greater than 2π or less than zero are coterminal with those already considered, the **period of the cosine is 2π**. The cosine curve is shown in Figure 5.17.

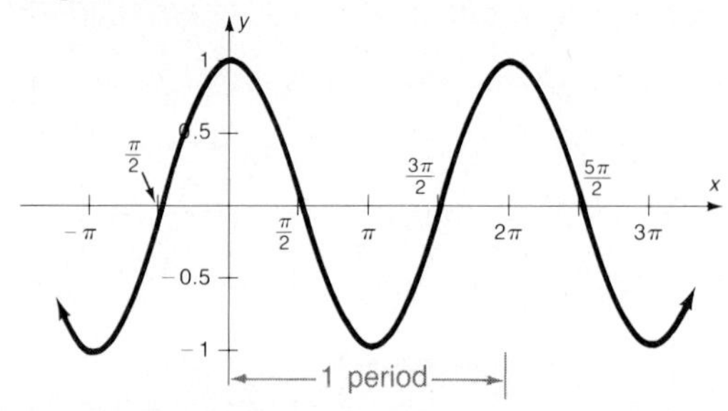

Figure 5.17 Graph of $y = \cos x$

By setting up a table of values and plotting points (the details are left as an exercise), notice that $y = \tan x$ does not exist at $\frac{\pi}{2}, \frac{3\pi}{2}$, or $\frac{\pi}{2} \pm n\pi$ for any integer n. The lines $x = \frac{\pi}{2}$, $x = \frac{3\pi}{2}, \ldots,$ $x = \frac{\pi}{2} \pm n\pi$ for which the tangent is not defined are vertical asymptotes. We now frame the tangent curve as in Figure 5.18.

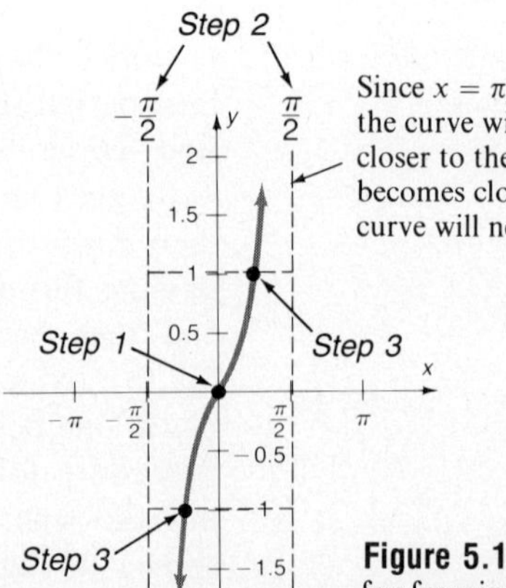

Figure 5.18 Procedure for framing a tangent curve

Procedure for framing a tangent curve:

1. Plot the midpoint $(0, 0)$.
2. Draw a pair of adjacent asymptotes—say $-\frac{\pi}{2}$ and $\frac{\pi}{2}$.
3. Halfway between the midpoint and the asymptotes, plot the points $(\frac{\pi}{4}, 1)$ and $(-\frac{\pi}{4}, -1)$.
4. The tangent curve is now framed. Draw the curve through the plotted points; it does not cross any of the asymptotes.

The tangent curve is indicated in Figure 5.19. Even though the curve repeats for values of x greater than 2π or less than zero, it also repeats after it has passed through an interval with length π. This result can be shown algebraically if we use the answers to Problems 61 and 62 of Section 5.3. Since $\sin(\theta + \pi) = -\sin\theta$ and $\cos(\theta + \pi) = -\cos\theta$,

$$\tan(\theta + \pi) = \frac{\sin(\theta + \pi)}{\cos(\theta + \pi)} = \frac{-\sin\theta}{-\cos\theta} = \frac{\sin\theta}{\cos\theta} = \tan\theta$$

Since $\tan(\theta + \pi) = \tan\theta$, then $\tan(\theta + n\pi) = \tan\theta$ for any integer n, and we see that the **tangent has a period of π.**

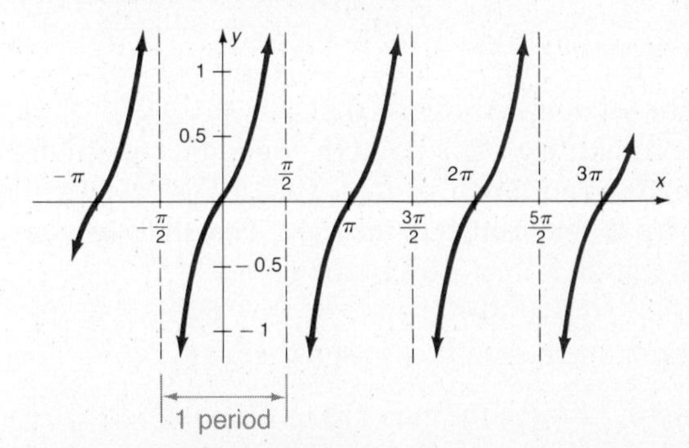

Figure 5.19 Graph of $y = \tan x$

The graph of the other three trigonometric functions can be done in the same fashion. Instead, however, we will make use of the reciprocal relationships and graph them, as shown in Example 1.

EXAMPLE 1 Sketch $y = \sec x$ by first sketching the reciprocal $y = \cos x$.

Solution Begin by sketching the reciprocal, $y = \cos x$ (dotted curve in Figure 5.20). Wherever $\cos x = 0$, $\sec x$ is undefined; draw asymptotes at these places. Now plot points by finding the reciprocals of the ordinates of points previously plotted. When

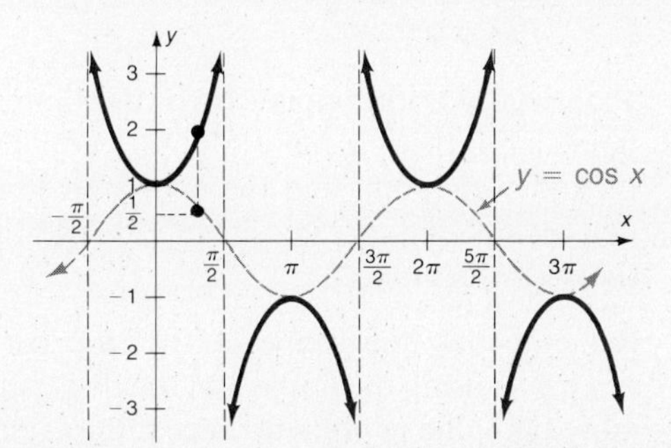

Figure 5.20 Graph of $y = \sec x$

$y = \cos x = \frac{1}{2}$, for example, the reciprocal is

$$y = \sec x = \frac{1}{\cos x} = \frac{1}{\frac{1}{2}} = 2$$

The completed graph is shown in Figure 5.20. ■

The graphs of the other reciprocal trigonometric functions are shown in Figure 5.21.

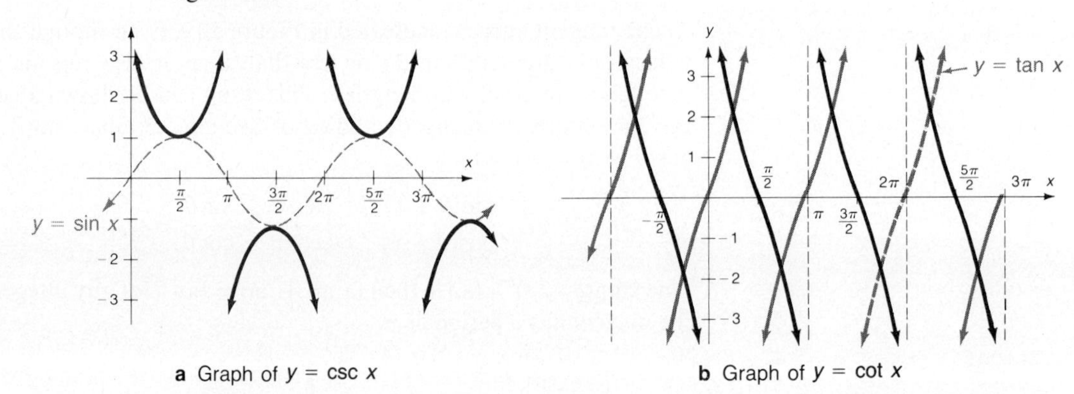

Figure 5.21 a Graph of $y = \csc x$ b Graph of $y = \cot x$

In Section 2.4 we saw that $y - k = f(x - h)$ can be sketched by translating the coordinate axes to a point (h, k) and then graphing the related function $y = f(x)$ on this new coordinate system. Thus if $f(x) = \sin x$, then $f(x - h) = \sin(x - h)$ is a sine curve shifted h units to the right. This shifting is called a **phase shift** but, in this book, we will treat it as a **translation**.

EXAMPLE 2 Graph one period of $y = \sin(x + \frac{\pi}{2})$.

Solution *Step 1* Frame the curve as in Figure 5.22.

a. Plot $(h, k) = (-\frac{\pi}{2}, 0)$.

b. The period of the sine curve is 2π, and it has a high point up 1 unit and a low point down 1 unit.

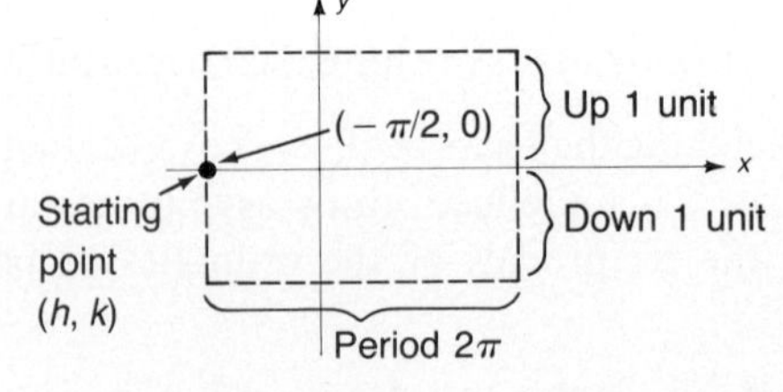

Figure 5.22 Framing the curve: this step is the same regardless of whether you are graphing a sine or a cosine

Step 2 Plot the five critical points (two endpoints, the midpoint, and two quarter-points). For the sine curve, plot the endpoint (h, k) and use the frame to plot the other endpoint and the midpoint. For the quarter-points, remember that the sine curve is "up–down"; use the frame to plot the quarter-points as shown in Figure 5.23.

Step 3 Remembering the shape of the sine curve, sketch one period of $y = \sin(x + \frac{\pi}{2})$ using the frame and the five critical points. If you want to show more than one period, just repeat the same pattern.

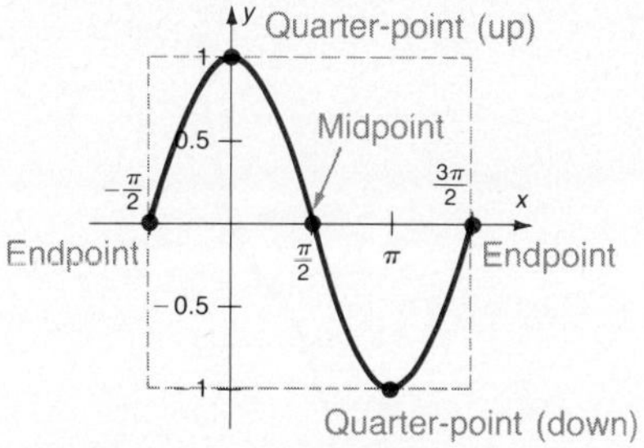

Figure 5.23 Graph of one period of $y = \sin(x + \frac{\pi}{2})$

Notice from Figure 5.23 that the graph of $y = \sin(x + \frac{\pi}{2})$ is the same as the graph of $y = \cos x$. Thus

$$\sin\left(x + \frac{\pi}{2}\right) = \cos x$$

EXAMPLE 3 Graph one period of $y - 2 = \cos(x - \frac{\pi}{6})$.

Solution *Step 1* Frame the curve as shown in Figure 5.24. Notice that $(h, k) = (\frac{\pi}{6}, 2)$.

Step 2 Plot the five critical points. For the cosine curve the left and right endpoints are at the upper corners of the frame; the midpoint is at the bottom of the frame; the quarter-points are on a line through the middle of the frame.

Step 3 Draw one period of the curve, as shown in Figure 5.24.

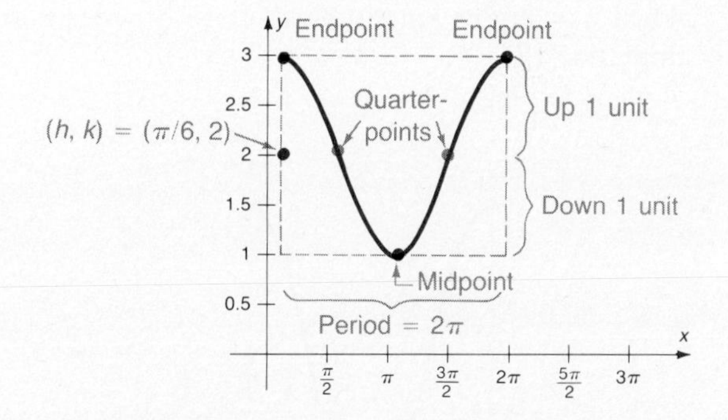

Figure 5.24 Graph of one period of $y - 2 = \cos(x - \frac{\pi}{6})$

EXAMPLE 4 Sketch one period of $y + 3 = \tan(x + \frac{\pi}{3})$.

Solution *Step 1* Frame the curve as shown in Figure 5.25. Notice that $(h, k) = (-\frac{\pi}{3}, -3)$, and remember that the period of the tangent is π.

Step 2 For the tangent curve, (h, k) is the midpoint of the frame. The endpoints, which are each a distance of one-half the period from the midpoint, determine the location of the asymptotes. The top and bottom of the frame are one unit from (h, k). Locate the quarter-points at the top and bottom of the frame as shown in Figure 5.25.

Step 3 Sketch one period of the curve as shown in Figure 5.25. Remember that the tangent curve is not contained entirely within the frame.

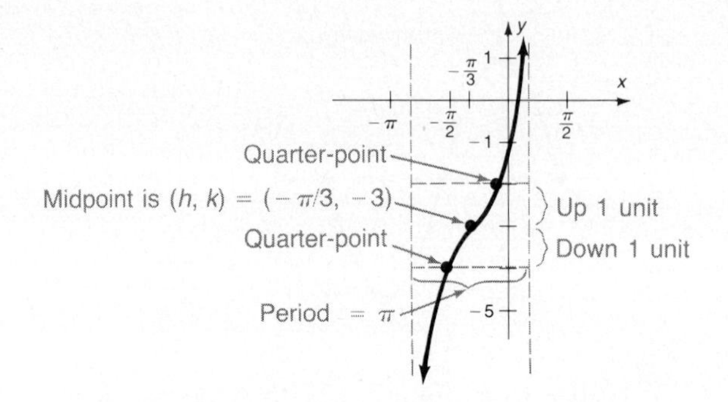

Figure 5.25 Graph of one period of $y + 3 = \tan(x + \frac{\pi}{3})$

■

We will now discuss two additional changes for the function defined by $y = f(x)$. The first, $y = af(x)$, changes the scale on the y axis; the second, $y = f(bx)$, changes the scale on the x axis.

For a function $y = af(x)$, it is clear that the y value is a times the corresponding value of $f(x)$, which means that $f(x)$ is stretched or shrunk in the y direction by the factor of a. For example, if $y = f(x) = \cos x$, then $y = 3f(x) = 3\cos x$ is the graph of $\cos x$ that has been stretched so that the high point is at 3 units and the low point is at negative 3 units. In general, given

$$y = af(x)$$

where f represents a trigonometric function, $2|a|$ gives the height of the frame for f. To graph $y = 3\cos x$, frame the cosine curve using an amplitude of 3 rather than 1 (see Figure 5.26). For the sine and cosine curves, $|a|$ is the **amplitude** of the function. When $a = 1$, the amplitude is 1, so $y = \cos x$ and $y = \sin x$ are said to have amplitude 1.

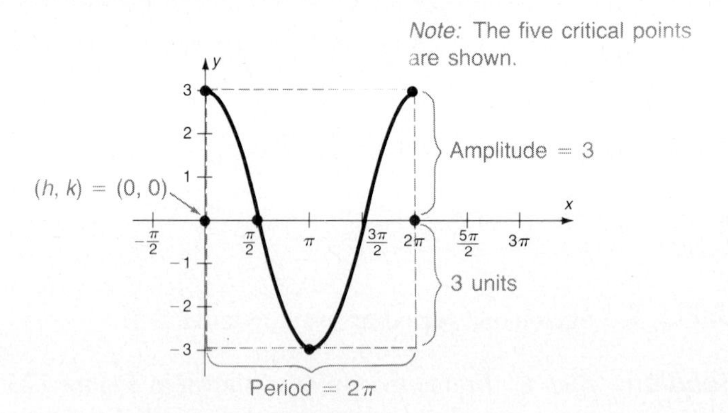

Figure 5.26 Graph of one period of $y = 3\cos x$

For a function $y = f(bx)$, $b > 0$, b affects the scale on the x axis. Recall that $y = \sin x$ has a period of 2π ($f(x) = \sin x$, so $b = 1$). A function $y = \sin 2x$ ($f(x) = \sin x$ and $f(2x) = \sin 2x$) must complete one period as $2x$ varies from zero to 2π. This means that one period is completed as x varies from zero to π. (Remember that for each value of x the result is doubled *before* we find the sine of that number.) In general, the period of $y = \sin bx$ is $\frac{2\pi}{b}$ and the period of $y = \cos bx$ is $\frac{2\pi}{b}$. Since the period of $y = \tan x$ is π, however, $y = \tan bx$ has a period of $\frac{\pi}{b}$. Therefore, when framing the curve, use $\frac{2\pi}{b}$ for the sine and cosine and $\frac{\pi}{b}$ for the tangent.

EXAMPLE 5 Graph one period of $y = \sin 2x$.

Solution The period is $\frac{2\pi}{2} = \pi$; thus the endpoints of the frame are $(0, 0)$ and $(\pi, 0)$, as shown in Figure 5.27.

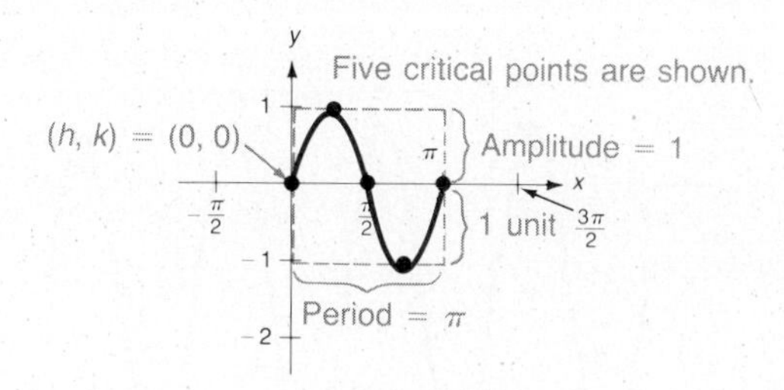

Figure 5.27 Graph of one period of $y = \sin 2x$

■

Summarizing all these results, we have the *general* cosine, sine, and tangent curves:

General Cosine, Sine, and Tangent Curves

$$y - k = a \cos b(x - h)$$
$$y - k = a \sin b(x - h)$$
$$y - k = a \tan b(x - h)$$

1. The origin has been translated or shifted to the point (h, k).
2. The height of the frame is $2|a|$.
3. The length of the frame is $\frac{2\pi}{b}$ for the cosine and sine curves and $\frac{\pi}{b}$ for the tangent curve.
4. The curves are sketched by translating the origin to the point (h, k) and then framing the curve to complete the graph.

EXAMPLE 6 Graph $y + 1 = 2 \sin \frac{2}{3}(x - \frac{\pi}{2})$.

Solution Notice that $(h, k) = (\frac{\pi}{2}, -1)$ and that the amplitude is 2; the period is $2\pi/(\frac{2}{3}) = 3\pi$. Now plot (h, k) and frame the curve. Then plot the five critical points (two endpoints, the midpoint, and two quarter-points). Finally, after sketching one period, draw the other periods as in Figure 5.28.

Figure 5.28 Graph of $y + 1 = 2 \sin \frac{2}{3}(x - \frac{\pi}{2})$; one period inside the frame is drawn first, and then the curve is extended outside the frame

■

EXAMPLE 7 Graph $y = 3\cos(2x + \frac{\pi}{2}) - 2$.

Solution Rewrite in standard form to obtain $y + 2 = 3\cos 2(x + \frac{\pi}{4})$
Notice that $(h, k) = (-\frac{\pi}{4}, -2)$; the amplitude is 3 and the period is $\frac{2\pi}{2} = \pi$. Plot (h, k) and frame the curve as shown in Figure 5.29.

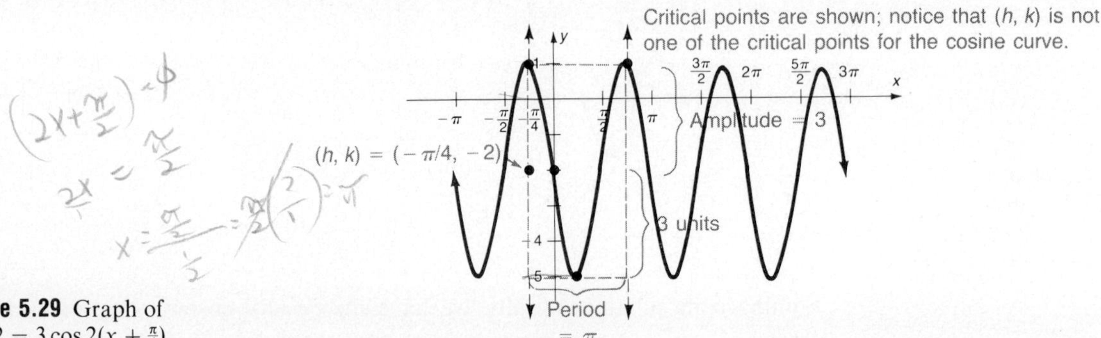

Figure 5.29 Graph of $y + 2 = 3\cos 2(x + \frac{\pi}{4})$

■

EXAMPLE 8 Graph $y - 2 = 3\tan\frac{1}{2}(x - \frac{\pi}{3})$.

Solution Notice that $(h, k) = (\frac{\pi}{3}, 2)$, $a = 3$, and the period is $\pi/(\frac{1}{2}) = 2\pi$. Plot (h, k) and frame the curve as shown in Figure 5.30.

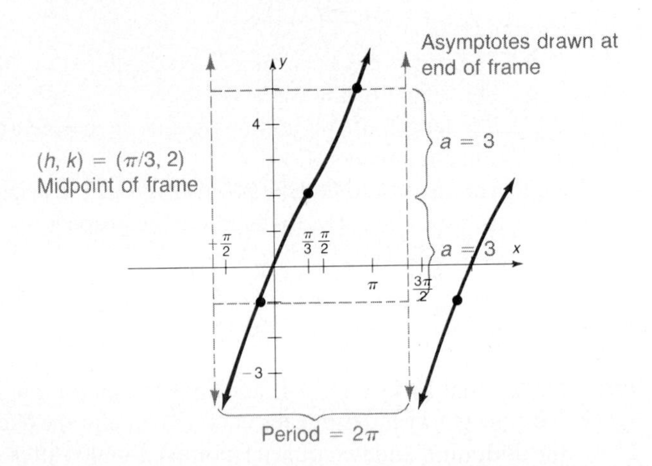

Figure 5.30 Graph of $y - 2 = 3\tan\frac{1}{2}(x - \frac{\pi}{3})$

■

PROBLEM SET 5.4

A

1. Complete the following table of values for $y = \cos x$:

x = angle	$\frac{2\pi}{3}$	$\frac{3\pi}{4}$	$\frac{5\pi}{6}$	$\frac{7\pi}{6}$	$\frac{5\pi}{4}$	$\frac{4\pi}{3}$	$\frac{7\pi}{4}$	$\frac{11\pi}{6}$
Quadrant; sign of $\cos x$								
$y = \cos x$								
y (approximate)								

Use this table, along with other values if necessary, to plot $y = \cos x$.

2. Complete a table of values like the one in Problem 1 for $y = \tan x$. Use this table, along with other values if necessary, to plot $y = \tan x$.

3. We have emphasized the fact that the sine function can be considered as a function of a real number, x. Instead of using units of π, graph the sine function by finding

additional values for the table shown here:

x = real number	0	1	2	3	4	5	6
$y = \sin x$	0	0.84	0.91	0.14	−0.76	−0.96	−0.28

Plot the ordered pairs, (0, 0), (1, 0.84), (2, 0.91), ..., and complete the graph of $y = \sin x$.

4. We have emphasized the fact that the cosine function can be considered as a function of a real number, x. Instead of using units of π, graph the cosine function by finding additional values for the table shown here:

x = real number	0	1	2	3	4	5	6
$y = \cos x$	1	0.54	−0.42	−0.99	−0.65	0.28	0.96

Plot the ordered pairs, (0, 1), (1, 0.54), (2, −0.42), ..., and complete the graph of $y = \cos x$.

Graph one period of each function given in Problems 5–19.

5. $y = \sin(x + \pi)$
6. $y = \cos(x + \frac{\pi}{2})$
7. $y = \tan(x + \frac{\pi}{3})$
8. $y = 3 \sin x$
9. $y = 2 \cos x$
10. $y = \frac{1}{2} \sin x$
11. $y = \sin 3x$
12. $y = \cos 2x$
13. $y = \cos \frac{1}{2}x$
14. $y = \tan(x - \frac{3\pi}{2})$
15. $y = \tan(x + \frac{\pi}{6})$
16. $y = \frac{1}{2} \sin x$
17. $y = \frac{1}{3} \tan x$
18. $y = 4 \tan x$
19. $y = 5 \sin x$

B

Graph one period of each function given in Problems 20–31.

20. $y - 2 = \sin(x - \frac{\pi}{2})$
21. $y + 1 = \cos(x + \frac{\pi}{3})$
22. $y - 3 = \tan(x + \frac{\pi}{6})$
23. $y - \frac{1}{2} = \frac{1}{2} \cos x$
24. $y - 1 = 2 \sin x$
25. $y + 2 = 3 \cos x$
26. $y - 1 = 2 \cos(x - \frac{\pi}{4})$
27. $y - 1 = \cos 2(x - \frac{\pi}{4})$
28. $y + 2 = 3 \sin(x + \frac{\pi}{6})$
29. $y + 2 = \sin 3(x + \frac{\pi}{6})$
30. $y = 1 + \tan 2(x - \frac{\pi}{4})$
31. $y + 2 = \tan(x - \frac{\pi}{4})$

Graph the curves given in Problems 32–46.

32. $y = \sin(4x + \pi)$
33. $y = \sin(3x + \pi)$
34. $y = \tan(2x - \frac{\pi}{2})$
35. $y = \tan(\frac{x}{2} + \frac{\pi}{3})$
36. $y = \frac{1}{2} \cos(x + \frac{\pi}{6})$
37. $y = \cos(\frac{1}{2}x + \frac{\pi}{12})$
38. $y = 3 \cos(3x + 2\pi) - 2$
39. $y = 4 \sin(\frac{1}{2}x + 2)$
40. $y = \sqrt{2} \cos(x - \sqrt{2}) - 1$
41. $y = \sqrt{3} \sin(\frac{1}{3}x - \sqrt{\frac{1}{3}})$
42. $y = 2 \sin(2\pi x)$
43. $y = 3 \cos(3\pi x)$
44. $y = 4 \tan(\frac{\pi x}{5})$
45. $y + 2 = \frac{1}{2} \cos(\pi x + 2\pi)$
46. $y - 3 = 3 \cos(2\pi x + 4)$

Use the technique of plotting points to graph the functions in Problems 47–55.

47. $y = \sec x$
48. $y = 2 \sec x$
49. $y = \csc x$
50. $y = \cot x$
51. $y = \csc 2x$
52. $y = 2 \cot x - 1$
53. $y = \sin x + \cos x$
54. $y = \sin 2x + \cos x$
55. $y = 2 \cos x + \sin 2x$

So far we have limited ourselves to $a > 0$. If $a < 0$, the curve is reflected through the x axis. Graph the curves in Problems 56–61.

56. $y = -\sin x$
57. $y = -\cos x$
58. $y = -\tan x$
59. $y = -3 \sin x$
60. $y = -2 \cos x$
61. $y = -\sin 3x$

C

62. ***Electrical Engineering*** The current I (in amperes) in a certain circuit is given by

$$I = 60 \cos(120\pi t - \pi)$$

where t is time in seconds. Graph this equation for $0 \le t \le \frac{1}{30}$.

63. ***Engineering*** Suppose a point P on a waterwheel with a 30-ft radius is d units from the water as shown in Figure 5.31. If it turns at 6 revolutions per minute, then

$$d = 29 + 30 \cos(\tfrac{\pi}{5}t - \pi)$$

Graph this equation for $0 \le t \le 20$.

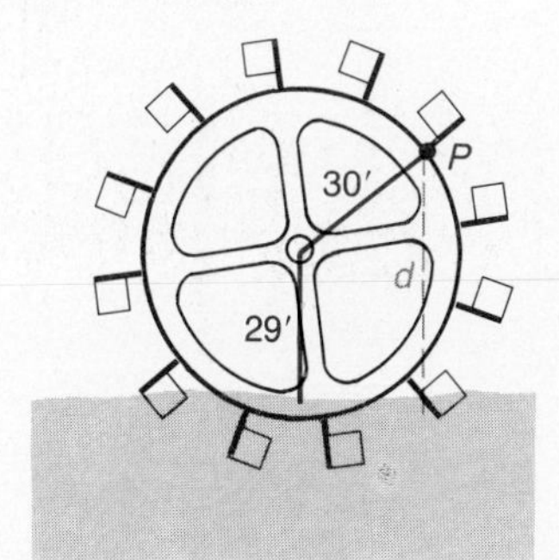

Figure 5.31

64. ***Space Science*** The distance that a certain satellite is north or south of the equator is given by

$$y = 3000 \cos(\tfrac{\pi}{60}t + \tfrac{\pi}{5})$$

where t is the number of minutes that have elapsed since liftoff.

a. Graph the equation for $0 \le t \le 120$.
b. What is the farthest distance that the satellite ever reaches north of the equator?
c. How long does it take to complete one orbit?

65. Plot points to graph $y = (\sin x)/x$.

5.5 INVERSE TRIGONOMETRIC FUNCTIONS

In Section 2.6 the notion of inverse functions was introduced. In this section, that idea is applied to the trigonometric functions. Recall that a function f must be one-to-one in order to have an inverse function. This means that the trigonometric functions do not have inverse functions. We can, however, restrict the domains of the trigonometric functions so that they become one-to-one. We will illustrate with the sine function. Figure 5.32 shows the graph of $y = \sin x$.

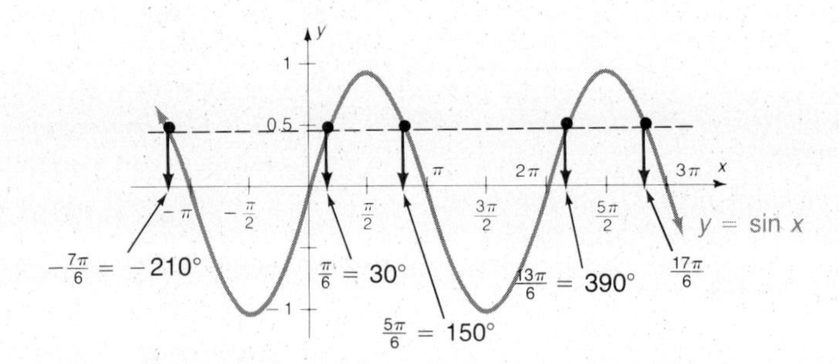

Figure 5.32 Graph of $y = \sin x$

Notice that if $y = \sin x$, then each x, say $\frac{5\pi}{6}$, is associated with exactly one y value, $\frac{1}{2}$ in this case. But it is *not* one-to-one because for a given y value, say $\frac{1}{2}$, there are infinitely many x values as shown in Figure 5.32.

Define a new function related to $y = \sin x$ but having the property that it is one-to-one. We do this simply by restricting its domain to the first and fourth quadrants. That is, define

$$y = \text{Sin}\, x \text{ so that } x \text{ is on the interval } \left[-\frac{\pi}{2}, \frac{\pi}{2}\right]$$

Notice the capital S in $y = \text{Sin}\, x$. $\text{Sin}\, x \neq \sin x$ because of their different domains. This function is shown in color in Figure 5.33a; notice that this function *is* one-to-one.

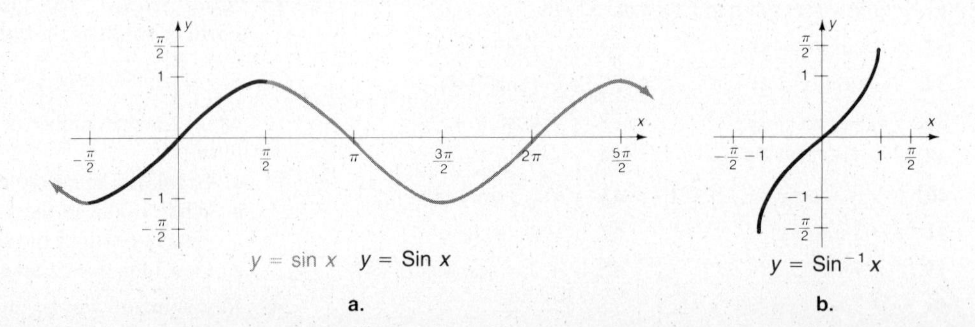

Figure 5.33 Comparison graphs of $y = \sin x$, $y = \text{Sin}\, x$, and $y = \text{Sin}^{-1} x$

Now we can define the inverse by using the notation introduced in Section 2.6. Remember: The inverse of $y = \operatorname{Sin} x$ is found by interchanging the x and y components to $x = \operatorname{Sin} y$.

Inverse Sine

> Thc **inverse sine function**, denoted by $\operatorname{Sin}^{-1} x$, is defined by
>
> $$y = \operatorname{Sin}^{-1} x \qquad \text{if and only if} \qquad x = \operatorname{Sin} y$$
>
> where $-1 \le x \le 1$ and $-\frac{\pi}{2} \le y \le \frac{\pi}{2}$.

EXAMPLE 1 Find $\operatorname{Sin}^{-1}(\frac{1}{2}\sqrt{3})$.

Solution Let $\theta = \operatorname{Sin}^{-1}(\frac{1}{2}\sqrt{3})$. (Remember: An inverse sine is an angle, so we denote it by θ.) Find the angle or real number θ with sine equal to $\frac{1}{2}\sqrt{3}$ so that $-\frac{\pi}{2} \le \theta \le \frac{\pi}{2}$. From the memorized table of exact values you know that $\sin(\frac{\pi}{3}) = \frac{1}{2}\sqrt{3}$. And since $\frac{\pi}{3}$ is between $-\frac{\pi}{2}$ and $\frac{\pi}{2}$, you have $\operatorname{Sin}^{-1}(\frac{1}{2}\sqrt{3}) = \frac{\pi}{3}$. ■

EXAMPLE 2 Find $\operatorname{Sin}^{-1}(-\frac{1}{2}\sqrt{3})$.

Solution You will find it easier to work with reference angles when finding inverse trigonometric functions. That is, because the table of exact values was memorized for the first quadrant, work with the reference angle. Let

$$\theta' = \operatorname{Sin}^{-1}(\tfrac{1}{2}\sqrt{3})$$

↑ reference angle; ↑ absolute value of the given number

$$\theta' = \frac{\pi}{3} \qquad \text{From Example 1}$$

Now place θ in the appropriate quadrant. The sine is negative in both the third and fourth quadrants, but you choose the fourth quadrant because of the restrictions on $y = \operatorname{Sin} x$. Thus θ is the fourth-quadrant angle with its reference angle $\theta' = \frac{\pi}{3}$. Therefore $\operatorname{Sin}^{-1}(-\frac{1}{2}\sqrt{3}) = -\frac{\pi}{3}$. ■

The other trigonometric functions are handled similarly:

Given function	Inverse	Other notations for inverse	
$y = \operatorname{Cos} x$	$x = \operatorname{Cos} y$	$y = \operatorname{Cos}^{-1} x$	$y = \operatorname{Arccos} x$
$y = \operatorname{Sin} x$	$x = \operatorname{Sin} y$	$y = \operatorname{Sin}^{-1} x$	$y = \operatorname{Arcsin} x$
$y = \operatorname{Tan} x$	$x = \operatorname{Tan} y$	$y = \operatorname{Tan}^{-1} x$	$y = \operatorname{Arctan} x$
$y = \operatorname{Sec} x$	$x = \operatorname{Sec} y$	$y = \operatorname{Sec}^{-1} x$	$y = \operatorname{Arcsec} x$
$y = \operatorname{Csc} x$	$x = \operatorname{Csc} y$	$y = \operatorname{Csc}^{-1} x$	$y = \operatorname{Arccsc} x$
$y = \operatorname{Cot} x$	$x = \operatorname{Cot} y$	$y = \operatorname{Cot}^{-1} x$	$y = \operatorname{Arccot} x$

These are the same.

Consider the graphs in Figure 5.34. We have to restrict each trigonometric function so it is one-to-one, but we also want to include all possible values in the range of the original function. For the sine curve, x was restricted so that $-\frac{\pi}{2} \leq x \leq \frac{\pi}{2}$. Then the inverse is the function

$$y = \text{Sin}^{-1} x \qquad \text{where } -\frac{\pi}{2} \leq y \leq \frac{\pi}{2}$$

The same restrictions (leaving out the values $-\frac{\pi}{2}$ and $\frac{\pi}{2}$) apply for the tangent and arctangent curves. For the cosine function, however, notice that by restricting x to the same interval you obtain only positive values for $y = \cos x$. Thus, to include the entire range of the cosine curve, x can be restricted so that $0 \leq x \leq \pi$. Then the inverse is the function

$$y = \text{Cos}^{-1} x \qquad \text{where } 0 \leq y \leq \pi$$

The cotangent function is restricted in almost the same way, and the results are summarized in the following definitions. The inverse functions are sketched in Figure 5.34.

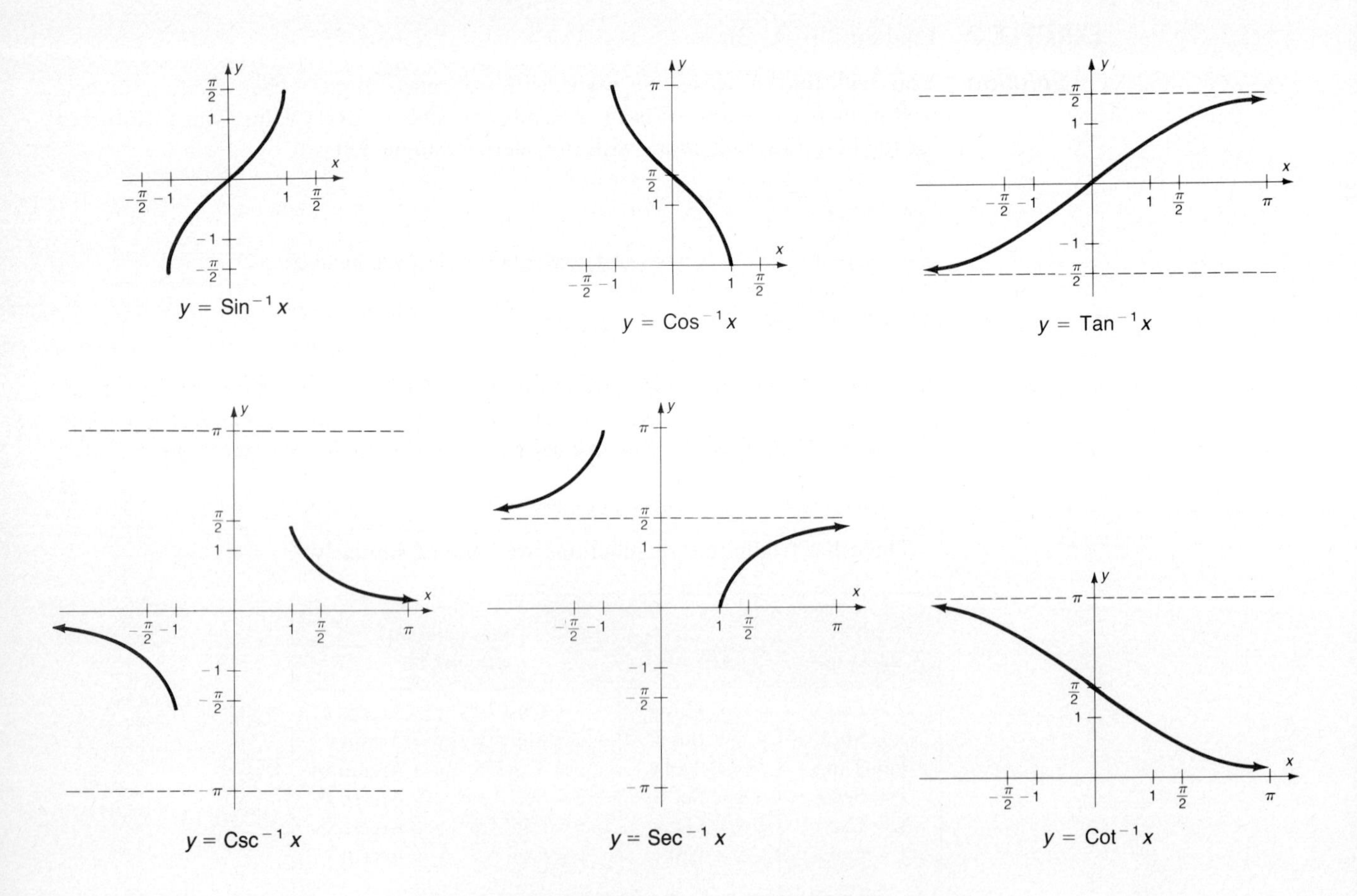

Figure 5.34 Graphs of inverse trigonometric functions

Inverse Trigonometric Functions

Inverse function	*Domain*	*Range*
$y = \text{Arccos}\, x$ or $y = \text{Cos}^{-1} x$	$-1 \le x \le 1$	$0 \le y \le \pi$
$y = \text{Arcsin}\, x$ or $y = \text{Sin}^{-1} x$	$-1 \le x \le 1$	$-\frac{\pi}{2} \le y \le \frac{\pi}{2}$
$y = \text{Arctan}\, x$ or $y = \text{Tan}^{-1} x$	All reals	$-\frac{\pi}{2} < y < \frac{\pi}{2}$
$y = \text{Arccot}\, x$ or $y = \text{Cot}^{-1} x$	All reals	$0 < y < \pi$
$y = \text{Arcsec}\, x$ or $y = \text{Sec}^{-1} x$	$x \le -1; x \ge 1$	$0 \le y \le \pi; y \ne \frac{\pi}{2}$
$y = \text{Arccsc}\, x$ or $y = \text{Csc}^{-1} x$	$x \le -1; x \ge 1$	$-\frac{\pi}{2} \le y \le \frac{\pi}{2}; y \ne 0$

EXAMPLE 3 Find Arctan 1.

Solution Let Arctan $1 = \theta$. You are looking for an angle θ with tangent equal to 1. Since this is an exact value, you know that $\theta = \frac{\pi}{4}$ or 45°. ■

EXAMPLE 4 Find $\text{Arccot}(-\sqrt{3})$.

Solution Let $\text{Arccot}(-\sqrt{3}) = \theta$. Find θ so that $\cot\theta = -\sqrt{3}$; the reference angle is 30°, and the cotangent is negative in Quadrants II and IV. Since Arccot x is defined in Quadrant II but not in Quadrant IV, $\text{Arccot}(-\sqrt{3}) = \frac{5\pi}{6}$ or 150°. ■

EXAMPLE 5 Find Arcsin(−0.4695).

Solution Let $\text{Arcsin}(-0.4695) = \theta$. Find θ so that $\sin\theta = -0.4695$; from Table B.3 we find that the reference angle is 28°. Since Arcsin x is defined for values between −90° and 90°, $\text{Arcsin}(-0.4695) = -28°$. If you want to solve this problem on your calculator, check the algorithm used on your own calculator. You may have an [arc] button instead of an [inv] button, but in the text we will indicate inverse by showing an [inv] button. For this example:

PRESS: [.4695] [+/−] [inv] [sin]

The display: −28.00184535 is the decimal representation of the angle in degrees (if the calculator is set to degrees) or the angle in radians (if it is set to radians). ■

EXAMPLE 6 Find Arccot(−2.747).

Solution Let $\text{Arccot}(-2.747) = \theta$. You can use Table B.3 to find the reference angle 20°. Since $0 < \cot^{-1} x < \pi$, you must place this angle in Quadrant II, so $\theta = 160°$. Because calculators have no cotangent function, note that if $\cot y = x$ then $\tan y = \frac{1}{x}$. Thus

$$y = \cot^{-1} x \text{ and } y = \tan^{-1}\left(\frac{1}{x}\right) \qquad \text{so} \qquad \cot^{-1} x = \tan^{-1}\left(\frac{1}{x}\right)$$

This tells you that to find the inverse cotangent on a calculator you must first take the reciprocal of the given value and then complete the problem.

Another way of looking at this is to consider the keys pressed on a calculator to find cotangent. For example, find cot 30°:

PRESS: [30] [tan] [1/x] DISPLAY: 1.73205081

Thus to find the arccot 1.73205081, just reverse the steps:

PRESS: [1/x] [inv] [tan] DISPLAY: 30

For this example:

PRESS: [2.747] [1/x] [inv] [tan] DISPLAY: 20.00320032

This result is correct for inverse tangent since it is in Quadrant IV. However, inverse cotangent must be in Quadrant I or II, so it is necessary to add 180° (or π if working in radians) for the proper placement of the angle. ■

PRESS: [2.747] [+/−] [1/x] [inv] [tan] [+] [180] [=]
DISPLAY: 159.9967997

Thus, the answer is approximately 160°. ■

We can now summarize the calculator steps illustrated in Examples 5 and 6.

Inverse function	**Calculator** Enter the value of x, then press:
$y = \text{Arccos } x$	[inv] [cos]
$y = \text{Arcsin } x$	[inv] [sin]
$y = \text{Arctan } x$	[inv] [tan]
$y = \text{Arcsec } x$	[1/x] [inv] [cos]
$y = \text{Arccsc } x$	[1/x] [inv] [sin]
$y = \text{Arccot } x$	If $x > 0$: [1/x] [inv] [tan] If $x < 0$: [1/x] [inv] [tan] [+] [π] [=]*

* Use 180 instead of π if working in degrees.

EXAMPLE 7 Find θ using a calculator.

a. $\text{Arcsec}(-3) = \theta$ **b.** $\text{Arccsc } 7.5 = \theta$
c. $\text{Arccot } 2.4747 = \theta$ **d.** $\text{Arccot}(-4.852) = \theta$

Solution Once again, we remind you to make sure your calculator is in the proper mode. These are all in radian mode.

a. Arcsec$(-3) = \theta$.

PRESS: [3] [+/−] [1/x] [inv] [cos]   DISPLAY: 1.910633236

b. Arccsc $7.5 = \theta$.

PRESS: [7.5] [1/x] [inv] [sin]   DISPLAY: .1337315894

c. Arccot $2.4747 = \theta$. Since 2.4747 is positive,

PRESS: [2.4747] [1/x] [inv] [tan]   DISPLAY: .3840267299

d. Arccot(-4.852). Since -4.852 is negative,

PRESS: [4.852] [+/−] [1/x] [inv] [tan] [+] [π] [=]

DISPLAY: 2.938338095 ■

A final word of caution is in order regarding the inverse trigonometric functions, especially when you are using a calculator. Recall from Section 2.6 that

$$(f^{-1} \circ f)(x) = (f \circ f^{-1})(x) = x$$

In the context of this section, this means

$$\begin{aligned} \text{Cos}^{-1}(\text{Cos}\, x) = x \quad &\text{and} \quad \text{Cos}(\text{Cos}^{-1} x) = x \\ \text{Sin}^{-1}(\text{Sin}\, x) = x \quad &\text{and} \quad \text{Sin}(\text{Sin}^{-1} x) = x \\ \text{Tan}^{-1}(\text{Tan}\, x) = x \quad &\text{and} \quad \text{Tan}(\text{Tan}^{-1} x) = x \\ \text{Cot}^{-1}(\text{Cot}\, x) = x \quad &\text{and} \quad \text{Cot}(\text{Cot}^{-1} x) = x \end{aligned}$$

However, you should not forget the appropriate restrictions that are also part of the definition. Consider Examples 8 to 13.

EXAMPLE 8 $\text{Cos}^{-1}(\cos 2.2) = 2.2$

Calculator check: [2.2] [cos] [inv] [cos]   DISPLAY: 2.2

(Some calculators may show a display such as 2.1999997; this should be considered a proper check.) ■

EXAMPLE 9 $\text{Sin}^{-1}(\sin 2.2) \approx 0.9$ (Not 2.2)

The reason for this answer is that $\text{Sin}\, x$ is defined in Quadrants I and IV, whereas 2.2 is not an angle in these quadrants.

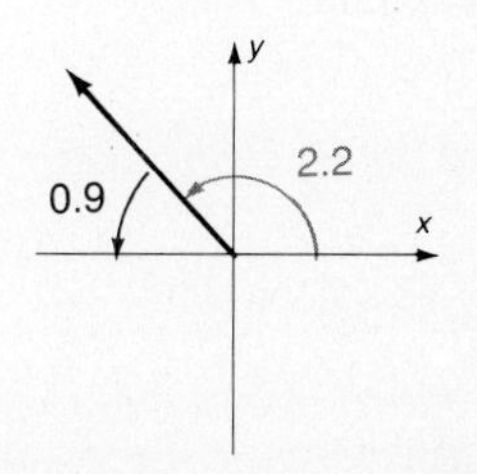

Since $\sin 2.2 \approx \sin 0.9$ (by the reduction principle) and since 0.9 is an angle in Quadrants I or IV,

$$\begin{aligned} \text{Sin}^{-1}(\sin 2.2) &\approx \text{Sin}^{-1}(\sin 0.9) \\ &= 0.9 \end{aligned}$$

Calculator check: [2.2] [sin] [inv] [sin]   DISPLAY: .94159265 ■

EXAMPLE 10 $\text{Arccos}(\cos 4) \approx 2.3$

Since the angle 4 radians is in Quadrant III, $\cos 4$ is negative. The Arccosine of a negative angle will be a Quadrant II angle. This is the Quadrant II angle having the

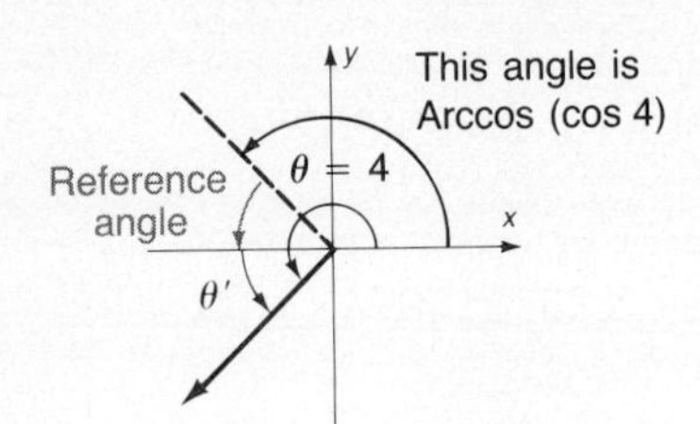

same reference angle as 4. The reference angle is $4 - \pi \approx 0.86$. In Quadrant II, π − reference angle ≈ 2.28. Therefore

$$\text{Arccos}(\cos 4) \approx 2.28$$

By calculator: 4 cos inv cos DISPLAY: 2.283185307 ■

EXAMPLE 11 $\sin(\text{Sin}^{-1} 0.463) = 0.463$ ■

EXAMPLE 12 $\sin(\text{Sin}^{-1} 2.463)$ is not defined since 2.463 is not between -1 and $+1$. ■

EXAMPLE 13 $\tan(\text{Tan}^{-1} 2.463) = 2.463$ ■

PROBLEM SET 5.5

A

In Problems 1–7, obtain the given angle from memory.

1. a. Arcsin 0 **b.** $\text{Tan}^{-1}(\frac{\sqrt{3}}{3})$
c. Arccot $\sqrt{3}$ **d.** Arccos 1

2. a. $\text{Cos}^{-1}(\frac{\sqrt{3}}{2})$ **b.** Arcsin $\frac{1}{2}$
c. $\text{Tan}^{-1} 1$ **d.** $\text{Sin}^{-1} 1$

3. a. Arctan $\sqrt{3}$ **b.** $\text{Cos}^{-1}(\frac{\sqrt{2}}{2})$
c. Arcsin$(\frac{1}{2}\sqrt{2})$ **d.** Arccot 1

4. a. Arcsin(-1) **b.** $\text{Cot}^{-1}(-1)$
c. Arcsin$(-\frac{\sqrt{3}}{2})$ **d.** $\text{Cos}^{-1}(-1)$

5. a. $\text{Cot}^{-1}(-\sqrt{3})$ **b.** Arctan(-1)
c. $\text{Sin}^{-1}(-\frac{1}{2}\sqrt{2})$ **d.** $\text{Cos}^{-1}(-\frac{1}{2})$

6. a. Arccos$(-\frac{\sqrt{2}}{2})$ **b.** $\text{Cot}^{-1}(-\frac{\sqrt{3}}{3})$
c. $\text{Sin}^{-1}(-\frac{1}{2})$ **d.** Arctan$(-\frac{\sqrt{3}}{3})$

7. a. $\text{Tan}^{-1} 0$ **b.** Arccot$(\frac{\sqrt{3}}{3})$
c. Arccos $\frac{1}{2}$ **d.** $\text{Sin}^{-1}(\frac{\sqrt{3}}{2})$

Use Table B.3 or a calculator to find the values (in radians) for the functions in Problems 8–25.

8. Arcsin 0.20846 **9.** $\text{Cos}^{-1} 0.83646$
10. Arctan 1.1156 **11.** $\text{Cot}^{-1}(-0.08097)$
12. $\text{Tan}^{-1}(-3.7712)$ **13.** Arccos(-0.94604)
14. $\text{Sin}^{-1} 0.75$ **15.** Arccos 0.25
16. Arctan 2 **17.** $\text{Tan}^{-1} 1.489$
18. $\text{Cot}^{-1} 3.451$ **19.** $\text{Sec}^{-1} 4.315$
20. $\text{Csc}^{-1} 5.791$ **21.** Arccsc 2.985
22. Arccot(-3) **23.** Arccot(-4)
24. Arctan(-2) **25.** Arcsec(-5)

Use Table B.3 or a calculator to find the values (in degrees) given in Problems 26–43.

26. $\text{Sin}^{-1} 0.3584$ **27.** $\text{Cos}^{-1} 0.3584$
28. Arccos 0.9455 **29.** $\text{Sin}^{-1}(-0.4695)$
30. $\text{Tan}^{-1} 2.050$ **31.** Arctan 1.036
32. $\text{Tan}^{-1}(-3.732)$ **33.** $\text{Cot}^{-1} 0.0875$
34. Arcsin(-0.9135) **35.** Arccot 0.7265
36. $\text{Cot}^{-1}(-0.3249)$ **37.** Arccot(-1.235)
38. $\text{Csc}^{-1} 2.816$ **39.** Arccot(-1)
40. Arccsc 3.945 **41.** Arccot(-2)
42. Arcsec(-6) **43.** Arctan(-3)

B

Simplify the expressions in Problems 44–58.

44. cot(Arccot 1) **45.** Arccos$[\cos(\frac{\pi}{6})]$
46. sin(Arcsin $\frac{1}{3}$) **47.** $\text{Tan}^{-1}[\tan(\frac{\pi}{15})]$
48. cos(Arccos $\frac{2}{3}$) **49.** Arcsin$[\sin(\frac{2\pi}{15})]$
50. Arccot(cot 35°) **51.** tan(Arctan 0.4163)
52. Arcsin(sin 4) **53.** Arccos(cos 5)
54. sin(Arcsin 0.7568) **55.** cos(Arccos 0.2836)
56. tan(Arctan 0.2910) **57.** Arctan(tan 2.5)
58. Arctan(Tan 2.5)

In Problems 59–61 graph the given pair of curves on the same axes.

59. $y = \text{Sin}\, x;\ y = \text{Sin}^{-1} x$ **60.** $y = \text{Cos}\, x;\ y = \text{Cos}^{-1} x$
61. $y = \text{Tan}\, x;\ y = \text{Tan}^{-1} x$

C

Graph the curves given in Problems 62–67.

62. $y + 2 = \text{Arctan}\, x$ **63.** $y - 1 = \text{Arcsin}\, x$
64. $y = 2\,\text{Cos}^{-1} x$ **65.** $y = 3\,\text{Sin}^{-1} x$
66. $y = \text{Arcsin}(x - 2)$ **67.** $y = \text{Arcsin}(x + 1)$

CHAPTER 5 SUMMARY

The material of this chapter is reviewed in the following list of objectives. After each objective there are some practice questions. For a sample test, select the first question of each set and check your answers with the answer section. For a sample test without answers, use the second question of each set. Additional practice is given by the other questions in each set. If you are having trouble with a particular type of problem, look back to that section for extra help.

5.1 ANGLES AND THE UNIT CIRCLE

Objective 1 Know the definition and equation of a unit circle. Know the definition and notation of an angle, including positive and negative angles. Know the Greek letters. Know what it means for an angle to be in standard position.

1. Name the Greek letter used for each angle: **a.** λ **b.** θ **c.** ϕ **d.** α **e.** β

Fill in the blanks.

2. A unit circle is ______________________________.
3. The equation of a unit circle is ______________________________.
4. An angle is in standard position if ______________________________.

Objective 2 Find angles coterminal with a given angle. Find the positive angle coterminal with the given angle and less than one revolution.

5. $-215°$ **6.** $\frac{11\pi}{3}$ **7.** $-\frac{5\pi}{6}$ **8.** $1000°$

Objective 3 Be familiar with the degree measure of an angle and be able to approximate the angle associated with a given degree measure without using any measuring devices. Draw the indicated angles.

9. $180°$ **10.** $120°$ **11.** $-30°$ **12.** $135°$

Objective 4 Be familiar with the radian measure of an angle and be able to approximate the angle associated with a given radian measure without using any measuring devices. Draw the indicated angles.

13. $\frac{\pi}{3}$ **14.** $\frac{\pi}{4}$ **15.** $\frac{5\pi}{6}$ **16.** 2

Objective 5 Change from radian to degree measure; know the commonly used degree and radian measure equivalences.

17. $\frac{3\pi}{2}$ **18.** 2 **19.** $-\frac{7\pi}{4}$ **20.** $\frac{5\pi}{6}$

Objective 6 Change from degree measure to radian measure; know the commonly used radian and degree measure equivalences. Use exact values when possible.

21. $300°$ **22.** $-45°$ **23.** $54°$ **24.** $-210°$

Objective 7 Know the arc length formula and be able to apply it.

25. The arc length formula is ______________ where s is the arc length, r is the ______________ and θ is ______________.
26. If the radius is 1 and the angle is 1, then what is the arc length?
27. If the minute hand on a clock is 15 cm long, how far does the tip move in 10 minutes?
28. A curve on a highway is laid out as the arc of a circle of radius 500 m. If the curve subtends a central angle of $18°$, what is the distance around this section of road? Give the exact answer and an answer rounded off to the nearest meter.

Objective 8 Find the reference angle θ for a given angle θ.

29. $300°$ **30.** $\frac{11\pi}{3}$ **31.** -4 **32.** $-215°$

5.2 TRIGONOMETRIC FUNCTIONS

Objective 9 Know the unit circle definition of the trigonometric functions. Fill in the blanks.

33. Let θ be ______________. Then the trigonometric functions are defined as follows:
34. ______________ ______________
35. ______________ ______________
36. ______________ ______________

Objective 10 Know the signs of the six trigonometric functions in each of the four quadrants. Name the function(s) which is (are) positive in the given quadrant.

37. I **38.** II **39.** III **40.** IV

Objective 11 Know that $\tan\theta = \dfrac{\sin\theta}{\cos\theta}$*; know the reciprocal relationships.*

41. If $\sin\theta = \frac{1}{2}$ and $\cos\theta = \frac{3}{4}$, then what is $\tan\theta$?
42. What is the reciprocal function of tangent?
43. What is the reciprocal function of cosine?
44. What is the reciprocal function of sine?

Objective 12 Be able to evaluate the trigonometric functions using tables or a calculator.

45. $\sec 23.4°$ **46.** $\cot 2.5$ **47.** $\csc 43.28°$ **48.** $\sin 7$

Objective 13 Know the ratio definition of the trigonometric functions.

49. If ____, then
50. ______________ ______________
51. ______________ ______________
52. ______________ ______________

Objective 14 Use the definition of the trigonometric functions to approximate their values for a given angle or for an angle passing through a given point. Assume that the terminal side passes through the given point.

53. $(5, -12)$ **54.** $(3, -4)$ **55.** $(-5, 2)$ **56.** $(4, 5)$

5.3 VALUES OF THE TRIGONOMETRIC FUNCTIONS

Objective 15 Know and be able to derive the table of exact values. Complete the table.

	Function	0	$\frac{\pi}{6}$	$\frac{\pi}{4}$	$\frac{\pi}{3}$	$\frac{\pi}{2}$	π	$\frac{3\pi}{2}$
57.	$\cos\theta$	a.	b.	c.	d.	e.	f.	g.
58.	$\csc\theta$	a.	b.	c.	d.	e.	f.	g.
59.	$\tan\theta$	a.	b.	c.	d.	e.	f.	g.
60.	$\sin\theta$	a.	b.	c.	d.	e.	f.	g.

Objective 16 Use the reduction principle, along with the table of exact values, to evaluate certain trigonometric functions.

61. $\cos(-\frac{5\pi}{3})$ **62.** $\sin(\frac{11\pi}{6})$ **63.** $\tan 135°$ **64.** $\cos(-210°)$

Objective 17 Use the reduction principle, along with tables, to approximate values of the trigonometric functions.

65. $\csc 43.28°$ **66.** $\sin 9$ **67.** $\sec 23.4°$ **68.** $\cot 2.5$

5.4 GRAPHS OF THE TRIGONOMETRIC FUNCTIONS

Objective 18 Graph the trigonometric functions, or variations, by plotting points.

69. $y = 2\cot\theta$
70. $y = 2\sec\theta$
71. $y = \frac{1}{2}\csc\theta$
72. $y = 2\cos\theta + \sin 2\theta$

Objective 19 Sketch $y = \cos x$, $y = \sin x$, *and* $y = \tan x$ *from memory; know their periods and amplitudes.*

73. $y = \sin x$ **74.** $y = \cos x$ **75.** $y = \tan x$ **76.** Fill in the blanks:

Function	Period	Amplitude
$\sin x$	**a.** ______	**b.** ______
$\cos x$	**c.** ______	**d.** ______
$\tan x$	**e.** ______	**f.** ______

Objective 20 Graph the general cosine, sine, and tangent curves.

77. $y = 2\cos\frac{2}{3}x$
78. $y = \cos(x + \frac{\pi}{4})$
79. $y - 2 = \sin(x - \frac{\pi}{6})$
80. $y = \tan(x - \frac{\pi}{3}) - 2$

5.5 INVERSE TRIGONOMETRIC FUNCTIONS

Objective 21 Know the definition of the inverse cosine, sine, tangent, and cotangent functions, especially the range of each. Fill in the blanks.

	Inverse Function	*Domain*	*Range*
81.	$y = \text{Arctan}\, x$	All reals	______________
82.	$y = \text{Cos}^{-1} x$	$-1 \le x \le 1$	______________
83.	$y = \text{Cot}^{-1} x$	All reals	______________
84.	$y = \text{Arcsin}\, x$	$-1 \le x \le 1$	______________

Objective 22 Evaluate inverse cosine, sine, tangent, and cotangent functions using exact values.

85. $\text{Arcsin}\,\frac{1}{2}$ **86.** $\text{Cot}^{-1}(\frac{1}{3}\sqrt{3})$ **87.** $\text{Arccos}(\frac{\sqrt{3}}{2})$ **88.** $\text{Tan}^{-1}(-\sqrt{3})$

Objective 23 Evaluate inverse cosine, sine, tangent, and cotangent functions using tables or a calculator. Answer in radians.

89. Arcsin 0.3140 **90.** Arccos(−0.6494) **91.** Arctan 3.271 **92.** Arccot 2

Objective 24 Graph the inverse cosine, sine, and tangent functions.

93. $y = \text{Sin}^{-1} x$ **94.** $y = \text{Arccos}\, x$ **95.** $y = \text{Arctan}\, x$ **96.** $y - 1 = \text{Sin}^{-1} x$

Objective 25 Simplify a function and its inverse function.

97. Arccos(cos 1) **98.** Arcsin(sin 1) **99.** Arcsin(cos 2) **100.** Arcsin(sin 2)

6

Analytic Trigonometry

Sesostris . . . made a division of the soil of Egypt among the inhabitants. . . . If the river carried away any portion of man's lot, . . . the king sent persons to examine, and determine by measurement the exact extent of the loss. . . . From this practice, I think, geometry first came to be known in Egypt, whence it passed into Greece.

HERODOTUS

Benjamin Banneker
(1731–1806)

* Optional Section

Analytic trigonometry has its roots in geometry and in the practical application of surveying. Surveying is the technique of locating points on or near the surface of the earth. The Egyptians are credited with being the first to do surveying. The annual flooding of the Nile and its constant destruction of property markers led the Egyptians to the principles of surveying. The first recorded measurements to determine the size of the earth were also made in Egypt when Eratosthenes measured a meridian arc in 230 B.C. (See Problem 62 of Section 5.1.) A modern-day example of a surveyor's dream is the city of Washington, D.C. The center of the city is the Capitol, and the city is divided into four sections; Northwest, Northeast, Southwest, and Southeast. These sections are separated by North, South, and East Capitol streets, which all converge at the Capitol. The initial surveying of this city was done by a group of mathematicians and surveyors who included the first distinguished Black mathematician, Benjamin Banneker.

Benjamin Banneker, born in 1731, was the first Black to be recognized as a mathematician and astronomer. He exhibited unusual talent in mathematics and, with little formal education, produced an almanac and invented a clock.

Banneker was born free in a troubled time of Black slavery. The right to vote was his by birth. It is important to note that this right to vote was denied him as well as other free Black Americans in 1803 in his native state of Maryland. Yet, history must acknowledge him as the first American Black man of science.

CHAPTER OVERVIEW Analytic trigonometry refers to trigonometric identities and the relationships of the trigonometric functions. The eight fundamental identities introduced in Section 6.2 are essential to your understanding of more advanced mathematics. Optional Section 6.6 uses trigonometry to find roots of equations and the last two sections introduce you to the idea of solving a triangle. The 22 objectives are listed on pages 274–278. The second Extended Application, about solar power, follows the chapter summary.

6.1 TRIGONOMETRIC EQUATIONS

Section 5.5 introduced notation for inverse functions. In this section we will use a similar notation to solve equations.

If $\cos x = \frac{1}{2}$, then write $x = \cos^{-1}(\frac{1}{2})$ to mean that x is *any* angle or real number whose cosine is $\frac{1}{2}$. Note the use of the small letter on $\cos^{-1}(\frac{1}{2})$ rather than the capital C we used for the inverse cosine function. The procedure for finding $\cos^{-1}(\frac{1}{2})$ relies on knowing $\text{Cos}^{-1}(\frac{1}{2})$. These steps are summarized:

1. Find $\text{Cos}^{-1}|y|$; this will give you the reference angle. It can be found by using the table of exact values, Appendix Table B3, or a calculator.
2. Find the principal values; use the sign of y to determine the proper quadrant placement. Use reference angles for finding the values of x less than one revolution that satisfy the equation.

   For cosine:
   - y positive: Quadrants I and IV
   - y negative: Quadrants II and III

   For sine:
   - y positive: Quadrants I and II
   - y negative: Quadrants III and IV

   For tangent:
   - y positive: Quadrants I and III
   - y negative: Quadrants II and IV

   Use:

Sine pos	All pos
Tangent pos	Cosine pos

3. For the entire solution, use the period of the function:

   For cosine and sine: add multiples of 2π or $360°$
   For tangent: add multiples of π or $180°$

EXAMPLE 1 Solve $\cos\theta = \frac{1}{2}$.

Solution *Step 1* $\theta' = \text{Cos}^{-1}|\frac{1}{2}| = 60°$ or $\frac{\pi}{3}$.

Step 2 $\frac{1}{2}$ is positive; cosine is positive in Quadrants I and IV; find the angles less than one revolution whose reference angle is $60°$ or $\frac{\pi}{3}$.

	Degrees	*Radians*
Quadrant I:	$60°$	$\frac{\pi}{3}$
Quadrant IV:	$300°$	$\frac{5\pi}{3}$

Step 3 Add multiples (let k be any integer):

$$\theta = \begin{cases} 60^\circ + 360^\circ k \\ 300^\circ + 360^\circ k \end{cases} \quad \text{or} \quad \theta = \begin{cases} \dfrac{\pi}{3} + 2k\pi \\ \dfrac{5\pi}{3} + 2k\pi \end{cases}$$

The solution is infinite. To check, select *any* integral value of k, say $k = 5$. From the solution, $300^\circ + 360^\circ(5) = 2100^\circ$. If this is a solution, it must satisfy the given equation: $\cos 2100^\circ = \frac{1}{2}$, which checks. ■

EXAMPLE 2 Solve $\cos\theta = -\frac{1}{2}$.

Solution $\theta' = \text{Cos}^{-1}|-\frac{1}{2}| = 60^\circ$ or $\frac{\pi}{3}$; this time the reference angle is in Quadrants II and III (since cosine is negative):

$$\theta = \begin{cases} 120^\circ + 360^\circ k \\ 240^\circ + 360^\circ k \end{cases} \quad \text{or} \quad \theta = \begin{cases} \dfrac{2\pi}{3} + 2k\pi \\ \dfrac{4\pi}{3} + 2k\pi \end{cases}$$

■

In trigonometry it is customary to delete the step in which multiples of 2π (for cosine and sine) or π (for tangent) are added to the principal values. In the examples that follow, you need only find the solutions less than one revolution. Give your solutions in radians unless specifically asked for degrees. The reason for this is that radians occur in calculus more often than degrees.

EXAMPLE 3 Solve $\sin\theta = -0.4446$ for $0^\circ \le \theta < 360^\circ$ (in degrees).

Solution **a.** From the table, $\theta' \approx 26.4^\circ$. Sine is negative in Quadrants III and IV:

206.4°, 333.6°

b. By calculator: [0.4446] [inv] [sin]

gives the reference angle 26.397749720; when placed in the proper quadrants, the solution is

206.3977497°, 333.6022503°

Quadrant III: $180^\circ + \theta'$ — Quadrant IV: $360^\circ - \theta'$ ■

EXAMPLE 4 Solve $\tan\theta = -0.66956$ for $0 \le \theta < 2\pi$.

Solution **a.** From the table, $\theta' \approx 0.59$. (Work in radians if degrees are not specified.) Tangent is negative in Quadrants II and IV:

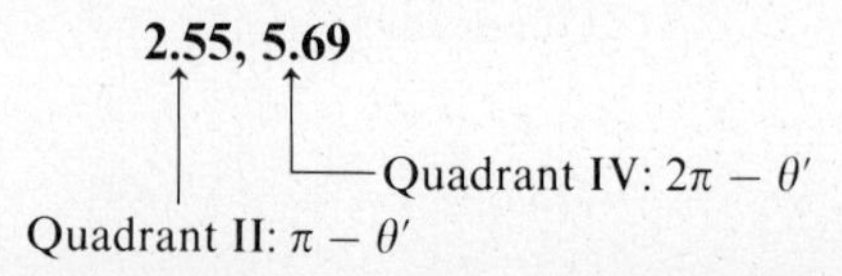

b. By calculator: .66956 | inv | tan

gives the reference angle 0.59000301; when placed in the proper quadrants, the solution is

$$\underset{\pi - \theta'}{\mathbf{2.5515896}}, \quad \underset{2\pi - \theta}{\mathbf{5.6931823}}$$

■

If the angle is a multiple of θ, the following inequalities hold true:

$$\begin{array}{ll} 0^\circ \le \theta < 360^\circ & 0 \le \theta < 2\pi \\ 0^\circ \le 2\theta < 720^\circ & 0 \le 2\theta < 4\pi \\ 0^\circ \le 3\theta < 1080^\circ & 0 \le 3\theta < 6\pi \\ 0^\circ \le 4\theta < 1440^\circ & 0 \le 4\theta < 8\pi \\ \vdots & \vdots \end{array}$$

EXAMPLE 5 Solve $\sin 2\theta = \frac{1}{2}$ for $0 \le \theta < 2\pi$.

Solution Since $0 \le \theta < 2\pi$, solve for 2θ such that $0 \le 2\theta < 4\pi$. Thus $\theta' = \frac{\pi}{6}$.

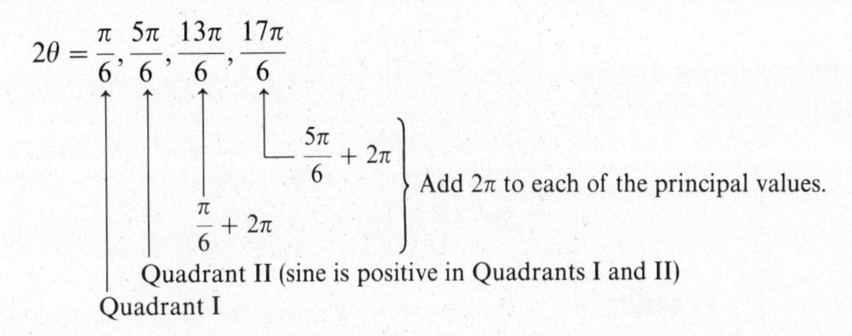

Mentally solve each equation for θ:

$$2\theta = \frac{\pi}{6} \qquad 2\theta = \frac{5\pi}{6} \qquad 2\theta = \frac{13\pi}{6} \qquad 2\theta = \frac{17\pi}{6}$$

As you solve these equations, notice that in each case θ is between zero and 2π. The solution is

$$\boldsymbol{\frac{\pi}{12}, \frac{5\pi}{12}, \frac{13\pi}{12}, \frac{17\pi}{12}}$$

■

EXAMPLE 6 Solve $\cos 3\theta = 1.2862$.

Solution The solution is **empty** since it is not true that $-1 \le \cos 3\theta < 1$, which is why $\cos 3\theta \ne 1.2862$. ■

EXAMPLE 7 Solve $\cos 3\theta = -0.68222$ for $0 \le \theta < 2\pi$.

Solution By table, $\theta' \approx 0.82$; 3θ is in Quadrants II and III and $0 \le 3\theta < 6\pi$.

$$3\theta \approx 2.32, 3.96, 8.60, 10.24, 14.89, 16.53$$

Divide by 3 to find the solution:

0.77, 1.32, 2.87, 3.41, 4.96, 5.51

By calculator: [.68222] [inv] [cos]

The reference angle is 0.82000165. To two decimal places the solution is the same as shown above. ■

Remember,

The unknown is x.

$$\cos(2x + 1) = 0$$

The angle is $2x + 1$.
The function is cosine.

The steps in solving a trigonometric equation are now given.

Procedure for Solving Trigonometric Equations

1. Solve for a single trigonometric function. You may use identities, factoring, or the Quadratic Formula.
2. Solve for the angle. You will use the definition of the inverse trigonometric functions for this step.
3. Solve for the unknown.

EXAMPLE 8 Solve $2\cos\theta\sin\theta = \sin\theta$ for $0 \le \theta < 2\pi$.

Solution This problem is solved by factoring:

$$2\cos\theta\sin\theta - \sin\theta = 0$$
$$\sin\theta(2\cos\theta - 1) = 0$$
$$\sin\theta = 0 \qquad 2\cos\theta - 1 = 0$$
$$\theta = 0, \pi \qquad \cos\theta = \frac{1}{2}$$
$$\theta = \frac{\pi}{3}, \frac{5\pi}{3}$$

Solution: $\mathbf{0, \pi, \frac{\pi}{3}, \frac{5\pi}{3}}$ ■

EXAMPLE 9 Solve $2\sin^2\theta = 1 + 2\sin\theta$ for $0 \le \theta < 2\pi$.

Solution This problem is solved by the Quadratic Formula: If $ax^2 + bx + c = 0, a \ne 0$, then

$$x = \frac{-b \pm \sqrt{b^2 - 4ac}}{2a}$$

Since $2\sin^2\theta - 2\sin\theta - 1 = 0$, let $x = \sin\theta$ in the Quadratic Formula:

$$\sin\theta = \frac{2 \pm \sqrt{4 - 4(2)(-1)}}{2(2)}$$

$$= \frac{1 \pm \sqrt{3}}{2}$$

$$\approx 1.366, \ -0.366025$$

(1.366: Reject since $-1 \le \sin\theta \le 1$.)

Solve $\sin\theta \approx -0.366025$ by using Table B.3 or a calculator to find a reference angle of 0.3747. Since the sine is negative in Quadrants III and IV, the solutions are $\pi + 0.3747 \approx 3.5163$ and $2\pi - 0.3747 \approx 5.9085$. To four decimal places: **3.5163, 5.9085.** ■

PROBLEM SET 6.1

A

Solve each of the equations in Problems 1–12. Use exact values.

1. $\sin x = \frac{1}{2}$ $(0 \le x < \frac{\pi}{2})$ **2.** $\sin x = \frac{1}{2}$ $(0 \le x < 90°)$

3. $\sin x = -\frac{1}{2}$ $(-\frac{\pi}{2} \le x \le \frac{\pi}{2})$

4. $\sin x = -\frac{1}{2}$ $(0 \le x \le \pi)$

5. $\sin x = \frac{1}{2}$ **6.** $\sin x = -\frac{1}{2}$

7. $\cos x = \frac{1}{2}$ $(0 \le x < \frac{\pi}{2})$ **8.** $\cos x = \frac{1}{2}$ $(0 \le x < 90°)$

9. $\cos x = -\frac{1}{2}$ $(-\frac{\pi}{2} \le x \le \frac{\pi}{2})$

10. $\cos x = -\frac{1}{2}$ $(0 \le x \le \pi)$

11. $\cos x = \frac{1}{2}$ **12.** $\cos x = -\frac{1}{2}$

Solve each of the equations in Problems 13–27 for exact values such that $0 \le x < 2\pi$.

13. $\cos 2x = \frac{1}{2}$ **14.** $\cos 3x = \frac{1}{2}$

15. $\cos 2x = -\frac{1}{2}$ **16.** $\sin 2x = \frac{\sqrt{2}}{2}$

17. $\sin 2x = -\frac{\sqrt{3}}{2}$ **18.** $\sin 3x = \frac{\sqrt{2}}{2}$

19. $\tan 3x = 1$ **20.** $\tan 3x = -1$

21. $\sec 2x = -\frac{2\sqrt{3}}{3}$ **22.** $(\sin x)(\cos x) = 0$

23. $(\sec x)(\tan x) = 0$ **24.** $(\sin x)(\cot x) = 0$

25. $(\cot x)(\cos x) = 0$

26. $(\csc x - 2)(2\cos x - 1) = 0$

27. $(\sec x - 2)(2\sin x - 1) = 0$

B

Solve each of the equations in Problems 28–41 for $0 \le x < 2\pi$. Use exact values where possible, but state approximate answers to four decimal places.

28. $\tan^2 x = \sqrt{3}\tan x$ **29.** $\tan^2 x = \tan x$

30. $\sin^2 x = \frac{1}{2}$ **31.** $\cos^2 x = \frac{1}{2}$

32. $3\sin x\cos x = \sin x$ **33.** $2\cos x\sin x = \sin x$

34. $\sin^2 x - \sin x - 2 = 0$ **35.** $\cos^2 x - 1 - \cos x = 0$

36. $4\cot^2 x - 8\cot x + 3 = 0$ **37.** $\tan^2 x - 3\tan x + 1 = 0$

38. $\sec^2 x - \sec x - 1 = 0$ **39.** $\csc^2 x - \csc x - 1 = 0$

40. $\cos x + 1 = \sqrt{3}$ **41.** $\sin x + 1 = \sqrt{3}$

Solve each of the equations in Problems 42–55 for $0 \le x < 2\pi$ correct to two decimal places.

42. $2\cos 2x\sin 2x = \sin 2x$ **43.** $\sin 2x + 2\cos x\sin 2x = 0$

44. $\cos 3x + 2\sin 2x\cos 3x = 0$

45. $\sin 2x + 1 = \sqrt{3}$ **46.** $\cos 3x - 1 = \sqrt{2}$

47. $\tan 2x + 1 = \sqrt{3}$ **48.** $1 - 2\sin^2 x = \sin x$

49. $2\cos^2 x - 1 = \cos x$ **50.** $2\cos x\sin x + \cos x = 0$

51. $1 - \sin x = 1 - 2\sin^2 x$ **52.** $\cos x = 2\sin x\cos x$

53. $\sin^2 3x + \sin 3x + 1 = 1 - \sin^2 3x$

54. $3\sin 4x - \sin 3x + \sqrt{3} = 2\sin 2x - \sin 3x + 3\sin 4x$

55. $3\cos x - \cos 3x + 2\cos 5x = 1 + 3\cos x - \cos 3x$

Find all solutions in radian measure of the equations given in Problems 56–61.

56. $\tan x = -\sqrt{3}$ **57.** $\cos x = -\frac{\sqrt{3}}{2}$ **58.** $\sin x = -\frac{\sqrt{2}}{2}$

59. $\sin x = 0.3907$ **60.** $\cos x = 0.2924$ **61.** $\tan x = 1.376$

C

62. ***Sound Waves*** A tuning fork vibrating at 264 Hz ($f = 264$) with an amplitude of .0050 cm produces C on the musical scale and can be described by an equation of the form

$$y = .0050 \sin 528\pi x$$

Find the smallest positive value of x (correct to four decimal places) for which $y = .0020$.

63. ***Electrical*** In a certain electric circuit, the electromotive force V in volts and the time t (in seconds) are related by an equation of the form

$$V = \cos 2\pi t$$

Find the smallest positive value for t (correct to three decimal places) for which $V = .400$.

64. ***Space Science*** The orbit of a certain satellite alternates above and below the equator according to the equation

$$y = 4000 \sin\left(\frac{\pi}{45}t + \frac{5\pi}{18}\right)$$

where t is the time in minutes and y is the distance (in kilometers) from the equator. Find the times at which the satellite crosses the equator during the first hour and a half (that is, for $0 \leq t \leq 90$).

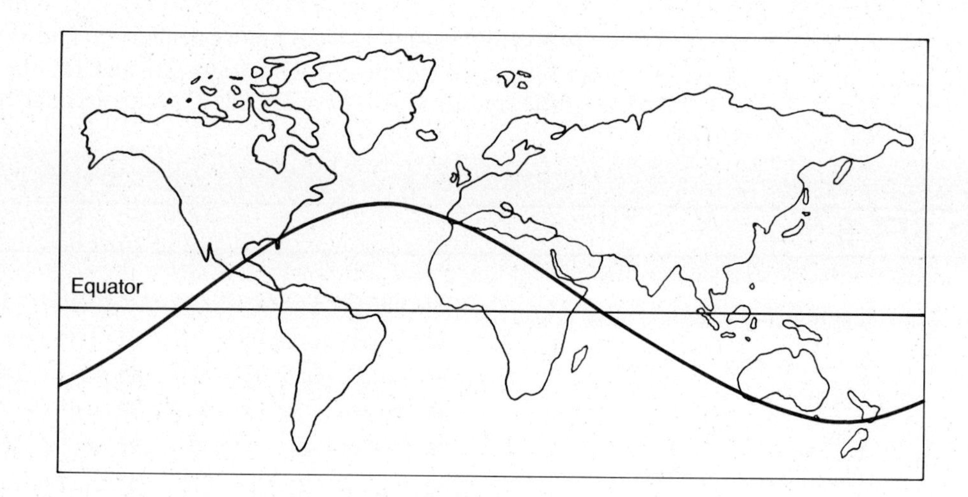

6.2 FUNDAMENTAL IDENTITIES

In the last section we solved some trigonometric equations. In this section we focus on trigonometric identities. The procedure for proving identities is quite different from the procedure for solving equations. Compare the following two examples from algebra:

SOLVE: $2x + 3x = 5$

SOLUTION: $2x + 3x = 5$ Given

$5x = 5$ Combining similar terms

$x = 1$ Multiply both sides by $\frac{1}{5}$

PROVE: $2x + 3x = 5x$

SOLUTION: $2x + 3x = (2 + 3)x$ Distributive property

$= 5x$ Closure

$2x + 3x = 5x$ Transitive property

Notice that when *solving* the equation, $2x + 3x = 5$ was given and *used as a starting point*. On the other hand, when *proving* the identity, $2x + 3x = 5x$ *could not be used as a starting point*. Indeed $2x + 3x = 5x$ was the *last step*, not the first step.

The reason for this difference in procedure is apparent if you look at the addition and multiplication principles:

Addition principle: If $a = b$, then $a + c = b + c$.

Multiplication principle: If $a = b$, then $ac = bc$.

In both cases you must *know* that $a = b$ before you can use the addition or multiplication principles. If you are asked to *prove* that $a = b$, you cannot assume $a = b$ to work the problem. You must begin with what is known to be true and *end* with the given identity.

All our work with trigonometric identities is ultimately based on eight basic identities called the **fundamental identities**. Notice that these identities are classified into three categories and numbered for later reference. Values of θ that cause division by zero are excluded.

Fundamental Identities

RECIPROCAL IDENTITIES:

1. $\sec\theta = \dfrac{1}{\cos\theta}$ 2. $\csc\theta = \dfrac{1}{\sin\theta}$ 3. $\cot\theta = \dfrac{1}{\tan\theta}$

RATIO IDENTITIES:

4. $\tan\theta = \dfrac{\sin\theta}{\cos\theta}$ 5. $\cot\theta = \dfrac{\cos\theta}{\sin\theta}$

PYTHAGOREAN IDENTITIES:

6. $\sin^2\theta + \cos^2\theta = 1$ 7. $1 + \tan^2\theta = \sec^2\theta$ 8. $1 + \cot^2\theta = \csc^2\theta$

The proofs of these identities follow directly from the definitions of the trigonometric functions. Some proofs were given in the last chapter, but they are repeated here and in Problem Set 6.2.

Let θ be an angle in standard position with point $P(x, y)$ on the terminal side a distance of r from the origin, with $r \neq 0$.

Proof of Identity 1 By definition, $\cos\theta = \frac{x}{r}$; thus

$$\frac{1}{\cos\theta} = \frac{1}{x/r}$$

$$= 1 \cdot \frac{r}{x} \qquad \text{Division of fractions}$$

$$= \frac{r}{x} \qquad \text{Multiplication of fractions}$$

$$= \sec\theta \qquad \text{By definition of } \sec\theta$$

Therefore $\dfrac{1}{\cos\theta} = \sec\theta$. □

Identities 2 and 3 are proved in precisely the same way and are left as problems.

Proof of Identity 4

$$\frac{\sin\theta}{\cos\theta} = \frac{y/r}{x/r} \quad \text{By definition of } \sin\theta \text{ and } \cos\theta$$

$$= \frac{y}{r}\cdot\frac{r}{x} \quad \text{Division of fractions}$$

$$= \frac{y}{x} \quad \text{Multiplication and simplification of fractions}$$

$$= \tan\theta \quad \text{By definition of } \tan\theta$$

Therefore $\dfrac{\sin\theta}{\cos\theta} = \tan\theta$. □

The proof of Identity 5 is similar to the proof of Identity 4 and is left as a problem. For Identities 6, 7, and 8, begin with the Pythagorean Theorem (which is why these are called the *Pythagorean identities*):

$$x^2 + y^2 = r^2 \quad \text{By the Pythagorean Theorem}$$

To prove Identity 6, divide both sides by r^2; for Identity 7 divide by x^2; and for Identity 8, divide by y^2. We will show the details for Identity 6 and leave Identities 7 and 8 as problems.

Proof of Identity 6

$$x^2 + y^2 = r^2 \quad \text{Pythagorean Theorem}$$

$$\frac{x^2}{r^2} + \frac{y^2}{r^2} = \frac{r^2}{r^2} \quad \text{Dividing both sides by } r^2\ (r \neq 0)$$

$$\left(\frac{x}{r}\right)^2 + \left(\frac{y}{r}\right)^2 = 1 \quad \text{Properties of exponents}$$

$$(\cos\theta)^2 + (\sin\theta)^2 = 1 \quad \text{Definition of } \cos\theta \text{ and } \sin\theta$$

$$\sin^2\theta + \cos^2\theta = 1 \quad \text{Commutative property}$$ □

EXAMPLE 1 Write all six trigonometric functions in terms of $\sin\theta$.

Solution

a. $\sin\theta = \sin\theta$

b. $\cos\theta = \pm\sqrt{1 - \sin^2\theta}$ From Identity 6

c. $\tan\theta = \dfrac{\sin\theta}{\cos\theta}$ Identity 4

$= \dfrac{\sin\theta}{\pm\sqrt{1 - \sin^2\theta}}$ From part b

d. $\cot\theta = \dfrac{1}{\tan\theta}$ Identity 5

$= \dfrac{\pm\sqrt{1 - \sin^2\theta}}{\sin\theta}$ From part c

e. $\csc\theta = \dfrac{1}{\sin\theta}$ From Identity 2

f. $\sec\theta = \dfrac{1}{\cos\theta}$ From Identity 1

$= \dfrac{1}{\pm\sqrt{1-\sin^2\theta}}$ From part b ■

The $\pm$ sign we have been using, as in

$$\cos\theta = \pm\sqrt{1-\sin^2\theta}$$

means that $\cos\theta$ is positive for some values of θ and negative for other values of θ. The plus or the minus sign is chosen by determining the proper quadrant, as shown in Example 2.

EXAMPLE 2 Given $\sin\theta = \frac{3}{5}$ and $\tan\theta < 0$, find the other functions of θ.

Solution Since the tangent is negative and the sine is positive, the quadrant is II. Thus

$$\begin{aligned}\cos\theta &= -\sqrt{1-\sin^2\theta} \quad \text{Since the cosine is negative in Quadrant II}\\ &= -\sqrt{1-(\tfrac{3}{5})^2}\\ &= -\sqrt{1-\tfrac{9}{25}}\\ &= -\sqrt{\tfrac{16}{25}}\\ &= -\tfrac{4}{5}\end{aligned}$$

Also,

$$\tan\theta = \frac{\sin\theta}{\cos\theta} = \frac{3/5}{-4/5} = -\frac{3}{4}$$

Using the reciprocal identities, $\cot\theta = -\frac{4}{3}$, $\sec\theta = -\frac{5}{4}$, and $\csc\theta = \frac{5}{3}$. ■

PROBLEM SET 6.2

A

1. State from memory the eight fundamental identities.

In Problems 2–9, state the quadrant or quadrants in which θ may lie to make the expression true.

2. $\sin\theta = \sqrt{1-\cos^2\theta}$

3. $\sin\theta = -\sqrt{1-\cos^2\theta}$

4. $\sec\theta = -\sqrt{1+\tan^2\theta}$

5. $\sec\theta = \sqrt{1+\tan^2\theta}$

6. $\csc\theta = \sqrt{1+\cot^2\theta}$; $\tan\theta < 0$

7. $\cos\theta = -\sqrt{1-\sin^2\theta}$; $\sin\theta > 0$

8. $\tan\theta = \sqrt{\sec^2\theta - 1}$; $\cos\theta < 0$

9. $\csc\theta = \sqrt{1+\cot^2\theta}$; $\cos\theta > 0$

Write each of the expressions in Problems 10–18 as a single trigonometric function of some angle by using one of the eight fundamental identities.

10. $\dfrac{\sin 50^\circ}{\cos 50^\circ}$

11. $\dfrac{\cos(A+B)}{\sin(A+B)}$

12. $\dfrac{1}{\sec 75^\circ}$

13. $\dfrac{1}{\cot(\frac{\pi}{15})}$

14. $\tan 42^\circ \cos 42^\circ$

15. $\cot\frac{\pi}{8}\sin\frac{\pi}{8}$

16. $1-\cos^2 18^\circ$

17. $-\sqrt{1-\sin^2 127^\circ}$

18. $\sec^2(\frac{\pi}{6}) - 1$

Evaluate the expressions in Problems 19–28 by using one of the eight fundamental identities.

19. $\cos 128° \sec 128°$

20. $\sin^2 \frac{\pi}{3} + \cos^2 \frac{\pi}{3}$

21. $\sec^2 \frac{\pi}{6} - \tan^2 \frac{\pi}{6}$

22. $\cot^2 45° - \csc^2 45°$

23. $\tan^2 135° - \sec^2 135°$

24. $\csc 85° \sin 85°$

25. Prove that $\csc\theta = 1/\sin\theta$.

26. Prove that $\cot\theta = 1/\tan\theta$.

27. Prove that $1 + \tan^2\theta = \sec^2\theta$.

28. Prove that $1 + \cot^2\theta = \csc^2\theta$.

B

In Problems 29–33, write all the trigonometric functions in terms of the given function.

29. $\cos\theta$ **30.** $\tan\theta$ **31.** $\cot\theta$ **32.** $\sec\theta$

33. $\csc\theta$

In Problems 34–45, find the other functions of θ using the given information.

34. $\cos\theta = \frac{3}{5}$; $\tan\theta > 0$

35. $\cos\theta = \frac{3}{5}$; $\csc\theta < 0$

36. $\cos\theta = \frac{5}{13}$; $\tan\theta < 0$

37. $\cos\theta = \frac{5}{13}$; $\tan\theta > 0$

38. $\tan\theta = \frac{5}{12}$; $\sin\theta > 0$

39. $\tan\theta = \frac{5}{12}$; $\sin\theta < 0$

40. $\sin\theta = \frac{2}{3}$; $\sec\theta > 0$

41. $\sin\theta = \frac{2}{3}$; $\sec\theta < 0$

42. $\sec\theta = \frac{\sqrt{34}}{5}$; $\tan\theta < 0$

43. $\sec\theta = \frac{\sqrt{34}}{5}$; $\tan > 0$

44. $\csc\theta = -\frac{\sqrt{10}}{3}$; $\cos\theta > 0$

45. $\csc\theta = -\frac{\sqrt{10}}{3}$; $\cos\theta < 0$

Simplify the expressions in Problems 46–54 using only sines, cosines, and the fundamental identities.

46. $\dfrac{1 - \sin^2\theta}{\cos\theta}$

47. $\dfrac{1 - \cos^2\theta}{\sin\theta}$

48. $\dfrac{\sin\theta}{\cos\theta} + \dfrac{\cos\theta}{\sin\theta}$

49. $\dfrac{1}{1 + \cos\theta} + \dfrac{1}{1 - \cos\theta}$

50. $\dfrac{\dfrac{\sin\theta}{\cos\theta} + \dfrac{\cos\theta}{\sin\theta}}{\dfrac{1}{\sin\theta\cos\theta}}$

51. $\sin\theta + \dfrac{\cos^2\theta}{\sin\theta}$

52. $\dfrac{\cos\theta + \dfrac{\sin^2\theta}{\cos\theta}}{\sin\theta}$

53. $\dfrac{\sin\theta - \dfrac{\cos^2\theta}{\sin\theta}}{\cos\theta}$

54. $\dfrac{\dfrac{\cos^4\theta}{\sin^2\theta} + \cos^2\theta}{\dfrac{\cos^2\theta}{\sin^2\theta}}$

Reduce the expressions in Problems 55–62 so that they involve only sines and cosines, and then simplify.

55. $\sin\theta + \cot\theta$

56. $\sec\theta + \tan\theta$

57. $\dfrac{\tan\theta + \cot\theta}{\sec\theta\csc\theta}$

58. $\dfrac{\sec\theta + \csc\theta}{\tan\theta\cot\theta}$

59. $\sec^2\theta + \tan^2\theta$

60. $\csc^2\theta + \cot^2\theta$

61. $(\cot\theta - \sec\theta)(\sin\theta\cos\theta)$

62. $(\tan\theta - \csc\theta)(\cos\theta\sin\theta)$

6.3 PROVING IDENTITIES

In the last section we considered eight fundamental identities, which are used to simplify and change the form of a variety of trigonometric expressions. Suppose you are given a trigonometric equation such as

$$\tan\theta + \cot\theta = \sec\theta\csc\theta$$

and are asked to show that it is an identity. You must be careful not to treat this problem as though it were an algebraic equation. When asked to prove an identity, do *not* start with the given expression, since you cannot assume it is true. You should *begin* with what you know is true and *end* with the given identity. There are three ways to proceed:

1. Reduce the left-hand side to the right-hand side by using algebra and the fundamental identities.
2. Reduce the right-hand side to the left-hand side.
3. Reduce both sides independently to the same expression.

EXAMPLE 1 Prove that $\tan\theta + \cot\theta = \sec\theta\csc\theta$.

Solution Begin with either the left- or right-hand side:

$$\begin{aligned}\tan\theta + \cot\theta &= \frac{\sin\theta}{\cos\theta} + \frac{\cos\theta}{\sin\theta}\\ &= \frac{\sin^2\theta + \cos^2\theta}{\cos\theta\sin\theta}\\ &= \frac{1}{\cos\theta\sin\theta}\end{aligned}$$

This is algebraically simplified, so return to the other side and begin anew:

$$\begin{aligned}\sec\theta\csc\theta &= \frac{1}{\cos\theta}\cdot\frac{1}{\sin\theta}\\ &= \frac{1}{\cos\theta\sin\theta}\end{aligned}$$

This too is simplified, but notice that the simplified forms for both the left and right sides are the same. Therefore

$$\tan\theta + \cot\theta = \sec\theta\csc\theta$$ ■

Usually it is easier to begin with the more complicated side and try to reduce it to the simpler side. If both sides seem equally complex, you might change all the functions to sines and cosines and then simplify.

EXAMPLE 2 Prove that $2\csc^2\theta = \dfrac{1}{1+\cos\theta} + \dfrac{1}{1-\cos\theta}$.

Solution Begin with the more complicated side:

$$\begin{aligned}\frac{1}{1+\cos\theta} + \frac{1}{1-\cos\theta} &= \frac{(1-\cos\theta)+(1+\cos\theta)}{(1+\cos\theta)(1-\cos\theta)}\\ &= \frac{2}{1-\cos^2\theta}\\ &= \frac{2}{\sin^2\theta}\\ &= 2\csc^2\theta\end{aligned}$$ ■

EXAMPLE 3 Prove that $\dfrac{\sec 2\lambda + \cot 2\lambda}{\sec 2\lambda} = 1 + \csc 2\lambda - \sin 2\lambda$.

Solution Begin with the left-hand side. When working with a fraction consisting of a single function as a denominator, it is often helpful to separate the fraction into the sum of several fractions:

$$\begin{aligned}\frac{\sec 2\lambda + \cot 2\lambda}{\sec 2\lambda} &= \frac{\sec 2\lambda}{\sec 2\lambda} + \frac{\cot 2\lambda}{\sec 2\lambda}\\ &= 1 + \cot 2\lambda \cdot \frac{1}{\sec 2\lambda}\\ &= 1 + \frac{\cos 2\lambda}{\sin 2\lambda} \cdot \cos 2\lambda\\ &= 1 + \frac{\cos^2 2\lambda}{\sin 2\lambda}\\ &= 1 + \frac{1 - \sin^2 2\lambda}{\sin 2\lambda}\\ &= 1 + \frac{1}{\sin 2\lambda} - \frac{\sin^2 2\lambda}{\sin 2\lambda}\\ &= 1 + \csc 2\lambda - \sin 2\lambda\end{aligned}$$ ■

EXAMPLE 4 Prove that $\dfrac{\cos\theta}{1 - \sin\theta} = \dfrac{1 + \sin\theta}{\cos\theta}$.

Solution Sometimes, when there is a binomial in the numerator or denominator, the identity can be proved by multiplying one side by 1, where 1 is written in the form of the conjugate of the binomial. When changing one side, keep a sharp eye on the other side, since it often gives a clue about what to do. Thus in this example we can multiply the numerator and denominator of the left-hand side by $1 + \sin\theta$:

$$\begin{aligned}\frac{\cos\theta}{1 - \sin\theta} &= \frac{\cos\theta}{1 - \sin\theta} \cdot \frac{\mathbf{1 + \sin\theta}}{\mathbf{1 + \sin\theta}}\\ &= \frac{\cos\theta(1 + \sin\theta)}{1 - \sin^2\theta}\\ &= \frac{\cos\theta(1 + \sin\theta)}{\cos^2\theta}\\ &= \frac{1 + \sin\theta}{\cos\theta}\end{aligned}$$ ■

EXAMPLE 5 Prove that $\dfrac{\sec^2 2\theta - \tan^2 2\theta}{\tan 2\theta + \sec 2\theta} = \dfrac{\cos 2\theta}{1 + \sin 2\theta}$.

Solution Sometimes the identity can be proved by factoring:

$$\begin{aligned}\frac{\sec^2 2\theta - \tan^2 2\theta}{\tan 2\theta + \sec 2\theta} &= \frac{(\sec 2\theta + \tan 2\theta)(\sec 2\theta - \tan 2\theta)}{\tan 2\theta + \sec 2\theta}\\ &= \sec 2\theta - \tan 2\theta\\ &= \frac{1}{\cos 2\theta} - \frac{\sin 2\theta}{\cos 2\theta}\end{aligned}$$

$$= \frac{1 - \sin 2\theta}{\cos 2\theta}$$

$$= \frac{1 - \sin 2\theta}{\cos 2\theta} \cdot \frac{\mathbf{1 + \sin 2\theta}}{\mathbf{1 + \sin 2\theta}}$$

$$= \frac{1 - \sin^2 2\theta}{\cos 2\theta(1 + \sin 2\theta)}$$

$$= \frac{\cos^2 2\theta}{\cos 2\theta(1 + \sin 2\theta)}$$

$$= \frac{\cos 2\theta}{1 + \sin 2\theta}$$ ■

EXAMPLE 6 Prove that $\dfrac{-2 \sin\theta \cos\theta}{1 - \sin\theta - \cos\theta} = 1 + \sin\theta + \cos\theta$.

Solution Sometimes, when there is a fraction on one side, the identity can be proved by multiplying the other side by 1 written so that the desired denominator is obtained. Thus, for this example,

$$1 + \sin\theta + \cos\theta$$

$$= (1 + \sin\theta + \cos\theta) \cdot \frac{\mathbf{1 - \sin\theta - \cos\theta}}{\mathbf{1 - \sin\theta - \cos\theta}}$$

$$= \frac{(1 + \sin\theta + \cos\theta)(1 - \sin\theta - \cos\theta)}{1 - \sin\theta - \cos\theta}$$

$$= \frac{1 - \sin\theta - \cos\theta + \sin\theta - \sin^2\theta - \sin\theta\cos\theta + \cos\theta - \cos\theta\sin\theta - \cos^2\theta}{1 - \sin\theta - \cos\theta}$$

$$= \frac{1 - (\sin^2\theta + \cos^2\theta) - 2\sin\theta\cos\theta}{1 - \sin\theta - \cos\theta}$$

$$= \frac{-2\sin\theta\cos\theta}{1 - \sin\theta - \cos\theta}$$ ■

In summary, there is no single method that is best for proving identities. However, the following hints should help:

Procedures for Proving Identities

1. If one side contains one function only, write all the trigonometric functions on the other side in terms of that function.
2. If the denominator of a fraction consists of only one function, break up the fraction.
3. Simplify by combining fractions.
4. Factoring is sometimes helpful.
5. Change all trigonometric functions to sines and cosines and simplify.
6. Multiply by the conjugate of either the numerator or the denominator.
7. If there are squares of functions, look for alternate forms of the Pythagorean identities.
8. Avoid the introduction of radicals.

PROBLEM SET 6.3

A

Prove that the equations in Problems 1–38 are identities.

1. $\sin\theta = \sin^3\theta + \cos^2\theta\sin\theta$
2. $\sec\theta = \sec\theta\sin^2\theta + \cos\theta$
3. $\tan\theta = \cot\theta\tan^2\theta$
4. $\dfrac{\sin\theta\cos\theta + \sin^2\theta}{\sin\theta} = \cos\theta + \sin\theta$
5. $\tan^2\theta - \sin^2\theta = \tan^2\theta\sin^2\theta$
6. $\cot^2\theta\cos^2\theta = \cot^2\theta - \cos^2\theta$
7. $\tan A + \cot A = \sec A\csc A$
8. $\cot A = \csc A\sec A - \tan A$
9. $\sin x + \cos x = \dfrac{\sec x + \csc x}{\csc x\sec x}$
10. $\dfrac{\cos\gamma + \tan\gamma\sin\gamma}{\sec\gamma} = 1$
11. $\dfrac{1 - \sec^2 t}{\sec^2 t} = -\sin^2 t$
12. $\dfrac{1 + \cot^2 t}{\cot^2 t} = \sec^2 t$
13. $(\sec\theta - \cos\theta)^2 = \tan^2\theta - \sin^2\theta$
14. $\dfrac{\sin\theta}{\csc\theta} + \dfrac{\cos\theta}{\sec\theta} = 1$
15. $1 - \sin 2\theta = \dfrac{1 - \sin^2 2\theta}{1 + \sin 2\theta}$
16. $\dfrac{1 - \tan^2 3\theta}{1 - \tan 3\theta} = 1 + \tan 3\theta$
17. $\sin\lambda = \dfrac{\sin^2\lambda + \sin\lambda\cos\lambda + \sin\lambda}{\sin\lambda + \cos\lambda + 1}$
18. $\dfrac{1 + \cot 2\lambda\sec 2\lambda}{\tan 2\lambda + \sec 2\lambda} = \cot 2\lambda$
19. $\sin 2\alpha\cos 2\alpha(\tan 2\alpha + \cot 2\alpha) = 1$
20. $(\sin\beta - \cos\beta)^2 + (\sin\beta + \cos\beta)^2 = 2$
21. $\csc 3\beta - \cos 3\beta\cot 3\beta = \sin 3\beta$
22. $\dfrac{1 + \cot^2 A}{1 + \tan^2 A} = \cot^2 A$
23. $\dfrac{\sin^2 B - \cos^2 B}{\sin B + \cos B} = \sin B - \cos B$
24. $\dfrac{\tan^2\gamma - \cot^2\gamma}{\tan\gamma + \cot\gamma} = \tan\gamma - \cot\gamma$
25. $\tan^2 2\gamma + \sin^2 2\gamma + \cos^2 2\gamma = \sec^2 2\gamma$
26. $\cot^2 C + \cos^2 C + \sin^2 C = \csc^2 C$
27. $\dfrac{\tan\theta + \cot\theta}{\sec\theta\csc\theta} = 1$
28. $\dfrac{\tan\theta - \cot\theta}{\sec\theta\csc\theta} = \sin^2\theta - \cos^2\theta$
29. $1 + \sin^2\lambda = 2 - \cos^2\lambda$
30. $2 - \sin^2 3\lambda = 1 + \cos^2 3\lambda$
31. $\dfrac{\sin\alpha}{\tan\alpha} + \dfrac{\cos\alpha}{\cot\alpha} = \cos\alpha + \sin\alpha$
32. $\dfrac{1}{1 + \cos 2\alpha} + \dfrac{1}{1 - \cos 2\alpha} = 2\csc^2 2\alpha$
33. $\sec\beta + \cos\beta = \dfrac{2 - \sin^2\beta}{\cos\beta}$
34. $2\sin^2 3\beta - 1 = 1 - 2\cos^2 3\beta$
35. $\dfrac{\tan 2\theta + \cot 2\theta}{\sec 2\theta} = \csc 2\theta$
36. $\dfrac{\tan 3\theta + \cot 3\theta}{\csc 3\theta} = \sec 3\theta$
37. $\dfrac{\sec\lambda + \tan^2\lambda}{\sec\lambda} = 1 + \sec\lambda - \cos\lambda$
38. $\dfrac{\sin 2\lambda}{\tan 2\lambda} + \dfrac{\cos 2\lambda}{\cot 2\lambda} = \cos 2\lambda + \sin 2\lambda$

B

Prove that the equations in Problems 39–60 are identities.

39. $\dfrac{1 + \tan C}{1 - \tan C} = \dfrac{\sec^2 C + 2\tan C}{2 - \sec^2 C}$
40. $(\cot x + \csc x)^2 = \dfrac{\sec x + 1}{\sec x - 1}$
41. $\dfrac{\sin^3 x - \cos^3 x}{\sin x - \cos x} = 1 + \sin x\cos x$
42. $\dfrac{\tan^3 t - \cot^3 t}{\tan t - \cot t} = \sec^2 t + \cot^2 t$
43. $\dfrac{1 - \cos\theta}{1 + \cos\theta} = \left(\dfrac{1 - \cos\theta}{\sin\theta}\right)^2$
44. $\dfrac{(\sec^2\gamma + \tan^2\gamma)^2}{\sec^4\gamma - \tan^4\gamma} = 1 + 2\tan^2\gamma$
45. $\dfrac{(\cos^2\gamma - \sin^2\gamma)^2}{\cos^4\gamma - \sin^4\gamma} = 2\cos^2\gamma - 1$
46. $(\sec 2\theta + \csc 2\theta)^2 = \dfrac{1 + 2\sin 2\theta\cos 2\theta}{\cos^2 2\theta\sin^2 2\theta}$

47. $\dfrac{1}{\sec\theta + \tan\theta} = \sec\theta - \tan\theta$

48. $\csc\theta + \cot\theta = \dfrac{1}{\csc\theta - \cot\theta}$

49. $\sec^2 2\lambda + \csc^2 2\lambda = \csc^2 2\lambda \sec^2 2\lambda$

50. $\dfrac{1 + \tan^3\theta}{1 + \tan\theta} = \sec^2\theta - \tan\theta$

51. $\dfrac{1 - \sec^3\theta}{1 - \sec\theta} = \tan^2\theta + \sec\theta + 2$

52. $\dfrac{\cos^2\theta - \cos\theta\csc\theta}{\cos^2\theta\csc\theta - \cos\theta\csc^2\theta} = \sin\theta$

53. $\dfrac{\tan^2\theta - 2\tan\theta}{2\tan\theta - 4} = \dfrac{1}{2}\tan\theta$

54. $\dfrac{\tan\theta}{\cot\theta} - \dfrac{\cot\theta}{\tan\theta} = \sec^2\theta - \csc^2\theta$

55. $\sqrt{(3\cos\theta - 4\sin\theta)^2 + (3\sin\theta + 4\cos\theta)^2} = 5$

56. $\dfrac{\cos\theta + \cos^2\theta}{\cos\theta + 1} = \dfrac{\cos\theta\sin\theta + \cos^2\theta}{\sin\theta + \cos\theta}$

57. $\sec^2\lambda - \csc^2\lambda = (2\sin^2\lambda - 1)(\sec^2\lambda + \csc^2\lambda)$

58. $2\csc A = 2\csc A - \cot A\cos A + \cos^2 A\csc A$

59. $\dfrac{\csc\theta + 1}{\cot^2\theta + \csc\theta + 1} = \dfrac{\sin^2\theta + \sin\theta\cos\theta}{\sin\theta + \cos\theta}$

60. $\dfrac{\cos^4\theta - \sin^4\theta}{(\cos^2\theta - \sin^2\theta)^2} = \dfrac{\cos\theta}{\cos\theta + \sin\theta} + \dfrac{\sin\theta}{\cos\theta - \sin\theta}$

C

Prove that the equations in Problems 61–72 are identities.

61. $(\cos\alpha - \cos\beta)^2 + (\sin\alpha - \sin\beta)^2$
$= 2 - 2(\cos\alpha\cos\beta + \sin\alpha\sin\beta)$

62. $(\sec\alpha + \sec\beta)^2 - (\tan\alpha - \tan\beta)^2$
$= 2 + 2(\sec\alpha\sec\beta + \tan\alpha\tan\beta)$

63. $\tan A + \cot B = (\sin A\sin B + \cos A\cos B)\sec A\csc B$

64. $\sec A + \csc B = (\cos A\sin^2 B + \sin B\cos^2 A)\sec^2 A\csc^2 B$

65. $(\sin A\cos A\cos\beta + \sin B\cos B\cos A)\sec A\sec B$
$= \sin A + \sin B$

66. $(\cos A\cos B\tan A + \sin A\sin B\cot B)\csc A\sec B = 2$

67. $\sin\theta + \cos\theta + 1 = \dfrac{2\sin\theta\cos\theta}{\sin\theta + \cos\theta - 1}$

68. $\dfrac{2\tan^2\theta + 2\tan\theta\sec\theta}{\tan\theta + \sec\theta - 1} = \tan\theta + \sec\theta + 1$

69. $\dfrac{\csc\theta + 1}{\csc\theta - 1} - \dfrac{\sec\theta - \tan\theta}{\sec\theta + \tan\theta} = 4\tan\theta\sec\theta$

70. $\dfrac{\cos\theta + \sin\theta}{\cos\theta - \sin\theta} + \dfrac{\cot\theta - 1}{\cot\theta + 1} = \dfrac{-2}{\sin^2\theta - \cos^2\theta}$

71. $\dfrac{\cos\theta + 1}{\cos\theta - 1} + \dfrac{1 - \sec\theta}{1 + \sec\theta} = -2\cot^2\theta - 2\csc^2\theta$

72. $\dfrac{\sin\theta}{1 - \cos\theta} + \dfrac{\cos\theta}{1 - \sin\theta} = (1 + \sin\theta + \cos\theta)(\sec\theta\csc\theta)$

6.4 ADDITION LAWS

When proving identities, it is sometimes necessary to simplify the functional value of the sum or difference of two angles. If α and β represent any two angles,

$$\cos(\alpha - \beta) \neq \cos\alpha - \cos\beta$$

For example, if $\alpha = 60^\circ$ and $\beta = 30^\circ$ then

$$\cos(60^\circ - 30^\circ) = \cos 30^\circ = \frac{\sqrt{3}}{2} \qquad \text{and} \qquad \cos 60^\circ - \cos 30^\circ = \frac{1}{2} - \frac{\sqrt{3}}{2} = \frac{1 - \sqrt{3}}{2}$$

Thus $\cos(60^\circ - 30^\circ) \neq \cos 60^\circ - \cos 30^\circ$.

In this section, we discuss twelve more identities. First we consider $\cos(\alpha - \beta)$ expanded in terms of trigonometric functions of single angles. We shall later list this as Identity 16 but it is convenient to discuss it first because it provides the cornerstone for building a great many additional identities.

Difference of Angles Identity

$$\cos(\alpha - \beta) = \cos\alpha\cos\beta + \sin\alpha\sin\beta$$

Proof Find the length of any chord in a unit circle with a corresponding arc intercepted by the central angle θ, where θ is in standard position. Let A be the point $(1, 0)$ and P be the point on the intersection of the terminal side of angle θ and the unit circle. This means that the coordinates of P are $(\cos\theta, \sin\theta)$. Now find the length of the chord AP (see Figure 6.1) by using the distance formula:

$$\begin{aligned} AP &= \sqrt{(1 - \cos\theta)^2 + (0 - \sin\theta)^2} \\ &= \sqrt{1 - 2\cos\theta + \cos^2\theta + \sin^2\theta} \\ &= \sqrt{1 - 2\cos\theta + 1} \\ &= \sqrt{2 - 2\cos\theta} \end{aligned}$$

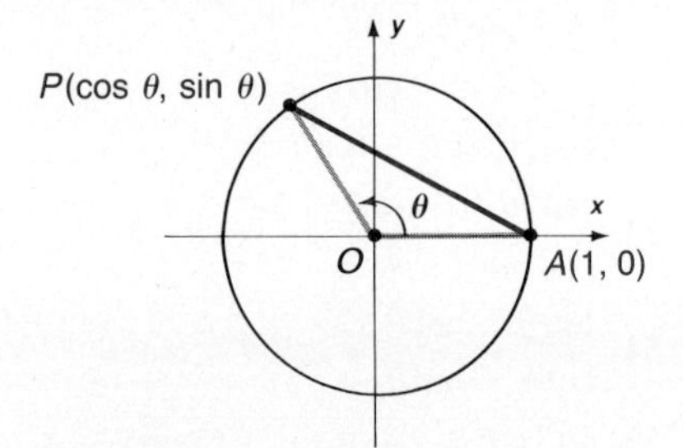

Figure 6.1 Length of a chord determined by an angle θ

Next apply this result to a chord determined by any two angles α and β, as shown in Figure 6.2. Let P_α and P_β be the points on the unit circle determined by the angles α and β, respectively. By the previous result,

$$P_\alpha P_\beta = \sqrt{2 - 2\cos(\alpha - \beta)}$$

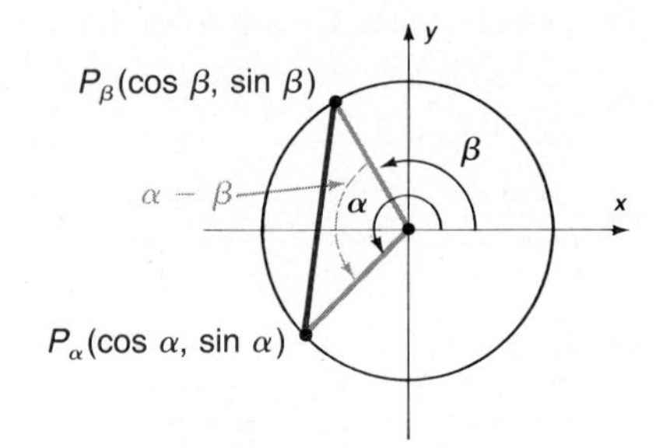

Figure 6.2 Distance between P_α and P_β

But you could also have found this distance directly via the distance formula:

$$\begin{aligned} P_\alpha P_\beta &= \sqrt{(\cos\beta - \cos\alpha)^2 + (\sin\beta - \sin\alpha)^2} \\ &= \sqrt{\cos^2\beta - 2\cos\alpha\cos\beta + \cos^2\alpha + \sin^2\beta - 2\sin\alpha\sin\beta + \sin^2\alpha} \\ &= \sqrt{(\cos^2\beta + \sin^2\beta) + (\cos^2\alpha + \sin^2\alpha) - 2(\cos\alpha\cos\beta + \sin\alpha\sin\beta)} \\ &= \sqrt{2 - 2(\cos\alpha\cos\beta + \sin\alpha\sin\beta)} \end{aligned}$$

Finally, equate these quantities since they both represent the distance between P_α and P_β:

$$\begin{aligned} \sqrt{2 - 2\cos(\alpha - \beta)} &= \sqrt{2 - 2(\cos\alpha\cos\beta + \sin\alpha\sin\beta)} \\ 2 - 2\cos(\alpha - \beta) &= 2 - 2(\cos\alpha\cos\beta + \sin\alpha\sin\beta) \\ -2\cos(\alpha - \beta) &= -2(\cos\alpha\cos\beta + \sin\alpha\sin\beta) \\ \boldsymbol{\cos(\alpha - \beta)} &= \boldsymbol{\cos\alpha\cos\beta + \sin\alpha\sin\beta} \end{aligned}$$

□

You can use this identity to find the exact values of functions of angles that are multiples of 15° as shown by Example 1.

EXAMPLE 1

$$\begin{aligned}\cos 345^\circ &= \cos 15^\circ && \text{Reduction principle}\\ &= \cos(45^\circ - 30^\circ) && \text{Since } 45^\circ - 30^\circ = 15^\circ\\ &= \cos 45^\circ \cos 30^\circ + \sin 45^\circ \sin 30^\circ\\ &= \frac{1}{2}\sqrt{2}\cdot\frac{1}{2}\sqrt{3} + \frac{1}{2}\sqrt{2}\cdot\frac{1}{2} && \text{Using exact values}\\ &= \frac{\sqrt{6}}{4} + \frac{\sqrt{2}}{4}\\ &= \frac{\sqrt{6}+\sqrt{2}}{4}\end{aligned}$$

■

Even though this identity is helpful for making evaluations (as in the preceding example), its real value lies in the fact that it is true for *any* choice of α and β. By making some particular choices for α and β, we find several useful special cases of this identity.

EXAMPLE 2 Prove $\cos(\frac{\pi}{2} - \theta) = \sin\theta$.

Solution This proof is based on the identity

$$\cos(\alpha - \beta) = \cos\alpha\cos\beta + \sin\alpha\sin\beta$$

Let $\alpha = \frac{\pi}{2}$ and $\beta = \theta$:

$$\begin{aligned}\cos\left(\frac{\pi}{2} - \theta\right) &= \cos\frac{\pi}{2}\cos\theta + \sin\frac{\pi}{2}\sin\theta\\ &= 0\cdot\cos\theta + 1\cdot\sin\theta\\ &= \sin\theta\end{aligned}$$

■

Example 2 is one of three identities known as the **cofunction identities**. (Remember that Identities 1–8 are the fundamental identities discussed in Section 6.2.)

Cofunction Identities

For any real number (or angle) θ,

9. $\cos\left(\frac{\pi}{2} - \theta\right) = \sin\theta$ 10. $\sin\left(\frac{\pi}{2} - \theta\right) = \cos\theta$ 11. $\tan\left(\frac{\pi}{2} - \theta\right) = \cot\theta$

The proof of Identity 9 was done in Example 2, and the proof of Identity 10 depends on Identity 9, and is shown below.

Proof of Identity 10

$$\begin{aligned}\cos\theta &= \cos\left[\frac{\pi}{2} - \left(\frac{\pi}{2} - \theta\right)\right]\\ &= \sin\left(\frac{\pi}{2} - \theta\right) && \text{This is Identity 9.}\end{aligned}$$

Therefore $\sin\left(\frac{\pi}{2} - \theta\right) = \cos\theta$. □

Identities involving the tangent are usually proved after proving similar identities for cosine and sine. The fundamental identity $\tan\theta = \sin\theta/\cos\theta$ is applied first, allowing you then to use the appropriate identities for cosine and sine. This process is illustrated with the following proof.

Proof of Identity 11

$$\tan\left(\frac{\pi}{2} - \theta\right) = \frac{\sin(\frac{\pi}{2} - \theta)}{\cos(\frac{\pi}{2} - \theta)}$$

$$= \frac{\cos\theta}{\sin\theta}$$

$$= \cot\theta$$

□

The cofunction identities allow us to change a trigonometric function to the cofunction of its complement.

EXAMPLE 3 Write each function in terms of its cofunction.

a. $\sin 28°$ **b.** $\cos 43°$ **c.** $\cot 9°$ **d.** $\sin\frac{\pi}{6}$

Solution

a. $\sin 28° = \cos(90° - 28°) = \cos 62°$

b. $\cos 43° = \sin(90° - 43°) = \sin 47°$

c. $\cot 9° = \tan(90° - 9°) = \tan 81°$

d. $\sin\frac{\pi}{6} = \cos(\frac{\pi}{2} - \frac{\pi}{6}) = \cos\frac{2\pi}{6} = \cos\frac{\pi}{3}$

■

Suppose the given angle in Example 3 is larger than 90°; for example,

$$\cos 125° = \sin(90° - 125°)$$
$$= \sin(-35°)$$

This result can be further simplified using the following **opposite-angle identities**.

Opposite-Angle Identities

For any real number (or angle) θ,

12. $\cos(-\theta) = \cos\theta$ 13. $\sin(-\theta) = -\sin\theta$ 14. $\tan(-\theta) = -\tan\theta$

Proof of Identity 12 Let $\alpha = 0$ and $\beta = \theta$ in the identity $\cos(\alpha - \beta) = \cos\alpha\cos\beta + \sin\alpha\sin\beta$.

$$\cos(0 - \theta) = \cos 0\cos\theta + \sin 0\sin\theta$$
$$= 1 \cdot \cos\theta + 0 \cdot \sin\theta$$
$$= \cos\theta$$

But, if you simplify directly,

$$\cos(0 - \theta) = \cos(-\theta)$$

Therefore $\cos(-\theta) = \cos\theta$ □

The proofs of Identities 13 and 14 are left as problems.

EXAMPLE 4 Write each as a function of a positive angle or number.

a. $\cos(-19^\circ)$ **b.** $\sin(-19^\circ)$ **c.** $\tan(-2)$

Solution **a.** $\cos(-19^\circ) = \cos 19^\circ$ **b.** $\sin(-19^\circ) = -\sin 19^\circ$ **c.** $\tan(-2) = -\tan 2$ ■

EXAMPLE 5 Write the given functions in terms of their cofunctions.

a. $\cos 125^\circ$ **b.** $\sin 102^\circ$ **c.** $\tan 2.5$

Solution **a.**
$$\begin{aligned}\cos 125^\circ &= \sin(90^\circ - 125^\circ)\\ &= \sin(-35^\circ)\\ &= -\sin 35^\circ\end{aligned}$$

b.
$$\begin{aligned}\sin 102^\circ &= \cos(90^\circ - 102^\circ)\\ &= \cos(-12^\circ)\\ &= \cos 12^\circ\end{aligned}$$

c.
$$\begin{aligned}\cot 2.5 &= \tan(\tfrac{\pi}{2} - 2.5)\\ &\approx \tan(-0.9292)\\ &= -\tan 0.9292\end{aligned}$$
■

You may also need to use opposite-angle identities together with other identities. For example, you know from algebra that

$$a - b \qquad \text{and} \qquad b - a$$

are opposites. This means that $a - b = -(b - a)$. In trigonometry you often see angles like $\frac{\pi}{2} - \theta$ and want to write $\theta - \frac{\pi}{2}$. This means that $\frac{\pi}{2} - \theta = -(\theta - \frac{\pi}{2})$. In particular,

$$\begin{aligned}\cos\left(\frac{\pi}{2} - \theta\right) &= \cos\left[-\left(\theta - \frac{\pi}{2}\right)\right]\\ &= \cos\left(\theta - \frac{\pi}{2}\right) \qquad \text{By Identity 12}\end{aligned}$$

EXAMPLE 6 Write the given functions using the opposite-angle identities.

a. $\sin(\frac{\pi}{2} - \theta)$ **b.** $\tan(\frac{\pi}{2} - \theta)$ **c.** $\cos(\pi - \theta)$

Solution **a.**
$$\begin{aligned}\sin(\tfrac{\pi}{2} - \theta) &= \sin[-(\theta - \tfrac{\pi}{2})]\\ &= -\sin(\theta - \tfrac{\pi}{2})\end{aligned}$$

b.
$$\begin{aligned}\tan(\tfrac{\pi}{2} - \theta) &= \tan[-(\theta - \tfrac{\pi}{2})]\\ &= -\tan(\theta - \tfrac{\pi}{2})\end{aligned}$$

c.
$$\begin{aligned}\cos(\pi - \theta) &= \cos[-(\theta - \pi)]\\ &= \cos(\theta - \pi)\end{aligned}$$
■

The procedure for graphing $y = -\cos\theta$ is identical to the procedure for graphing $y = \cos\theta$, except that, after the frame is drawn, the endpoints are at the bottom of the frame instead of at the top of the frame, and the midpoint is at the top, as shown in Figure 6.3.

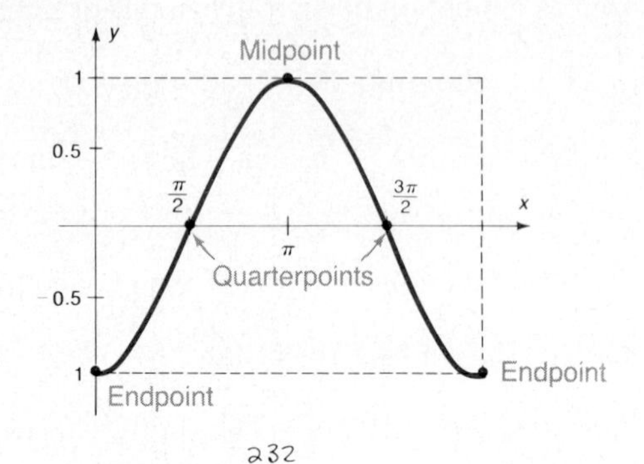

Figure 6.3 One period of $y = -\cos\theta$

EXAMPLE 7 Graph $y = \sin(-\theta)$.

Solution First use an opposite-angle identity, if necessary: $\sin(-\theta) = -\sin\theta$. To graph $y = -\sin\theta$, build the frame as before; the endpoints and midpoints are the same, but the quarter-points are reversed as shown in Figure 6.4.

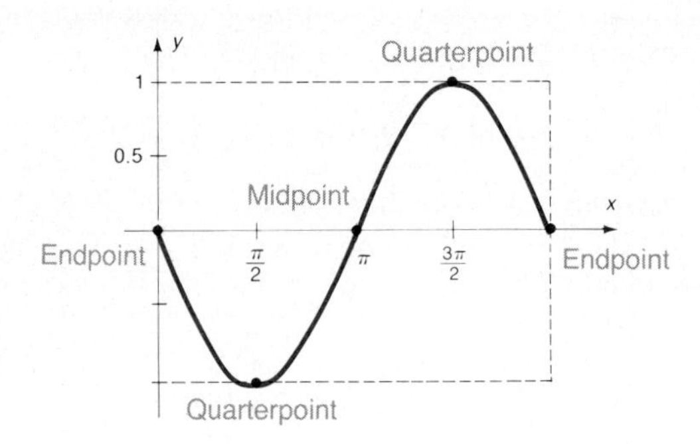

Figure 6.4 One period of $y = -\sin\theta$

■

The difference of angles identity proved at the beginning of this section is one of six identities known as the **addition laws**. Since subtraction can easily be written as a sum, the designation *addition laws* refers to both addition and subtraction.

Addition Laws

15. $\cos(\alpha + \beta) = \cos\alpha\cos\beta - \sin\alpha\sin\beta$
16. $\cos(\alpha - \beta) = \cos\alpha\cos\beta + \sin\alpha\sin\beta$
17. $\sin(\alpha + \beta) = \sin\alpha\cos\beta + \cos\alpha\sin\beta$
18. $\sin(\alpha - \beta) = \sin\alpha\cos\beta - \cos\alpha\sin\beta$
19. $\tan(\alpha + \beta) = \dfrac{\tan\alpha + \tan\beta}{1 - \tan\alpha\tan\beta}$
20. $\tan(\alpha - \beta) = \dfrac{\tan\alpha - \tan\beta}{1 + \tan\alpha\tan\beta}$

EXAMPLE 8 Write $\cos(\frac{2\pi}{3} + \theta)$ as a function of θ only.

Solution Use Identity 15:

$$\cos\left(\frac{2\pi}{3} + \theta\right) = \cos\frac{2\pi}{3}\cos\theta - \sin\frac{2\pi}{3}\sin\theta$$

$$= \left(-\frac{1}{2}\right)\cos\theta - \left(\frac{\sqrt{3}}{2}\right)\sin\theta$$ Substitute exact values where possible.

$$= -\frac{1}{2}(\cos\theta + \sqrt{3}\sin\theta)$$ Simplify. ■

EXAMPLE 9 Evaluate

$$\frac{\tan 18^\circ - \tan 40^\circ}{1 + \tan 18^\circ \tan 40^\circ}$$

using Table B.3 or a calculator.

Solution You can do a lot of arithmetic, or use Identity 20:

$$\frac{\tan 18^\circ - \tan 40^\circ}{1 + \tan 18^\circ \tan 40^\circ} = \tan(18^\circ - 40^\circ)$$

$$= \tan(-22^\circ)$$

$$= -\tan 22^\circ$$ Identity 14

$$\approx -0.4040$$ By Table B.3 or a calculator ■

We will conclude this section by proving some of the addition laws; others are left as problems.

Proof of Identity 15

$$\cos(\alpha + \beta) = \cos[\alpha - (-\beta)]$$

$$= \cos\alpha\cos(-\beta) + \sin\alpha\sin(-\beta)$$ By Identity 16, proved earlier

$$= \cos\alpha\cos\beta - \sin\alpha\sin\beta$$ By Identities 12 and 13

□

Identity 16 is the main cosine identity proved at the beginning of this section. It is numbered and included here for the sake of completeness.

Proof of Identity 17

$$\sin(\alpha + \beta) = \cos\left[\frac{\pi}{2} - (\alpha + \beta)\right]$$

$$= \cos\left[\left(\frac{\pi}{2} - \alpha\right) - \beta\right]$$

$$= \cos\left(\frac{\pi}{2} - \alpha\right)\cos\beta + \sin\left(\frac{\pi}{2} - \alpha\right)\sin\beta$$

$$= \sin\alpha\cos\beta + \cos\alpha\sin\beta$$

□

To prove Identity 18, replace β by $-\beta$ in Identity 17; the details are left as an exercise.

Proof of Identity 19

$$\tan(\alpha+\beta) = \frac{\sin(\alpha+\beta)}{\cos(\alpha+\beta)}$$

$$= \frac{\sin\alpha\cos\beta + \cos\alpha\sin\beta}{\cos\alpha\cos\beta - \sin\alpha\sin\beta}$$

$$= \frac{\sin\alpha\cos\beta + \cos\alpha\sin\beta}{\cos\alpha\cos\beta - \sin\alpha\sin\beta} \cdot \frac{\dfrac{1}{\cos\alpha\cos\beta}}{\dfrac{1}{\cos\alpha\cos\beta}} \quad \text{Multiply by 1.}$$

$$= \frac{\dfrac{\sin\alpha\cos\beta}{\cos\alpha\cos\beta} + \dfrac{\cos\alpha\sin\beta}{\cos\alpha\cos\beta}}{\dfrac{\cos\alpha\cos\beta}{\cos\alpha\cos\beta} - \dfrac{\sin\alpha\sin\beta}{\cos\alpha\cos\beta}}$$

$$= \frac{\tan\alpha + \tan\beta}{1 - \tan\alpha\tan\beta} \qquad \square$$

The proof of Identity 20 is left as an exercise.

PROBLEM SET 6.4

A

Change each of the expressions in Problems 1–9 to functions of θ only.

1. $\cos(30^\circ + \theta)$ **2.** $\sin(\theta - 45^\circ)$ **3.** $\tan(45^\circ + \theta)$

4. $\cos(\frac{\pi}{3} - \theta)$ **5.** $\cos(\theta - \frac{\pi}{4})$ **6.** $\sin(\frac{2\pi}{3} + \theta)$

7. $\cos(\theta + \theta)$ **8.** $\sin(\theta + \theta)$ **9.** $\tan(\theta + \theta)$

Write each function in Problems 10–15 in terms of its cofunction by using the cofunction identities.

10. $\cos 15^\circ$ **11.** $\sin 38^\circ$ **12.** $\cot 41^\circ$

13. $\sin\frac{\pi}{6}$ **14.** $\cos\frac{5\pi}{6}$ **15.** $\cot\frac{2\pi}{3}$

Write the functions in Problems 16–21 as functions of positive angles.

16. $\cos(-18^\circ)$ **17.** $\tan(-49^\circ)$ **18.** $\sin(-41^\circ)$

19. $\sin(-31^\circ)$ **20.** $\cos(-39^\circ)$ **21.** $\tan(-24^\circ)$

Evaluate each of the expressions in Problems 22–27. You may use tables or a calculator.

22. $\sin 158^\circ \cos 92^\circ - \cos 158^\circ \sin 92^\circ$

23. $\cos 114^\circ \cos 85^\circ + \sin 114^\circ \sin 85^\circ$

24. $\cos 30^\circ \cos 48^\circ - \sin 30^\circ \sin 48^\circ$

25. $\sin 18^\circ \cos 23^\circ + \cos 18^\circ \sin 23^\circ$

26. $\dfrac{\tan 32^\circ + \tan 18^\circ}{1 - \tan 32^\circ \tan 18^\circ}$ **27.** $\dfrac{\tan 59^\circ - \tan 25^\circ}{1 + \tan 59^\circ \tan 25^\circ}$

B

Using the identities of this section, find the exact values of the sine, cosine, and tangent of each of the angles given in Problems 28–33.

28. 15° **29.** -15°

30. 195° **31.** 75°

32. 165° **33.** 105°

34. Write each as a function of $\theta - \frac{\pi}{3}$.
a. $\cos(\frac{\pi}{3} - \theta)$ **b.** $\sin(\frac{\pi}{3} - \theta)$ **c.** $\tan(\frac{\pi}{3} - \theta)$

35. Write each as a function of $(\theta - \frac{2\pi}{3})$.
a. $\cos(\frac{2\pi}{3} - \theta)$ **b.** $\sin(\frac{2\pi}{3} - \theta)$ **c.** $\tan(\frac{2\pi}{3} - \theta)$

36. Write each as a function of $\alpha - \beta$.
a. $\cos(\beta - \alpha)$ **b.** $\sin(\beta - \alpha)$ **c.** $\tan(\beta - \alpha)$

Use the opposite-angle identities, if necessary, to graph the functions in Problems 37–46.

37. $y = -2\cos\theta$ **38.** $y = -3\sin\theta$

39. $y = \tan(-\theta)$ **40.** $y = \cos(-2\theta)$

41. $y = \sin(-3\theta)$ **42.** $y - 1 = \sin(2 - \theta)$

43. $y - 2 = \cos(\pi - \theta)$ **44.** $y - 3 = \sin(\pi - \theta)$

45. $y - 1 = \tan(\frac{\pi}{6} - \theta)$ **46.** $y + 2 = \cos(3 - \theta)$

Prove the identities in Problems 47–60.

47. $\sin(\alpha - \beta) = \sin\alpha\cos\beta - \cos\alpha\sin\beta$

48. $\tan(\alpha - \beta) = \dfrac{\tan\alpha - \tan\beta}{1 + \tan\alpha\tan\beta}$

49. $\cot(\alpha + \beta) = \dfrac{\cot\alpha\cot\beta - 1}{\cot\beta + \cot\alpha}$

50. $\cot(\alpha - \beta) = \dfrac{\cot\alpha\cot\beta + 1}{\cot\beta - \cot\alpha}$

51. $\dfrac{\cos 5\theta}{\sin\theta} - \dfrac{\sin 5\theta}{\cos\theta} = \dfrac{\cos 6\theta}{\sin\theta\cos\theta}$

52. $\dfrac{\sin 6\theta}{\sin 3\theta} - \dfrac{\cos 6\theta}{\cos 3\theta} = \sec 3\theta$

53. $\sec(\frac{\pi}{2} - \theta) = \csc\theta$

54. $\cot(\frac{\pi}{2} - \theta) = \tan\theta$

55. $\csc(\frac{\pi}{2} - \theta) = \sec\theta$

56. $\cot(-\theta) = -\cot\theta$

57. $\sec(-\theta) = \sec\theta$

58. $\csc(-\theta) = -\csc\theta$

59. $\sin(\alpha + \beta)\cos\beta - \cos(\alpha + \beta)\sin\beta = \sin\alpha$

60. $\cos(\alpha - \beta)\cos\beta - \sin(\alpha - \beta)\sin\beta = \cos\alpha$

C

Prove the identities in Problems 61–67.

61. $\dfrac{\tan(\alpha + \beta) - \tan\beta}{1 + \tan(\alpha + \beta)\tan\beta} = \tan\alpha$

62. $\dfrac{\sin(\theta + h) - \sin\theta}{h} = \cos\theta\left(\dfrac{\sin h}{h}\right) - \sin\theta\left(\dfrac{1 - \cos h}{h}\right)$

63. $\dfrac{\cos(\theta + h) - \cos\theta}{h} = -\sin\theta\left(\dfrac{\sin h}{h}\right) - \cos\theta\left(\dfrac{1 - \cos h}{h}\right)$

64. $\sin(\alpha + \beta + \gamma) = \sin\alpha\cos\beta\cos\gamma + \cos\alpha\sin\beta\cos\gamma + \cos\alpha\cos\beta\sin\gamma - \sin\alpha\sin\beta\sin\gamma$

65. $\cos(\alpha + \beta + \gamma) = \cos\alpha\cos\beta\cos\gamma - \cos\alpha\sin\beta\sin\gamma - \sin\alpha\cos\beta\sin\gamma - \sin\alpha\sin\beta\cos\gamma$

66. $\tan(\alpha + \beta + \gamma) = \dfrac{\tan\alpha + \tan\beta + \tan\gamma - \tan\alpha\tan\beta\tan\gamma}{1 - \tan\beta\tan\gamma - \tan\alpha\tan\gamma - \tan\alpha\tan\beta}$

67. $\cot(\alpha + \beta + \gamma) = \dfrac{\cot\alpha\cot\beta\cot\gamma - \cot\alpha - \cot\beta - \cot\gamma}{\cot\beta\cot\gamma + \cot\alpha\cot\gamma + \cot\alpha\cot\beta - 1}$

6.5 DOUBLE-ANGLE AND HALF-ANGLE IDENTITIES

Two additional, special cases of the addition laws of Section 6.4 are now considered. The first is that of the **double-angle identities**.

Double-Angle Identities

21. $\cos 2\theta = \cos^2\theta - \sin^2\theta$
$= 2\cos^2\theta - 1$
$= 1 - 2\sin^2\theta$

22. $\sin 2\theta = 2\sin\theta\cos\theta$

23. $\tan 2\theta = \dfrac{2\tan\theta}{1 - \tan^2\theta}$

Proof Use the addition laws where $\alpha = \theta$ and $\beta = \theta$.

$$\begin{aligned}\cos 2\theta = \cos(\theta + \theta) &= \cos\theta\cos\theta - \sin\theta\sin\theta\\ &= \boldsymbol{\cos^2\theta - \sin^2\theta}\\ &= \cos^2\theta - (1 - \cos^2\theta)\\ &= \boldsymbol{2\cos^2\theta - 1}\\ &= 2(1 - \sin^2\theta) - 1\\ &= \boldsymbol{1 - 2\sin^2\theta}\end{aligned}$$

$$\begin{aligned}\sin 2\theta = \sin(\theta + \theta) &= \sin\theta\cos\theta + \cos\theta\sin\theta\\ &= \boldsymbol{2\sin\theta\cos\theta}\end{aligned}$$

$$\begin{aligned}\tan 2\theta = \tan(\theta + \theta) &= \frac{\tan\theta + \tan\theta}{1 - \tan\theta\tan\theta}\\ &= \boldsymbol{\frac{2\tan\theta}{1 - \tan^2\theta}}\end{aligned}$$

□

EXAMPLE 1 $\cos 100x = \cos(2 \cdot 50x) = \cos^2 50x - \sin^2 50x$ ■

EXAMPLE 2 $\sin 120^\circ = \sin 2(60^\circ) = 2 \sin 60^\circ \cos 60^\circ$ ■

EXAMPLE 3

$$\begin{aligned}
\cos 3\theta = \cos(2\theta + \theta) &= \cos 2\theta \cos \theta - \sin 2\theta \sin \theta \\
&= (\cos^2 \theta - \sin^2 \theta)\cos \theta - (2 \sin \theta \cos \theta) \sin \theta \\
&= \cos^3 \theta - \sin^2 \theta \cos \theta - 2 \sin^2 \theta \cos \theta \\
&= \cos^3 \theta - 3 \sin^2 \theta \cos \theta \\
&= \cos^3 \theta - 3(1 - \cos^2 \theta) \cos \theta \\
&= \cos^3 \theta - 3 \cos \theta + 3 \cos^3 \theta \\
&= 4 \cos^3 \theta - 3 \cos \theta
\end{aligned}$$ ■

EXAMPLE 4 Evaluate $\dfrac{2 \tan \frac{\pi}{16}}{1 - \tan^2 \frac{\pi}{16}}$.

Solution Notice this is the right-hand side of Identity 23 so it is the same as $\tan(2 \cdot \frac{\pi}{16}) = \tan \frac{\pi}{8}$. Now use tables or a calculator to find $\tan \frac{\pi}{8} \approx 0.4142$. ■

EXAMPLE 5 If $\cos \theta = \frac{3}{5}$ and θ is in Quadrant IV, find $\cos 2\theta$, $\sin 2\theta$, and $\tan 2\theta$.

Solution Since $\cos 2\theta = 2 \cos^2 \theta - 1$,

$$\begin{aligned}
\cos 2\theta &= 2\left(\frac{3}{5}\right)^2 - 1 \\
&= 2\left(\frac{9}{25}\right) - 1 \\
&= -\frac{7}{25}
\end{aligned}$$

For the other functions of 2θ, you need to know $\sin \theta$. Begin with the fundamental identity relating cosine and sine:

$$\begin{aligned}
\sin^2 \theta &= 1 - \cos^2 \theta \\
\sin \theta &= -\sqrt{1 - \cos^2 \theta} && \text{Negative since the sine is negative in Quadrant IV} \\
&= -\sqrt{1 - \left(\frac{3}{5}\right)^2} \\
&= -\sqrt{1 - \frac{9}{25}} \\
&= -\sqrt{\frac{16}{25}} \\
&= -\frac{4}{5}
\end{aligned}$$

Now,

$$\begin{aligned}\sin 2\theta &= 2\sin\theta\cos\theta\\ &= 2\left(-\frac{4}{5}\right)\left(\frac{3}{5}\right)\\ &= -\frac{24}{25}\end{aligned}$$

Finally,

$$\begin{aligned}\tan 2\theta &= \frac{\sin 2\theta}{\cos 2\theta}\\ &= \frac{-\frac{24}{25}}{-\frac{7}{25}}\\ &= \frac{24}{7}\end{aligned}$$

■

Identity 21 leads us to the second important special case of the addition laws, called the **half-angle identities**. We wish to solve $\cos 2\alpha = 2\cos^2\alpha - 1$ for $\cos^2\alpha$.

$$\begin{aligned}2\cos^2\alpha - 1 &= \cos 2\alpha\\ 2\cos^2\alpha &= 1 + \cos 2\alpha\\ \cos^2\alpha &= \frac{1 + \cos 2\alpha}{2}\end{aligned}$$

Now, if $\alpha = \frac{1}{2}\theta$, then $2\alpha = \theta$ and

$$\cos^2\frac{1}{2}\theta = \frac{1 + \cos\theta}{2}$$

If $\frac{1}{2}\theta$ is in Quadrant I or IV, then

$$\cos\frac{1}{2}\theta = \sqrt{\frac{1 + \cos\theta}{2}}$$

If $\frac{1}{2}\theta$ is in Quadrant II or III, then

$$\cos\frac{1}{2}\theta = -\sqrt{\frac{1 + \cos\theta}{2}}$$

These results are summarized by writing

$$\cos\frac{1}{2}\theta = \pm\sqrt{\frac{1 + \cos\theta}{2}}$$

However, *you must be careful.* The sign + or − is chosen according to which quadrant $\frac{1}{2}\theta$ is in. The formula requires either + or −, but not both. This use of $\pm$ is different from the use of $\pm$ in algebra. For example, when using $\pm$ in the Quadratic Formula, we are indicating *two* possible correct roots. In this trigonometric identity we will obtain *one* correct value depending on the quadrant of $\frac{1}{2}\theta$.

For the sine, solve $\cos 2\alpha = 1 - 2\sin^2\alpha$ for $\sin^2\alpha$.

$$\cos 2\alpha = 1 - 2\sin^2\alpha$$
$$2\sin^2\alpha = 1 - \cos 2\alpha$$
$$\sin^2\alpha = \frac{1 - \cos 2\alpha}{2}$$

Replace $\alpha = \frac{1}{2}\theta$, and

$$\sin^2\frac{1}{2}\theta = \frac{1 - \cos\theta}{2}$$

or

$$\sin\frac{1}{2}\theta = \pm\sqrt{\frac{1 - \cos\theta}{2}}$$

where the sign depends on the quadrant of $\frac{1}{2}\theta$. If $\frac{1}{2}\theta$ is in Quadrant I or II, you use $+$; if it is in Quadrant III or IV, you use $-$.

Finally, to find the half-angle identity for the tangent, write

$$\tan\frac{1}{2}\theta = \frac{\sin\frac{1}{2}\theta}{\cos\frac{1}{2}\theta}$$

$$= \frac{\pm\sqrt{\dfrac{1 - \cos\theta}{2}}}{\pm\sqrt{\dfrac{1 + \cos\theta}{2}}}$$

$$= \pm\sqrt{\frac{1 - \cos\theta}{1 + \cos\theta}}$$

$$= \pm\sqrt{\frac{1 - \cos\theta}{1 + \cos\theta}\cdot\frac{1 - \cos\theta}{1 - \cos\theta}}$$

$$= \pm\sqrt{\frac{(1 - \cos\theta)^2}{\sin^2\theta}}$$

Remember that $1 - \cos^2\theta = \sin^2\theta$.

$$= \frac{1 - \cos\theta}{\sin\theta}$$

Notice that $1 - \cos\theta$ is positive. Also, since $\tan\frac{1}{2}\theta$ and $\sin\theta$ have the same sign regardless of the quadrant of θ, the desired result follows.

You can also show that $\tan\frac{1}{2}\theta = \dfrac{\sin\theta}{1 + \cos\theta}$.

Half-Angle Identities

24. $\cos\frac{1}{2}\theta = \pm\sqrt{\dfrac{1 + \cos\theta}{2}}$

25. $\sin\frac{1}{2}\theta = \pm\sqrt{\dfrac{1 - \cos\theta}{2}}$

26. $\tan\frac{1}{2}\theta = \dfrac{1 - \cos\theta}{\sin\theta} = \dfrac{\sin\theta}{1 + \cos\theta}$

To help you remember the correct sign between the first two half-angle identities, remember "*sinus-minus*"—the sine is minus.

EXAMPLE 6 Find the exact value of $\cos\frac{9\pi}{8}$.

Solution

$$\cos\frac{9\pi}{8} = \cos\left(\frac{1}{2}\cdot\frac{9\pi}{4}\right) = -\sqrt{\frac{1+\cos\frac{9\pi}{4}}{2}}$$

Choose a negative sign, since $\frac{9\pi}{8}$ is in Quadrant III and the cosine is negative in this quadrant.

$$= -\sqrt{\frac{1+\cos\frac{\pi}{4}}{2}}$$

$$= -\sqrt{\frac{1+\frac{\sqrt{2}}{2}}{2}}$$

$$= -\sqrt{\frac{2+\sqrt{2}}{4}}$$

$$= -\frac{1}{2}\sqrt{2+\sqrt{2}}$$ ■

EXAMPLE 7 If $\cot 2\theta = \frac{3}{4}$, find $\cos\theta$, $\sin\theta$, and $\tan\theta$, where 2θ is in Quadrant I.

Solution You need to find $\cos 2\theta$ so that you can use it in the half-angle identities. To do this, first find $\tan 2\theta$:

$$\tan 2\theta = \frac{1}{\cot 2\theta} = \frac{4}{3}$$

Next, find $\sec 2\theta$:

$$\sec 2\theta = \pm\sqrt{1+\tan^2 2\theta}$$ From Identity 7

$$= \sqrt{1+\frac{16}{9}}$$ It is positive because 2θ is in Quadrant I.

$$= \frac{5}{3}$$

Finally, $\cos 2\theta$ is the reciprocal of $\sec 2\theta$: $\cos 2\theta = \frac{3}{5}$

Next use the half-angle identities:

$$\cos\theta = \pm\sqrt{\frac{1+\cos 2\theta}{2}} \quad\text{and}\quad \sin\theta = \pm\sqrt{\frac{1-\cos 2\theta}{2}}$$

Do you see that θ is one-half of 2θ in these formulas?

$$\cos\theta = +\sqrt{\frac{1+\frac{3}{5}}{2}} \qquad \sin\theta = +\sqrt{\frac{1-\frac{3}{5}}{2}}$$

Positive value chosen because θ is in Quadrant I.

$$\cos\theta = \frac{2}{\sqrt{5}} \qquad \sin\theta = \frac{1}{\sqrt{5}}$$

$$\begin{aligned}\tan\theta &= \frac{\sin\theta}{\cos\theta}\\ &= \frac{1/\sqrt{5}}{2/\sqrt{5}}\\ &= \frac{1}{2}\end{aligned}$$

■

EXAMPLE 8 Prove that $\sin\theta = \dfrac{2\tan\frac{1}{2}\theta}{1+\tan^2\frac{1}{2}\theta}$.

Solution When proving identities involving functions of different angles, you should write all the trigonometric functions in the problems as functions of a single angle.

$$\begin{aligned}\frac{2\tan\frac{1}{2}\theta}{1+\tan^2\frac{1}{2}\theta} &= \frac{2\dfrac{\sin\frac{1}{2}\theta}{\cos\frac{1}{2}\theta}}{\sec^2\frac{1}{2}\theta}\\ &= 2\frac{\sin\frac{1}{2}\theta}{\cos\frac{1}{2}\theta}\cdot\cos^2\tfrac{1}{2}\theta\\ &= 2\sin\tfrac{1}{2}\theta\cos\tfrac{1}{2}\theta\\ &= \sin\theta\end{aligned}$$

From Identity 22 (double-angle identity) ■

It is sometimes convenient, or even necessary, to write a trigonometric sum as a product or a product as a sum. We conclude this section with eight additional identities.

Product Identities

27. $2\cos\alpha\cos\beta = \cos(\alpha-\beta) + \cos(\alpha+\beta)$
28. $2\sin\alpha\sin\beta = \cos(\alpha-\beta) - \cos(\alpha+\beta)$
29. $2\sin\alpha\cos\beta = \sin(\alpha+\beta) + \sin(\alpha-\beta)$
30. $2\cos\alpha\sin\beta = \sin(\alpha+\beta) - \sin(\alpha-\beta)$

The proof of the product identities involves a system of equations, and will therefore be delayed until Chapter 7.

EXAMPLE 9 Write $2\sin 3\sin 1$ as the sum of two functions.

Solution Use Identity 28, where $\alpha = 3$ and $\beta = 1$:

$$\begin{aligned}2\sin 3\sin 1 &= \cos(3-1) - \cos(3+1)\\ &= \cos 2 - \cos 4\end{aligned}$$

■

EXAMPLE 10 Write $\sin 40° \cos 12°$ as the sum of two functions.

Solution Use Identity 29 where $\alpha = 40°$ and $\beta = 12°$:

$$\begin{aligned} 2\sin 40° \cos 12° &= \sin(40° + 12°) + \sin(40° - 12°) \\ &= \sin 52° + \sin 28° \end{aligned}$$

But what about the coefficient 2? Since you know that the preceding is an *equation* that is true, you can divide both sides by 2 to obtain:

$$\sin 40° \cos 12° = \frac{1}{2}(\sin 52° + \sin 28°)$$ ■

By making appropriate substitutions and again using systems (as shown in Chapter 7) identities 27–30 can be rewritten in a form known as the **sum identities**.

Sum Identities

31. $\cos x + \cos y = 2\cos\left(\frac{x+y}{2}\right)\cos\left(\frac{x-y}{2}\right)$

32. $\cos x - \cos y = -2\sin\left(\frac{x+y}{2}\right)\sin\left(\frac{x-y}{2}\right)$

33. $\sin x + \sin y = 2\sin\left(\frac{x+y}{2}\right)\cos\left(\frac{x-y}{2}\right)$

34. $\sin x - \sin y = 2\sin\left(\frac{x-y}{2}\right)\cos\left(\frac{x+y}{2}\right)$

EXAMPLE 11 Write $\sin 35° + \sin 27°$ as a product.

Solution $x = 35°$, $y = 27°$, and

$$\frac{x+y}{2} = \frac{35° + 27°}{2} = 31°; \qquad \frac{x-y}{2} = 4°$$

Therefore, $\sin 35° + \sin 27° = 2\sin 31° \cos 4°$. ■

You sometimes will use these product and sum identities to prove other identities.

EXAMPLE 12 Prove $\dfrac{\sin 7\gamma + \sin 5\gamma}{\cos 7\gamma - \cos 5\gamma} = -\cot\gamma$.

Solution

$$\begin{aligned} \frac{\sin 7\gamma + \sin 5\gamma}{\cos 7\gamma - \cos 5\gamma} &= \frac{2\sin\left(\frac{7\gamma+5\gamma}{2}\right)\cos\left(\frac{7\gamma-5\gamma}{2}\right)}{-2\sin\left(\frac{7\gamma+5\gamma}{2}\right)\sin\left(\frac{7\gamma-5\gamma}{2}\right)} \\ &= \frac{2\sin 6\gamma\cos\gamma}{-2\sin 6\gamma\sin\gamma} \\ &= -\frac{\cos\gamma}{\sin\gamma} \\ &= -\cot\gamma \end{aligned}$$ ■

PROBLEM SET 6.5

A

Use the double-angle or half-angle identities to evaluate each of Problems 1–9 using exact values.

1. $2\cos^2 22.5^\circ - 1$
2. $\dfrac{2\tan\frac{\pi}{8}}{1-\tan^2\frac{\pi}{8}}$
3. $\sqrt{\dfrac{1-\cos 60^\circ}{2}}$
4. $\cos^2 15^\circ - \sin^2 15^\circ$
5. $1 - 2\sin^2 90^\circ$
6. $-\sqrt{\dfrac{1-\cos 420^\circ}{2}}$
7. $\sin 22.5^\circ$
8. $\cos\frac{\pi}{8}$
9. $\tan 22.5^\circ$

In each of Problems 10–15, find the exact values of cosine, sine, and tangent of 2θ.

10. $\sin\theta = \frac{3}{5}$; θ in Quadrant I
11. $\sin\theta = \frac{5}{13}$; θ in Quadrant II
12. $\tan\theta = -\frac{5}{12}$; θ in Quadrant IV
13. $\tan\theta = -\frac{3}{4}$; θ in Quadrant II
14. $\cos\theta = \frac{5}{9}$; θ in Quadrant I
15. $\cos\theta = -\frac{5}{13}$; θ in Quadrant III

In each of Problems 16–21, find the exact values of cosine, sine, and tangent of $\frac{1}{2}\theta$.

16. $\sin\theta = \frac{3}{5}$; θ in Quadrant I
17. $\sin\theta = \frac{5}{13}$; θ in Quadrant II
18. $\tan\theta = -\frac{5}{12}$; θ in Quadrant IV
19. $\tan\theta = -\frac{3}{4}$; θ in Quadrant II
20. $\cos\theta = \frac{5}{9}$; θ in Quadrant I
21. $\cos\theta = -\frac{5}{13}$; θ in Quadrant III

Write each of the expressions in Problems 22–33 as the sum of two functions.

22. $2\cos 75^\circ \cos 35^\circ$
23. $2\cos 46^\circ \cos 18^\circ$
24. $2\sin 35^\circ \sin 24^\circ$
25. $2\sin 53^\circ \cos 24^\circ$
26. $\sin 70^\circ \sin 88^\circ$
27. $\cos 53^\circ \cos 70^\circ$
28. $\sin 41^\circ \cos 19^\circ$
29. $\cos 115^\circ \sin 200^\circ$
30. $\sin 225^\circ \sin 300^\circ$
31. $\sin 2\theta \sin 5\theta$
32. $\cos\theta \cos 3\theta$
33. $\cos 3\theta \sin 2\theta$

Write each of the expressions in Problems 34–45 as a product of two functions.

34. $\sin 43^\circ + \sin 63^\circ$
35. $\sin 22^\circ - \sin 6^\circ$
36. $\cos 81^\circ - \cos 79^\circ$
37. $\cos 78^\circ + \cos 25^\circ$
38. $\sin 215^\circ + \sin 300^\circ$
39. $\cos 25^\circ - \cos 100^\circ$
40. $\sin x - \sin 2x$
41. $\cos 5x - \cos 3x$
42. $\sin x + \sin 2x$
43. $\cos 3\theta + \cos 2\theta$
44. $\cos 5y + \cos 9y$
45. $\sin 6z - \sin 9z$

B

In each of Problems 46–51, find $\cos\theta$, $\sin\theta$, *and* $\tan\theta$ *when* θ *is in Quadrant I and* $\cot 2\theta$ *is given.*

46. $\cot 2\theta = -\frac{3}{4}$
47. $\cot 2\theta = 0$
48. $\cot 2\theta = \frac{1}{\sqrt{3}}$
49. $\cot 2\theta = -\frac{1}{\sqrt{3}}$
50. $\cot 2\theta = -\frac{4}{3}$
51. $\cot 2\theta = \frac{4}{3}$

52. ***Aviation*** An airplane flying faster than the speed of sound is said to have a speed greater than Mach 1. The Mach number is the ratio of the speed of the plane to the speed of sound and is denoted by M. When a plane flies faster than the speed of sound, a sonic boom is heard, created by sound waves that form a cone with a vertex angle θ, as shown in Figure 6.5. It can be shown that, if $M > 1$, then

$$\sin\frac{\theta}{2} = \frac{1}{M}$$

   a. If $\theta = \frac{\pi}{6}$, find the Mach number to the nearest tenth.
   b. Find the exact Mach number for part *a*.

Figure 6.5 Pattern of sound waves creating a sonic boom

Prove each of the identities in Problems 53–61.

53. $\sin \alpha = 2 \sin \frac{\alpha}{2} \cos \frac{\alpha}{2}$

54. $\cos 4\theta = \cos^2 2\theta - \sin^2 2\theta$

55. $\sin 2\theta = \dfrac{2 \tan \theta}{1 + \tan^2 \theta}$

56. $\tan \dfrac{3}{2}\beta = \dfrac{2 \tan \frac{3\beta}{4}}{1 - \tan^2 \frac{3\beta}{4}}$

57. $\tan \dfrac{1}{2}\theta = \dfrac{1 - \cos \theta}{\sin \theta}$

58. $\tan \dfrac{1}{2}\theta = \dfrac{\sin \theta}{1 + \cos \theta}$

59. $\dfrac{\sin 5\theta + \sin 3\theta}{\cos 5\theta + \cos 3\theta} = \tan 4\theta$

60. $\dfrac{\cos 5w + \cos w}{\cos w - \cos 5w} = \dfrac{\cot 2w}{\tan 3w}$

61. $\dfrac{\cos 3\theta - \cos \theta}{\sin \theta - \sin 3\theta} = \tan 2\theta$

C

Prove each of the identities in Problems 62–66.

62. $\sin 3\theta = 3 \sin \theta - 4 \sin^3 \theta$

63. $\tan \frac{B}{2} = \csc B - \cot B$

64. $\sin 4\theta = 4 \sin \theta \cos \theta - 8 \sin^3 \theta \cos \theta$

65. $\frac{1}{2} \cot x - \frac{1}{2} \tan x = \cot 2x$

66. $\cos^4 \theta = \frac{1}{8}(3 + 4 \cos 2\theta + \cos 4\theta)$

6.6 DE MOIVRE'S THEOREM*

Consider a graphical representation of a complex number. To give a graphic representation of complex numbers, such as

$$2 + 3i, \qquad -i, \qquad -3 - 4i, \qquad 3i, \qquad -2 + \sqrt{2}i, \qquad \frac{3}{2} - \frac{5}{2}i,$$

a two-dimensional coordinate system is used. The horizontal axis represents the **real axis** and the vertical axis is the **imaginary axis**, so that $a + bi$ is represented by the ordered pair (a, b). Remember that a and b represent real numbers, so we plot (a, b) in the usual manner, as shown in Figure 6.6. The coordinate system in Figure 6.6 is called the **complex plane** or the **Gaussian plane**, in honor of Karl Friedrich Gauss.

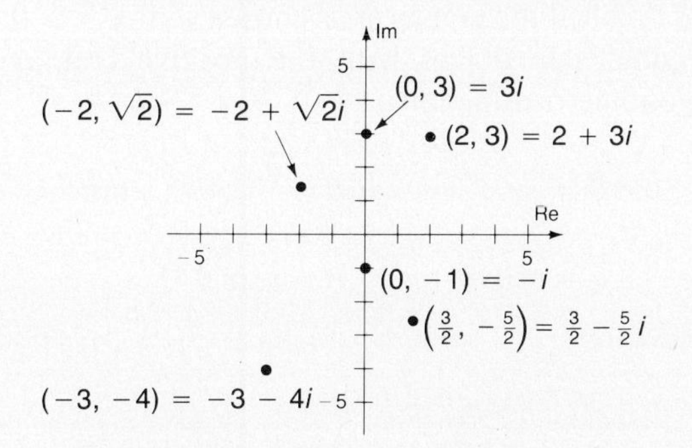

Figure 6.6 Complex plane

The **absolute value** of a complex number z is, graphically, the distance between z and the origin (just as it is for real numbers). The absolute value of a complex number is also called the **modulus**. The distance formula leads to the following definition.

* This is an optional section which requires complex numbers from Section 1.3.

Absolute Value, or Modulus, of a Complex Number

If $z = a + bi$, then the *absolute value*, or *modulus*, of z is denoted by $|z|$ and defined by

$$|z| = \sqrt{a^2 + b^2}$$

EXAMPLE 1 Find the absolute value.

a. $3 + 4i$

$$\text{Absolute value: } |3 + 4i| = \sqrt{3^2 + 4^2} = \sqrt{25} = \mathbf{5}$$

b. $-2 + \sqrt{2}i$

$$\text{Absolute value: } |-2 + \sqrt{2}i| = \sqrt{4 + 2} = \sqrt{\mathbf{6}}$$

c. -3

$$\text{Absolute value: } |-3 + 0i| = \sqrt{9 + 0} = \mathbf{3}$$

This example shows that the definition of absolute value for complex numbers is consistent with the definition of absolute value given for real numbers. ■

The form $a + bi$ is called the **rectangular form**, but another useful representation uses trigonometry. Consider the graphical representation of a complex number $a + bi$, as shown in Figure 6.7. Let r be the distance from the origin to (a, b) and let θ be the angle the segment makes with the real axis. Then

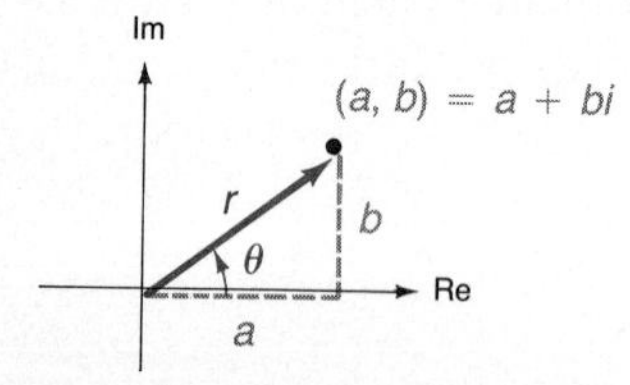

Figure 6.7 Trigonometric form and rectangular form of a complex number

$$r = \sqrt{a^2 + b^2}$$

and θ, called the **argument**, is chosen so that it is the smallest nonnegative angle the terminal side makes with the positive real axis. From the definition of the trigonometric functions,

$$\cos\theta = \frac{a}{r} \qquad \sin\theta = \frac{b}{r} \qquad \tan\theta = \frac{b}{a}$$

$$a = r\cos\theta \qquad b = r\sin\theta$$

Therefore

$$a + bi = r\cos\theta + ir\sin\theta = r(\cos\theta + i\sin\theta)$$

Sometimes $r(\cos\theta + i\sin\theta)$ is abbreviated by

$$r(\cos\theta + i\sin\theta)$$
$$\downarrow\downarrow \quad \downarrow\downarrow\ \downarrow$$
$$r(\text{c} \quad i\,\text{s} \quad \theta) \qquad \text{or } r\,\text{cis}\,\theta$$

Trigonometric Form of a Complex Number

> The **trigonometric form** of a complex number $z = a + bi$ is
>
> $r(\cos\theta + i\sin\theta) = r\operatorname{cis}\theta$
>
> where $r = \sqrt{a^2 + b^2}$; $\tan\theta = b/a$ if $a \neq 0$; $a = r\cos\theta$; $b = r\sin\theta$. This representation is unique for $0 \leq \theta < 360°$ for all z except $0 + 0i$.

The placement of θ in the proper quadrant is an important consideration because there are two values of $0 \leq \theta < 360°$ that will satisfy the relationship

$$\tan\theta = \frac{b}{a}$$

For example, compare the following:

1. $-1 + i$ $\quad a = -1, b = 1 \quad \tan\theta = \dfrac{1}{-1}$ or $\tan\theta = -1$

2. $1 - i$ $\quad a = 1, b = -1 \quad \tan\theta = \dfrac{-1}{1}$ or $\tan\theta = -1$

Notice the same trigonometric equation for both complex numbers, even though $-1 + i$ is in Quadrant II and $1 - i$ is in Quadrant IV. This consideration of quadrants is even more important when you are doing the problem on a calculator, since the proper sequence of steps for this example is

[1] [+/−] [inv] [tan]

This gives the result $-45°$, which is not true for either example since $0 \leq \theta < 360°$. The entire process can be dealt with quite simply if you let θ' be the reference angle for θ. Then find the reference angle

$$\theta' = \tan^{-1}\left|\frac{b}{a}\right|$$

After you know the reference angle and the quadrant, it is easy to find θ. For these examples,

$\theta' = \tan^{-1}|-1|$ $\quad$ On a calculator: [1] [inv] [tan]

$= 45°$

For Quadrant II, $\theta = 135°$; for Quadrant IV, $\theta = 315°$.

$-1 + i$ is in Quadrant III (plot it), so $\theta = 135°$; $1 - i$ is in Quadrant IV, so $\theta = 315°$.

EXAMPLE 2 Change the complex numbers to trigonometric form.

a. $1 - \sqrt{3}i$

$a = 1$ and $b = -\sqrt{3}$; the number is in Quadrant IV.

$$r = \sqrt{1^2 + (-\sqrt{3})^2} = \sqrt{4} = 2 \qquad \theta' = \tan^{-1}\left|\frac{-\sqrt{3}}{1}\right| = 60°$$

On a calculator: [3] [$\sqrt{x}$] [inv] [tan]

The reference angle is 60°; in Quadrant IV: $\theta = 300°$.

Thus $\mathbf{1 - \sqrt{3}i = 2\operatorname{cis}300°}$.

b. $6i$

$a = 0$ and $b = 6$; notice that $\tan\theta$ is not defined for $\theta = 90°$. By inspection, $6i = 6 \operatorname{cis} 90°$.

c. $4.310 + 5.516i$

$a = 4.310$ and $b = 5.516$; the number is in Quadrant I.

$$r = \sqrt{(4.310)^2 + (5.516)^2} \approx \sqrt{49} = 7 \qquad \theta = \tan^{-1}\left|\frac{5.516}{4.310}\right| \approx \tan^{-1}(1.2798) \approx 52°$$

Thus $\mathbf{4.310 + 5.516}\boldsymbol{i} \approx \mathbf{7 \operatorname{cis} 52°}$. ■

EXAMPLE 3 Change the complex numbers to rectangular form.

a. $4 \operatorname{cis} 330°$

$r = 4$ and $\theta = 330°$

$$a = 4\cos 330° = 4\left(\frac{\sqrt{3}}{2}\right) = 2\sqrt{3} \qquad b = 4\sin 330° = 4\left(-\frac{1}{2}\right) = -2$$

Thus $\mathbf{4 \operatorname{cis} 330° = 2\sqrt{3} - 2}\boldsymbol{i}$.

b. $5(\cos 38° + i\sin 38°)$

$r = 5$ and $\theta = 38°$

$$a = 5\cos 38° \approx 3.94 \qquad b = 5\sin 38° \approx 3.08$$

Thus $\mathbf{5(\cos 38° + }\boldsymbol{i}\mathbf{\sin 38°) \approx 3.94 + 3.08}\boldsymbol{i}$. ■

The great advantage of trigonometric form over rectangular form is the ease with which you can multiply and divide complex numbers.

Products and Quotients of Complex Numbers in Trigonometric Form

Let $z_1 = r_1 \operatorname{cis} \theta_1$ and $z_2 = r_2 \operatorname{cis} \theta_2$ be nonzero complex numbers. Then

$$z_1 z_2 = r_1 r_2 \operatorname{cis}(\theta_1 + \theta_2) \qquad \frac{z_1}{z_2} = \frac{r_1}{r_2} \operatorname{cis}(\theta_1 - \theta_2)$$

Proof

$$\begin{aligned} z_1 z_2 &= (r_1 \operatorname{cis} \theta_1)(r_2 \operatorname{cis} \theta_2) \\ &= [r_1(\cos\theta_1 + i\sin\theta_1)][r_2(\cos\theta_2 + i\sin\theta_2)] \\ &= r_1 r_2(\cos\theta_1 + i\sin\theta_1)(\cos\theta_2 + i\sin\theta_2) \\ &= r_1 r_2(\cos\theta_1\cos\theta_2 + i\cos\theta_1\sin\theta_2 + i\sin\theta_1\cos\theta_2 - \sin\theta_1\sin\theta_2) \\ &= r_1 r_2[(\cos\theta_1\cos\theta_2 - \sin\theta_1\sin\theta_2) + i(\cos\theta_1\sin\theta_2 + \sin\theta_1\cos\theta_2)] \\ &= r_1 r_2[\cos(\theta_1 + \theta_2) + i\sin(\theta_1 + \theta_2)] \\ &= r_1 r_2 \operatorname{cis}(\theta_1 + \theta_2) \end{aligned}$$

The proof of the quotient form is similar and is left as a problem. □

EXAMPLE 4 Simplify:

a. $5 \text{ cis } 38° \cdot 4 \text{ cis } 75° = 5 \cdot 4 \text{ cis}(38° + 75°)$
$= 20 \text{ cis } 113°$

b. $\sqrt{2} \text{ cis } 188° \cdot 2\sqrt{2} \text{ cis } 310° = 4 \text{ cis } 498°$
$= 4 \text{ cis } 138°$

c. $(2 \text{ cis } 48°)^3 = (2 \text{ cis } 48°)(2 \text{ cis } 48°)^2$
$= (2 \text{ cis } 48°)(4 \text{ cis } 96°)$
$= 8 \text{ cis } 144°$

Notice that this result is the same as $(2 \text{ cis } 48°)^3 = 2^3 \text{ cis}(3 \cdot 48°) = 8 \text{ cis } 144°$ ■

EXAMPLE 5 Find $\dfrac{15(\cos 48° + i \sin 48°)}{5(\cos 125° + i \sin 125°)}$.

Solution

$$\frac{15 \text{ cis } 48°}{5 \text{ cis } 125°} = 3 \text{ cis}(48° - 125°)$$
$$= 3 \text{ cis}(-77°)$$
$$= 3 \text{ cis } 283°$$

Remember that arguments should be between 0 and 360°. ■

EXAMPLE 6 Simplify $(1 - \sqrt{3}i)^5$.

Solution First change to trigonometric form:

$a = 1$; $b = -\sqrt{3}$; Quadrant IV

$r = \sqrt{1 + 3}$ $\quad \theta' = \tan^{-1}|-\frac{\sqrt{3}}{1}|$
$= 2$ $\quad = 60°$
$\theta = 300°$ (Quadrant IV)

$$(1 - \sqrt{3}i)^5 = (2 \text{ cis } 300°)^5$$
$$= 2^5 \text{ cis}(5 \cdot 300°)$$
$$= 32 \text{ cis } 1500°$$
$$= 32 \text{ cis } 60°$$

If you want the answer in rectangular form, you can now change back:

$a = 32 \cos 60°$ $\quad b = 32 \sin 60°$
$= 32\left(\frac{1}{2}\right)$ $\quad = 32\left(\frac{1}{2}\sqrt{3}\right)$
$= 16$ $\quad = 16\sqrt{3}$

Thus $(1 - \sqrt{3}i)^5 = 16 + 16\sqrt{3}i$. ■

As you can see from Example 6, multiplication in trigonometric form extends quite nicely to any positive integral power in a result called *De Moivre's Theorem.* This theorem is proved by mathematical induction (Chapter 8).

De Moivre's Theorem

If n is a natural number, then

$$(r\operatorname{cis}\theta)^n = r^n \operatorname{cis} n\theta$$

for a complex number $r\operatorname{cis}\theta = r(\cos\theta + i\sin\theta)$.

Although De Moivre's Theorem is useful for powers as illustrated by Example 6, its real usefulness is in finding the complex roots of numbers. Recall from algebra that $\sqrt[n]{r} = r^{1/n}$ is used to denote the principal nth root of r. However, $r^{1/n}$ is only *one* of the nth roots of r. How do you find *all* nth roots of r? To find the principal root, you can use a calculator or logarithms along with the following theorem, which follows directly from De Moivre's Theorem.

nth Root Theorem

If n is any positive integer, then the nth roots of $r\operatorname{cis}\theta$ are given by

$$\sqrt[n]{r}\operatorname{cis}\left(\frac{\theta + 360°k}{n}\right) \quad \text{or} \quad \sqrt[n]{r}\operatorname{cis}\left(\frac{\theta + 2\pi k}{n}\right)$$

for $k = 0, 1, 2, 3, \ldots, n-1$.

The proof of this theorem is left as a problem.

EXAMPLE 7 Find the square roots of $-\frac{9}{2} + \frac{9}{2}\sqrt{3}i$.

Solution First change to trigonometric form:

$$r = \sqrt{\left(-\frac{9}{2}\right)^2 + \left(\frac{9}{2}\sqrt{3}\right)^2} = \sqrt{\frac{81}{4} + \frac{81\cdot 3}{4}} = \sqrt{81\left(\frac{1}{4} + \frac{3}{4}\right)} = 9$$

$$\theta' = \tan^{-1}\left|\frac{\frac{9}{2}\sqrt{3}}{-\frac{9}{2}}\right| = \tan^{-1}(\sqrt{3}) = 60°$$

$$\theta = 120° \qquad \text{(Quadrant II)}$$

By the nth root theorem, the square roots of $9\operatorname{cis}120°$ are

$$9^{1/2}\operatorname{cis}\left(\frac{120° + 360°k}{2}\right) = 3\operatorname{cis}(60° + 180°k)$$

$$k = 0: \quad 3\operatorname{cis}60° = \mathbf{\frac{3}{2} + \frac{3}{2}\sqrt{3}i}$$

$$k = 1: \quad 3\operatorname{cis}240° = \mathbf{-\frac{3}{2} - \frac{3}{2}\sqrt{3}i}$$

All other integral values of k repeat one of the previously found roots. For example,

$$k = 2: \quad 3\operatorname{cis}420° = \frac{3}{2} + \frac{3}{2}\sqrt{3}i$$

Check:

$$\left(\frac{3}{2}+\frac{3}{2}\sqrt{3}i\right)^2 = \frac{9}{4}+\frac{9}{2}\sqrt{3}i+\frac{9}{4}\cdot 3i^2 \qquad \left(-\frac{3}{2}-\frac{3}{2}\sqrt{3}i\right)^2 = \frac{9}{4}+\frac{9}{2}\sqrt{3}i+\frac{9}{4}\cdot 3i^2$$

$$= -\frac{9}{2}+\frac{9}{2}\sqrt{3}i \qquad\qquad = -\frac{9}{2}+\frac{9}{2}\sqrt{3}i$$

■

EXAMPLE 8 Find the fifth roots of 32.

Solution Begin by writing 32 in trigonometric form: $32 = 32 \operatorname{cis} 0^\circ$. The fifth roots are found by

$$32^{1/5} \operatorname{cis}\left(\frac{0^\circ + 360^\circ k}{5}\right) = 2 \operatorname{cis} 72^\circ k$$

$k = 0$: $2 \operatorname{cis} 0^\circ = 2$ ← The first root, which is located so that its argument is θ/n, is called the **principal nth root.**

$k = 1$: $2 \operatorname{cis} 72^\circ = 0.6180 + 1.9021i$

$k = 2$: $2 \operatorname{cis} 144^\circ = -1.6180 + 1.1756i$

$k = 3$: $2 \operatorname{cis} 216^\circ = -1.6180 - 1.1756i$

$k = 4$: $2 \operatorname{cis} 288^\circ = 0.6180 - 1.9021i$

All other integral values for k repeat those listed here. ■

If all the fifth roots of 32 are represented graphically, as shown in Figure 6.8, notice that they all lie on a circle of radius 2 and are equally spaced.

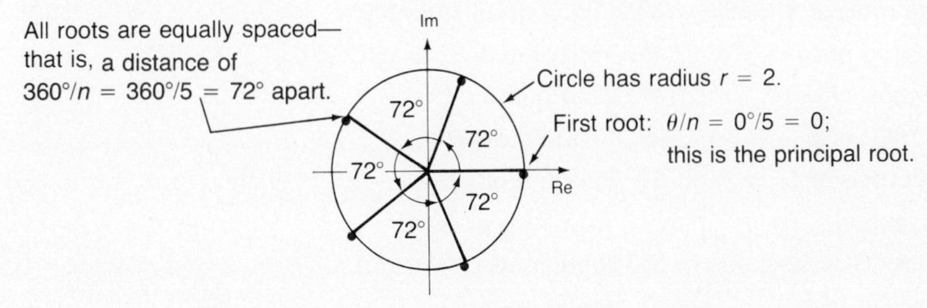

Figure 6.8 Graphical representation of the fifth roots of 32

If n is a positive integer, then the nth roots of a complex number $a + bi = r \operatorname{cis} \theta$ are equally spaced on the circle of radius r centered at the origin.

PROBLEM SET 6.6

A

Plot the complex numbers given in Problems 1–3. Find the modulus of each.

1. a. $3 + i$ **b.** $7 - i$ **c.** $3 + 2i$ **d.** $-3 - 2i$

2. a. $-1 + 3i$ **b.** $2 + 4i$ **c.** $5 + 6i$ **d.** $2 - 5i$

3. a. $-2 + 5i$ **b.** $-5 + 4i$ **c.** $4 - i$ **d.** $-1 + i$

Plot and then change to trigonometric form each of the numbers in Problems 4–15.

4. $1 + i$ **5.** $1 - i$ **6.** $\sqrt{3} - i$ **7.** $\sqrt{3} + i$

8. $1 - \sqrt{3}i$ **9.** $-1 - \sqrt{3}i$ **10.** 1 **11.** 5

12. $-4i$ **13.** $5.7956 - 1.5529i$

14. $-0.6946 + 3.9392i$ **15.** $1.5321 - 1.2856i$

Plot and then change to rectangular form each of the numbers in Problems 16–27. Use exact values whenever possible.

16. $2(\cos 45^\circ + i\sin 45^\circ)$

17. $3(\cos 60^\circ + i\sin 60^\circ)$

18. $4(\cos 315^\circ + i\sin 315^\circ)$

19. $5\operatorname{cis}(\frac{4\pi}{3})$

20. $\operatorname{cis}(\frac{5\pi}{6})$

21. $5\operatorname{cis}(\frac{3\pi}{2})$

22. $4\operatorname{cis} 30^\circ$

23. $2\operatorname{cis} \pi$

24. $10\operatorname{cis} 65^\circ$

25. $8\operatorname{cis} 24^\circ$

26. $6\operatorname{cis} 247^\circ$

27. $9\operatorname{cis} 190^\circ$

B

Perform the indicated operations in Problems 28–39.

28. $2\operatorname{cis} 60^\circ \cdot 3\operatorname{cis} 150^\circ$

29. $3\operatorname{cis} 48^\circ \cdot 5\operatorname{cis} 92^\circ$

30. $4(\cos 65^\circ + i\sin 65^\circ)\cdot 12(\cos 87^\circ + i\sin 87^\circ)$

31. $\dfrac{5(\cos 315^\circ + i\sin 315^\circ)}{2(\cos 48^\circ + i\sin 48^\circ)}$

32. $\dfrac{8\operatorname{cis} 30^\circ}{4\operatorname{cis} 15^\circ}$

33. $\dfrac{12\operatorname{cis} 250^\circ}{4\operatorname{cis} 120^\circ}$

34. $(2\operatorname{cis} 50^\circ)^3$

35. $(3\operatorname{cis} 60^\circ)^4$

36. $(\cos 210^\circ + i\sin 210^\circ)^5$

37. $(2 - 2i)^4$

38. $(1 + i)^6$

39. $(\sqrt{3} - i)^8$

Find the indicated roots of the numbers in Problems 40–57. Leave your answers in trigonometric form.

40. Square roots of $16\operatorname{cis} 100^\circ$

41. Cube roots of $8\operatorname{cis} 240^\circ$

42. Fourth roots of $81\operatorname{cis} 88^\circ$

43. Fifth roots of $32\operatorname{cis} 200^\circ$

44. Cube roots of $64\operatorname{cis} 216^\circ$

45. Fifth roots of $32\operatorname{cis} 160^\circ$

46. Cube roots of -1

47. Cube roots of 27

48. Cube roots of 8

49. Fourth roots of i

50. Fourth roots of $1 + i$

51. Fourth roots of $-1 - i$

52. Sixth roots of -64

53. Sixth roots of $64i$

54. Ninth roots of 1

55. Ninth roots of $-1 + i$

56. Tenth roots of i

57. Tenth roots of 1

Find the indicated roots of the numbers in Problems 58–63. Leave your answers in rectangular form. Show the roots graphically.

58. Cube roots of 1

59. Fourth roots of 1

60. Cube roots of -8

61. Cube roots of $4\sqrt{3} - 4i$

62. Square roots of $\dfrac{25}{2} - \dfrac{25\sqrt{3}}{2}i$

63. Fourth roots of $12.2567 + 10.2846i$

C

64. Find the cube roots of $(4\sqrt{2} + 4\sqrt{2}i)^2$.

65. Find the fifth roots of $(-16 + 16\sqrt{3}i)^3$.

66. Solve $x^5 - 1 = 0$.

67. Solve $x^4 + x^3 + x^2 + x + 1 = 0$.

68. Prove that

$$\frac{r_1 \operatorname{cis} \theta_1}{r_2 \operatorname{cis} \theta_2} = \frac{r_1}{r_2}\operatorname{cis}(\theta_1 - \theta_2)$$

69. Prove that

$$\left[\sqrt[n]{r}\operatorname{cis}\left(\frac{\theta + 360^\circ k}{n}\right)\right]^n = r\operatorname{cis}\theta$$

70. If $z_1 = a + bi$ and $z_2 = c + di$, show that
$|z_1 + z_2| \le |z_1| + |z_2|$.
This relationship is called the *triangle inequality*.

71. If

$$\cos\theta = 1 - \frac{\theta^2}{2!} + \frac{\theta^4}{4!} - \frac{\theta^6}{6!} + \cdots + \frac{(-1)^n\theta^{2n}}{(2n)!} + \cdots$$

$$\sin\theta = \theta - \frac{\theta^3}{3!} + \frac{\theta^5}{5!} - \frac{\theta^7}{7!} + \cdots + \frac{(-1)^n\theta^{2n+1}}{(2n+1)!} + \cdots$$

and

$$e^{i\theta} = 1 + (i\theta) + \frac{(i\theta)^2}{2!} + \frac{(i\theta)^3}{3!} + \frac{(i\theta)^4}{4!} + \cdots + \frac{(i\theta)^n}{n!} + \cdots$$

show that $e^{i\theta} = \cos\theta + i\sin\theta$. This equation is called *Euler's formula*.

72. Using Problem 71, show that $e^{i\pi} = -1$.

6.7 RIGHT TRIANGLES

One of the most important uses of trigonometry is in solving triangles. Recall from geometry that every **triangle** has three sides and three angles, which are called the six *parts* of the triangle. We say that a **triangle is solved** if all six parts are known. Typically three parts will be given, or known, and you will want to find the other

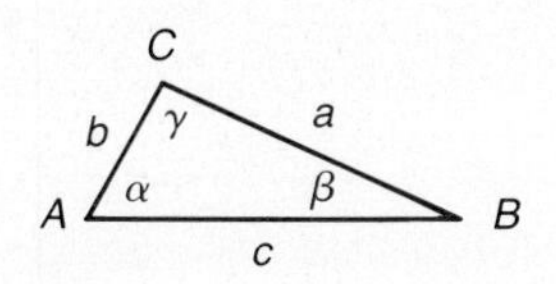

Figure 6.9 Correctly labeled triangle

three parts. Label a triangle as shown in Figure 6.9. The vertices are labeled A, B, and C, with the sides opposite those vertices a, b, and c, respectively. The angles are labeled α, β, and γ, respectively. In this section the examples are limited to right triangles in which γ denotes the right angle and c the **hypotenuse**.

According to the definition of the trigonometric functions, the angle under consideration must be in standard position. This requirement is sometimes inconvenient, so we use that definition to create a special case that applies to any acute angle θ of a right triangle. Notice that in Figure 6.9, θ might be α or β, but it would not be γ since γ is not an acute angle. Also notice that the hypotenuse is one of the sides of both acute angles. The other side making up the angle is called the **adjacent side**. Thus side a is adjacent to β and side b is adjacent to α. The third side of the triangle (the one not making up the angle) is called the **opposite side**. Thus side a is opposite α and side b is opposite β.

Right-Triangle Definition of the Trigonometric Functions

If θ is an acute angle in a right triangle, then

$$\cos\theta = \frac{\text{adjacent side}}{\text{hypotenuse}}$$

$$\sin\theta = \frac{\text{opposite side}}{\text{hypotenuse}}$$

$$\tan\theta = \frac{\text{opposite side}}{\text{adjacent side}}$$

The other trigonometric functions are the reciprocals of these relationships.

We can now use this definition to solve some given triangles.

EXAMPLE 1 Solve the triangle with $a = 50$, $\alpha = 35°$, and $\gamma = 90°$. (*Note:* $\alpha = 35°$ means the measure of angle α is 35°.)

Solution

$\boldsymbol{\alpha = 35°}$ Given

$\boldsymbol{\beta = 55°}$ Since $\alpha + \beta = 90°$ for any right triangle with right angle at C

$\boldsymbol{\gamma = 90°}$ Given

$\boldsymbol{a = 50}$ Given

$$b\text{: } \tan 35° = \frac{50}{b}$$

$$b = \frac{50}{\tan 35°}$$

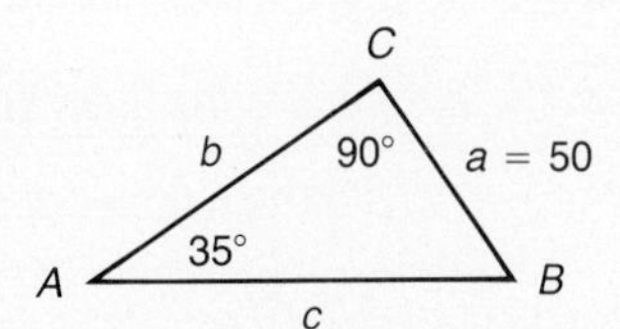

By table:

$$b = \frac{50}{\tan 35°} \approx \frac{50}{0.7002}$$ From Table B.3, $\tan 35° \approx 0.7002$

$$\approx 71.4082$$

or

$$b = 50 \cot 35^\circ \qquad \text{This avoids division.}$$
$$\approx 50(1.4281) \qquad \text{From Table B.3}$$
$$= 71.405$$

By calculator with algebraic logic:

[50] [÷] [35] [tan] [=]   DISPLAY: 71.40740034

By calculator with RPN logic:

[50] [ENTER] [35] [tan] [÷]   DISPLAY: 71.40740034

Notice that some of the answers above differ. However, to two significant digits, **$b = 71$**.

$$c\text{: } \sin 35^\circ = \frac{50}{c}$$

$$c = \frac{50}{\sin 35^\circ}$$

$$\approx 87.1723 \qquad \text{Use a calculator or Table B.3.}$$

To two significant digits, **$c = 87$**. ■

Appendix A explains how to determine the correct number of significant digits.

EXAMPLE 2 Solve the triangle with $a = 32$, $b = 58$, and $\gamma = 90^\circ$.

Solution α: $\tan\alpha = \dfrac{32}{58}$ or $\alpha = \text{Tan}^{-1}\left(\dfrac{32}{58}\right)$

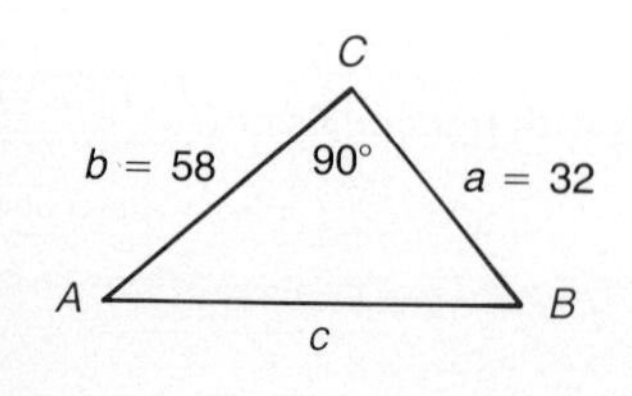

1. By table: $\frac{32}{58} \approx 0.5517$, so from Table B.3, $\alpha \approx 28.9^\circ$.
2. By calculator with algebraic logic:

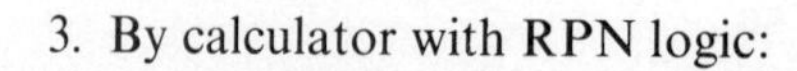

[32] [÷] [58] [=] [inv] [tan]   DISPLAY: 28.88658177

3. By calculator with RPN logic:

[32] [ENTER] [58] [÷] [arc] [tan]   DISPLAY: 28.88658176

To the nearest degree,

$\alpha = 29^\circ$
$\beta = 61^\circ$ $\quad \beta = 90^\circ - \alpha$
$\gamma = 90^\circ$ $\quad$ Given
$a = 32$ $\quad$ Given
$b = 58$ $\quad$ Given

c: To find c by table:

$$\sin \alpha = \frac{32}{c}$$

$$c = \frac{32}{\sin \alpha}$$

$$\approx \frac{32}{0.4833} \qquad \text{By table, } \sin 28.9^\circ = .4833$$

$$\approx 66.2139$$

To find c by calculator, use the Pythagorean Theorem:

$$c = \sqrt{a^2 + b^2}$$

Algebraic logic: [32] [x^2] [+] [58] [x^2] [=] [$\sqrt{x}$] DISPLAY: 66.24198065

RPN logic: [32] [ENTER] [×] [58] [ENTER] [×] [+] [$\sqrt{x}$]

DISPLAY: 66.24198067

To two significant digits, **$c = 66$.** ■

As you can see, there are many ways to solve a triangle; the method you choose will depend on the accuracy you want and the type of table or calculator you have. In the rest of this chapter, it is assumed that you have access to a calculator.

The solution of right triangles is necessary in a variety of situations. The first one we will consider concerns an observer looking at an object. The **angle of depression** is the acute angle measured down from the horizontal line to the line of sight whereas the **angle of elevation** is the acute angle measured up from a horizontal line to the line of sight. These ideas are illustrated in Examples 3 and 4.

EXAMPLE 3 The angle of elevation of a tree from a point on the ground 42 m from its base is 33°. Find the height of the tree.

Solution Let θ = angle of elevation and h = height of tree. Then

$$\tan \theta = \frac{h}{42}$$

$$h = 42 \tan 33^\circ$$

$$\approx 42(0.6494)$$

$$\approx 27.28$$

h

θ

42

The tree is 27 m tall. ■

EXAMPLE 4 The distance across a canyon can be determined from an airplane. Suppose the angles of depression to the two sides of the canyon are 43° and 55° as shown in Figure 6.10. If the altitude of the plane is 20,000 ft, how far is it across the canyon?

Solution Label parts x, y, θ, and ϕ as shown in Figure 6.10.

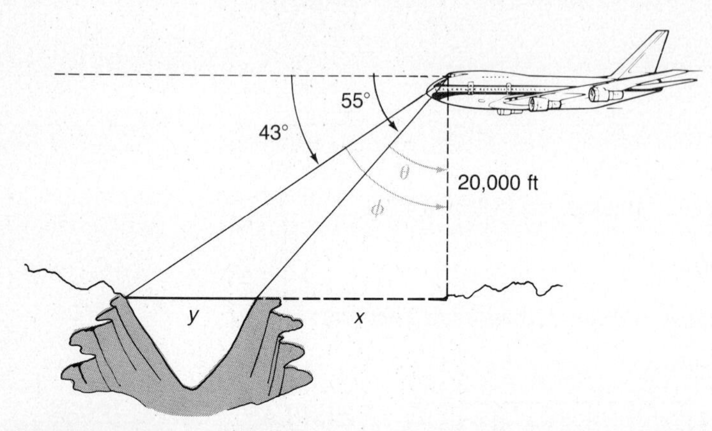

Figure 6.10 Determining the distance across a canyon from the air

$$\theta = 90^\circ - 55^\circ = 35^\circ \quad \text{and} \quad \phi = 90^\circ - 43^\circ = 47^\circ$$

First find x:

$$\tan 35^\circ = \frac{x}{20{,}000}$$

$$x = 20{,}000 \tan 35^\circ \approx 14{,}004$$

Next find $x + y$:

$$\tan 47^\circ = \frac{x + y}{20{,}000}$$

$$x + y = 20{,}000 \tan 47^\circ \approx 21{,}447$$

Thus

$$y \approx 21{,}447 - 14{,}004 = 7443$$

Remember, when you are working these problems on your calculator, do not work with the rounded results here, but with the entire accuracy possible with your calculator. You should round only once and that is with your final answer.

To two significant digits, **the distance across the canyon is 7400 ft.** ■

A second application of the solution of right triangles involves the **bearing** of a line, which is defined as an acute angle made with a north–south line. When giving the bearing of a line, first write N or S to determine whether to measure the angle from the north or the south side of a point on the line. Then give the measure of the angle followed by E or W, denoting on which side of the north–south line the angle is to be measured. Some examples are shown in Figure 6.11.

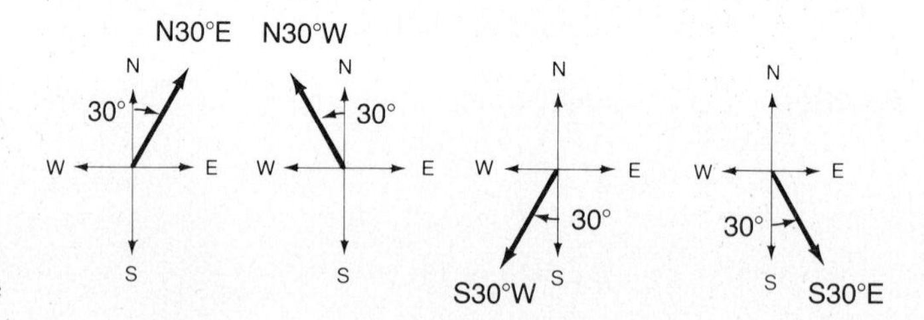

Figure 6.11 Bearing of an angle

EXAMPLE 5 To find the width AB of a canyon, a surveyor measures 100 m from A in the direction of N42.6°W to locate point C. The surveyor then determines that the bearing of CB is N73.5°E. Find the width of the canyon if point B is situated so that $\angle BAC = 90.0^\circ$.

Solution Let $\theta = \angle BCA$ in Figure 6.12.

$$\angle BCE' = 16.5^\circ \qquad \text{Complementary angles}$$
$$\angle ACS' = 42.6^\circ \qquad \text{Alternate interior angles}$$
$$\angle E'CA = 47.4^\circ \qquad \text{Complementary angles}$$
$$\theta = \angle BCA = \angle BCE' + \angle E'CA$$
$$= 16.5^\circ + 47.4^\circ$$
$$= 63.9^\circ$$

Since $\tan\theta = \dfrac{AB}{AC}$,

$$AB = AC\tan\theta$$
$$= 100\tan\theta$$
$$= 100\tan 63.9^\circ$$
$$\approx 204.125$$

The canyon is 204 m across.

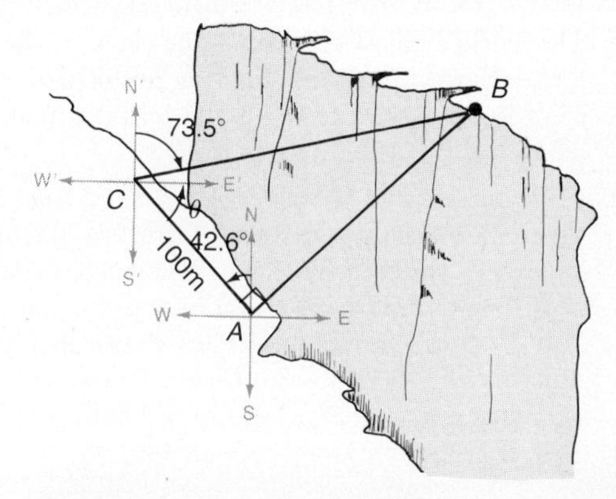

Figure 6.12 Surveying problem ■

PROBLEM SET 6.7

A

Solve the right triangles ($\gamma = 90^\circ$) in Problems 1–30.

1. $a = 80$; $\beta = 60^\circ$
2. $a = 18$; $\beta = 30^\circ$
3. $a = 9.0$; $\beta = 45^\circ$
4. $b = 37$; $\alpha = 65^\circ$
5. $b = 15$; $\alpha = 37^\circ$
6. $b = 50$; $\alpha = 53^\circ$
7. $a = 69$; $c = 73$
8. $b = 23$; $c = 45$
9. $b = 13$; $c = 22$
10. $a = 68$; $b = 83$
11. $a = 24$; $b = 29$
12. $a = 12$; $b = 8.0$
13. $a = 29$; $\alpha = 76^\circ$
14. $a = 93$; $\alpha = 26^\circ$
15. $a = 49$; $\beta = 45^\circ$
16. $b = 13$; $\beta = 65^\circ$
17. $b = 90$; $\beta = 13^\circ$
18. $b = 47$; $\beta = 108^\circ$
19. $b = 82$; $\alpha = 50^\circ$
20. $c = 28.3$; $\alpha = 69.2^\circ$
21. $c = 36$; $\alpha = 6^\circ$
22. $\beta = 57.4^\circ$; $a = 70.0$
23. $\alpha = 56.00^\circ$; $b = 2350$
24. $\beta = 23^\circ$; $a = 9000$
25. $b = 3100$; $c = 3500$
26. $\beta = 16.4^\circ$; $b = 2580$
27. $\alpha = 42^\circ$; $b = 350$
28. $b = 3200$; $c = 7700$
29. $b = 4100$; $c = 4300$
30. $c = 75.4$; $\alpha = 62.5^\circ$

B

31. ***Surveying*** The angle of elevation of a building from a point on the ground 30 m from its base is 38°. Find the height of the building.

32. ***Surveying*** Repeat Problem 31 for 150 ft instead of 30 m.

33. ***Surveying*** From a cliff 150 m above the shoreline, the angle of depression of a ship is 37°. Find the distance of the ship from a point directly below the observer.

34. ***Surveying*** Repeat Problem 33 for 450 ft instead of 150 m.

35. From a police helicopter flying at 1000 ft, a stolen car is sighted at an angle of depression of 71°. Find the distance of the car from a point directly below the helicopter.

36. Repeat Problem 35 for an angle of depression of 64° instead of 71°.

37. ***Surveying*** To find the east–west boundary of a piece of land, a surveyor must divert his path from point C on the boundary by proceeding due south for 300 ft to point A. Point B, which is due east of point C, is now found to be in the direction of N49°E from point A. What is the distance CB?

38. ***Surveying*** To find the distance across a river that runs east–west, a surveyor locates points P and Q on a north–south line on opposite sides of the river. She then paces out 150 ft from Q due east to a point R. Next she determines that the bearing of RP is N58°W. How far is it across the river?

39. A 16-ft ladder on level ground is leaning against a house. If the angle of elevation of the ladder is 52°, how far above the ground is the top of the ladder?

40. How far is the base of the ladder in Problem 39 from the house?

41. If the ladder in Problem 39 is moved so that the bottom is 9 ft from the house, what will be the angle of elevation?

42. Find the height of the Barrington Space Needle if the angle of elevation at 1000 ft from a point on the ground directly below the top is 58.15°.

43. The world's tallest chimney is the stack of the International Nickel Company. Find its height if the angle of elevation at 1000 ft from a point on the ground directly below the top of the stack is 51.36°.

44. In the movie *Close Encounters of the Third Kind*, there was a scene in which the star, Richard Dreyfuss, was approaching Devil's Tower in Wyoming. He could have determined his distance from Devil's Tower by first stopping at a point P and estimating the angle P, as shown in Figure 6.13. After moving 100 m toward Devil's Tower, he could have estimated the angle N, as shown in Figure 6.13. How far away from Devil's Tower is point N?

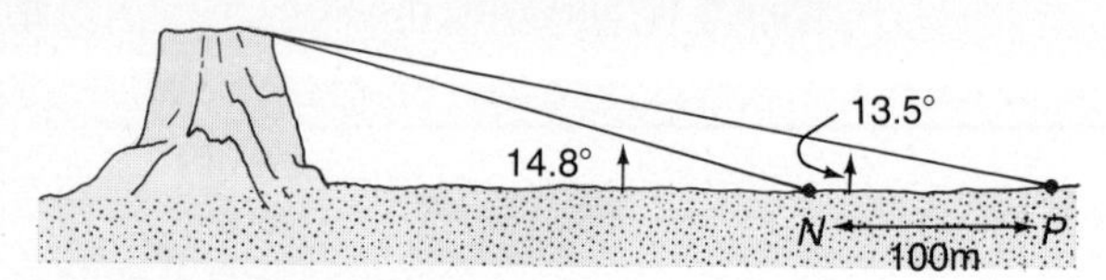

Figure 6.13 Procedure for determining the height of Devil's Tower

45. Determine the height of Devil's Tower in Problem 44.

46. ***Surveying*** To find the boundary of a piece of land, a surveyor must divert his path from a point A on the boundary for 500 ft in the direction S50°E. He then determines that the bearing of a point B located directly south of A is S40°W. Find the distance AB.

47. ***Surveying*** To find the distance across a river, a surveyor locates points P and Q on either side of the river. Next she measures 100 m from point Q in the direction S35°E to point R. Then she determines that point P is now in the direction of N25.0°E from point R and that $\angle PQR$ is a right angle. Find the distance across the river.

48. ***Surveying*** If the Empire State Building and the Sears Tower were situated 1000 ft apart, the angle of depression from the top of the Sears Tower to the top of the Empire State Building would be 11.53°, and the angle of depression to the foot of the Empire State Building would be 55.48°. Find the heights of the buildings.

49. ***Surveying*** On the top of the Empire State Building is a television tower. From a point 1000 ft from a point on the ground directly below the top of the tower, the angle of elevation to the bottom of the tower is 51.34° and to the top of the tower is 55.81°. What is the length of the television tower?

50. ***Physics*** A wheel 5.00 ft in diameter rolls up a 15.0° incline. What is the height of the center of the wheel above the base of the incline after the wheel has completed one revolution?

51. ***Physics*** What is the height of the center of the wheel in Problem 50 after three revolutions?

52. ***Astronomy*** If the distance from the earth to the sun is 92.9 million mi and the angle formed between Venus, the earth, and the sun (as shown in Figure 6.14) is 47.0°, find the distance from the sun to Venus.

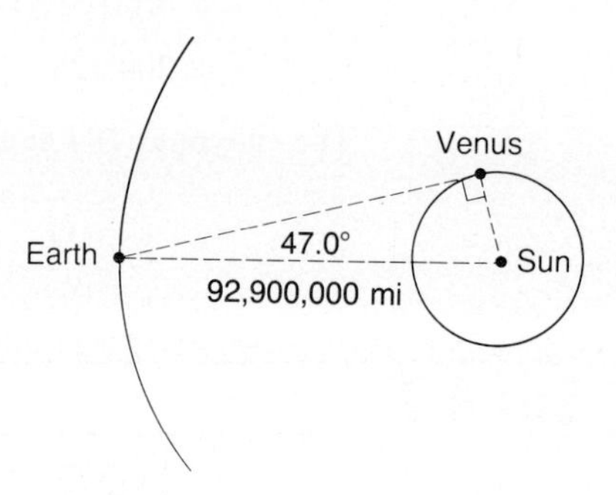

Figure 6.14

53. ***Astronomy*** Use the information in Problem 52 to find the distance from the earth to Venus.

C

54. The largest ground area covered by any office building is that of the Pentagon in Arlington, Virginia. If the radius of the circumscribed circle is 783.5 ft, find the length of one side of the Pentagon.

55. Use the information in Problem 54 to find the radius of the circle inscribed in the Pentagon.

56. ***Surveying*** To determine the height of the building shown in Figure 6.15, we select a point P and find that the angle of elevation is 59.64°. We then move out a distance of 325.4 ft (on a level plane) to point Q and find that the angle of elevation is now 41.32°. Find the height h of the building.

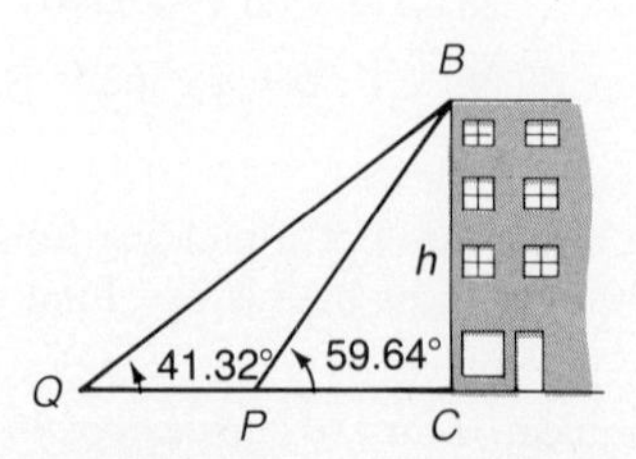

Figure 6.15

57. Using Figure 6.15, let the angle of elevation at P be α and at Q be β, and let the distance from P to Q be d. If h is the height of the building, show that

$$h = \frac{d \sin\alpha \sin\beta}{\sin(\alpha - \beta)}$$

58. A 6.0-ft person is casting a shadow of 4.2 ft. What time of the morning is it if the sun rose at 6:15 A.M. and is directly overhead at 12:15 P.M.?

59. How long will the shadow of the person in Problem 58 be at 8:00 A.M.?

60. ***Surveying*** From the top of a tower 100 ft high, the angles of depression to two landmarks on the plane upon which the tower stands are 18.5° and 28.4°.

a. Find the distance between the landmarks when they are on the same side of the tower.

b. Find the distance between the landmarks when they are on opposite sides of the tower.

61. Show that in every right triangle the value of c lies between $(a + b)/\sqrt{2}$ and $a + b$.

6.8 OBLIQUE TRIANGLES

Oblique triangles are triangles with no right angles. We will approach the solution of oblique triangles by studying the possible combinations of given information. In general, given three parts of a triangle, you will want to find the remaining three parts. But can you do so given *any* three parts? Consider the possibilities:

1. SSS: By SSS we mean that you are given three sides and want to find the three angles.
2. SAS: You are given two sides and an included angle.
3. AAA: You are given three angles.
4. ASA or AAS: You are given two angles and a side.
5. SSA: You are given two sides and the angle opposite one of them.

We will consider these possibilities one at a time.

SSS

To solve a triangle given SSS, it is necessary for the sum of the lengths of the two smaller sides to be greater than the length of the largest side. (Otherwise, there is no solution.) In this case, use a generalization of the Pythagorean Theorem called the **Law of Cosines:**

Law of Cosines

In triangle ABC labeled as shown in Figure 6.16,

$$c^2 = a^2 + b^2 - 2ab\cos\gamma$$

Notice that for a right triangle, $\gamma = 90°$. This means that

$$c^2 = a^2 + b^2 - 2ab\cos 90° \qquad \text{or} \qquad c^2 = a^2 + b^2$$

since $\cos 90° = 0$. This last equation is simply the Pythagorean Theorem.

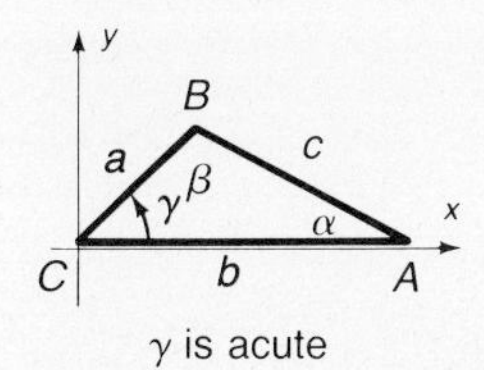

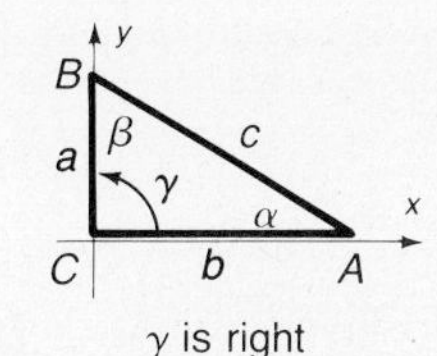

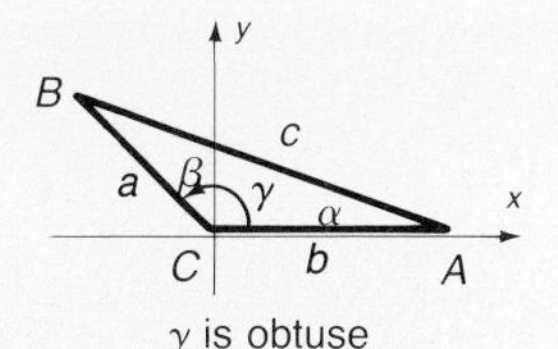

Figure 6.16 Examples of triangles

Proof Let γ be an angle in standard position with A on the positive x axis as shown in Figure 6.16. The coordinates of the vertices are as follows:

$C(0,0)$	Since C is in standard position
$A(b,0)$	Since A is on the x axis a distance of b units from the origin
$B(a\cos\gamma, a\sin\gamma)$	Let B be at (x, y); then by definition of the trigonometric functions, $\cos\gamma = x/a$ and $\sin\gamma = y/a$. Thus $x = a\cos\gamma$ and $y = a\sin\gamma$.

Use the distance formula for the distance between $A(b,0)$ and $B(a\cos\gamma, a\sin\gamma)$:

$$\begin{aligned} c^2 &= (a\cos\gamma - b)^2 + (a\sin\gamma - 0)^2 \\ &= a^2\cos^2\gamma - 2ab\cos\gamma + b^2 + a^2\sin^2\gamma \\ &= a^2(\cos^2\gamma + \sin^2\gamma) + b^2 - 2ab\cos\gamma \\ &= a^2 + b^2 - 2ab\cos\gamma \end{aligned}$$

□

By letting A and B, respectively, be in standard position, it can also be shown that

$$a^2 = b^2 + c^2 - 2bc\cos\alpha$$

and

$$b^2 = a^2 + c^2 - 2ac\cos\beta$$

To find the angles when you are given three sides, solve for α, β, or γ.

Law of Cosines (Alternate Forms)

$$a^2 = b^2 + c^2 - 2bc\cos\alpha \qquad \cos\alpha = \frac{b^2 + c^2 - a^2}{2bc}$$

$$b^2 = a^2 + c^2 - 2ac\cos\beta \qquad \cos\beta = \frac{a^2 + c^2 - b^2}{2ac}$$

$$c^2 = a^2 + b^2 - 2ab\cos\gamma \qquad \cos\gamma = \frac{a^2 + b^2 - c^2}{2ab}$$

EXAMPLE 1 What is the smallest angle of a triangular patio whose sides measure 25, 18, and 21 ft?

Solution If γ represents the smallest angle, then c (the side opposite γ) must be the smallest side, so $c = 18$. Then:

$$\cos\gamma = \frac{a^2 + b^2 - c^2}{2ab}$$

$$= \frac{25^2 + 21^2 - 18^2}{2(25)(21)}$$

Use this number and trig tables if you have only a four-function calculator.

$$\gamma = \cos^{-1}\left(\frac{25^2 + 21^2 - 18^2}{2(25)(21)}\right)$$

By calculator with algebraic logic:

[25] [x^2] [+] [21] [x^2] [−] [18] [x^2] [=] [÷] [2] [÷] [25] [÷] [21] [=] [inv] [cos]

DISPLAY: 45.03565072

By calculator with RPN logic:

[25] [ENTER] [×] [21] [ENTER] [×] [+] [18] [ENTER] [×] [−] [2] [÷] [25] [÷]
[21] [÷] [arc] [cos]

DISPLAY: 45.03565071

To two significant digits, the answer is **45°**. ■

SAS The second possibility listed for solving oblique triangles is that of being given two sides and an included angle. It is necessary that the given angle be less than 180°. Again use the Law of Cosines for this possibility, as shown by Example 2.

EXAMPLE 2 Find c when $a = 52.0$, $b = 28.3$, and $\gamma = 28.5°$.

Solution By the Law of Cosines:

$$\begin{aligned} c^2 &= a^2 + b^2 - 2ab\cos\gamma \\ &= (52.0)^2 + (28.3)^2 - 2(52.0)(28.3)\cos 28.5° \end{aligned}$$

By calculator:

$$\begin{aligned} c^2 &\approx 918.355474 \\ c &\approx 30.30438044 \end{aligned}$$

To three significant digits, $\boldsymbol{c} = \mathbf{30.3}$. ■

If you do not have a calculator available, you should use the Law of Tangents for this problem, since the Law of Cosines does not lend itself to logarithmic calculations. The Law of Tangents is given in Problem 63 of Problem Set 6.8.

AAA The third case supposes that three angles are given. However, from what you know of similar triangles (see Figure 6.17), you can conclude that the triangle cannot be solved unless you know the length of at least one side.

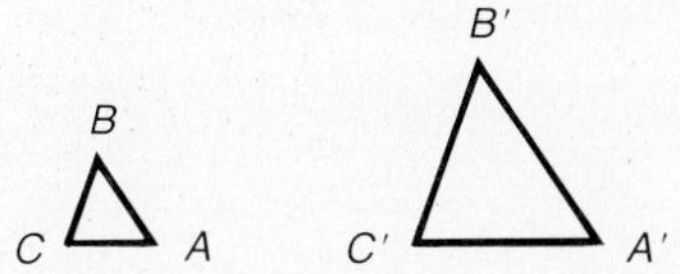

Figure 6.17 Similar triangles—similar triangles have the same shape but not necessarily the same size; that is, corresponding angles of similar triangles have equal measure

ASA or AAS Case 4 supposes that two angles and a side are given. For a triangle to be formed, the sum of the two given angles must be less than 180° and the given side must be greater than zero. If you know two angles, you can easily find the third angle since the sum of the three angles is 180°. The Law of Cosines is not sufficient in this case because you need at least two sides for the application of that law. We state and prove a result called the **Law of Sines**.

Law of Sines

In any $\triangle ABC$,

$$\frac{\sin\alpha}{a} = \frac{\sin\beta}{b} = \frac{\sin\gamma}{c}$$

The equation in the Law of Sines means that you can use any of the following equations:

$$\frac{\sin\alpha}{a} = \frac{\sin\beta}{b}$$

$$\frac{\sin\alpha}{a} = \frac{\sin\gamma}{c}$$

$$\frac{\sin\beta}{b} = \frac{\sin\gamma}{c}$$

Proof Consider any oblique triangle, as shown in Figure 6.18. Let h = height of the triangle with base CA. Then

$$\sin\alpha = \frac{h}{c} \qquad \text{and} \qquad \sin\gamma = \frac{h}{a}$$

Solving for h,

$$h = c\sin\alpha \qquad \text{and} \qquad h = a\sin\gamma$$

Thus

$$c\sin\alpha = a\sin\gamma$$

Dividing by ac,

$$\frac{\sin\alpha}{a} = \frac{\sin\gamma}{c}$$

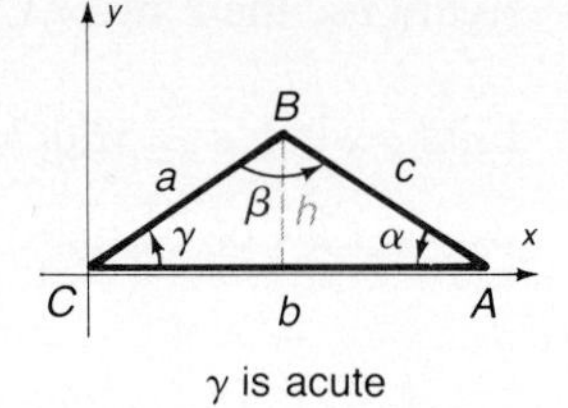

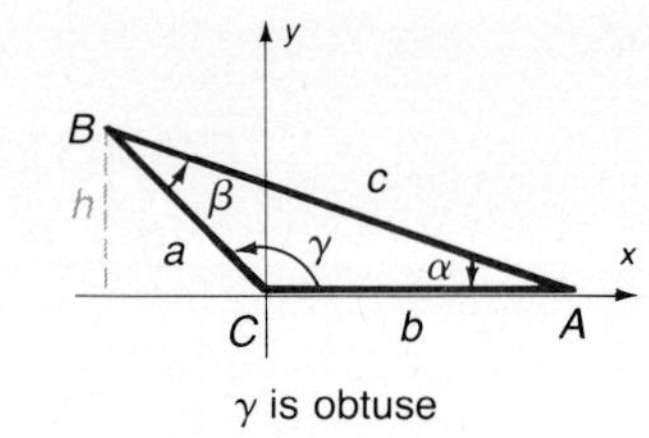

Figure 6.18 Oblique triangles for law of sines

Repeat these steps for the height of the same triangle with base AB:

$$\frac{\sin\alpha}{a} = \frac{\sin\beta}{b}$$

□

EXAMPLE 3 Solve the triangle in which $a = 20$, $\alpha = 38°$, and $\beta = 121°$.

Solution

$\boldsymbol{\alpha = 38°}$ Given

$\boldsymbol{\beta = 121°}$ Given

$\boldsymbol{\gamma = 21°}$ Since $\alpha + \beta + \gamma = 180°$, then $\gamma = 180° - 38° - 121° = 21°$

$\boldsymbol{a = 20}$ Given

$\boldsymbol{b = 28}$ Use the Law of Sines:

$$\frac{\sin 38°}{20} = \frac{\sin 121°}{b}$$

Then

$$b = \frac{20\sin 121°}{\sin 38°} \qquad \text{Use tables or a calculator.}$$

$$\approx \frac{20(0.8572)}{0.6157} \qquad \text{Use logarithms or a calculator.}$$

$$\approx 27.85 \qquad \text{Give answer to two significant digits.}$$

$c = 12$ Use the Law of Sines:

$$\frac{\sin 38^\circ}{20} = \frac{\sin 21^\circ}{c}$$

Then

$$c \approx \frac{20 \sin 21^\circ}{\sin 38^\circ}$$

$$\approx 11.64$$

Notice that the answers are given to two significant digits (see boldface). ■

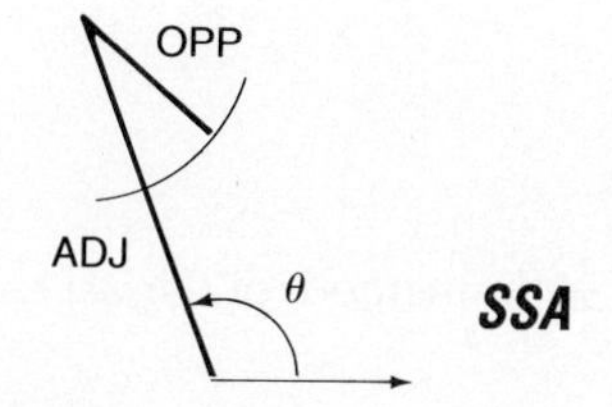

Figure 6.19 $\theta > 90^\circ$, OPP $\leq$ ADJ

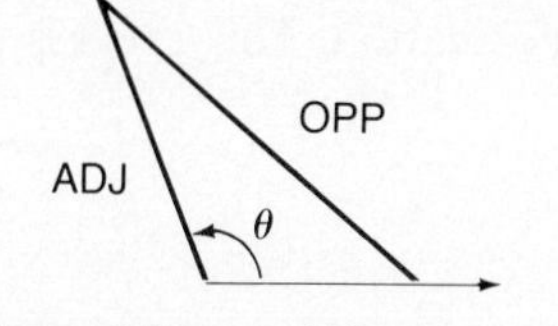

Figure 6.20 $\theta > 90^\circ$, OPP $>$ ADJ

SSA The remaining case of solving oblique triangles is case 5, in which two sides and an angle that is not an included angle are given. Call the given angle θ (which may be α, β, or γ depending on the problem). Since you are given SSA, one of the given sides must not be one of the sides of θ. Call this side "OPP." The given side that is one of the sides of θ is called "ADJ."

1. Suppose that $\theta > 90^\circ$. There are two possibilities:

i. OPP $\leq$ ADJ

No triangle is formed (see Figure 6.19).

ii. OPP $>$ ADJ

One triangle is formed (see Figure 6.20). Use the Law of Sines as shown in Example 4.

EXAMPLE 4 Let $a = 3.0$, $b = 2.0$, and $\alpha = 110^\circ$. Solve the triangle.

$\alpha = 110^\circ$ Given

$\beta = 39^\circ$ Work shown at right $\qquad \dfrac{\sin \alpha}{a} = \dfrac{\sin \beta}{b}$

$\gamma = 31^\circ$ $\gamma = 180^\circ - 110^\circ - 39^\circ$ $\qquad \dfrac{\sin 110^\circ}{3} = \dfrac{\sin \beta}{2}$

$a = 3.0$ Given $\qquad \sin \beta = \frac{2}{3} \sin 110^\circ$

$\approx \frac{2}{3}(0.9397)$

$b = 2.0$ Given $\qquad \approx 0.6265$

$c = 1.7$ Use the Law of Sines: $\qquad \beta \approx 38.79$

$$\frac{\sin 110^\circ}{3} = \frac{\sin \gamma}{c}$$

$$c = \frac{3 \sin \gamma}{\sin 110^\circ}$$

$$\approx 1.654$$

■

Do not work with rounded results. When finding c in Example 4, notice that if you work with

$$\gamma \approx 31.2104436^\circ$$

you obtain $c \approx 1.654$, or 1.7 to two significant digits. If you work with $\gamma \approx 31°$, however, you obtain $c \approx 1.644$, or 1.6 to two significant digits. This means that you should round only when stating answers.

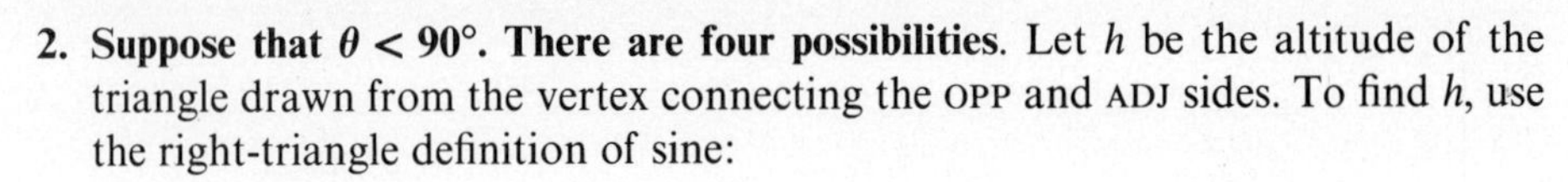

2. **Suppose that $\theta < 90°$. There are four possibilities.** Let h be the altitude of the triangle drawn from the vertex connecting the OPP and ADJ sides. To find h, use the right-triangle definition of sine:

$$h = (\text{ADJ})\sin\theta$$

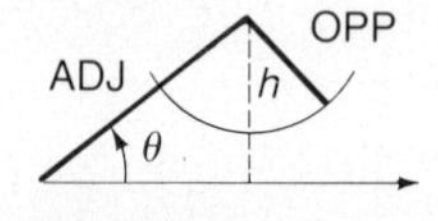

Figure 6.21 $\theta < 90°$, OPP $< h <$ ADJ

   i. **OPP $< h <$ ADJ**
   No triangle is formed (see Figure 6.21).

   ii. **OPP $= h <$ ADJ**
   A right triangle is formed (see Figure 6.22). Use the methods of the last section to solve the triangle.

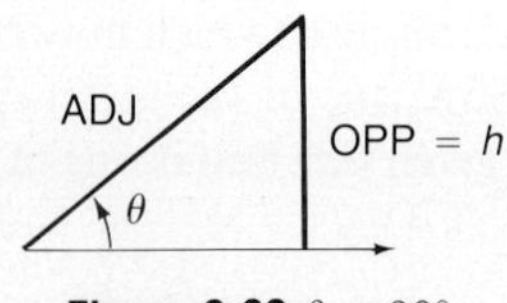

Figure 6.22 $\theta < 90°$, $h <$ OPP $<$ ADJ

   iii. **$h <$ OPP $<$ ADJ**
   This situation is called the **ambiguous case** and is really the only special case you must watch for. All the other cases can be determined from the calculations without special consideration. Notice from Figure 6.23 that two *different* triangles are formed with the given information. This process for finding both solutions is shown in Example 5.

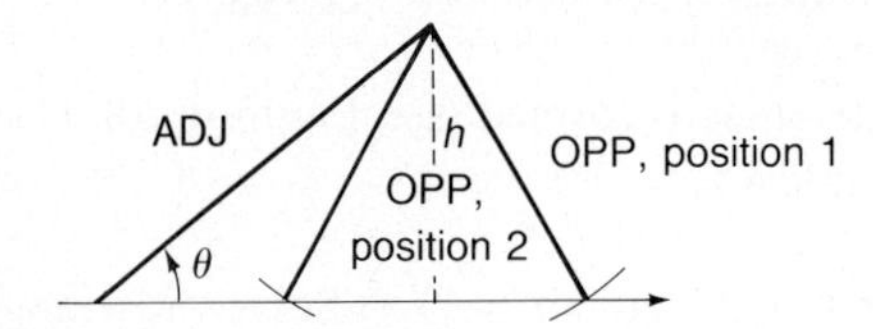

Figure 6.23 The ambiguous case: $\theta < 90°$, $h <$ OPP $<$ ADJ

EXAMPLE 5 Solve the triangle with $a = 1.50$, $b = 2.00$, and $\alpha = 40.0°$.

Solution

$$\frac{\sin\alpha}{a} = \frac{\sin\beta}{b}$$

$$\frac{\sin 40°}{1.5} = \frac{\sin\beta}{2}$$

$$\sin\beta = \frac{2\sin 40°}{1.5}$$

$$\approx \frac{4}{3}(0.6428)$$

$$\approx 0.8570$$

$$\beta \approx 59.0°$$

Figure 6.24

But from Figure 6.24 you can see that this is only the acute-angle solution. For the obtuse-angle solution—call it β'—find

$$\beta' = 180° - \beta$$

$$\approx 121°$$

Finish the problem by working two calculations, which are presented side by side.

Solution 1:		*Solution 2:*	
$\alpha = 40.0^\circ$	Given	$\alpha = 40.0^\circ$	Given
$\beta = 59.0^\circ$	See above	$\beta' = 121^\circ$	See above
$\gamma = 81.0^\circ$	$\gamma = 180^\circ - \alpha - \beta$	$\gamma' = 19.0^\circ$	$\gamma' = 180^\circ - \alpha - \beta'$
$a = 1.50$	Given	$a = 1.50$	Given
$b = 2.00$	Given	$b = 2.00$	Given
$c = 2.30$	$\dfrac{\sin\alpha}{a} = \dfrac{\sin\gamma}{c}$	$c' = 0.76$	$\dfrac{\sin\alpha}{a} = \dfrac{\sin\gamma'}{c'}$
	$c = \dfrac{1.5\sin\gamma}{\sin 40^\circ}$		$c' = \dfrac{1.5\sin\gamma'}{\sin 40^\circ}$
	≈ 2.3049		≈ 0.7597

■

iv. OPP $\geq$ ADJ

One triangle is formed as shown in Example 6.

EXAMPLE 6 Solve the triangle given by $a = 3.0$, $b = 2.0$, and $\alpha = 40^\circ$ (Figure 6.25).

Solution

$$\frac{\sin\alpha}{a} = \frac{\sin\beta}{b}$$

$$\frac{\sin 40^\circ}{3} = \frac{\sin\beta}{2}$$

$$\sin\beta = \frac{2}{3}\sin 40^\circ$$

$$\approx \frac{2}{3}(0.6428)$$

$$\approx 0.4285$$

$$\beta \approx 25.374$$

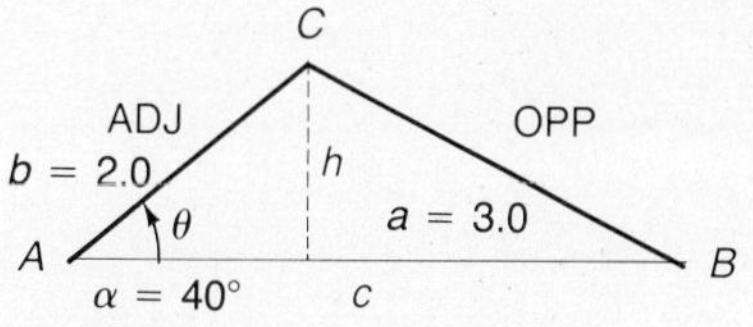

Figure 6.25 $\theta < 90^\circ$, OPP $\geq$ ADJ

$\alpha = 40^\circ$	Given
$\beta = 25^\circ$	See work shown above.
$\gamma = 115^\circ$	$\gamma = 180^\circ - \alpha - \beta$
$a = 3.0$	Given
$b = 2.0$	Given
$c = 4.2$	Since $\dfrac{\sin 40^\circ}{3} = \dfrac{\sin\gamma}{c}$
	$c = \dfrac{3\sin\gamma}{\sin 40^\circ}$
	≈ 4.2427

■

The most important skill to be learned from this section is the ability to select the proper trigonometric law when given a problem. A review of various types of problems may be helpful:

Summary for solving triangles

To Solve an Oblique Triangle *ABC*

Given	Conditions on given information	Law to use for solution
1. **SSS**	*a.* The sum of the lengths of the two smaller sides is less than or equal to the length of the larger side.	No solution
	b. The sum of the lengths of the two smaller sides is greater than the length of the larger side.	**Law of Cosines**
2. **SAS**	*a.* The angle is greater than or equal to 180°.	No solution
	b. The angle is less than 180°.	**Law of Cosines**
3. **AAA**		**No solution**
4. **ASA** or **AAS**	*a.* The sum of the angles is greater than or equal to 180°.	No solution
	b. The sum of the angles is less than 180°.	**Law of Sines**
5. **SSA**	Let θ be the given angle with adjacent (ADJ) and opposite (OPP) sides given; the height h is found by $h = (\text{ADJ})\sin\theta$	
	a. $\theta > 90°$	
	i. OPP $\le$ ADJ	No solution
	ii. OPP $>$ ADJ	**Law of Sines**
	b. $\theta < 90°$	
	i. OPP $< h <$ ADJ	No solution
	ii. OPP $= h <$ ADJ	Right-triangle solution
	iii. $h <$ OPP $<$ ADJ	***Ambiguous case:*** Use Law of Sines to find two solutions
	iv. OPP $\ge$ ADJ	**Law of Sines**

Remember: *When given two sides and an angle that is not an included angle, check to see whether one side is between the height of the triangle and the length of the other side. If it is, there will be two solutions.*

EXAMPLE 7 An airplane is 100 km N40°E of a loran station and is traveling due west at 240 kph. How long will it be (to the nearest minute) *before* the plane is 90 km from the loran station?

Solution You are given SSA, so you should check the other conditions for the ambiguous case. Let θ be angle $= 50°$ (see Figure 6.26); 100 km is the side adjacent to θ; 90 km is the side opposite to θ.

$$h = (\text{ADJ})\sin\theta$$
$$= 100 \sin 50°$$
$$\approx 76.6$$
$$h < \text{OPP} < \text{ADJ}$$
$$76.6 < 90 < 100$$

which is the ambiguous case.

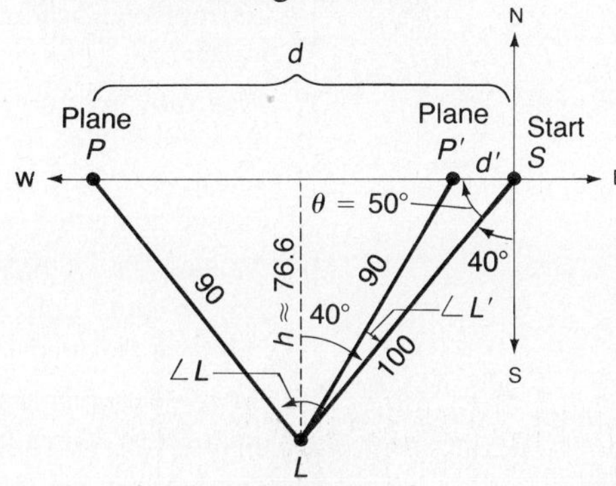

Figure 6.26

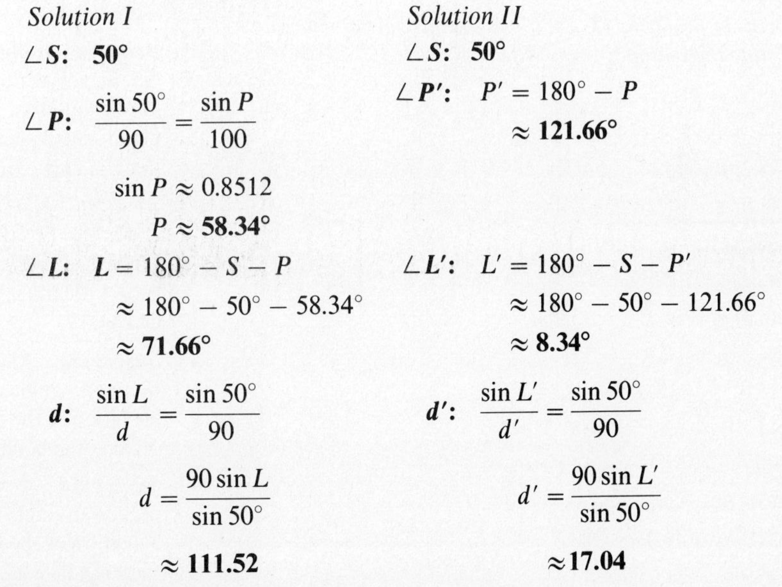

Solution I

$\angle S$: **50°**

$\angle P$: $\dfrac{\sin 50°}{90} = \dfrac{\sin P}{100}$

$\sin P \approx 0.8512$

$P \approx \mathbf{58.34°}$

$\angle L$: $L = 180° - S - P$

$\approx 180° - 50° - 58.34°$

$\approx \mathbf{71.66°}$

d: $\dfrac{\sin L}{d} = \dfrac{\sin 50°}{90}$

$d = \dfrac{90 \sin L}{\sin 50°}$

$\approx \mathbf{111.52}$

Solution II

$\angle S$: **50°**

$\angle P'$: $P' = 180° - P$

$\approx \mathbf{121.66°}$

$\angle L'$: $L' = 180° - S - P'$

$\approx 180° - 50° - 121.66°$

$\approx \mathbf{8.34°}$

d': $\dfrac{\sin L'}{d'} = \dfrac{\sin 50°}{90}$

$d' = \dfrac{90 \sin L'}{\sin 50°}$

$\approx \mathbf{17.04}$

At 240 kph, the times are

$$\frac{111.52}{240} \approx 0.464668 \quad \text{and} \quad \frac{17.04}{240} \approx 0.070989$$

To convert these to minutes, multiply by 60 and round to the nearest minute. The times are 28 min and 4 min. ■

PROBLEM SET 6.8

A

Solve $\triangle ABC$ in Problems 1–14. If the triangle does not have a solution, state the reason.

1. $a = 7.0$; $b = 8.0$; $c = 2.0$

2. $a = 18$; $b = 25$; $\gamma = 30°$

3. $b = 14$; $c = 12$; $\alpha = 82°$

4. $a = 3.0$; $b = 4.0$; $\alpha = 125°$

5. $a = 5.0$; $b = 7.0$; $\alpha = 75°$

6. $a = 5.0$; $b = 4.0$; $\alpha = 125°$

7. $a = 5.0$; $b = 4.0$; $\alpha = 80°$

8. $a = 7.0$; $b = 9.0$; $\alpha = 52°$

9. $a = 10.2$; $b = 11.8$; $\alpha = 47.0°$

10. $a = 4.0$; $b = 5.0$; $\alpha = 56°$

11. $a = 4.5$; $b = 5.0$; $\alpha = 56°$

12. $a = 7.0$; $b = 5.0$; $\alpha = 56°$

13. $a = 78$; $c = 14$; $\alpha = 103°$

14. $a = 28$; $c = 10$; $\alpha = 112°$

B

Solve $\triangle ABC$ in Problems 15–28. If the triangle does not have a solution, state the reason.

15. $a = 38$; $b = 41$; $c = 25$

16. $a = 45$; $b = 92$; $c = 41$

17. $a = 38.2$; $b = 14.8$; $\gamma = 48.2°$

18. $a = 41.0$; $\alpha = 45.2°$; $\beta = 21.5°$

19. $a = 26$; $b = 71$; $c = 88$

20. $\alpha = 48°$; $\beta = 105°$; $\gamma = 27°$

21. $a = 14.2$; $b = 16.3$; $\gamma = 35.0°$

22. $a = 14.2$; $b = 16.3$; $\gamma = 135.0°$

23. $\beta = 15.0°$; $\gamma = 18.0°$; $b = 23.5$

24. $b = 45.7$; $\alpha = 82.3°$; $\beta = 61.5°$

25. $b = 82.5$; $c = 52.2$; $\gamma = 32.1°$

26. $a = 151$; $b = 234$; $c = 416$

27. $a = 21.3$; $b = 49.0$; $\alpha = 23.1°$

28. $b = 45.3$; $c = 11.9$; $\beta = 14.1°$

29. Prove that $a^2 = b^2 + c^2 - 2bc \cos \alpha$.

30. Prove that $b^2 = a^2 + c^2 - 2ac \cos \beta$.

31. Given $\triangle ABC$, show that $(\sin \alpha)/a = (\sin \beta)/b$.

32. Given $\triangle ABC$, show that $(\sin \beta)/b = (\sin \gamma)/c$.

33. Given $\triangle ABC$, show that $(\sin \alpha)/a = (\sin \gamma)/c$.

34. Given $\triangle ABC$, show that $(\sin \alpha)/\sin \beta = a/b$.

35. ***Navigation*** Two boats leave a dock at the same time; one travels S15° W at 8 kph, and the other travels N28° W at 12 kph. How far apart are they in 3 hr?

36. ***Aviation*** Two airplanes leave an airport at the same time; one travels at 180 mph on a heading of 280° (measured clockwise from north), while the other travels at 260 mph on a heading of 35°. How far apart are the planes in 1 hr?

37. ***Aviation*** At noon a plane leaves an airport and flies on a heading of 61° (measured clockwise from north) at 200 mph. When will the plane be 35 mi from a point located 50 mi due east of the airport?

38. ***Navigation*** A boat leaves a dock at noon and travels N53° W at 10 kph. When will the boat be 12 km from a lighthouse located 15 km due west of the dock?

39. If two sides of a parallelogram are 50 cm and 70 cm with a diagonal of 40 cm, find the angles of the parallelogram.

40. Two sides of a parallelogram make an angle of 47°. If the lengths of the sides are 18 m and 24 m, what is the length of each diagonal?

41. ***Aviation*** Buffalo, New York, is approximately 475 km N9° E of Washington, D.C., and New York City is N49° W of Washington, D.C. How far is Buffalo from New York City if the distance from New York City to Washington, D.C., is approximately 350 km?

42. ***Aviation*** New Orleans, Louisiana, is approximately 1800 km S56° E of Denver, Colorado, and Chicago, Illinois, is N76° E of Denver. How far is Chicago from New Orleans if Denver is approximately 1500 km from Chicago?

43. ***Surveying*** A tree stands vertically on a hillside that makes an angle of 14° with the horizontal. If the angle of elevation of the tree 58 ft down the hill from the base is 52°, what is the height of the tree?

44. ***Surveying*** A tree stands vertically on a hillside that makes an angle of 5° with the horizontal. If the angle of elevation of the tree at exactly 50 ft down the hill from the base is 43°, what is the height of the tree?

45. ***Ballistics*** An artillery observer must determine the distance to a target at point T (see Figure 6.27). He knows that the target is 5.20 mi from point I on a nearby island. He also knows that he (at point H) is 4.30 mi from point I. If $\angle HIT$ is 68.4°, how far is he from the target?

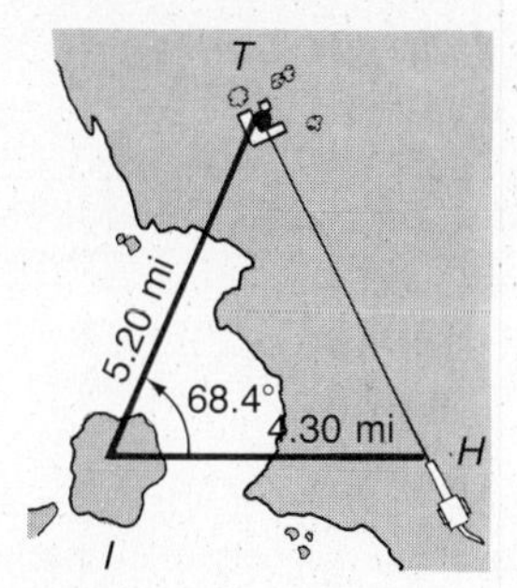

Figure 6.27

46. ***Consumer*** A buyer is interested in purchasing a triangular lot with vertices LOT, but the marker at point L has been lost. The deed indicates that TO is 453 ft and LO is 112 ft and the angle at L is 82.6°. What is the distance from L to T?

47. ***Navigation*** A UFO is sighted by people in two cities 2.300 mi apart. The UFO is between the two cities and in the same vertical plane. The angle of elevation of the UFO from the first city is 10.48° and from the second it is 40.79°. At what altitude is the UFO flying? What is the actual distance of the UFO from each city?

48. ***Engineering*** At 500 ft in the direction that the Tower of Pisa is leaning, the angle of elevation is 20.24°. If the tower leans at an angle of 5.45° from the vertical, what is the length of the tower?

49. ***Engineering*** What is the angle of elevation of the leaning Tower of Pisa (described in Problem 48) if you measure from a point 500 ft in the direction exactly opposite from the way it is leaning?

50. ***Navigation*** From a blimp, the angle of depression to the top of the Eiffel Tower is 23.2° and to the bottom it is 64.6° (see Figure 6.28). After flying over the tower at the same height and to a distance of 1000 ft from the first location, you determine that the angle of depression to the top of the tower is now 31.4°. What is the height of the Eiffel Tower given that these measurements are in the same vertical plane?

51. ***Surveying*** The world's longest deepwater jetty is at Le Havre, France. Since access to the jetty is restricted, it was necessary for me to calculate its length by noting that it forms an angle of 85.0° with the shoreline. After pacing out 1000 ft along the line making an 85.0° angle with the jetty, I calculated the angle to the end of the jetty to be 83.6°. What is the length of the jetty?

52. ***Surveying*** The world's longest pier is at Hasa, Saudi Arabia. The length of the pier can be determined by

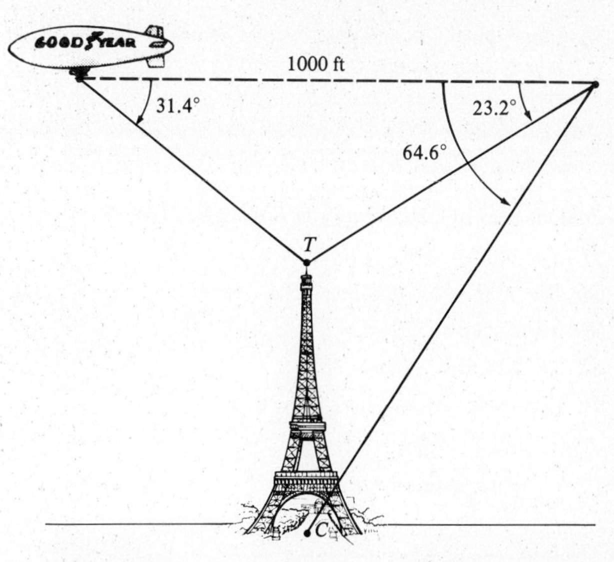

Figure 6.28

measuring out a distance of 1.00 mi along a line on shore that makes an angle of 83.4° with the foot of the pier. At that point, the angle to the end of the pier is 88.05°. How long is the pier?

53. ***Consumer*** A level lot has dimensions as shown in Figure 6.29. What is the total cost of treating the area for poison oak if the fee is \$50 per acre (1 acre = 43,560 ft^2)?

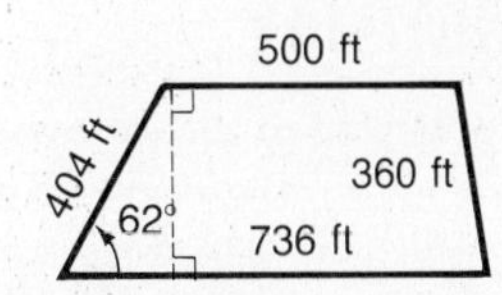

Figure 6.29

C

54. Show that the area K of a triangle ABC can be found by

$$K = \frac{1}{2}bc \sin \alpha = \frac{1}{2}ac \sin \beta = \frac{1}{2}ab \sin \gamma$$

when two sides and an included angle are known.

55. Show that the area K of a triangle ABC can be found by

$$K = \frac{a^2 \sin \beta \sin \gamma}{2 \sin \alpha}$$

when three angles and one side are known.

56. Use Problem 54 and the Law of Cosines to show that the area K of a triangle can be found by

$$K = \sqrt{s(s-a)(s-b)(s-c)}$$

where $s = \frac{1}{2}(a+b+c)$.

Find the area of each triangle in Problems 57–60.

57. $a = 14.2$; $b = 16.3$; $\gamma = 35.0°$

58. $B = 15.0°$; $C = 18.0°$; $b = 23.5$

59. $b = 82.5$; $c = 52.2$; $\alpha = 32.1°$

60. $a = 124$; $b = 325$; $c = 351$

61. Using the Law of Sines, show that

$$\frac{\sin\alpha - \sin\beta}{\sin\alpha + \sin\beta} = \frac{a-b}{a+b}$$

62. Using Problem 61 and the formulas for the sum and difference of sines, show that

$$\frac{2\cos\frac{1}{2}(\alpha+\beta)\sin\frac{1}{2}(\alpha-\beta)}{2\sin\frac{1}{2}(\alpha+\beta)\cos\frac{1}{2}(\alpha-\beta)} = \frac{a-b}{a+b}$$

63. Using Problem 62, prove the Law of Tangents:

$$\frac{\tan\frac{1}{2}(\alpha-\beta)}{\tan\frac{1}{2}(\alpha+\beta)} = \frac{a-b}{a+b}$$

Other forms of the Law of Tangents are

$$\frac{\tan\frac{1}{2}(\alpha-\gamma)}{\tan\frac{1}{2}(\alpha+\gamma)} = \frac{a-c}{a+c}$$

and

$$\frac{\tan\frac{1}{2}(\beta-\gamma)}{\tan\frac{1}{2}(\beta+\gamma)} = \frac{b-c}{b+c}$$

64. *Newton's Formula* involves all six parts of a triangle. It is not useful in solving a triangle, but it is helpful in checking results. Show that

$$\frac{a+b}{c} = \frac{\cos\frac{1}{2}(\alpha-\beta)}{\sin\frac{1}{2}\gamma}$$

65. Show that the radius R of a circumscribed circle of $\triangle ABC$ satisfies the equations

$$R = \frac{a}{2\sin\alpha} = \frac{b}{2\sin\beta} = \frac{c}{2\sin\gamma}$$

66. Show that the radius r of an inscribed circle of $\triangle ABC$ satisfies the equation

$$r = \sqrt{\frac{(s-a)(s-b)(s-c)}{s}}$$

where $s = \frac{1}{2}(a+b+c)$.

CHAPTER 6 SUMMARY

The material of this chapter is reviewed in the following list of objectives. After each objective there are some practice questions. For a sample test, select the first question of each set and check your answers with the answer section. For a sample test without answers, use the second question of each set. Additional practice is given by the other questions in each set. If you are having trouble with a particular type of problem, look back to that section for extra help.

6.1 TRIGONOMETRIC EQUATIONS

Objective 1 Solve first-degree trigonometric equations. Solve for $0 \le \theta < 2\pi$.

1. $3\tan 2\theta - \sqrt{3} = 0$

2. $\sin^2\theta + 2\cos\theta = 1 + 3\cos\theta + \sin^2\theta$

3. $2\sin 3\theta = \sqrt{2}$

4. $3\sec 2\theta - 2 = 0$

Objective 2 Solve second-degree trigonometric equations by factoring or by using the Quadratic Formula. Solve for $0 \le \theta < 2\pi$.

5. $4\cos^2\theta = 1$

6. $3\cos^2\theta = 1 + \cos\theta$

7. $\frac{1}{2}\cos^2\theta = 1$

8. $\tan^2 2\theta = 3\tan 2\theta$

6.2 FUNDAMENTAL IDENTITIES

Objective 3 State and prove the eight fundamental identities.

9. State the reciprocal identities.

10. State the ratio identities.

11. State the Pythagorean identities.

12. Prove one of the Pythagorean identities.

Objective 4 Use the fundamental identities to find the other trigonometric functions if you are given the value of one function and want to find the others.

13. Find the other trigonometric functions so that $\sin\delta = \frac{3}{5}$ when $\tan\delta < 0$.
14. Find the other trigonometric functions so that $\cos\beta = \frac{3}{5}$ and $\tan\beta < 0$.
15. Find the other trigonometric functions so that $\sin\omega = -\frac{3}{5}$ when $\tan\omega > 0$.
16. Find the other trigonometric functions so that $\cos\omega = -\frac{3}{5}$ when $\tan\omega < 0$.

Objective 5 Algebraically simplify expressions involving the trigonometric functions. Leave your answer in terms of sines and cosines.

17. $\dfrac{\sin\theta}{\cos\theta} + \dfrac{1}{\sin\theta}$

18. $\dfrac{1}{\sin\theta + \cos\theta} + \dfrac{1}{\sin\theta - \cos\theta}$

19. $\tan^2\theta + \sec^2\theta$

20. $\dfrac{\tan\theta + \cot\theta}{\sec\theta}$

6.3 PROVING IDENTITIES

Objective 6 Prove identities using algebraic simplification and the eight fundamental identities.

21. $\dfrac{\csc^2\alpha}{1 + \cot^2\alpha} = 1$

22. $\dfrac{1 + \tan^2\theta}{\csc\theta} = \sec\theta\tan\theta$

23. $\dfrac{\cos\theta}{\sec\theta} - \dfrac{\sin\theta}{\cot\theta} = \dfrac{\cos\theta\cot\theta - \tan\theta}{\csc\theta}$

24. $\tan\beta + \sec\beta = \dfrac{1 + \csc\beta}{\cos\beta\csc\beta}$

Objective 7 Prove a given identity by using various "tricks of the trade."

25. $\dfrac{\sin^2\theta - \cos^2\theta}{\sin\theta + \cos\theta} = \sin\theta - \cos\theta$

26. $(\tan\theta + \cot\theta)^2 = \sec^2\theta + \csc^2\theta$

27. $\cos^2 x \tan x \csc x \sec x = 1$

28. $\dfrac{1}{\sin\theta + \cos\theta} + \dfrac{1}{\sin\theta - \cos\theta} = \dfrac{2\sin\theta}{\sin^4\theta - \cos^4\theta}$

6.4 ADDITION LAWS

Objective 8 Use the cofunction identities.

29. Write $\sin 38^\circ$ in terms of its cofunction.
30. Write $\tan\frac{\pi}{8}$ in terms of its cofunction.
31. Write $\cos 0.456$ in terms of its cofunction.
32. Prove $\tan(\frac{\pi}{2} - \theta) = \cot\theta$

Objective 9 Use the opposite-angle identities.

33. Write $\cos(\frac{\pi}{6} - \theta)$ as a function of $(\theta - \frac{\pi}{6})$.
34. Write $\tan(1 - \theta)$ as a function of $(\theta - 1)$.
35. Graph $y - 1 = 2\sin(\frac{\pi}{6} - \theta)$.
36. Graph $y + 1 = 2\cos(2 - \theta)$.

Objective 10 Use the addition laws.

37. Write $\cos(\theta - 30^\circ)$ as a function of $\cos\theta$ and $\sin\theta$.
38. Find the exact value of $\sin 105^\circ$.
39. Evaluate $\dfrac{\tan 23^\circ - \tan 85^\circ}{1 + \tan 23^\circ \tan 85^\circ}$.
40. Prove $\cos(\alpha - \beta) = \cos\alpha\cos\beta + \sin\alpha\sin\beta$.

6.5 DOUBLE-ANGLE AND HALF-ANGLE IDENTITIES

Objective 11 Use the double-angle identities. Let θ be a positive angle.

41. Evaluate $\dfrac{2\tan\frac{\pi}{6}}{1 - \tan^2\frac{\pi}{6}}$ using exact values.

42. If $\cos\theta = -\frac{4}{5}$ and 2θ is in Quadrant IV, find the exact value of $\cos 2\theta$.
43. If $\cos\theta = \frac{4}{5}$ and 2θ is in Quadrant IV, find the exact value of $\sin 2\theta$.
44. If $\cos\theta = -\frac{4}{5}$ and 2θ is in Quadrant IV, find the exact value of $\tan 2\theta$.

Objective 12 Use the half-angle identities.

45. Evaluate $-\sqrt{\dfrac{1+\cos 240^\circ}{2}}$ using exact values.
46. If $\cot 2\theta = -\frac{4}{3}$, find the exact value of $\cos\theta$ (θ in Quadrant I).
47. If $\cot 2\theta = \frac{4}{3}$, find the exact value of $\sin\theta$ (θ in Quadrant I).
48. If $\cot 2\theta = -\frac{4}{3}$, find the exact value of $\tan\theta$ (θ in Quadrant I).

Objective 13 Use the product and sum identities.

49. Write $\sin 3\theta \cos\theta$ as a sum.
50. Write $\sin 40^\circ - \sin 65^\circ$ as a product.
51. Write $\sin(x+h) - \sin x$ as a product.
52. Prove $\dfrac{\sin 5\theta + \sin 3\theta}{\cos 5\theta - \cos 3\theta} = -\cot\theta$.

*6.6 DE MOIVRE'S THEOREM

Objective 14 Change rectangular-form complex numbers to trigonometric form.

53. $7 - 7i$
54. $-3i$
55. $\frac{7}{2}\sqrt{3} - \frac{7}{2}i$
56. $2 + 3i$

Objective 15 Change trigonometric-form complex numbers to rectangular form.

57. $4(\cos\frac{7\pi}{4} + i\sin\frac{7\pi}{4})$
58. $2 \operatorname{cis} 150^\circ$
59. $5 \operatorname{cis} 270^\circ$
60. $5 \operatorname{cis} 25^\circ$

Objective 16 Multiply and divide complex numbers. Perform the operations and leave your answer in the form indicated.

61. $(\sqrt{12} - 2i)^4$; rectangular
62. $\dfrac{(3+3i)(\sqrt{3}-i)}{1+i}$; rectangular
63. $\dfrac{2 \operatorname{cis} 158^\circ \cdot 4 \operatorname{cis} 212^\circ}{(2 \operatorname{cis} 312^\circ)^3}$; trigonometric
64. $2i(-1+i)(-2+2i)$; trigonometric

Objective 17 State De Moivre's Theorem and the nth Root Theorem, and find all roots of a complex number.

65. State the nth Root Theorem.
66. State De Moivre's Theorem.
67. Find and plot the square roots of $\frac{7}{2}\sqrt{3} - \frac{7}{2}i$. Leave your answer in rectangular form.
68. Find the cube roots of i. Leave your answer in trigonometric form.

6.7 RIGHT TRIANGLES

Objective 18 Know the right triangle definition of the trigonometric functions and solve right triangles.

69. State the right triangle definition of the trigonometric functions.
70. Solve $\triangle ABC$ where $a = 7.3$, $c = 15$, and $\gamma = 90^\circ$.
71. Solve $\triangle ABC$ where $b = 678$, $\beta = 55.0^\circ$, and $\gamma = 90.0^\circ$.
72. Solve $\triangle ABC$ where $a = 3.0$, $b = 4.0$, and $c = 5.0$.

---

* Optional section

6.8 OBLIQUE TRIANGLES

Objective 19 Know the laws of cosines and sines, as well as the proof for each.

73. State the law of cosines.
74. State the law of sines.
75. Prove the law of cosines.
76. Prove the law of cosines.

Objective 20 Know when to apply the laws of cosines and sines, as well as recognizing the ambiguous case. Complete the table for triangle ABC labeled in the usual fashion.

	To find	Known	Procedure	Solution
	β	a, b, α	$\dfrac{\sin\alpha}{a} = \dfrac{\sin\beta}{b}$	$\beta = \sin^{-1}\left(\dfrac{b\sin\alpha}{a}\right)$
77.	α	a, b, c		
78.	α	a, b, β		
79.	α	a, β, γ		
80.	b	a, α, β		

Objective 21 Solve oblique triangles.

81. $a = 34$, $c = 61$, $\beta = 58^\circ$
82. $a = 6.8$, $b = 12.2$, $c = 21.5$
83. $b = 34$, $c = 21$, $\gamma = 16^\circ$
84. $b = 4.6$, $\alpha = 108^\circ$, $\gamma = 38^\circ$

Objective 22 Solve applied problems using either right triangles or oblique triangles.

85. A mine shaft is dug into the side of a sloping hill. The shaft is dug horizontally for 485 ft. Next a turn is made so that the angle of elevation of the second shaft is 58.0°, thus forming a 58° angle between the shafts. The shaft is then continued for 382 ft before exiting, as shown in Figure 6.30. How far is it along a straight line from the entrance to the exit, assuming that all tunnels are in a single plane? If the slope of the hill follows the line from the entrance to the exit, what is the angle of elevation from the entrance to the exit?

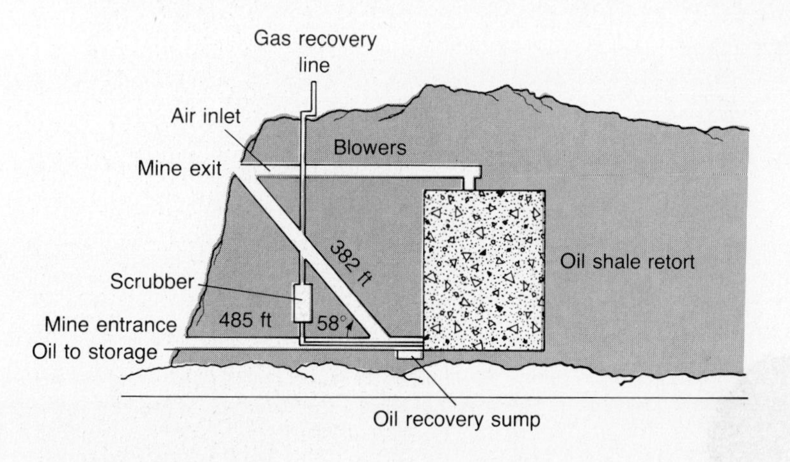

Figure 6.30 Determining the exit of a mine shaft

86. Ferndale is 7 mi N50°W of Fortuna. If I leave Fortuna at noon and travel due west at 2 mph, when will I be exactly 6 mi from Ferndale?

87. To measure the span of the Rainbow Bridge in Utah, a surveyor selected two points, P and Q, on either end of the bridge. From point Q, the surveyor measured 500 ft in the direction N38.4°E to point R. Point P was then determined to be in the direction S67.5°W. What is the span of the Rainbow Bridge if all the preceding measurements are in the same plane and $\angle PQR$ is a right angle?

88. When viewing Angel Falls (the world's highest waterfall) from Observation Platform A, located on the same level as the bottom of the falls, we calculate the angle of elevation to the top of the falls to be 69.30°. From Observation Platform B, which is located on the same level exactly 1000 ft from the first observation point, we calculate the angle of elevation to the top of the falls to be 52.90°. How high are the falls?

Solar Power

Energy from the sun provides enough power to heat water or even a home

BY NEALE LESLIE A revolutionary solar-powered water heater that aligns itself facing directly into the sun now is available on the market.

Unlike flat plate collectors which have to be installed in a southward direction, usually on roofs of the buildings being served, this device can be installed on any part of a north or south-facing roof—or any portion of the yard—as long as the sun's rays reach it.

This device, called the SAV, manufactured by SAV Solar Systems Inc. of Los Angeles, follows the sun. It is both a parabolic reflector and storage tank combined.

The tank—riding piggyback above the parabolic mirror—is a collector itself. It is coated with black chrome, one of the most superior materials known for absorbing heat. Actually, it absorbs 95 percent of all incoming radiation, and only emits 10 percent of the heat collected. The result: A trapping and holding of 85 percent of the solar radiation.

The tank is further encased in cylindrical double glazing made of clear plastic, with one inch of air space between the tank and inner glazing and another inch of air space between the inner and outer glazing.

This provides a "glass house" heating and insulating effect. At the same time the tank is tilted to the most direct angle to the sun, an angle the homeowner can adjust to conform to his latitude and the change of the season.

The tank, designed by Jon Makeever, SAV Solar Systems president, permits the sun's heat to set up "thermosyphonic" circulation patterns, greatly increasing the efficiency of the unit.

Makeever also incorporated a system of parabolic surfaces concentrating the energy of the sun directly onto the solar collector and storage tank. And working in conjunction with this is his tracking system which causes the entire mechanism, including the tank, to follow the sun across the sky.

Makeever chose Freon 12, available anywhere in refrigeration service shops, as the agent to provide the power for the tracking mechanism. The Freon 12 is energized by the sun. It is maintenance free, requires no wires, no electricity, timers or other apparatus. It is nonexplosive, providing enough pressure to operate a couple of hydraulic pistons, but not enough to burst any of the tracker's components. The Freon 12 has to be replaced once every four years.

The unit, approximately 4 by 4 feet, provides up to 98 percent more heat per collector area than flat plate systems used in a test, Makeever said.

One unit would supply the hot water needs of a typical family, but three would be needed for space heating a house. Empty, the unit weighs 140 pounds; filled another 100, but way under the minimal weight standard for roofs.

The unit is said to be adaptable to any part of a sloping or flat roof, and capable of withstanding wind velocities up to 120 m.p.h.

Retail price of the unit is about $2,500 installed, available through plumbing, heating and air conditioning firms and solar equipment dealers.

SOURCE: Courtesy of *The Press Democrat*, Santa Rosa, California, July 29, 1979. Reprinted by permission.

The efficient use of a solar collector requires knowledge of the length of daylight and the angle of the sun throughout the year at the location of the collector. Table 6.1 on page 280 gives the times of sunrise and sunset for various latitudes. Figure 6.31 shows a graph of the sunrise and sunset times for latitude 35°N. These are definitely not graphs of sine or cosine functions. If you plot the length of daylight (see Problem 7), however, you will obtain a curve that is nearly a sine curve.

If you wish to use Table 6.1 for your own town's latitude and it is not listed, you can use a procedure called **linear interpolation**. Your local Chamber of Commerce will probably be able to tell you your town's latitude. The latitude for Santa Rosa, California, is about 38°N, which falls between 35° and 40° on the table (see Figure 6.32).

Table 6.1
Times of sunrise and sunset for various latitudes

Date	Time of Sunrise: 20°N. Latitude (Hawaii) h m	30°N. Latitude (New Orleans) h m	35°N. Latitude (Albuquerque) h m	40°N. Latitude (Philadelphia) h m	45°N. Latitude (Minneapolis) h m	60°N. Latitude (Alaska) h m	Time of Sunset: 20°N. Latitude (Hawaii) h m	30°N. Latitude (New Orleans) h m	35°N. Latitude (Albuquerque) h m	40°N. Latitude (Philadelphia) h m	45°N. Latitude (Minneapolis) h m	60°N. Latitude (Alaska) h m
Jan. 1	6 35	6 56	7 08	7 22	7 38	9 03	17 31	17 10	16 58	16 44	16 28	15 03
Jan. 15	6 38	6 57	7 08	7 20	7 35	8 48	17 41	17 22	17 12	16 59	16 44	15 31
Jan. 30	6 36	6 52	7 01	7 11	7 23	8 19	17 51	17 35	17 27	17 17	17 05	16 09
Feb. 14	6 30	6 41	6 48	6 55	7 03	7 42	17 59	17 48	17 42	17 34	17 26	16 48
Mar. 1	6 20	6 26	6 30	6 34	6 39	6 59	18 05	17 59	17 56	17 52	17 47	17 27
Mar. 16	6 08	6 09	6 10	6 11	6 11	6 15	18 10	18 09	18 08	18 08	18 07	18 04
Mar. 31	5 55	5 51	5 49	5 46	5 43	5 29	18 14	18 18	18 20	18 23	18 26	18 41
Apr. 15	5 42	5 34	5 28	5 23	5 16	4 44	18 18	18 27	18 32	18 38	18 45	19 18
Apr. 30	5 32	5 18	5 11	5 02	4 51	4 01	18 23	18 37	18 44	18 53	19 04	19 55
May 15	5 24	5 07	4 57	4 45	4 31	3 23	18 29	18 46	18 56	19 08	19 22	20 31
May 30	5 20	5 00	4 48	4 34	4 18	2 53	18 35	18 55	19 07	19 21	19 38	21 04
June 14	5 20	4 58	4 45	4 30	4 13	2 37	18 40	19 02	19 15	19 30	19 48	21 24
June 29	5 23	5 02	4 49	4 34	4 16	2 40	18 43	19 05	19 18	19 33	19 51	21 26
July 14	5 29	5 08	4 56	4 43	4 26	3 01	18 43	19 03	19 15	19 29	19 45	21 09
July 29	5 34	5 17	5 07	4 55	4 41	3 33	18 39	18 56	19 06	19 17	19 31	20 38
Aug. 13	5 39	5 26	5 18	5 09	4 59	4 09	18 30	18 43	18 51	19 00	19 10	19 59
Aug. 28	5 43	5 35	5 29	5 24	5 17	4 45	18 19	18 27	18 33	18 38	18 45	19 16
Sept. 12	5 47	5 43	5 40	5 38	5 35	5 20	18 06	18 10	18 12	18 14	18 17	18 31
Sept. 27	5 50	5 51	5 51	5 52	5 53	5 55	17 52	17 51	17 50	17 49	17 49	17 45
Oct. 12	5 54	6 00	6 03	6 07	6 11	6 31	17 39	17 33	17 29	17 26	17 21	17 01
Oct. 22	5 57	6 06	6 12	6 18	6 25	6 56	17 32	17 22	17 17	17 11	17 04	16 32
Nov. 6	6 04	6 18	6 26	6 35	6 45	7 35	17 23	17 10	17 02	16 52	16 42	15 52
Nov. 21	6 12	6 30	6 40	6 52	7 05	8 12	17 19	17 02	16 51	16 40	16 26	15 19
Dec. 6	6 22	6 42	6 54	7 07	7 23	8 44	17 20	17 00	16 48	16 35	16 19	14 58
Dec. 21	6 30	6 52	7 04	7 18	7 35	9 02	17 26	17 05	16 52	16 38	16 21	14 54

Table courtesy of U.S. Naval Observatory. This table of sunrise and sunset may be used in any year of the 20th century with an error not exceeding two minutes and generally less than one minute. It may also be used anywhere in the vicinity of the stated latitude with an additional error of less than one minute for each 9 miles.

Angle	*Time*	
35°	7:08	Known, from Table 6.1
38°	x	Unknown; this is the latitude of the town for which the times of sunrise and sunset are not available.
40°	7:22	Known, from Table 6.1

Use the known information and a proportion to find x:

$$\mathbf{5}\left[\begin{array}{ll} \mathbf{3}\big[\,35^\circ & 7{:}08 \\ \quad\; 38^\circ & x \\ \quad\; 40^\circ & 7{:}22 \end{array}\right]\mathbf{14}$$

Since 38° is $\frac{3}{5}$ of the way between 35° and 40°, the sunrise time will be $\frac{3}{5}$ of the way between 7:08 and 7:22.

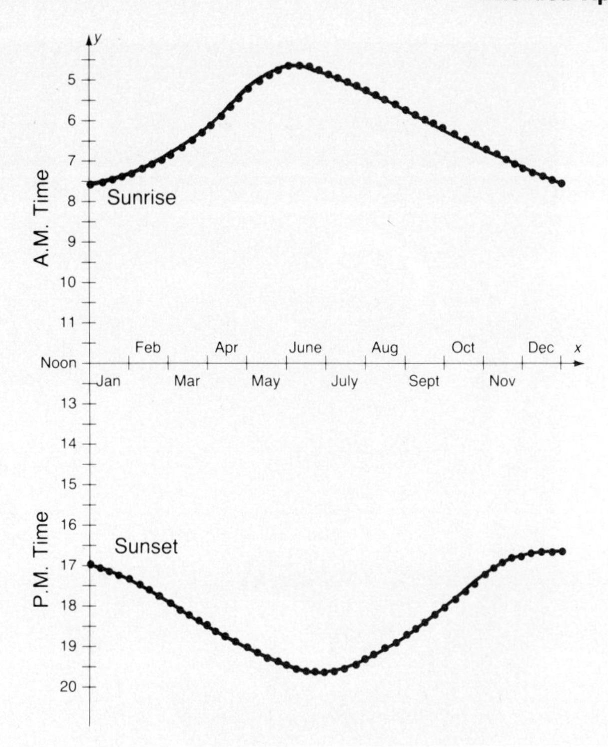

Figure 6.31 Sunrise and sunset times for latitude 35°N

$$\frac{3}{5} = \frac{x}{14}$$

$$x = \frac{3}{5}(14)$$

$$= 8.4 \qquad \text{To the nearest minute this is :08.}$$

The sunrise time is about 7:08 + :08 = 7:16.

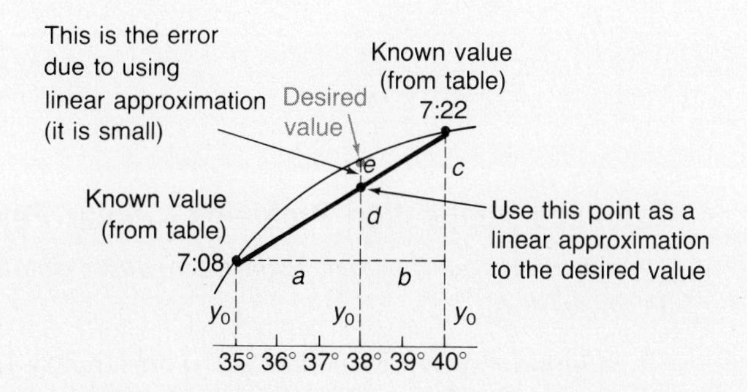

Figure 6.32 Linear interpolation

This solar tank riding piggyback above a parabolic mirror is called the SAV. It is pictured with its designer, Jon Makeever, president of SAV Solar Systems. Solar collectors have recently become a big business in the U.S. (Courtesy of *The Press Democrat*.)

One type of solar collector requires the construction of a central cylinder. The construction of a template for cutting a 45° angle in a cardboard tube to make a rotating mirror is described in the May 1978 issue of *Byte* magazine. The template is shown in Figure 6.33, and the details of construction are given in Problem 12.

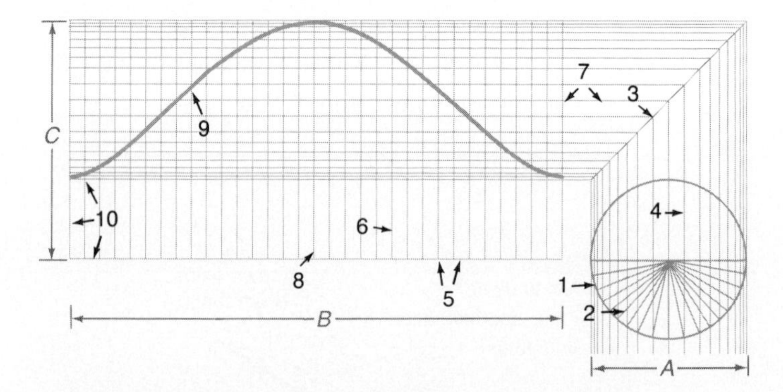

Figure 6.33 Construction of a template for cutting a 45° angle in a cardboard tube to make a rotating mirror

Extended Application Problems—Solar Power

Use linear interpolation and Table 6.1 to approximate the times requested in each of Problems 1–6.

1. Sunset for Santa Rosa (lat. 38°N) on January 1.
2. Sunrise for Tampa, Florida (lat. 28°N), on June 29.

3. Sunset for Tampa, Florida (lat. 28°N), on June 29.

4. Sunset for Winnipeg, Canada (lat. 50°N), on October 12.

5. Sunrise for Winnipeg, Canada (lat. 50°N), on October 12.

6. Sunrise for Juneau, Alaska (lat. 58°N), on March 1.

7. Plot the length of daylight in Chattanooga, Tennessee (lat. 35°N).

8. Plot the time of sunrise and sunset for Seward, Alaska (lat. 60°).

9. Plot the length of daylight for Seward, Alaska (lat. 60°N).

10. Plot the sunrise and sunset for your town's latitude by drawing a graph similar to that in Figure 6.31.

11. Plot the length of daylight for your town from the data in Problem 10.

12. Carry out the following steps from *Byte* magazine (May 1978, p. 129) to cut a 45° angle in a tube of diameter A, circumference B, and finished height C:*

1. Draw a circle the same size as the outside diameter of the tube.
2. Divide the circle into 32 equal parts of $11\frac{1}{4}$ in. each.
3. Draw a 45° angle above the circle.
4. Carry the points at the outside of the circle straight up to the 45° angle.
5. Find the circumference of the circle, draw a straight line, and divide it into 32 equal parts.
6. Carry lines from those divisions straight upward.
7. Bring lines straight across from the intersections at the 45° angle line and intersect with the vertical lines.
8. Starting at the centerline, mark points of intersection.
9. Fill in between points of intersection.
10. Cut out pattern and wrap around tube, lining up straight circumference side of pattern with (cut) end of tube square.
11. Carefully trace curved line end of pattern onto tube.
12. Remove pattern from tube and save for making black paper cover for tube.
13. Cut along traced line on tube with an X-acto (or similar) knife.
14. Tape large sheet of sandpaper to a table top or other flat surface.
15. Sand end of tube until flat.
16. Tube is now ready for application of Mylar mirror surface.

---

* From "How to Multiply in a Wet Climate," by J. Bryant and M. Swasdee, May, 1978, p. 129. Copyright © Byte Publications, Inc. Reprinted by permission.

7

Systems and Matrices

I have so many ideas that may perhaps be of some use in time if others more penetrating than I go deeply into them some day and join the beauty of their minds to the labor of mine.

LEIBNIZ

Gottfried Wilhelm von Leibniz (1646–1716)

Gottfried Leibniz has been described as a true example of a "Universal genius." He was a professional diplomat and historian, besides being an important philosopher and a great mathematician. By the time he was 20, he had mastered most of the important works on mathematics and had set the stage for modern logic. At 30, he invented the calculus, and at 36 he invented a reckoning machine, a forerunner of the computer. Throughout his life he was searching for a *Universal Mathematics* in which he aimed to create "a general method in which all truths of the reason would be reduced to a kind of calculation." Because of his vast talent, his search for this universal mathematics led him to many applied problems. In this chapter we investigate an important tool in solving a wide variety of applied problems. That tool is systems of equations and inequalities. Systems of equations were solved by the Babylonians as early as 1600 B.C., but the first rule for solving a certain set of n simultaneous linear equations in n unknowns was given by the Greek scholar Thymridas about 350 B.C. By 1683 the Japanese mathematician Seki Kōwa had updated an old Chinese method of solving simultaneous linear equations that involved rearrangements of the rods similar to the way we simplify determinants in this chapter. It was Leibniz, however, who originated the notation of determinants. Leibniz' later years were dimmed by a bitter controversy with Isaac Newton concerning whether he had discovered the calculus independently of Newton. When he died, in 1716, his funeral was attended only by his secretary. Today we ascribe the independent discovery of calculus to both Leibniz and Newton.

CHAPTER OVERVIEW This chapter introduces a very important tool in mathematics, namely the matrix. We are motivated by our desire to solve systems of equations, but we soon learn that matrices have many other applications and uses. This chapter concludes with an introduction to a very important new branch of mathematics, linear programming. There are 14 specific objectives in this chapter, which are listed on pages 335–338.

7.1 SYSTEMS OF EQUATIONS

Solving systems of equations is a procedure that arises throughout mathematics. We will begin our study of this topic by considering an arbitrary system of two equations with two variables:

$$\begin{cases} a_{11}x_1 + a_{12}x_2 = b_1 \\ a_{21}x_1 + a_{22}x_2 = b_2 \end{cases}$$

The notation may seem strange, but it will prove to be useful. The variables are x_1 and x_2; the constants are $a_{11}, a_{12}, a_{21}, a_{22}, b_1$, and b_2. By a **system** of two equations with two variables, we mean any two equations in those variables. The **simultaneous solution** of a system is the intersection of the solution sets of the individual equations. The brace is used to show that this intersection is desired. If all the equations in a system are linear, it is called a **linear system**.

You solved linear systems with two variables in elementary algebra, so we will begin by reviewing the methods used there. Then we will generalize first to nonlinear systems and finally to more complicated linear systems.

Since the graph of each equation in a system of linear equations in two variables is a line, the solution set is the intersection of two lines. In two dimensions, two lines must be related to each other in one of three possible ways:

1. They intersect at a single point.
2. The graphs are parallel lines. In this case, the solution set is empty and the system is called *inconsistent*. In general, any system that has an empty solution set is referred to as an **inconsistent system**.
3. The graphs are the same line. In this case, there are infinitely many points in the solution set and any solution of one equation is also a solution of the other. Such a system is called **dependent**. This word is also used to describe nonlinear systems.

EXAMPLE 1 Relate the system

$$\begin{cases} 2x - 3y = -8 \\ x + y = 6 \end{cases}$$

to the general system

$$\begin{cases} a_{11}x_1 + a_{12}x_2 = b_1 \\ a_{21}x_1 + a_{22}x_2 = b_2 \end{cases}$$

and solve by graphing.

Solution The variables are $x_1 = x$ and $x_2 = y$. The subscripts in a_{11}, a_{12}, a_{21}, and a_{22} are called double subscripts and indicate *position*. That is, a_{12} should not be read "*a* sub twelve" but rather "a sub one two." It represents the second constant in the first equation:

$$a_{12}$$

equation number ↗ ↖ constant number in equation

This effort in using notation may seem unnecessarily complicated when dealing with only two equations and two variables, but we want to develop a notation that can be used with many variables and many equations. For this example,

$$a_{11} = 2 \qquad a_{12} = -3 \qquad b_1 = -8$$
$$a_{21} = 1 \qquad a_{22} = 1 \qquad b_2 = 6$$

The solution is $x = 2$, $y = 4$, or **(2, 4)** as shown in Figure 7.1. ■

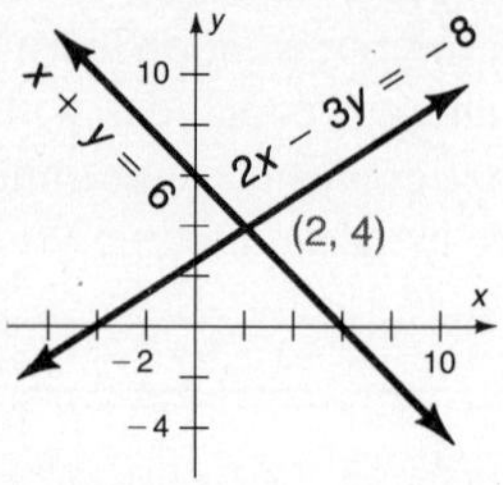

Figure 7.1 Graph of the system
$\begin{cases} 2x - 3y = -8 \\ x + y = 6 \end{cases}$

EXAMPLE 2 Relate the system

$$\begin{cases} 2x - 3y = -8 \\ 4x - 6y = 0 \end{cases}$$

to the general system and solve by graphing.

Solution The variables are $x_1 = x$, $x_2 = y$, and the constants are

$$a_{11} = 2 \qquad a_{12} = -3 \qquad b_1 = -8$$
$$a_{21} = 4 \qquad a_{22} = -6 \qquad b_2 = 0$$

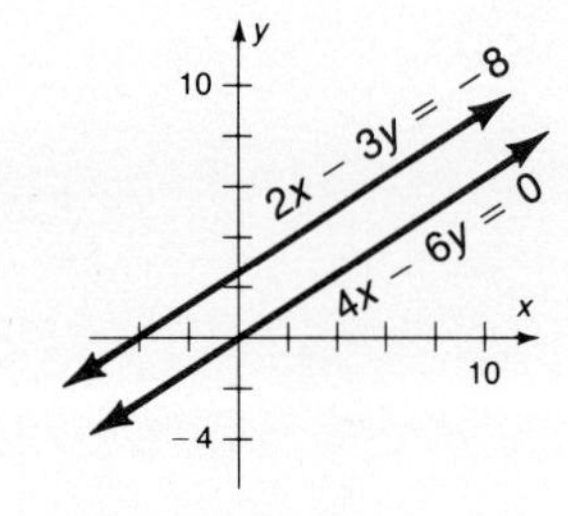

Figure 7.2 Graph of the system
$\begin{cases} 2x - 3y = -8 \\ 4x - 6y = 0 \end{cases}$

The graphs of the lines are parallel since there is no point of intersection (Figure 7.2). This is an **inconsistent system.** ■

EXAMPLE 3 Relate the system

$$\begin{cases} 2x - 3y = -8 \\ y = \dfrac{2}{3}x + \dfrac{8}{3} \end{cases}$$

to the general system and solve by graphing.

Solution To relate to the general system, both equations must be algebraically in the usual form; the second equation needs to be rewritten:

$$y = \frac{2}{3}x + \frac{8}{3}$$
$$3y = 2x + 8$$
$$-2x + 3y = 8$$

This means that the variables are $x_1 = x$, $x_2 = y$, and the constants are

$$a_{11} = 2 \qquad a_{12} = -3 \qquad b_1 = -8$$
$$a_{21} = -2 \qquad a_{22} = 3 \qquad b_2 = 8$$

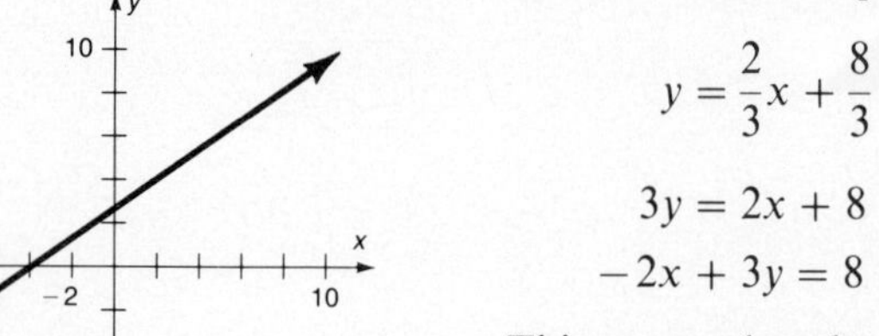

Figure 7.3 Graph of the system
$\begin{cases} 2x - 3y = -8 \\ y = \frac{2}{3}x + \frac{8}{3} \end{cases}$

The equations represent the same line as shown in Figure 7.3. This is a **dependent system.** ■

EXAMPLE 4 Solve the system $\begin{cases} x - y = 3 \\ 6x - y = x^2 + 7 \end{cases}$

Solution This is a nonlinear system. The graphs of $y = x - 3$ and $y = -x^2 + 6x - 7$ are shown in Figure 7.4. Remember that to graph this second-degree equation you need to complete the square:

$$y = -x^2 + 6x - 7$$

$$y + 7 - 9 = -(x^2 - 6x + 9)$$

$\frac{1}{2}$ of (-6) squared

negative 1 times 9 added to both sides

$$y - 2 = -(x - 3)^2$$

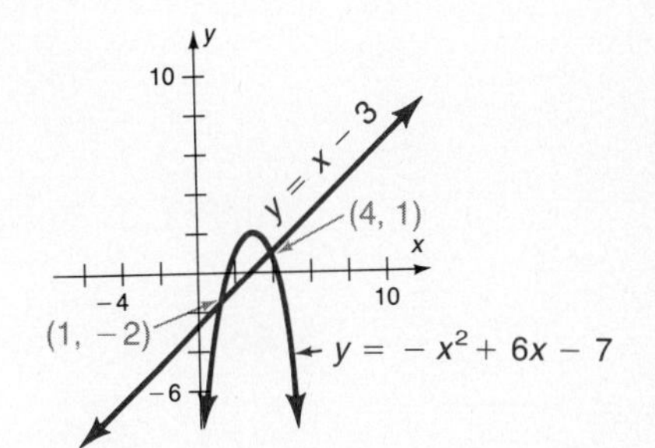

Figure 7.4 Graph of $\begin{cases} x - y = 3 \\ 6x - y = x^2 + 7 \end{cases}$

This is a parabola that opens downward with vertex at (3, 2). By inspection, the solution is **(4, 1) and (1, −2)**.

To check an answer, make sure that every member of the solution set satisfies all the equations of the system:

Check (4, 1): $x - y = 4 - 1 = 3$ (checks)

$6x - y = 6(4) - 1 = 23$ and $x^2 + 7 = (4)^2 + 7 = 23$

checks

Check (1, −2): $x - y = 1 - (-2) = 3$ (checks)

$6x - y = 6(1) - (-2) = 8$ and $x^2 + 7 = (1)^2 + 7 = 8$

checks

Both (4, 1) and (1, −2) check. ■

The graphing method can give solutions only as accurate as the graphs you can draw, and consequently it is inadequate for most applications. There is therefore a need for more efficient methods.

In general, given a system, the procedure is to write a simpler equivalent system. Two systems are said to be **equivalent** if they have the same solution set. In this chapter we will limit ourselves to finding only real roots. There are several ways to go about writing equivalent systems. The first nongraphical method we will consider comes from the substitution property of real numbers and leads to a **substitution method** for solving systems.

Substitution Method for Solving Systems of Equations

1. *Solve* one of the equations for one of the variables.
2. *Substitute* the expression that you obtain into the other equation.
3. *Solve* the resulting equation in a single variable for the value of that variable.
4. *Substitute* that value into either of the original equations to determine the value of the other variable.
5. *State* the solution.

EXAMPLE 5 Solve $\begin{cases} 2p + 3q = 5 \\ q = -2p + 7 \end{cases}$ by substitution.

Solution Since $q = -2p + 7$, substitute $-2p + 7$ for q in the other equation:

$$\begin{aligned} 2p + \quad 3q \quad &= 5 \\ 2p + 3(\mathbf{-2p + 7}) &= 5 \\ 2p - 6p + 21 &= 5 \\ -4p &= -16 \\ p &= 4 \end{aligned}$$

Substitute 4 for p in either of the given equations:

$$\begin{aligned} q &= -2p + 7 \\ &= -2(\mathbf{4}) + 7 \\ &= -1 \end{aligned}$$

The solution is $(p, q) = (4, -1)$. ■

EXAMPLE 6 Solve $\begin{cases} x - y + 7 = 0 \\ y = x^2 + 4x + 3 \end{cases}$

Solution Solve one of the equations for one of the variables. The second equation is solved for y, so substitute $x^2 + 4x + 3$ for y in the first equation:

$$\begin{aligned} x - \quad y \quad + 7 &= 0 \\ x - (\mathbf{x^2 + 4x + 3}) + 7 &= 0 \\ -x^2 - 3x + 4 &= 0 \\ x^2 + 3x - 4 &= 0 \\ (x + 4)(x - 1) &= 0 \\ x &= -4, 1 \end{aligned}$$

If $x = -4$, then

$$\begin{aligned} y &= (-4)^2 + 4(-4) + 3 \\ &= 3 \end{aligned}$$

If $x = 1$, then

$$\begin{aligned} y &= 1^2 + 4(1) + 3 \\ &= 8 \end{aligned}$$

Solution: $\mathbf{(-4, 3)}$ **and** $\mathbf{(1, 8)}$. ■

EXAMPLE 7 Solve $\begin{cases} x^2 + 2xy = 0 \\ x^2 - 5xy = 14 \end{cases}$

Solution You want to solve the first equation for y, but that requires $x \neq 0$. If $x = 0$, then the first equation is satisfied but the second is not. Thus you can say that $x \neq 0$. Then

$$y = \frac{-x^2}{2x} = \frac{-x}{2}$$

Substitute into the second equation:

$$\begin{aligned} x^2 - \quad 5x\,y \quad &= 14 \\ x^2 - 5x\left(\frac{-x}{2}\right) &= 14 \\ 2x^2 + 5x^2 &= 28 \\ 7x^2 &= 28 \\ x^2 &= 4 \\ x^2 - 4 &= 0 \\ (x-2)(x+2) &= 0 \\ x &= 2, -2 \end{aligned}$$

If $x = 2$, then $y = \frac{-2}{2} = -1$; if $x = -2$, then $y = \frac{-(-2)}{2} = 1$. Thus the solution is **(2, −1) and (−2, 1).** ■

EXAMPLE 8 Solve $\begin{cases} y = x^2 - x - 1 \\ 4x = 7 + 2y - y^2 \end{cases}$

Solution Substitute the first equation into the second one:

$$\begin{aligned} 4x &= 7 + \quad 2y \quad - \quad y^2 \\ 4x &= 7 + 2(\mathbf{x^2 - x - 1}) - (\mathbf{x^2 - x - 1})^2 \\ 4x &= 7 + 2x^2 - 2x - 2 - x^4 + 2x^3 + x^2 - 2x - 1 \\ 4x &= 4 + 3x^2 - 4x - x^4 + 2x^3 \\ x^4 &- 2x^3 - 3x^2 + 8x - 4 = 0 \end{aligned}$$

The possible rational roots are: $\pm 1, \pm 2, \pm 4$.

	1	−2	−3	8	−4
1	1	−1	−4	4	0
1	1	0	−4	0	

The depressed equation is

$$\begin{aligned} x^2 - 4 &= 0 \\ (x-2)(x+2) &= 0 \\ x &= 2, -2 \end{aligned}$$

The roots of this fourth-degree equation are $x = 1, 2,$ and -2.

If $x = 1$, then $y = 1^2 - 1 - 1 = -1$.

If $x = 2$, then $y = 2^2 - 2 - 1 = 1$.

If $x = -2$, then $y = (-2)^2 - (-2) - 1 = 5$.

The solution is **(1, −1), (2, 1), and (−2, 5).** ■

A third method for solving systems is called the **linear combination method**. It involves substitution and the idea that if equal quantities are added to equal quantities, the resulting equation is equivalent to the original system. In general, such addition will not simplify matters unless the numerical coefficients of one or more terms are opposites. However, you can often force them to be opposites by multiplying one or both of the given equations by nonzero constants.

Linear Combination Method for Solving Systems of Equations

1. *Multiply* one or both of the equations by a constant or constants, so that the coefficients of one of the variables become opposites.
2. *Add* corresponding members of the equations to obtain a new equation in a single variable.
3. *Solve* the derived equation for that variable.
4. *Substitute* the value of the found variable into either of the original equations and solve for the second variable.
5. *State* the solution.

EXAMPLE 9 Solve $\begin{cases} 3x + 5y = -2 \\ 2x + 3y = 0 \end{cases}$

Solution Multiply both sides of the first equation by 2 and both sides of the second equation by -3. This procedure, denoted as shown below, forces the coefficients of x to be opposites:

$$\begin{matrix} \mathbf{2} \\ \mathbf{-3} \end{matrix} \begin{cases} 3x + 5y = -2 \\ 2x + 3y = 0 \end{cases}$$

This means you should add the equations of the system.

$$+\begin{cases} 6x + 10y = -4 \\ -6x - 9y = 0 \end{cases}$$

$$[6x + (-6x)] + [10y + (-9y)] = -4 + 0 \leftarrow \text{This step should be done mentally.}$$

$$y = -4$$

If $y = -4$, then $2x + 3y = 0$ means $2x + 3(-4) = 0$, or $x = 6$. The solution is **(6, −4)**. ■

EXAMPLE 10 Solve $\begin{cases} y^2 - 5xy = 3 \\ 5xy + 4 = y^2 \end{cases}$

Solution

$$\begin{matrix} \mathbf{-1} \\ \; \end{matrix} \begin{cases} y^2 - 5xy = 3 \\ y^2 - 5xy = 4 \end{cases}$$

$$+\begin{cases} -y^2 + 5xy = -3 \\ y^2 - 5xy = 4 \end{cases}$$

$$0 = 1$$

This is an **inconsistent system**. ■

PROBLEM SET 7.1

A

Solve the systems in Problems 1–9 by graphing.

1. $\begin{cases} y = 3x - 7 \\ y = -2x + 8 \end{cases}$

2. $\begin{cases} x - y = -1 \\ 3x - y = 5 \end{cases}$

3. $\begin{cases} y = \frac{2}{3}x - 7 \\ 2x + 3y = 3 \end{cases}$

4. $\begin{cases} y = \frac{3}{5}x + 2 \\ 3x - 5y = -10 \end{cases}$

5. $\begin{cases} 2x - 3y = 9 \\ y = \frac{2}{3}x - 3 \end{cases}$

6. $\begin{cases} x + y = 0 \\ y = -x^2 - 6x - 4 \end{cases}$

7. $\begin{cases} x + y = 4 \\ y = x^2 + 6x + 14 \end{cases}$

8. $\begin{cases} y = x^2 + 8x + 11 \\ x + y = -7 \end{cases}$

9. $\begin{cases} y = x^2 - 4x \\ y = x^2 - 4x + 8 \end{cases}$

Solve the systems in Problems 10–18 by substitution. Relate each system to the general system

$$\begin{cases} a_{11}x_1 + a_{12}x_2 = b_1 \\ a_{21}x_1 + a_{22}x_2 = b_2 \end{cases}$$

10. $\begin{cases} a = 3b - 7 \\ a = -2b + 8 \end{cases}$

11. $\begin{cases} s + t = 1 \\ 3s + t = -5 \end{cases}$

12. $\begin{cases} m = \frac{2}{3}n - 7 \\ 2n + 3m = 3 \end{cases}$

13. $\begin{cases} v = \frac{3}{5}u + 2 \\ 3u - 5v = 10 \end{cases}$

14. $\begin{cases} 2p - 3q = 9 \\ q = \frac{2}{3}p - 3 \end{cases}$

15. $\begin{cases} 3t_1 + 5t_2 = 1541 \\ t_2 = 2t_1 + 160 \end{cases}$

16. $\begin{cases} \alpha = -7\beta - 3 \\ 2\alpha + 5\beta = 3 \end{cases}$

17. $\begin{cases} \gamma = 3\delta - 4 \\ 5\gamma - 4\delta = -9 \end{cases}$

18. $\begin{cases} \theta + 3\phi = 0 \\ \theta = 5\phi + 16 \end{cases}$

Solve the systems in Problems 19–27 by linear combinations. Relate each system to the general system

$$\begin{cases} a_{11}x_1 + a_{12}x_2 = b_1 \\ a_{21}x_1 + a_{22}x_2 = b_2 \end{cases}$$

19. $\begin{cases} c + d = 2 \\ 2c - d = 1 \end{cases}$

20. $\begin{cases} 2s_1 + s_2 = 10 \\ 5s_1 - 2s_2 = 16 \end{cases}$

21. $\begin{cases} 3q_1 - 4q_2 = 3 \\ 5q_1 + 3q_2 = 5 \end{cases}$

22. $\begin{cases} 9x + 3y = 5 \\ 3x + 2y = 2 \end{cases}$

23. $\begin{cases} 7x + y = 5 \\ 14x - 2y = -2 \end{cases}$

24. $\begin{cases} 2x + 3y = 1 \\ 3x - 2y = 0 \end{cases}$

25. $\begin{cases} \alpha + \beta = 12 \\ \alpha - 2\beta = -4 \end{cases}$

26. $\begin{cases} 2\gamma - 3\delta = 16 \\ 5\gamma + 2\delta = 21 \end{cases}$

27. $\begin{cases} 2\theta + 5\phi = 7 \\ 3\theta + 4\phi = 0 \end{cases}$

B

Solve the systems in Problems 28–45 for x and y by any method. Limit your answers to the set of real numbers.

28. $\begin{cases} 5x + 4y = 5 \\ 15x - 2y = 8 \end{cases}$

29. $\begin{cases} 3x + 2y = 1 \\ 6x + 4y = 2 \end{cases}$

30. $\begin{cases} 4x - 2y = -28 \\ y = \frac{1}{2}x + 5 \end{cases}$

31. $\begin{cases} 12x - 5y = -39 \\ y = 2x + 9 \end{cases}$

32. $\begin{cases} y = 2x - 1 \\ y = -3x - 9 \end{cases}$

33. $\begin{cases} y = \frac{2}{3}x - 5 \\ y = -\frac{4}{3}x + 7 \end{cases}$

34. $\begin{cases} x + y = \alpha \\ x - y = \beta \end{cases}$

35. $\begin{cases} x - y = \gamma \\ x - 2y = \delta \end{cases}$

36. $\begin{cases} x + y = 2\alpha \\ x - y = 2\beta \end{cases}$

37. $\begin{cases} x + y = a \\ x - y = b \end{cases}$

38. $\begin{cases} x + 2y = a \\ x - 3y = b \end{cases}$

39. $\begin{cases} 2x - y = c \\ 3x + y = d \end{cases}$

40. $\begin{cases} x^2 - 10x = -5 - 2y \\ y = x + 1 \end{cases}$

41. $\begin{cases} y = 2x + 6 \\ 2x^2 + 8x = -2 - 3y \end{cases}$

42. $\begin{cases} 3x^2 + 4y^2 = 12 \\ x^2 + y^2 = -8 \end{cases}$

43. $\begin{cases} 3x^2 + 4y^2 = 19 \\ x^2 + y^2 = 5 \end{cases}$

44. $\begin{cases} x^2 + y^2 = 25 \\ y^2 = 5 - x \end{cases}$

45. $\begin{cases} x^2 - 12y - 1 = 0 \\ x^2 - 4y^2 - 9 = 0 \end{cases}$

46. Use the system

$$\begin{cases} \cos(x + y) = \cos x \cos y - \sin x \sin y \\ \cos(x - y) = \cos x \cos y + \sin x \sin y \end{cases}$$

to prove the identity

$$2 \cos x \cos y = \cos(x + y) + \cos(x - y)$$

47. Use the system

$$\begin{cases} \sin(x + y) = \sin x \cos y + \cos x \sin y \\ \sin(x - y) = \sin x \cos y - \cos x \sin y \end{cases}$$

to prove the identity

$$2 \sin x \cos y = \sin(x + y) + \sin(x - y)$$

48. Use the system in Problem 46 to prove the identity

$$2 \sin x \sin y = \cos(x - y) - \cos(x + y)$$

49. Use the system in Problem 47 to prove the identity

$$2 \cos x \sin y = \sin(x + y) - \sin(x - y)$$

50. Find two numbers whose sum is nine and whose product is eighteen.

51. Find two numbers whose difference is 4 and whose product is -3.

52. Find the lengths of the legs of a right triangle whose area is 60 ft^2 and whose hypotenuse is 17 ft.

53. Find the length and width of a rectangle whose area is 60 ft^2 with a diagonal of length 13 ft.

C

Solve the systems in Problems 54–63 for x and y. For Problems 54–56 give only those solutions between 0 and 6.

54. $\begin{cases} \sin x + \cos y = 1 \\ \sin x - \cos y = 0 \end{cases}$

55. $\begin{cases} \cos x - \sin y = 1 \\ \cos x + \sin y = 0 \end{cases}$

56. $\begin{cases} x + 3y = \cos^2 60° \\ x + \ \ y = -\sin^2 60° \end{cases}$

57. $\begin{cases} x^2 - xy + y^2 = 21 \\ xy - y^2 = 15 \end{cases}$

58. $\begin{cases} x^2 - xy + y^2 = 3 \\ x^2 + y^2 = 6 \end{cases}$

59. $\begin{cases} x^2 + 2xy = 8 \\ x^2 - 4y^2 = 8 \end{cases}$

60. $\begin{cases} \frac{3}{4} + \frac{4}{y} = 3 \\ \frac{9}{x} - \frac{2}{y} = 2 \end{cases}$

61. $\begin{cases} \frac{2}{x-1} - \frac{5}{y+2} = 12 \\ \frac{4}{x-1} - \frac{2}{y+2} = -12 \end{cases}$

62. $\begin{cases} y^2 - 5xy = 1 \\ y^2 - 4xy = 2y \end{cases}$

63. $\begin{cases} x^2 - 3xy = 2 \\ x^2 - 4xy = 3x \end{cases}$

64. If the parabola $(x - h)^2 = y - k$ passes through the points $(-2, 6)$ and $(-4, 2)$, find (h, k).

65. If the parabola $-3(x - h)^2 = y - k$ passes through the points $(2, 5)$ and $(-1, -4)$, find (h, k).

66. The Babylonians knew how to solve systems of equations as early as 1600 B.C. One tablet, called the Yale Tablet, shows a system equivalent to

$$\begin{cases} xy = 600 \\ (x + y)^2 - 150(x - y) = 100 \end{cases}$$

Find a solution for this system correct to the nearest tenth.

67. The Louvre Tablet from the Babylonian civilization is dated at about 1500 B.C. It shows a system equivalent to

$$\begin{cases} xy = 1 \\ x + y = a \end{cases}$$

Solve this system for x and y in terms of a.

7.2 CRAMER'S RULE

As you encounter systems more difficult than those considered in the last section, you will need more efficient methods for solving them. The first new method we will consider is a solution by *formula* in a procedure known as **Cramer's Rule**. Consider

$$\begin{cases} a_{11}x_1 + a_{12}x_2 = b_1 \\ a_{21}x_1 + a_{22}x_2 = b_2 \end{cases}$$

Solve this system by using linear combinations.

$$\begin{matrix} \boldsymbol{a_{22}} \\ \boldsymbol{-a_{12}} \end{matrix} \begin{cases} a_{11}x_1 + a_{12}x_2 = b_1 \\ a_{21}x_1 + a_{22}x_2 = b_2 \end{cases}$$

$$+\begin{cases} a_{11}a_{22}x_1 + a_{12}a_{22}x_2 = b_1a_{22} \\ -a_{12}a_{21}x_1 - a_{12}a_{22}x_2 = -b_2a_{12} \end{cases}$$

$$(a_{11}a_{22} - a_{12}a_{21})x_1 = b_1a_{22} - b_2a_{12}$$

If $a_{11}a_{22} - a_{12}a_{21} \neq 0$, then

$$x_1 = \frac{b_1a_{22} - b_2a_{12}}{a_{11}a_{22} - a_{12}a_{21}}$$

Similarly, solving for x_2,

$$x_2 = \frac{b_2a_{11} - b_1a_{21}}{a_{11}a_{22} - a_{12}a_{21}}$$

The difficulty with using this formula is that it is next to impossible to remember. To help with this matter, the following definition is given.

Determinant of Order 2

The **determinant** of order 2 is denoted by

$$\begin{vmatrix} a & b \\ c & d \end{vmatrix}$$

and is defined to be the real number $ad - bc$.

EXAMPLE 1 Evaluate the given determinants.

a. $\begin{vmatrix} 1 & 2 \\ 3 & 4 \end{vmatrix} = (1)(4) - (2)(3)$
$= -2$

b. $\begin{vmatrix} 3 & -2 \\ 4 & 1 \end{vmatrix} = 3 + 8$
$= 11$

c. $\begin{vmatrix} \sin\theta & \cos\theta \\ -\cos\theta & \sin\theta \end{vmatrix} = \sin^2\theta + \cos^2\theta$
$= 1$

d. $\begin{vmatrix} a_{11} & a_{12} \\ a_{21} & a_{22} \end{vmatrix} = a_{11}a_{22} - a_{12}a_{21}$

e. $\begin{vmatrix} b_1 & a_{12} \\ b_2 & a_{22} \end{vmatrix} = b_1 a_{22} - b_2 a_{12}$

f. $\begin{vmatrix} a_{11} & b_1 \\ a_{21} & b_2 \end{vmatrix} = b_2 a_{11} - b_1 a_{21}$ ■

The solution to the general system

$$\begin{cases} a_{11}x_1 + a_{12}x_2 = b_1 \\ a_{21}x_1 + a_{22}x_2 = b_2 \end{cases}$$

can now be stated using determinant notation:

$$x_1 = \frac{\begin{vmatrix} b_1 & a_{12} \\ b_2 & a_{22} \end{vmatrix}}{\begin{vmatrix} a_{11} & a_{12} \\ a_{21} & a_{22} \end{vmatrix}} \qquad x_2 = \frac{\begin{vmatrix} a_{11} & b_1 \\ a_{21} & b_2 \end{vmatrix}}{\begin{vmatrix} a_{11} & a_{12} \\ a_{21} & a_{22} \end{vmatrix}}$$

Notice that the denominator of both variables is the same. This determinant is called the **determinant of the coefficients** for the system and is denoted by $|D|$. The numerators are also found by looking at $|D|$. For x_1, replace the first column of $|D|$ by the constant numbers b_1 and b_2, respectively. Denote this by $|D_1|$. Similarly, let $|D_2|$ be the determinant formed by replacing the coefficients of x_2 in $|D|$ by the constant numbers b_1 and b_2, respectively. The solution to the system can now be stated with a result called *Cramer's Rule*.

Cramer's Rule (Two Unknowns)

Let $|D|$ be the determinant of the coefficients ($|D| \neq 0$) of a system of linear equations, and let $|D_1|$ and $|D_2|$ be the determinants where the coefficients of x_1 and x_2 are replaced respectively by b_1 and b_2. Then

$$x_1 = \frac{|D_1|}{|D|} \quad \text{and} \quad x_2 = \frac{|D_2|}{|D|}$$

EXAMPLE 2 Solve $\begin{cases} 2x - 3y = -8 \\ x + y = 6 \end{cases}$ using Cramer's Rule.

$$x = \frac{\begin{vmatrix} -8 & -3 \\ 6 & 1 \end{vmatrix}}{\begin{vmatrix} 2 & -3 \\ 1 & 1 \end{vmatrix}} = \frac{-8 + 18}{2 + 3} = \frac{10}{5} = 2 \qquad y = \frac{\begin{vmatrix} 2 & -8 \\ 1 & 6 \end{vmatrix}}{5} = \frac{12 + 8}{5} = \frac{20}{5} = 4$$

The solution is **(2, 4)**. ■

One advantage of Cramer's Rule is the ease with which its form can be used when working with more complicated solutions.

EXAMPLE 3 Solve $\begin{cases} 14x - 3y = 1 \\ 5x + 7y = -2 \end{cases}$ using Cramer's Rule.

Solution

$$x = \frac{\begin{vmatrix} 1 & -3 \\ -2 & 7 \end{vmatrix}}{\begin{vmatrix} 14 & -3 \\ 5 & 7 \end{vmatrix}} = \frac{7 - 6}{98 + 15} = \mathbf{\frac{1}{113}}$$

$$y = \frac{\begin{vmatrix} 14 & 1 \\ 5 & -2 \end{vmatrix}}{113} = \frac{-28 - 5}{113} = \mathbf{\frac{-33}{113}}$$ ■

EXAMPLE 4 Solve $\begin{cases} 2x + 3y = 9 \\ y = -\frac{2}{3}x + 1 \end{cases}$ using Cramer's Rule.

Solution You must first put both equations into the usual form:

$$\begin{cases} 2x + 3y = 9 \\ 2x + 3y = 3 \end{cases}$$

$$x = \frac{\begin{vmatrix} 9 & 3 \\ 3 & 3 \end{vmatrix}}{\begin{vmatrix} 2 & 3 \\ 2 & 3 \end{vmatrix}} = \frac{27 - 9}{6 - 6} = \frac{18}{0}$$

Since division by zero is not defined, **Cramer's Rule fails.** ■

Notice, however, from Example 4 that if $a_{11}a_{22} - a_{12}a_{21} = 0$ and the numerators are not zero, the system is inconsistent. To show this, note that

$$a_{11}a_{22} = a_{12}a_{21}$$

since $a_{11}a_{22} - a_{12}a_{21} = 0$. Divide both sides by $a_{12}a_{22}$. (If $a_{12}a_{22} = 0$, then $a_{12}a_{21} = 0$ and the problem would be trivial.) Assume, therefore, that $a_{12}a_{22} \neq 0$:

$$\frac{a_{11}}{a_{12}} = \frac{a_{21}}{a_{22}}$$

Now multiply both sides by -1:

$$-\frac{a_{11}}{a_{12}} = -\frac{a_{21}}{a_{12}}$$

But $-a_{11}/a_{12}$ is the slope of the line represented by the first equation of the system, and $-a_{21}/a_{22}$ is the slope of the line represented by the second equation. This means that if $a_{11}a_{22} - a_{12}a_{21} = 0$, then the slopes of the lines are the same. Thus the lines are either parallel or coincident. Next we show that they are not coincident by considering the numerators:

$$b_1a_{22} - b_2a_{12} \neq 0 \qquad\qquad a_{11}b_2 - a_{21}b_1 \neq 0$$
$$b_1a_{22} \neq b_2a_{12} \qquad\qquad a_{11}b_2 \neq a_{21}b_1$$
$$\frac{b_1}{b_2} \neq \frac{a_{12}}{a_{22}} \qquad\qquad \frac{b_1}{b_2} \neq \frac{a_{11}}{a_{21}}$$

This shows that the lines differ in at least one point, but since they are parallel or coincident they must be parallel. Therefore, if Cramer's Rule gives a solution of the form where the denominator is zero and the numerators are not zero, the system is *inconsistent*. (That is, the lines are parallel.) If Cramer's Rule yields a form $\frac{0}{0}$, the system is *dependent*.

PROBLEM SET 7.2

A

Evaluate the determinants given in Problems 1–15.

1. $\begin{vmatrix} 3 & 1 \\ 2 & 4 \end{vmatrix}$ **2.** $\begin{vmatrix} 6 & 3 \\ 2 & 2 \end{vmatrix}$ **3.** $\begin{vmatrix} -2 & 4 \\ 1 & 3 \end{vmatrix}$

4. $\begin{vmatrix} -5 & -3 \\ 4 & 2 \end{vmatrix}$ **5.** $\begin{vmatrix} 6 & -3 \\ 2 & 5 \end{vmatrix}$ **6.** $\begin{vmatrix} -3 & 2 \\ -4 & 5 \end{vmatrix}$

7. $\begin{vmatrix} 6 & 8 \\ -2 & -1 \end{vmatrix}$ **8.** $\begin{vmatrix} 6 & -3 \\ 4 & -2 \end{vmatrix}$ **9.** $\begin{vmatrix} 8 & -3 \\ 4 & -5 \end{vmatrix}$

10. $\begin{vmatrix} \sqrt{3} & \sqrt{5} \\ \sqrt{5} & \sqrt{3} \end{vmatrix}$ **11.** $\begin{vmatrix} \sqrt{2} & 3 \\ 1 & \sqrt{2} \end{vmatrix}$ **12.** $\begin{vmatrix} \pi & 3 \\ 0 & 1 \end{vmatrix}$

13. $\begin{vmatrix} \sin\theta & -\cos\theta \\ \cos\theta & \sin\theta \end{vmatrix}$ **14.** $\begin{vmatrix} \tan\theta & 1 \\ -1 & \tan\theta \end{vmatrix}$

15. $\begin{vmatrix} \csc\theta & 1 \\ 1 & \csc\theta \end{vmatrix}$

Solve the systems in Problems 16–30 by using Cramer's Rule.

16. $\begin{cases} x + y = 1 \\ 3x + y = -5 \end{cases}$ **17.** $\begin{cases} 2s + t = 7 \\ 3s + t = 12 \end{cases}$

18. $\begin{cases} c + d = 2 \\ 2c - d = 1 \end{cases}$ **19.** $\begin{cases} 2x + 3y = 1 \\ 3x - 2y = 0 \end{cases}$

20. $\begin{cases} 2s_1 + 3s_2 = 3 \\ 5s_1 - 2s_2 = 1 \end{cases}$ **21.** $\begin{cases} 3q_1 - 4q_2 = 3 \\ 5q_1 + 3q_2 = 4 \end{cases}$

22. $\begin{cases} 4x + 3y = 5 \\ 3x + 2y = 2 \end{cases}$ **23.** $\begin{cases} 2x - 3y = 6 \\ 5x + 2y = 3 \end{cases}$

24. $\begin{cases} y = \frac{2}{3}x - 7 \\ 2x + 3y = 3 \end{cases}$ **25.** $\begin{cases} y = \frac{3}{5}x + 2 \\ 3x - 5y = 10 \end{cases}$

26. $\begin{cases} 2x - 3y = 9 \\ y = \frac{2}{3}x - 3 \end{cases}$ **27.** $\begin{cases} 2p - 3q = 9 \\ q = \frac{2}{3}p - 3 \end{cases}$

28. $\begin{cases} y = \frac{2}{3}x + 5 \\ x + 7y = 1 \end{cases}$ **29.** $\begin{cases} 3x + 5y + 6 = 0 \\ y = -\frac{5}{3}x + 2 \end{cases}$

30. $\begin{cases} 2x - 3y - 4 = 0 \\ y = \frac{3}{2}x - 8 \end{cases}$

B

Solve the systems in Problems 31–51 for (x, y) by using Cramer's Rule.

31. $\begin{cases} 2x - 3y - 5 = 0 \\ 3x - 5y + 2 = 0 \end{cases}$ **32.** $\begin{cases} y = \frac{2}{3}x + 3 \\ y = -\frac{3}{4}x - 5 \end{cases}$

33. $\begin{cases} y = \frac{1}{2}x - 4 \\ y = \frac{2}{3}x + 5 \end{cases}$ **34.** $\begin{cases} x + y = a \\ x - y = b \end{cases}$

35. $\begin{cases} 2x + 3y = a \\ x - 5y = b \end{cases}$ **36.** $\begin{cases} 2x + y = \alpha \\ x - 3y = \beta \end{cases}$

37. $\begin{cases} ax + by = 1 \\ bx + ay = 0 \end{cases}$ **38.** $\begin{cases} ax + by = 0 \\ bx + ay = 1 \end{cases}$

39. $\begin{cases} cx + dy = s \\ ex + fy = t \end{cases}$

40. $\begin{cases} y = m_1x + b_1 \\ y = m_2x + b_2 \end{cases}$

41. $\begin{cases} ax + by = \alpha \\ cx + dy = \beta \end{cases}$

42. $\begin{cases} ax - by = \gamma \\ cx + dy = \delta \end{cases}$

43. $\begin{cases} 0.12x + 0.06y = 108 \\ x + y = 1000 \end{cases}$

44. $\begin{cases} 0.12x + 0.06y = 210 \\ x + y = 2000 \end{cases}$

45. $\begin{cases} 0.12x + 0.06y = 228 \\ x + y = 2000 \end{cases}$

46. $\begin{cases} 0.12x + 0.06y = 1140 \\ x + y = 10{,}000 \end{cases}$

47. $\begin{cases} q + d = 147 \\ 0.25q + 0.1d = 24.15 \end{cases}$

48. $\begin{cases} x + y = 10 \\ 0.4x + 0.9y = 0.5 \end{cases}$

49. $\begin{cases} 2x + 3y = \cos^2 45^\circ \\ x - y = -\sin^2 45^\circ \end{cases}$

50. $\begin{cases} 3x - y = \sec^2 30^\circ \\ x - y = \tan^2 30^\circ \end{cases}$

51. $\begin{cases} x + 3y = \csc^2 60^\circ \\ x - 2y = \cot^2 60^\circ \end{cases}$

For Problems 52–56 let $|A| = \begin{vmatrix} a & b \\ c & d \end{vmatrix}$

52. Show that $\begin{vmatrix} c & d \\ a & b \end{vmatrix} = -|A|$

53. Show that $\begin{vmatrix} a & b \\ a + c & b + d \end{vmatrix} = |A|$

54. Show that $\begin{vmatrix} a & a + b \\ c & c + d \end{vmatrix} = |A|$

55. Show that for any constant k, $\begin{vmatrix} ak & bk \\ c & d \end{vmatrix} = k|A|$

56. Show that $\begin{vmatrix} a & b \\ c + ka & d + kb \end{vmatrix} = \begin{vmatrix} a + kc & b + kd \\ c & d \end{vmatrix} = |A|$

C

Solve the systems in Problems 57–60 by using Cramer's Rule.

57. $\begin{cases} 3x^2 + 4y^2 = 16 \\ x^2 + y^2 = 5 \end{cases}$

58. $\begin{cases} 13x^2 + 12y^2 = 169 \\ x^2 + y^2 = 12 \end{cases}$

59. $\begin{cases} \dfrac{3}{x} + \dfrac{4}{y} = 3 \\ \dfrac{9}{x} - \dfrac{2}{y} = 2 \end{cases}$

60. $\begin{cases} \dfrac{2}{x - 1} - \dfrac{5}{y + 2} = 12 \\ \dfrac{4}{x - 1} - \dfrac{2}{y + 2} = -12 \end{cases}$

7.3 PROPERTIES OF DETERMINANTS

The strength of Cramer's Rule is that it can be applied to linear systems of n equations with n unknowns. However, to apply Cramer's Rule to these systems, some additional notation and definitions are required. Since the number of rows and columns in a determinant is the same, we call the size of either the **order** or **dimension** of the determinant. The determinants of the last section were of order 2. Consider a determinant of order 3:

$$\begin{vmatrix} a_{11} & a_{12} & a_{13} \\ a_{21} & a_{22} & a_{23} \\ a_{31} & a_{32} & a_{33} \end{vmatrix}$$

3 rows

3 columns

The entry a_{ij} refers to the entry in the ith row and jth column. That is, a_{23} refers to the entry in the second row and third column. The **minor** of an element in a determinant is the (smaller) determinant that remains after deleting all entries in its row and column.

EXAMPLE 1 Consider

$$\begin{vmatrix} 2 & -4 & 0 \\ 1 & -3 & -1 \\ 6 & 5 & 3 \end{vmatrix}$$

a. The a_{13} entry is 0; the a_{31} entry is 6; and the a_{23} entry is -1.

b. The minor of -3 is

$$\begin{vmatrix} 2 & -4 & 0 \\ 1 & \textcircled{-3} & -1 \\ 6 & 5 & 3 \end{vmatrix} \quad \text{or} \quad \begin{vmatrix} 2 & 0 \\ 6 & 3 \end{vmatrix} = 6$$

Delete shaded portions.

c. The minor of 0 is

$$\begin{vmatrix} 1 & -3 \\ 6 & 5 \end{vmatrix} = 5 + 18 = 23$$

d. The minor of 2 is

$$\begin{vmatrix} -3 & -1 \\ 5 & 3 \end{vmatrix} = -9 + 5 = -4$$

■

The **cofactor** of an entry a_{ij} is $(-1)^{i+j}$ times the minor of the a_{ij} entry. This says that if the sum of the row and column number is even, the cofactor is the same as the minor. If the sum of the row and column of an entry is odd, the cofactor of that entry is the opposite of its minor.

EXAMPLE 2 Consider

$$\begin{vmatrix} 2 & -4 & 0 \\ 1 & -3 & -1 \\ 6 & 5 & 3 \end{vmatrix}$$

a. The cofactor of 1 is

$$-\underbrace{\begin{vmatrix} -4 & 0 \\ 5 & 3 \end{vmatrix}}_{\text{minor}} = -(-12 - 0) = 12$$

Notice that this sign is independent of whether the entry is positive or negative or whether the minor is positive or negative.

b. The cofactor of -3 is

$$\underset{\text{sign}}{+}\underbrace{\begin{vmatrix} 2 & 0 \\ 6 & 3 \end{vmatrix}}_{\text{minor}} = 6$$

c. The cofactor of -1 is

$$-\begin{vmatrix} 2 & -4 \\ 6 & 5 \end{vmatrix} = -(10 + 24) = -34$$

■

EXAMPLE 3 Consider

$$\begin{vmatrix} 1 & 32 & -3 & 4 & 5 \\ -6 & 7 & 8 & 9 & 10 \\ 11 & -10 & -9 & -8 & -7 \\ -6 & -5 & -4 & 3 & \textcircled{2} \\ 1 & 0 & 3 & 5 & 12 \end{vmatrix}$$

The a_{45} entry is 2 and its cofactor is

$$-\underbrace{\begin{vmatrix} 1 & 32 & -3 & 4 \\ -6 & 7 & 8 & 9 \\ 11 & -10 & -9 & -8 \\ 1 & 0 & 3 & 5 \end{vmatrix}}_{\text{minor}}$$

sign: Row number is 4; column number is 5; the sum is 9, which is an odd number, so use the negative sign. ■

Determinant of Order n

> A determinant of order n is a real number whose value is the sum of the products obtained by multiplying each element of a row (or column) by its cofactor.

EXAMPLE 4 Expand $\begin{vmatrix} 1 & -2 & -5 \\ \mathbf{2} & \mathbf{-1} & \mathbf{0} \\ -4 & 5 & 6 \end{vmatrix}$ about the second row.

Solution

$$-(2)\begin{vmatrix} -2 & -5 \\ 5 & 6 \end{vmatrix} + (-1)\begin{vmatrix} 1 & -5 \\ -4 & 6 \end{vmatrix} - (0)\begin{vmatrix} 1 & -2 \\ -4 & 5 \end{vmatrix}$$

signs of cofactors

$= -2(-12 + 25) - (6 - 20) - 0$

$= -26 - (-14)$

$= \mathbf{-12}$ ■

EXAMPLE 5 Expand the determinant of Example 4 about the first column.

Solution

$$+(1)\begin{vmatrix} -1 & 0 \\ 5 & 6 \end{vmatrix} - (2)\begin{vmatrix} -2 & -5 \\ 5 & 6 \end{vmatrix} + (-4)\begin{vmatrix} -2 & -5 \\ -1 & 0 \end{vmatrix}$$

$= (-6 - 0) - 2(-12 + 25) - 4(0 - 5)$

$= -6 - 2(13) + 20$

$= \mathbf{-12}$ ■

Notice from Examples 4 and 5 that the same value is obtained regardless of the row or column chosen. You are asked to prove this for order 3 in the problem set.

The method of evaluating determinants by rows or columns as shown in Examples 4 and 5 is not very efficient for higher-order determinants. The following theorem considerably simplifies the work in evaluating determinants.

Determinant Reduction Theorem

> If $|A'|$ is a determinant obtained from a determinant $|A|$ by multiplying any row by a constant k and adding the result to any other row (entry by entry), then $|A'| = |A|$. The same result holds for columns.

Proof We will prove this theorem for the determinant of order 2; in the problems you are asked to prove it for order 3. For order 2, let

$$|A| = \begin{vmatrix} a & b \\ c & d \end{vmatrix} = ad - bc$$

Multiply row 1 by k and add it to row 2:

$$\begin{aligned} |A'| = \begin{vmatrix} a & b \\ c + ak & d + bk \end{vmatrix} &= a(d + bk) - b(c + ak) \\ &= ad + abk - bc - abk \\ &= ab - bc \\ &= |A| \end{aligned}$$

Thus $|A'| = |A|$. Next multiply row 2 by k and add it to row 1 to obtain the same result. □

EXAMPLE 6 Expand

$$\begin{vmatrix} 1 & -2 & -5 \\ 2 & -1 & 0 \\ -4 & 5 & 6 \end{vmatrix}$$

by using the determinant reduction theorem.

Solution *You want to obtain some row or column with two zeros.* Add twice the second column to the first column:

$$\begin{vmatrix} 1 & -2 & -5 \\ 2 & -1 & 0 \\ -4 & 5 & 6 \end{vmatrix} = \begin{vmatrix} -3 & -2 & -5 \\ 0 & -1 & 0 \\ 6 & 5 & 6 \end{vmatrix}$$

× 2 These columns are unchanged.

Now expand about the second row (do not even bother to write down the products that are zero):

positive understood

$$\begin{aligned} &= (-1)\begin{vmatrix} -3 & -5 \\ 6 & 6 \end{vmatrix} \\ &= (-1)(-18 + 30) \\ &= \mathbf{-12} \end{aligned}$$

■

EXAMPLE 7 Expand

$$\begin{vmatrix} 2 & -3 & 2 & 5 & 0 \\ 4 & 2 & -1 & 4 & 0 \\ 5 & 1 & 0 & -2 & 0 \\ 6 & 2 & 3 & 6 & 0 \\ 3 & 4 & 6 & 1 & -2 \end{vmatrix}$$

by using the determinant reduction theorem.

Solution First notice that all the entries in the fifth column except one are zero, so begin by expanding about the fifth column:

$$+(-2)\begin{vmatrix} 2 & -3 & 2 & 5 \\ 4 & 2 & -1 & 4 \\ 5 & 1 & 0 & -2 \\ 6 & 2 & 3 & 6 \end{vmatrix}$$

Row 5, column 5; 5 + 5 is even, so this sign is positive.

Next obtain a row or column with all entries zero except one; row 3 or column 3 seem to be likely candidates. We will work toward obtaining zeros in column 3:

$$(-2)\begin{vmatrix} 2 & -3 & 2 & 5 \\ 4 & 2 & -1 & 4 \\ 5 & 1 & 0 & -2 \\ 6 & 2 & 3 & 6 \end{vmatrix} \times 2 = (-2)\begin{vmatrix} 10 & 1 & 0 & 13 \\ 4 & 2 & -1 & 4 \\ 5 & 1 & 0 & -2 \\ 6 & 2 & 3 & 6 \end{vmatrix} \times 3$$

$$= (-2)\begin{vmatrix} 10 & 1 & 0 & 13 \\ 4 & 2 & -1 & 4 \\ 5 & 1 & 0 & -2 \\ 18 & 8 & 0 & 18 \end{vmatrix} = (-2)\left[-(-1)\begin{vmatrix} 10 & 1 & 13 \\ 5 & 1 & -2 \\ 18 & 8 & 18 \end{vmatrix}\right]$$

$\times(-5)$

Finished when all entries but one are zero; now reduce the order of the determinant.

Row 2, column 3; 2 + 3 is odd.

$$= (-2)\begin{vmatrix} 5 & 1 & 13 \\ 0 & 1 & -2 \\ -22 & 8 & 18 \end{vmatrix} = (-2)\begin{vmatrix} 5 & 1 & 15 \\ 0 & 1 & 0 \\ -22 & 8 & 34 \end{vmatrix}$$

$\times 2$

$$= (-2)\left[+(1)\begin{vmatrix} 5 & 15 \\ -22 & 34 \end{vmatrix}\right]$$

Row 2, column n; 2 + 2 is even.

$$= -2(170 + 330) = -2(500) = \mathbf{-1000}$$

We can now state Cramer's Rule for n linear equations with n unknowns. Consider the general n by n system

$$\begin{cases} a_{11}x_1 + a_{12}x_2 + a_{13}x_3 + \cdots + a_{1n}x_n = b_1 \\ a_{21}x_1 + a_{22}x_2 + a_{23}x_3 + \cdots + a_{2n}x_n = b_2 \\ a_{31}x_1 + a_{32}x_2 + a_{33}x_3 + \cdots + a_{3n}x_n = b_3 \\ \vdots \\ a_{n1}x_1 + a_{n2}x_2 + a_{n3}x_3 + \cdots + a_{nn}x_n = b_n \end{cases}$$

The unknowns are $x_1, x_2, x_3, \ldots, x_n$; a_{ij} are the coefficients and $b_1, b_2, b_3, \ldots, b_n$ are the constants. When written in this form, the system is said to be in **standard form**.

Cramer's Rule

> Let $|D|$ be the determinant of the coefficients ($|D| \neq 0$) of a system of n linear equations with n unknowns. Let $|D_i|$ be the determinant where the coefficients of x_i have been replaced by the constants $b_1, b_2, b_3, \ldots, b_n$, respectively. Then
>
> $$x_i = \frac{|D_i|}{|D|}$$

As before, if $|D_i|$ and $|D|$ are zero, the system is dependent; if at least one of the $|D_i|$ is not zero when $|D|$ is zero, the system is inconsistent.

EXAMPLE 8 Solve

$$\begin{cases} x - 2y - 5z = -12 \\ 2x - y = 7 \\ 5y + 6z = 4x + 1 \end{cases}$$

Solution Rewrite the system in standard form:

$$\begin{cases} x - 2y - 5z = -12 \\ 2x - y \qquad = 7 \\ -4x + 5y + 6z = 1 \end{cases}$$

Be sure to consider "missing terms" as terms with zero coefficient.

$$|D| = \begin{vmatrix} 1 & -2 & -5 \\ 2 & -1 & 0 \\ -4 & 5 & 6 \end{vmatrix} = -12 \quad \text{From Example 6}$$

$$|D_1| = \begin{vmatrix} -12 & -2 & -5 \\ 7 & -1 & 0 \\ 1 & 5 & 6 \end{vmatrix} = \begin{vmatrix} -26 & -2 & -5 \\ 0 & -1 & 0 \\ 36 & 5 & 6 \end{vmatrix} = (-1)\begin{vmatrix} -26 & -5 \\ 36 & 6 \end{vmatrix}$$

$\times 7$

$$= -(-156 + 180)$$
$$= -24$$

$$|D_2| = \begin{vmatrix} 1 & -12 & -5 \\ 2 & 7 & 0 \\ -4 & 1 & 6 \end{vmatrix} \times 1 = \begin{vmatrix} 1 & -12 & -5 \\ 2 & 7 & 0 \\ -3 & -11 & 1 \end{vmatrix} \times 5$$

$$= \begin{vmatrix} -14 & -67 & 0 \\ 2 & 7 & 0 \\ -3 & -11 & 1 \end{vmatrix} = \begin{vmatrix} -14 & -67 \\ 2 & 7 \end{vmatrix} = -98 + 134 = 36$$

$$|D_3| = \begin{vmatrix} 1 & -2 & -12 \\ 2 & -1 & 7 \\ -4 & 5 & 1 \end{vmatrix} = \begin{vmatrix} -3 & -2 & -26 \\ 0 & -1 & 0 \\ 6 & 5 & 36 \end{vmatrix} = (-1)\begin{vmatrix} -3 & -26 \\ 6 & 36 \end{vmatrix}$$

$\times 2 \quad \times 7$

$$= -(-108 + 156) = -48$$

Thus

$$x = \frac{|D_1|}{|D|} = \frac{-24}{-12} = 2 \qquad y = \frac{|D_2|}{|D|} = \frac{36}{-12} = -3 \qquad z = \frac{|D_3|}{|D|} = \frac{-48}{-12} = 4$$

The solution is $(x, y, z) = (2, -3, 4)$. ■

PROBLEM SET 7.3

A

Evaluate the determinants in Problems 1–15. If you want to check, evaluate the determinant by using a different row or column.

1. $\begin{vmatrix} 3 & 0 & 0 \\ 2 & 1 & 4 \\ 3 & 6 & -1 \end{vmatrix}$

2. $\begin{vmatrix} 1 & -2 & 3 \\ 0 & 4 & 0 \\ 3 & -1 & -3 \end{vmatrix}$

3. $\begin{vmatrix} 4 & -2 & 6 \\ 3 & 1 & 4 \\ 0 & 0 & -2 \end{vmatrix}$

4. $\begin{vmatrix} 0 & 1 & 1 \\ -3 & 2 & 4 \\ 0 & -2 & -3 \end{vmatrix}$

5. $\begin{vmatrix} 1 & -2 & 3 \\ -2 & 0 & 4 \\ 3 & 0 & 5 \end{vmatrix}$

6. $\begin{vmatrix} 3 & 1 & 0 \\ -2 & 4 & 0 \\ -3 & 5 & -4 \end{vmatrix}$

7. $\begin{vmatrix} 1 & 1 & 1 \\ 1 & 3 & 2 \\ 1 & -2 & 1 \end{vmatrix}$

8. $\begin{vmatrix} 4 & 1 & 1 \\ 4 & 3 & 2 \\ 7 & -2 & 1 \end{vmatrix}$

9. $\begin{vmatrix} 1 & 4 & 1 \\ 1 & 4 & 2 \\ 1 & 7 & 1 \end{vmatrix}$

10. $\begin{vmatrix} -3 & -1 & 1 \\ 14 & 2 & -3 \\ 12 & 3 & -1 \end{vmatrix}$

11. $\begin{vmatrix} 2 & -3 & 1 \\ 1 & 14 & -3 \\ 3 & 12 & -1 \end{vmatrix}$

12. $\begin{vmatrix} 2 & -1 & -3 \\ 1 & 3 & 14 \\ 3 & 3 & 12 \end{vmatrix}$

13. $\begin{vmatrix} 2 & 4 & 3 \\ -2 & 3 & -2 \\ 4 & 3 & 5 \end{vmatrix}$

14. $\begin{vmatrix} 6 & 3 & -3 \\ 2 & 0 & 5 \\ 3 & 5 & -2 \end{vmatrix}$

15. $\begin{vmatrix} 4 & 8 & 5 \\ 3 & 2 & 3 \\ 5 & 5 & 4 \end{vmatrix}$

B

Evaluate the determinants in Problems 16–30.

16. $\begin{vmatrix} 5 & 2 & 6 & -11 \\ -3 & 0 & 3 & 1 \\ 4 & 0 & 0 & 6 \\ 5 & 0 & 0 & -1 \end{vmatrix}$

17. $\begin{vmatrix} 4 & 3 & 2 & 1 \\ -5 & 0 & 0 & 0 \\ 11 & -4 & 0 & 0 \\ 9 & 6 & 3 & -5 \end{vmatrix}$

18. $\begin{vmatrix} 7 & 2 & -5 & 3 \\ 1 & 0 & -3 & 2 \\ 4 & 0 & 1 & -5 \\ 0 & 0 & 3 & 0 \end{vmatrix}$

19. $\begin{vmatrix} 2 & 1 & -1 & 3 \\ 4 & 0 & 0 & 0 \\ 2 & 1 & -2 & 3 \\ 1 & 4 & 3 & 5 \end{vmatrix}$

20. $\begin{vmatrix} 6 & 3 & 0 & -2 \\ 4 & 3 & 4 & -1 \\ 1 & 2 & 0 & 5 \\ 3 & -2 & 0 & 5 \end{vmatrix}$

21. $\begin{vmatrix} 1 & 6 & 2 & -1 \\ 0 & -5 & 0 & 0 \\ 3 & 4 & 5 & -3 \\ 2 & 1 & 4 & 0 \end{vmatrix}$

22. $\begin{vmatrix} 2 & 1 & 2 & 4 \\ 3 & -1 & 2 & 5 \\ -3 & 2 & 3 & -4 \\ -3 & 2 & 8 & -4 \end{vmatrix}$

23. $\begin{vmatrix} 3 & 1 & -1 & 2 \\ 4 & 0 & 3 & 0 \\ 2 & 4 & 3 & -3 \\ 6 & 1 & 4 & 0 \end{vmatrix}$

24. $\begin{vmatrix} 2 & 1 & 3 & 1 \\ 6 & 3 & -3 & 2 \\ 2 & 0 & 5 & 1 \\ 3 & 5 & -2 & -1 \end{vmatrix}$

25. $\begin{vmatrix} 0 & -3 & 5 & 6 & -1 \\ 0 & 0 & 1 & 1 & 3 \\ 5 & -3 & 8 & -5 & 1 \\ 0 & 0 & 2 & 0 & 0 \\ 0 & 0 & 4 & -1 & 2 \end{vmatrix}$

26. $\begin{vmatrix} 3 & 1 & -3 & -2 & 5 \\ -1 & 5 & 1 & 0 & 1 \\ 5 & 0 & 3 & 0 & 0 \\ 4 & 0 & 0 & 0 & 0 \\ 2 & 2 & 4 & 0 & -1 \end{vmatrix}$

27. $\begin{vmatrix} -3 & 4 & 5 & 8 & -9 \\ 0 & 0 & 0 & 0 & 6 \\ 3 & 0 & 0 & -1 & 4 \\ 0 & 0 & 4 & 0 & 9 \\ -3 & 0 & 0 & 1 & 8 \end{vmatrix}$

28. $\begin{vmatrix} 3 & 4 & 1 & -2 & 0 \\ 1 & -2 & 3 & 0 & 1 \\ 0 & 4 & -2 & 1 & 0 \\ 5 & 0 & -3 & 0 & 0 \\ 2 & 1 & -2 & 2 & 0 \end{vmatrix}$

29. $\begin{vmatrix} 2 & -1 & 0 & 1 & -1 \\ 3 & 0 & 0 & 2 & 0 \\ 2 & 1 & 3 & 1 & 3 \\ 0 & 0 & 0 & -2 & 0 \\ 1 & 3 & -1 & 2 & 1 \end{vmatrix}$

30. $\begin{vmatrix} 2 & 1 & 3 & 0 & 0 \\ 6 & 1 & 5 & 2 & -1 \\ 1 & 4 & -5 & 9 & 3 \\ 3 & 2 & 5 & 7 & 2 \\ 2 & -3 & -2 & 4 & 6 \end{vmatrix}$

Use Cramer's Rule to solve the systems in Problems 31–40.

31. $\begin{cases} x + y + z = 6 \\ 2x - y + z = 3 \\ x - 2y - 3z = 6 \end{cases}$

32. $\begin{cases} 2x - y + z = 3 \\ x - 3y + 2z = 7 \\ x - y - z = -1 \end{cases}$

33. $\begin{cases} x + y + z = 4 \\ x + 3y + 2z = 4 \\ x - 2y + z = 7 \end{cases}$

34. $\begin{cases} x + y + z = 3 \\ 2x + z = -1 \\ y = 5 \end{cases}$

35. $\begin{cases} x + y + z = 3 \\ 3y - z = -11 \\ x = 4 \end{cases}$

36. $\begin{cases} 2x + y + z = -5 \\ 3x - y = -9 \\ z = -4 \end{cases}$

37. $\begin{cases} 2x + 2y + 3z = 1 \\ 2x - z = -11 \\ 3y + 2z = 6 \end{cases}$

38. $\begin{cases} x + 2y + z = 1 \\ x - 3y - 2z = 2 \\ 3x - 2y + z = 3 \end{cases}$

39. $\begin{cases} 2x - y + z = 4 \\ 3x - 2y + 2z = 3 \\ x - y + 3z = 2 \end{cases}$

40. $\begin{cases} 5x - 3y + 2z = 10 \\ 4x + 2y - 3z = 4 \\ 3x + y + 4z = -8 \end{cases}$

Use the results of Problem 51 to find the equations of the lines passing through the points given in Problems 41–43.

41. $(-3, -2), (1, -3)$ **42.** $(4, -5), (7, -8)$

43. $(1, 5), (-2, -3)$

Use the results of Problem 52 to find the areas of the triangles with vertices given in Problems 44–46.

44. $(1, 1), (-2, -3), (11, -3)$ **45.** $(-3, 12), (5, 6), (-3, -9)$

46. $(-8, 0), (12, 10), (4, -5)$

Solve the systems in Problems 47–50.

47. $\begin{cases} w + x - y + z = 7 \\ 2w + y - 3z = 1 \\ 2x - z + w = 4 \\ y - w + z = -4 \end{cases}$

48. $\begin{cases} 3t - u + x = 20 \\ x - 2t - u = 0 \\ 3u - 2x + 5t = 1 \end{cases}$

49. $\begin{cases} s + 3t - 2u = 4 \\ u + 2x - t = -5 \\ x - u - t = 0 \\ s - 2x = -1 \end{cases}$

50. $\begin{cases} 2x - y + w - v = -4 \\ 3x + 2w = 0 \\ x + y + 3z + w + 3v = 5 \\ -2w = -6 \\ x + 3y - z + 2w + v = 10 \end{cases}$

C

51. Prove that

$$\begin{vmatrix} x & y & 1 \\ x_1 & y_1 & 1 \\ x_2 & y_2 & 1 \end{vmatrix} = 0$$

is the equation of a line passing through (x_1, y_1) and (x_2, y_2).

52. Prove that the area of a triangle with vertices at (x_1, y_1), (x_2, y_2), and (x_3, y_3) is the absolute value of

$$\frac{1}{2}\begin{vmatrix} x_1 & y_1 & 1 \\ x_2 & y_2 & 1 \\ x_3 & y_3 & 1 \end{vmatrix}$$

53. If r_1, r_2, r_3, and r_4 are the fourth roots of 1, show that

$$\begin{vmatrix} r_1 & r_2 & r_3 & r_4 \\ r_2 & r_3 & r_4 & r_1 \\ r_3 & r_4 & r_1 & r_2 \\ r_4 & r_1 & r_2 & r_3 \end{vmatrix} = 0$$

For Problems 54–60, let

$$|A| = \begin{vmatrix} a_{11} & a_{12} & a_{13} \\ a_{21} & a_{22} & a_{23} \\ a_{31} & a_{32} & a_{33} \end{vmatrix}$$

54. Show that you obtain the same result if you expand along the first or third row.

55. Show that you obtain the same result if you expand along the second row or second column.

56. Prove the determinant reduction theorem for $|A|$. (Without loss of generality, prove it by multiplying the first row by k and adding it to the second row.)

57. Prove that $|A| = 0$ if two rows (say the first and second rows) are identical.

58. Prove that $|A| = 0$ if two columns (say the first and second columns) are identical.

59. Prove that if two rows of $|A|$ are interchanged (say the first and third), the resulting determinant is $-|A|$.

60. Prove that

$$k|A| = \begin{vmatrix} a_{11} & a_{12} & a_{13} \\ ka_{21} & ka_{22} & ka_{23} \\ a_{31} & a_{32} & a_{33} \end{vmatrix}$$

7.4 ALGEBRA OF MATRICES

Instead of considering the determinant of the coefficients, we will now arrange the coefficients of a system of linear equations into a rectangular **array** of numbers. Such an array of numbers is called a **matrix**. The number of columns and rows of a matrix need not be the same; but if they are, the matrix is called a **square matrix**. The **order** or **dimension** of a matrix is given by an expression $\boldsymbol{m \times n}$ (pronounced "m by n"), where m is the number of rows and n is the number of columns.

EXAMPLE 1

$$A = [5 \quad 2 \quad 1] \qquad B = [4 \quad 8 \quad -5]$$

$$C = \begin{bmatrix} 7 & 3 & 2 \\ 5 & -4 & -3 \end{bmatrix} \qquad D = \begin{bmatrix} 4 & -2 & 1 \\ -3 & 3 & -1 \\ 2 & 4 & -1 \end{bmatrix}$$

$$E = \begin{bmatrix} 3 & 4 & -1 \\ 2 & 0 & 5 \\ -4 & 2 & 3 \end{bmatrix} \qquad G = [g_{ij}]_{m,n}$$

Note that A and B have order 1×3; C has order 2×3; D and E have order 3×3—these are square **matrices** (plural form of *matrix*). Since m and n are the same for square matrices, we often refer to them as having order n: D and E have order 3.

Matrix G is an arbitrary m by n matrix; it is shorthand notation for

$$\begin{bmatrix} g_{11} & g_{12} & g_{13} & \cdots & g_{1n} \\ g_{21} & g_{22} & g_{23} & \cdots & g_{2n} \\ g_{31} & g_{32} & g_{33} & \cdots & g_{3n} \\ & & \vdots & & \\ g_{m1} & g_{m2} & g_{m3} & \cdots & g_{mn} \end{bmatrix}$$

■

Capital letters are generally used to denote matrices and the same lowercase letters for entries of that matrix. Also, don't confuse matrices, determinants, and their notation. A matrix M is an *array* of real numbers whereas the determinant $|M|$ is itself a single real number. The entries of a matrix are always shown enclosed in square brackets, whereas determinants are enclosed by vertical lines.

As in the last chapter with complex numbers, we need to state a definition of equality along with the fundamental matrix operations.

Matrix Operations

EQUALITY: $M = N$ if and only if matrices M and N have the same dimension and $m_{ij} = n_{ij}$ for all i and j.

ADDITION: $M + N = P$ if and only if M and N have the same dimension and $p_{ij} = m_{ij} + n_{ij}$ for all i and j.

MULTIPLICATION OF A MATRIX AND A REAL NUMBER: $cM = Mc = [cm_{ij}]$ for any real number c.

SUBTRACTION: $M - N = P$ if and only if matrices M and N have the same dimension and $p_{ij} = m_{ij} - n_{ij}$ for all i and j.

MULTIPLICATION: $MN = P$ if and only if the number of columns of matrix M is the same as the number of rows of matrix N and

$$p_{ij} = m_{i1}n_{1j} + m_{i2}n_{2j} + m_{i3}n_{3j} + \cdots + m_{in}n_{nj}$$

The matrix P has the same number of rows as M and the same number of columns as N.

All of these definitions, except multiplication, are straightforward, so we will consider multiplication separately after Example 2. If an addition or multiplication cannot be performed because of the order of the given matrices, the matrices are said to be not **conformable**.

EXAMPLE 2 Let A, B, C, D, and E be the matrices defined in Example 1.

a.
$$\begin{aligned} A + B &= [5 \quad 2 \quad 1] + [4 \quad 8 \quad -5] \\ &= [5 + 4 \quad 2 + 8 \quad 1 + (-5)] \\ &= [9 \quad 10 \quad -4] \end{aligned}$$

b. $A + C$ is not defined because A and C are not conformable.

c. $E + D = \begin{bmatrix} 3 & 4 & -1 \\ 2 & 0 & 5 \\ -4 & 2 & 3 \end{bmatrix} + \begin{bmatrix} 4 & -2 & 1 \\ -3 & 3 & -1 \\ 2 & 4 & -1 \end{bmatrix}$

$= \begin{bmatrix} 7 & 2 & 0 \\ -1 & 3 & 4 \\ -2 & 6 & 2 \end{bmatrix}$ Add entry by entry.

d. $(-5)C = (-5)\begin{bmatrix} 7 & 3 & 2 \\ 5 & -4 & -3 \end{bmatrix}$

$= \begin{bmatrix} -35 & -15 & -10 \\ -25 & 20 & 15 \end{bmatrix}$

e. $2A - 3B = 2[5 \quad 2 \quad 1] + (-3)[4 \quad 8 \quad -5]$

$= [10 \quad 4 \quad 2] + [-12 \quad -24 \quad 15]$

$= [-2 \quad -20 \quad 17]$ ■

You will need to study the definition of matrix multiplication carefully to understand the process. Consider the following specific example:

$$M = \begin{bmatrix} m_{11} & m_{12} & m_{13} & m_{14} \\ m_{21} & m_{22} & m_{23} & m_{24} \end{bmatrix} = \begin{bmatrix} \mathbf{1} & \mathbf{2} & \mathbf{3} & \mathbf{4} \\ \mathbf{5} & \mathbf{6} & \mathbf{7} & \mathbf{8} \end{bmatrix}$$

$$N = \begin{bmatrix} n_{11} & n_{12} & n_{13} \\ n_{21} & n_{22} & n_{23} \\ n_{31} & n_{32} & n_{33} \\ n_{41} & n_{42} & n_{43} \end{bmatrix} = \begin{bmatrix} -3 & 1 & -2 \\ 0 & -1 & 5 \\ -4 & 3 & -1 \\ 2 & 3 & -2 \end{bmatrix}$$

The matrix N and its entries are shown in color so you can keep track of them.

$$MN = \begin{bmatrix} m_{11} & m_{12} & m_{13} & m_{14} \\ m_{21} & m_{22} & m_{23} & m_{24} \end{bmatrix} \begin{bmatrix} n_{11} & n_{12} & n_{13} \\ n_{21} & n_{22} & n_{23} \\ n_{31} & n_{32} & n_{33} \\ n_{41} & n_{42} & n_{43} \end{bmatrix} = \underbrace{\begin{bmatrix} p_{11} & p_{12} & p_{13} \\ p_{21} & p_{22} & p_{23} \end{bmatrix}}_{\text{These are the entries of the product.}}$$

These must be the same for multiplication to be conformable.

Order of M is 2×4 Order of N is 4×3

Order of P is 2×3; this is the order of the product.

$$MN = \begin{bmatrix} 1 & 2 & 3 & 4 \\ 5 & 6 & 7 & 8 \end{bmatrix} \begin{bmatrix} -3 & 1 & -2 \\ 0 & -1 & 5 \\ -4 & 3 & -1 \\ 2 & 3 & -2 \end{bmatrix} = \begin{bmatrix} p_{11} & p_{12} & p_{13} \\ p_{21} & p_{22} & p_{23} \end{bmatrix}$$

(Row 1 and Row 2 of M are marked; Column 1, Column 2, Column 3 of N are marked.)

Entry in Product Row	Column	Notation
1	1	p_{11}
1	2	p_{12}
1	3	p_{13}
2	1	p_{21}
2	2	p_{22}
2	3	p_{23}

$$p_{11} = m_{11}n_{11} + m_{12}n_{21} + m_{13}n_{31} + m_{14}n_{41} = 1(-3) + 2(0) + 3(-4) + 4(2) = -7$$

$$p_{12} = m_{11}n_{12} + m_{12}n_{22} + m_{13}n_{32} + m_{14}n_{42} = 1(1) + 2(-1) + 3(3) + 4(3) = 20$$

$$p_{13} = m_{11}n_{13} + m_{12}n_{23} + m_{13}n_{33} + m_{14}n_{43} = 1(-2) + 2(5) + 3(-1) + 4(-2) = -3$$

$$p_{21} = m_{21}n_{11} + m_{22}n_{21} + m_{23}n_{31} + m_{24}n_{41} = 5(-3) + 6(0) + 7(-4) + 8(2) = -27$$

$$p_{22} = m_{21}n_{12} + m_{22}n_{22} + m_{23}n_{32} + m_{24}n_{42} = 5(1) + 6(-1) + 7(3) + 8(3) = 44$$

$$p_{23} = m_{21}n_{13} + m_{22}n_{23} + m_{23}n_{33} + m_{24}n_{43} = 5(-2) + 6(5) + 7(-1) + 8(-2) = -3$$

This example makes the process seem very lengthy, but it has been written out so that you could study the *process*. The way it would look in your work is shown in Example 3*a*.

EXAMPLE 3 Find the products.

a. $\begin{bmatrix} 1 & 2 & 3 & 4 \\ 5 & 6 & 7 & 8 \end{bmatrix} \begin{bmatrix} -3 & 1 & -2 \\ 0 & -1 & 5 \\ -4 & 3 & -1 \\ 2 & 3 & -2 \end{bmatrix}$

$$= \begin{bmatrix} 1(-3) + 2(0) + 3(-4) + 4(2) & 1(1) + 2(-1) + 3(3) + 4(3) & 1(-2) + 2(5) + 3(-1) + 4(-2) \\ 5(-3) + 6(0) + 7(-4) + 8(2) & 5(1) + 6(-1) + 7(3) + 8(3) & 5(-2) + 6(5) + 7(-1) + 8(-2) \end{bmatrix}$$

$$= \begin{bmatrix} -7 & 20 & -3 \\ -27 & 44 & -3 \end{bmatrix}$$

b. $\begin{bmatrix} 3 & -1 & 4 \\ 2 & 1 & 0 \\ -1 & 3 & 2 \end{bmatrix} \begin{bmatrix} 5 & 1 & -1 \\ 2 & 3 & -2 \\ 0 & 3 & 4 \end{bmatrix}$

$$= \begin{bmatrix} 3(5) + (-1)(2) + 4(0) & 3(1) + (-1)(3) + 4(3) & 3(-1) + (-1)(-2) + 4(4) \\ 2(5) + (1)(2) + 0(0) & 2(1) + (1)(3) + 0(3) & 2(-1) + (1)(-2) + 0(4) \\ (-1)(5) + 3(2) + 2(0) & (-1)(1) + 3(3) + 2(3) & (-1)(-1) + 3(-2) + 2(4) \end{bmatrix} = \begin{bmatrix} 13 & 12 & 15 \\ 12 & 5 & -4 \\ 1 & 14 & 3 \end{bmatrix}$$ ■

Systems of equations can be written in the form of matrix multiplication, as shown by Example 4.

EXAMPLE 4 Let

$$A = \begin{bmatrix} 1 & 2 & 3 \\ 4 & -1 & 5 \\ 3 & 2 & -1 \end{bmatrix} \quad X = \begin{bmatrix} x \\ y \\ z \end{bmatrix} \quad B = \begin{bmatrix} 3 \\ 16 \\ 5 \end{bmatrix}$$

What is $AX = B$?

Solution

$$AX = \begin{bmatrix} x + 2y + 3z \\ 4x - y + 5z \\ 3x + 2y - z \end{bmatrix}$$

so $AX = B$ is a matrix equation representing the system

$$\begin{cases} x + 2y + 3z = 3 \\ 4x - y + 5z = 16 \\ 3x - 2y - z = 5 \end{cases}$$

■

Properties for an algebra of matrices can also be developed. The $m \times n$ **zero matrix**, denoted by 0, is the matrix with m rows and n columns in which each entry is 0. The **identity matrix**, denoted by I_n, is the square matrix with n rows and n columns consisting of a 1 in each position on the **main diagonal** (entries $m_{11}, m_{22}, m_{33}, \ldots$) and zeros elsewhere:

$$I_2 = \begin{bmatrix} 1 & 0 \\ 0 & 1 \end{bmatrix} \quad I_3 = \begin{bmatrix} 1 & 0 & 0 \\ 0 & 1 & 0 \\ 0 & 0 & 1 \end{bmatrix} \quad I_4 = \begin{bmatrix} 1 & 0 & 0 & 0 \\ 0 & 1 & 0 & 0 \\ 0 & 0 & 1 & 0 \\ 0 & 0 & 0 & 1 \end{bmatrix}$$

The **additive inverse** of a matrix M is denoted by $-M$ and is defined by $(-1)M$; the **multiplicative inverse** of M is denoted by M^{-1} if it exists. (*Note:* $M^{-1} \neq 1/M$.) The following list summarizes the properties of matrices. Assume that matrices M, N, and P all have order n, which forces them to be conformable for the given operations.

TABLE 7.1 Properties of matrices

Property	Addition	Multiplication
Commutative	$M + N = N + M$	$MN \neq NM$ Matrix multiplication is not commutative.
Associative	$(M + N) + P = M + (N + P)$	$(MN)P = M(NP)$
Identity	$M + 0 = 0 + M = M$	$I_n M = M I_n = M$ For a square matrix M of order n.
Inverse	$M + (-M) = (-M) + M = 0$	$M(M^{-1}) = (M^{-1})M = I_n$ For a square matrix M for which M^{-1} exists.
Distributive	$M(N + P) = MN + MP$ and $(N + P)M = NM + PM$	

These properties are straightforward except for the inverse property. There seem to be two unanswered questions. Given a square matrix M, when does M^{-1} exist? And if it exists, how do you find it?

If a square matrix M has an inverse, it is called **nonsingular**. If

$$M = \begin{bmatrix} a_{11} & a_{12} \\ a_{21} & a_{22} \end{bmatrix}$$

and it has an inverse

$$M^{-1} = \begin{bmatrix} w & x \\ y & z \end{bmatrix}$$

then $MM^{-1} = I_2$ if and only if $|M| \neq 0$. To prove this, simply multiply MM^{-1} and solve the resulting systems for w, x, y, and z to find

$$w = \frac{a_{22}}{|M|} \qquad x = \frac{-a_{12}}{|M|} \qquad y = \frac{-a_{21}}{|M|} \qquad z = \frac{a_{11}}{|M|}$$

You are asked to do this in the problems. Thus we see that the existence of the inverse of a matrix depends on whether the value of the associated determinant is zero.

The procedure for finding an inverse depends on **elementary row operations** and on equivalent matrices. We say that a matrix M **is equivalent to** a matrix N, written $M \sim N$, if M can be changed into N by one or more of the following elementary row operations.

Elementary Row Operations

There are three elementary row operations:

1. Interchange any two rows.
2. Multiply all the elements of a row by the same nonzero real number.
3. Multiply all the elements of a row by a real number and add the product to the corresponding entry of another row.

If M has order n, write $[M \,\vdots\, I_n]$ and carry out the following procedure using only elementary row operations.

Procedure for Finding an Inverse Matrix

Step 1 Obtain a 1 in the first position on the main diagonal. If this is not possible (the first column has all zeros), the matrix does not have an inverse.

Step 2 Use the 1 in the first position on the main diagonal as a pivot to obtain zeros for all the entries in the first column under the 1.

Step 3 Obtain a 1 in the second position on the main diagonal (if possible).

Step 4 Use the 1 in the second position on the main diagonal as a pivot to obtain zeros for all the entries in the second column under the 1.

Step 5 Obtain a 1 in the third position on the main diagonal (if possible). Use this to obtain zeros for all entries below this 1. Continue this procedure until the main diagonal of M contains all 1's and all entries below those 1's are zeros.

Step 6 Use the 1 in the last position on the main diagonal to obtain zeros in all entries in the last column above the 1. Continue this process until all the entries above all the 1's on the diagonal of M are zero.

Step 7 The matrix in the position of I is the inverse of M.

EXAMPLE 5 Find the inverse, if possible, of the matrix

$$\begin{bmatrix} 1 & 2 \\ 1 & 4 \end{bmatrix}$$

Solution

$$\left[\begin{array}{cc:cc} 1 & 2 & 1 & 0 \\ 1 & 4 & 0 & 1 \end{array}\right] \sim \left[\begin{array}{cc:cc} 1 & 2 & 1 & 0 \\ 0 & 2 & -1 & 1 \end{array}\right]$$ Multiply row 1 by -1 and add to row 2.

$$\sim \left[\begin{array}{cc:cc} 1 & 2 & 1 & 0 \\ 0 & 1 & -\frac{1}{2} & \frac{1}{2} \end{array}\right]$$ Multiply row 2 by $\frac{1}{2}$.

$$\sim \left[\begin{array}{cc:cc} 1 & 0 & 2 & -1 \\ 0 & 1 & -\frac{1}{2} & \frac{1}{2} \end{array}\right]$$ Multiply row 2 by -2 and add to row 1.

The inverse is on the right side of the dotted line:

$$\begin{bmatrix} 2 & -1 \\ -\frac{1}{2} & \frac{1}{2} \end{bmatrix} = \frac{1}{2}\begin{bmatrix} 4 & -2 \\ -1 & 1 \end{bmatrix}$$ ■

EXAMPLE 6 Find the inverse of the matrix

$$A = \begin{bmatrix} 0 & 1 & 2 \\ 2 & -1 & 1 \\ -1 & 1 & 0 \end{bmatrix}$$

if it exists.

Solution

$$\left[\begin{array}{ccc:ccc} 0 & 1 & 2 & 1 & 0 & 0 \\ 2 & -1 & 1 & 0 & 1 & 0 \\ -1 & 1 & 0 & 0 & 0 & 1 \end{array}\right]$$

$$\sim \left[\begin{array}{ccc:ccc} -1 & 1 & 0 & 0 & 0 & 1 \\ 2 & -1 & 1 & 0 & 1 & 0 \\ 0 & 1 & 2 & 1 & 0 & 0 \end{array}\right]$$ Interchange row 1 and row 3.

$$\sim \left[\begin{array}{ccc:ccc} 1 & -1 & 0 & 0 & 0 & -1 \\ 2 & -1 & 1 & 0 & 1 & 0 \\ 0 & 1 & 2 & 1 & 0 & 0 \end{array}\right]$$ Multiply row 1 by -1.

$$\sim \left[\begin{array}{ccc:ccc} 1 & -1 & 0 & 0 & 0 & -1 \\ 0 & 1 & 1 & 0 & 1 & 2 \\ 0 & 1 & 2 & 1 & 0 & 0 \end{array}\right]$$ Multiply row 1 by -2 and add to row 2.

$$\sim\left[\begin{array}{ccc|ccc} 1 & -1 & 0 & 0 & 0 & -1 \\ 0 & 1 & 1 & 0 & 1 & 2 \\ 0 & 0 & 1 & 1 & -1 & -2 \end{array}\right]$$ Multiply row 2 by -1 and add to row 3.

$$\sim\left[\begin{array}{ccc|ccc} 1 & -1 & 0 & 0 & 0 & -1 \\ 0 & 1 & 0 & -1 & 2 & 4 \\ 0 & 0 & 1 & 1 & -1 & -2 \end{array}\right]$$ Multiply row 3 by -1 and add to row 2.

$$\sim\left[\begin{array}{ccc|ccc} 1 & 0 & 0 & -1 & 2 & 3 \\ 0 & 1 & 0 & -1 & 2 & 4 \\ 0 & 0 & 1 & 1 & -1 & -2 \end{array}\right]$$ Add row 2 to row 1.

$$A^{-1} = \begin{bmatrix} -1 & 2 & 3 \\ -1 & 2 & 4 \\ 1 & -1 & -2 \end{bmatrix}$$

■

In Example 4 we saw how a system of equations can be written in matrix form. We can now see how to solve a system of linear equations by using the inverse. Consider a system of n linear equations with n unknowns whose matrix of coefficients A has an inverse A^{-1}:

$AX = B$	Given system
$(A^{-1})AX = A^{-1}B$	Multiply both sides by A^{-1}
$(A^{-1}A)X = A^{-1}B$	Associative property
$I_nX = A^{-1}B$	Inverse property
$X = A^{-1}B$	Identity property

This says that to solve a system, simply multiply A^{-1} and B to find X.

EXAMPLE 7 Solve the system

$$\begin{cases} y + 2z = 0 \\ 2x - y + z = -1 \\ y - x = 1 \end{cases}$$

Solution Write in matrix form:

$$A = \begin{bmatrix} 0 & 1 & 2 \\ 2 & -1 & 1 \\ -1 & 1 & 0 \end{bmatrix} \quad X = \begin{bmatrix} x \\ y \\ z \end{bmatrix} \quad B = \begin{bmatrix} 0 \\ -1 \\ 1 \end{bmatrix}$$

From Example 6,

$$A^{-1} = \begin{bmatrix} -1 & 2 & 3 \\ -1 & 2 & 4 \\ 1 & -1 & -2 \end{bmatrix}$$

Thus

$$X = A^{-1}B = \begin{bmatrix} -1 & 2 & 3 \\ -1 & 2 & 4 \\ 1 & -1 & -2 \end{bmatrix} \begin{bmatrix} 0 \\ -1 \\ 1 \end{bmatrix} = \begin{bmatrix} 0-2+3 \\ 0-2+4 \\ 0+1-2 \end{bmatrix} = \begin{bmatrix} 1 \\ 2 \\ -1 \end{bmatrix}$$

Therefore $x = 1$, $y = 2$, and $z = -1$. ■

The method of solving a system by using the inverse matrix is very efficient if you know the inverse. Unfortunately, *finding* the inverse for one system is usually more work than using another method to solve the system. However, there are certain applications that yield the same system over and over, and the only thing to change is the constants. In this case the inverse method is worthwhile. And, finally, computers can often find approximations for inverse matrices quite easily, so this method might be appropriate if one is programming a computer.

PROBLEM SET 7.4

A

In Problems 1–20, find the indicated matrices if possible.

$$A = \begin{bmatrix} 1 & 2 \\ 4 & 0 \\ -1 & 3 \\ 2 & 1 \end{bmatrix} \qquad B = \begin{bmatrix} 4 & 2 \\ -1 & 3 \end{bmatrix}$$

$$C = \begin{bmatrix} 1 & 0 & 0 & 0 \\ 0 & 1 & 0 & 0 \\ 0 & 0 & 1 & 0 \\ 0 & 0 & 0 & 1 \end{bmatrix}$$

$$D = \begin{bmatrix} 4 & 1 & 3 & 6 \\ -1 & 0 & -2 & 3 \end{bmatrix} \qquad E = \begin{bmatrix} 1 & 0 & 2 \\ 3 & -1 & 2 \\ 4 & 1 & 0 \end{bmatrix}$$

$$F = \begin{bmatrix} 1 & 4 & 0 \\ 3 & -1 & 2 \\ -2 & 1 & 5 \end{bmatrix} \qquad G = \begin{bmatrix} 8 & 1 & 6 \\ 3 & 5 & 7 \\ 4 & 9 & 2 \end{bmatrix}$$

1. $E + F$
2. EF
3. EG
4. $EF + EG$
5. $E(F + G)$
6. $2E - G$
7. FG
8. GF
9. $(EF)G$
10. $E(FG)$
11. AB
12. BA
13. B^2
14. CA
15. BD
16. DB
17. $(B + C)A$
18. $BA + CA$
19. C^3
20. CD

B

Find the inverse of each matrix in Problems 21–29, if it exists.

21. $\begin{bmatrix} 4 & -7 \\ -1 & 2 \end{bmatrix}$

22. $\begin{bmatrix} 8 & 6 \\ -2 & 4 \end{bmatrix}$

23. $\begin{bmatrix} 1 & 3 \\ 2 & 0 \end{bmatrix}$

24. $\begin{bmatrix} 1 & 0 & 2 \\ 2 & 1 & 0 \\ 0 & -2 & 9 \end{bmatrix}$

25. $\begin{bmatrix} 6 & 1 & 20 \\ 1 & -1 & 0 \\ 0 & 1 & 3 \end{bmatrix}$

26. $\begin{bmatrix} 4 & 1 & 0 \\ 2 & -1 & 4 \\ -3 & 2 & 1 \end{bmatrix}$

27. $\begin{bmatrix} 1 & 0 & 0 & 1 \\ 0 & 2 & 0 & 0 \\ 0 & 0 & 0 & 1 \\ 2 & 0 & 1 & 0 \end{bmatrix}$

28. $\begin{bmatrix} 0 & 1 & 2 & 0 \\ 0 & 0 & 0 & 1 \\ 1 & 1 & 3 & 0 \\ 2 & 4 & 0 & 0 \end{bmatrix}$

29. $\begin{bmatrix} 1 & 2 & 0 & 0 \\ 0 & 0 & 1 & 0 \\ 1 & 3 & 0 & 1 \\ 4 & 0 & 0 & 2 \end{bmatrix}$

Solve the systems in Problems 30–61 by solving the corresponding matrix equation with an inverse if possible. Problems 30–35 use the inverse found in Problem 21.

30. $\begin{cases} 4x - 7y = -2 \\ -x + 2y = 1 \end{cases}$

31. $\begin{cases} 4x - 7y = -65 \\ -x + 2y = 18 \end{cases}$

32. $\begin{cases} 4x - 7y = 48 \\ -x + 2y = -13 \end{cases}$

33. $\begin{cases} 4x - 7y = 2 \\ -x + 2y = 3 \end{cases}$

34. $\begin{cases} 4x - 7y = 5 \\ -x + 2y = 4 \end{cases}$

35. $\begin{cases} 4x - 7y = -3 \\ -x + 2y = 8 \end{cases}$

Problems 36–41 use the inverse found in Problem 22.

36. $\begin{cases} 8x + 6y = 12 \\ -2x + 4y = -14 \end{cases}$

37. $\begin{cases} 8x + 6y = 16 \\ -2x + 4y = 18 \end{cases}$

38. $\begin{cases} 8x + 6y = -6 \\ -2x + 4y = -26 \end{cases}$

39. $\begin{cases} 8x + 6y = -28 \\ -2x + 4y = 18 \end{cases}$

40. $\begin{cases} 8x + 6y = -26 \\ -2x + 4y = 12 \end{cases}$

41. $\begin{cases} 8x + 6y = -36 \\ -2x + 4y = -2 \end{cases}$

Problems 42–47 all use the same inverse.

42. $\begin{cases} 2x + 3y = 9 \\ x - 6y = -3 \end{cases}$

43. $\begin{cases} 2x + 3y = 2 \\ x - 6y = 16 \end{cases}$

44. $\begin{cases} 2x + 3y = 2 \\ x - 6y = -14 \end{cases}$

45. $\begin{cases} 2x + 3y = 9 \\ x - 6y = 42 \end{cases}$

46. $\begin{cases} 2x + 3y = -22 \\ x - 6y = 49 \end{cases}$

47. $\begin{cases} 2x + 3y = 12 \\ x - 6y = -24 \end{cases}$

Problems 48–53 use the inverse found in Problem 24.

48. $\begin{cases} x + 2z = 7 \\ 2x + y = 16 \\ -2y + 9z = -3 \end{cases}$

49. $\begin{cases} x + 2z = 4 \\ 2x + y = 0 \\ -2y + 9z = 19 \end{cases}$

50. $\begin{cases} x + 2z = 7 \\ 2x + y = 0 \\ -2y + 9z = 31 \end{cases}$

51. $\begin{cases} x + 2z = 7 \\ 2x + y = 1 \\ -2y + 9z = 28 \end{cases}$

52. $\begin{cases} x + 2z = 12 \\ 2x + y = 0 \\ -2y + 9z = 10 \end{cases}$

53. $\begin{cases} x + 2z = 5 \\ 2x + y = 8 \\ -2y + 9z = 9 \end{cases}$

Problems 54–56 use the inverse found in Problem 25.

54. $\begin{cases} 6x + y + 20z = 27 \\ x - y = 0 \\ y + 3z = 4 \end{cases}$

55. $\begin{cases} 6x + y + 20z = 14 \\ x - y = 1 \\ y + 3z = 1 \end{cases}$

56. $\begin{cases} 6x + y + 20z = 11 \\ x - y = 5 \\ y + 3z = -3 \end{cases}$

Problems 57–59 use the inverse found in Problem 26.

57. $\begin{cases} 4x + y = 6 \\ 2x - y + 4z = 12 \\ -3x + 2y + z = 4 \end{cases}$

58. $\begin{cases} 4x + y = 7 \\ 2x - y + 4z = -11 \\ -3x + 2y + z = -12 \end{cases}$

59. $\begin{cases} 4x + y = -10 \\ 2x - y + 4z = 20 \\ -3x + 2y + z = 20 \end{cases}$

Problems 60–62 use the inverse found in Problem 29.

60. $\begin{cases} x + 2y = 5 \\ z = 3 \\ x + 3y + w = 9 \\ 4x + 2w = 8 \end{cases}$

61. $\begin{cases} x + 2y = 0 \\ z = -4 \\ x + 3y + w = 4 \\ 4x + 2w = -2 \end{cases}$

62. $\begin{cases} x + 2y = 7 \\ z = -7 \\ x + 3y + w = 16 \\ 4x + 2w = -4 \end{cases}$

C

63. Let

$$M = \begin{bmatrix} a_{11} & a_{12} \\ a_{21} & a_{22} \end{bmatrix}$$

Show that if $|M| \neq 0$, then the inverse exists.

64. Let M be defined as in Problem 64. Show that if the inverse exists, then $|M| \neq 0$.

65. If M and N are nonsingular, then MN is nonsingular. Show that the inverse of MN is $(MN)^{-1} = N^{-1}M^{-1}$.

66. If M is nonsingular, show that if $MN = MP$ then $N = P$.

67. Prove that if M is nonsingular and $MN = M$, then $N = I_3$ for square matrices of order 3.

7.5 MATRIX SOLUTION OF SYSTEMS

Now that we have seen a variety of different methods for solving systems of linear equations, we will conclude our study of this topic by introducing the most general method yet: the **Gauss–Jordan method**. It is based on the method of solving a system by linear combinations but uses matrix notation to increase the efficiency of this method when working with more complicated systems. Instead of augmenting the matrix of the coefficients of the system as you did when finding the inverse, simply augment A_n with the constant terms. Next perform elementary row operations to transform A_n to I_n. The solution of the system will then be obvious, as shown by Examples 1 and 2.

EXAMPLE 1 Solve the system

$$\begin{cases} x + 2y - z = 0 \\ 2x + 3y - 2z = 3 \\ -x - 4y + 3z = -2 \end{cases}$$

Solution

$$\left[\begin{array}{rrr|r} 1 & 2 & -1 & 0 \\ 2 & 3 & -2 & 3 \\ -1 & -4 & 3 & -2 \end{array}\right] \sim \left[\begin{array}{rrr|r} 1 & 2 & -1 & 0 \\ 0 & -1 & 0 & 3 \\ 0 & -2 & 2 & -2 \end{array}\right]$$

Add -2 times the first row to the second row and add the first row to the third row.

$$\sim \left[\begin{array}{rrr|r} 1 & 2 & -1 & 0 \\ 0 & 1 & 0 & -3 \\ 0 & -2 & 2 & -2 \end{array}\right]$$

Multiply the second row by -1.

$$\sim \left[\begin{array}{rrr|r} 1 & 2 & -1 & 0 \\ 0 & 1 & 0 & -3 \\ 0 & 0 & 2 & -8 \end{array}\right]$$

Add two times the second row to the third row.

$$\sim \left[\begin{array}{rrr|r} 1 & 2 & -1 & 0 \\ 0 & 1 & 0 & -3 \\ 0 & 0 & 1 & -4 \end{array}\right]$$

Multiply the third row by $\frac{1}{2}$.

$$\sim \left[\begin{array}{rrr|r} 1 & 2 & 0 & -4 \\ 0 & 1 & 0 & -3 \\ 0 & 0 & 1 & -4 \end{array}\right]$$

Add the third row to the first row.

$$\sim \left[\begin{array}{rrr|r} 1 & 0 & 0 & 2 \\ 0 & 1 & 0 & -3 \\ 0 & 0 & 1 & -4 \end{array}\right]$$

Multiply the second row by -2 and add it to the first row.

This is equivalent to the system

$$\begin{aligned} 1 \cdot x + 0 \cdot y + 0 \cdot z &= 2 \\ 0 \cdot x + 1 \cdot y + 0 \cdot z &= -3 \\ 0 \cdot x + 0 \cdot y + 1 \cdot z &= -4 \end{aligned} \qquad \text{or} \qquad \begin{cases} x = 2 \\ y = -3 \\ z = -4 \end{cases}$$

The solution is now obvious: $\mathbf{(x, y, z) = (2, -3, -4)}$. ■

EXAMPLE 2 Solve the system

$$\begin{cases} w - x + 2y + 3z = -6 \\ 2w + x - y + 2z = -4 \\ w - 3x + y - z = 0 \\ w + 2x + 3y + 4z = -1 \end{cases}$$

Solution We will show the steps without mentioning which of the elementary row operations are being applied. Study this example carefully to make sure you can follow each step:

$$\left[\begin{array}{rrrr|r} 1 & -1 & 2 & 3 & -6 \\ 2 & 1 & -1 & 2 & -4 \\ 1 & -3 & 1 & -1 & 0 \\ 1 & 2 & 3 & 4 & -1 \end{array}\right] \sim \left[\begin{array}{rrrr|r} 1 & -1 & 2 & 3 & -6 \\ 0 & 3 & -5 & -4 & 8 \\ 0 & -2 & -1 & -4 & 6 \\ 0 & 3 & 1 & 1 & 5 \end{array}\right]$$

$$\sim \left[\begin{array}{rrrr|r} 1 & -1 & 2 & 3 & -6 \\ 0 & 1 & -6 & -8 & 14 \\ 0 & -2 & -1 & -4 & 6 \\ 0 & 3 & 1 & 1 & 5 \end{array}\right] \sim \left[\begin{array}{rrrr|r} 1 & -1 & 2 & 3 & -6 \\ 0 & 1 & -6 & -8 & 14 \\ 0 & 0 & -13 & -20 & 34 \\ 0 & 0 & 19 & 25 & -37 \end{array}\right]$$

$$\sim \left[\begin{array}{rrrr|r} 1 & -1 & 2 & 3 & -6 \\ 0 & 1 & -6 & -8 & 14 \\ 0 & 0 & -13 & -20 & 34 \\ 0 & 0 & 6 & 5 & -3 \end{array}\right] \sim \left[\begin{array}{rrrr|r} 1 & -1 & 2 & 3 & -6 \\ 0 & 1 & -6 & -8 & 14 \\ 0 & 0 & -1 & -10 & 28 \\ 0 & 0 & 6 & 5 & -3 \end{array}\right]$$

$$\sim \left[\begin{array}{rrrr|r} 1 & -1 & 2 & 3 & -6 \\ 0 & 1 & -6 & -8 & 14 \\ 0 & 0 & 1 & 10 & -28 \\ 0 & 0 & 6 & 5 & -3 \end{array}\right] \sim \left[\begin{array}{rrrr|r} 1 & -1 & 2 & 3 & -6 \\ 0 & 1 & -6 & -8 & 14 \\ 0 & 0 & 1 & 10 & -28 \\ 0 & 0 & 0 & -55 & 165 \end{array}\right]$$

$$\sim \left[\begin{array}{rrrr|r} 1 & -1 & 2 & 3 & -6 \\ 0 & 1 & -6 & -8 & 14 \\ 0 & 0 & 1 & 10 & -28 \\ 0 & 0 & 0 & 1 & -3 \end{array}\right] \sim \left[\begin{array}{rrrr|r} 1 & -1 & 2 & 0 & 3 \\ 0 & 1 & -6 & 0 & -10 \\ 0 & 0 & 1 & 0 & 2 \\ 0 & 0 & 0 & 1 & -3 \end{array}\right]$$

$$\sim \left[\begin{array}{rrrr|r} 1 & -1 & 0 & 0 & -1 \\ 0 & 1 & 0 & 0 & 2 \\ 0 & 0 & 1 & 0 & 2 \\ 0 & 0 & 0 & 1 & -3 \end{array}\right] \sim \left[\begin{array}{rrrr|r} 1 & 0 & 0 & 0 & 1 \\ 0 & 1 & 0 & 0 & 2 \\ 0 & 0 & 1 & 0 & 2 \\ 0 & 0 & 0 & 1 & -3 \end{array}\right]$$

Thus **$w = 1$, $x = 2$, $y = 2$, and $z = -3$.** ■

Our discussion concludes with two applied problems.

EXAMPLE 3 A rancher has to mix three types of feed for her cattle. The following analysis shows the amounts per bag (100 lb) of grain:

Grain	Protein	Carbohydrates	Sodium
A	7 lb	88 lb	1 lb
B	6 lb	90 lb	1 lb
C	10 lb	70 lb	2 lb

How many bags of each type of grain should she mix to provide 71 lb of protein, 854 lb of carbohydrates, and 12 lb of sodium?

Solution Let a, b, and c be the number of bags of grains A, B, and C respectively that are needed for the mixture. Then:

Grain	Protein	Carbohydrates	Sodium
A	$7a$	$88a$	a
B	$6b$	$90b$	b
C	$10c$	$70c$	$2c$
Total	71	854	12

Thus

$$\begin{cases} 7a + 6b + 10c = 71 \\ 88a + 90b + 70c = 854 \\ a + b + 2c = 12 \end{cases}$$

$$\left[\begin{array}{ccc|c} 7 & 6 & 10 & 71 \\ 88 & 90 & 70 & 854 \\ 1 & 1 & 2 & 12 \end{array}\right] \sim \left[\begin{array}{ccc|c} 1 & 1 & 2 & 12 \\ 88 & 90 & 70 & 854 \\ 7 & 6 & 10 & 71 \end{array}\right]$$

$$\sim \left[\begin{array}{ccc|c} 1 & 1 & 2 & 12 \\ 0 & 2 & -106 & -202 \\ 0 & -1 & -4 & -13 \end{array}\right] \sim \left[\begin{array}{ccc|c} 1 & 1 & 2 & 12 \\ 0 & 1 & -53 & -101 \\ 0 & -1 & -4 & -13 \end{array}\right]$$

$$\sim \left[\begin{array}{ccc|c} 1 & 1 & 2 & 12 \\ 0 & 1 & -53 & -101 \\ 0 & 0 & -57 & -114 \end{array}\right] \sim \left[\begin{array}{ccc|c} 1 & 1 & 2 & 12 \\ 0 & 1 & -53 & -101 \\ 0 & 0 & 1 & 2 \end{array}\right]$$

$$\sim \left[\begin{array}{ccc|c} 1 & 1 & 0 & 8 \\ 0 & 1 & 0 & 5 \\ 0 & 0 & 1 & 2 \end{array}\right] \sim \left[\begin{array}{ccc|c} 1 & 0 & 0 & 3 \\ 0 & 1 & 0 & 5 \\ 0 & 0 & 1 & 2 \end{array}\right]$$

Mix three bags of grain A, five bags of grain B, and two bags of grain C. ■

EXAMPLE 4 Suppose the equation of a certain parabola has the form

$$y = ax^2 + bx + c$$

Find the equation of the parabola passing through $(-1, -6)$, $(2, 9)$, and $(-2, -3)$.

Solution If a curve passes through a point, then the coordinates of that point must satisfy the equation

$$\begin{aligned} (-1, -6)&: \quad -6 = a(-1)^2 + b(-1) + c \\ (2, 9)&: \quad 9 = a(2)^2 + b(2) + c \\ (-2, -3)&: \quad -3 = a(-2)^2 + b(-2) + c \end{aligned}$$

In standard form the system is

$$\begin{cases} a - b + c = -6 \\ 4a + 2b + c = 9 \\ 4a - 2b + c = -3 \end{cases}$$

$$\left[\begin{array}{ccc:c} 1 & -1 & 1 & -6 \\ 4 & 2 & 1 & 9 \\ 4 & -2 & 1 & -3 \end{array}\right] \sim \left[\begin{array}{ccc:c} 1 & -1 & 1 & -6 \\ 0 & 4 & 0 & 12 \\ 4 & -2 & 1 & -3 \end{array}\right] \sim \left[\begin{array}{ccc:c} 1 & -1 & 1 & -6 \\ 0 & 1 & 0 & 3 \\ 0 & 2 & -3 & 21 \end{array}\right]$$

$$\sim \left[\begin{array}{ccc:c} 1 & -1 & 1 & -6 \\ 0 & 1 & 0 & 3 \\ 0 & 0 & -3 & 15 \end{array}\right] \sim \left[\begin{array}{ccc:c} 1 & -1 & 1 & -6 \\ 0 & 1 & 0 & 3 \\ 0 & 0 & 1 & -5 \end{array}\right] \sim \left[\begin{array}{ccc:c} 1 & -1 & 0 & -1 \\ 0 & 1 & 0 & 3 \\ 0 & 0 & 1 & -5 \end{array}\right]$$

$$\sim \left[\begin{array}{ccc:c} 1 & 0 & 0 & 2 \\ 0 & 1 & 0 & 3 \\ 0 & 0 & 1 & -5 \end{array}\right]$$

Thus $a = 2$, $b = 3$, and $c = -5$, so the equation of the parabola is

$$y = 2x^2 + 3x - 5$$

■

PROBLEM SET 7.5

A

Given the matrices in Problems 1–6, perform elementary row operations to obtain a 1 in the row 1, column 1 position.

1. $\left[\begin{array}{ccc:c} 3 & 1 & 2 & 1 \\ 0 & 2 & 4 & 5 \\ 1 & 3 & -4 & 9 \end{array}\right]$ **2.** $\left[\begin{array}{ccc:c} -2 & 3 & 5 & 9 \\ 1 & 0 & 2 & -8 \\ 0 & 1 & 0 & 5 \end{array}\right]$

3. $\left[\begin{array}{ccc:c} 2 & 4 & 10 & -12 \\ 6 & 3 & 4 & 6 \\ 10 & -1 & 0 & 1 \end{array}\right]$ **4.** $\left[\begin{array}{ccc:c} 5 & 20 & 15 & 6 \\ 7 & -5 & 3 & 2 \\ 12 & 0 & 1 & 4 \end{array}\right]$

5. $\left[\begin{array}{ccc:c} 5 & 6 & -3 & 4 \\ 4 & 1 & 9 & 2 \\ 7 & 6 & 1 & 3 \end{array}\right]$ **6.** $\left[\begin{array}{ccc:c} 4 & 8 & 5 & 9 \\ 3 & 2 & 1 & -4 \\ -2 & 5 & 0 & 1 \end{array}\right]$

Given the matrices in Problems 7–12, perform elementary row operations to obtain zeros under the one in the first column.

7. $\left[\begin{array}{ccc:c} 1 & 2 & -3 & 0 \\ 0 & 3 & 1 & 4 \\ 2 & 5 & 1 & 6 \end{array}\right]$ **8.** $\left[\begin{array}{ccc:c} 1 & 3 & -5 & 6 \\ -3 & 4 & 1 & 2 \\ 0 & 5 & 1 & 3 \end{array}\right]$

9. $\left[\begin{array}{ccc:c} 1 & 2 & 4 & 1 \\ -2 & 5 & 0 & 2 \\ -4 & 5 & 1 & 3 \end{array}\right]$ **10.** $\left[\begin{array}{ccc:c} 1 & 5 & 3 & 2 \\ 2 & 3 & -1 & 4 \\ 3 & 2 & 1 & 0 \end{array}\right]$

11. $\left[\begin{array}{cccc:c} 1 & 4 & -1 & 3 & 3 \\ -3 & 4 & 6 & 4 & 0 \\ 5 & 1 & 9 & 1 & -2 \\ 1 & 0 & 2 & 0 & 0 \end{array}\right]$

12. $\left[\begin{array}{cccc:c} 1 & 8 & 5 & 2 & -1 \\ 2 & 0 & 7 & 5 & -1 \\ 6 & -2 & 7 & 7 & 5 \\ 3 & 1 & 0 & 0 & 0 \end{array}\right]$

Given the matrices in Problems 13–18, perform elementary row operations to obtain a one in the second row, second column without changing the entries in the first column.

13. $\left[\begin{array}{ccc:c} 1 & 3 & 5 & 2 \\ 0 & 2 & 6 & -8 \\ 0 & 3 & 4 & 1 \end{array}\right]$ **14.** $\left[\begin{array}{ccc:c} 1 & 5 & -3 & 5 \\ 0 & 3 & 9 & -15 \\ 0 & 2 & 1 & 5 \end{array}\right]$

15. $\left[\begin{array}{ccc:c} 1 & 4 & -1 & 6 \\ 0 & 5 & 1 & 3 \\ 0 & 4 & 6 & 5 \end{array}\right]$ **16.** $\left[\begin{array}{ccc:c} 1 & 3 & -2 & 0 \\ 0 & 4 & 2 & 9 \\ 0 & 3 & 6 & 1 \end{array}\right]$

17. $\left[\begin{array}{ccc:c} 1 & 3 & -2 & 4 \\ 0 & 5 & 1 & 3 \\ 0 & 7 & 9 & 2 \end{array}\right]$ **18.** $\left[\begin{array}{ccc:c} 1 & 3 & -2 & 0 \\ 0 & 4 & 2 & 9 \\ 0 & 10 & -3 & 4 \end{array}\right]$

Given the matrices in Problems 19–21, perform elementary row operations to obtain a zero (or zeros) under the one in the second column without changing the entries in the first column.

19. $\left[\begin{array}{ccc|c} 1 & 5 & -3 & 2 \\ 0 & 1 & 4 & 5 \\ 0 & 3 & 4 & 2 \end{array}\right]$

20. $\left[\begin{array}{cccc|c} 1 & 6 & -3 & 4 & 1 \\ 0 & 1 & 7 & 3 & 0 \\ 0 & 3 & 4 & 0 & -2 \\ 0 & -2 & 3 & 1 & 0 \end{array}\right]$

21. $\left[\begin{array}{cccc|c} 1 & 7 & 6 & 6 & 2 \\ 0 & 1 & 9 & 2 & 1 \\ 0 & -4 & 6 & 1 & 1 \\ 0 & 5 & 8 & 10 & 3 \end{array}\right]$

Given the matrices in Problems 22–24, perform elementary row operations to obtain a one in the third row, third column without changing the entries in the first two columns.

22. $\left[\begin{array}{ccc|c} 1 & 3 & 4 & 5 \\ 0 & 1 & -3 & 6 \\ 0 & 0 & 5 & 10 \end{array}\right]$ **23.** $\left[\begin{array}{ccc|c} 1 & -3 & 4 & -5 \\ 0 & 1 & 3 & 6 \\ 0 & 0 & 8 & 12 \end{array}\right]$

24. $\left[\begin{array}{ccc|c} 1 & -2 & 5 & 6 \\ 0 & 1 & 9 & 1 \\ 0 & 0 & -3 & 5 \end{array}\right]$

Given the matrices in Problems 25–27, perform elementary row operations to obtain zeros above the one in the third column without changing the entries in the first or second columns.

25. $\left[\begin{array}{ccc|c} 1 & 3 & -1 & 5 \\ 0 & 1 & 2 & 6 \\ 0 & 0 & 1 & 4 \end{array}\right]$ **26.** $\left[\begin{array}{ccc|c} 1 & 6 & -3 & -2 \\ 0 & 1 & 4 & 5 \\ 0 & 0 & 1 & 3 \end{array}\right]$

27. $\left[\begin{array}{cccc|c} 1 & 6 & 3 & 0 & -2 \\ 0 & 1 & -2 & 0 & 1 \\ 0 & 0 & 1 & 0 & 0 \\ 0 & 0 & 0 & 1 & -6 \end{array}\right]$

Given the matrices in Problems 28–30, perform elementary row operations to obtain a zero above the one in the second column and interpret the solution to the system if the variables are x, y, and z. For example,

$$\left[\begin{array}{ccc|c} 1 & 3 & 0 & 5 \\ 0 & 1 & 0 & 2 \\ 0 & 0 & 1 & 3 \end{array}\right] \sim \left[\begin{array}{ccc|c} 1 & 0 & 0 & -1 \\ 0 & 1 & 0 & 2 \\ 0 & 0 & 1 & 3 \end{array}\right]$$

so the solution is $(x, y, z) = (-1, 2, 3)$.

28. $\left[\begin{array}{ccc|c} 1 & 3 & 0 & 4 \\ 0 & 1 & 0 & -5 \\ 0 & 0 & 1 & 3 \end{array}\right]$ **29.** $\left[\begin{array}{ccc|c} 1 & -8 & 0 & 3 \\ 0 & 1 & 0 & 4 \\ 0 & 0 & 1 & -21 \end{array}\right]$

30. $\left[\begin{array}{ccc|c} 1 & -4 & 0 & 3 \\ 0 & 1 & 0 & -\frac{8}{3} \\ 0 & 0 & 1 & \frac{3}{5} \end{array}\right]$

Solve the systems given in Problems 31–45 by using the Gauss–Jordan method of solution.

31. $\begin{cases} 4x - 7y = -2 \\ -x + 2y = 1 \end{cases}$ **32.** $\begin{cases} 4x - 7y = -5 \\ 2y - x = 1 \end{cases}$

33. $\begin{cases} 8x + 6y = 17 \\ y - x = \frac{1}{2} \end{cases}$ **34.** $\begin{cases} 8x + 6y = 14 \\ 4y - 2x = 24 \end{cases}$

35. $\begin{cases} y = -\frac{2}{3}x + 3 \\ x - 6y = -3 \end{cases}$ **36.** $\begin{cases} y = \frac{1}{2}x + 1 \\ x + 3y = 18 \end{cases}$

37. $\begin{cases} x + y + z = 6 \\ 2x - y + z = 3 \\ x - 2y - 3z = -12 \end{cases}$ **38.** $\begin{cases} 2x - y + z = 3 \\ x - 3y + 2z = 7 \\ x - y - z = -1 \end{cases}$

39. $\begin{cases} x + y + z = 4 \\ x + 3y + 2z = 4 \\ x - 2y + z = 7 \end{cases}$ **40.** $\begin{cases} x + 2z = 13 \\ 2x + y = 8 \\ -2y + 9z = 41 \end{cases}$

41. $\begin{cases} x + 2z = 7 \\ x + y = 11 \\ -2y + 9z = -3 \end{cases}$ **42.** $\begin{cases} 4x + y + 2z = 7 \\ x + 2y = 0 \\ 3x - y - z = 7 \end{cases}$

43. $\begin{cases} 6x + y + 20z = 27 \\ x - y = 0 \\ y + z = 2 \end{cases}$ **44.** $\begin{cases} 2x - y + 4z = 13 \\ 3x + 6y = 0 \\ 2y - 3z = 3 + 3x \end{cases}$

45. $\begin{cases} 3x - 2y + z = 5 \\ 5x - 3y = 24 \\ 2y + z = -5 \end{cases}$

B

Solve the systems in Problems 46–54 by using the Gauss–Jordan method of solution.

46. $\begin{cases} 2x + 2y + 3z = 1 \\ 2x - z = -11 \\ 3y + 2z = 6 \end{cases}$ **47.** $\begin{cases} x + 2y + z = 1 \\ x - 3y - 2z = 2 \\ 3x - 2y + z = 3 \end{cases}$

48. $\begin{cases} 2x - y + z = 4 \\ 3x - 2y + 2z = 3 \\ x - y + 3z = 2 \end{cases}$ **49.** $\begin{cases} 2x + 10y - 6z = 28 \\ 5x - 3y + z = -14 \\ x + 5y - 3z = 14 \end{cases}$

50. $\begin{cases} x + 5y - 3z = 2 \\ 2x + 2y + z = -1 \\ 3x - y + 5z = 3 \end{cases}$ **51.** $\begin{cases} x + y + z = 1 \\ 2x - y + z = 2 \\ 3x + 2y - 3z = 3 \end{cases}$

52. $\begin{cases} x + 2y = 5 \\ z = 3 \\ w + x + 3y = 6 \\ 2w + 4x = 2 \end{cases}$

53. $\begin{cases} x + y + z + w = -2 \\ 2x + y + w = 2 \\ w + 3x + z = 5 \\ 2x + y + 3z = -5 \end{cases}$

54. $\begin{cases} 2x + y + z + w = 3 \\ x - y - z + 2w = -3 \\ x + 3y + 2z + 4w = -12 \\ x - y - z + w = 1 \end{cases}$

Suppose the equation of a certain parabola has the form $y = ax^2 + bx + c$. Find the equation of the parabola passing through the points given in Problems 55–57.

55. (0, 5), (−1, 2), (3, 26)

56. (1, −2), (−2, −14), (3, −4)

57. (4, −4), (5, −5), (7, −1)

C

58. ***Agriculture*** In order to control a certain type of crop disease, it is necessary to use 23 gal of chemical *A* and 34 gal of chemical *B*. The dealer can order commercial spray I, each container of which holds 5 gal of chemical *A* and 2 gal of chemical *B*, and commercial spray II, each container of which holds 2 gal of chemical *A* and 7 gal of chemical *B*. How many containers of each type of commercial spray should be used to attain exactly the right proportion of chemicals needed?

59. ***Business*** In order to manufacture a certain alloy, it is necessary to use 33 kg (kilograms) of metal *A* and 56 kg of metal *B*. It is cheaper for the manufacturer if she buys and mixes an alloy, each bar of which contains 3 kg of metal *A* and 5 kg of metal *B*, along with another alloy, each bar of which contains 4 kg of metal *A* and 7 kg of metal *B*. How much of the two alloys should she use to produce the alloy desired?

60. ***Business*** A candy maker mixes chocolate, milk, and coconut to produce three kinds of candy—I, II, and III—with the following proportions:

I: 7 lb chocolate, 5 gal milk, 1 oz almonds
II: 3 lb chocolate, 2 gal milk, 2 oz almonds
III: 4 lb chocolate, 3 gal milk, 3 oz almonds

If 67 lb of chocolate, 48 gal of milk, and 32 oz of almonds are available, how much of each kind of candy can be produced?

61. ***Business*** Using the data from Problem 60, how much of each type of candy can be produced with 62 lb of chocolate, 44 gal of milk, and 32 oz of almonds?

7.6 SYSTEMS OF INEQUALITIES

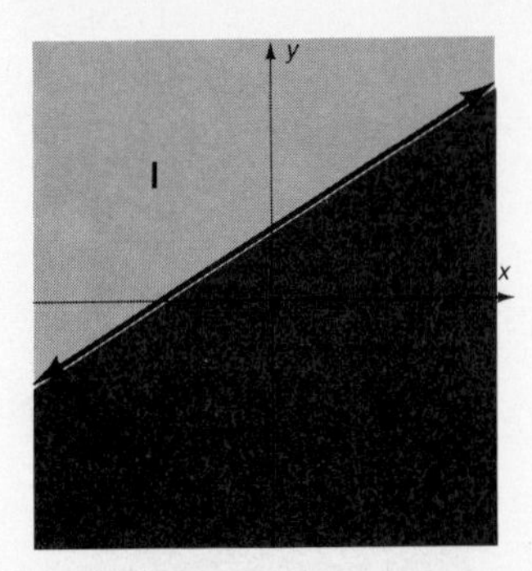

Figure 7.5 Half-planes

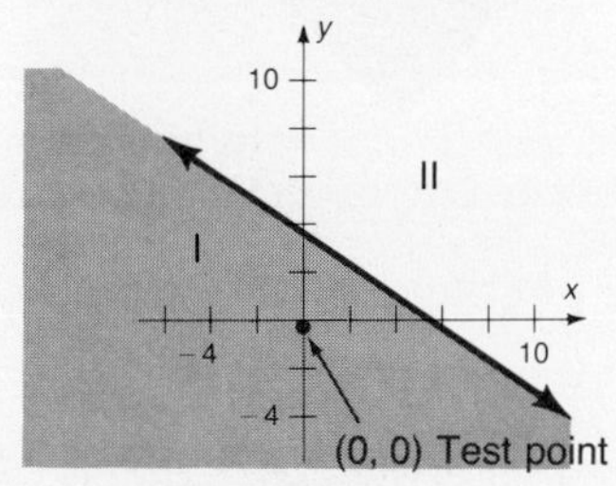

Figure 7.6

In previous sections we have discussed the simultaneous solution of a system of equations. In this section we will discuss the simultaneous solution of a **system of linear inequalities**. Let us begin by considering some preliminary ideas. The **solution** for an inequality in *x* and *y* is defined as an ordered pair that when substituted for (x, y) makes the inequality true. To **solve an inequality** means to find the set of all solutions, which is usually an infinite set. We therefore usually represent the solution graphically so that the **graph of the inequality** is the graph of all solutions of that inequality.

Graphing a linear inequality is similar to graphing a linear equation. A line divides the plane into three regions as shown in Figure 7.5. Regions I and II in Figure 7.5 are called **open half-planes**. Thus the three regions determined by the line are the open half-planes labeled I and II and the set of points on the line. The line is called the **boundary** of each open half-plane. An open half-plane, along with its boundary, is called a **closed half-plane**.

Every linear inequality with one or two variables determines an associated linear equation that is the boundary for the solution set of the inequality. For example, $2x + 3y \le 12$ has a boundary line $2x + 3y = 12$. To graph the solution set of an inequality, begin by graphing the boundary line as shown in Figure 7.6. Next decide whether the solution is half-plane I or half-plane II. To do this, **choose *any* point not on the boundary**. For example, choose the point (0, 0) in Figure 7.6—this choice, if not on the boundary, is usually the best because of the ease of the arithmetic

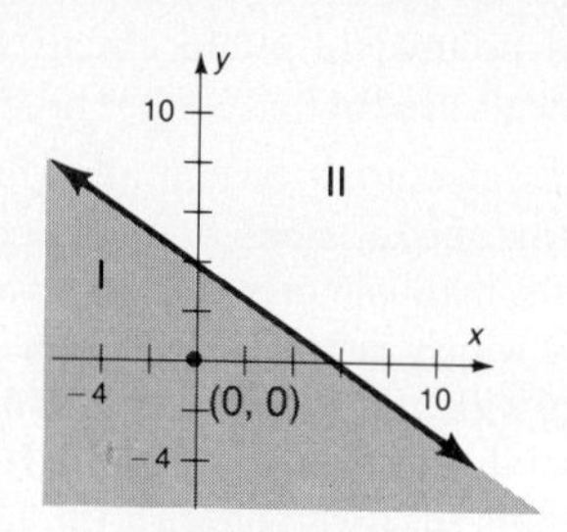

Figure 7.7 Graph of $2x + 3y \le 12$

involved. Notice from Figure 7.6 that the point $(0, 0)$ is in half-plane I. If $(0, 0)$ makes the inequality true, then the solution is the half-plane containing $(0, 0)$; if $(0, 0)$ makes the inequality false, then the solution set is the half-plane not containing $(0, 0)$. Checking by substituting $(0, 0)$ into $2x + 3y \le 12$, you have

$$2(\mathbf{0}) + 3(\mathbf{0}) \le 12 \qquad \text{True}$$

Therefore the solution set is the area shown as half-plane I. This is the shaded portion of Figure 7.7. Notice that the solution set is the closed half-plane I, since it includes the boundary. This is shown on the graph by using a solid line for the boundary. If the boundary is not included (when the inequality symbols are $<$ or $>$), a dotted line is used to indicate the boundary.

EXAMPLE 1 Graph $150x - 75y > 1875$.

Solution Graph the boundary line $150x - 75y = 1875$:

$$y = 2x - 25$$

Check some point, say $(0, 0)$, not on the boundary and test:

$$150(\mathbf{0}) - 75(\mathbf{0}) > 1875 \qquad \text{False}$$

The solution is the half-plane not including $(0, 0)$. It is the shaded portion of Figure 7.8.

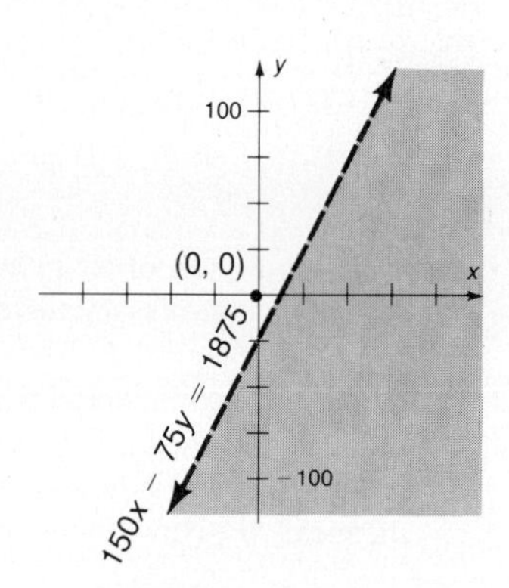

Figure 7.8 Graph of $150x - 75y > 1875$ ■

Many curves divide the plane into two regions with the curve serving as a boundary. The test point method illustrated for lines works efficiently in many different settings, as illustrated by Examples 2 and 3.

EXAMPLE 2 Graph $y \ge |x + 3| + 2$.

Solution Begin by graphing the boundary:

$$y = |x + 3| + 2$$

We recognize this as the *form* $y = |x|$ translated to the point $(-3, 2)$. This is drawn as a solid curve in Figure 7.9. Now this curve divides the plane into two regions, so plot a test point $(0, 0)$ and check the truth or falsity in the given inequality:

$$y \ge |x + 3| + 2$$
$$\mathbf{0} \ge |\mathbf{0} + 3| + 2$$
$$0 \ge 5 \qquad \text{False}$$

Since it is false, shade the region *not* containing the test point $(0, 0)$ as shown in Figure 7.9.

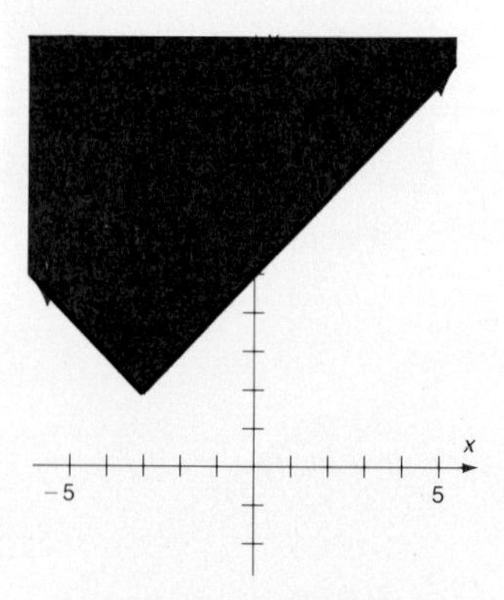

Figure 7.9 Graph of $y \ge |x + 3| + 2$ ■

EXAMPLE 3 Sketch the graph $y > x^2 + 2x + 3$.

Solution Consider the associated equation

$$y = x^2 + 2x + 3$$

or $y - 2 = (x + 1)^2$, which is a parabola that has vertex at $(-1, 2)$ and opens upward. Sketch this boundary equation as in Figure 7.10. In this case, the boundary is dotted because it is not included. Plot a test point $(0, 0)$ and check the inequality:

$$y > x^2 + 2x + 3$$
$$\mathbf{0} > \mathbf{0}^2 + 2 \cdot \mathbf{0} + 3 \qquad \text{False}$$

Since it is false, $(0, 0)$ is not in the solution set. Shade the appropriate region on the graph.

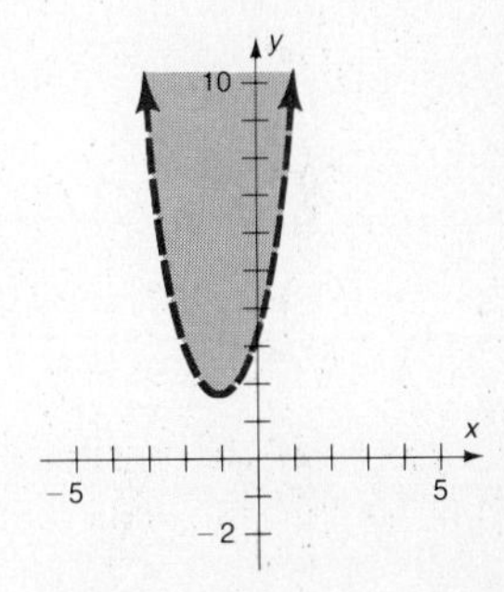

Figure 7.10 Graph of $y > x^2 + 2x + 3$ ■

The solution of a **system of inequalities** refers to the intersection of the solutions of the individual inequalities in the system. The procedure for graphing the solution of a system of inequalities is to graph the solution of the individual inequalities and then shade in the intersection.

EXAMPLE 4 Graph the solution of the system

$$\begin{cases} y > -20x + 100 \\ x \geq 0 \\ y \geq 0 \\ x \leq 8 \end{cases}$$

Solution *Step 1:* Graph $y > -20x + 100$.

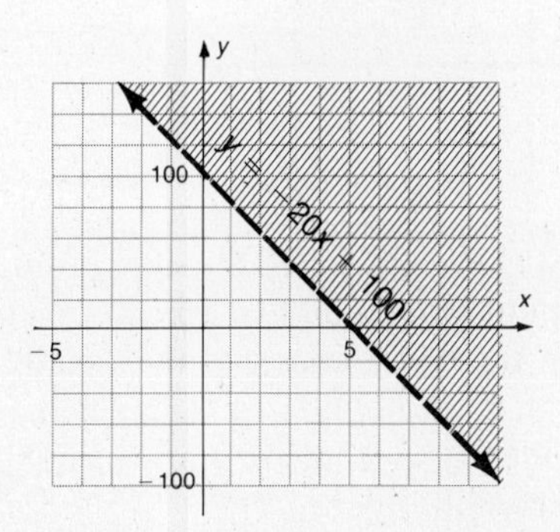

Step 2: Graph $x \geq 0$.

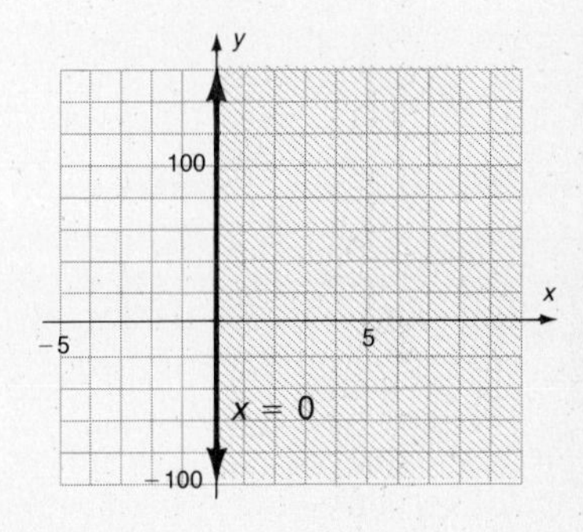

Step 3: Combine Steps 1 and 2. The intersection of the two graphs is the part that is shaded darker.

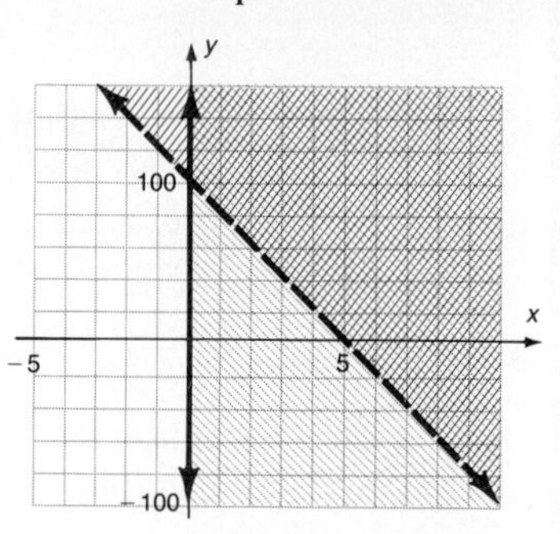

Step 4: Graph $y \geq 0$.

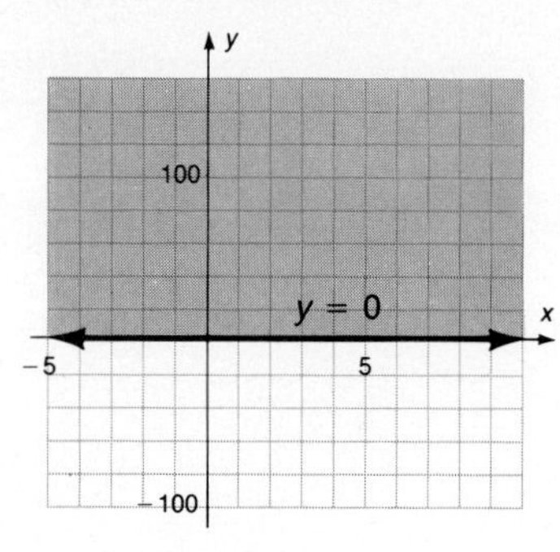

Step 5: Look at the intersection of Steps 3 and 4.

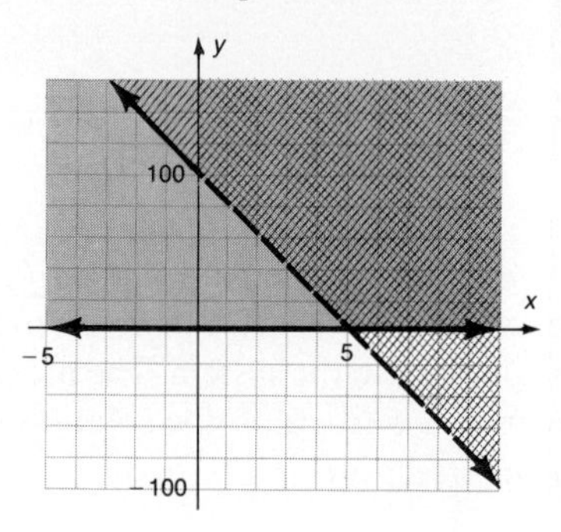

Step 6: Graph $x \leq 8$.

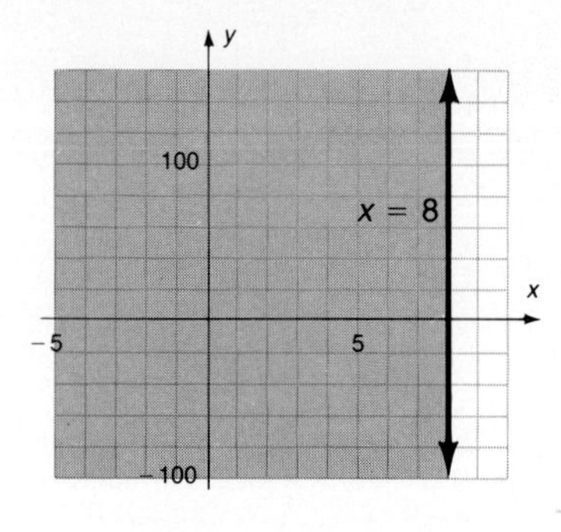

Step 7: Look at the intersection of Steps 5 and 6.

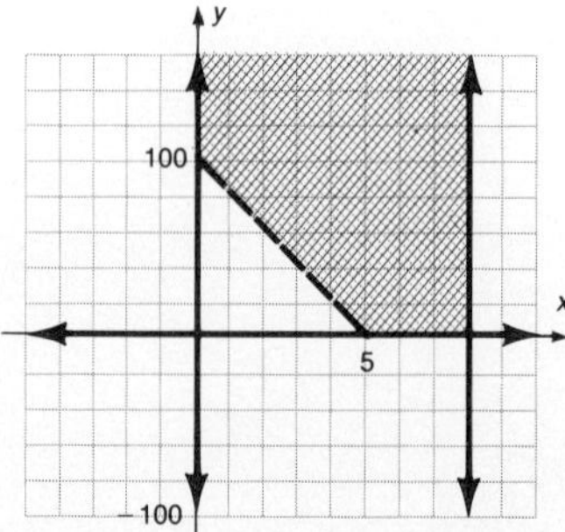

The way we have illustrated the steps of this solution was for explanation only; your work will not look at all like this. In practice, show the individual steps with little arrows on the boundaries and then shade in the intersection only after you have drawn in all of the boundaries. This device replaces the use of a lot of shading, which can be confusing if there are many inequalities in the system. The work for Example 4 is shown in Figure 7.11. ■

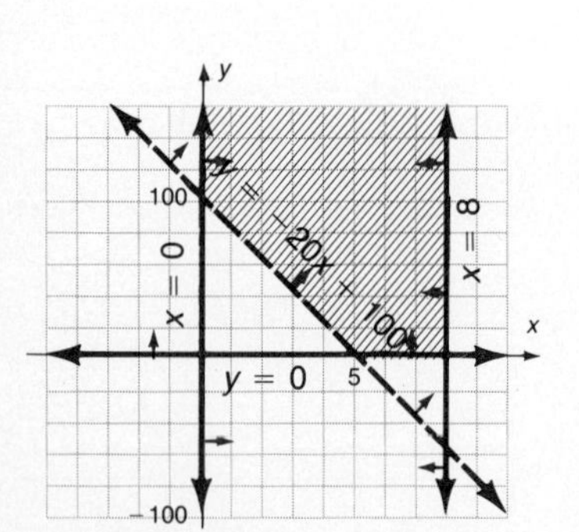

Figure 7.11 Graph of the solution of a system of inequalities

EXAMPLE 5 Graph the solution of the system

$$\begin{cases} 2x + y \le 3 \\ x - y > 5 \\ x \ge 0 \\ y > -10 \end{cases}$$

Solution The graphs of the individual inequalities and their intersections are shown in Figure 7.12. Notice the use of arrows to show the solutions of the individual inequalities.

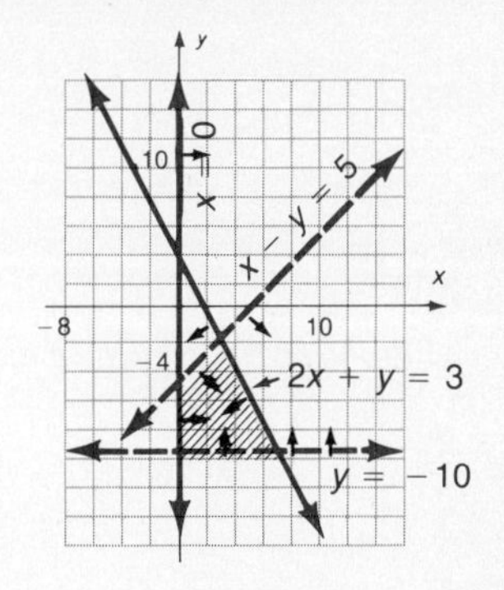

Figure 7.12 ■

PROBLEM SET 7.6

A

Graph the solution for each of the linear inequalities given in Problems 1–30.

1. $y \ge 2x - 3$
2. $y \ge 3x + 2$
3. $y \ge 4x + 5$
4. $y \le \frac{1}{2}x + 2$
5. $y \le \frac{2}{3}x - 4$
6. $y \le \frac{4}{5}x - 3$
7. $x - y > 3$
8. $2x + y < -3$
9. $3x + y < 4$
10. $x + 2y > 4$
11. $x - 3y > 12$
12. $3x - 4y > 8$
13. $y \le 6$
14. $x \ge 2$
15. $y < -2$
16. $y > 0$
17. $x < 0$
18. $x < 8$
19. $y \ge |x|$
20. $y < |x|$
21. $y < |x + 1|$
22. $y - 3 > |x + 1|$
23. $y + 1 \le |x - 3|$
24. $y < |x - 1| + 2$
25. $y \le x^2$
26. $y \ge (x - 3)^2$
27. $y - 1 \ge (x + 1)^2$
28. $y < x^2 - 6x + 7$
29. $x^2 - 4x + y + 5 < 0$
30. $x^2 + 6x + y + 7 > 0$

B

Graph the solution of each system given in Problems 31–57.

31. $\begin{cases} x \ge 0 \\ y \ge 0 \end{cases}$
32. $\begin{cases} x \ge 0 \\ y \le 0 \end{cases}$
33. $\begin{cases} x \le 0 \\ y \le 0 \end{cases}$
34. $\begin{cases} y \ge 0 \\ x < 8 \\ y < 5 \end{cases}$
35. $\begin{cases} x \ge 0 \\ y \ge 0 \\ x \le 5 \\ y \le 6 \end{cases}$
36. $\begin{cases} x \ge 0 \\ y \ge 0 \\ x < 500 \\ y < 1000 \end{cases}$
37. $\begin{cases} 2x + y > 3 \\ 3x - y < 2 \end{cases}$
38. $\begin{cases} y \le 3x - 4 \\ y \ge -2x + 5 \end{cases}$
39. $\begin{cases} 3x - 2y \ge 6 \\ 2x + 3y \le 6 \end{cases}$
40. $\begin{cases} y - 5 \le 0 \\ y \ge 0 \end{cases}$
41. $\begin{cases} x - 10 \le 0 \\ x \ge 0 \end{cases}$
42. $\begin{cases} y - 25 \le 0 \\ y \ge 0 \end{cases}$
43. $\begin{cases} -10 \le x \\ x \le 6 \\ 3 < y \\ y < 8 \end{cases}$
44. $\begin{cases} -5 < x \\ 3 \ge x \\ 5 > y \\ -2 \le y \end{cases}$
45. $\begin{cases} -5 < x \\ x \le 2 \\ -4 \le y \\ y < 9 \end{cases}$
46. $\begin{cases} x \ge 0 \\ y \ge 0 \\ x + y \le 9 \\ 2x - 3y \ge -6 \\ x - y \le 3 \end{cases}$
47. $\begin{cases} x \ge 0 \\ y \ge 0 \\ x + y \le 8 \\ y \le 4 \\ x \le 6 \end{cases}$
48. $\begin{cases} x \ge 0 \\ y \ge 0 \\ 2x + y \ge 8 \\ y \le 5 \\ x - y \le 2 \\ 3x - y \ge 5 \end{cases}$
49. $\begin{cases} 2x - 3y + 30 \ge 0 \\ 3x - 2y + 20 \le 0 \\ x \le 0 \\ y \ge 0 \end{cases}$
50. $\begin{cases} 2x + 3y \le 30 \\ 3x + 2y \ge 20 \\ x \ge 0 \\ y \ge 0 \end{cases}$
51. $\begin{cases} x + y - 10 \le 0 \\ x + y + 4 \ge 0 \\ x - y \le 6 \\ y - x \le 4 \end{cases}$
52. $\begin{cases} 3y = x^2 + 21 \\ y \le 10 \end{cases}$
53. $\begin{cases} 3y + 16x = x^2 + 85 \\ y \le 10 \end{cases}$
54. $\begin{cases} 3y + x^2 = 2x + 25 \\ -1 \le x \le 2 \end{cases}$

55. $\begin{cases} 4y + 6x + 3 = x^2 \\ |x - 3| \le 4 \end{cases}$

56. $\begin{cases} 3y + x^2 + 23 = 14x \\ |x - 7| \le 2 \end{cases}$

57. $\begin{cases} 3x^2 + 14 = 2y + 12x \\ |x - 2| \le 2 \end{cases}$

C

In Problems 58–63 sketch the systems of inequalities.

58. $\begin{cases} y \ge x^2 \\ 5x - 4y + 26 \ge 0 \end{cases}$

59. $\begin{cases} y < 4 - x^2 \\ y > \frac{1}{2}x \end{cases}$

60. $\begin{cases} y \ge \frac{1}{2}(x - 2)^2 \\ y - 4 < \frac{1}{4}(x - 2)^2 \end{cases}$

61. $\begin{cases} y - 2 \ge \frac{1}{3}(x - 3)^2 \\ y - 4 \le -\frac{1}{3}(x - 3)^2 \end{cases}$

62. $\begin{cases} x \ge (y + 2)^2 \\ y + 1 \ge \frac{1}{2}(x - \frac{3}{2})^2 \end{cases}$

63. $\begin{cases} x + 1 \ge (y - 5)^2 \\ y - 3 \ge (x + 1)^2 \end{cases}$

7.7 LINEAR PROGRAMMING*

This section introduces you to the topic of linear programming. Linear programming is a branch of mathematics that can be applied when you are interested in maximizing or minimizing a linear function. It was developed during World War II, and today it is used in a variety of applications, such as maximizing profits, minimizing costs, finding the most efficient shipping schedules, minimizing waste, securing the proper mix of ingredients, controlling inventories, and finding the most efficient assignment of personnel. In this section we apply the ideas of systems of inequalities to solve certain types of linear programming problems graphically. In later courses you will learn how to apply the processes of linear programming to more advanced applications.

We begin with some terminology. A set of points S is called a **convex set** if, for *any* two points P and Q in S, the entire segment $\overline{PQ}$ is in S. Some examples are shown in Figure 7.13.

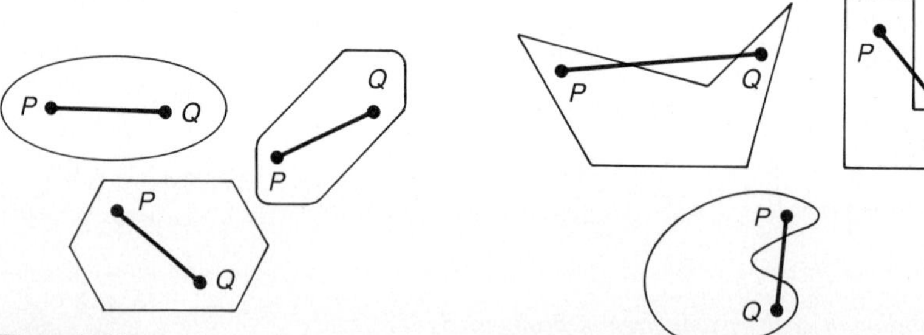

Figure 7.13
(a) Convex sets (b) Sets that are not convex

The applications of linear programming can be generalized as follows. Suppose you are given a function $c_1x + c_2y$ in two variables x and y with constants c_1 and c_2, and the problem is to find the maximum or minimum value of this function. This function is called the **objective function**. This objective function is subject to certain limitations called **constraints**. In linear programming these constraints are specified

* Optional section

by a system of linear inequalities whose solution forms a convex set S. Each point of the set S is called a **feasible solution**, and a point at which the objective function takes on a maximum or a minimum value is called an **optimum solution**. The proof of the following Linear Programming Theorem is beyond the scope of this course, so it is stated without proof.

Linear Programming Theorem

A linear function in two variables.

$$c_1x + c_2y$$

defined over a convex set S whose sides are line segments takes on its maximum value at a corner point of S and its minimum value at a corner point of S. If S is unbounded there may or may not be an optimum value, but if there is, then it must occur at a corner point.

In summary, to solve a linear programming problem:

1. Find the objective function (the quantity to be maximized or minimized).
2. Graph the constraints defined by a system of linear inequalities; the simultaneous solution is called the set S.
3. Find the corners of S; this may require the solution of a system of two equations with two unknowns, one for each corner.
4. Find the value of the objective function for the coordinates of each corner point. The largest value is the maximum; the smallest value is the minimum.

EXAMPLE 1 Maximize $C = 4x + 5y$, subject to:

$$\begin{cases} 2x + 5y \le 25 \\ 6x + 5y \le 45 \\ x \ge 0 \\ y \ge 0 \end{cases}$$

Solution The constraints give a set of feasible solutions; graph this system of inequalities as was illustrated in Section 7.6. This region is shown in Figure 7.14. To solve the linear

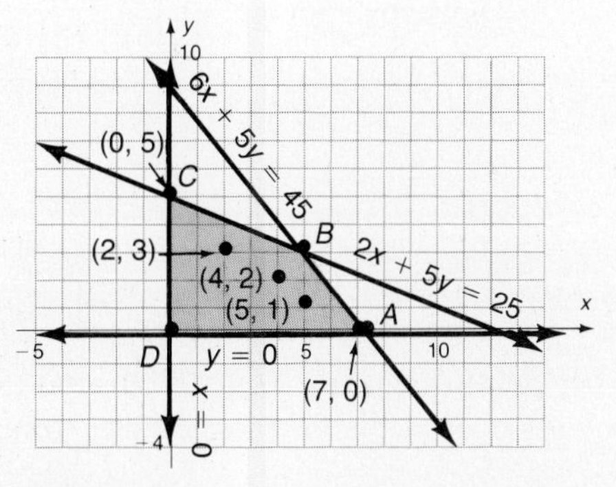

Figure 7.14

programming problem we must now find the feasible solution which makes the objective function as large as possible. Consider the following table of possible solutions.

Feasible Solution (A point in the solution set of the system.)	Objective Function $C = 4x + 5y$
(2, 3)	$4(2) + 5(3) = 8 + 15 = 23$
(4, 2)	$4(4) + 5(2) = 16 + 10 = 26$
(5, 1)	$4(5) + 5(1) = 20 + 5 = 25$
(7, 0)	$4(7) + 5(0) = 28 + 0 = 28$
(0, 5)	$4(0) + 5(5) = 0 + 25 = 25$

The point from the table that makes the objective function the largest is (7, 0). But is this the largest for all feasible solutions? How about (6, 1)? or (5, 3)? The Linear Programming Theorem tells us that the maximum value will occur at a corner point. The corner points are labeled A, B, C, and D on Figure 7.14. Some corner points can usually be found by inspection. In this example we see $D = (0, 0)$ and $C = (0, 5)$. Other corner points may require some work with the boundary *lines*. Use the equations of the boundaries, not the inequalities giving the regions, to find the points of intersection.

Point A: System $\begin{cases} y = 0 \\ 6x + 5y = 45 \end{cases}$

Solve by substitution: $6x + 5(0) = 45$

$$x = \frac{45}{6} = \frac{15}{2}$$

Point B: System $\begin{cases} 2x + 5y = 25 \\ 6x + 5y = 45 \end{cases}$

Solve by adding: $\begin{cases} -2x - 5y = -25 \\ 6x + 5y = 45 \end{cases}$

$$4x = 20$$
$$x = 5$$

If $x = 5$, then $2(5) + 5y = 25$

$$5y = 15$$
$$y = 3$$

The corner points are: (0, 0), (0, 5), $(\frac{15}{2}, 0)$, and (5, 3). The maximum value of $C = 4x + 5y$ will occur at one of these points. Compare the following table with the previous one.

Corner Points	Objective Function $C = 4x + 5y$	
$(0, 0)$	$4(0) + 5(0) = 0$	
$(0, 5)$	$4(0) + 5(5) = 25$	
$(\frac{15}{2}, 0)$	$4(\frac{15}{2}) + 5(0) = 30$	
$(5, 3)$	$4(5) + 5(3) = 35$	← Maximum

Look for the maximum value of C on *this* list; it is 35 so **the maximum value of C is 35.** ■

EXAMPLE 2 A farmer has 100 acres on which to plant two crops: corn or wheat. To produce these crops, there are certain expenses:

	Item	Cost per Acre
Corn:	seed	\$ 12
	fertilizer	\$ 58
	planting/care/harvesting	\$ 50
	TOTAL	\$120
Wheat:	seed	\$ 40
	fertilizer	\$ 80
	planting/care/harvesting	\$ 90
	TOTAL	\$210

After the harvest, the farmer must store the crops awaiting proper market conditions. Each acre yields an average of 110 bushels of corn or 30 bushels of wheat. The limitations of resources are as follows:

Available capital:	\$15,000
Available storage facilities:	4000 bushels

If the net profit (after all expenses have been subtracted) per bushel of corn is \$1.30 and for wheat is \$2.00, how should the farmer plant the 100 acres to maximize the profits?

Solution First, you might try to solve this problem using your intuition. If you plant all 100 acres with wheat the production is $30 \times 100 = 3000$ bushels for a net profit of $3000 \times 2 = \$6000$. But the farmer could not plant the entire 100 acres because wheat costs \$210 per acre for a total of \$21,000 (210×100) and there is only \$15,000 available. On the other hand, if the farmer plants 100 acres of corn the total cost is $\$120 \times 100 = \$12{,}000$. For this option, the net profit is $110 \times \$1.30 \times 100 = \$14{,}300$, but the production is $110 \times 100 = 11{,}000$ bushels, which exceeds the available storage capacity of 4000 bushels. Clearly a mix is necessary.

To formulate a mathematical model, begin by letting

x = number of acres to be planted in corn

y = number of acres to be planted in wheat

There are certain limitations or constraints.

$\mathbf{x \geq 0}$ The number of acres planted cannot be negative. *These constraints will apply in almost every model even though they are not explicitly stated as*

$\mathbf{y \geq 0}$ *part of the given problem.*

$\mathbf{x + y \leq 100}$ The amount of available land is 100 acres. Why not $x + y = 100$? It might be more profitable for the farmer to leave some land out of production. That is, it is not *necessary* to plant all the land.

$120x$ = expenses for planting the corn

$210y$ = expenses for planting the wheat

$\mathbf{120x + 210y \leq 15{,}000}$ The total expenses cannot exceed \$15,000; this is the *available capital.*

$110x$ = yield of acreage planted in corn

$30y$ = yield of acreage planted in wheat

$\mathbf{110x + 30y \leq 4000}$ The total yield cannot exceed the storage capacity of 4000 bushels.

Summary of constraints (in boldface above):

$$\begin{cases} x \geq 0 \\ y \geq 0 \\ x + y \leq 100 \\ 120x + 210y \leq 15{,}000 \\ 110x + 30y \leq 4000 \end{cases}$$

Now, let P = total profit. The farmer wants to maximize the profit, P.

$$\begin{aligned} \text{profit from the corn} &= \text{value} \times \text{amount} \\ &= 1.30 \times 110x \\ &= 143x \end{aligned}$$

$$\begin{aligned} \text{profit from the wheat} &= \text{value} \times \text{amount} \\ &= 2.00 \times 30y \\ &= 60y \end{aligned}$$

$$\begin{aligned} P &= \text{profit from the corn} + \text{profit from the wheat} \\ &= 143x + 60y \end{aligned}$$

The linear programming model is stated as follows:

Maximize: $P = 143x + 60y$

Subject to: $\begin{cases} x \geq 0 \\ y \geq 0 \\ x + y \leq 100 \\ 120x + 210y \leq 15{,}000 \\ 110x + 30y \leq 4000 \end{cases}$

First graph the set of feasible solutions by graphing the system of inequalities (see Figure 7.15).

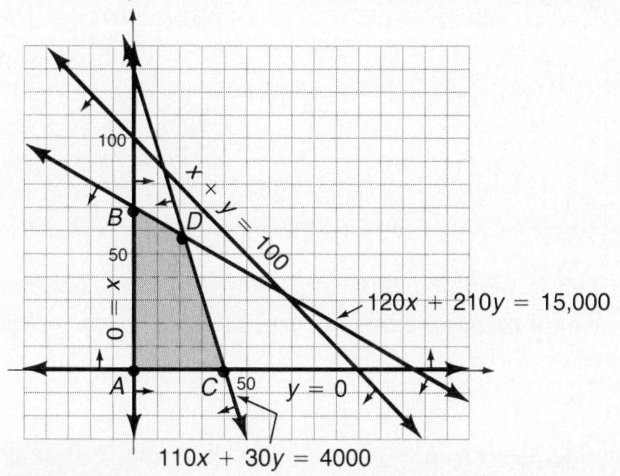

Figure 7.15

Next, find the corner points:

$\boldsymbol{A = (0, 0)}$ By inspection

$$B: \begin{cases} 120x + 210y = 15{,}000 \\ x = 0 \end{cases}$$

$$120(0) + 210y = 15{,}000$$

$$y = \frac{15{,}000}{210}$$

$$= \frac{500}{7}$$

$$\boldsymbol{B = \left(0, \frac{500}{7}\right)}$$

$$C: \begin{cases} 110x + 30y = 4000 \\ y = 0 \end{cases}$$

$$110x + 30(0) = 4000$$

$$110x = 4000$$

$$x = \frac{400}{11}$$

$$\boldsymbol{C = \left(\frac{400}{11}, 0\right)}$$

$$D: \quad -7\begin{cases} 110x + 30y = 400 \\ 120x + 210y = 15{,}000 \end{cases} \qquad \begin{cases} -770x - 210y = -28{,}000 \\ 120x + 210y = 15{,}000 \end{cases}$$

$$-650x = -13{,}000$$

$$x = 20$$

$$110(20) + 30y = 4000$$

$$30y = 1800$$

$$y = 60$$

$$\boldsymbol{D = (20, 60)}$$

Use the Linear Programming Theorem and check the corner points.

Corner Point	Objective Function $P = 143x + 60y$
$(0, 0)$	$143(0) + 60(0) = 0$
$\left(0, \frac{500}{7}\right)$	$143(0) + 60\left(\frac{500}{7}\right) \approx 4286$
$\left(\frac{400}{11}, 0\right)$	$143\left(\frac{400}{11}\right) + 60(0) = 5200$
$(20, 60)$	$143(20) + 60(60) = 6460$

The maximum value of P is at $(20, 60)$. **This means to maximize profit the farmer should plant 20 acres in corn, plant 60 acres in wheat, and leave 20 acres unplanted.** ■

Notice from the graph in Example 2 that some of the constraints could be eliminated from the problem and everything else would remain unchanged. For example, the boundary $x + y = 100$ was not necessary in finding the maximum value of P. Such a condition is said to be a **superfluous constraint**. It is not uncommon to have superfluous constraints in a linear programming problem. Suppose, however, that the farmer in Example 2 contracted to have the grain stored at a neighboring farm and now the contract calls for *at least* 4000 bushels to be stored. This change from $110x + 30y \leq 4000$ to $110x + 30y \geq 4000$ *now* makes the condition $x + y \leq 100$ important to the solution of the problem. Therefore, you must be careful about superfluous constraints even though they do not affect the solution at the present time.

The next example is solved more succinctly to show you the way your work will probably look.

EXAMPLE 3 The Sticky Widget Company makes two types of widgets: regular and deluxe. Each widget is produced at a station consisting of a machine and a person who finishes the widgets by hand. The regular widget requires 2 hr of machine time and 1 hr of finishing time. The deluxe widget requires 3 hr of machine time and 5 hr of finishing time. The profit on the regular widget is \$25; on the deluxe widget it is \$30. If the workday is 8 hr, how many of each type of widget should be produced at each station to maximize the profit?

Solution Let x = number of regular widgets produced

y = number of deluxe widgets produced

$$\begin{cases} \left.\begin{aligned} x \geq 0 \\ y \geq 0 \end{aligned}\right\} & \text{The number of widgets must be nonnegative.} \\ 2x + 3y \leq 8 & \text{The machine workday is not more than 8 hr.} \\ x + 5y \leq 8 & \text{The person's workday is not more than 8 hr.} \end{cases}$$

The set S is shown in color in Figure 7.16. The corner points are found by considering the piecewise simultaneous solution of the equations of the boundary

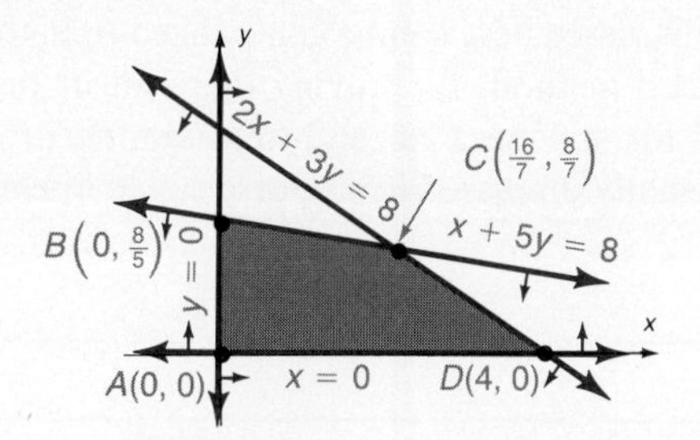

Figure 7.16

lines. You can find these intersection points by inspection or by following the methods of this chapter.

Corner Point	System to Solve	Solution	Objective Function $25x+30y$	Maximum or Minimum
A	$\begin{cases} x=0 \\ y=0 \end{cases}$	$(0,0)$	$25(0)+30(0)=0$	minimum
B	$\begin{cases} x=0 \\ x+5y=8 \end{cases}$	$(0,\frac{8}{5})$	$25(0)+30(\frac{8}{5})=48$	
C	$\begin{cases} 2x+3y=8 \\ x+5y=8 \end{cases}$	$(\frac{16}{7},\frac{8}{7})$	$25(\frac{16}{7})+30(\frac{8}{7})\approx 91.43$	
D	$\begin{cases} y=0 \\ 2x+3y=8 \end{cases}$	$(4,0)$	$25(4)+30(0)=100$	maximum

Profits are maximized if only regular widgets are produced. The company should produce four regular widgets per day at each station. ■

EXAMPLE 4 The Sticky Widget Company of the preceding example changes its operating procedure so that the profit on a regular widget is \$20 and that on a deluxe widget is \$35. All the constraints of the previous example are the same. How many of each type of widget should be produced to maximize the profits?

Solution The objective function to be maximized is $20x + 35y$. Check the values at each of the corner points:

A: $20(0) + 35(0) = 0$ minimum

B: $20(0) + 35\left(\frac{8}{5}\right) = 56$

C: $20\left(\frac{16}{7}\right) + 35\left(\frac{8}{7}\right) \approx 85.71$ maximum

D: $20(4) + 35(0) = 80$

The profits are maximized if the company produces 16 regular widgets and 8 deluxe widgets at each station every 7 working days. ■

The method discussed here can be generalized to higher dimensions by means of more sophisticated methods for solving these linear programming problems. The general method for solving a linear programming problem is called the *simplex method* and is usually discussed in a course called linear programming.

PROBLEM SET 7.7

A

For Problems 1–6, decide whether the given point is a feasible solution for the constraints

$$\begin{cases} x \geq 0 \\ y \geq 0 \\ 3x + 2y \leq 10 \\ 2x + 4y \leq 8 \end{cases}$$

1. $(1, 3)$ **2.** $(2, 1)$ **3.** $(1, 2)$ **4.** $(-1, 4)$
5. $(2, 2)$ **6.** $(0, 4)$

For Problems 7–12, decide whether the given point is a corner point for the constraints

$$\begin{cases} x \geq 0 \\ y \geq 0 \\ 2x + 3y \geq 120 \\ 2x + y \geq 80 \end{cases}$$

7. $(0, 0)$ **8.** $(0, 80)$ **9.** $(80, 0)$ **10.** $(60, 0)$
11. $(30, 20)$ **12.** $(20, 30)$

Find the corner points for each set of feasible solutions given in Problems 13–24.

13. $\begin{cases} x \geq 0 \\ y \geq 0 \\ 2x + y \leq 12 \\ x + 2y \leq 9 \end{cases}$ **14.** $\begin{cases} x \geq 0 \\ y \geq 0 \\ 2x + 5y \leq 20 \\ 2x + y \leq 12 \end{cases}$

15. $\begin{cases} x \geq 0 \\ y \geq 0 \\ 3x + 2y \leq 12 \\ x + 2y \leq 8 \end{cases}$ **16.** $\begin{cases} x \geq 0 \\ y \geq 0 \\ x \leq 10 \\ y \leq 8 \\ 3x + 2y \geq 12 \end{cases}$

17. $\begin{cases} x \geq 0 \\ y \geq 0 \\ x + y \leq 8 \\ y \leq 4 \\ x \leq 6 \end{cases}$ **18.** $\begin{cases} x \geq 0 \\ y \geq 0 \\ x + y \geq 6 \\ -2x + y \geq -16 \\ y \leq 9 \end{cases}$

19. $\begin{cases} x \geq 0 \\ y \geq 0 \\ 3x + 2y \leq 8 \\ x + 5y \leq 8 \end{cases}$ **20.** $\begin{cases} x \geq 0 \\ y \geq 0 \\ x \leq 8 \\ y \geq 2 \\ x + y \leq 10 \\ x \leq 3y \end{cases}$

21. $\begin{cases} x \geq 0 \\ y \geq 0 \\ 10x + 5y \geq 200 \\ 2x + 5y \geq 100 \\ 3x + 4y \geq 120 \end{cases}$ **22.** $\begin{cases} x \geq 0 \\ y \geq 0 \\ x + y \leq 9 \\ 2x - 3y \geq -6 \\ x - y \leq 3 \end{cases}$

23. $\begin{cases} x \geq 0 \\ y \geq 0 \\ 2x + y \geq 8 \\ y \leq 5 \\ x - y \leq 2 \\ 3x - 2y \geq 5 \end{cases}$ **24.** $\begin{cases} x \geq 0 \\ y \geq 0 \\ 2x + y \geq 8 \\ x - 2y \leq 7 \\ x - y \geq -3 \\ x \leq 9 \end{cases}$

B

Find the optimum value for each objective function given in Problems 25–44.

25. Maximize $W = 30x + 20y$ subject to the constraints of Problem 13.

26. Maximize $P = 40x + 10y$ subject to the constraints of Problem 13.

27. Maximize $T = 100x + 10y$ subject to the constraints of Problem 14.

28. Maximize $V = 20x + 30y$ subject to the constraints of Problem 14.

29. Maximize $P = 100x + 100y$ subject to the constraints of Problem 15.

30. Maximize $T = 30x + 10y$ subject to the constraints of Problem 15.

31. Minimize $C = 24x + 12y$ subject to the constraints of Problem 16.

32. Minimize $K = 6x + 18y$ subject to the constraints of Problem 16.

33. Maximize $F = 2x - 3y$ subject to the constraints of Problem 17.

34. Maximize $P = 5x + y$ subject to the constraints of Problem 17.

35. Minimize $I = 90x + 20y$ subject to the constraints of Problem 18.

36. Minimize $T = 400x + 100y$ subject to the constraints of Problem 18.

37. Maximize $P = 23x + 46y$ subject to the constraints of Problem 18.

38. Minimize $C = 12x + 15y$ subject to the constraints of Problem 18.

39. Maximize $P = 6x + 3y$ subject to the constraints of Problem 22.

40. Maximize $T = x + 6y$ subject to the constraints of Problem 22.

41. Minimize $X = 5x + 3y$ subject to the constraints of Problem 23.

42. Minimize $A = 2x - 3y$ subject to the constraints of Problem 23.

43. Minimize $K = 140x + 250y$ subject to the constraints of Problem 24.

44. Minimize $C = 640x - 130y$ subject to the constraints of Problem 24.

Write a linear programming model including the objective function and the set of constraints for Problems 45–51. DO NOT SOLVE, but be sure to define all your variables.

45. ***Allocation of Resources in Manufacturing*** The Wadsworth Widget Company manufactures two types of widgets: regular and deluxe. Each widget is produced at a station consisting of a machine and a person who finishes the widgets by hand. The regular widget requires 3 hr of machine time and 2 hr of finishing time. The deluxe widget requires 2 hr of machine time and 4 hr of finishing time. The profit on the regular widget is \$25; on the deluxe widget it is \$30. If the workday is 8 hours, how many of each type of widget should be produced at each station per day in order to maximize the profit?

46. ***Diet Problem*** A convalescent hospital wishes to provide, at a minimum cost, a diet that has a minimum of 200 g of carbohydrates, 100 g of protein, and 120 g of fats per day. These requirements can be met with two foods:

Food	Carbohydrates	Protein	Fats
A	10 g	2 g	3 g
B	5 g	5 g	4 g

If food *A* costs \$0.29 per gram and food *B* \$0.15 per gram how many grams of each food should be purchased for each patient per day in order to meet the minimum requirements at the lowest cost?

47. ***Investment Application*** Brown Bros., Inc. is an investment company doing an analysis of the pension fund for a certain company. The fund has a maximum of \$10 million to invest in two places: no more than \$8 million in stocks yielding 12%, and at least \$2 million in long-term bonds yielding 8%. The stock-to-bond investment ratio cannot be more than 3 to 1. How should Brown Bros. advise their client so that the investments yield the maximum yearly return?

48. ***Office Management*** An office manager decides to purchase computers for each office worker. Apple IIe computers cost \$900 each and will provide 64K storage capability and will take up 5 sq ft of desk space. IBM PC computers will cost \$1200 each, will provide 128K storage, and will take up 3 sq ft of desk space. Proper utilization of space requires that a total of not more than 165 sq ft be used for this new equipment. If there is \$36,000 available to spend on computers, how many of each type of computer should the office manager purchase if she wishes to maximize storage capacity?

49. ***Allocation of Resources*** Foley's Motel has 200 rooms and a restaurant that seats 50 people. Experience shows that 40% of the commercial guests and 20% of the other guests eat in the restaurant. Suppose there is a net profit of \$4.50 per day from each commercial guest and \$3.50 per day from each other guest. Find the number of commercial and other guests needed in order to maximize the net profits assuming exactly one guest per room and that the capacity of the restaurant will not be exceeded.

50. ***Operating Costs*** Karlin Enterprises manufactures two electronic games. Standing orders require that at least 24,000 space battle and 5000 football games be produced. The company has two factories: The Gainesville plant can produce 600 space battle games and 100 football games per day; the Sacramento plant can produce 300 space battle games and 100 football games per day. If the Gainesville plant costs \$20,000 per day to operate and the Sacramento factory costs \$15,000 per day, find the number of days per month each factory should operate to minimize the cost. (Assume that each month has 30 days.)

51. ***Production Problem*** The Alco Company manufactures two products, Alpha and Beta. Each product must pass through two processing operations, and all materials are introduced at the first operation. Alco may produce either one product exclusively or various combinations of both products subject to the following constraints:

	First Process	Second Process	Profit per Unit
Hours required to produce one unit of:			
Alpha	1 hr	1 hr	$5.00
Beta	3 hr	2 hr	$8.00
Total capacity in hours per day:	1200 hr	1000 hr	

A shortage of technical labor has limited Alpha production to no more than 700 units per day. There are no constraints on the production of Beta other than the hour constraints in the schedule shown above. How many of each product should be manufactured in order to maximize the profit?

Solve the linear programming problems in Problems 52–60.

52. ***Allocation of Resources in Production*** Suppose the net profit per bushel of corn in Example 2 increased to $2.00 and the net profit per bushel of wheat dropped to $1.50. Maximize the profit if the other conditions in the example remain the same.

53. ***Allocation of Resources in Production*** Suppose the farmer in Example 2 contracted to have the grain stored at a neighboring farm and the contract calls for at least 4000 bushels to be stored. How many acres should be planted in corn and how many in wheat to maximize profits if the other conditions in Example 2 remain the same?

54. ***Allocation of Resources in Production*** A farmer has 500 acres on which to plant two crops, corn and wheat. It costs $120 per acre to produce corn and $60 per acre to produce wheat and there is $24,000 available to pay for this year's production. If the yields per acre are 100 bushels of corn and 40 bushels of wheat and the farmer has contracted to store at least 18,000 bushels, how much should the farmer plant to maximize profits when the net profit is $1.20 per bushel for corn and $2.50 per bushel for wheat?

55. ***Allocation of Resources in Manufacturing*** The Wadsworth Widget Company manufactures two types of widgets: regular and deluxe. Each widget is produced at a station consisting of a machine and a person who finishes the widgets by hand. The regular widget requires 2 hr of machine time and 1 hr of finishing time. The deluxe widget requires 3 hr of machine time and 5 hr of finishing time. The profit on the regular widget is $25; on the deluxe widget it is $30. If the workday is 8 hours, how many of each type of widget would be produced at each station per day in order to maximize the profit?

C

56. ***Allocation of Resources in Manufacturing*** The Thompson Company manufactures two industrial products, standard ($45 profit per item) and economy ($30 profit per item). These items are built using machine time and manual labor. The standard product requires 3 hr of machine time and 2 hr of manual labor. The economy model requires 3 hr of machine time and no manual labor. If the week's supply of manual labor is limited to 800 hr and machine time to 15,000 hr, how much of each type of product should be produced each week in order to maximize the profit?

57. ***Diet Problem*** A convalescent hospital wishes to provide at a minimum cost, a diet that has at least 200 g of carbohydrates, 100 g of protein, and 120 g of fats per day. These requirements can be met with two foods:

Food	Carbohydrates	Protein	Fats
A	10 g	2 g	3 g
B	5 g	5 g	4 g

If food *A* costs $0.29 per gram and food *B* is $0.15 per gram how many grams of each food should be purchased for each patient per day in order to meet the minimum requirements at the lowest cost?

58. ***Diet Problem*** The following carbohydrate information is given on the side of the following cereal boxes (for 1 oz of cereal with 1/2 cup of whole milk):

	Starch and Related Carbohydrates	Sucrose and Other Sugars
Kellogg's Corn Flakes	23 g	7 g
Post Honeycombs	14 g	17 g

What is the minimum cost in order to receive at least 322 g starch and 119 g sucrose by consuming these two cereals if Corn Flakes cost $0.07 per ounce and Honeycombs cost $0.19 per ounce?

59. ***Investment Application*** Brown Bros., Inc. is an investment company doing an analysis of the pension fund for a certain company. The fund has a maximum of $10 million to invest in two places: no more than $8 million in stocks yielding 12%, and at least $2 million in long-term bonds yielding 8%. The stock-to-bond investment ratio cannot be more than 3 to 1. How should Brown Bros. advise their client so that the investments yield the maximum yearly return?

60. ***Investment Application*** Your broker tells you of two investments she thinks are worthwhile. She advises a new issue of Pertec stock which should yield 20% over the next year, and then to balance your account she advises Campbell Municipal Bonds with a 10% yearly yield. The stock-to-bond ratio should be no less than 3 to 1. If you have no more than $100,000 to invest and do not want to invest more than $70,000 in Pertec or less than $20,000 in bonds, how much should be invested in each to maximize your return?

CHAPTER 7 SUMMARY

The material of this chapter is reviewed in the following list of objectives. After each objective there are some practice questions. For a sample test, select the first question of each set and check your answers with the answer section. For a sample test without answers, use the second question of each set. Additional practice is given by the other questions in each set. If you are having trouble with a particular type of problem, look back to that section for extra help.

7.1 SYSTEMS OF EQUATIONS

Objective 1 Solve a system of equations by graphing.

1. $\begin{cases} 2x - y = -8 \\ y = \frac{3}{5}x + 1 \end{cases}$ **2.** $\begin{cases} y = \frac{1}{3}x - 5 \\ 2x - 4y - 12 = 0 \end{cases}$

3. $\begin{cases} 2x - 3y + 21 = 0 \\ 5x + 4y - 5 = 0 \end{cases}$ **4.** $\begin{cases} y = x^2 - 6x + 5 \\ x + y - 5 = 0 \end{cases}$

Objective 2 Solve a system of equations by substitution.

5. $\begin{cases} 4x + 3y = -18 \\ y = -\frac{2}{3}x - 2 \end{cases}$ **6.** $\begin{cases} 5y - 3x = 0 \\ y = x + 2 \end{cases}$

7. $\begin{cases} x + 2y = 26 \\ 5x - 2y = -122 \end{cases}$ **8.** $\begin{cases} y = x^2 + 4x - 3 \\ x - y + 1 = 0 \end{cases}$

Objective 3 Solve a system of equations by linear combinations.

9. $\begin{cases} 2x + 3y = 6 \\ 3x + 2y = -1 \end{cases}$ **10.** $\begin{cases} 5x - 2y = 30 \\ 3x - 2y = -2 \end{cases}$

11. $\begin{cases} 4x + 3y = -7 \\ 2x - 5y = 55 \end{cases}$ **12.** $\begin{cases} 3x^2 - 2y^2 = 9 \\ x^2 + y^2 = 8 \end{cases}$

7.2 CRAMER'S RULE

Objective 4 Solve a system of two equations in two unknowns by using Cramer's Rule.

13. $\begin{cases} 3x - y = 2 \\ x + 5y = -3 \end{cases}$ **14.** $\begin{cases} 5x - 3y = 2 \\ 2x + 4y = 5 \end{cases}$

15. $\begin{cases} 3x + 4y = 0 \\ 6x + 8y = 0 \end{cases}$ **16.** $\begin{cases} y = m_1x + b_1 \\ y = m_2x + b_2 \end{cases}$

7.3 PROPERTIES OF DETERMINANTS

Objective 5 Evaluate determinants.

17. $\begin{vmatrix} 1 & 3 & -2 \\ 4 & 5 & 1 \\ 3 & -2 & 4 \end{vmatrix}$ **18.** $\begin{vmatrix} 3 & 1 & -2 \\ 4 & 4 & 0 \\ 2 & -3 & 1 \end{vmatrix}$

19. $\begin{vmatrix} 3 & -2 & 0 \\ 2 & 5 & 8 \\ 5 & 3 & 5 \end{vmatrix}$

20. $\begin{vmatrix} 0 & 3 & 0 & 0 \\ 3 & 4 & 0 & 2 \\ 3 & 8 & -4 & 1 \\ 1 & -5 & 0 & -1 \end{vmatrix}$

Objective 6 Solve a system of n unknowns by using Cramer's Rule.

21. $\begin{cases} 3x - 2y + z = 9 \\ 2x + 5y - 3z = 17 \\ x - 3y + 2z = -2 \end{cases}$

22. $\begin{cases} 3x + y - 2z = 2 \\ 4x + 4y = 8 \\ 2x - 3y + z = 0 \end{cases}$

23. $\begin{cases} 2x + z = 6 + y \\ 3x + z = 8 + 2y \\ x + y = 11 - 3z \end{cases}$

24. $\begin{cases} w + 2z = 7 \\ x - 3y = 5 \\ z - 4w = -1 \\ 2w + y = 5 \end{cases}$

7.4 ALGEBRA OF MATRICES

Objective 7 Perform matrix operations. Let

$$A = \begin{bmatrix} 3 & -2 & 1 \\ -2 & 1 & 4 \\ 1 & -3 & 1 \end{bmatrix} \quad B = \begin{bmatrix} 2 & 1 & 0 \\ -1 & 7 & 3 \\ 2 & -3 & 5 \end{bmatrix} \quad \text{and } C = \begin{bmatrix} 2 & 4 & -2 \\ 1 & -1 & 2 \end{bmatrix}$$

Find, if possible:

25. $A + B$ **26.** $A(B + C)$ **27.** AB **28.** $C(BA)$

Objective 8 Find the inverse of a matrix (if it exists).

29. $\begin{bmatrix} 1 & -2 \\ -3 & 7 \end{bmatrix}$ **30.** $\begin{bmatrix} 4 & -3 \\ 5 & 4 \end{bmatrix}$ **31.** $\begin{bmatrix} 1 & 1 \\ 1 & 1 \end{bmatrix}$ **32.** $\begin{bmatrix} 1 & 2 & 2 \\ 1 & 3 & 2 \\ 1 & 2 & 3 \end{bmatrix}$

Objective 9 Solve a system of equations by using an inverse matrix. The matrices

$$\begin{bmatrix} -1 & 1 & -1 \\ 2 & -1 & 2 \\ 2 & -1 & 1 \end{bmatrix} \quad \text{and} \quad \begin{bmatrix} 1 & 0 & 1 \\ 2 & 1 & 0 \\ 0 & 1 & -1 \end{bmatrix}$$

are inverses. Use this information to solve the systems in Problems 33–36.

33. $\begin{cases} -x + y - z = 6 \\ 2x - y + 2z = -5 \\ 2x - y + z = -5 \end{cases}$

34. $\begin{cases} y = x + z \\ 2x + 2z = y + 1 \\ 2x + z = y + 3 \end{cases}$

35. $\begin{cases} x + z = 11 \\ 2x + y = 14 \\ y - z = -5 \end{cases}$

36. $\begin{cases} x + z = -4 \\ 2x + y = -4 \\ y - z = 1 \end{cases}$

7.5 MATRIX SOLUTION OF SYSTEMS

Objective 10 Solve a system of equations by using the Gauss–Jordan method.

37. $\begin{cases} 2x + 3y - z = -2 \\ x + y - 5z = -3 \\ 5x - 7y - 10z = 60 \end{cases}$

38. $\begin{cases} x - 3y + z = 11 \\ 2x + y - z = 0 \\ x - 2y + 4z = 25 \end{cases}$

39. $\begin{cases} 2x - y + z = -3 \\ 3x + y - 2z = 11 \\ 5x - 2y + 3z = -8 \end{cases}$

40. $\begin{cases} w + 3x = -5 \\ x + y = -4 \\ 2x - 3z = -7 \\ z + y = -1 \end{cases}$

7.6 SYSTEMS OF INEQUALITIES

Objective 11 Graph a linear inequality in two variables.

41. $3x - 2y + 16 \geq 9$

42. $y < x - 1$

43. $25x - 5y + 120 > 0$

44. $y \geq x^2 - 8x + 19$

Objective 12 Graph the solutions of a system of inequalities.

45. $\begin{cases} 2x + 7y \geq 420 \\ 2x + 2y \leq 500 \\ x > 50 \\ y < 100 \end{cases}$

46. $\begin{cases} y \geq 0 \\ 3x + 2y > -3 \\ x - y < 0 \end{cases}$

47. $\begin{cases} x - y - 8 \leq 0 \\ x - y + 4 \geq 0 \\ x + y \geq -5 \\ y + x \leq 4 \end{cases}$

48. $\begin{cases} 2x + 3y \leq 600 \\ x + 2y \leq 360 \\ 3x - 2y \leq 480 \\ x \geq 0 \\ y \geq 0 \end{cases}$

*7.7 LINEAR PROGRAMMING

Objective 13 Solve a linear programming problem.

49. Maximize $5x + 7y$ subject to:

$$\begin{cases} x \geq 0 \\ y \geq 0 \\ 4x + y \geq 7 \\ x + 2y \leq 12 \\ x \leq 7 \end{cases}$$

50. Minimize $8x + 5y$ subject to:

$$\begin{cases} x \geq 0 \\ y \geq 0 \\ 2x + y \geq 8 \\ x - y \leq 2 \end{cases}$$

51. Maximize $5x + 4y$ subject to:

$$\begin{cases} 2x + y \leq 420 \\ 2x + 2y \leq 500 \\ 2x + 3y \leq 600 \\ x \geq 0 \\ y \geq 0 \end{cases}$$

52. Maximize $x + y$ subject to:

$$\begin{cases} x \geq 0 \\ y \geq 0 \\ 12x + 150y \leq 1200 \\ 6x + 200y \leq 1200 \\ 16x + 50y \leq 800 \end{cases}$$

Objective 14 Solve applied problems involving systems of equations and linear programming.

53. ***Chemistry*** Two commercial preparations, I and II, contain two ingredients, A and B, in the following proportions:

	A	B
I	70%	30%
II	40%	60%

How many grams of each preparation must be mixed to obtain 60 g of a preparation that contains the ingredients in equal parts?

* This problem is from the optional section.

54. ***Business*** A manufacturer of auto accessories uses three basic parts A, B, and C in its two products I and II, in the following proportions:

	A	B	C
I	2	1	1
II	2	2	1

The inventory shows 1250 of A, 900 of B, and 2000 of C on hand. How many of each product should be manufactured from the parts available to give the largest total number of products?

* 55. ***Medicine*** A hospital wishes to provide for its patients a diet that has a minimum of 100 g of carbohydrates, 60 g of proteins, and 40 g of fats per day. These requirements can be met with two foods:

Food	Carbohydrates	Proteins	Fats
A	6 g	3 g	1 g
B	2 g	2 g	2 g

It is also important to minimize costs, and food A costs $0.14 per gram and food B costs $0.06 per gram. How many grams of each food should be bought for each patient per day in order to meet the minimum daily requirements at the lowest cost?

*56. ***Utilization of Sales Force*** Bradbury Bros. Realty plans to open several new branch offices, which will employ either four or six people, and has $1,275,000 in capital for this expansion. A four-person branch requires an initial cash outlay of $175,000, and a six-person branch, $200,000. Bradbury Bros. has also decided not to hire more than 32 new people and will not open more than 10 branches. How many of each type of branch should be opened to maximize cash inflow if it is expected to be $50,000 for four-person branches and $65,000 for six-person branches?

* This problem is from the optional section.

8

Additional Topics in Algebra

Sir Isaac Newton (1642–1727)

Analysis and natural philosophy owe their most important discoveries to this fruitful means, which is called "induction." Newton was indebted to it for his theorem of the binomial and the principle of universal gravity.

PIERRE-SIMON LAPLACE
A Philosophical Essay on Probabilities

Isaac Newton was one of the greatest mathematicians of all time. However, when he was a schoolboy at Grantham he was last in his class until he wanted to beat a bully both physically and mentally. He challenged the bully to a fight and won; then he set about to beat the bully in school and did not stop until he was at the top of his class. He entered Trinity College, Cambridge when he was 18 and remained there until 1696 as student or professor. In 1665 the university closed because of bubonic plague, and Newton had to stay home for a year. During this year he interested himself in various physical questions, and among other things he formulated the basis of his theory of gravitation and invented the calculus. Leibniz, who was featured at the beginning of Chapter 7, invented the calculus at about the same time. The Binomial Theorem of this chapter was proved for real exponents by Newton in 1665. His statement of this theorem is awkward by today's standards, but remember that he found it by a laborious trial-and-error procedure. The following statement of the Binomial Theorem is attributed to Newton in D. E. Smith's *Source Book in Mathematics* (1929), pp. 224–228:

$$\overline{P + PQ}\Big|\frac{m}{n} = P\frac{m}{n} + \frac{m}{n}AQ + \frac{m-n}{2n}BQ + \frac{m-2n}{3n}CQ + \frac{m-3n}{4n}DQ + \text{etc.}$$

CHAPTER OVERVIEW Two of the most far-reaching results in algebra are presented in this chapter, mathematical induction and the binomial theorem. The last three sections, which deal with sequences and series, should be considered together as a unit. There are 9 objectives in this chapter, which are listed on pages 370–371.

8.1 MATHEMATICAL INDUCTION

Mathematical induction is an important method of proof in mathematics. Let us begin with a simple example. Suppose you want to know the sum of the first n odd integers. You could begin by looking for a pattern:

$$\begin{aligned} &1 &&= 1 \\ &1 + 3 &&= 4 \\ &1 + 3 + 5 &&= 9 \\ &1 + 3 + 5 + 7 &&= 16 \\ &1 + 3 + 5 + 7 + 9 &&= 25 \end{aligned}$$

Do you see a pattern here? It appears that the sum of the first n odd numbers is n^2 since $1 = 1^2, 4 = 2^2, 9 = 3^2, 16 = 4^2$, and so on. Now *prove deductively* that

$$1 + 3 + 5 + \cdots + \underbrace{(2n - 1)}_{n\text{th odd number}} = n^2$$

is true for all positive integers n. How can you proceed? Use a method called **mathematical induction**, which is used to prove certain propositions about the positive integers. The proposition is denoted by $P(n)$. In this case, $P(n)$ is the proposition

$$P(n)\!: \quad 1 + 3 + 5 + \cdots + (2n - 1) = n^2$$

Then

$$\begin{aligned} &P(1)\!: && 1 = 1^2 \\ &P(2)\!: && 1 + 3 = 2^2 \\ &P(3)\!: && 1 + 3 + 5 = 3^2 \\ &P(4)\!: && 1 + 3 + 5 + 7 = 4^2 \\ &\quad\vdots && \quad\vdots \\ &P(100)\!: && 1 + 3 + 5 + \cdots + 199 = 100^2 \\ &P(x - 1)\!: && 1 + 3 + 5 + \cdots + (2x - 3) = (x - 1)^2 \\ &P(x)\!: && 1 + 3 + 5 + \cdots + (2x - 1) = x^2 \\ &P(x + 1)\!: && 1 + 3 + 5 + \cdots + (2x + 1) = (x + 1)^2 \end{aligned}$$

We want to show that $P(n)$ is true for *all* n when $n = 1, n = 2, \ldots$ (n a positive integer). Let S denote the set of positive integers for which $P(n)$ is true. If we can show that 1 is in S and that if k is in S then $k + 1$ is in S, we know that all positive integers are in S. This leads to a statement of the **Principle of Mathematical Induction**.

Principle of Mathematical Induction (PMI)

> If a given proposition $P(n)$ is true for $P(1)$ and if the truth of $P(k)$ implies the truth of $P(k + 1)$, then $P(n)$ is true for all positive integers.

This sets up the following procedure for proof by mathematical induction.

1. **Prove $P(1)$ is true.**
2. **Assume $P(k)$ is true.**
3. **Prove $P(k + 1)$ is true.**
4. **Conclude that $P(n)$ is true for all positive integers n.**

Students often have a certain uneasiness when they first use the Principle of Mathematical Induction as a method of proof. Suppose this principle is used with a stack of large dominoes set up as shown in the cartoon. How can the man in the cartoon be certain of knocking over all the dominoes?

1. He would have to be able to knock over the first one.
2. He would have to have the dominoes arranged so that if the kth domino falls, then the next one, the $(k + 1)$st, will also fall. That is, each domino is set up so that if it falls it causes the next one to fall. We have set up a kind of chain reaction here. The first domino falls; this knocks over the next one (the second domino); the second one knocks over the next one (the third domino); the third one knocks over the next one; and so on. This continues until all the dominoes are knocked over.

EXAMPLE 1 Prove that $1 + 3 + 5 + \cdots + (2n - 1) = n^2$ is true for all positive integers n.

Solution Proof:

Step 1 Prove $P(1)$ true: $1 = 1^2$ is true.

Step 2 Assume $P(k)$ true: $1 + 3 + 5 + \cdots + (2k - 1) = k^2$.

Step 3 Prove $P(k + 1)$ true.
To prove:

$$1 + 3 + 5 + \cdots + [2(k + 1) - 1] = (k + 1)^2$$

or

$$1 + 3 + 5 + \cdots + (2k + 1) = (k + 1)^2 \quad \textit{Since } 2(k + 1) - 1 = 2k + 2 - 1 = 2k + 1$$

Statements	Reasons
1. $1 + 3 + 5 + \cdots + (2k - 1) = k^2$	1. By hypothesis (step 2)
2. $1 + 3 + 5 + \cdots + (2k - 1) + (2k + 1) = k^2 + (2k + 1)$	2. Add $(2k + 1)$ to both sides
3. $1 + 3 + 5 + \cdots + (2k - 1) + (2k + 1) = k^2 + 2k + 1$	3. Associative
4. $1 + 3 + 5 + \cdots + (2k - 1) + (2k + 1) = (k + 1)^2$	4. Factoring (distributive)
5. $1 + 3 + 5 + \cdots + (2k - 1) + [2(k + 1) - 1] = (k + 1)^2$	5. $2k + 1 = 2k + 2 - 1$ $= 2(k + 1) - 1$

Step 4 The proposition $P(n)$ is true for all positive integers n by PMI (the Principle of Mathematical Induction). ■

Remember: A single example showing that a proposition is false serves as a counterexample to disprove the proposition.

EXAMPLE 2 Prove or disprove that $2 + 4 + 6 + \cdots + 2n = n(n + 1)$ is true for all positive integers n.

Solution *Step 1* Prove $P(1)$ true: $2 \stackrel{?}{=} 1(1 + 1)$
$2 = 2$; it is true

Step 2 Assume $P(k)$.
Hypothesis: $2 + 4 + 6 + \cdots + 2k = k(k + 1)$

Step 3 Prove $P(k + 1)$.
To prove: $2 + 4 + 6 + \cdots + 2(k + 1) = (k + 1)(k + 2)$

Statements	Reasons
1. $2 + 4 + 6 + \cdots + 2k = k(k + 1)$	1. Hypothesis (step 2)
2. $2 + 4 + 6 + \cdots + 2k + 2(k + 1)$ $= k(k + 1) + 2(k + 1)$	2. Add $2(k + 1)$ to both sides
3. $= (k + 1)(k + 2)$	3. Factoring

Step 4 The proposition $P(n)$ is true for all positive integers n by PMI. ■

Mathematical induction does not apply only to propositions involving sums of terms, as shown by the next examples.

EXAMPLE 3 Prove or disprove that $n^3 + 2n$ is divisible by 3 for all positive integers n.

Solution Proof: An integer is divisible by 3 if it has a factor of 3.

Step 1 Prove $P(1)$: $1^3 + 2 \cdot 1 = 3$, which is divisible by 3; thus, $P(1)$ is true.

Step 2 Assume $P(k)$.
Hypothesis: $k^3 + 2k$ is divisible by 3

Step 3 Prove $P(k + 1)$.
To prove: $(k + 1)^3 + 2(k + 1)$ is divisible by 3

Statements	Reasons
1. $(k + 1)^3 + 2(k + 1)$ $= k^3 + 3k^2 + 3k + 1 + 2k + 2$	1. Distributive, associative, and commutative axioms
2. $= (3k^2 + 3k + 3) + (k^3 + 2k)$	2. Commutative and associative
3. $= 3(k^2 + k + 1) + (k^3 + 2k)$	3. Distributive
4. $3(k^2 + k + 1)$ is divisible by 3	4. Definition of divisibility by 3
5. $k^3 + 2k$ is divisible by 3	5. Hypothesis
6. $(k + 1)^3 + 2(k + 1)$ is divisible by 3	6. Both terms are divisible by 3; therefore the sum is divisible by 3

Step 4 The proposition $P(n)$ is true for all positive integers n by PMI. ■

EXAMPLE 4 Prove or disprove that $n + 1$ is prime for all positive integers n.

Solution Proof:

Step 1 Prove $P(1)$: $1 + 1 = 2$ is a prime

Step 2 Assume $P(k)$.
Hypothesis: $k + 1$ is a prime

Step 3 Prove $P(k + 1)$.
To prove: $(k + 1) + 1$ is a prime; it is not possible since $(k + 1) + 1 = k + 2$, which is not prime whenever k is an even positive integer.

Step 4 Any conclusion? You cannot conclude that it is false—only that induction does not work. But it is in fact false, and a counterexample is found by letting $n = 3$; then $n + 1 = 4$ is not prime. ■

Example 4 shows that even though you made an assumption in step 2, it is not going to change the conclusion. If the proposition you are trying to prove is not true, making the assumption in step 2 that it is true will *not* enable you to prove it true. Another common mistake of students working with mathematical induction for the first time is illustrated by Example 5.

EXAMPLE 5 Prove that $1 \cdot 2 \cdot 3 \cdot 4 \cdots n < 0$ for all positive integers n.

Solution Students often slip into the habit of skipping either the first or second step in a proof by mathematical induction. This is dangerous. It is important to check every step. Suppose a careless person did not verify the first step. Then the results could be as follows:

Step 2 Assume $P(k)$.
Hypothesis: $1 \cdot 2 \cdot 3 \cdots k < 0$

Step 3 Prove $P(k+1)$.
To prove: $1 \cdot 2 \cdot 3 \cdots k \cdot (k+1) < 0$
Proof: $1 \cdot 2 \cdot 3 \cdots k < 0$ by hypothesis; $k+1$ is positive since k is a positive integer; therefore

$$\underbrace{1 \cdot 2 \cdot 3 \cdots k}_{\text{negative}} \cdot \underbrace{(k+1)}_{\text{positive}} < 0$$

Step 3 is proved.

Step 4 The proposition is not true for all positive integers, since the first step, $1 < 0$, does not hold. ■

PROBLEM SET 8.1

A

If $P(n)$ represents each statement given in Problems 1–10, state and prove or disprove $P(1)$.

1. $5 + 9 + 13 + \cdots + (4n+1) = n(2n+3)$
2. $3 + 9 + 15 + \cdots + (6n-3) = 3n^2$
3. $2^2 + 4^2 + 6^2 + \cdots + (2n)^2 = \dfrac{2n(n+1)(2n+1)}{3}$
4. $1^3 + 2^3 + 3^3 + \cdots + n^3 = \dfrac{n^2(n+1)^2}{4}$
5. $\cos(\theta + n\pi) = (-1)^n \cos\theta$
6. $\sin(\frac{\pi}{4} + n\pi) = (-1)^n(\frac{\sqrt{2}}{2})$
7. $n^2 + n$ is even
8. $n^3 - n + 3$ is divisible by 3 (Steps in proof may vary.)
9. $\left(\dfrac{2}{3}\right)^{n+1} < \left(\dfrac{2}{3}\right)^n$
10. $1 + 2n \le 3^n$

If $P(n)$ represents each statement given in Problems 11–20, state $P(k)$ and $P(k+1)$.

11. $5 + 9 + 13 + \cdots + (4n+1) = n(2n+3)$
12. $3 + 9 + 15 + \cdots + (6n-3) = 3n^2$
13. $2^2 + 4^2 + 6^2 + \cdots + (2n)^2 = \dfrac{2n(n+1)(2n+1)}{3}$
14. $1^3 + 2^3 + 3^3 + \cdots + n^3 = \dfrac{n^2(n+1)^2}{4}$
15. $\cos(\theta + n\pi) = (-1)^n \cos\theta$
16. $\sin(\frac{\pi}{4} + n\pi) = (-1)^n(\frac{\sqrt{2}}{2})$
17. $n^2 + n$ is even
18. $n^3 - n + 3$ is divisible by 3
19. $\left(\dfrac{2}{3}\right)^{n+1} < \left(\dfrac{2}{3}\right)^n$
20. $1 + 2n \le 3^n$

In each of Problems 21–35, prove that the given formula is true for all positive integers n.

21. $1 + 2 + 3 + \cdots + n = \dfrac{n(n+1)}{2}$
22. $1 + 4 + 7 + \cdots + (3n-2) = \dfrac{n(3n-1)}{2}$
23. $5 + 9 + 13 + \cdots + (4n+1) = n(2n+3)$
24. $3 + 9 + 15 + \cdots + (6n-3) = 3n^2$

25. $2 + 7 + 12 + \cdots + (5n - 3) = \dfrac{n(5n-1)}{2}$

26. $1^2 + 2^2 + 3^2 + \cdots + n^2 = \dfrac{n(n+1)(2n+1)}{6}$

27. $1^2 + 3^2 + 5^2 + \cdots + (2n-1)^2 = \dfrac{n(2n-1)(2n+1)}{3}$

28. $1^3 + 2^3 + 3^3 + \cdots + n^3 = \dfrac{n^2(n+1)^2}{4}$

29. $2^2 + 4^2 + 6^2 + \cdots + (2n)^2 = \dfrac{2n(n+1)(2n+1)}{3}$

30. $1 \cdot 2 + 2 \cdot 3 + 3 \cdot 4 + \cdots + n(n+1) = \dfrac{n(n+1)(n+2)}{3}$

31. $1 \cdot 3 + 2 \cdot 4 + 3 \cdot 5 + \cdots + n(n+2) = \dfrac{n(n+1)(2n+7)}{6}$

32. $3 + 3^2 + \cdots + 3^n = \dfrac{3^{n+1} - 3}{2}$

33. $1 + 5 + 5^2 + \cdots + 5^n = \dfrac{5^{n+1} - 1}{4}$

34. $1 + r + r^2 + \cdots + r^n = \dfrac{r^{n+1} - 1}{r - 1}$

35. $\log(a_1 a_2 \cdots a_n) = \log a_1 + \log a_2 + \cdots + \log a_n$ for $n \geq 2$ (Assume all the a_i are positive real numbers.)

B

Define $b^{n+1} = b^n \cdot b$ and $b^0 = 1$. Use this definition to prove the properties of exponents in Problems 36–39 for all positive integers n.

36. $b^m \cdot b^n = b^{m+n}$

37. $(b^m)^n = b^{mn}$

38. $(ab)^n = a^n b^n$

39. $\left(\dfrac{a}{b}\right)^n = \dfrac{a^n}{b^n}$

Prove that the statements in Problems 40–53 are true for every positive integer n.

40. $\cos(\theta + n\pi) = (-1)^n \cos\theta$

41. $\sin(\frac{\pi}{4} + n\pi) = (-1)^n(\frac{\sqrt{2}}{2})$

42. $\cos(\frac{\pi}{3} + n\pi) = \dfrac{(-1)^n}{2}$

43. $n^2 + n$ is even

44. $n^5 - n$ is divisible by 5

45. $n(n+1)(n+2)$ is divisible by 6

46. $n^3 - n + 3$ is divisible by 3

47. $10^{n+1} + 3 \cdot 10^n + 5$ is divisible by 9

48. $(1 + n)^2 \geq 1 + n^2$

49. $2^n > n$

50. $\left(\dfrac{2}{3}\right)^{n+1} < \left(\dfrac{2}{3}\right)^n$

51. $1 + 2n \leq 3^n$

52. $\dfrac{1}{2} + \dfrac{1}{3} + \dfrac{1}{4} + \dfrac{1}{5} + \cdots + \dfrac{1}{n+1} < n$

53. $1 + 2 + 3 + \cdots + n < \dfrac{(2n+1)^2}{8}$

54. Prove the generalized distributive property: $a(b_1 + b_2 + \cdots + b_n) = ab_1 + ab_2 + \cdots + ab_n$.

C

55. Notice the following:

$$1^3 = 1^2$$
$$1^3 + 2^3 = 3^2$$
$$1^3 + 2^3 + 3^3 = 6^2$$
$$1^3 + 2^3 + 3^3 + 4^3 = 10^2$$

Make a conjecture based on this pattern and then prove your conjecture.

56. Notice the following:

$$1 = 1$$
$$1 + 4 = 5$$
$$1 + 4 + 7 = 12$$
$$1 + 4 + 7 + 10 = 22$$

Make a conjecture based on this pattern and then prove your conjecture.

57. Notice the following:

$$2 = 2$$
$$2 + 2 \cdot 3 = 8$$
$$2 + (2 \cdot 3) + (2 \cdot 3^2) = 26$$
$$2 + (2 \cdot 3) + (2 \cdot 3^2) + (2 \cdot 3^3) = 80$$

Make a conjecture based on this pattern and then prove your conjecture.

58. Notice the following:

$$(-1)^1 = -1$$
$$(-1)^1 + (-1)^2 = 0$$
$$(-1)^1 + (-1)^2 + (-1)^3 = -1$$
$$(-1)^1 + (-1)^2 + (-1)^3 + (-1)^4 = 0$$

Make a conjecture based on this pattern and then prove your conjecture.

59. Prove $\sin x + \sin^2 x + \cdots + \sin^n x = \dfrac{\sin^{n+1} x - \sin x}{\sin x - 1}$ for all positive integers n.

60. Prove $e^x + e^{2x} + \cdots + e^{nx} = \dfrac{e^{(n+1)x} - e^x}{e^x - 1}$ for all positive integers n.

61. Use mathematical induction to prove De Moivre's Theorem: $(r \operatorname{cis} \theta)^n = r^n \operatorname{cis}(n\theta)$ for every positive integer n.

8.2 BINOMIAL THEOREM

In mathematics it is frequently necessary to expand $(a + b)^n$. If n is very large, direct calculation is rather tedious so we try to find an easy pattern that will not only help us find $(a + b)^n$ but will also allow us to find any given term in that expansion.

Consider the powers of $(a + b)$, which are found by direct multiplication:

$$\begin{aligned}
(a+b)^0 &= 1\\
(a+b)^1 &= \mathbf{1}\cdot a + \mathbf{1}\cdot b\\
(a+b)^2 &= \mathbf{1}\cdot a^2 + \mathbf{2}\cdot ab + \mathbf{1}\cdot b^2\\
(a+b)^3 &= \mathbf{1}\cdot a^3 + \mathbf{3}\cdot a^2b + \mathbf{3}\cdot ab^2 + \mathbf{1}\cdot b^3\\
(a+b)^4 &= \mathbf{1}\cdot a^4 + \mathbf{4}\cdot a^3b + \mathbf{6}\cdot a^2b^2 + \mathbf{4}\cdot ab^3 + \mathbf{1}\cdot b^4\\
(a+b)^5 &= \mathbf{1}\cdot a^5 + \mathbf{5}\cdot a^4b + \mathbf{10}\cdot a^3b^2 + \mathbf{10}\cdot a^2b^3 + \mathbf{5}\cdot ab^4 + \mathbf{1}\cdot b^5\\
&\vdots
\end{aligned}$$

Ignore the coefficients and focus your attention only on the variables:

$$\begin{array}{llllll}
(a+b)^1: & a & b & & & \\
(a+b)^2: & a^2 & ab & b^2 & & \\
(a+b)^3: & a^3 & a^2b & ab^2 & b^3 & \\
(a+b)^4: & a^4 & a^3b & a^2b^2 & ab^3 & b^4\\
 & & & \vdots & &
\end{array}$$

Do you see a pattern? As you read from left to right, the powers of a decrease and the powers of b increase. Notice that the sum of the exponents for each term is the same as the original exponent:

$$(a+b)^n:\quad a^nb^0 \quad a^{n-1}b^1 \quad a^{n-2}b^2 \cdots a^{n-r}b^r \cdots a^2b^{n-2} \quad a^1b^{n-1} \quad a^0b^n$$

Next consider the coefficients:

$$\begin{array}{lccccccccccc}
(a+b)^0: & & & & & & 1 & & & & & \\
(a+b)^1: & & & & & 1 & & 1 & & & & \\
(a+b)^2: & & & & 1 & & 2 & & 1 & & & \\
(a+b)^3: & & & 1 & & 3 & & 3 & & 1 & & \\
(a+b)^4: & & 1 & & 4 & & 6 & & 4 & & 1 & \\
(a+b)^5: & 1 & & 5 & & 10 & & 10 & & 5 & & 1\\
 & & & & & & \vdots & & & & &
\end{array}$$

Do you see this pattern? This arrangement of numbers is called **Pascal's triangle**. The rows and columns of Pascal's triangle are usually numbered as shown in Figure 8.1.

Notice that the rows of Pascal's triangle are numbered to correspond to the exponent n on $(a + b)^n$. There are many relationships associated with this pattern,

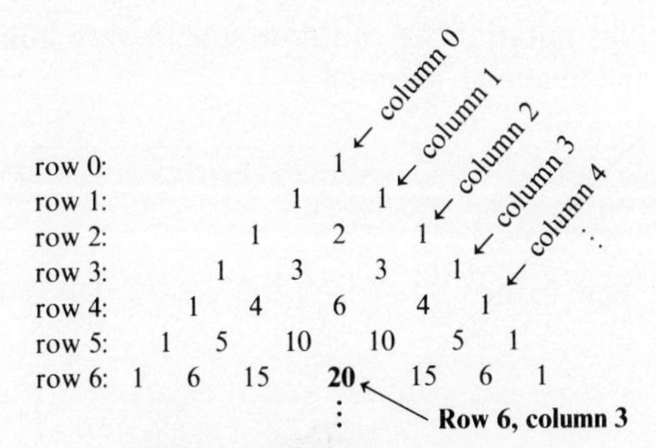

Figure 8.1 Pascal's triangle

but we are concerned with an expression representing the entries in the pattern. Do you see how to generate additional rows of the triangle?

1. Each row begins and ends with a 1.
2. Notice that we began counting the rows with row 0. This is because after row 0, the second entry in the row is the same as the row number. Thus row 7 begins 1 7 ....
3. The triangle is symmetric about the middle. This means that the entries of each row are the same at the beginning and the end. Thus row 7 ends with ... 7 1. (This property is proved in Problem 58.)
4. To find new entries, we can simply add the two entries just above in the preceding row. Thus row 7 is found by looking at row 6:

Row 6: 1 6 15 20 15 6 1

Row 7: 1 7 21 35 35 21 7 1

(This property is proved in Problem 57.)

Write $\binom{n}{r}$ to represent the element in the nth row and rth column of the triangle. Therefore:

$$\binom{0}{0} = 1$$

$$\binom{1}{0} = 1 \quad \binom{1}{1} = 1$$

$$\binom{2}{0} = 1 \quad \binom{2}{1} = 2 \quad \binom{2}{2} = 1$$

$$\binom{3}{0} = 1 \quad \binom{3}{1} = 3 \quad \binom{3}{2} = 3 \quad \binom{3}{3} = 1$$

$$\vdots$$

Using this notation we can now state a very important theorem in mathematics called the **Binomial Theorem**:

Binomial Theorem

For any positive integer n,

$$(a+b)^n = \binom{n}{0}a^n + \binom{n}{1}a^{n-1}b + \binom{n}{2}a^{n-2}b^2 + \cdots + \binom{n}{r}a^{n-r}b^r + \cdots$$
$$+ \binom{n}{n-2}a^2b^{n-2} + \binom{n}{n-1}ab^{n-1} + \binom{n}{n}b^n$$

The proof of the Binomial Theorem is by mathematical induction and the procedure is lengthy, so we leave the proof for the problem set. You are led through the steps and are asked to fill in the details in Problems 60–63.

EXAMPLE 1 Find $(x+y)^8$.

Solution Use Pascal's triangle to obtain the coefficients in the expansion. Thus

$$(x+y)^8 = x^8 + 8x^7y + 28x^6y^2 + 56x^5y^3 + 70x^4y^4 + 56x^3y^5 + 28x^2y^6 + 8xy^7 + y^8$$ ■

EXAMPLE 2 Find $(x-2y)^4$.

Solution In this example, $a = x$ and $b = -2y$ and the coefficients are in Pascal's triangle.

$$(x-2y)^4 = 1 \cdot x^4 + 4 \cdot x^3(-2y)^1 + 6 \cdot x^2(-2y)^2 + 4 \cdot x(-2y)^3 + 1 \cdot (-2y)^4$$
$$= x^4 - 8x^3y + 24x^2y^2 - 32xy^3 + 16y^4$$ ■

If the power of the binomial is very large, then Pascal's triangle is not efficient, so the next step is to find a formula for $\binom{n}{r}$. This formula is found by using a notation called **factorial notation**.

Factorial Notation

The symbol

$$n! = 1 \cdot 2 \cdot 3 \cdots (n-1) \cdot n$$

is called ***n* factorial** (n a natural number). Also, we define $0! = 1$ and $1! = 1$.

EXAMPLE 3

$0! = 1$

$1! = 1$

$2! = 1 \cdot 2 = 2$

$3! = 1 \cdot 2 \cdot 3 = 6$

$4! = 1 \cdot 2 \cdot 3 \cdot 4 = 24$

$5! = 1 \cdot 2 \cdot 3 \cdot 4 \cdot 5 = 120$

$$6! = 1 \cdot 2 \cdot 3 \cdot 4 \cdot 5 \cdot 6 = 720$$
$$7! = 1 \cdot 2 \cdot 3 \cdot 4 \cdot 5 \cdot 6 \cdot 7 = 5040$$
$$8! = 1 \cdot 2 \cdot 3 \cdot \cdots \cdot 7 \cdot 8 = 40{,}320$$
$$9! = 1 \cdot 2 \cdot 3 \cdot \cdots \cdot 8 \cdot 9 = 362{,}880$$
$$10! = 1 \cdot 2 \cdot 3 \cdot \cdots \cdot 9 \cdot 10 = 3{,}628{,}800$$

■

In finding the values of Example 3, notice that $3! = 3 \cdot 2!$, $4! = 4 \cdot 3!$, $5! = 5 \cdot 4!, \ldots$; in general $n! = n(n-1)!$

EXAMPLE 4 Evaluate the given expressions.

a. $5! - 4! = 120 - 24$
$= 96$

b. $(5 - 4)! = 1!$
$= 1$

c. $\dfrac{8!}{4!} = \dfrac{8 \cdot 7 \cdot 6 \cdot 5 \cdot \not{4} \cdot \not{3} \cdot \not{2} \cdot \not{1}}{\not{4} \cdot \not{3} \cdot \not{2} \cdot \not{1}}$
$= 8 \cdot 7 \cdot 6 \cdot 5$
$= 1680$

d. $\left(\dfrac{8}{4}\right)! = 2!$
$= 2$

e. $\dfrac{10!}{8!} = \dfrac{10 \cdot 9 \cdot \not{8!}}{\not{8!}}$
$= 90$

Notice that
$10! = 10 \cdot 9!$
$= 10 \cdot 9 \cdot 8!$
$= 10 \cdot 9 \cdot 8 \cdot 7!$
and so on.

■

Now we can state a formula for $\binom{n}{r}$ using this notation.

Binomial Coefficient

> The symbol $\binom{n}{r}$ is defined for integers r and n such that $0 \le r \le n$.
>
> $$\binom{n}{r} = \frac{n!}{r!(n-r)!}$$
>
> is called the **binomial coefficient *n*, *r*.**

EXAMPLE 5 Find $\binom{3}{2}$ both by Pascal's triangle and by formula.

Solution Row 3, column 2; entry in Pascal's triangle is 3. By formula,

$$\binom{3}{2} = \frac{3!}{2!(3-2)!}$$
$$= \frac{3!}{2!1!}$$
$$= 3$$

■

EXAMPLE 6 Find $\binom{6}{4}$ both by Pascal's triangle and by formula.

Solution Row 6, column 4; entry is 15. By formula,

$$\begin{aligned}\binom{6}{4} &= \frac{6!}{4!(6-4)!}\\ &= \frac{6!}{4!2!}\\ &= \frac{6\cdot 5\cdot 4!}{2\cdot 1\cdot 4!}\\ &= 15\end{aligned}$$

■

EXAMPLE 7 Evaluate the given expressions. (You would not use Pascal's triangle for these.)

a. $$\begin{aligned}\binom{52}{2} &= \frac{52!}{2!(52-2)!}\\ &= \frac{52!}{2!50!}\\ &= \frac{52\cdot 51\cdot 50!}{2\cdot 1\cdot 50!}\\ &= 1326\end{aligned}$$

b. $$\begin{aligned}\binom{n}{n} &= \frac{n!}{n!(n-n)!}\\ &= \frac{n!}{n!0!}\\ &= 1\end{aligned}$$

c. $$\begin{aligned}\binom{n}{n-1} &= \frac{n!}{(n-1)![n-(n-1)]!}\\ &= \frac{n!}{(n-1)!1!}\\ &= \frac{n(n-1)!}{(n-1)!}\\ &= n\end{aligned}$$

■

EXAMPLE 8 Find $(a+b)^{15}$.

Solution The power is rather large, so use the formula to find the coefficients.

$$\begin{aligned}(a+b)^{15} &= \binom{15}{0}a^{15} + \binom{15}{1}a^{14}b + \binom{15}{2}a^{13}b^2 + \cdots + \binom{15}{14}ab^{14} + \binom{15}{15}b^{15}\\ &= \frac{15!}{0!15!}a^{15} + \frac{15!}{1!14!}a^{14}b + \frac{15!}{2!13!}a^{13}b^2 + \cdots + \frac{15!}{14!1!}ab^{14} + \frac{15!}{15!0!}b^{15}\\ &= a^{15} + 15a^{14}b + 105a^{13}b^2 + \cdots + 15ab^{14} + b^{15}\end{aligned}$$

■

EXAMPLE 9 Find the coefficient of the term x^2y^{10} in the expansion of $(x + 2y)^{12}$.

Solution $n = 12$, $k = 10$, $a = x$, and $b = 2y$; thus

$$\binom{12}{10}x^2(2y)^{10} = \frac{12!}{10!2!}(2)^{10}x^2y^{10}$$

The coefficient is $66(1024) = 67{,}584$. ■

PROBLEM SET 8.2

A

Evaluate the expressions in Problems 1–24.

1. $4! - 2!$

2. $5! - 3!$

3. $(4 - 2)!$

4. $(6 - 3)!$

5. $\frac{9!}{7!}$

6. $\frac{10!}{6!}$

7. $\frac{12!}{10!}$

8. $\frac{10!}{4!6!}$

9. $\frac{12!}{3!(12 - 3)!}$

10. $\frac{15!}{5!(15 - 5)!}$

11. $\frac{20!}{3!(20 - 3)!}$

12. $\frac{52!}{3!(52 - 3)!}$

13. $\binom{8}{1}$

14. $\binom{5}{4}$

15. $\binom{8}{2}$

16. $\binom{52}{3}$

17. $\binom{8}{3}$

18. $\binom{7}{4}$

19. $\binom{8}{4}$

20. $\binom{5}{5}$

21. $\binom{52}{2}$

22. $\binom{10}{1}$

23. $\binom{1000}{1}$

24. $\binom{574}{2}$

B

In Problems 25–40, expand by using Pascal's triangle or by formula.

25. $(a + b)^6$

26. $(a + b)^7$

27. $(x + 3)^3$

28. $(x - 3)^4$

29. $(x + y)^5$

30. $(x - y)^6$

31. $(x + 2)^5$

32. $(x - 2)^6$

33. $(x + y)^4$

34. $(2x + 3y)^4$

35. $(\frac{1}{2}x + y^3)^3$

36. $(x^{-2} + y^{-2})^4$

37. $(x^{1/2} + y^{1/2})^4$

38. $(1 + x)^{10}$

39. $(1 - x)^8$

40. $(1 - 2y)^6$

Find the first four terms in the expressions given in Problems 41–49.

41. $(a + b)^{10}$

42. $(a + b)^{12}$

43. $(a + b)^{14}$

44. $(x - y)^{15}$

45. $(x + 2y)^{16}$

46. $(x + \sqrt{2})^8$

47. $(x - 2y)^{12}$

48. $(1 - 0.03)^{13}$

49. $(1 - 0.02)^{12}$

50. Find the coefficient of a^5b^6 in $(a - b)^{11}$.

51. Find the coefficient of $a^{14}b$ in $(a^2 - 2b)^8$.

52. Find the coefficient of $a^{10}b^4$ in $(a + b)^{14}$.

53. Find the coefficient of $x^{10}y^2$ in $(2x^2 + \sqrt{y})^9$.

C

54. What is the constant term in the expansion of $(9x^{-1} + x^2/3)^6$?

55. What is the constant term in the expansion of $(4y^{-2} + y^3/2)^5$?

56. Show that

$$\binom{n}{0} + \binom{n}{1} + \binom{n}{2} + \cdots + \binom{n}{n-1} + \binom{n}{n} = 2^n$$

This says that the sum of the entries of the nth row of Pascal's triangle is 2^n.

57. Show that

$$\binom{n-1}{r-1} + \binom{n-1}{r} = \binom{n}{r}$$

This says that to find any entry in Pascal's triangle (except the first and last), simply add the entries above.

58. Show that

$$\binom{n}{r} = \binom{n}{n-r}$$

This says that Pascal's triangle is symmetric.

59. Use the formula $\binom{n}{k} = \dfrac{n!}{k!(n-k)!}$ to prove

$$\binom{k}{r} + \binom{k}{r-1} = \binom{k+1}{r}$$

Problems 60–63 will lead you through the induction proof of the Binomial Theorem. Let n be any positive integer. To prove:

$$(a+b)^n = \binom{n}{0}a^n + \binom{n}{1}a^{n-1}b + \binom{n}{2}a^{n-2}b^2 + \cdots$$

$$+ \binom{n}{r}a^{n-r}b^r + \cdots + \binom{n}{n-2}a^2b^{n-2}$$

$$+ \binom{n}{n-1}ab^{n-1} + \binom{n}{n}b^n$$

60. Prove it true for $n = 1$.

61. Assume it true for $n = k$. Fill in the statement of the hypothesis.

62. Prove it true for $n = k + 1$.
To prove:

$$(a+b)^{k+1} =$$

$$a^{k+1} + \cdots + \left[\binom{k}{r} + \binom{k}{r-1}\right]a^{k-r+1}b^r + \cdots + b^{k+1}$$

Hint: To prove this you will need to use the results of Problem 59.

63. Tie together Problems 60–62 to complete the proof of the Binomial Theorem.

64. What is wrong with the following "proof," which results in $1 = 2$?

$$(a+b)^n = a^n + na^{n-1}b + \frac{n(n-1)}{2!}a^{n-2}b^2$$

$+ \cdots + nab^{n-1} + b^n$ From the Binomial Theorem

Let $n = 0$. Then

$$(a+b)^0 = a^0 + 0 + 0 + \cdots + 0 + b^0$$

By substitution and zero multiplication

$$1 = 1 + 0 + 0 + \cdots + 0 + 1$$

By definition of zero exponent

$$1 = 2$$

8.3

SEQUENCES, SERIES, AND SUMMATION NOTATION

Patterns and proof are two of the cornerstones upon which mathematics is founded. In this chapter these two ideas are tied together. Consider a function whose domain is the set of counting numbers. Remember that the set of counting numbers is $N = \{1, 2, 3, 4, \ldots, n, \ldots\}$.

Infinite Sequence

An **infinite sequence** is a function s with a domain that consists of the set of counting numbers. The number $s(1)$ is called the *first term* of the sequence, $s(2)$ the *second term*, and $s(n)$ the *nth term* or *general term* of the sequence.

For convenience we sometimes refer to an infinite sequence simply as a **sequence** or **progression**. Sometimes we talk of a **finite sequence**, which means that the domain is the finite set $\{1, 2, 3, \ldots, n-1, n\}$ for some natural number n.

Two special types of sequences are studied in this chapter. The first is an **arithmetic sequence**. In an arithmetic sequence, there is a common difference between successive terms. That is, if any term is subtracted from the next term, the result is always the same, and this number is called the **common difference**.

EXAMPLE 1 Show that 1, 4, 7, 10, 13, ____, . . . is an arithmetic sequence and find the missing term.

Solution Look for a common difference by subtracting each term from the succeeding term:

$$4 - 1 = 3$$
$$7 - 4 = 3$$
$$10 - 7 = 3$$
$$13 - 10 = 3$$

↑ The common difference is 3 (be sure to find this for *each* of the *given* terms).

$$x - 13 = 3$$

The common difference is 3; x is the missing term.

Thus $x = 16$ is the next term (which is found by solving the equation). In order to view this as a function whose domain is the set of counting numbers, you will need to wait until we discuss the general term of the sequence in the next section. ■

The second type of sequence is called a **geometric sequence**. In a geometric sequence, there is a common ratio between successive terms. If any term is divided into the next term, the result is always the same, and this number is called the **common ratio**.

EXAMPLE 2 Show that 2, 4, 8, 16, 32, ____, . . . is a geometric sequence and find the missing term.

Solution

$$\frac{4}{2} = 2$$

$$\frac{8}{4} = 2$$

$$\frac{16}{8} = 2$$

$$\frac{32}{16} = 2$$

↑ The common ratio is 2.

$$\frac{x}{32} = 2 \qquad \text{(where } x \text{ is the next term)}$$

Thus

$$x = 64$$ ← The next term ■

There are other sequences, as shown by Example 3, that are neither arithmetic nor geometric.

EXAMPLE 3 Show that 1, 1, 2, 3, 5, 8, 13, ____, ____, ____, . . . is neither arithmetic nor geometric. Find the missing terms.

Solution First check to see if it is arithmetic: $1 - 1 = 0$; $2 - 1 = 1$; there is no *common* difference. Next check to see if it is geometric: $\frac{1}{1} = 1$; $\frac{2}{1} = 2$; there is no *common* ratio. Look for another pattern:

$$1 + 1 = 2$$
$$1 + 2 = 3$$
$$2 + 3 = 5$$
$$3 + 5 = 8$$
$$5 + 8 = 13$$

It looks like the pattern is obtained by adding the two preceding terms:

$$8 + 13 = \mathbf{21}$$
$$13 + 21 = \mathbf{34}$$
$$21 + 34 = \mathbf{55}$$

The missing terms are 21, 34, and 55. ■

A new notation is generally used when working with sequences. Remember that the domain is the set of counting numbers, so a sequence could be defined by

$$s(n) = 3n - 2 \quad \text{where } n \in \{1, 2, 3, \ldots\}$$

Thus

$$s(1) = 3(1) - 2 = 1$$
$$s(2) = 3(2) - 2 = 4$$
$$s(3) = 3(3) - 2 = 7$$
$$\vdots$$

Instead of writing $s(1)$, however, the notation s_1 is used; in place of $s(2)$, s_2; in place of $s(n)$, s_n. Thus s_{15} means the fifteenth term of the sequence. It is found in the same fashion as though the notation $s(15)$ were used:

$$s_{15} = 3(15) - 2$$
$$= 43$$

EXAMPLE 4 Find the first four terms of $s_n = 26 - 6n$.

Solution

$$s_1 = 26 - 6(1) = 20$$
$$s_2 = 26 - 6(2) = 14$$
$$s_3 = 26 - 6(3) = 8$$
$$s_4 = 26 - 6(4) = 2$$

The sequence is **20, 14, 8, 2,**. . . . It is an arithmetic sequence. ■

EXAMPLE 5 Find the first four terms of $s_n = (-2)^n$.

Solution

$s_1 = (-2)^1 = -2$
$s_2 = (-2)^2 = 4$
$s_3 = (-2)^3 = -8$
$s_4 = (-2)^4 = 16$ The sequence is **−2, 4, −8, 16,**.... It is a geometric sequence. ■

EXAMPLE 6 Find the first four terms of $s_n = s_{n-1} + s_{n-2}$, $n \geq 3$, where $s_1 = 1$ and $s_2 = 2$.

Solution

$s_1 = 1$ Given
$s_2 = 2$ Given
$s_3 = s_2 + s_1$
$= 2 + 1$ By substitution
$= 3$
$s_4 = s_3 + s_2$
$= 3 + 2$
$= 5$ The sequence is **1, 2, 3, 5,**.... This sequence is neither arithmetic nor geometric. ■

EXAMPLE 7 Find the first four terms of $s_n = 2n$.

Solution $s_1 = 2, s_2 = 4, s_3 = 6, s_4 = 8$ The sequence is **2, 4, 6, 8,**.... ■

EXAMPLE 8 Find the first four terms of $s_n = 2n + (n-1)(n-2)(n-3)(n-4)$.

Solution

$s_1 = 2(1) + 0 = 2$
$s_2 = 2(2) + 0 = 4$
$s_3 = 2(3) + 0 = 6$
$s_4 = 2(4) + 0 = 8$ The sequence is **2, 4, 6, 8,**.... ■

If you are given a general term, you can find a unique sequence. However, Examples 7 and 8 show that if only a finite number of successive terms is known and no general term is given, then a *unique* general term cannot be given. That is, if we are given the sequence

2, 4, 6, 8, ____

the next term is probably 10 (if we are thinking of the general term of Example 7), but it *may* be something different. In Example 8, $s_1 = 2$, $s_2 = 4$, $s_3 = 6$, $s_4 = 8$, and

$$\begin{aligned} s_5 &= 2(5) + (5-1)(5-2)(5-3)(5-4) \\ &= 10 + (4)(3)(2)(1) \\ &= 34 \end{aligned}$$

In general, you are looking for the simplest general term; nevertheless, you must remember that answers are not unique *unless the general term is given.*

It is sometimes necessary to consider the sum of the terms of a sequence. This sum is called a *series*.

Finite Series

> The indicated sum of the terms of a finite sequence $s_1, s_2, s_3, \ldots, s_n$ is called a **finite series** and is denoted by
>
> $$S_n = s_1 + s_2 + s_3 + \cdots + s_n$$

EXAMPLE 9 Let $s_n = 26 - 6n$ from Example 4. Find S_4.

Solution

$$\begin{aligned} S_4 &= s_1 + s_2 + s_3 + s_4 \\ &= 20 + 14 + 8 + 2 \qquad \text{From Example 4} \\ &= 44 \end{aligned}$$

■

EXAMPLE 10 Let $s_n = (-1)^n n^2$. Find S_3.

Solution $s_1 = (-1)^1(1)^2 = -1$; $s_2 = (-1)^2(2)^2 = 4$; $s_3 = (-1)^3(3)^2 = -9$. Now find S_3:

$$\begin{aligned} S_3 &= s_1 + s_2 + s_3 \\ &= (-1) + 4 + (-9) \\ &= -6 \end{aligned}$$

■

The terms of the sequence in Example 10 alternate in sign:

$$-1, 4, -9, 16, \ldots$$

A factor of $(-1)^n$ or $(-1)^{n+1}$ in the general term will cause the sign of the terms to alternate, creating a series called an **alternating series**.

The next two sections discuss more efficient ways of finding the general term of a sequence, as well as more efficient ways of finding the sum of n terms of these sequences. We will use the following **summation notation** to simplify the way we express sums.

Consider the function $s_k = 2k$ with the domain $N = \{1, 2, 3, 4\}$. The sum of the terms of this finite arithmetic sequence is the sum of the series

$$2 + 4 + 6 + 8$$

k	$s_k = 2k$
1	2
2	4
3	6
4	8

as shown in the table in the margin. Now denote this sum by using the symbol Σ (called **sigma**) as follows:

This is the last natural number in the domain.

$$\sum_{k=1}^{4} 2k = 2 + 4 + 6 + 8$$

This is the function being evaluated; it is the general term of the sequence.

This is the first natural number in the domain.

This variable is called the **index of summation**.

This symbol means to evaluate the function for each number in the domain and *add* the resulting terms.

Thus

$$\sum_{k=1}^{4} 2k = 2 + 4 + 6 + 8 = 20$$

EXAMPLE 11 Let $s_k = 2k - 1$ and $N = \{3, 4, 5, 6\}$. Evaluate $\sum_{k=3}^{6} (2k + 1)$.

Solution

	k	$s_k = 2k + 1$
First natural number in domain; $k = 3$ →	3	7
	4	9
	5	11
Last natural number in domain; $k = 6$ →	6	13

Thus (Σ means to add these values)

$$\sum_{k=3}^{6} (2k + 1) = \underset{k=3}{7} + \underset{k=4}{9} + \underset{k=5}{11} + \underset{k=6}{13} = 40$$

This is obtained by letting $k = 3$ and evaluating $(2k + 1)$. ■

EXAMPLE 12 Expand $\sum_{k=3}^{n} \frac{1}{2^k}$

Solution

$$\sum_{k=3}^{n} \frac{1}{2^k} = \underset{k=3}{\frac{1}{8}} + \underset{k=4}{\frac{1}{16}} + \underset{k=5}{\frac{1}{32}} + \cdots + \underset{k=n}{\frac{1}{2^n}}$$ ■

EXAMPLE 13 Write the sum of the arithmetic series $S_n = a + (a + d) + (a + 2d) + \cdots + (a + nd)$ using sigma notation.

Solution

$$S_n = \sum_{k=1}^{n} [a + (k - 1)d]$$ ■

EXAMPLE 14 Write the sum of the finite geometric series $S_n = a + ar + ar^2 + \cdots + ar^{n-1}$ using sigma notation.

Solution

$$S_n = \sum_{k=1}^{n} ar^{k-1}$$ ■

PROBLEM SET 8.3

A

For the sequences in Problems 1–16, answer the following questions. The answers are not necessarily unique.

a. *Classify each as arithmetic, geometric, or neither.*

b. *If arithmetic, state d; if geometric, state r; if neither, state a pattern in your own words.*

c. *Supply the missing term.*

1. 2, 5, 8, 11, 14, ____

2. 1, 2, 1, 1, 2, 1, 1, 1, 2, 1, 1, 1, 1, ____

3. 3, 6, 12, 24, 48, ____

4. 5, −15, 45, −135, 405, ____

5. 100, 99, 97, 94, 90, ____

6. 1, 1, 2, 3, 5, 8, 13, ____

7. p, pq, pq^2, pq^3, pq^4, ____

8. 97, 86, 75, 64, ____

9. 8, 12, 18, 26, ____

10. $5^5, 5^4, 5^3, 5^2$, ____

11. 2, 5, 2, 5, 5, 2, 5, 5, 5, ____

12. 5, −5, −15, −25, −35, ____

13. $1, \frac{1}{2}, \frac{1}{3}, \frac{2}{3}, \frac{1}{4}, \frac{3}{4}, \frac{1}{5}, \frac{2}{5}, \frac{3}{5}, \frac{4}{5}, \frac{1}{6}$, ____

14. $\frac{4}{3}, 2, 3, 4\frac{1}{2}$, ____

15. 1, 8, 27, 64, 125, ____

16. 2, 8, 18, 32, ____

Find the first three terms of the sequence with the nth term given in Problems 17–28.

17. $s_n = 4n - 3$

18. $s_n = -3 + 3n$

19. $s_n = \dfrac{10}{2^{n-1}}$

20. $s_n = a + nd$

21. $s_n = ar^{n-1}$

22. $s_n = \dfrac{n-1}{n+1}$

23. $s_n = (-1)^n$

24. $s_n = (-1)^n(n+1)$

25. $s_n = 1 + \dfrac{1}{n}$

26. $s_n = \dfrac{n+1}{n}$

27. $s_n = 2$

28. $s_n = -5$

Evaluate the expressions in Problems 29–40.

29. $\sum_{k=2}^{6} k$

30. $\sum_{m=1}^{4} m^2$

31. $\sum_{n=0}^{6} (2n+1)$

32. $\sum_{p=1}^{6} 2p$

33. $\sum_{k=2}^{5} (10 - 2k)$

34. $\sum_{k=2}^{5} (100 - 5k)$

35. $\sum_{k=1}^{5} (-2)^{k-1}$

36. $\sum_{k=0}^{4} 3(-2)^k$

37. $\sum_{k=0}^{3} 2(3^k)$

38. $\sum_{k=1}^{3} (-1)^k(k^2+1)$

39. $\sum_{k=1}^{10} [1^k + (-1)^k]$

40. $\sum_{k=0}^{5} [2^k + (-2)^k]$

B

41. Find the fifteenth term of the sequence in Problem 17.

42. Find the 102nd term of the sequence in Problem 18.

43. Find the tenth term of the sequence in Problem 19.

44. Find the twentieth term of the sequence in Problem 24.

45. Find the third term of the sequence $(-1)^{n+1}5^{n+1}$.

46. Find the second term of the sequence $(-1)^{n-1}7^{n-1}$.

47. Find the first five terms of the sequence where $s_1 = 2$ and $s_n = 3s_{n-1}, n \geq 2$.

48. Find the first five terms of the sequence where $s_1 = 3$ and $s_n = \frac{1}{3}s_{n-1}, n \geq 2$.

49. Find the first five terms of the sequence where $s_1 = 1$, $s_2 = 1$, and $s_n = s_{n-1} + s_{n-2}, n \geq 3$.

50. Find the first five terms of the sequence where $s_1 = 1$, $s_2 = 2$, and $s_n = s_{n-1} + s_{n-2}, n \geq 3$.

51. Write $\frac{1}{2} + \frac{1}{4} + \frac{1}{8} + \cdots + \frac{1}{2^r} + \cdots + \frac{1}{128}$ using summation notation.

52. Write $2 + 4 + 6 + \cdots + 2n + \cdots + 100$ using summation notation.

53. Write $1 + 6 + 36 + 216 + 1296$ using summation notation. The rth term is 6^{r-1}.

54. Write $5 + 15 + 45 + 135 + 405$ using summation notation. The rth term is $5 \cdot 3^{r-1}$.

Classify the situations in Problems 55–59 as arithmetic or geometric. Do NOT answer the questions.

55. Suppose that a teacher obtains a job with a starting salary of \$15,000 and receives a \$500 raise every year thereafter. What will the teacher's salary be in 10 years?

56. Suppose that an autoworker obtains a job with a starting salary of \$20,000 and receives an 8% raise every year thereafter. What will the autoworker's salary be in 10 years?

57. Suppose that a teacher obtains a job with a starting salary of \$15,000 and receives a $3\frac{1}{3}$% raise every year thereafter. What will the teacher's salary be in 10 years?

58. A grocery clerk must stack 30 cases of canned fruit, each containing 24 cans. He decides to display the cans by stacking them in a pyramid where each row after the bottom row contains one less can. Is it possible to use all the cans and end up with a top row of only one can?

59. Suppose that a chain letter asks you to send copies to 10 of your friends. If everyone carries out the directions and sends the chain letter, and if nobody receives more than one chain letter, how many people will be involved in 10 mailings?

60. Compare Problems 55 and 57. Since $3\frac{1}{3}$% of \$15,000 is \$500 do you think that the total amount earned in 10 years will be the same for both of these problems? If not, which one do you feel will be the larger, and why?

C

61. **a.** Write out $\sum_{j=1}^{r} a_j b_j$ without summation notation.

b. Let $b_j = k$ and show that $\sum_{j=1}^{r} ka_j = k\sum_{j=1}^{r} a_j$.

c. Let $a_j = 1$ and show that $\sum_{j=1}^{r} k = kr$.

62. Show that $\sum_{k=1}^{n} (a_k + b_k) = \sum_{k=1}^{n} a_k + \sum_{k=1}^{n} b_k$.

Find the next term for the sequences in Problems 63–65. Use any rule you can defend.

63. 1, 3, 4, 7, 11, 18, 29, ____

64. 225, 625, 1225, 2025, ____

65. 8, 5, 4, 9, 1, ____

8.4 ARITHMETIC SEQUENCES AND SERIES

This section focuses on arithmetic sequences and series. For clarity of notation, we will use s_n and S_n for general sequences that might be arithmetic, geometric, or neither. However, if you wish to denote the terms of a sequence that you know is arithmetic, use $a_1, a_2, a_3, \ldots, a_n$. Recall that an arithmetic sequence has a common difference d, which by definition is

$$d = a_n - a_{n-1}$$

for every $n > 1$. If you write this as

$$a_n = a_{n-1} + d$$

you can find successive terms by substitution.

If $n = 2$, then

$$a_2 = a_1 + d$$

If $n = 3$, then

$$\begin{aligned} a_3 &= a_2 + d \\ &= (a_1 + d) + d && \text{Substitute } a_1 + d \text{ for } a_2. \\ &= a_1 + 2d \end{aligned}$$

If $n = 4$, then

$$\begin{aligned} a_4 &= a_3 + d \\ &= (a_1 + 2d) + d && \text{Substitute} \\ &= a_1 + 3d \end{aligned}$$

Look for a pattern:

$$a_4 = a_1 + 3d$$

↑ n (under a_4); ↑ 1 less than n (under 3)

This pattern leads to the following formula, which can be proved by mathematical induction.

General Term of an Arithmetic Sequence

> The **general term of an arithmetic sequence** $a_1, a_2, a_3, \ldots, a_n, \ldots$ with common difference d is
>
> $$a_n = a_1 + (n - 1)d$$
>
> for $n \geq 1$.

EXAMPLE 1 Find the general term of the arithmetic sequence 18, 14, 10, 6, . . .

Solution $a_1 = 18$; $d = 14 - 18 = -4$; thus

$$\begin{aligned} a_n &= a_1 + (n-1)d \\ &= 18 + (n-1)(-4) \\ &= 18 - 4n + 4 \\ &= 22 - 4n \end{aligned}$$

■

EXAMPLE 2 If $a_5 = 14$, $a_{10} = 34$, find d.

Solution Use the formula $a_n = a_1 + (n-1)d$:

$a_5 = a_1 + 4d$ or, since $a_5 = 14$, $14 = a_1 + 4d$

$a_{10} = a_1 + 9d$ or, since $a_{10} = 34$, $34 = a_1 + 9d$

This can be written as the system

$$\begin{cases} a_1 + 4d = 14 \\ a_1 + 9d = 34 \end{cases}$$

Multiply the first equation by -1 and add:

$$5d = 20$$
$$d = 4$$

■

What is the total number of blocks shown in Figure 8.2?

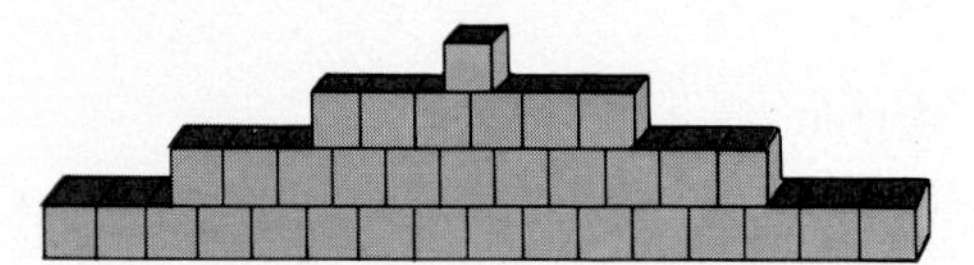

Figure 8.2 How many blocks?

The number of blocks in the successive rows form an arithmetic sequence: 1, 6, 11, 16. The total number of blocks is the sum

$$\sum_{j=1}^{4} (-4 + 5j)$$

The indicated sum of an arithmetic sequence is called an **arithmetic series**. Let A_n denote the arithmetic series

$$\sum_{k=1}^{n} a_k = a_1 + a_2 + a_3 + \cdots + a_n$$

EXAMPLE 3
a. How many blocks are in the stack shown in Figure 8.2?
b. How many blocks are in 10 rows of a stack of blocks similar to the one shown in Figure 8.2?

Solution **a.** $A_1 = a_1 = 1$

$A_2 = a_1 + a_2 = 1 + 6 = 7$

$A_3 = a_1 + a_2 + a_3 = 1 + 6 + 11 = 18$

$A_4 = a_1 + a_2 + a_3 + a_4 = 1 + 6 + 11 + 16 = 34$

This is the number of blocks in Figure 8.2.

b. $A_{10} = a_1 + a_2 + \cdots + a_9 + a_{10}$

$= 1 + 6 + 11 + 16 + 21 + 26 + 31 + 36 + 41 + 46$

$= 235$ blocks in a stack 10 rows high ■

In Example 3, we found A_{10} by brute force addition. This method would not be practical for large n, however, so it is desirable to find a general formula for A_n as we did for a_n. Consider the following method for finding A_{10} of Example 3:

$$A_{10} = 1 + 6 + 11 + 16 + 21 + 26 + 31 + 36 + 41 + 46$$

and

$$A_{10} = 46 + 41 + 36 + 31 + 26 + 21 + 16 + 11 + 6 + 1$$

Add these equations term by term:

$$A_{10} + A_{10} = (1 + 46) + (6 + 41) + (11 + 36) + \cdots + (41 + 6) + (46 + 1)$$
$$2A_{10} = 47 + 47 + 47 + \cdots + 47 + 47$$

Notice that the sums of all the numbers within parentheses are equal to the sum of the first and last terms. Instead of doing a lot of addition, the result can be found by multiplication. In this example $n = 10$, so the number of terms (without directly counting them) is 10. Thus

$$2A_{10} = 10(47)$$

(10: Number of terms; 47: Sum of first and last terms)

$$A_{10} = \frac{10(47)}{2}$$
$$= 10\left(\frac{47}{2}\right)$$

($\frac{47}{2}$: Average of first and last terms)

$$= 235$$

Arithmetic Series

For an arithmetic sequence $a_1, a_2, a_3, \ldots, a_n$, with common difference d, the sum of the arithmetic series

$$\sum_{k=1}^{n} a_k = a_1 + a_2 + a_3 + \cdots + a_n$$

is

$$A_n = n\left(\frac{a_1 + a_n}{2}\right) \quad \text{or, equivalently,} \quad A_n = \frac{n}{2}[2a_1 + (n - 1)d]$$

Proof

$$A_n = a_1 + a_2 + a_3 + \cdots + a_{n-1} + a_n$$
$$A_n = a_n + a_{n-1} + \cdots + a_3 + a_2 + a_1$$

Add these equations term by term:

$$2A_n = (a_1 + a_n) + (a_2 + a_{n-1}) + \cdots + (a_{n-1} + a_2) + (a_n + a_1)$$

All the quantities enclosed by parentheses are equal because

$$\begin{aligned}
a_1 + a_n &= a_1 + [a_1 + (n-1)d] \\
&= \mathbf{2a_1 + (n-1)d} \\
a_2 + a_{n-1} &= (a_1 + d) + [a_1 + (n-2)d] \\
&= \mathbf{2a_1 + (n-1)d} \\
a_3 + a_{n-2} &= (a_1 + 2d) + [a_1 + (n-3)d] \\
&= \mathbf{2a_1 + (n-1)d} \\
&\vdots
\end{aligned}$$

Since there is a total of n such sums,

$$2A_n = n[2a_1 + (n-1)d]$$

$$A_n = \frac{n}{2}[2a_1 + (n-1)d]$$

For the other part of the formula, replace $2a_1 + (n-1)d$ with $a_1 + a_n$ because

$$\begin{aligned}
a_1 + a_n &= a_1 + [a_1 + (n-1)d] \\
&= 2a_1 + (n-1)d
\end{aligned}$$

Therefore

$$A_n = n\left(\frac{a_1 + a_n}{2}\right)$$

□

EXAMPLE 4 Find A_{100} for the sequence of Example 3.

Solution $a_1 = 1$; $n = 100$; $d = 6 - 1 = 5$; thus

$$\begin{aligned}
A_{100} &= \frac{100}{2}[2(1) + 99(5)] \\
&= 50[2 + 495] \\
&= \mathbf{24{,}850}
\end{aligned}$$

■

EXAMPLE 5 Find A_{10} where $a_1 = 6$ and $a_7 = -18$.

Solution First find d:

$$a_n = a_1 + (n-1)d$$

$$a_7 = a_1 + (7-1)d \qquad \text{Replace } n \text{ by 7.}$$

$$-18 = 6 + 6d \qquad \text{Substitute the given values of } a_7 \text{ and } a_1.$$

$$-24 = 6d$$

$$-4 = d$$

Then find A_{10}:

$$A_n = \frac{n}{2}[2a_1 + (n-1)d]$$

$$\begin{aligned} A_{10} &= \frac{10}{2}[2(6) + (10-1)(-4)] \\ &= 5[12 - 36] \\ &= \mathbf{-120} \end{aligned}$$

■

PROBLEM SET 8.4

A

Write out the first four terms of the arithmetic sequences in Problems 1–12.

1. $a_1 = 5, d = 4$
2. $a_1 = -5, d = 3$
3. $a_1 = 85, d = 3$
4. $a_1 = 50, d = -10$
5. $a_1 = 100, d = -5$
6. $a_1 = 20, d = -4$
7. $a_1 = -\frac{5}{2}, d = \frac{1}{2}$
8. $a_1 = \frac{2}{3}, d = -\frac{5}{3}$
9. $a_1 = \sqrt{12}, d = \sqrt{3}$
10. $a_1 = 5, d = x$
11. $a_1 = x, d = y$
12. $a_1 = m, d = 2m$

Find a_1 and d for the arithmetic sequences in Problems 13–21.

13. $5, 8, 11, \ldots$
14. $5, 5, 5, \ldots$
15. $6, 11, 16, \ldots$
16. $35, 46, 57, \ldots$
17. $-8, -1, 6, \ldots$
18. $-1, 1, 3, \ldots$
19. $x, 2x, 3x, \ldots$
20. $x + \sqrt{3}, x + \sqrt{12}, x + \sqrt{27}, \ldots$
21. $x - 5b, x - 3b, x - b, \ldots$

B

Find an expression for the general term of each arithmetic sequence in Problems 22–30.

22. $5, 8, 11, \ldots$
23. $5, 5, 5, \ldots$
24. $6, 11, 16, \ldots$
25. $35, 46, 57, \ldots$
26. $-8, -1, 6, \ldots$
27. $-1, 1, 3, \ldots$
28. $x, 2x, 3x, \ldots$
29. $x + \sqrt{3}, x + \sqrt{12}, x + \sqrt{27}, \ldots$
30. $x - 5b, x - 3b, x - b, \ldots$

Find the indicated quantity for each arithmetic sequence in Problems 31–45.

31. $a_1 = 6, d = 5; a_{20}$
32. $a_1 = 35, d = 11; a_{10}$
33. $a_1 = -20, d = 5; a_{10}$
34. $a_1 = 35, d = 11; A_{10}$
35. $a_1 = -7, d = -2; A_{100}$
36. $a_1 = 15, d = -4; A_{50}$
37. $a_1 = -5, a_{30} = -63; d$
38. $a_1 = 4, a_6 = 24; d$
39. $a_1 = -13, a_{10} = 5; d$
40. $a_1 = 4, a_6 = 24; A_{15}$
41. $a_1 = -5, a_{30} = -63; A_{10}$
42. $a_1 = 110, a_{11} = 0; d$
43. $a_5 = 27, a_{10} = 47; d$
44. $a_4 = 36, a_5 = 60; d$
45. $a_3 = 36, a_5 = 60; d$

46. **a.** How many blocks are shown in Figure 8.3?
b. How many blocks would there be in 100 rows of a stack of blocks like the one shown in Figure 8.3?

Figure 8.3 How many blocks?

47. Find the sum of the first 20 terms of the arithmetic sequence with first term 100 and common difference 50.

48. Find the sum of the first 50 terms of the arithmetic sequence with first term −15 and common difference 5.

49. Find the sum of the even integers between 41 and 99.

50. Find the sum of the odd integers between 100 and 80.

51. Find the sum of the first n odd integers.

52. Find the sum of the first n even integers.

53. A sequence $s_1, s_2, \ldots, s_n$ is a **harmonic sequence** if its reciprocals form an arithmetic sequence. Which of the following are harmonic sequences?
a. $1, \frac{1}{2}, \frac{1}{3}, \frac{1}{4}, \frac{1}{5}, \ldots$
b. $\frac{1}{2}, \frac{1}{5}, \frac{1}{8}, \frac{1}{11}, \frac{1}{14}, \ldots$

c. $2, \frac{2}{3}, \frac{2}{5}, \frac{2}{7}, \ldots$
d. $\frac{1}{5}, -\frac{1}{5}, -\frac{1}{15}, -\frac{1}{25}, \ldots$
e. $\frac{3}{4}, \frac{1}{2}, \frac{1}{3}, \frac{2}{9}, \ldots$

54. Consider the arithmetic sequence a_1, x, a_3. The number x can be found as follows:

$$+\begin{cases} x = a_1 + d \\ x = a_3 - d \end{cases}$$

$$2x = a_1 + a_3 \quad \text{By adding}$$

$$x = \frac{a_1 + a_3}{2}$$

Here x is called the **arithmetic mean** between a_1 and a_3. Find the arithmetic mean between each of the given pairs of numbers.
a. 1, 8 **b.** 1, 7 **c.** $-5, 3$ **d.** 80, 88 **e.** 40, 56

55. Find the arithmetic mean (see Problem 54) between each of the given pairs of numbers.
a. 4, 20 **b.** 4, 15 **c.** $\frac{1}{2}, \frac{1}{3}$ **d.** $-10, -2$ **e.** $-\frac{2}{3}, \frac{4}{5}$

C

56. Suppose you were hired for a job paying \$21,000 per year and were given the following options:

Option A: annual salary increase of \$1440
Option B: semiannual salary increase of \$360

Which is the better option?

57. Repeat Problem 56 for the following options:

Option C: quarterly salary increase of \$90
Option D: monthly salary increase of \$10

58. What are the differences in the amounts earned in the first year from options A to D in Problems 56 and 57?

59. Repeat Problem 58 for the first 2 years.

60. Write the arithmetic series for the total amount of money earned in 10 years under the following options described in Problem 56.
a. option A **b.** option B

8.5 GEOMETRIC SEQUENCES AND SERIES

As with arithmetic sequences, we will denote the terms of a geometric sequence by using a special notation. Let $g_1, g_2, g_3, \ldots, g_n$ be the terms of a geometric sequence. To find the general term of a geometric sequence, remember that the common ratio is r. Thus

$$g_2 = rg_1$$
$$g_3 = rg_2 = r(rg_1) = r^2 g_1$$
$$g_4 = rg_3 = r(r^2 g_1) = r^3 g_1$$
$$\vdots$$
$$g_n = g_1 r^{n-1}$$

General Term of a Geometric Sequence

For a geometric sequence $g_1, g_2, g_3, \ldots, g_n$ with a common ratio r,

$$g_n = g_1 r^{n-1}$$

for every $n \geq 1$.

EXAMPLE 1 Find the general term of the geometric sequence 50, 100, 200, ...

Solution

$$g_1 = 50 \qquad r = \frac{100}{50} = 2$$

$$\begin{aligned} g_n &= 50(2)^{n-1} \\ &= 2 \cdot 5^2 \cdot 2^{n-1} \\ &= 2^n \cdot 5^2 \end{aligned}$$

■

EXAMPLE 2 Find r when $g_1 = 20$ and $g_5 = 200$.

Solution Use $g_n = g_1 r^{n-1}$:

$$g_5 = 20r^4$$
$$200 = 20r^4$$
$$10 = r^4$$
$$r = \sqrt[4]{10}$$

■

Suppose you receive the following letter:

> Dear Friend,
>
> This is a chain letter…
>
> Copy this letter six times and send it to six of your friends. In twenty days, you will have good luck.
>
> If you break this chain, you will have bad luck!…

Consider the number of people that could become involved with this chain letter if we assume that everyone carries out their task and does not break the chain. The first mailing would consist of six letters with seven people involved.

$$1 + 6 = 7$$

The second mailing would involve 43 people since the second mailing of 36 letters (each of the six people receiving a letter send out six more letters) is added to the total:

$$1 + 6 + 36 = 43$$

The number of letters in each successive mailing is a number of a geometric sequence:

1st mailing: $6 = 6$
2nd mailing: $6^2 = 36$
3rd mailing: $6^3 = 216$
4th mailing: $6^4 = 1{,}296$
⋮
10th mailing: $6^{10} = 60{,}466{,}176$
11th mailing: $6^{11} = 362{,}797{,}056$

By the eleventh mailing, more letters would have to be sent than there are people in the United States! The number of letters in only two more mailings would exceed the number of men, women, and children in the whole world.

How many people are involved in 11 mailings assuming that no person receives a letter more than once? To answer this question, consider the series associated with the geometric sequence. The G_n represents the sum of the first n terms of the geometric series. In this problem, then, you need to find G_{11}:

$$G_{11} = 1 + 6 + 36 + 216 + \cdots + 60{,}466{,}176 + 362{,}797{,}056$$

You could add these on your calculator, but that would take a long time; instead write these numbers by using exponents:

$$G_{11} = 1 + 6 + 6^2 + 6^3 + \cdots + 6^{10} + 6^{11}$$

Next multiply both sides by 6:

$$6G_{11} = 6 + 6^2 + 6^3 + 6^4 + \cdots + 6^{11} + 6^{12}$$

Finally, consider $G_{11} - 6G_{11}$:

$$G_{11} - 6G_{11} = 1 - 6^{12}$$

$$-5G_{11} = 1 - 6^{12}$$

$$G_{11} = \frac{1 - 6^{12}}{-5} \quad \text{or} \quad \frac{1}{5}(6^{12} - 1)$$

This number is easy to find on your calculator. But, more important, it leads to a procedure for finding G_n in general, where G_n represents the geometric series

$$\sum_{k=1}^{n} g_k = g_1 + g_2 + g_3 + \cdots + g_n$$

The next step is to find a formula for G_n:

$$G_n = g_1 + g_1 r + g_1 r^2 + g_1 r^3 + \cdots + g_1 r^{n-1}$$

Multiply both sides by r:

$$rG_n = g_1 r + g_1 r^2 + g_1 r^3 + g_1 r^4 + \cdots + g_1 r^n$$

Notice that, except for the first and last terms, all the terms in the expressions for G_n and rG_n are the same, so that

$$G_n - rG_n = g_1 - g_1 r^n$$

Now solve for G_n:

$$(1 - r)G_n = g_1(1 - r^n)$$

$$G_n = \frac{g_1(1 - r^n)}{1 - r} \quad (r \neq 1)$$

Geometric Series

For a geometric series $\sum_{k=1}^{n} g_k = g_1 + g_2 + g_3 + \cdots + g_n$ with common ratio $r \neq 1$,

$$G_n = \frac{g_1(1 - r^n)}{1 - r}$$

EXAMPLE 3 Find the sum of the first five terms of a geometric series with $g_1 = -15$ and $r = 2$.

Solution

$$G_n = \frac{g_1(1 - r^n)}{1 - r} = \frac{-15(1 - 2^n)}{1 - 2} = \frac{-15}{-1}(1 - 2^n) = 15(1 - 2^n)$$

$$G_5 = 15(1 - 2^5) = 15(1 - 32) = -465$$ ■

EXAMPLE 4 Find

$$\sum_{k=1}^{10} \left(\frac{1}{2}\right)^k = \left(\frac{1}{2}\right)^1 + \left(\frac{1}{2}\right)^2 + \left(\frac{1}{2}\right)^3 + \left(\frac{1}{2}\right)^4 + \cdots + \left(\frac{1}{2}\right)^{10}$$

$$= \frac{1}{2} + \frac{1}{4} + \frac{1}{8} + \frac{1}{16} + \cdots + \frac{1}{1024}$$

This is a geometric series with $g_1 = \frac{1}{2}$ and $r = \frac{1}{2}$.

$$\sum_{k=1}^{10} \left(\frac{1}{2}\right)^k = G_{10} = \frac{\frac{1}{2}[1 - (\frac{1}{2})^{10}]}{1 - \frac{1}{2}}$$

$$= 1 - \left(\frac{1}{2}\right)^{10}$$

$$= 1 - \frac{1}{1024}$$

$$= \frac{1023}{1024}$$ ■

The geometric series presented above is finite. Consider now an infinite geometric series. Remember:

1. $s_1, s_2, s_3, \ldots$ is a ***sequence***.
2. $s_1 + s_2 + s_3 + \cdots + s_n$ is a ***series*** denoted by S_n.
3. If we consider $s_1 + s_2 + s_3 + \cdots + s_n$, then S_n is called the ***n*th *partial sum***.
4. Now consider $S_1, S_2, S_3, \ldots$. This is a ***sequence of partial sums***.

Consider the sequence of partial sums for Example 4.

$$G_1 = \frac{1}{2}$$

$$G_2 = \frac{1}{2} + \frac{1}{4} = \frac{3}{4}$$

$$G_3 = \frac{1}{2} + \frac{1}{4} + \frac{1}{8} = \frac{7}{8}$$

$$\vdots$$

$$G_{10} = \frac{1023}{1024}$$ This result was found in Example 4.

It appears that the partial sums are getting closer to 1 as n becomes large. We *can* find the sum of an infinite geometric sequence. Consider

$$G_n = \frac{g_1(1 - r^n)}{1 - r}$$
$$= \frac{g_1 - g_1 r^n}{1 - r}$$
$$= \frac{g_1}{1 - r} - \frac{g_1}{1 - r} r^n$$

Now g_1, r, and $1 - r$ are fixed numbers. If $|r| < 1$, then $r^n \to 0$ as $n \to \infty$ and thus

$$G_n \to \frac{g_1}{1 - r} \text{ as } n \to \infty$$

because the second term is approaching zero. This is not a proof, of course, but it does lead to the following result, which can be proved in a calculus course.

Sum of an Infinite Geometric Series

If $g_1, g_2, g_3, \ldots, g_n$ is an infinite geometric sequence with a common ratio r such that $|r| < 1$, then its sum is denoted by G and found by

$$G = \frac{g_1}{1 - r}$$

If $|r| \geq 1$, the infinite geometric series has no sum.

EXAMPLE 5 Find the sum of the series $100 + 50 + 25 + \cdots$ if possible.

Solution Since $g_1 = 100$ and $r = \frac{1}{2}$, then $G = \dfrac{100}{(1 - \frac{1}{2})} = 200.$ ■

EXAMPLE 6 Find the sum of the series $-5 + 10 - 20 + \cdots$ if possible.

Solution $g_1 = -5$ and $r = -2$; since $|r| \geq 1$, this infinite series does not have a sum. ■

EXAMPLE 7 The repeating decimal $0.\overline{72} = 0.72727272\ldots$ is a rational number and can therefore be written as the quotient of two integers. Find this representation by using a geometric series.

Solution

$$0.727272\ldots = 0.72 + 0.0072 + 0.000072 + \cdots$$
$$= 0.72 + 0.72(0.01) + 0.72(0.0001) + \cdots$$
$$= 0.72 + 0.72(0.01) + 0.72(0.01)^2 + \cdots$$

Now $g_1 = 0.72$ and $r = 0.01$; then

$$G = \frac{0.72}{1 - 0.01} \quad \text{Since } |r| < 1$$
$$= \frac{0.72}{0.99} \quad \text{Remember: You want to write this out as a fraction, not as a decimal.}$$
$$= \frac{72/100}{99/100}$$
$$= \frac{72}{99}$$
$$= \frac{8}{11}$$

■

PROBLEM SET 8.5

A

Write out the first three terms of the geometric sequences in Problems 1–9 with first term g_1 and common ratio r.

1. $g_1 = 5, r = 3$

2. $g_1 = -12, r = 3$

3. $g_1 = 1, r = -2$

4. $g_1 = 1, r = 2$

5. $g_1 = -15, r = \frac{1}{5}$

6. $g_1 = 625, r = -\frac{1}{5}$

7. $g_1 = 8, r = x$

8. $g_1 = a, r = \frac{1}{2}$

9. $g_1 = x, r = y$

Find g_1 and r for the geometric sequences in Problems 10–15.

10. $3, 6, 12, \ldots$

11. $7, 14, 28, \ldots$

12. $1, \frac{1}{2}, \frac{1}{4}, \ldots$

13. $100, 50, 25, \ldots$

14. $x, x^2, x^3, \ldots$

15. $xyz, xy, \ldots$

Find the general term for each of the geometric sequences in Problems 16–21.

16. $3, 6, 12, \ldots$

17. $7, 14, 28, \ldots$

18. $1, \frac{1}{2}, \frac{1}{4}, \ldots$

19. $100, 50, 25, \ldots$

20. $x, x^2, x^3, \ldots$

21. $xyz, xy, \ldots$

Find the sum, if possible, of the infinite geometric series in Problems 22–27.

22. $1 + \frac{1}{2} + \frac{1}{4} + \cdots$

23. $1000 + 500 + 250 + \cdots$

24. $100 + 50 + 25 + \cdots$

25. $-20 + 10 - 5 + \cdots$

26. $-45 - 15 - 5 - \cdots$

27. $-216 - 36 - 6 - \cdots$

B

Find the indicated quantities in Problems 28–39 for the given geometric sequences.

28. $g_1 = 6, r = 3; g_5$

29. $g_1 = 100, r = \frac{1}{10}; g_{10}$

30. $g_1 = 6, r = 3; G_5$

31. $g_1 = 7, g_8 = 896; r$

32. $g_1 = 1, r = 10; G_{10}$

33. $g_1 = 3, r = \frac{1}{5}; G_3$

34. $g_1 = \frac{1}{3}, r = \frac{1}{3}; G$

35. $g_1 = \frac{1}{4}, r = \frac{1}{4}; G$

36. $g_1 = 1, r = 0.08; G$

37. $\sum_{k=1}^{4} \left(\frac{1}{10}\right)^k$

38. $\sum_{k=1}^{4} \left(\frac{1}{3}\right)^k$

39. $\sum_{k=1}^{4} \left(\frac{1}{4}\right)^k$

Represent each repeating decimal in Problems 40–51 as the quotient of two integers by considering an infinite geometric series.

40. $0.\overline{4}$

41. $0.\overline{5}$

42. $0.\overline{9}$

43. $0.\overline{27}$

44. $0.\overline{18}$

45. $0.\overline{45}$

46. $0.\overline{418}$

47. $0.\overline{218}$

48. $0.\overline{123}$

49. $2.\overline{45}$

50. $5.03\overline{1}$

51. $2.2\overline{534}$

52. ***Social Science*** Suppose that a chain letter asks you to send copies to 10 of your friends. If everyone carries out the directions and sends the chain letter, and if nobody receives more than one chain letter, how many people will be involved in five mailings? Count yourself, the 10 letters you mail, the letters they mail, and so forth.

53. ***Social Science*** According to the 1980 census, the U.S. population is 226,504,825. If everyone follows the directions in the chain letter mentioned in Problem 52, how many mailings would be necessary to include the *entire* U.S. population?

54. A new type of Superball advertises that it will rebound to nine-tenths of its original height. If it is dropped from a height of 10 ft, how far will the ball travel before coming to rest?

55. Repeat Problem 54 for a ball that rebounds to two-thirds of its original height.

56. ***Business*** Suppose that a piece of machinery costing \$10,000 depreciates 20% of its present value each year. That is, the first year \$10,000(0.20) = \$2000 is depreciated. The second year's depreciation is

$$\$8000(0.20) = \$1600$$

since the value for the second year is \$10,000 − \$2000 = \$8000. The third year's depreciation is

$$\$6400(0.20) = \$1280$$

If the depreciation is calculated this way indefinitely, what is the total depreciation?

57. ***Business*** Winnie Winner wins \$100 in a pie-baking contest run by the Hi-Do Pie Co. The company gives Winnie the \$100. However, the tax collector wants 20% of the \$100. Winnie pays the tax. But then she realizes that she didn't really win a \$100 prize and tells her story to the Hi-Do Co. The friendly Hi-Do Co. gives Winnie the \$20 she paid in taxes. Unfortunately, the tax collector now wants 20% of the \$20. She pays the tax again and then goes back to the Hi-Do Co. with her story. Assume that this can go on indefinitely. How much money does the Hi-Do Co. have to give Winnie so that she will really win \$100? How much does she pay in taxes?

58. Consider the geometric sequence g_1, x, g_3. The number x can be found by considering

$$\frac{x}{g_1} = r \quad \text{and} \quad \frac{g_3}{x} = r$$

Thus

$$\frac{x}{g_1} = \frac{g_3}{x} \quad \text{so} \quad x^2 = g_1 g_3$$

This equation has two solutions:

$$x = \sqrt{g_1 g_3} \quad \text{and} \quad x = -\sqrt{g_1 g_3}$$

If g_1 and g_3 are both positive, then $\sqrt{g_1 g_3}$ is called the **geometric mean** of g_1 and g_3. If g_1 and g_3 are both negative, then $-\sqrt{g_1 g_3}$ is called the **geometric mean**. Find the geometric mean of each of the given pairs of numbers:

a. 1, 8 **b.** 2, 8 **c.** $-5, -3$
d. $-10, -2$ **e.** 4, 20

Find the sum of the infinite geometric series in Problems 59–62.

59. $2 + \sqrt{2} + 1 + \cdots$

60. $3 + \sqrt{3} + 1 + \cdots$

61. $(1 + \sqrt{2}) + 1 + (-1 + \sqrt{2}) + \cdots$

62. $(\sqrt{2} - 1) + 1 + (\sqrt{2} + 1) + \cdots$

63. Find three distinct numbers with a sum equal to 9 so that these numbers form an arithmetic sequence and their squares form a geometric sequence.

64. Square $ABCD$ has sides of length 1. Square $EFGH$ is formed by connecting the midpoints of the sides of the first square, as shown in Figure 8.4. Assume that the pattern of shaded regions in the square is continued indefinitely. What is the area of the shaded regions?

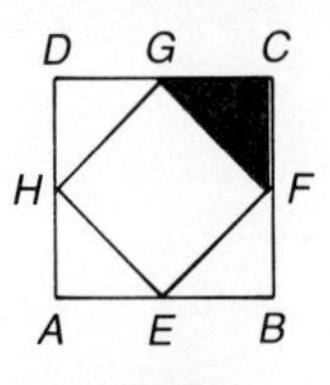

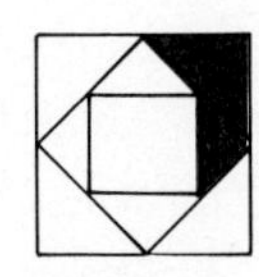

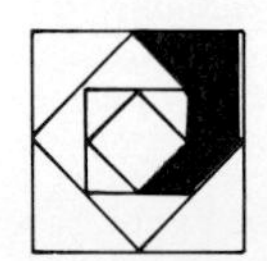

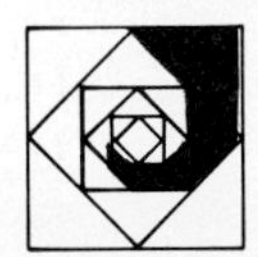

Figure 8.4 Find the area of the shaded regions.

65. Repeat Problem 64 for a square whose sides have length a.

CHAPTER 8 SUMMARY

The material of this chapter is reviewed in the following list of objectives. After each objective there are some practice questions. For a sample test, select the first question of each set and check your answers with the answer section. For a sample test without answers, use the second question of each set. Additional practice is given by the other questions in each set. If you are having trouble with a particular type of problem, look back to that section for extra help.

8.1 MATHEMATICAL INDUCTION

Objective 1 State the principle of mathematical induction and prove propositions using this principle.

1. State the principle of mathematical induction.

2. If $4 + 8 + 12 + \cdots + 4n = 2n(n + 1)$, then state the proposition $P(1)$ and prove it is true.

3. State the propositions $P(k)$ and $P(k + 1)$ for the proposition given in Problem 2.

4. Prove that $4 + 8 + 12 + \cdots + 4n = 2n(n + 1)$ for all positive integers n.

8.2 BINOMIAL THEOREM

Objective 2 Use Pascal's triangle to expand binomials. Expand:

5. $(a + b)^5$ **6.** $(x - y)^5$ **7.** $(2x + y)^5$ **8.** $(3x^2 - y^3)^5$

Objective 3 Simplify expressions containing factorial and binomial coefficient notations.

9. $\dfrac{52!}{5!\,47!}$ **10.** $\dbinom{8}{4}!$ **11.** $\dbinom{8}{4}$ **12.** $\dbinom{p}{q}$

Objective 4 State the binomial theorem, expand binomials by formula, and find particular terms in a binomial expansion.

13. State the binomial formula.

14. Write out the first four terms in the expansion of $(a + b)^{18}$.

15. Write the rth term in the expansion of $(x - y)^{15}$.
16. What is the coefficient of x^8y^4 in the expansion of $(x + 2y)^{12}$?

8.3 SEQUENCES, SERIES, AND SUMMATION NOTATION

Objective 5 Classify a sequence as arithmetic, geometric, or neither. If it is arithmetic or geometric find d, r, and the general term; if it is neither, find the pattern and give the next two terms.

17. 1, 11, 21, 31,...
18. 1, 11, 121, 1331,...
19. 1, 11, 111, 1111,...
20. 54, 18, 6, 2,...

8.4, 8.5 ARITHMETIC AND GEOMETRIC SEQUENCES AND SERIES

Objective 6 Use summation notation with series. Evaluate:

21. $\sum_{k=0}^{3} 3^k$
22. $\sum_{k=1}^{4} 5$
23. $\sum_{k=1}^{10} 2(3)^{k-1}$
24. $\sum_{k=1}^{100} [5 + (k - 1)4]$

Objective 7 Find the sum of n terms of an arithmetic or geometric sequence. Find the sum of the first 10 terms of the following sequences.

25. 1, 11, 21, 31,...
26. 1, 11, 121, 1331,...
27. 54, 18, 6, 2,...
28. 5, 5, 5, 5,...

Objective 8 Work with the following formulas for sequences and series:

Arithmetic Sequence

$$a_n = a_1 + (n - 1)d$$

Geometric Sequence

$$g_n = g_1r^{n-1}$$

Arithmetic Series

$$A_n = n\left(\frac{a_1 + a_n}{2}\right)$$

$$A_n = \frac{n}{2}[2a_1 + (n - 1)d]$$

Geometric Series

$$G_n = \frac{g_1(1 - r^n)}{1 - r}$$

$$G = \frac{g_1}{1 - r}, |r| < 1$$

Find the indicated quantities for the given sequences:

29. $a_1 = 2, a_{10} = 20; d, A_{10}$
30. $a_2 = 5, d = 13; a_n$
31. $g_1 = 5, r = 2; g_{10}, G_5$
32. $a_1 = 50, d = -5; a_{10}, A_5$

Objective 9 Find the sum of an infinite geometric series including the fractional representation of a repeating decimal.

33. Find the sum of the infinite series $1000 + 500 + 250 + \cdots$.
34. Find the sum of the infinite series $\frac{1}{8} + \frac{1}{16} + \frac{1}{32} + \cdots$.
35. Find $2.\overline{18}$ as the quotient of two integers by considering an infinite geometric series.
36. Find $3.1\overline{6}$ as the quotient of two integers by considering an infinite geometric series.

9

Analytic Geometry-Conic Sections

René Descartes (1596–1665)

It is impossible not to feel stirred at the thought of the emotions of men at certain historic moments of adventure and discovery—Columbus when he first saw the Western shore, Pizarro when he stared at the Pacific Ocean, Franklin when the electric spark came from the string of his kite, Galileo when he first turned his telescope to the heavens. Such moments are also granted to students in the abstract regions of thought, and high among them must be placed the morning when Descartes lay in bed and invented the method of coordinate geometry.

A. N. WHITEHEAD
An Introduction to Mathematics

In this text we have seen many ideas which revolutionized mathematics and the nature of thought. None is more profound than the powerful idea of associating ordered pairs of real numbers with points in a plane, thereby making possible a correspondence between curves in the plane and equations in two variables. This idea stems from the French mathematician René Descartes. The rectangular coordinate system is sometimes called the Cartesian coordinate system to honor him. There is a legend about how this idea came to Descartes. Because of his frail health he was accustomed to staying in bed as long as he wished. He thought of this coordinate system while he was lying in bed watching a fly crawl around on the ceiling of his room. He noticed that the path of the fly could be described if he knew the relation connecting the fly's distances from the walls. However, 1800 years earlier the Greek mathematician Apollonius (ca. 262–190 B.C.) made a thorough investigation of the conic sections introduced in this chapter. He considered himself a rival of Archimedes and was the first person to use the words *parabola*, *ellipse*, and *hyperbola.* His methods were expounded in an eight-volume work called *Conics*, and his work was so modern it is sometimes judged to be an analytic geometry preceding Descartes.

CHAPTER OVERVIEW Although many calculus books include analytic geometry, they often assume that students have a rudimentary knowledge of the fundamentals, in particular of conic sections. For that reason, a thorough introduction to the conic sections is presented in this chapter. Section 9.4 on rotations may easily be omitted. There are 10 objectives for this chapter, which are listed on pages 408–409. The final Extended Application, on planetary orbits, follows the Chapter Summary.

9.1 PARABOLAS

In Chapter 2 we looked at quadratic functions of the form

$$y = ax^2 + bx + c \qquad (a \neq 0)$$

The graph of this quadratic equation is a parabola, but not all parabolas can be represented by this equation, because not all parabolas are graphs of functions. Consider the general second-degree equation

$$Ax^2 + Bxy + Cy^2 + Dx + Ey + F = 0$$

for any constants A, B, C, D, E, and F. If $A = B = C = 0$, then the equation is not quadratic but linear (first degree); and if at least one of A, B, or C is not zero, then the equation is quadratic.

Historically, second-degree equations in two variables were first considered in a geometric context and were called **conic sections** because the curves they represent can be described as the intersections of a double-napped right circular cone and a plane. There are three general ways a plane can intersect a cone, as shown in Figure 9.1. (Several special cases are discussed later.)

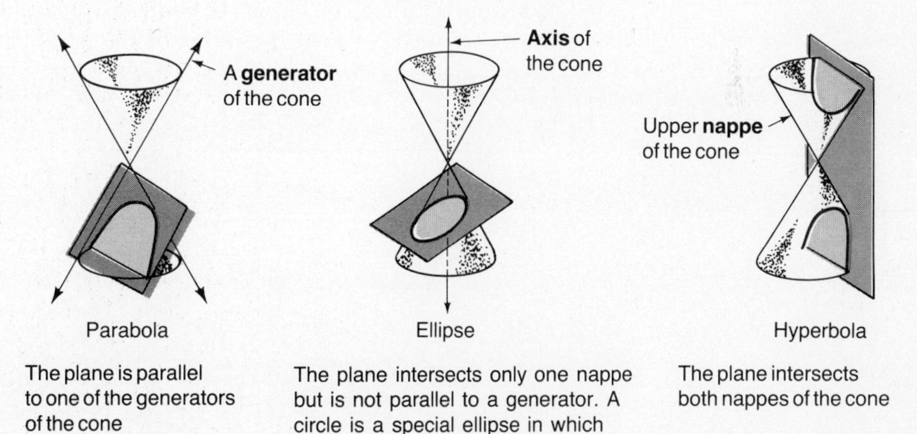

Figure 9.1 Conic sections

Reconsider the parabola; this time take a different geometric viewpoint:

Parabola

A **parabola** is a set of all points in the plane equidistant from a given point (called the **focus**) and a given line (called the **directrix**).

To obtain the graph of a parabola from this definition, you can use the special type of graph paper shown in Figure 9.2, where F is the focus and L is the directrix.

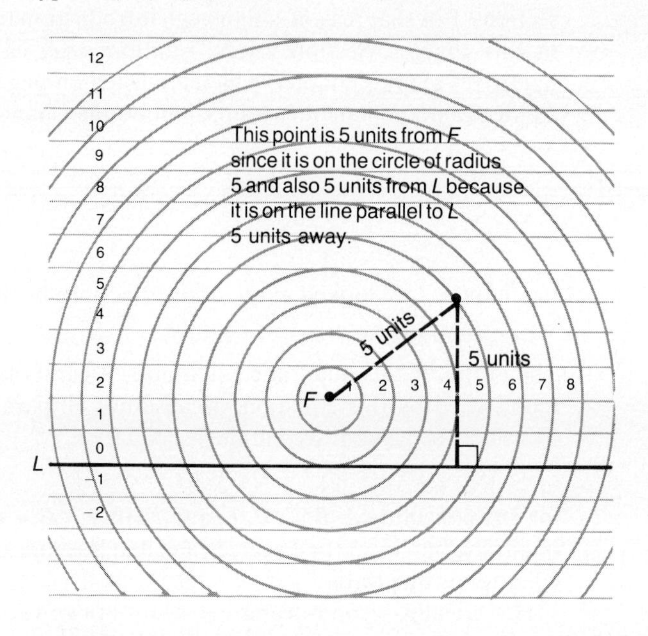

Figure 9.2 Parabola graph paper

To sketch a parabola using the definition, let F be any point and let L be any line, as shown in Figure 9.2. Plot points in the plane equidistant from the focus and the directrix. Draw a line through the focus and perpendicular to the directrix. This line is called the **axis** of the parabola. Let V be the point on this line halfway between the focus and the directrix. This is the point of the parabola nearest to both the focus and the directrix. It is called the **vertex** of the parabola. Plot other points equidistant from F and L as shown in Figure 9.3.

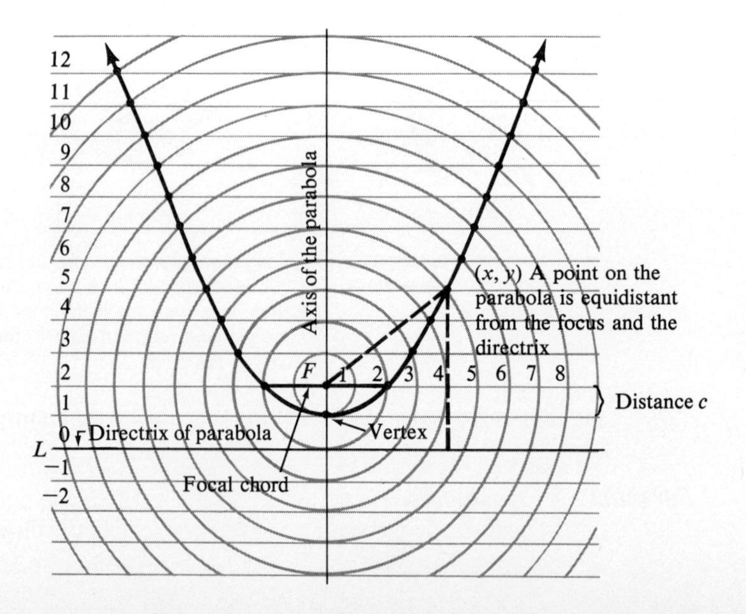

Figure 9.3 Parabola graphed from definition

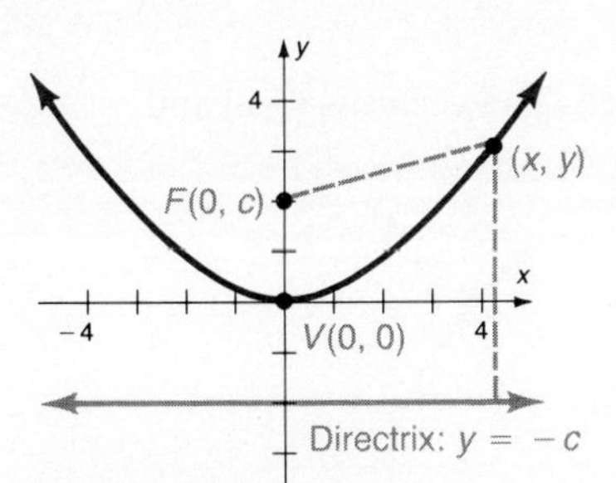

Figure 9.4 Graph of the parabola $x^2 = 4cy$

In Figure 9.3, let c be the distance from the vertex to the focus. Notice that the distance from the vertex to the directrix is also c. Consider the segment that passes through the focus perpendicular to the axis and with endpoints on the parabola. This segment has length $4c$ and is called the **focal chord**.

To obtain the equation of a parabola, first consider a special case—a parabola with focus $F(0, c)$ and directrix $y = -c$, where c is any positive number. This parabola must have its vertex at the origin (remember that the vertex is halfway between the focus and the directrix) and must open upward as shown in Figure 9.4.

Let (x, y) be any point on the parabola. Then, from the definition of a parabola,

$$\text{Distance from } (x, y) \text{ to } (0, c) = \text{distance from } (x, y) \text{ to directrix}$$

or

$$\sqrt{(x-0)^2 + (y-c)^2} = y + c$$

Square both sides:

$$x^2 + (y-c)^2 = (y+c)^2$$
$$x^2 + y^2 - 2cy + c^2 = y^2 + 2cy + c^2$$
$$x^2 = 4cy$$

This is the equation of the parabola with vertex $(0, 0)$ and directrix $y = -c$.

You can repeat this argument for parabolas that have vertex at the origin and open downward, to the left, and to the right to obtain the results summarized below. A positive number c, the distance from the focus to the vertex, is assumed given. These are called the **standard-form parabola equations** with vertex $(0, 0)$.

Standard-Form Equations for Parabolas with Vertex (0, 0)

Parabola	Focus	Directrix	Vertex	Equation
Opens *upward*	$(0, c)$	$y = -c$	$(0, 0)$	$\mathbf{x^2 = 4cy}$
Opens *downward*	$(0, -c)$	$y = c$	$(0, 0)$	$\mathbf{x^2 = -4cy}$
Opens *right*	$(c, 0)$	$x = -c$	$(0, 0)$	$\mathbf{y^2 = 4cx}$
Opens *left*	$(-c, 0)$	$x = c$	$(0, 0)$	$\mathbf{y^2 = -4cx}$

EXAMPLE 1 Graph $x^2 = 8y$

Solution This equation represents a parabola that opens upward. The vertex is $(0, 0)$, and notice by inspection that

$$4c = 8$$
$$c = 2$$

Thus the focus is $(0, 2)$. After plotting the vertex $V(0, 0)$ and the focus $F(0, 2)$, the only question is the width of the parabola. In Chapter 2 you plotted a couple of points or found the y intercept. The method in these examples is more useful and efficient for determining the graph of a parabola. In this chapter we will determine the width of the parabola by using the focal chord. Remember that the **length of the focal chord is 4c**, so that in this case it is 8. Do you see that in each case $4c$ is the absolute value of the coefficient of the first-degree term in the standard-form equations? Since a parabola is symmetric with respect to its axis, draw a segment of length 8 with the midpoint at F. Using these three points (the vertex and the endpoints of the focal chord), sketch the parabola as shown in Figure 9.5. ■

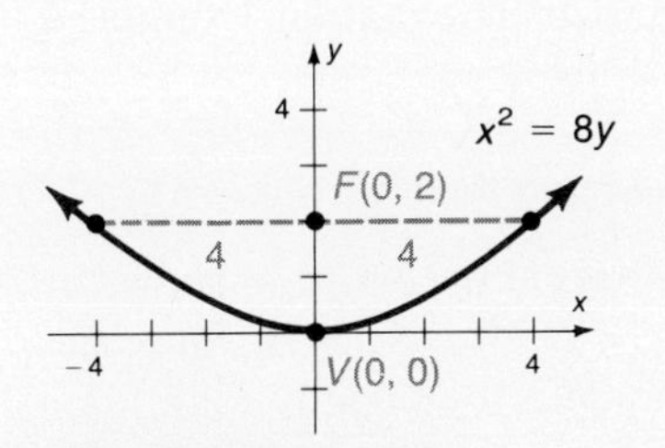

Figure 9.5 Graph of the parabola $x^2 = 8y$

EXAMPLE 2 Graph $y^2 = -12x$.

Solution This equation represents a parabola that opens left. The vertex is (0, 0) and

$$4c = 12$$
$$c = 3$$

(recall that c is positive), so the focus is $(-3, 0)$. The length of the focal chord is 12 and the parabola is drawn as in Figure 9.6.

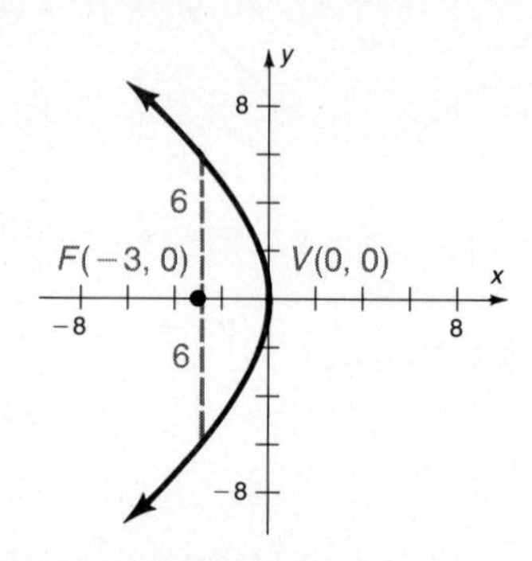

Figure 9.6 Graph of the parabola $y^2 = -12x$ ■

EXAMPLE 3 Graph $2y^2 - 5x = 0$.

Solution You might first put the equation into standard form by solving for the second-degree term:

$$y^2 = \frac{5}{2}x$$

The vertex is (0, 0), and

$$4c = \frac{5}{2}$$

$$c = \frac{5}{8}$$

Thus the parabola opens to the right, the focus is $(\frac{5}{8}, 0)$, and the length of the focal chord is $\frac{5}{2}$, as shown in Figure 9.7.

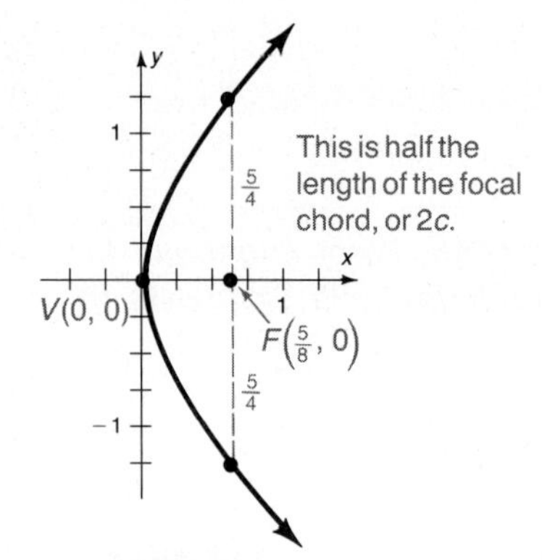

Figure 9.7 Graph of the parabola $2y^2 - 5x = 0$ ■

There are two basic types of problems you will need to solve concerning each curve in analytic geometry:

1. Given the equation, draw the graph; this is what you did in Examples 1 to 3.
2. Given the graph (or information about the graph), write the equation. Example 4 is a problem of this type.

EXAMPLE 4 Find the equation of the parabola with directrix $y = 4$ and focus at $F(0, -4)$.

Solution This curve is a parabola that opens downward with vertex at the origin, as shown in Figure 9.8. The value for c is found by inspection: $c = 4$. Thus $4c = 16$. Since the equation is of the form $x^2 = -4cy$, the desired equation is (by substitution)

$$x^2 = -16y$$

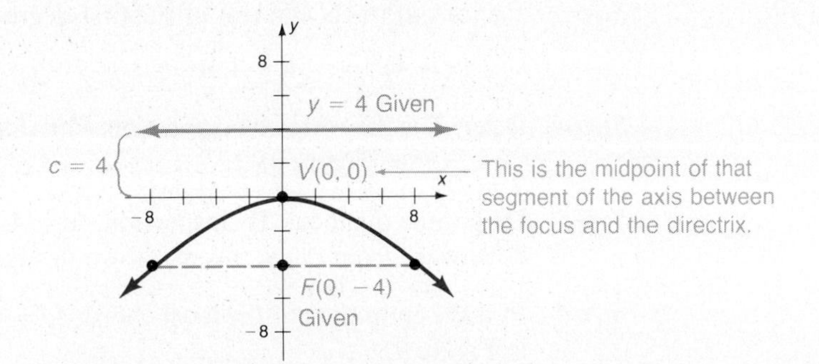

Figure 9.8 Graph of the parabola with focus at $(0, -4)$ and directrix $y = 4$

■

The types of parabolas we have been considering are quite limited, since we have assumed that the vertex is at the origin and the directrix is parallel to one of the coordinate axes. Suppose, however, that you are given a parabola with vertex at (h, k) and a directrix parallel to one of the coordinate axes. In Section 2.4 we showed that you can translate the axes to (h, k) by a substitution:

$$x' = x - h$$
$$y' = y - k$$

Therefore the **standard-form parabola equations** with vertex (h, k) can be summarized by the following table.

Standard-Form Equations for Translated Parabolas

Parabola	Focus	Directrix	Vertex	Equation
Opens upward	$(h, k+c)$	$y=k-c$	(h, k)	$(x-h)^2=4c(y-k)$
Opens downward	$(h, k-c)$	$y=k+c$	(h, k)	$(x-h)^2=-4c(y-k)$
Opens right	$(h+c, k)$	$x=h-c$	(h, k)	$(y-k)^2=4c(x-h)$
Opens left	$(h-c, k)$	$x=h+c$	(h, k)	$(y-k)^2=-4c(x-h)$

For the rest of the material in this chapter you will need to remember the procedure for **completing the square**. Examples 5 and 6 should provide enough detail to refresh your memory; if you need additional review, see Section 2.4, pages 62–67.

EXAMPLE 5 Sketch $x^2 + 4y + 8x + 4 = 0$.

Solution *Step 1* Associate together the terms involving the variable that is squared:

$$x^2 + 8x = -4y - 4$$

Step 2 Complete the square for the variable that is squared.

Coefficient is 1; if it is not, divide both sides by this coefficient.

$$x^2 + 8x + \left(\frac{1}{2}\cdot 8\right)^2 = -4y - 4 + \left(\frac{1}{2}\cdot 8\right)^2$$

Take one-half of this coefficient, square it, and add it to both sides.

$$x^2 + 8x + \mathbf{16} = -4y - 4 + \mathbf{16}$$
$$(x + 4)^2 = -4y + 12$$

Step 3 Factor out the coefficient of the first-degree term:

$$(x + 4)^2 = -4(y - 3)$$

Step 4 Determine the vertex by inspection. Plot (h, k); in this example, the vertex is $(-4, 3)$. (See Figure 9.9.)

Step 5 Determine the focus. By inspection, $4c = 4$, $c = 1$, and the parabola opens downward from the vertex as shown in Figure 9.9.

Step 6 Plot the endpoints of the focal chord; $4c = 4$. Draw the parabola as shown in Figure 9.9.

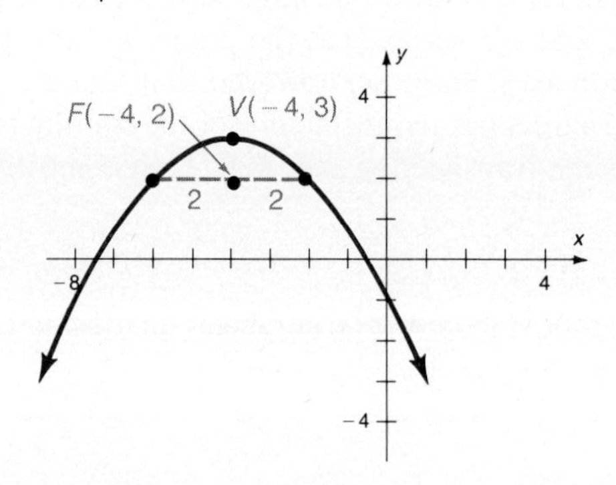

Figure 9.9 Graph of $x^2 + 4y + 8x + 4 = 0$

■

EXAMPLE 6 Sketch $2y^2 + 6y + 5x + 10 = 0$.

Solution

$$2y^2 + 6y = -5x - 10$$

$$y^2 + 3y = -\frac{5}{2}x - 5$$

Divide both sides by 2 so the leading coefficient is 1; then complete the square.

$$y^2 + 3y + \frac{9}{4} = -\frac{5}{2}x - 5 + \frac{9}{4}$$

$$\left(y + \frac{3}{2}\right)^2 = -\frac{5}{2}x - \frac{11}{4}$$

$$\left(y + \frac{3}{2}\right)^2 = -\frac{5}{2}\left(x + \frac{11}{10}\right)$$

The last step is to factor out the coefficient of the first-degree term.

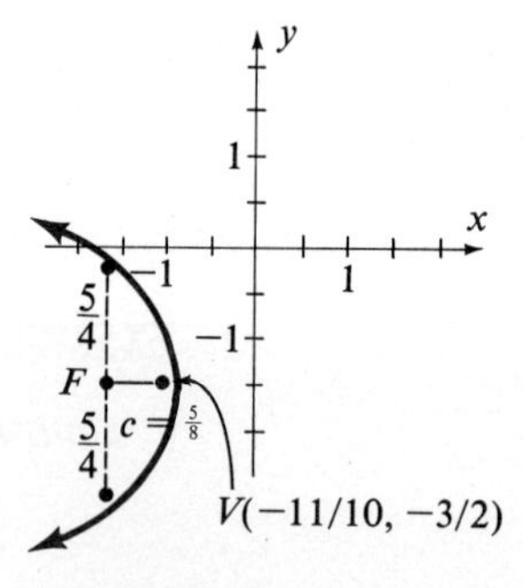

Figure 9.10 Graph of $2y^2 + 6y + 5x + 10 = 0$

The vertex is $(-\frac{11}{10}, -\frac{3}{2})$ and $4c = \frac{5}{2}$, $c = \frac{5}{8}$. Sketch the curve as shown in Figure 9.10.

■

EXAMPLE 7 Find the equation of the parabola with focus at $(4, -3)$ and directrix the line $x + 2 = 0$.

Solution Sketch the given information as shown in Figure 9.11. The vertex is $(1, -3)$ since it must be equidistant from F and the directrix. Note that $c = 3$. Thus substitute into the equation

$$(y - k)^2 = 4c(x - h)$$

since the parabola opens to the right.
The desired equation is

$$(y + 3)^2 = 12(x - 1)$$

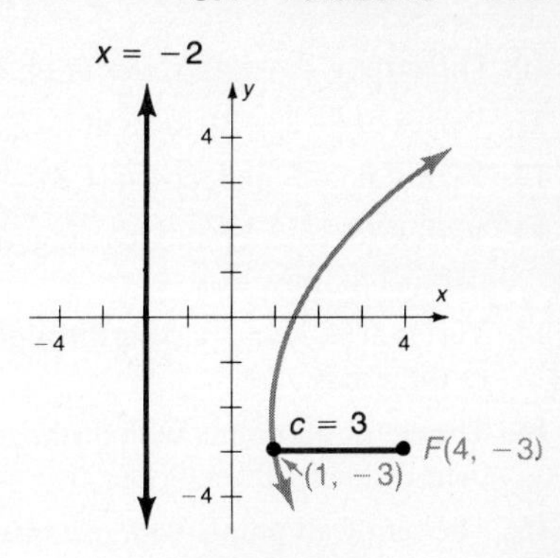

Figure 9.11 ■

By sketching the focus and directrix, it is easy to find c and the vertex by inspection. The sketch of the curve is not necessary in order to find this equation.

EXAMPLE 8 Find the equation of the parabola with vertex at $(2, -1)$, axis parallel to the y axis, and passing through $(3, 2)$.

Solution Sketch the given information as shown in Figure 9.12. The parabola opens upward and thus has the form

$$(x - h)^2 = 4c(y - k)$$

Since the vertex is $(2, -1)$,

$$(x - 2)^2 = 4c(y + 1)$$

Also, since it passes through $(3, 2)$, this point must satisfy the equation

$$(3 - 2)^2 = 4c(2 + 1)$$

Solving for c,

$$c = \frac{1}{12}$$

Therefore the desired equation is

$$(x - 2)^2 = \frac{1}{3}(y + 1)$$

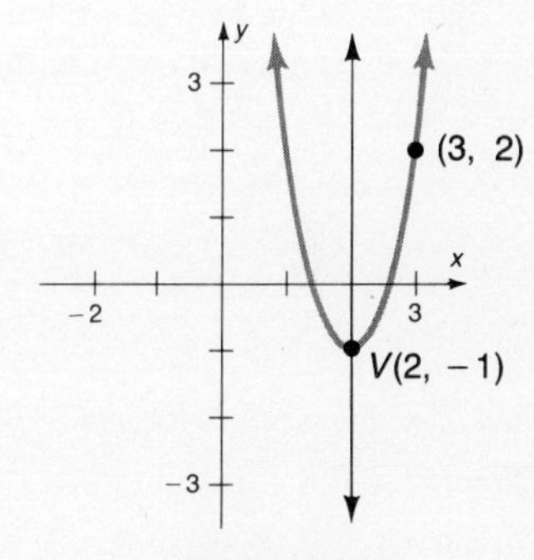

Figure 9.12 ■

PROBLEM SET 9.1

A

In Problems 1–6, use the definition of a parabola to sketch a parabola such that the distance between the focus and the directrix is the given number. (See Figure 9.2.)

1. 4 **2.** 6 **3.** 8 **4.** 10

5. 3 **6.** 7

Sketch the parabola satisfying the conditions given in Problems 7–16.

7. Directrix $x = 0$; focus at $(5, 0)$

8. Directrix $y = 0$; focus at $(0, -3)$

9. Directrix $x - 3 = 0$; vertex at $(-1, 2)$

10. Directrix $y + 4 = 0$; vertex at $(4, -1)$
11. Vertex at $(-2, -3)$; focus at $(-2, 3)$
12. Vertex at $(-3, 4)$; focus at $(1, 4)$
13. Vertex at $(-3, 2)$ and passing through $(-2, -1)$; axis parallel to the y axis
14. Vertex at $(4, 2)$ and passing through $(-3, -4)$; axis parallel to the x axis
15. The set of all points with distances from $(4, 3)$ that equal their distances from $(0, 3)$
16. The set of all points with distances from $(4, 3)$ that equal their distances from $(-2, 1)$

Sketch the curves given by the equations in Problems 17–40. Label the focus F and the vertex V.

17. $y^2 = 8x$
18. $y^2 = -12x$
19. $y^2 = -20x$
20. $4x^2 = 10y$
21. $3x^2 = -12y$
22. $2x^2 = -4y$
23. $2x^2 + 5y = 0$
24. $5y^2 + 15x = 0$
25. $3y^2 - 15x = 0$
26. $4y^2 + 3x = 12$
27. $5x^2 + 4y = 20$
28. $4x^2 + 3y = 12$
29. $(y - 1)^2 = 2(x + 2)$
30. $(y + 3)^2 = 3(x - 1)$
31. $(x + 2)^2 = 2(y - 1)$
32. $(x - 1)^2 = 3(y + 3)$
33. $(x - 1) = -2(y + 2)$
34. $(x + 3) = -3(y - 1)$
35. $y^2 + 4x - 3y + 1 = 0$
36. $y^2 - 4x + 10y + 13 = 0$
37. $2y^2 + 8y - 20x + 148 = 0$
38. $x^2 + 9y - 6x + 18 = 0$
39. $9x^2 + 6x + 18y - 23 = 0$
40. $9x^2 + 6y + 18x - 23 = 0$

B

Find the equation of each curve in Problems 41–50. You sketched these curves in Problems 7–16.

41. Directrix $x = 0$; focus at $(5, 0)$
42. Directrix $y = 0$; focus at $(0, -3)$
43. Directrix $x - 3 = 0$; vertex at $(-1, 2)$
44. Directrix $y + 4 = 0$; vertex at $(4, -1)$
45. Vertex at $(-2, -3)$; focus at $(-2, 3)$
46. Vertex at $(-3, 4)$; focus at $(1, 4)$
47. Vertex at $(-3, 2)$ and passing through $(-2, -1)$; axis parallel to the y axis
48. Vertex at $(4, 2)$ and passing through $(-3, -4)$; axis parallel to the x axis
49. The set of all points with distances from $(4, 3)$ that equal their distances from $(0, 3)$
50. The set of all points with distances from $(4, 3)$ that equal their distances from $(-2, 1)$

51. ***Physics*** If the path of a baseball is parabolic and is 200 ft wide at the base and 50 ft high in the vertex, write the equation that gives the path of the baseball if the origin is the point of departure for the ball.
52. ***Engineering*** A parabolic archway has the dimensions shown in Figure 9.13. Find the equation of the parabolic portion.

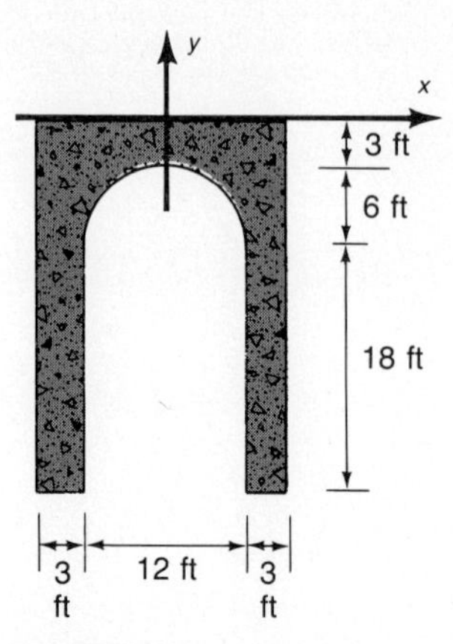

Figure 9.13 A parabolic archway

53. ***Engineering*** A radar antenna is constructed so that a cross section along its axis is a parabola with the receiver at the focus. Find the focus if the antenna is 12 m across and its depth is 4 m. See Figure 9.14.

a Radar antenna

Figure 9.14 A three-dimensional model of a parabola is a **parabolic reflector** (or **parabolic mirror**). A radar antenna serves as an example. If a source of light is placed at the focus of the mirror, the light rays will reflect from the mirror as rays parallel to the axis (as in an automobile headlamp). The radar antenna works in reverse—parallel incoming rays are focused at a single location. (Courtesy of Wide World Photos, Inc.)

b Dimensions for radar antenna in Problem 53

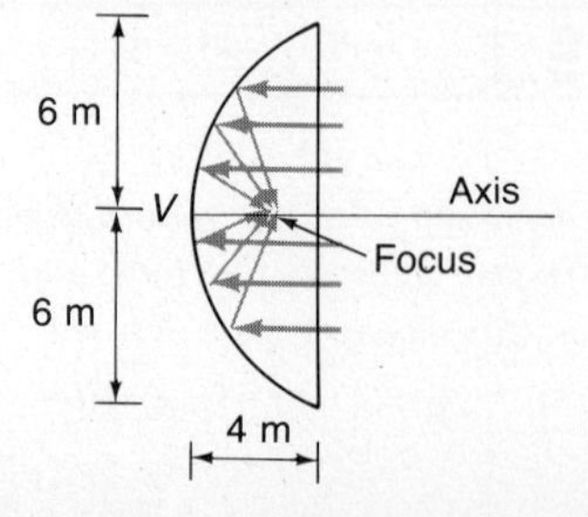

54. ***Engineering*** If the diameter of the parabolic reflector in Problem 53 is 16 cm and the depth is 8 cm, find the focus. (See Figure 9.14.)

C

Find the points of intersection (if any) for each line and parabola in Problems 55–58. Show the result both algebraically and geometrically.

55. $\begin{cases} y = 2x + 10 \\ y = x^2 + 4x + 7 \end{cases}$

56. $\begin{cases} y = -2x + 4 \\ y = x^2 - 12x + 25 \end{cases}$

57. $\begin{cases} 2x + y - 7 = 0 \\ y^2 - 6y - 4x + 17 = 0 \end{cases}$

58. $\begin{cases} 3x + 2y - 5 = 0 \\ y^2 + 4y + 3x - 4 = 0 \end{cases}$

59. ***Sports*** Phil Lee, a physical education expert, has made a study and determined that a woman who runs the 100-m dash reaches her peak at age 20. Her time T, for a 100-m dash at a particular age A, is

$$T = c_1(A - 20)^2 + c_2$$

for constants c_1 and c_2. If Wyomia Tyus ran the race when she was 16 years old in 11.4 sec and when she was 20 in 11.0 sec, predict her time for running the race when she is 40 years old. Graph this function.

60. ***Sports*** According to another physical education expert, Jim Kintzi, the peak age for a woman runner is 24. Using this information, answer the questions posed in Problem 59.

61. Derive the equation of a parabola with $F(0, -c)$, where c is a positive number and the directrix is the line $y = c$.

62. Derive the equation of a parabola with $F(c, 0)$, where c is a positive number and the directrix is the line $x = -c$.

63. Derive the equation of a parabola with $F(-c, 0)$, where c is a positive number and the directrix is the line $x = c$.

64. Show that the length of the focal chord for the parabola $y^2 = 4cx$ is $4c$ $(c > 0)$.

65. Let L: $Ax + By + C = 0$ be any nonvertical line and let $P(x_0, y_0)$ be any point not on the line.
    a. Find the slope of L.
    b. Let L' be a line through P perpendicular to L. Find the slope of L'.
    c. Find the equation of L'.
    d. Let Q be the point of intersection of L and L'. Find Q.
    e. Find the distance between P and Q to show that the distance from a point to a line is

$$d = \frac{|Ax_0 + By_0 + C|}{\sqrt{A^2 + B^2}}$$

66. Find the equation of the parabola with focus at $(4, -3)$ and directrix $x - y + 3 = 0$. (*Hint:* Use the definition of a parabola and the formula derived in Problem 65.)

67. Find the equation of the parabola with focus at $(3, -5)$ and directrix $12x - 5y + 4 = 0$. (*Hint:* Use the definition of a parabola and the formula derived in Problem 65.)

9.2 ELLIPSES

The second conic considered in this chapter is called an *ellipse*.

Ellipse

An **ellipse** is the set of all points in the plane such that, for each point on the ellipse, the sum of its distances from two fixed points is a constant.

The fixed points are called the **foci** (plural of **focus**). To see what an ellipse looks like, we will use the special type of graph paper shown in Figure 9.15a, where F_1 and F_2 are the foci.

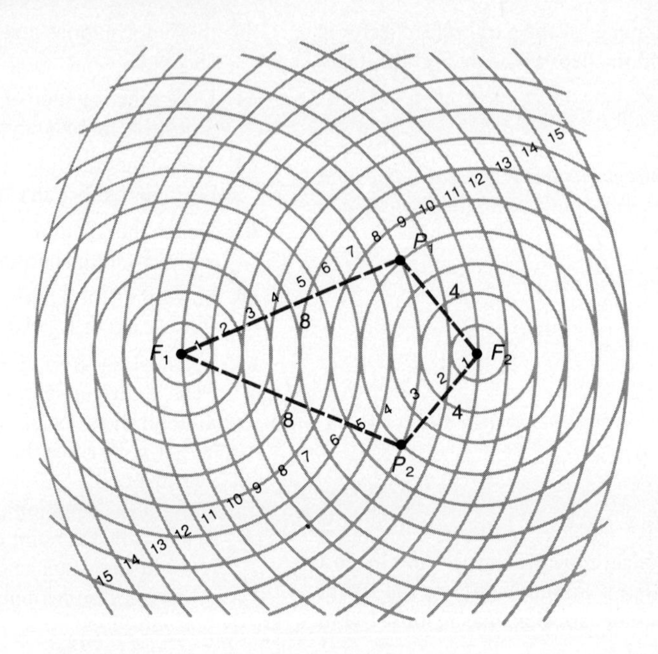

Figure 9.15a Ellipse graph paper

Let the given constant be 12. Plot all the points in the plane so that the sum of their distances from the foci is 12. If a point is 8 units from F_1, for example, then it is 4 units from F_2 and you can plot the points P_1 and P_2. The completed graph of this ellipse is shown in Figure 9.15b.

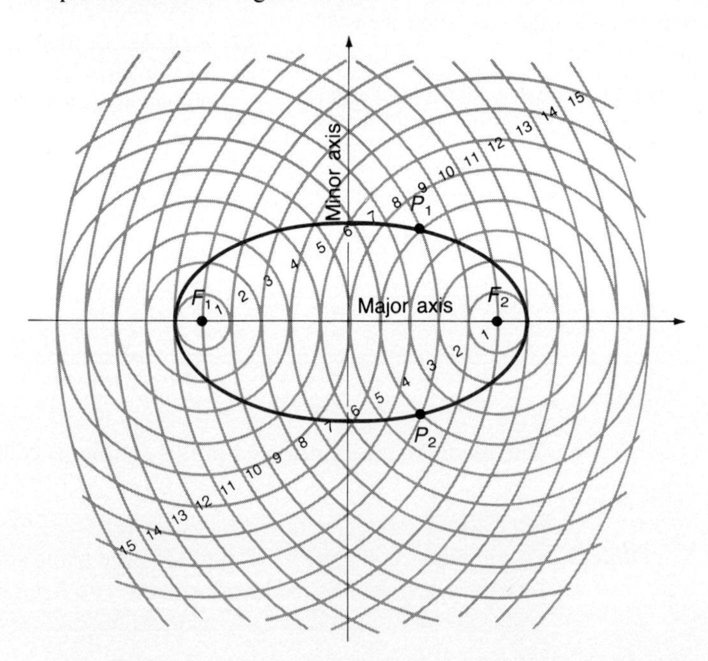

Figure 9.15b Graphing an ellipse

The line passing through F_1 and F_2 is called the **major axis**. The **center** is the midpoint of the segment $\overline{F_1F_2}$. The line passing through the center perpendicular to the major axis is called the **minor axis**. The ellipse is symmetric with respect to both the major and minor axes.

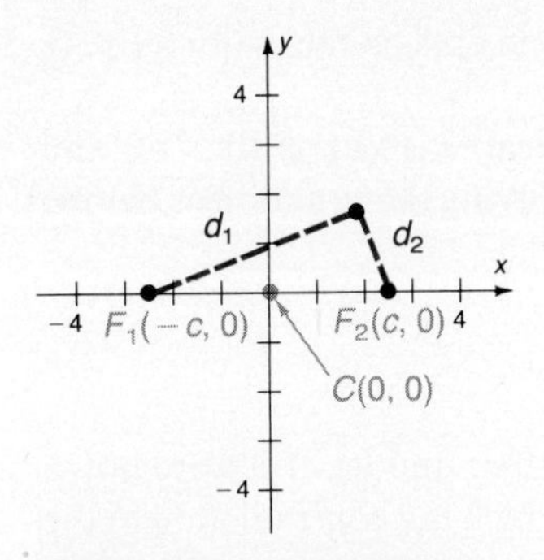

Figure 9.16 Developing the equation of an ellipse by using the definition

To find the equation of an ellipse, first consider a special case where the center is at the origin. Let the distance from the center to a focus be the positive number c; that is, let $F_1(-c, 0)$ and $F_2(c, 0)$ be the foci and let the constant distance be $2a$, as shown in Figure 9.16.

Notice that the center of the ellipse in Figure 9.16 is $(0, 0)$. Let $P(x, y)$ be any point on the ellipse, and use the distance formula and the definition of an ellipse to derive the equation of this ellipse:

$$d_1 + d_2 = 2a$$

or

$$\sqrt{(x+c)^2 + (y-0)^2} + \sqrt{(x-c)^2 + (y-0)^2} = 2a$$

Simplifying,

$$\sqrt{(x+c)^2 + y^2} = 2a - \sqrt{(x-c)^2 + y^2} \qquad \text{Isolate one radical.}$$

$$(x+c)^2 + y^2 = 4a^2 - 4a\sqrt{(x-c)^2 + y^2} + (x-c)^2 + y^2 \qquad \text{Square both sides.}$$

$$x^2 + 2cx + c^2 + y^2 = 4a^2 - 4a\sqrt{(x-c)^2 + y^2} + x^2 - 2cx + c^2 + y^2$$

$$4a\sqrt{(x-c)^2 + y^2} = 4a^2 - 4cx$$

$$\sqrt{(x-c)^2 + y^2} = a - \frac{c}{a}x \qquad \text{Since } a \neq 0\text{, divide by } 4a.$$

$$(x-c)^2 + y^2 = \left(a - \frac{c}{a}x\right)^2 \qquad \text{Square both sides again.}$$

$$x^2 - 2cx + c^2 + y^2 = a^2 - 2cx + \frac{c^2}{a^2}x^2$$

$$x^2 + y^2 = a^2 - c^2 + \frac{c^2}{a^2}x^2$$

$$x^2 - \frac{c^2}{a^2}x^2 + y^2 = a^2 - c^2$$

$$\left(1 - \frac{c^2}{a^2}\right)x^2 + y^2 = a^2 - c^2$$

$$\frac{a^2 - c^2}{a^2}x^2 + y^2 = a^2 - c^2$$

$$\frac{x^2}{a^2} + \frac{y^2}{a^2 - c^2} = 1 \qquad \text{Divide both sides by } a^2 - c^2.$$

Let $b^2 = a^2 - c^2$; then

$$\frac{x^2}{a^2} + \frac{y^2}{b^2} = 1$$

If $x = 0$, the y intercepts are obtained:

$$\frac{y^2}{b^2} = 1$$

$$y = \pm b$$

If $y = 0$, the x intercepts are obtained: $x = \pm a$. The intercepts on the major axis are called the **vertices** of the ellipse.

The equation of the ellipse with major axis vertical, $F_1(0, c)$, $F_2(0, -c)$, and constant distance $2a$ is found in a similar fashion. Simplifying the equation as before,

$$\frac{y^2}{a^2} + \frac{x^2}{b^2} = 1$$

where $b^2 = a^2 - c^2$.

Notice that in both cases a^2 must be larger than both c^2 and b^2. If it were not, a square number would be equal to a negative number, which is a contradiction in the set of real numbers.

Standard-Form Equations for Ellipses with Center (0, 0)

Ellipse	Foci	Constant distance	Center	Equation
Horizontal	$(-c, 0)$, $(c, 0)$	$2a$	$(0, 0)$	$\frac{x^2}{a^2} + \frac{y^2}{b^2} = 1$
Vertical	$(0, c)$, $(0, -c)$	$2a$	$(0, 0)$	$\frac{y^2}{a^2} + \frac{x^2}{b^2} = 1$
where $b^2 = a^2 - c^2$ or $c^2 = a^2 - b^2$				

EXAMPLE 1 Sketch $\frac{x^2}{9} + \frac{y^2}{4} = 1$.

Solution The center of the ellipse is $(0, 0)$. The x intercepts are ± 3 (these are the vertices) and the y intercepts are ± 2. Sketch the ellipse as shown in Figure 9.17.

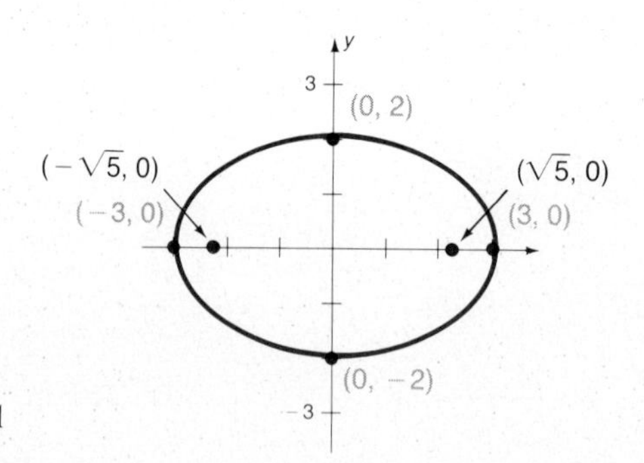

Figure 9.17 Graph of $\frac{x^2}{9} + \frac{y^2}{4} = 1$

The foci can also be found, since

$$c^2 = a^2 - b^2$$
$$c^2 = 9 - 4$$
$$c = \pm\sqrt{5}$$

■

EXAMPLE 2 Sketch $\frac{x^2}{4} + \frac{y^2}{9} = 1$.

Solution Here $a^2 = 9$ and $b^2 = 4$, which is an ellipse with major axis vertical. The x intercepts are ± 2 and the y intercepts are ± 3 (these are the vertices). The sketch is shown in Figure 9.18.

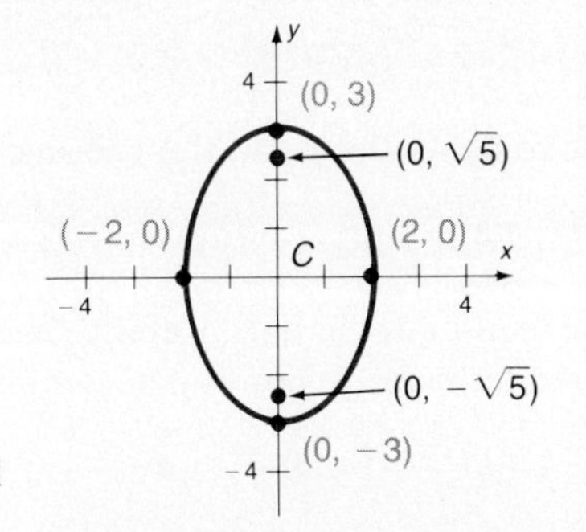

Figure 9.18 Graph of $\frac{x^2}{4} + \frac{y^2}{9} = 1$

The foci are found by

$$c^2 = a^2 - b^2$$
$$c^2 = 9 - 4$$
$$c = \pm\sqrt{5}$$

Remember: The foci are always on the major axis, so plot $(0, \sqrt{5})$ and $(0, -\sqrt{5})$ for the foci. ■

If the center of the ellipse is at (h, k), the equations can be written in terms of a translation as shown by the following table.

Standard-Form Equations for Translated Ellipses

Ellipse	Center	Foci	Intercepts: Major axis	Intercepts: Minor axis	Equation
Horizontal	(h, k)	$(h + c, k)$ $(h - c, k)$	$(h - a, k)$ $(h + a, k)$	$(h, k + b)$ $(h, k - b)$	$\frac{(x-h)^2}{a^2} + \frac{(y-k)^2}{b^2} = 1$
Vertical	(h, k)	$(h, k + c)$ $(h, k - c)$	$(h, k + a)$ $(h, k - a)$	$(h - b, k)$ $(h + b, k)$	$\frac{(y-k)^2}{a^2} + \frac{(x-h)^2}{b^2} = 1$

The segment from the center to a vertex on the major axis is called a **semimajor axis** and has length a; the segment from the center to an intercept on the minor axis is called a **semiminor axis** and has length b.

EXAMPLE 3 Graph $\frac{(x-3)^2}{25} + \frac{(y-1)^2}{16} = 1$.

Solution *Step 1* Plot the center (h, k). By inspection, the center of this ellipse is $(3, 1)$. This becomes the center of a new translated coordinate system. The vertices and foci are now measured with reference to the new origin at $(3, 1)$.

Step 2 Plot the x' and y' intercepts. These are ± 5 and ± 4, respectively. Remember to measure these distances from $(3, 1)$ as shown in Figure 9.19.

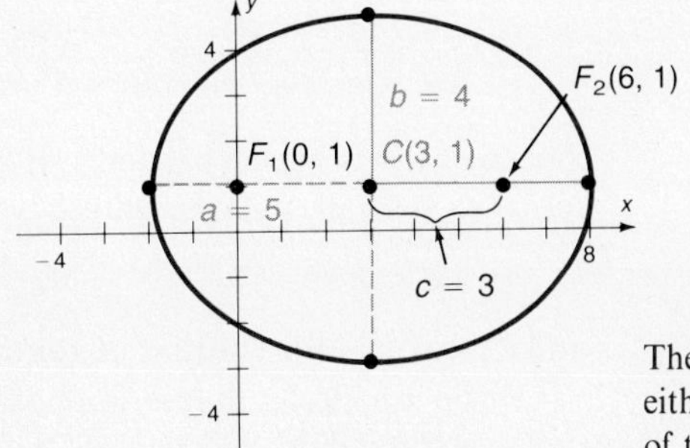

Figure 9.19 Graph of $\frac{(x-3)^2}{25} + \frac{(y-1)^2}{16} = 1$

The foci are found by

$$c^2 = a^2 - b^2$$
$$= 25 - 16$$
$$= \pm 9$$

The distance from the center to either focus is 3, so the coordinates of the foci are $(6, 1)$ and $(0, 1)$. ■

EXAMPLE 4 Graph $3x^2 + 4y^2 + 24x - 16y + 52 = 0$.

Solution *Step 1* Associate together the x and the y terms:

$$(3x^2 + 24x) + (4y^2 - 16y) = -52$$

Step 2 Complete the squares in both x and y. This requires that the coefficients of the squared terms be 1. You can accomplish this by factoring:

$$\mathbf{3}(x^2 + 8x \qquad) + \mathbf{4}(y^2 - 4y \qquad) = -52$$

Next complete the square for both x and y, being sure to add the same number to both sides:

Add 16 to both sides.

$$\mathbf{3}(x^2 + 8x + \mathbf{16}) + \mathbf{4}(y^2 - 4y + \mathbf{4}) = -52 + \mathbf{48} + \mathbf{16}$$

Add 48 to both sides.

Step 3 Factor:

$$3(x + 4)^2 + 4(y - 2)^2 = 12$$

Step 4 Divide both sides by 12:

$$\frac{(x + 4)^2}{4} + \frac{(y - 2)^2}{3} = 1$$

Step 5 Plot the center (h, k). By inspection, you can see the center is $(-4, 2)$. The vertices are at ± 2, and the length of the semiminor axis is $\sqrt{3}$, as shown in Figure 9.20. ■

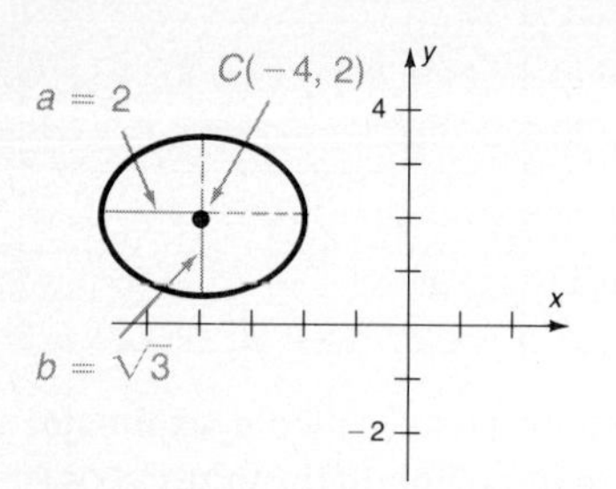

Figure 9.20 Graph of $3x^2 + 4y^2 + 24x - 16y + 52 = 0$

We have seen that some ellipses are more circular and some more flat than others. A measure of the amount of flatness of an ellipse is called its **eccentricity**, which is defined as

$$\epsilon = \frac{c}{a}$$

Notice that

$$\epsilon = \frac{c}{a} = \frac{\sqrt{a^2 - b^2}}{a} = \sqrt{\frac{a^2 - b^2}{a^2}} = \sqrt{1 - \left(\frac{b}{a}\right)^2}$$

Since $c < a$, ϵ is between 0 and 1. If $a = b$, then $\epsilon = 0$ and the conic is a **circle**. If the ratio b/a is small, then the ellipse is very flat. Thus, for an ellipse,

$$0 \leq \epsilon < 1$$

and ϵ measures the amount of roundness of the ellipse.

Consider a circle; that is, suppose $a = b$. In this case, let $r = a = b$ and call this distance the **radius**. You can see that a circle is a special case of an ellipse.

Standard-Form Equation of a Circle

> The equation of a **circle** with center at (h, k) and radius r is
>
> $$(x-h)^2 + (y-k)^2 = r^2$$

EXAMPLE 5 Graph $x^2 + y^2 + 6x - 14y + 22 = 0$.

Solution Complete the square in x and y:

$$(x^2 + 6x \quad\ ) + (y^2 - 14y \quad\ ) = -22$$
$$(x^2 + 6x + \mathbf{9}) + (y^2 - 14y + \mathbf{49}) = -22 + \mathbf{9} + \mathbf{49}$$
$$(x+3)^2 + (y-7)^2 = 36$$

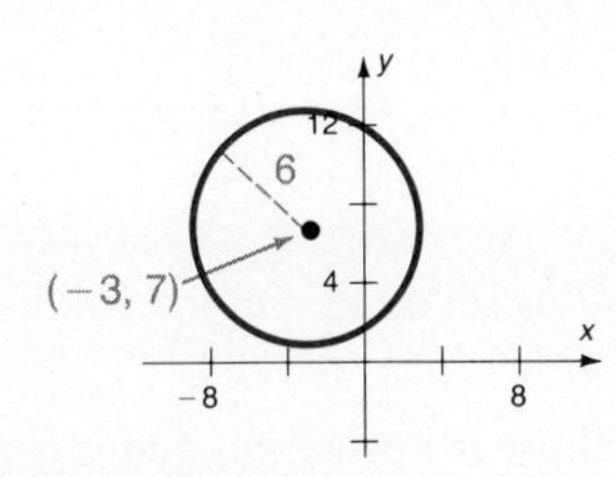

Figure 9.21 Graph of $x^2 + y^2 + 6x - 14y + 22 = 0$

This is a circle with center at $(-3, 7)$ and radius 6, as shown in Figure 9.21. ■

EXAMPLE 6 Find the equation of the circle passing through the points $(-3, 4)$, $(4, 5)$, and $(1, -4)$. Sketch the graph.

Solution Now $(x-h)^2 + (y-k)^2 = r^2$ can be written as $x^2 + y^2 + C_1x + C_2y + C_3 = 0$ for some constants C_1, C_2, and C_3. Since the points lie on the circle, they satisfy the equation. Thus

$$\begin{aligned} (-3,4)&: \quad -3C_1 + 4C_2 + C_3 = -25 \\ (4,5)&: \quad 4C_1 + 5C_2 + C_3 = -41 \\ (1,-4)&: \quad C_1 - 4C_2 + C_3 = -17 \end{aligned}$$

This is a system of three linear equations in three unknowns that can be solved simultaneously by using the methods given in Chapter 7. (We will not show the details of this solution here.) After several steps you will find that $C_1 = -2$, $C_2 = -2$, and $C_3 = -23$. Now substitute these values into the equation of the circle:

$$x^2 + y^2 - 2x - 2y - 23 = 0$$
$$(x^2 - 2x + 1) + (y^2 - 2y + 1) = 23 + 1 + 1$$
$$(x-1)^2 + (y-1)^2 = 25$$

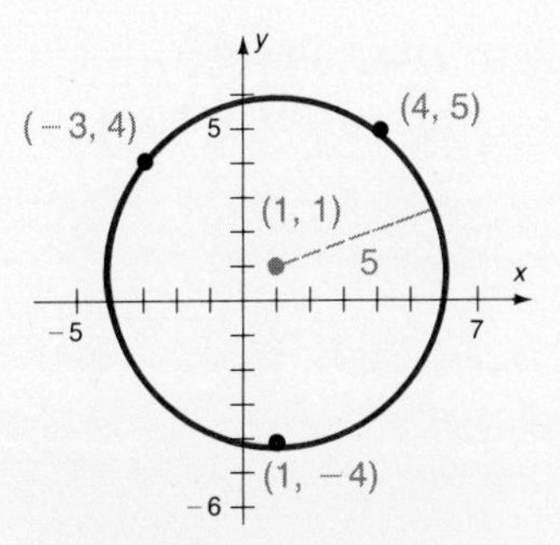

Figure 9.22 Graph of the circle passing through $(-3, 4)$, $(4, 5)$, and $(1, -4)$

The graph is shown in Figure 9.22. ■

EXAMPLE 7 Find the equation of the ellipse with vertices at $(3, 2)$ and $(3, -4)$ and foci at $(3, \sqrt{5} - 1)$ and $(3, -\sqrt{5} - 1)$.

Solution By inspection, the ellipse is vertical, and it is centered at $(3, -1)$, where $a = 3$ and $c = \sqrt{5}$ (see Figure 9.23). Thus

$$c^2 = a^2 - b^2$$
$$5 = 9 - b^2$$
$$b^2 = 4$$

The equation is

$$\frac{(y + 1)^2}{9} + \frac{(x - 3)^2}{4} = 1$$

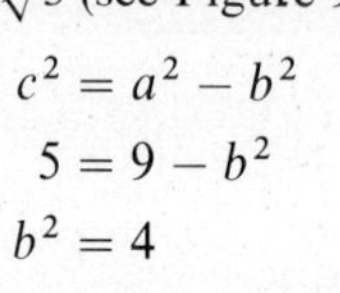

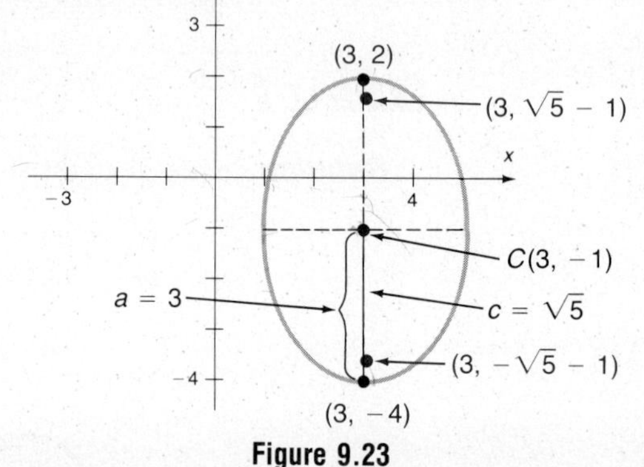

Figure 9.23

■

EXAMPLE 8 Find the equation of the set of all points with the sum of distances from $(-3, 2)$ and $(5, 2)$ equal to 16.

Solution By inspection, the ellipse is horizontal and is centered at $(1, 2)$. You are given

$$2a = 16 \qquad \text{This is the sum of the distances.}$$
$$a = 8$$

and $c = 4$—the distance from the center $(1, 2)$ to a focus $(5, 2)$. Thus

$$c^2 = a^2 - b^2$$
$$16 = 64 - b^2$$
$$b^2 = 48$$

The equation is

$$\frac{(x - 1)^2}{64} + \frac{(y - 2)^2}{48} = 1$$

■

EXAMPLE 9 Find the equation of the ellipse with foci at $(-3, 6)$ and $(-3, 2)$ with $\epsilon = \frac{1}{5}$.

Solution By inspection, the ellipse is vertical and is centered at $(-3, 4)$ with $c = 2$. Since

$$\epsilon = \frac{c}{a} = \frac{1}{5}$$

and $c = 2$, then

$$\frac{2}{a} = \frac{1}{5}$$

which implies $a = 10$. Just because

$$\frac{c}{a} = \frac{1}{5}$$

you cannot assume that $c = 1$ and $a = 5$; all you know is that the reduced *ratio* of c to a is $\frac{1}{5}$. Also,

$$c^2 = a^2 - b^2$$
$$4 = 100 - b^2$$
$$b^2 = 96$$

Thus the equation is

$$\frac{(y-4)^2}{100} + \frac{(x+3)^2}{96} = 1$$

■

PROBLEM SET 9.2

A

Sketch the curves in Problems 1–18.

1. $\frac{x^2}{4} + \frac{y^2}{9} = 1$
2. $\frac{x^2}{25} + \frac{y^2}{36} = 1$
3. $x^2 + \frac{y^2}{9} = 1$
4. $4x^2 + 9y^2 = 36$
5. $25x^2 + 16y^2 = 400$
6. $36x^2 + 25y^2 = 900$
7. $3x^2 + 2y^2 = 6$
8. $4x^2 + 3y^2 = 12$
9. $5x^2 + 10y^2 = 7$
10. $(x-2)^2 + (y+3)^2 = 25$
11. $(x+4)^2 + (y-2)^2 = 49$
12. $(x-1)^2 + (y-1)^2 = \frac{1}{4}$
13. $\frac{(x+3)^2}{81} + \frac{(y-1)^2}{49} = 1$
14. $\frac{(x-3)^2}{16} + \frac{(y-2)^2}{9} = 1$
15. $\frac{(x+2)^2}{25} + \frac{(y+4)^2}{9} = 1$
16. $3(x+1)^2 + 4(y-1)^2 = 12$
17. $10(x-5)^2 + 6(y+2)^2 = 60$
18. $5(x+2)^2 + 3(y+4)^2 = 60$

Sketch the curves in Problems 19–30.

19. The set of points 6 units from the point $(4, 5)$
20. The set of points 3 units from the point $(-2, 3)$
21. The set of points 6 units from $(-1, -4)$
22. The set of points such that the sum of the distances from $(-6, 0)$ and $(6, 0)$ is 20
23. The set of points such that the sum of the distances from $(0, 4)$ and $(0, -4)$ is 10
24. The set of points such that the sum of the distances from $(-4, 1)$ and $(2, 1)$ is 10
25. The ellipse with vertices at $(0, 7)$ and $(0, -7)$ and foci at $(0, 5)$ and $(0, -5)$
26. The ellipse with vertices at $(4, 3)$ and $(4, -5)$ and foci at $(4, 2)$ and $(4, -4)$
27. The ellipse with vertices $(-6, 3)$ and $(4, 3)$ and foci at $(-4, 3)$ and $(2, 3)$
28. The ellipse with foci at $(-4, -3)$ and $(2, -3)$ with eccentricity $\frac{4}{5}$
29. The circle passing through $(2, 2)$, $(-2, -6)$, and $(5, 1)$
30. The ellipse passing through $(5, 2)$ and $(3, \sqrt{5})$ with axes along the coordinate axes

B

Find the equations of the curves in Problems 31–42. These are the curves you sketched in Problems 19–30.

31. The set of points 6 units from the point $(4, 5)$
32. The set of points 3 units from the point $(-2, 3)$
33. The set of points 5 units from $(-1, -4)$
34. The set of points such that the sum of the distances from $(-6, 0)$ and $(6, 0)$ is 20
35. The set of points such that the sum of the distances from $(0, 4)$ and $(0, -4)$ is 10
36. The set of points such that the sum of the distances from $(-4, 1)$ and $(2, 1)$ is 10
37. The ellipse with vertices at $(0, 7)$ and $(0, -7)$ and foci at $(0, 5)$ and $(0, -5)$
38. The ellipse with vertices at $(4, 3)$ and $(4, -5)$ and foci at $(4, 2)$ and $(4, -4)$
39. The ellipse with vertices $(-6, 3)$ and $(4, 3)$ and foci at $(-3, 3)$ and $(2, 3)$
40. The ellipse with foci at $(-4, -3)$ and $(2, -3)$ and eccentricity $\frac{4}{5}$
41. The circle passing through $(2, 2)$, $(-2, -6)$, and $(5, 1)$
42. The ellipse passing through $(5, 2)$ and $(3, \sqrt{5})$ with axes along the coordinate axes

Sketch the curves in Problems 43–54.

43. $x^2 + 4x + y^2 + 6y - 12 = 0$

44. $9x^2 + 4y^2 - 18x + 16y - 11 = 0$

45. $16x^2 + 9y^2 + 96x - 36y + 36 = 0$

46. $y^2 + 6y + 25x + 159 = 0$

47. $3x^2 + 4y^2 + 2x - 8y + 4 = 0$

48. $144x^2 + 72y^2 - 72x + 48y - 7 = 0$

49. $y^2 + 4x^2 + 2y - 8x + 1 = 0$

50. $4y^2 + x^2 - 16y + 4x - 8 = 0$

51. $x^2 + y^2 - 10x - 14y - 70 = 0$

52. $x^2 + y^2 - 4x + 10y + 15 = 0$

53. $4x^2 + y^2 + 24x + 4y + 16 = 0$

54. $x^2 + 9y^2 - 4x - 18y - 14 = 0$

55. Derive the equation of the ellipse with foci at $(0, c)$ and $(0, -c)$ and constant distance $2a$. Let $b^2 = a^2 - c^2$. Show all your work.

56. Derive the equation of a circle with center (h, k) and radius r by using the distance formula.

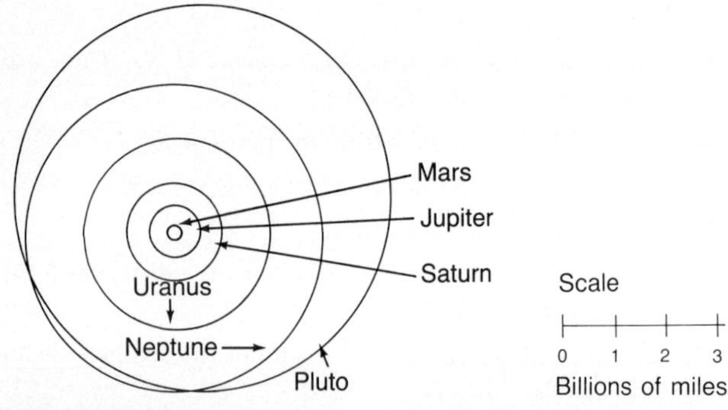

Figure 9.24 Planetary orbits: The orbits of planets or satellites serve as an example of ellipses. See the Extended Application at the end of this chapter for a discussion of this application.

The following is needed for Problems 57–59. The orbit of a planet can be described by

$$\frac{x^2}{a^2} + \frac{y^2}{b^2} = 1$$

with the sun at one focus. The orbit is commonly identified by its ***major axis*** *$2a$ and its eccentricity ϵ.*

$$\epsilon = \frac{c}{a} = \sqrt{1 - \frac{b^2}{a^2}} \qquad (a \geq b > 0)$$

where c is the distance between the center and a focus.

57. The orbits of the planets can be described as standard-form ellipses with the sun at one focus (see Figure 9.24). The point at which a planet is farthest from the sun is called **aphelion**, and the point at which it is closest is called **perihelion**. (For a more detailed discussion of planetary orbits, see the Extended Application at the end of this chapter.) If the major axis of the earth's orbit is 186,000,000 miles and its eccentricity is $\frac{1}{62}$, how far is the earth from the sun at aphelion and at perihelion?

58. If the planet Mercury is 28 million miles from the sun at perihelion, and the eccentricity of its orbit is $\frac{1}{5}$, how long is the major axis of Mercury's orbit? (See Problem 57.)

59. The moon's orbit is elliptical with the earth at one focus. The point at which the moon is farthest from the earth is called **apogee**, and the point at which it is closest is called **perigee**. If the moon is 199,000 miles from the earth at apogee and the major axis of its orbit is 378,000 miles, what is the eccentricity of the moon's orbit?

C

60. ***Engineering*** A stone tunnel is to be constructed such that the opening is a semielliptic arch as shown in Figure 9.25. It is necessary to know the height at 4-ft intervals from the center. That is, how high is the tunnel at 4, 8, 12, 16, and 20 ft from the center? (Answer to the nearest tenth of a foot.)

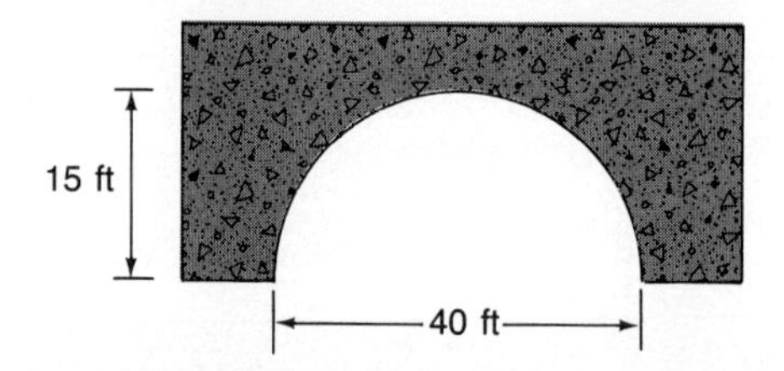

Figure 9.25 Semielliptic arch

61. Find the points of intersection (if any) of the ellipse $16(x - 2)^2 + 9(y + 1)^2 = 144$ and the line $4x - 3y - 23 = 0$. Graph both equations.

62. Find the points of intersection (if any) of the circle $x^2 + y^2 + 4x + 6y + 4 = 0$ and the parabola $x^2 + 4x + 8y + 4 = 0$. Graph both equations.

63. If we are given an ellipse with foci at $(-c, 0)$ and $(c, 0)$ and vertices at $(-a, 0)$ and $(a, 0)$, we define the *directrices* of the ellipse as the lines $x = a/\epsilon$ and $x = -a/\epsilon$. Show that an ellipse is the set of all points with distances from $F(c, 0)$ equal to ϵ times their distances from the line $x = a/\epsilon$ $(a > 0, c > 0)$.

64. A line segment through a focus parallel to a directrix (see Problem 63) and cut off by the ellipse is called the *focal chord.* Show that the length of the focal chord of the following ellipse is $2b^2/a$:

$$\frac{x^2}{a^2} + \frac{y^2}{b^2} = 1$$

9.3 HYPERBOLAS

The last of the conic sections to be considered has a definition similar to that of the ellipse.

Hyperbola

A **hyperbola** is the set of all points in the plane such that, for each point on the hyperbola, the difference of its distances from two fixed points is a constant.

The fixed points are called the **foci**. A hyperbola with foci at F_1 and F_2, where the given constant is 8, is shown in Figure 9.26.

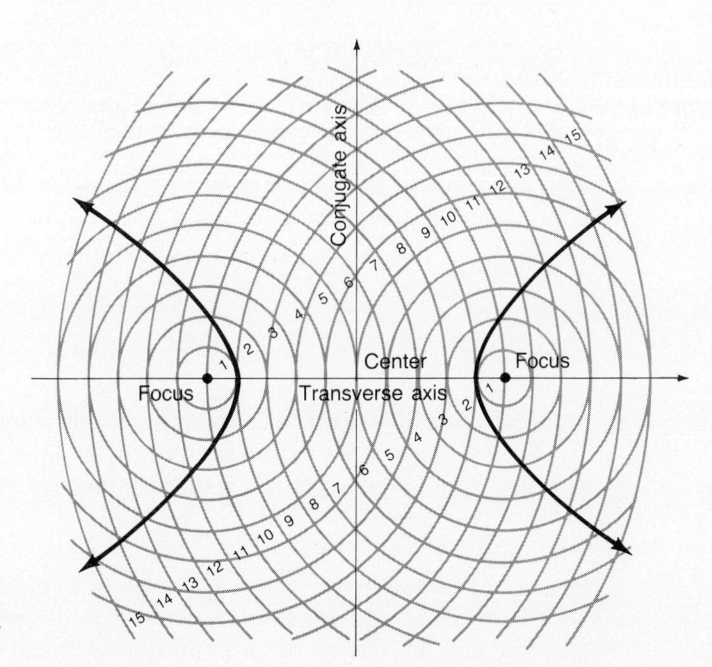

Figure 9.26 Graph of a hyperbola from the definition

The line passing through the foci is called the **transverse axis**. The **center** is the midpoint of the segment connecting the foci. The line passing through the center perpendicular to the transverse axis is called the **conjugate axis**. The hyperbola is symmetric with respect to both the transverse and the conjugate axes.

If you use the definition, you can derive the equation for a hyperbola with foci at $(-c, 0)$ and $(c, 0)$ with constant distance $2a$. If (x, y) is any point on the curve, then

$$\left|\sqrt{(x+c)^2+(y-0)^2}-\sqrt{(x-c)^2+(y-0)^2}\right|=2a$$

The procedure for simplifying this expression is the same as that shown for the ellipse, so the details are left as a problem (see Problems 59 and 60). After several steps, you should obtain

$$\frac{x^2}{a^2}-\frac{y^2}{c^2-a^2}=1$$

If $b^2=c^2-a^2$, then

$$\frac{x^2}{a^2}-\frac{y^2}{b^2}=1$$

which is the standard-form equation. Notice that $c^2=a^2-b^2$ for the ellipse and that $c^2=a^2+b^2$ for the hyperbola. For the ellipse it is necessary that $a^2>b^2$, but for the hyperbola there is no restriction on the relative sizes for a and b.

Repeat the argument for a hyperbola with foci $(0,c)$ and $(0,-c)$, and you will obtain the other standard-form equation for a hyperbola with a vertical transverse axis.

Standard-Form Equations for the Hyperbola with Center (0, 0)

Hyperbola	Foci	Constant distance	Center	Equation
Horizontal	$(-c,0),(c,0)$	$2a$	$(0,0)$	$\frac{x^2}{a^2}-\frac{y^2}{b^2}=1$
Vertical	$(0,c),(0,-c)$	$2a$	$(0,0)$	$\frac{y^2}{a^2}-\frac{x^2}{b^2}=1$

where $b^2=c^2-a^2$ or $c^2=a^2+b^2$

As with the other conics, we will sketch a hyperbola by determining some information about the curve directly from the equation by inspection. The points of intersection of the hyperbola with the transverse axis are called the **vertices**. For

$$\frac{x^2}{a^2}-\frac{y^2}{b^2}=1 \quad \text{and} \quad \frac{y^2}{a^2}-\frac{x^2}{b^2}=1$$

notice that the vertices occur at $(a,0)$, $(-a,0)$ and $(0,a)$, $(0,-a)$, respectively. The number $2a$ is the **length of the transverse axis**. The hyperbola does not intersect the conjugate axis, but if you plot the points $(0,b)$, $(0,-b)$ and $(-b,0)$, $(b,0)$, respectively, you determine a segment on the conjugate axis called the **length of the conjugate axis**.

EXAMPLE 1 Sketch $\frac{x^2}{4}-\frac{y^2}{9}=1$.

Solution The center of the hyperbola is $(0, 0)$, $a = 2$, and $b = 3$. Plot the vertices at ± 2, as shown in Figure 9.27. The transverse axis is along the x axis and the conjugate axis is along the y axis. Plot the length of the conjugate axis at ± 3. We call the points the **pseudovertices**, since the curve does not actually pass through these points.

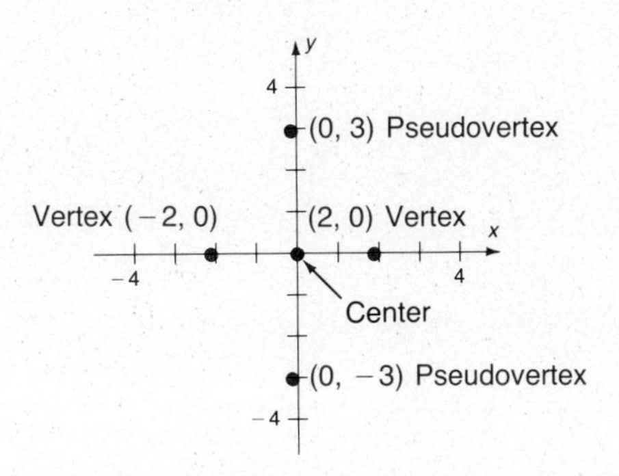

Figure 9.27

Next draw lines through the vertices and pseudovertices parallel to the axes of the hyperbola. These lines form what we will call the **central rectangle**. The diagonal lines passing through the corners of the central rectangle are **slant asymptotes** for the hyperbola, as shown in Figure 9.28; they aid in sketching the hyperbola.

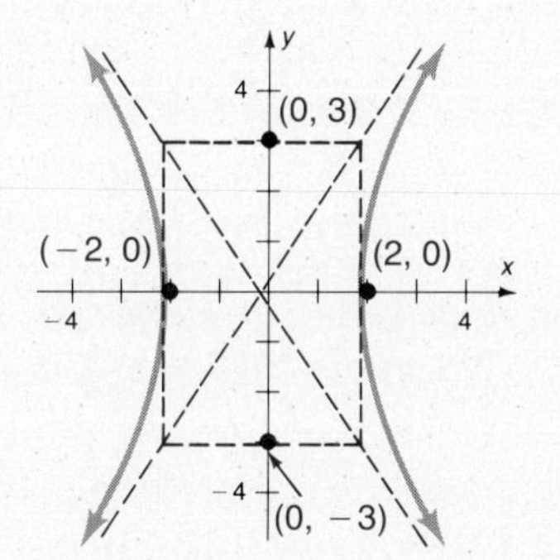

Figure 9.28 Graph of $\frac{x^2}{4} - \frac{y^2}{9} = 1$ ■

For the general hyperbola given by the equation

$$\frac{x^2}{a^2} - \frac{y^2}{b^2} = 1$$

the equations of the slant asymptotes described in Example 1 are

$$y = \frac{b}{a}x \quad \text{and} \quad y = -\frac{b}{a}x$$

To justify this result, solve the equation of the hyperbola for y:

$$\frac{y^2}{b^2} = \frac{x^2}{a^2} - 1$$

$$y^2 = b^2\left(\frac{x^2 - a^2}{a^2}\right)$$

$$y = \pm\frac{b}{a}\sqrt{x^2 - a^2}$$

Now write

$$y = \pm\frac{b}{a}\sqrt{x^2\left(1 - \frac{a^2}{x^2}\right)}$$

$$= \pm\frac{bx}{a}\sqrt{1 - \frac{a^2}{x^2}}$$

in order to see that, as $|x| \to \infty$,

$$\sqrt{1 - \frac{a^2}{x^2}} \to 1$$

So, as $|x|$ becomes large,

$$y \to \pm\frac{b}{a}x$$

If the center of the hyperbola is (h, k), the following equations for a hyperbola are obtained:

Standard-Form Equations for Translated Hyperbolas

Hyperbola	Center	Foci	Vertices	Pseudo-vertices	Equations
Horizontal	(h, k)	$(h + c, k)$ $(h - c, k)$	$(h - a, k)$ $(h + a, k)$	$(h, k + b)$ $(h, k - b)$	$\frac{(x-h)^2}{a^2} - \frac{(y-k)^2}{b^2} = 1$
Vertical	(h, k)	$(h, k + c)$ $(h, k - c)$	$(h, k + a)$ $(h, k - a)$	$(h - b, k)$ $(h + b, k)$	$\frac{(y-k)^2}{a^2} - \frac{(x-h)^2}{b^2} = 1$

EXAMPLE 2 Sketch $16x^2 - 9y^2 - 128x - 18y + 103 = 0$.

Solution Complete the square in both x and y:

$$(16x^2 - 128x) + (-9y^2 - 18y) = -103$$

$$16(x^2 - 8x \qquad) - 9(y^2 + 2y \qquad) = -103$$

$$16(x^2 - 8x + 16) - 9(y^2 + 2y + 1) = -103 + 256 - 9$$

$$16(x - 4)^2 - 9(y + 1)^2 = 144$$

$$\mathbf{\frac{(x-4)^2}{9} - \frac{(y+1)^2}{16} = 1}$$

The graph is shown in Figure 9.29.

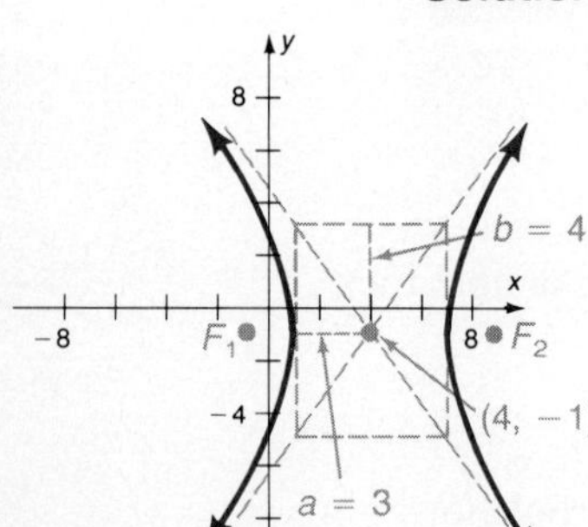

Figure 9.29 Sketch of $16x^2 - 9y^2 - 128x - 18y + 103 = 0$

■

EXAMPLE 3 Find the equation of the hyperbola with vertices at $(2,4)$ and $(2,-2)$ and foci at $(2,6)$ and $(2,-4)$.

Solution Plot the given points as shown in Figure 9.30. Notice that the center of the hyperbola is $(2,1)$ since it is the midpoint of the segment connecting the foci. Also, $c=5$ and $a=3$. Since

$$c^2 = a^2 + b^2$$

you have

$$25 = 9 + b^2$$

$$b^2 = 16$$

and the equation is

$$\frac{(y-1)^2}{9} - \frac{(x-2)^2}{16} = 1$$

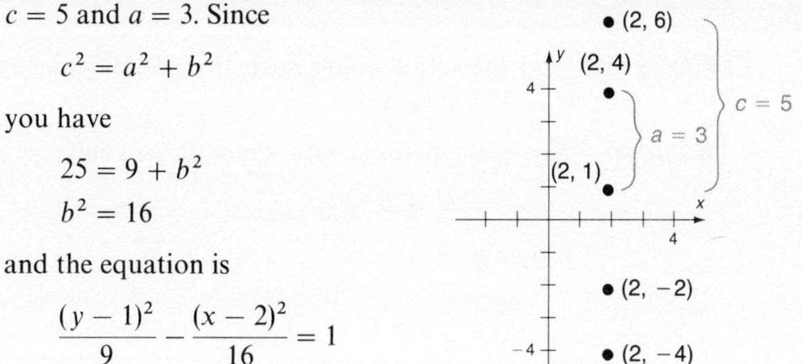

Figure 9.30

■

The eccentricity of the hyperbola and parabola is defined by the same equation that was used for the ellipse, namely

$$\epsilon = \frac{c}{a}$$

Remember that for the ellipse, $0 \le \epsilon < 1$; however, for the hyperbola $c > a$ so $\epsilon > 1$, and for the parabola $c = a$ so $\epsilon = 1$.

EXAMPLE 4 Find the equation of the hyperbola with foci at $(-3,2)$ and $(5,2)$ and with eccentricity $\frac{3}{2}$.

Solution The center of the hyperbola is $(1,2)$ and $c=4$. Also, since

$$\epsilon = \frac{c}{a} = \frac{3}{2}$$

you have

$$\frac{4}{a} = \frac{3}{2}$$

$$a = \frac{8}{3}$$

Since $c^2 = a^2 + b^2$,

$$16 = \frac{64}{9} + b^2$$

$$b^2 = \frac{80}{9}$$

Thus the equation is

$$\frac{(x-1)^2}{\frac{64}{9}} - \frac{(y-2)^2}{\frac{80}{9}} = 1 \quad \text{or} \quad \frac{9(x-1)^2}{64} - \frac{9(y-2)^2}{80} = 1$$

■

EXAMPLE 5 Find the set of points such that the difference of their distances from $(6, 2)$ and $(6, -5)$ is always 3.

Solution This is a hyperbola with center $(6, -\frac{3}{2})$ and $c = \frac{7}{2}$. Also $2a = 3$, so $a = \frac{3}{2}$. Since

$$c^2 = a^2 + b^2$$

you have

$$\frac{49}{4} = \frac{9}{4} + b^2$$

$$b^2 = 10$$

The equation is

$$\frac{\left(y + \frac{3}{2}\right)^2}{\frac{9}{4}} - \frac{(x-6)^2}{10} = 1 \quad \text{or} \quad \frac{4\left(y + \frac{3}{2}\right)^2}{9} - \frac{(x-6)^2}{10} = 1$$

■

Conic Section Summary

We have now considered the graphs of equations of the form

$$Ax^2 + Bxy + Cy^2 + Dx + Ey + F = 0$$

Geometrically they represent the intersection of a plane and a cone, usually resulting in a line, parabola, ellipse, or hyperbola. However, there are certain positions of the plane that result in what are called **degenerate conics**. To visualize some of these degenerate conics, first consider Figure 9.31.

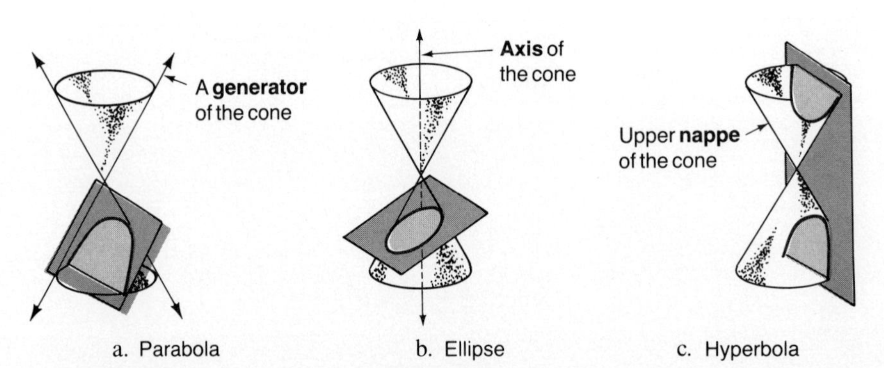

Figure 9.31 Conic sections

For a **degenerate parabola**, visualize the cone (see Figure 9.31a) situated so that one of its generators lies in the plane; a line results. For a **degenerate ellipse**, visualize the plane intersecting at the vertex of the upper and lower nappes (see Figure 9.31b); a point results. And finally, for a **degenerate hyperbola**, visualize the plane situated so that the axis of the cone lies in the plane (see Figure 9.31c); a pair of intersecting lines results.

EXAMPLE 6 Sketch $\frac{(x-2)^2}{4} + \frac{(y+3)^2}{9} = 0$

Solution There is only one point that satisfies this equation—namely $(2, -3)$. This is an example of a *degenerate ellipse.* Notice that except for the zero the equation has the "form of an ellipse." See Figure 9.32.

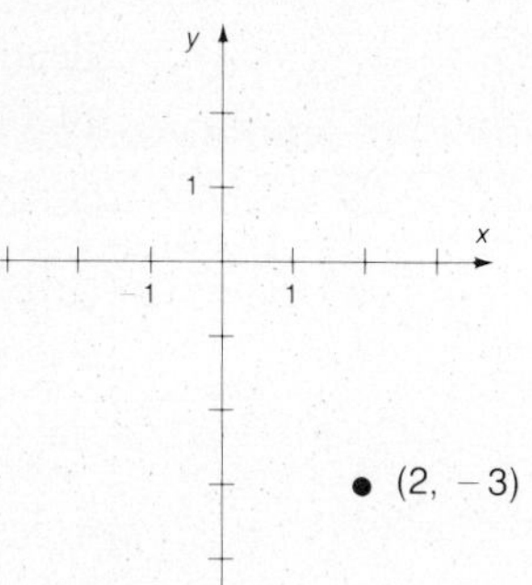

Figure 9.32 Graph of $\frac{(x-2)^2}{4} + \frac{(y+3)^2}{9} = 0$ ■

EXAMPLE 7 Sketch $\frac{x^2}{4} - \frac{y^2}{9} = 0.$

Solution This equation has the "form of a hyperbola," but because of the zero it cannot be put into standard form. You can, however, treat this as a factored form as described in Section 3.1:

$$\left(\frac{x}{2} - \frac{y}{3}\right)\left(\frac{x}{2} + \frac{y}{3}\right) = 0$$

$$\frac{x}{2} - \frac{y}{3} = 0 \quad \text{or} \quad \frac{x}{2} + \frac{y}{3} = 0$$

The graph (Figure 9.33) is a *degenerate hyperbola.*

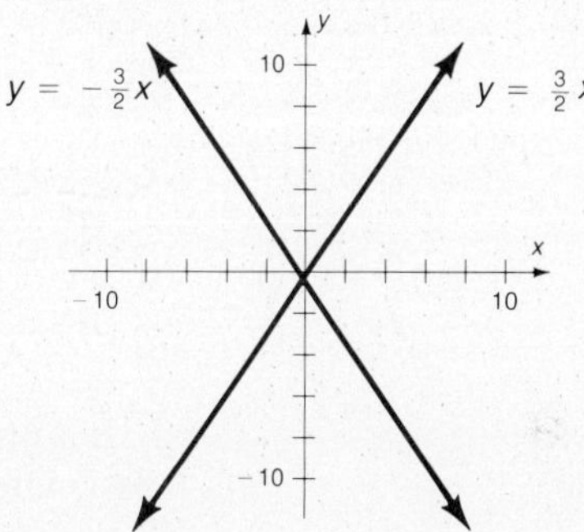

Figure 9.33 Graph of $\frac{x^2}{4} - \frac{y^2}{9} = 0$ ■

It is important to be able to recognize the curve by inspection of the equation before you begin. The first thing to notice is whether or not there is an xy term.

Up to now we have considered conics for which $B = 0$ (that is, no xy term). If $B = 0$, then:

Type of curve	Degree of equation	Degree in x	Degree in y	Relationship to the general equation $Ax^2+Bxy+Cy^2+Dx+Ey+F=0$
1. Line	First	First	First	$A = C = 0$
2. Parabola	Second	First	Second	$A=0$ and $C \neq 0$
		Second	First	$A \neq 0$ and $C = 0$
3. Ellipse	Second	Second	Second	A and C have same sign
4. Circle	Second	Second	Second	$A = C$
5. Hyperbola	Second	Second	Second	A and C have opposite signs

In the next section, we will graph conics for which $B \neq 0$ (that is, there is an xy term). However, we will begin in this section by identifying this type of conic. This identification consists of calculating $B^2 - 4AC$. (Remember, $b^2 - 4ac$ in the quadratic formula was called the discriminant.)

Type of curve	Relationship to general equation
1. Ellipse	$B^2 - 4AC < 0$
2. Parabola	$B^2 - 4AC = 0$
3. Hyperbola	$B^2 - 4AC > 0$

These tests do not distinguish degenerate cases. This means that the test may tell you the curve is an ellipse that may turn out to be a single point. Remember too that a circle is a special case of the ellipse. The expression $B^2 - 4AC$ is called the **discriminant**.

EXAMPLE 8 Identify each curve.

Solution

a. $x^2 + 4xy + 4y^2 = 9$
$B^2 - 4AC = 16 - 4(1)(4) = 0$; parabola

b. $2x^2 + 3xy + y^2 = 25$
$B^2 - 4AC = 9 - 4(2)(1) > 0$; hyperbola

c. $x^2 + xy + y^2 - 8x + 8y = 0$
$B^2 - 4AC = 1 - 4(1)(1) < 0$; ellipse

d. $xy = 5$
$B^2 - 4AC = 1 - 4(0)(0) > 0$; hyperbola ■

It is important to remember the standard-form equations of the conics and some basic information about these curves. The important ideas are summarized in the following conic summary.

Summary of Standard-Position Conics

	Parabola	Ellipse	Hyperbola
Definition	All points equidistant from a given point and a given line	All points with the sum of distances from two fixed points constant	All points with the difference of distances from two fixed points constant
Equations	Up: $x^2 = 4cy$ Down: $x^2 = -4cy$ Right: $y^2 = 4cx$ Left: $y^2 = -4cx$	$c^2 = a^2 - b^2$ Horizontal axis: $\frac{x^2}{a^2} + \frac{y^2}{b^2} = 1$ Vertical axis: $\frac{y^2}{a^2} + \frac{x^2}{b^2} = 1$	$c^2 = a^2 + b^2$ Horizontal axis: $\frac{x^2}{a^2} - \frac{y^2}{b^2} = 1$ Vertical axis: $\frac{y^2}{a^2} - \frac{x^2}{b^2} = 1$

	Parabola	Ellipse	Hyperbola
Recognition	Second-degree equation; linear in one variable, quadratic in the other variable	Second-degree equation; coefficients of x^2 and y^2 have same sign	Second-degree equation; coefficients of x^2 and y^2 have different signs
Eccentricity	$\epsilon = 1$	$0 \leq \epsilon < 1$	$\epsilon > 1$
Directrix	Perpendicular to axis c units from the vertex (one directrix)	Perpendicular to major axis $\pm a/\epsilon$ units from center (two directrices)	Perpendicular to transverse axis $\pm a/\epsilon$ units from center (two directrices)
Translations to the point (h, k): $x' = x - h$ and $y' = y - k$			

PROBLEM SET 9.3

A

Sketch the curves in Problems 1–21.

1. $x^2 - y^2 = 1$

2. $x^2 - y^2 = 4$

3. $y^2 - x^2 = 1$

4. $\dfrac{x^2}{4} - \dfrac{y^2}{9} = 1$

5. $\dfrac{x^2}{9} - \dfrac{y^2}{4} = 1$

6. $\dfrac{y^2}{9} - \dfrac{x^2}{4} = 1$

7. $\dfrac{x^2}{16} - \dfrac{y^2}{25} = 1$

8. $\dfrac{y^2}{16} - \dfrac{x^2}{25} = 1$

9. $\dfrac{x^2}{36} - \dfrac{y^2}{9} = 1$

10. $36y^2 - 25x^2 = 900$

11. $3x^2 - 4y^2 = 12$

12. $3y^2 = 4x^2 + 12$

13. $3x^2 - 4y^2 = 5$

14. $4y^2 - 4x^2 = 5$

15. $4y^2 - x^2 = 9$

16. $\dfrac{(x-2)^2}{4} - \dfrac{(y+3)^2}{16} = 1$

17. $\dfrac{(x+3)^2}{8} - \dfrac{(y-1)^2}{5} = 1$

18. $\dfrac{(y-1)^2}{6} - \dfrac{(x+2)^2}{8} = 1$

19. $\dfrac{(x-2)^2}{16} - \dfrac{(y+1)^2}{9} = 1$

20. $\dfrac{(y+2)^2}{25} - \dfrac{(x+1)^2}{16} = 1$

21. $\dfrac{(y+2)^2}{1/4} - \dfrac{(x-1)^2}{1/9} = 1$

In Problems 22–33, identify and sketch the curve.

22. $2x - y - 8 = 0$

23. $2x + y - 10 = 0$

24. $4x^2 - 16y = 0$

25. $\dfrac{(x-3)^2}{4} - \dfrac{(y+2)}{6} = 1$

26. $\dfrac{(x-3)^2}{9} - \dfrac{(y+2)^2}{25} = 1$

27. $\dfrac{x-3}{9} + \dfrac{y-2}{25} = 1$

28. $(x+3)^2 + (y-2)^2 = 0$

29. $9(x+3)^2 + 4(y-2) = 0$

30. $9(x+3)^2 - 4(y-2)^2 = 0$

31. $x^2 + 8(y-12)^2 = 16$

32. $x^2 + 64(y+4)^2 = 16$

33. $x^2 + y^2 - 3y = 0$

B

Find the equations of the curves in Problems 34–39.

34. The hyperbola with vertices at $(0, 5)$ and $(0, -5)$ and foci at $(0, 7)$ and $(0, -7)$

35. The set of points such that the difference of their distances from $(-6, 0)$ and $(6, 0)$ is 10

36. The hyperbola with foci at $(5, 0)$ and $(-5, 0)$ and eccentricity 5

37. The hyperbola with vertices at $(4, 4)$ and $(4, 8)$ and foci at $(4, 3)$ and $(4, 9)$

38. The set of points such that the difference of their distances from $(4, -3)$ and $(-4, -3)$ is 6

39. The hyperbola with vertices at $(-2, 0)$ and $(6, 0)$ passing through $(10, 3)$

Sketch the curves in Problems 40–47.

40. $5(x-2)^2 - 2(y+3)^2 = 10$

41. $4(x+4)^2 - 3(y+3)^2 = -12$

42. $3x^2 - 4y^2 = 12x + 80y + 88$

43. $9x^2 - 18x - 11 = 4y^2 + 16y$

44. $4y^2 - 8y + 4 = 3x^2 - 2x$

45. $x^2 - 4x + y^2 + 6y - 12 = 0$

46. $x^2 - y^2 = 2x + 4y - 3$

47. $3x^2 - 5y^2 + 18x + 10y - 8 = 0$

In Problems 48–57, identify and sketch the curve.

48. $4x^2 - 3y^2 - 24y - 112 = 0$

49. $y^2 - 4x + 2y + 21 = 0$

50. $9x^2 + 2y^2 - 48y + 270 = 0$

51. $x^2 + 4x + 12y + 64 = 0$

52. $y^2 - 6y - 4x + 5 = 0$

53. $100x^2 - 7y^2 + 98y - 368 = 0$

54. $x^2 + y^2 + 2x - 4y - 20 = 0$

55. $4x^2 + 12x + 4y^2 + 4y + 1 = 0$

56. $x^2 - 4y^2 - 6x - 8y - 11 = 0$

57. $9x^2 + 25y^2 - 54x - 200y + 256 = 0$

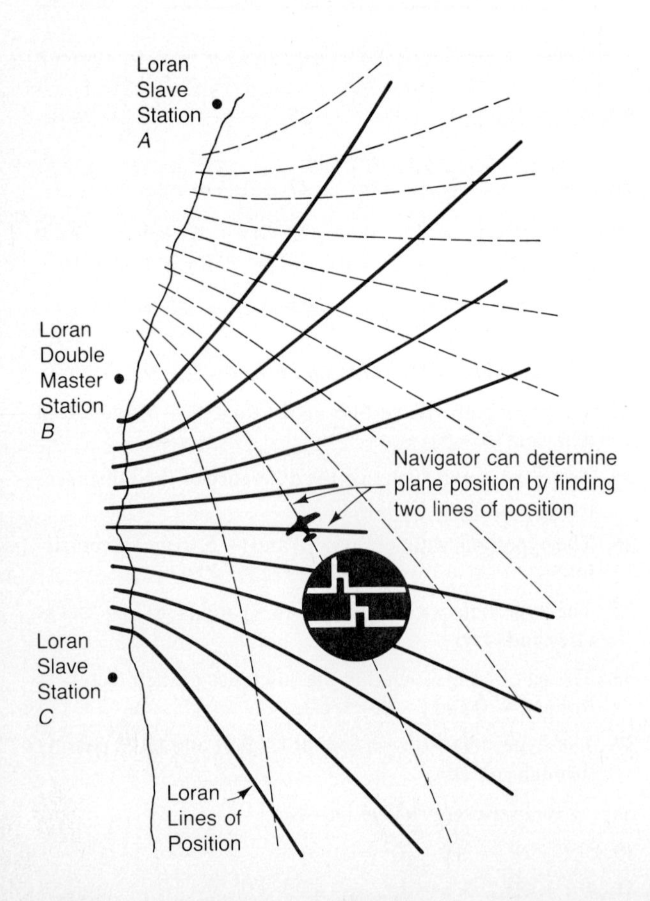

Hyperbolas are used in some techniques of navigation, and of satellite tracking, where the difference of distances from two fixed points is observed (a plane's distances from two fixed stations on the ground, a satellite's distances from two antennas). If A and B are land stations and the difference in distance from them is electronically observed on the plane, its location must be on a certain hyperbola with A and B as foci. Also, if the difference in distance from B and a third station C is observed, the plane must be on a certain hyperbola with B and C as foci; thus, its position is on the intersection of the two hyperbolic curves. The *two sets of hyperbolas serve as a coordinate system*, comparable to the Cartesian system, where two sets of straight lines intersect at right angles.

C

58. Consider a person A who fires a rifle at a distant gong B. Assuming that the ground is flat, where must you stand to hear the sound of the gun and the sound of the gong simultaneously? *Hint:* To answer this question, let x be the distance sound travels in the length of time it takes the bullet to travel from the gun to the gong. Show that the person who hears the sounds simultaneously must stand on a branch of a hyperbola (the one nearest the target) so that the difference of the distances from A to B is x.

59. Derive the equation of the hyperbola with foci at $(-c, 0)$ and $(c, 0)$ and constant distance $2a$. Let $b^2 = c^2 - a^2$. Show all your work.

60. Derive the equation of the hyperbola with foci at $(0, c)$ and $(0, -c)$ and constant distance $2a$. Let $b^2 = c^2 - a^2$. Show all your work.

61. Let d represent the vertical distance between the hyperbola in the first quadrant

$$y = \frac{b}{a}\sqrt{x^2 - a^2}$$

and the line

$$y = \frac{b}{a}x$$

in the first quadrant. Show that as $|x| \to \infty$, then $d \to 0$.

62. Show what happens to d as $|x| \to \infty$ if (x, y) is in Quadrant IV.

63. Given the hyperbola

$$\frac{x^2}{a^2} - \frac{y^2}{b^2} = 1$$

show that the length of the diagonal of the central rectangle of the hyperbola is $2c$.

64. Given the hyperbola

$$\frac{x^2}{a^2} - \frac{y^2}{b^2} = 1$$

we define the *directrices* of the hyperbola as the lines

$$x = \frac{a}{\epsilon} \quad \text{and} \quad x = -\frac{a}{\epsilon}$$

Show that the hyperbola is the set of all points with distances from $F(c, 0)$ that are equal to ϵ times their distances from the line $x = a/\epsilon$.

65. A line through a focus parallel to a directrix and cut off by the hyperbola is called the *focal chord.* Show that the length of the focal chord of the following hyperbola is $2b^2/a$:

$$\frac{x^2}{a^2} - \frac{y^2}{b^2} = 1$$

9.4 ROTATIONS*

In Section 2.4 we introduced the idea of a **translation** in order to simplify the equation of a curve by writing it relative to a new translated coordinate system. In this section we will consider the idea of a **rotation** in which the equation of a curve can be simplified by writing it in terms of a rotated coordinate system. Suppose the coordinate axes are rotated through an angle θ $(0 < \theta < 90°)$. The relationship between the old coordinates (x, y) and the new coordinates (x', y') can be found by considering Figure 9.34.

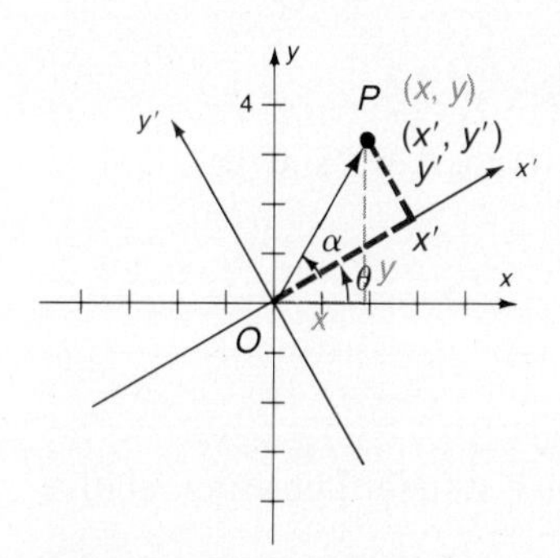

Figure 9.34 Rotation of axes

Let O be the origin and P be a point with coordinates (x, y) relative to the old coordinate system and (x', y') relative to the new rotated coordinate system. Let θ be the amount of rotation and let α be the angle between the x' axis and $|OP|$. Then, using the definition of sine and cosine,

$$x = |OP|\cos(\theta + \alpha) \qquad x' = |OP|\cos\alpha$$
$$y = |OP|\sin(\theta + \alpha) \qquad y' = |OP|\sin\alpha$$

Now

$$\begin{aligned} x = |OP|\cos(\theta + \alpha) &= |OP|\,[\cos\theta\cos\alpha - \sin\theta\sin\alpha] \\ &= |OP|\cos\theta\cos\alpha - |OP|\sin\theta\sin\alpha \\ &= (|OP|\cos\alpha)\cos\theta - (|OP|\sin\alpha)\sin\theta \\ &= x'\cos\theta - y'\sin\theta \end{aligned}$$

Also,

$$\begin{aligned} y = |OP|\sin(\theta + \alpha) &= |OP|\sin\theta\cos\alpha + |OP|\cos\theta\sin\alpha \\ &= x'\sin\theta + y'\cos\theta \end{aligned}$$

Therefore:

Rotation of Axes Formulas

$$x = x'\cos\theta - y'\sin\theta \qquad y = x'\sin\theta + y'\cos\theta$$

* Optional section

All the curves considered in this chapter can be characterized by the general second-degree equation

$$Ax^2 + Bxy + Cy^2 + Dx + Ey + F = 0$$

Notice that the xy term has not appeared before. The presence of this term indicates that the conic has been rotated. Thus in this section we assume that $B \neq 0$ and now need to determine the amount this conic has been rotated from standard position. That is, the new axes should be rotated the same amount as the given conic so that it will be in standard position after the rotation. To find out how much to rotate the axes, substitute

$$x = x'\cos\theta - y'\sin\theta \qquad y = x'\sin\theta + y'\cos\theta$$

into

$$Ax^2 + Bxy + Cy^2 + Dx + Ey + F = 0 \qquad (B \neq 0)$$

After a lot of simplifying you will obtain

$$\begin{aligned}(A\cos^2\theta &+ B\cos\theta\sin\theta + C\sin^2\theta)x'^2\\ &+ [B(\cos^2\theta - \sin^2\theta) + 2(C - A)\sin\theta\cos\theta]x'y'\\ &+ (A\sin^2\theta - B\sin\theta\cos\theta + C\cos^2\theta)y'^2 + (D\cos\theta + E\sin\theta)x'\\ &+ (-D\sin\theta + E\cos\theta)y' + F = 0\end{aligned}$$

This looks terrible, but it is still in the form

$$A'x'^2 + B'x'y' + C'y'^2 + D'x' + E'y' + F = 0$$

You want to choose θ so that $B' = 0$. This will give you a standard position relative to the new coordinate axes. That is,

$$B(\cos^2\theta - \sin^2\theta) + 2(C - A)\sin\theta\cos\theta = 0$$

$$B\cos 2\theta + (C - A)\sin 2\theta = 0$$

$$B\cos 2\theta = (A - C)\sin 2\theta$$

$$\frac{\cos 2\theta}{\sin 2\theta} = \frac{A - C}{B}$$

Using double-angle identities, $B \neq 0$, $\theta \neq 0$.

Simplifying, you obtain the following result:

Amount of Rotation Formula

$$\cot 2\theta = \frac{A - C}{B}$$

Notice that we required $0 < \theta < 90°$, so 2θ is in Quadrant I or Quadrant II. This means that if $\cot 2\theta$ is positive, then 2θ must be in Quadrant I; if $\cot 2\theta$ is negative, then 2θ is in Quadrant II.

In Examples 1 to 4, find the appropriate rotation so that the given curve will be in standard position relative to the rotated axes. Also find the x and y values in the new coordinate system.

EXAMPLE 1 $xy = 6$

Solution

$$\cot 2\theta = \frac{A - C}{B} = \frac{0 - 0}{1} = 0$$

Thus $2\theta = 90^\circ$ and $\theta = 45^\circ$

$$\mathbf{x = x' \cos\theta - y' \sin\theta}$$
$$x = x' \cos \mathbf{45^\circ} - y' \sin \mathbf{45^\circ} = x'\left(\frac{1}{\sqrt{2}}\right) - y'\left(\frac{1}{\sqrt{2}}\right) = \frac{1}{\sqrt{2}}(x' - y')$$

$$\mathbf{y = x' \sin\theta + y' \cos\theta}$$
$$y = x' \sin \mathbf{45^\circ} + y' \cos \mathbf{45^\circ} = x'\left(\frac{1}{\sqrt{2}}\right) + y'\left(\frac{1}{\sqrt{2}}\right) = \frac{1}{\sqrt{2}}(x' + y')$$

■

EXAMPLE 2 $7x^2 - 6\sqrt{3}xy + 13y^2 - 16 = 0$

Solution

$$\cot 2\theta = \frac{A - C}{B} = \frac{7 - 13}{-6\sqrt{3}} = \frac{1}{\sqrt{3}}$$

Thus $2\theta = 60^\circ$ and $\theta = 30^\circ$

$$\mathbf{x = x' \cos\theta - y' \sin\theta} = x'\left(\frac{\sqrt{3}}{2}\right) - y'\left(\frac{1}{2}\right) = \frac{1}{2}(\sqrt{3}x' - y')$$

$$\mathbf{y = x' \sin\theta + y' \cos\theta} = \frac{1}{2}(x' + \sqrt{3}y')$$

■

EXAMPLE 3 $x^2 - 4xy + 4y^2 + 5\sqrt{5}y - 10 = 0$

Solution

$$\cot 2\theta = \frac{1 - 4}{-4} = \frac{3}{4}$$

Since this is not an exact value for θ (as it was in Examples 1 and 2), you will need to use some trigonometric identities to find $\cos\theta$ and $\sin\theta$. If $\cot 2\theta = \frac{3}{4}$, then $\tan 2\theta = \frac{4}{3}$ and $\sec 2\theta = \sqrt{1 + (\frac{4}{3})^2} = \frac{5}{3}$. Then $\cos 2\theta = \frac{3}{5}$ and you can now apply the half-angle identities:

$$\cos\theta = \pm\sqrt{\frac{1 + \cos 2\theta}{2}} \qquad \sin\theta = \pm\sqrt{\frac{1 - \cos 2\theta}{2}}$$

$$\cos\theta = \sqrt{\frac{1 + (\frac{3}{5})}{2}} \qquad \sin\theta = \sqrt{\frac{1 - (\frac{3}{5})}{2}} \qquad \text{Positive because } \theta \text{ is in Quadrant I}$$

$$= \frac{2}{\sqrt{5}} \qquad = \frac{1}{\sqrt{5}}$$

To find the amount of rotation, use a calculator and one of the preceding equations to find $\theta \approx 26.6^\circ$. Finally, the rotation of axes formulas provides

$$\mathbf{x = x' \cos\theta - y' \sin\theta} = x'\left(\frac{\mathbf{2}}{\sqrt{\mathbf{5}}}\right) - y'\left(\frac{\mathbf{1}}{\sqrt{\mathbf{5}}}\right) = \frac{1}{\sqrt{5}}(2x' - y')$$

$$\mathbf{y = x' \sin\theta + y' \cos\theta} = x'\left(\frac{\mathbf{1}}{\sqrt{\mathbf{5}}}\right) + y'\left(\frac{\mathbf{2}}{\sqrt{\mathbf{5}}}\right) = \frac{1}{\sqrt{5}}(x' + 2y')$$

■

EXAMPLE 4 $10x^2 + 24xy + 17y^2 - 9 = 0$

Solution

$$\cot 2\theta = \frac{10 - 17}{24}$$
$$= -\frac{7}{24}$$

Since this is negative, 2θ must be in Quadrant II; this means that $\sec 2\theta$ is negative in the following sequence of identities: Since $\cot 2\theta = -\frac{7}{24}$, then $\tan 2\theta = -\frac{24}{7}$ and $\sec 2\theta = -\sqrt{1 + (-\frac{24}{7})^2} = -\frac{25}{7}$. Thus $\cos 2\theta = -\frac{7}{25}$, which gives

$$\cos\theta = \sqrt{\frac{1 + (-\frac{7}{25})}{2}} \qquad \sin\theta = \sqrt{\frac{1 - (-\frac{7}{25})}{2}}$$
$$= \frac{3}{5} \qquad\qquad = \frac{4}{5}$$

Using either of these equations and a calculator, you find that the rotation is $\theta \approx 53.1°$. The rotation of axes formulas provide

$$\mathbf{x = x'\cos\theta - y'\sin\theta} \qquad \mathbf{y = x'\sin\theta + y'\cos\theta}$$
$$= x'\left(\frac{3}{5}\right) - y'\left(\frac{4}{5}\right) \qquad = x'\left(\frac{4}{5}\right) + y'\left(\frac{3}{5}\right)$$
$$= \frac{1}{5}(3x' - 4y') \qquad = \frac{1}{5}(4x' + 3y')$$

■

Procedure for Sketching a Rotated Conic

1. Find the angle of rotation.
2. Find x and y in the new coordinate system.
3. Substitute the values found in step 2 into the given equation and simplify.
4. Sketch the resulting equation relative to the new x' and y' axes. You may have to complete the square if it is not centered at the origin.

EXAMPLE 5 Sketch $xy = 6$.

Solution From Example 1, the rotation is $\theta = 45°$ and

$$x = \frac{1}{\sqrt{2}}(x' - y') \qquad y = \frac{1}{\sqrt{2}}(x' + y')$$

Substitute these values into the original equation $xy = 6$:

$$\left[\frac{1}{\sqrt{2}}(x' - y')\right]\left[\frac{1}{\sqrt{2}}(x' + y')\right] = 6$$

Simplify (see Problem 1 for the details) to obtain

$$x'^2 - y'^2 = 12$$

$$\frac{x'^2}{12} - \frac{y'^2}{12} = 1$$

This curve is a hyperbola that has been rotated 45°. Draw the rotated axis and sketch this equation relative to the rotated axes. The result is shown in Figure 9.35.

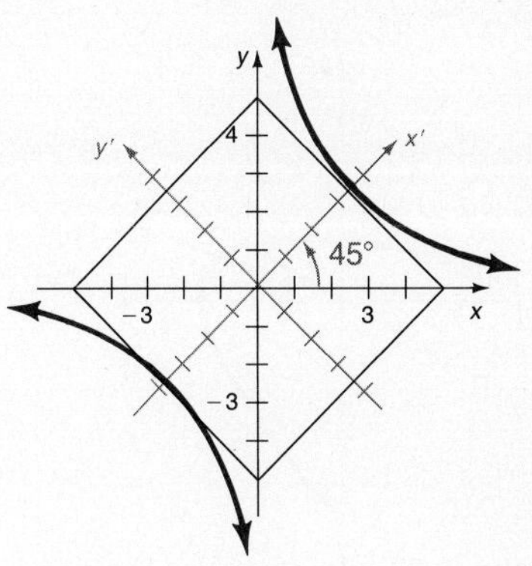

Figure 9.35 Graph of $xy = 6$ ■

EXAMPLE 6 Sketch $7x^2 - 6\sqrt{3}xy + 13y^2 - 16 = 0$.

Solution From Example 2, $\theta = 30°$ and

$$x = \frac{1}{2}(\sqrt{3}x' - y') \qquad y = \frac{1}{2}(x' + \sqrt{3}y')$$

Substitute into the original equation:

$$7\left(\frac{1}{2}\right)^2(\sqrt{3}x' - y')^2 - 6\sqrt{3}\left(\frac{1}{2}\right)(\sqrt{3}x' - y')\left(\frac{1}{2}\right)(x' + \sqrt{3}y')$$

$$+ 13\left(\frac{1}{2}\right)^2(x' + \sqrt{3}y')^2 - 16 = 0$$

Simplify (see Problem 2 for the details) to obtain

$$\frac{x'^2}{4} + \frac{y'^2}{1} = 1$$

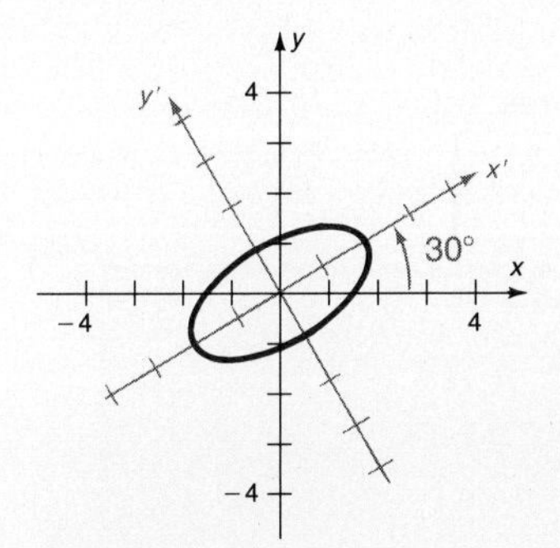

Figure 9.36 Graph of $7x^2 - 6\sqrt{3}xy + 13y^2 - 16 = 0$

This curve is an ellipse with a 30° rotation. The sketch is shown in Figure 9.36. ■

EXAMPLE 7 Sketch $x^2 - 4xy + 4y^2 + 5\sqrt{5}y - 10 = 0$.

Solution From Example 3, the rotation is $\theta \approx 26.6°$ and

$$x = \frac{1}{\sqrt{5}}(2x' - y') \qquad y = \frac{1}{\sqrt{5}}(x' + 2y')$$

Substitute

$$\frac{1}{5}(2x' - y')^2 - 4\left(\frac{1}{5}\right)(2x' - y')(x' + 2y') + 4\left(\frac{1}{5}\right)(x' + 2y')^2$$

$$+ 5\sqrt{5}\left(\frac{1}{\sqrt{5}}\right)(x' + 2y') - 10 = 0$$

Simplify (see Problem 3 for the details) to obtain

$$y'^2 + 2y' = -x' + 2$$

This curve is a parabola with a rotation of about 26.6°. Next complete the square to obtain

$$(y' + 1)^2 = -(x' - 3)$$

This sketch is shown in Figure 9.37. ■

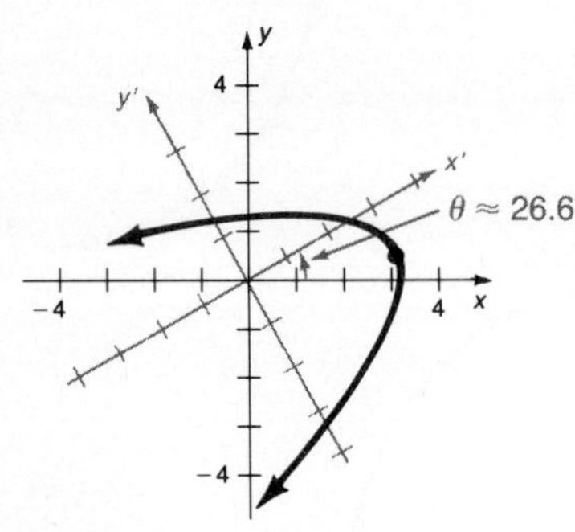

Figure 9.37 Graph of $x^2 - 4xy + 4y^2 + 5\sqrt{5}y - 10 = 0$

To graph the general second-degree equation

$$Ax^2 + Bxy + Cy^2 + Dx + Ey + F = 0$$

follow the steps shown in Figure 9.38.

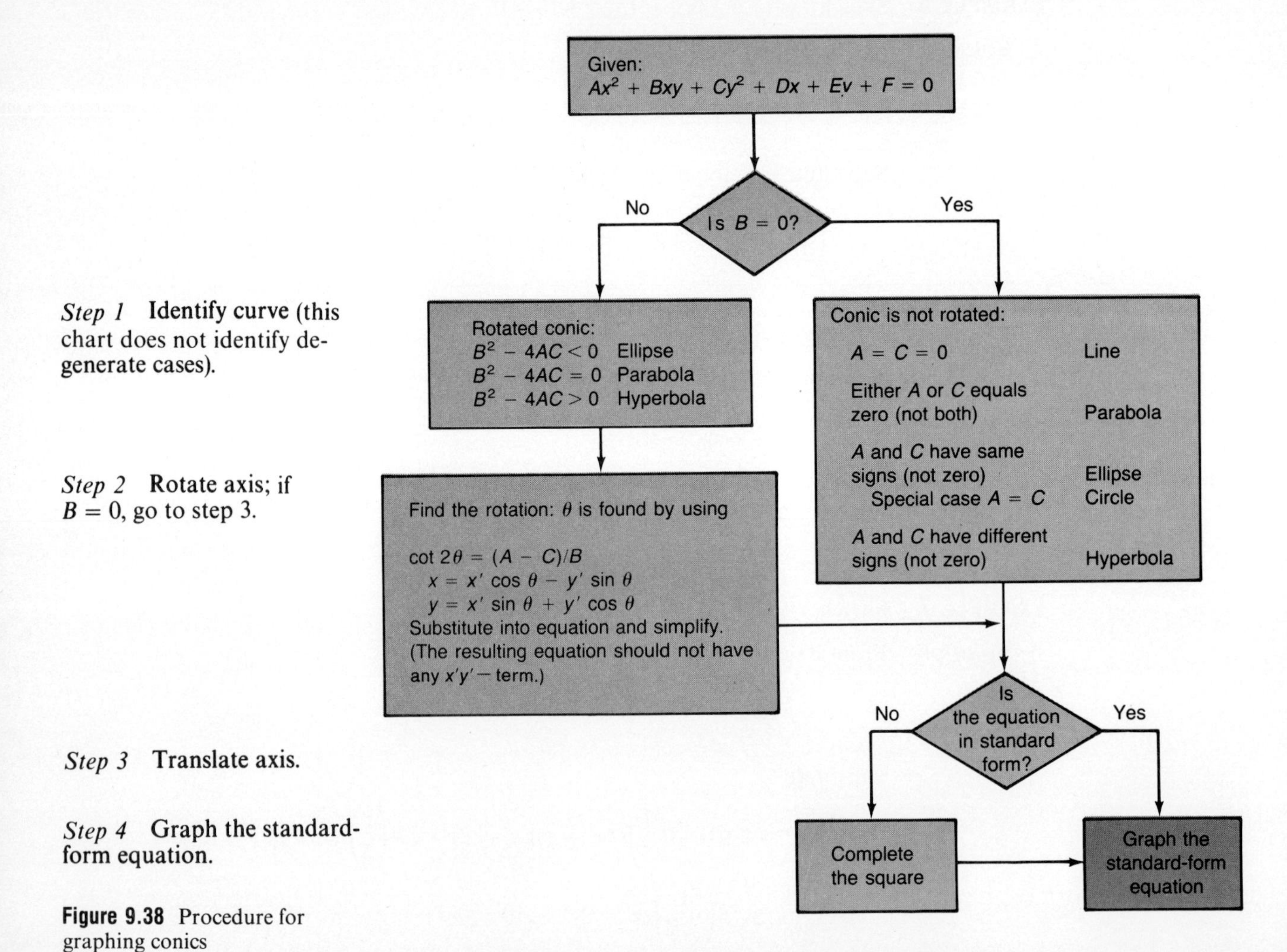

Figure 9.38 Procedure for graphing conics

PROBLEM SET 9.4

A

Because of the amount of arithmetic and algebra involved, many algebraic steps were left out of the examples of this section. In Problems 1–3, fill in the details left out of the indicated example.

1. Example 5 **2.** Example 6 **3.** Example 7

Identify the curves whose equations are given in Problems 4–21.

4. $xy = 10$

5. $xy = -1$

6. $xy = -4$

7. $13x^2 - 10xy + 13y^2 - 72 = 0$

8. $5x^2 - 26xy + 5y^2 + 72 = 0$

9. $x^2 + 4xy + 4y^2 + 10\sqrt{5}x = 9$

10. $5x^2 - 4xy + 8y^2 = 36$

11. $23x^2 + 26\sqrt{3}xy - 3y^2 - 144 = 0$

12. $3x^2 + 2\sqrt{3}xy + y^2 + 16x - 16\sqrt{3}y = 0$

13. $24x^2 + 16\sqrt{3}xy + 8y^2 - x + \sqrt{3}y - 8 = 0$

14. $3x^2 - 2\sqrt{3}xy + y^2 + 24x + 24\sqrt{3}y = 0$

15. $13x^2 - 6\sqrt{3}xy + 7y^2 + (16\sqrt{3} - 8)x + (-16 - 8\sqrt{3})y + 16 = 0$

16. $3x^2 - 10xy + 3y^2 - 32 = 0$

17. $5x^2 - 3xy + y^2 + 65x - 25y + 203 = 0$

18. $24x^2 + 16\sqrt{3}xy + 8y^2 - x + \sqrt{3}y - 8 = 0$

19. $5x^2 - 3xy + y^2 + 65x - 25y + 203 = 0$

20. $3x^2 - 10xy + 3y^2 - 32 = 0$

21. $13x^2 - 6\sqrt{3}xy + 7y^2 + (16\sqrt{3} - 8)x + (-16 - 8\sqrt{3})y + 16 = 0$

B

Find the appropriate rotation in Problems 22–35 so that the given curve will be in standard position relative to the rotated axes. Also find the x and y values in the new coordinate system by using the rotation of axes formulas.

22. $xy = 10$

23. $xy = -1$

24. $xy = -4$

25. $13x^2 - 10xy + 13y^2 - 72 = 0$

26. $5x^2 - 26xy + 5y^2 + 72 = 0$

27. $x^2 + 4xy + 4y^2 + 10\sqrt{5}x = 9$

28. $5x^2 - 4xy + 8y^2 = 36$

29. $23x^2 + 26\sqrt{3}xy - 3y^2 - 144 = 0$

30. $3x^2 + 2\sqrt{3}xy + y^2 + 16x - 16\sqrt{3}y = 0$

31. $24x^2 + 16\sqrt{3}xy + 8y^2 - x + \sqrt{3}y - 8 = 0$

32. $3x^2 - 2\sqrt{3}xy + y^2 + 24x + 24\sqrt{3}y = 0$

33. $13x^2 - 6\sqrt{3}xy + 7y^2 + (16\sqrt{3} - 8)x + (-16 - 8\sqrt{3})y + 16 = 0$

34. $3x^2 - 10xy + 3y^2 - 32 = 0$

35. $5x^2 - 3xy + y^2 + 65x - 25y + 203 = 0$

C

Sketch the curves in Problems 36–59.

36. $xy = 10$ **37.** $xy = -1$ **38.** $xy = -4$ **39.** $xy = 8$

40. $8x^2 - 4xy + 5y^2 = 36$

41. $13x^2 - 10xy + 13y^2 - 72 = 0$

42. $5x^2 - 26xy + 5y^2 + 72 = 0$

43. $x^2 + 4xy + 4y^2 + 10\sqrt{5}x = 9$

44. $5x^2 - 4xy + 8y^2 = 36$

45. $23x^2 + 26\sqrt{3}xy - 3y^2 - 144 = 0$

46. $3x^2 + 2\sqrt{3}xy + y^2 + 16x - 16\sqrt{3}y = 0$

47. $24x^2 + 16\sqrt{3}xy + 8y^2 - x + \sqrt{3}y - 8 = 0$

48. $3x^2 - 2\sqrt{3}xy + y^2 + 24x + 24\sqrt{3}y = 0$

49. $3x^2 - 10xy + 3y^2 - 32 = 0$

50. $x^2 + 2xy + y^2 + 12\sqrt{2}x - 6 = 0$

51. $10x^2 + 24xy + 17y^2 - 9 = 0$

52. $5x^2 - 3xy + y^2 + 65x - 25y + 203 = 0$

53. $3xy - 4y^2 + 18 = 0$

54. $17x^2 - 12xy + 8y^2 - 80 = 0$

55. $5x^2 - 8xy + 5y^2 - 9 = 0$

56. $16x^2 - 24xy + 9y^2 - 60x - 80y + 100 = 0$

57. $13x^2 - 6\sqrt{3}xy + 7y^2 + (16\sqrt{3} - 8)x + (-16 - 8\sqrt{3})y + 16 = 0$

58. $x^2 + 2\sqrt{3}xy + 3y^2 + 2\sqrt{3}x + 2y - 16 = 0$

59. $21x^2 + 10\sqrt{3}xy + 31y^2 - 72x - 16\sqrt{3}x - 72\sqrt{3}y + 16y + 16 = 0$

60. Let $Ax^2 + Bxy + Cy^2 + Dx + Ey + F = 0$. Show that $B^2 - 4AC = B'^2 - 4A'C'$ for any angle θ through which the axes may be rotated and that A', B', and C' are the values given on page 402. Use this fact to prove that (if the graph exists)

If $B^2 - 4AC = 0$, the graph is a parabola
If $B^2 - 4AC < 0$, the graph is an ellipse
If $B^2 - 4AC > 0$, the graph is a hyperbola

CHAPTER 9 SUMMARY

The material of this chapter is reviewed in the following list of objectives. After each objective there are some practice questions. For a sample test, select the first question of each set and check your answers with the answer section. For a sample test without answers, use the second question of each set. Additional practice is given by the other questions in each set. If you are having trouble with a particular type of problem, look back to that section for extra help.

9.1 PARABOLAS

Objective 1 Graph parabolas.

1. $x^2 = y$
2. $(y-1)^2 = 8(x+2)$
3. $8y^2 - x - 32y + 31 = 0$
4. $y^2 + 4x + 4y = 0$

Objective 2 Find the equations of parabolas given certain information about the graph.

5. Vertex at $(6, 3)$; directrix $x = 1$
6. Directrix $y - 3 = 0$; focus $(-3, -2)$
7. Vertex $(-3, 5)$; focus $(-3, -1)$
8. Vertex at $(4, 2)$ and passing through $(-3, -4)$; axis parallel to the y axis

9.2 ELLIPSES

Objective 3 Graph ellipses.

9. $25x^2 + 16y^2 = 400$
10. $5(x+3)^2 + 9(y-2)^2 = 45$
11. $x^2 + y^2 = 4x + 2y - 3$
12. $9x^2 + 16y^2 - 90x - 32y + 97 = 0$

Objective 4 Find the equations of ellipses given certain information about the graph.

13. The ellipse with the center at $(4, 1)$, a focus at $(5, 1)$, and a semimajor axis 2
14. The set of points such that the sum of the distances from $(-3, 4)$ and $(-7, 4)$ is 12
15. The set of points 8 units from the point $(-1, -2)$
16. The ellipse with foci at $(2, 3)$ and $(-1, 3)$ with eccentricity $\frac{3}{5}$.

9.3 HYPERBOLAS

Objective 5 Graph hyperbolas.

17. $x^2 - y^2 + x - y = 3$
18. $x(x - y) = y(y - x) - 1$
19. $5(x+2)^2 - 3(y+4)^2 = 60$
20. $12x^2 - 4y^2 + 24x - 8y + 4 = 0$

Objective 6 Find the equations of hyperbolas given certain information about the graph.

21. The set of points with the difference of distances from $(-3, 4)$ and $(-7, 4)$ equal to 2
22. The hyperbola with vertices at $(-3, 0)$ and $(3, 0)$ and foci at $(5, 0)$ and $(-5, 0)$
23. The hyperbola with vertices at $(0, -3)$ and $(0, 3)$ and eccentricity $\frac{5}{3}$
24. The hyperbola with vertices at $(-3, 1)$ and $(-5, 1)$ and foci at $(-4 - \sqrt{6}, 1)$ and $(-4 + \sqrt{6}, 1)$

Objective 7 Know the definition and standard-form equations for the conic section. State the appropriate standard-form equation.

25. Horizontal ellipse
26. Vertical hyperbola
27. Parabola opening right
28. Circle

Objective 8 Graph conic sections. Name the type of curve (by inspection) and then graph the curve.

29. $3x - 2y^2 - 4y + 7 = 0$
30. $\frac{x}{16} + \frac{y}{4} = 1$
31. $25x^2 + 9y^2 = 225$
32. $25(x-2)^2 + 25(y+1)^2 = 400$

*9.4 ROTATIONS

Objective 9 Use the rotation of axes formulas and the amount of rotation formula.

33. What is the rotation for the curve whose equation is $xy - 7 = 0$?

34. What is the rotation for the curve whose equation is
$5x^2 + 4xy + 5y^2 + 3x - 2y + 5 = 0$?

35. What is the rotation for the curve whose equation is $4x^2 + 4xy + y^2 + 3x - 2y + 7 = 0$?

36. Use the rotation formulas to rewrite the equation in Problem 33 so that there is no xy term.

Objective 10 Identify the conic by looking at its equation.

37. $xy + x^2 - 3x = 5$

38. $x^2 + y^2 + xy + 3x - y = 3$

39. $x^2 + 2xy + y^2 = 10$

40. $(x - 1)(y + 1) = 7$

* Optional section

A 'Planet X' Way Out There?

LIVERMORE, Calif. (UPI)—A scientist at the Lawrence Livermore Laboratory suggested Friday the existence of a "Planet X" three times as massive as Saturn and nearly six billion miles from earth.

The planet, far beyond Pluto which is currently the outermost of the nine known planets of the solar system, was predicted on sophisticated mathematical computations of the movements of Halley's Comet.

Joseph L. Brady, a Lawrence mathematician and an authority on the comet, reported the calculations in the Journal of the Astronomical Society of the Pacific.

Brady said he and his colleagues, Edna M. Carpenter and Francis H. McMahon, used a computer to process mathematical observations of the strange deviations in Halley's Comet going back to before Christ.

Lawrence officials said the existence of a 10th planet has been predicted before, but Brady is the first to predict its orbit, mass and position.

Brady said the planet was about 65 times as far from the sun as earth, which is about 93 million miles from the sun. From earth "Planet X" would be located in the constellation Casseiopeia on the border of the Milky Way.

The size and location of "Planet X" were proposed to account for mysterious deviations in the orbit of Halley's Comet. But the calculations subsequently were found to account for deviations in the orbits of two other reappearing comets, Olbers and Pons-Brooks, Brady said.

No contradiction between the proposed planet and the known orbits of comets and other planets has been found.

The prediction of unseen planets is not new. The location of Neptune was predicted in 1846 on the basis of deviations in the orbit of Uranus. Deviations in Neptune's orbit led to a prediction of Pluto's location in 1915.

Although no such deviations of Pluto have been found, Brady pointed out that since its discovery in 1930, Pluto has been observed through less than one-fourth of its revolution around the sun and a complete picture of its orbit is not available.

Brady said "Planet X" may be as elusive as Pluto was a half century ago. It took 15 years to find it from the time of its prediction.

"The proposed planet is located in the densely populated Milky Way where even a tiny area encompasses thousands of stars, many of which are brighter than we expect this planet to be," he said. "If it exists, it will be extremely difficult to find."

"Planet X," in its huge orbit, takes 600 years to complete a revolution around the sun, he said.

SOURCE: Courtesy of *The Press Democrat*, Santa Rosa, California, April 30, 1972. Reprinted by permission.

In 1986 Halley's Comet returned for our once-every-76-year view. By using mathematics and a knowledge of the conic sections, we can predict the comet's path as well as its exact arrival time (closest to Earth on April 10, 1986).

After the planet Uranus was discovered in 1781, its motion revealed gravitational perturbations caused by an unknown planet. Independent mathematical calculations by Urbain Leverrier and John Couch Adams predicted the position of this unknown planet—the discovery of Neptune in 1846 is one of the greatest triumphs of celestial mechanics and mathematics in the history of astronomy. A similar search lead to the discovery of Pluto in 1930. The article reproduced here is dated April 30, 1972, but at the time of this printing no planet has yet been found. Remember, though, that it took 15 years to find Pluto after the time of its prediction.

The orbits of the planets are elliptical in shape. If the sun is placed at one of the foci of a giant ellipse, the orbit of the earth is elliptical. The *perihelion* is the point where the planet comes closest to the sun; the *aphelion* is the farthest distance the planet travels from the sun. The eccentricity of a planet tells us the amount of roundness of that planet's orbit. The eccentricity of a circle is 0 and that of a parabola is 1. The eccentricity for each planet in our solar system is given here:

Planet	Eccentricity
Mercury	0.194
Venus	0.007
Earth	0.017
Mars	0.093
Jupiter	0.048
Saturn	0.056
Uranus	0.047
Neptune	0.009
Pluto	0.249

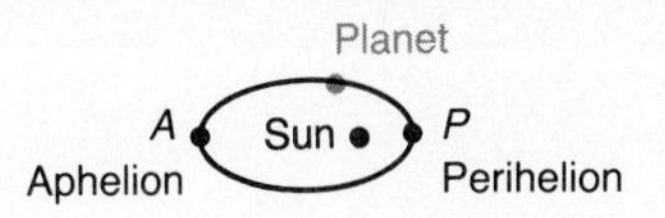

EXAMPLE 1 The orbit of the earth around the sun is elliptical with the sun at one focus. If the semimajor axis of this orbit is 9.3×10^7 mi and the eccentricity is about 0.017, determine the greatest and least distance of the earth from the sun (correct to two significant digits).

Solution We are given $a = 9.3 \times 10^7$ and $\epsilon = 0.17$. Now

$$\epsilon = \frac{c}{a}$$

$$0.017 = \frac{c}{9.3 \times 10^7}$$

$$c \approx 1.581 \times 10^6$$

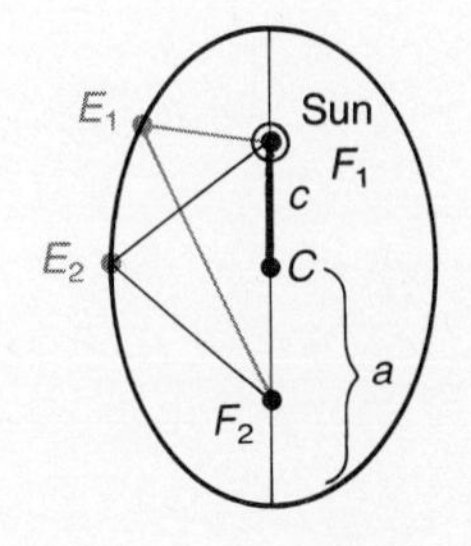

The greatest distance is $a + c \approx \mathbf{9.5 \times 10^7}$.
The least distance is $a - c \approx \mathbf{9.1 \times 10^7}$. ■

The orbit of a satellite can be calculated from Kepler's third law:

$$\frac{\text{Mass of (planet + satellite)}}{\text{Mass of (sun + planet + satellite)}} = \frac{(\text{semimajor axis of satellite orbit})^3}{(\text{semimajor axis of planet orbit})^3} \times \frac{(\text{period of planet})^2}{(\text{period of satellite})^2}$$

EXAMPLE 2 If the mass of the earth is 6.58×10^{21} tons, the sun 2.2×10^{27} tons, and the moon 8.1×10^{19} tons, calculate the orbit of the moon. Assume that the semimajor axis of the earth is 9.3×10^7 mi, the period of the earth 365.25 days, and the period of the moon 27.3 days.

Solution First solve the equation given as Kepler's third law for the unknown—the semimajor axis of the satellite orbit:

$$(\text{Semimajor axis of satellite orbit})^3$$

$$= \frac{\text{mass of (planet + satellite)}}{\text{mass of (sun + planet + satellite)}} \times \frac{(\text{period of satellite})^2}{(\text{period of planet})^2} \times (\text{semimajor axis of planet orbit})^3$$

Thus

$$(\text{Semimajor axis of satellite orbit})^3$$
$$= \frac{(6.58 \cdot 10^{21}) + (8.1 \cdot 10^{19})}{(2.2 \cdot 10^{27}) + (6.58 \cdot 10^{21}) + (8.1 \cdot 10^{19})} \times \frac{(27.3)^2}{(365.25)^2} \times (9.3 \cdot 10^7)^3$$
$$\approx \frac{3.993131141 \cdot 10^{48}}{2.934975261 \cdot 10^{32}}$$
$$\approx 1.360533151 \cdot 10^{16}$$

$$\text{Semimajor axis of satellite orbit} \approx 2.387278258 \cdot 10^5$$

The moon's orbit has a semimajor axis of about 239,000 mi. ■

Extended Application Problems—Planetary Orbits

1. The orbit of Mars about the sun is elliptical with the sun at one focus. If the semimajor axis of this orbit is 1.4×10^8 mi and the eccentricity is about 0.093, determine the greatest and least distance of Mars from the sun, correct to two significant digits.
2. The orbit of Venus about the sun is elliptical with the sun at one focus. If the semimajor axis of this orbit is 6.7×10^7 mi and the eccentricity is about 0.007, determine the greatest and least distance of Venus from the sun, correct to two significant digits.
3. The orbit of Neptune about the sun is elliptical with the sun at one focus. If the semimajor axis of this orbit is 3.66×10^9 mi and the eccentricity is about 0.009, determine the greatest and least distance of Neptune from the sun, correct to two significant digits.
4. If a Planet X has an elliptical orbit about the sun with a semimajor axis of 6.89×10^{10} mi and eccentricity of about 0.35, what is the least distance between Planet X and the sun?
5. If the mass of Mars is 7.05×10^{20} tons, its satellite Phobos 4.16×10^{12} tons, and the sun 2.2×10^{27} tons, calculate the length of the semimajor axis of Phobos. Assume that the semimajor axis of Mars is 1.4×10^8 mi, the period of Mars 693.5 days, and the period of Phobos 0.3 days.
6. Use the information in Problem 5 to find the approximate mass of the martian satellite Deimos if its elliptical orbit has a semimajor axis of 15,000 mi and a period of 1.26 days.
7. If the perihelion distance of Mercury is 2.9×10^7 mi and the aphelion distance is 4.3×10^7 mi, write the equation for the orbit of Mercury.
8. If the perihelion distance of Venus is 6.7234×10^7 mi and the aphelion distance is 6.8174×10^7 mi, write the equation for the orbit of Venus.
9. If the perihelion distance of Earth is 9.225×10^7 mi and the aphelion distance is 9.542×10^7 mi, write the equation for the orbit of Earth.
10. If the perihelion distance of Mars is 1.29×10^8 mi and the aphelion distance is 1.55×10^8 mi, write the equation for the orbit of Mars.

10

Additional Topics in Analytic Geometry

Charlotte Angas Scott
(1858–1931)

Indeed, mathematics, the indispensable tool of the sciences, defying the senses to follow its splendid flights, is demonstrating today, as it never has been demonstrated before, the supremacy of the pure reason.

NICHOLAS BUTLER
The Meaning of Education and Other Essays and Addresses

*** Optional Section**

The first prominent woman mathematician in America was Angas Scott. She is best known because of her lifelong work as an educator. She was educated in England but, soon after receiving her Ph.D., she came to America and developed the mathematics program at Bryn Mawr College. Today, it is difficult for us to understand the problems encountered by a woman in Scott's day who wished to receive a college education. Scott, for example, was permitted to take her final undergraduate examinations at Cambridge only informally. According to Karen Rappaport, "Scott tied for eighth place in the mathematics exam. Mathematics was an unprecedented area for a woman to excel in and the achievement attracted public attention, especially when her name was not mentioned at the official ceremony. When the official name was read, loud shouts of 'Scott of Girton' could be heard in the gallery. The public also responded to the slight. The February 7, 1880 issue of *Punch* stated: 'But when the academy doors are reopened to the Ladies let them be opened to their full worth. Let us not hear of any restrictions or exclusions from this or that function or privilege....' As a result of this support and a public petition, women were formally admitted to the Tripos Exams the following year but they were still not permitted Cambridge degrees. (That event did not occur until 1948.)" Mathematically, Scott was concerned with the study of specific algebraic curves of degree higher than two. She published thirty papers in the field of algebraic geometry, as well as a text, *Modern Analytic Geometry,* in 1894. In 1922, members of the American Mathematical Society and many former students organized a dinner to honor Charlotte Scott. Alfred North Whitehead came from England to give the main address, in which he stated: "A life's work such as that of Professor Charlotte Angas Scott is worth more to the world than many anxious efforts of diplomatists."

CHAPTER OVERVIEW In the first three sections of this chapter we introduce vectors, along with the basic vector operations. Vectors will be an important tool in your more advanced work with mathematics. The later part of this chapter is concerned with polar coordinates, polar-form curves, and parametric equations. The 20 objectives for this chapter are listed on pages 454–457.

10.1 VECTORS

Many applications of mathematics involve quantities that have *both* magnitude and direction, such as forces, velocities, accelerations, and displacements. Vectors are used to describe such quantities. A **vector** is a directed line segment specifying both a magnitude and a direction. The length of the vector represents the **magnitude** of the quantity being represented; the **direction** of the vector represents the direction of the quantity. Two vectors are **equal** if they have the same magnitude and direction.

Suppose we choose a point O in the plane and call it the origin. A vector is a directed line segment from O to a point $P(x, y)$ in the plane. This vector is denoted by $\overrightarrow{OP}$ or **v**. In the text we use **v**; in your work you will write $\vec{v}$. The magnitude of $\overrightarrow{OP}$ is denoted by $|\overrightarrow{OP}|$ or $|\mathbf{v}|$. The vector from O to O is called the zero vector **0**.

If **v** and **w** represent any two vectors having different (but not opposite) directions, then the **sum** or **resultant** is the vector drawn as the diagonal of the parallelogram having **v** and **w** as the adjacent sides, as shown in Figure 10.1. The vectors **v** and **w** are called **components**.

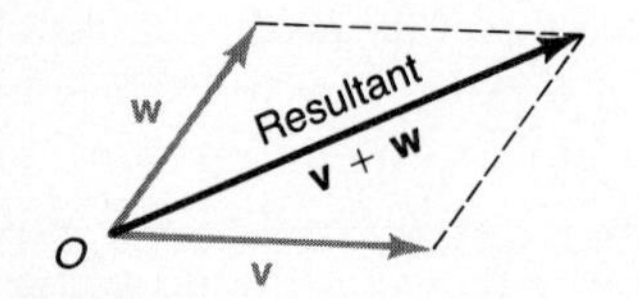

Figure 10.1 Resultant of vectors **v** and **w**

There are basically two types of vector problems dealing with addition of vectors. The first is to find the *resultant vector*. To do this you use a right triangle, the Law of Sines, or the Law of Cosines. The second problem is to *resolve* a vector into two component vectors. If the two vectors form a right angle, they are called **rectangular components**. You will usually resolve a vector into rectangular components.

EXAMPLE 1 Consider two forces, one with magnitude 3.0 in a N20°W direction and the other with magnitude 7.0 in a S50°W direction. Find the resultant vector.

Solution Sketch the given vectors and draw the parallelogram formed by these vectors. The diagonal is the resultant vector as shown in Figure 10.2. You can easily find $\theta = 110°$, but you really need an angle inside the shaded triangle. Use the property from geometry that adjacent angles in a parallelogram are supplementary (they add up to 180°). This tells you that $\phi = 70°$. Thus you know SAS, so you should use the Law of Cosines to find the magnitude $|\mathbf{v}|$:

$$|\mathbf{v}|^2 = 3^2 + 7^2 - 2(3)(7)\cos 70°$$

$$|\mathbf{v}| \approx 6.60569103$$

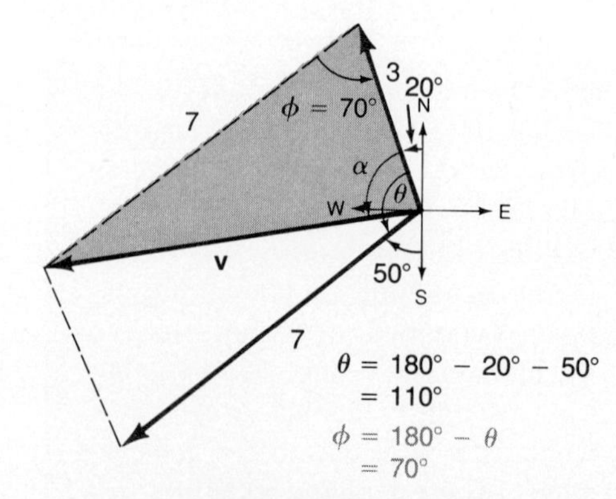

Figure 10.2

The direction of $\mathbf{v}$ can be found by using the Law of Sines to derive α:

$$\frac{\sin 70^\circ}{|\mathbf{v}|} = \frac{\sin \alpha}{7}$$

$$\sin \alpha = \frac{7}{|\mathbf{v}|} \sin 70^\circ$$

$$\approx 0.99578505$$

Thus $\alpha \approx 84.737565^\circ$. (Do not forget significant digits—see Appendix A.)

Since $20^\circ + 85^\circ = 105^\circ$, you can see that the direction of $\mathbf{v}$ should be measured from the south. Thus the magnitude of $\mathbf{v}$ is 6.6 and the direction is S75°W. ■

EXAMPLE 2 Suppose a vector $\mathbf{v}$ has a magnitude of 5.00 and a direction given by $\theta = 30.0^\circ$, where θ is the angle the vector makes with the positive x axis. Resolve this vector into horizontal and vertical components. (Do not forget significant digits—see Appendix A.)

Solution Let $\mathbf{v}_x$ be the horizontal component and $\mathbf{v}_y$ be the vertical component as shown in Figure 10.3. Then

$$\cos \theta = \frac{|\mathbf{v}_x|}{|\mathbf{v}|} \qquad \sin \theta = \frac{|\mathbf{v}_y|}{|\mathbf{v}|}$$

$$|\mathbf{v}_x| = |\mathbf{v}| \cos \theta \qquad \mathbf{v}_y = |\mathbf{v}| \sin \theta$$

$$= 5 \cos 30^\circ \qquad = 5 \sin 30^\circ$$

$$= \frac{5}{2}\sqrt{3} \qquad = \frac{5}{2}$$

$$\approx 4.33 \qquad = 2.50$$

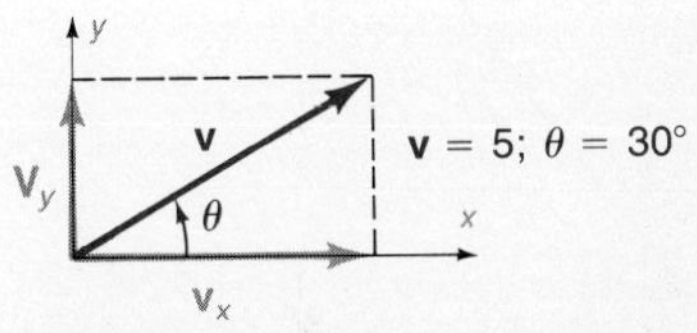

Figure 10.3 Resolving a vector

Thus $\mathbf{v}_x$ is a horizontal vector with magnitude 4.33 and $\mathbf{v}_y$ is a vertical vector with magnitude 2.50. ■

The process of resolving a vector can be simplified by using scalar multiplication and two special vectors. **Scalar multiplication** is the multiplication of a vector by a real number. It is called scalar multiplication because real numbers are sometimes called scalars. If c is a positive real number and $\mathbf{v}$ is a vector, then the vector $c\mathbf{v}$ is a vector representing the scalar multiplication of c and $\mathbf{v}$. It is defined geometrically as a vector in the same direction as $\mathbf{v}$ but with a magnitude c times the original magnitude of $\mathbf{v}$, as shown in Figure 10.4. If c is a negative real number, then the scalar multiplication results in a vector in exactly the opposite direction as $\mathbf{v}$ with length c times as long, as shown in Figure 10.4. If $c = 0$, then $c\mathbf{v}$ is the zero vector.

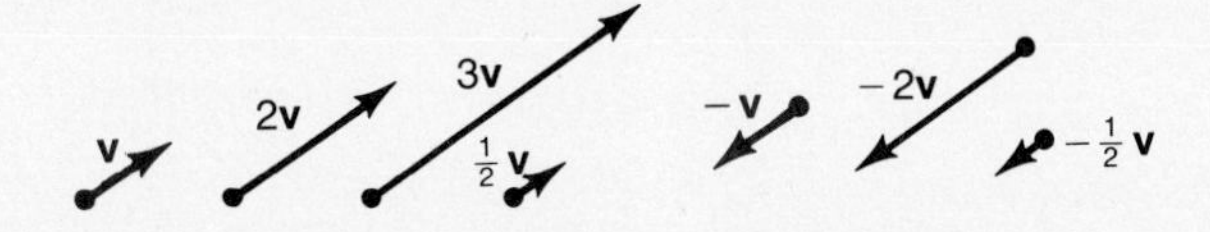

Figure 10.4 Examples of scalar multiplication

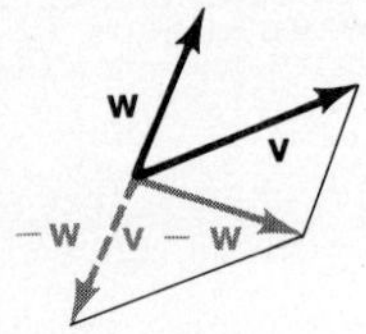

Scalar multiplication allows us to define **subtraction** for vectors:

$$\mathbf{v} - \mathbf{w} = \mathbf{v} + (-\mathbf{w})$$

Geometrically, subtraction is shown in Figure 10.5.

Figure 10.5 Vector subtraction

Two special vectors help us to treat vectors algebraically as well as geometrically:

Definition of the i and j Vectors

> **i** is the vector of unit length in the direction of the positive x axis.
> **j** is the vector of unit length in the direction of the positive y axis.

Example 2 shows how the vector **v** with magnitude 5.00 and $\theta = 30.0°$ can be resolved into rectangular components:

1. $\mathbf{v}_x$ (horizontal component) with magnitude 4.33
2. $\mathbf{v}_y$ (vertical component) with magnitude 2.50

However, since **i** has unit length and is horizontal and **j** has unit length and is vertical, we can write

$$\mathbf{v}_x = 4.33\mathbf{i} \quad \text{and} \quad \mathbf{v}_y = 2.50\mathbf{j}$$

Consequently,

$$\begin{aligned} \mathbf{v} &= \mathbf{v}_x + \mathbf{v}_y \\ &= 4.33\mathbf{i} + 2.50\mathbf{j} \end{aligned}$$

In general, any vector **v** can be written as

$$\mathbf{v} = a\mathbf{i} + b\mathbf{j}$$

where a and b are the magnitude of the horizontal and vertical components, respectively. This is called the **algebraic representation of a vector** and is shown in Figure 10.6.

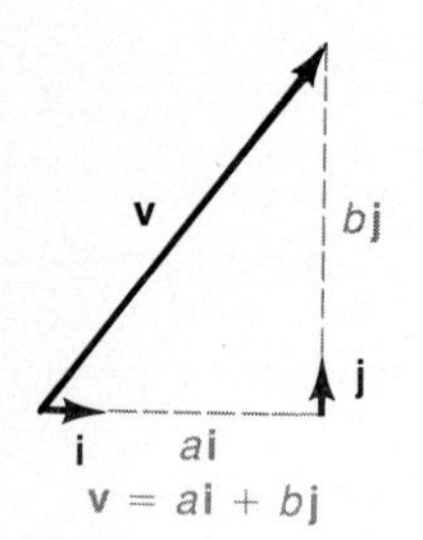

Figure 10.6 Algebraic representation of a vector

EXAMPLE 3 Find the algebraic representation for a vector **v** with magnitude 10 making an angle of 60° with the positive x axis.

Solution Figure 10.7 shows the general procedure for writing the algebraic representation of a vector when given the magnitude and direction of that vector.

$$|b\mathbf{j}| = b$$

$$|a\mathbf{i}| = a$$

$$\cos\theta = \frac{a}{|\mathbf{v}|} \qquad a = |\mathbf{v}|\cos\theta$$

$$\sin\theta = \frac{b}{|\mathbf{v}|} \qquad b = |\mathbf{v}|\sin\theta$$

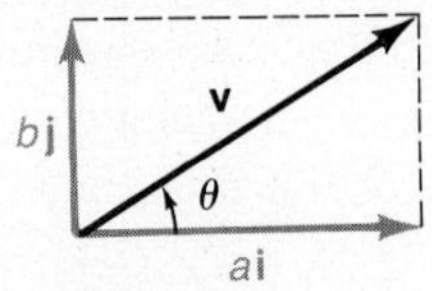

Figure 10.7
$\mathbf{v} = |\mathbf{v}|\cos\theta\mathbf{i} + |\mathbf{v}|\sin\theta\mathbf{j}$

From Figure 10.7,

$$a = |\mathbf{v}|\cos\theta \qquad b = |\mathbf{v}|\sin\theta$$
$$= 10\cos 60^\circ \qquad = 10\sin 60^\circ$$
$$= 5.0 \qquad \approx 8.7$$

Therefore

$$\mathbf{v} = 5.0\mathbf{i} + 8.7\mathbf{j}$$

■

EXAMPLE 4 Find the algebraic representation for a vector $\mathbf{v}$ with initial point $(4, -3)$ and endpoint $(-2, 4)$.

Solution Figure 10.8 shows the general procedure for writing the algebraic representation of a vector when given the endpoints of that vector.

$$a = x_2 - x_1 \quad \text{and} \quad b = y_2 - y_1$$

From Figure 10.8,

$$a = x_2 - x_1 \qquad b = y_2 - y_1$$
$$= -2 - 4 \qquad = 4 - (-3)$$
$$= -6 \qquad = 7$$

Thus $\mathbf{v} = -6\mathbf{i} + 7\mathbf{j}$

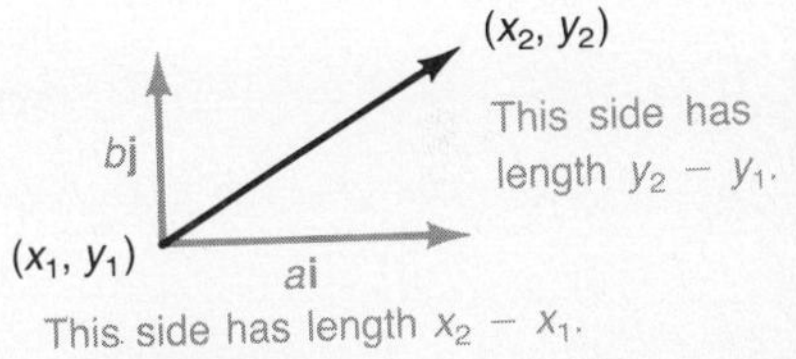

Figure 10.8
$\mathbf{v} = (x_2 - x_1)\mathbf{i} + (y_2 - y_1)\mathbf{j}$

■

EXAMPLE 5 Find the magnitude of the vector in Example 4.

Solution $\mathbf{v} = -6\mathbf{i} + 7\mathbf{j}$. Thus

$$|\mathbf{v}| = \sqrt{(-6)^2 + (7)^2}$$
$$= \sqrt{36 + 49}$$
$$= \sqrt{85}$$

■

Example 5 leads to the following general result.

Magnitude of a Vector

The **magnitude** of a vector $\mathbf{v} = a\mathbf{i} + b\mathbf{j}$ is given by

$$|\mathbf{v}| = \sqrt{a^2 + b^2}$$

The operations of addition, subtraction, and scalar multiplication can also be stated algebraically. Let $\mathbf{v} = a\mathbf{i} + b\mathbf{j}$ and $\mathbf{w} = c\mathbf{i} + d\mathbf{j}$. Then

$$\mathbf{v} + \mathbf{w} = (a + c)\mathbf{i} + (b + d)\mathbf{j}$$
$$\mathbf{v} - \mathbf{w} = (a - c)\mathbf{i} + (b - d)\mathbf{j}$$
$$c\mathbf{v} = ca\mathbf{i} + cb\mathbf{j}$$

EXAMPLE 6 Let $\mathbf{v} = 6\mathbf{i} + 4\mathbf{j}$ and $\mathbf{w} = -2\mathbf{i} + 3\mathbf{j}$.

Solution

a. $|\mathbf{v}| = \sqrt{6^2 + 4^2} = \sqrt{36 + 16} = 2\sqrt{13}$

b. $|\mathbf{w}| = \sqrt{(-2)^2 + 3^2} = \sqrt{4 + 9} = \sqrt{13}$

c. $\mathbf{v} + \mathbf{w} = (6 - 2)\mathbf{i} + (4 + 3)\mathbf{j} = 4\mathbf{i} + 7\mathbf{j}$

d. $\mathbf{v} - \mathbf{w} = (6 + 2)\mathbf{i} + (4 - 3)\mathbf{j} = 8\mathbf{i} + \mathbf{j}$

e. $-\mathbf{v} = (-1)6\mathbf{i} + (-1)4\mathbf{j} = -6\mathbf{i} - 4\mathbf{j}$

f. $-2\mathbf{w} = (-2)(-2)\mathbf{i} + (-2)(3)\mathbf{j} = 4\mathbf{i} - 6\mathbf{j}$ ■

You can see from Example 6 that the algebraic representation of a vector makes it easy to handle vectors and their operations. Another advantage of the algebraic representation is that it specifies a direction and a magnitude, but not a particular location. The directed line segment in Example 4 defined a given vector. However, there are infinitely many other vectors represented by the form $-6\mathbf{i} + 7\mathbf{j}$. Two of these are shown in Figure 10.9.

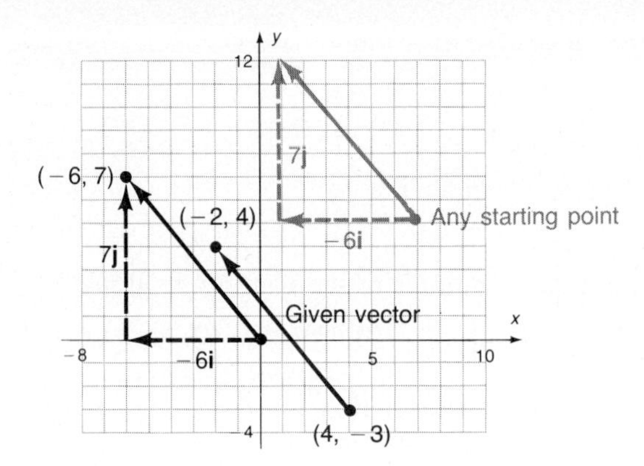

Figure 10.9 Some vectors represented by $\mathbf{v} = -6\mathbf{i} + 7\mathbf{j}$

In arithmetic the words *multiplication* and *product* are used to mean the same thing. When working with vectors, however, these words are used to denote different ideas. Recall that scalar multiplication does not tell how to multiply two vectors; instead it tells how to multiply a scalar and a vector. Now we define an operation called **scalar product** in order to multiply two vectors to obtain a number. It is called *scalar* product because a number or scalar is obtained as an answer. In Section 10.3 we will define another vector multiplication, called **vector product**, in which a vector is obtained as an answer.

Definition of Scalar Product

Let $\mathbf{v} = a\mathbf{i} + b\mathbf{j}$ and $\mathbf{w} = c\mathbf{i} + d\mathbf{j}$. Then the **scalar product**, written $\mathbf{v} \cdot \mathbf{w}$, is defined by

$$\mathbf{v} \cdot \mathbf{w} = ac + bd$$

Sometimes this product is called the *dot product*, from the form in which it is written.

EXAMPLE 7 Find the scalar product of the given vectors.

Solution **a.** If $\mathbf{v} = 2\mathbf{i} + 5\mathbf{j}$ and $\mathbf{w} = 6\mathbf{i} - 3\mathbf{j}$, then

$$\begin{aligned}\mathbf{v}\cdot\mathbf{w} &= 2(6) + 5(-3)\\ &= 12 - 15\\ &= -3\end{aligned}$$

b. If $\mathbf{v} = \cos 30°\mathbf{i} + \sin 30°\mathbf{j}$ and $\mathbf{w} = \cos 60°\mathbf{i} - \sin 60°\mathbf{j}$, then

$$\begin{aligned}\mathbf{v}\cdot\mathbf{w} &= \cos 30° \cos 60° - \sin 30° \sin 60°\\ &= \cos(30° + 60°)\\ &= 0\end{aligned}$$

c. If $\mathbf{v} = -\sqrt{3}\mathbf{i} + \sqrt{2}\mathbf{j}$ and $\mathbf{w} = 3\sqrt{3}\mathbf{i} + 5\sqrt{2}\mathbf{j}$, then

$$\begin{aligned}\mathbf{v}\cdot\mathbf{w} &= (-\sqrt{3})(3\sqrt{3}) + \sqrt{2}(5\sqrt{2})\\ &= -9 + 10\\ &= 1\end{aligned}$$

d. If $\mathbf{v} = 2\mathbf{i} - 3\mathbf{j}$ and $\mathbf{w} = 4\mathbf{i} + a\mathbf{j}$, then

$$\mathbf{v}\cdot\mathbf{w} = 8 - 3a$$

■

There is a very useful geometric property for scalar product that is apparent if we find an expression for the angle between two vectors:

Angle between Vectors

> The angle θ between vectors $\mathbf{v}$ and $\mathbf{w}$ is found by
>
> $$\cos\theta = \frac{\mathbf{v}\cdot\mathbf{w}}{|\mathbf{v}||\mathbf{w}|}$$

Proof Let $\mathbf{v} = a\mathbf{i} + b\mathbf{j}$ and $\mathbf{w} = c\mathbf{i} + d\mathbf{j}$ be drawn with their bases at the origin, as shown in Figure 10.10. Let x be the distance between the endpoints of the vectors. Then, by the Law of Cosines,

$$\begin{aligned}\cos\theta &= \frac{|\mathbf{v}|^2 + |\mathbf{w}|^2 - x^2}{2|\mathbf{v}||\mathbf{w}|}\\ &= \frac{(\sqrt{a^2 + b^2})^2 + (\sqrt{c^2 + d^2})^2 - (\sqrt{(a - c)^2 + (b - d)^2})^2}{2|\mathbf{v}||\mathbf{w}|}\\ &= \frac{a^2 + b^2 + c^2 + d^2 - (a^2 - 2ac + c^2 + b^2 - 2bd + d^2)}{2|\mathbf{v}||\mathbf{w}|}\\ &= \frac{2ac + 2bd}{2|\mathbf{v}||\mathbf{w}|}\\ &= \frac{ac + bd}{|\mathbf{v}||\mathbf{w}|}\\ &= \frac{\mathbf{v}\cdot\mathbf{w}}{|\mathbf{v}||\mathbf{w}|}\end{aligned}$$

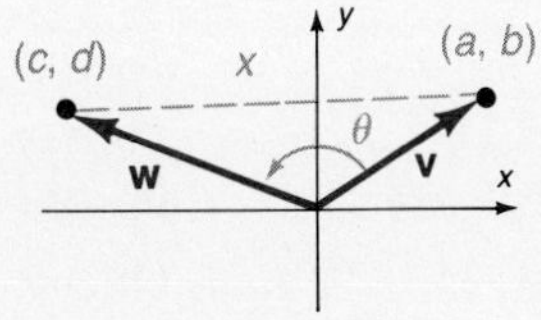

□

Figure 10.10 Finding the angle between two vectors

There is a useful geometric property of vectors whose directions differ by 90°. If you are dealing with lines, they are called **perpendicular lines**, but if they are vectors forming a 90° angle, they are called **orthogonal vectors**. Notice that if $\theta = 90°$, then $\cos 90° = 0$; therefore, from the angle between vectors formula,

$$0 = \frac{\mathbf{v} \cdot \mathbf{w}}{|\mathbf{v}||\mathbf{w}|}$$

If you multiply both sides by $|\mathbf{v}||\mathbf{w}|$, the result is the following important condition.

Orthogonal Vectors

Vectors $\mathbf{v}$ and $\mathbf{w}$ are orthogonal if and only if $\mathbf{v} \cdot \mathbf{w} = 0$.

EXAMPLE 8 Show that $\mathbf{v} = 3\mathbf{i} - 2\mathbf{j}$ and $\mathbf{w} = 6\mathbf{i} + 9\mathbf{j}$ are orthogonal.

Solution

$$\begin{aligned} \mathbf{v} \cdot \mathbf{w} &= 3(6) + (-2)(9) \\ &= 18 - 18 \\ &= 0 \end{aligned}$$

Since the scalar product is zero, the vectors are orthogonal. ■

EXAMPLE 9 Find a so that $\mathbf{v} = 3\mathbf{i} + a\mathbf{j}$ and $\mathbf{w} = \mathbf{i} - 2\mathbf{j}$ are orthogonal.

Solution

$$\mathbf{v} \cdot \mathbf{w} = 3 - 2a$$

If they are orthogonal, then

$$3 - 2a = 0$$

$$a = \frac{3}{2}$$ ■

PROBLEM SET 10.1

A

Find the algebraic representation for each vector given in Problems 1–18. $|\mathbf{v}|$ is the magnitude of the vector $\mathbf{v}$, and θ is the angle the vector makes with the positive x axis. A and B are the endpoints of the vector $\mathbf{v}$, and A is the base point. Draw each vector.

1. $|\mathbf{v}| = 12, \theta = 60°$
2. $|\mathbf{v}| = 8, \theta = 30°$
3. $|\mathbf{v}| = \sqrt{2}, \theta = 45°$
4. $|\mathbf{v}| = 9, \theta = 45°$
5. $|\mathbf{v}| = 7, \theta = 23°$
6. $|\mathbf{v}| = 5, \theta = 72°$
7. $|\mathbf{v}| = 4, \theta = 112°$
8. $|\mathbf{v}| = 10, \theta = 214°$
9. $A(4, 1), B(2, 3); \overrightarrow{AB}$
10. $A(-1, -3), B(4, 5); \overrightarrow{AB}$
11. $A(1, -2), B(-5, -7); \overrightarrow{AB}$
12. $A(6, -8), B(5, -2); \overrightarrow{AB}$
13. $A(-3, 2), B(5, -8); \overrightarrow{AB}$
14. $A(0, 0), B(-3, -4); \overrightarrow{AB}$
15. $A(7, 1), B(0, 0); \overrightarrow{AB}$
16. $A(2, 9), B(-5, 8); \overrightarrow{AB}$
17. $A(6, -1); B(-3, -7); \overrightarrow{AB}$
18. $A(-2, -8), B(-4, 7); \overrightarrow{AB}$

Find the magnitude of each of the vectors given in Problems 19–27.

19. $\mathbf{v} = 3\mathbf{i} + 4\mathbf{j}$
20. $\mathbf{v} = 5\mathbf{i} - 12\mathbf{j}$
21. $\mathbf{v} = 6\mathbf{i} - 7\mathbf{j}$
22. $\mathbf{v} = -3\mathbf{i} + 5\mathbf{j}$
23. $\mathbf{v} = -2\mathbf{i} + 2\mathbf{j}$
24. $\mathbf{v} = 5\mathbf{i} - 8\mathbf{j}$
25. $\mathbf{v} = \mathbf{i} - 3\mathbf{j}$
26. $\mathbf{v} = 2\mathbf{i} - \mathbf{j}$
27. $\mathbf{v} = 4\mathbf{i} + 5\mathbf{j}$

State whether the given pairs of vectors in Problems 28–33 are orthogonal.

28. $\mathbf{v} = 3\mathbf{i} - 2\mathbf{j}; \mathbf{w} = 6\mathbf{i} + 9\mathbf{j}$
29. $\mathbf{v} = 2\mathbf{i} + 3\mathbf{j}; \mathbf{w} = 6\mathbf{i} - 9\mathbf{j}$
30. $\mathbf{v} = 4\mathbf{i} - 5\mathbf{j}; \mathbf{w} = 8\mathbf{i} + 10\mathbf{j}$
31. $\mathbf{v} = 5\mathbf{i} + 4\mathbf{j}; \mathbf{w} = 8\mathbf{i} - 10\mathbf{j}$
32. $\mathbf{v} = 2\mathbf{i} - 3\mathbf{j}; \mathbf{w} = 3\mathbf{i} + 2\mathbf{j}$
33. $\mathbf{v} = \mathbf{i}; \mathbf{w} = \mathbf{j}$

B

34. ***Navigation*** A woman sets out in a rowboat heading due west and rows at 4.8 mph. The current is carrying the boat due south at 12 mph. What is the true course of the rowboat, and how fast is the boat traveling relative to the ground?

35. ***Aviation*** An airplane is headed due west at 240 mph. The wind is blowing due south at 43 mph. What is the true course of the plane, and how fast is it traveling across the ground?

36. ***Aviation*** An airplane is heading 215° with a velocity of 723 mph. How far south has it traveled in 1 hr?

37. ***Aviation*** An airplane is heading 43.0° with a velocity of 248 mph. How far east has it traveled in 2 hr?

In Problems 38–49, find $\mathbf{v} \cdot \mathbf{w}$, $|\mathbf{v}|$, $|\mathbf{w}|$, *and* $\cos\theta$, *where* θ *is the angle between* **v** *and* **w**.

38. $\mathbf{v} = 3\mathbf{i} + 4\mathbf{j}$, $\mathbf{w} = 5\mathbf{i} + 12\mathbf{j}$

39. $\mathbf{v} = 8\mathbf{i} - 6\mathbf{j}$, $\mathbf{w} = -5\mathbf{i} + 12\mathbf{j}$

40. $\mathbf{v} = 2\mathbf{i} + \sqrt{5}\mathbf{j}$, $\mathbf{w} = 3\sqrt{5}\mathbf{i} - 3\mathbf{j}$

41. $\mathbf{v} = 7\mathbf{i} - \sqrt{15}\mathbf{j}$, $\mathbf{w} = 2\sqrt{15}\mathbf{i} + 14\mathbf{j}$

42. $\mathbf{v} = -2\mathbf{i} + 3\mathbf{j}$, $\mathbf{w} = 6\mathbf{i} + 5\mathbf{j}$

43. $\mathbf{v} = 3\mathbf{i} + 9\mathbf{j}$, $\mathbf{w} = 2\mathbf{i} - 5\mathbf{j}$

44. $\mathbf{v} = \mathbf{i}$, $\mathbf{w} = \mathbf{i}$

45. $\mathbf{v} = \mathbf{j}$, $\mathbf{w} = \mathbf{j}$

46. $\mathbf{v} = \mathbf{i}$, $\mathbf{w} = -\mathbf{j}$

47. $\mathbf{v} = 5\mathbf{i} - \mathbf{j}$, $\mathbf{w} = 2\mathbf{i} + 3\mathbf{j}$

48. $\mathbf{v} = 4\mathbf{i} + 2\mathbf{j}$, $\mathbf{w} = 3\mathbf{i} - \mathbf{j}$

49. $\mathbf{v} = \mathbf{i} + \mathbf{j}$, $\mathbf{w} = \mathbf{i}$

In Problems 50–55, find the angle θ *to the nearest degree,* $0^\circ \le \theta \le 180^\circ$, *between the vectors* **v** *and* **w**.

50. $\mathbf{v} = \frac{1}{2}\mathbf{i} + \frac{\sqrt{3}}{2}\mathbf{j}$, $\mathbf{w} = \frac{1}{2}\mathbf{i} + \frac{1}{2}\mathbf{j}$

51. $\mathbf{v} = \sqrt{2}\mathbf{i} - \sqrt{2}\mathbf{j}$, $\mathbf{w} = \frac{\sqrt{3}}{2}\mathbf{i} + \frac{1}{2}\mathbf{j}$

52. $\mathbf{v} = \mathbf{j}$, $\mathbf{w} = \frac{1}{2}\mathbf{i} - \frac{\sqrt{3}}{2}\mathbf{j}$

53. $\mathbf{v} = -\mathbf{i}$, $\mathbf{w} = -2\sqrt{2}\mathbf{i} + 2\sqrt{2}\mathbf{j}$

54. $\mathbf{v} = 2\mathbf{i} + 3\mathbf{j}$, $\mathbf{w} = -\mathbf{i} + 4\mathbf{j}$

55. $\mathbf{v} = -3\mathbf{i} + 2\mathbf{j}$, $\mathbf{w} = 6\mathbf{i} + 9\mathbf{j}$

In Problems 56–58, find a number a so that the given vectors are orthogonal.

56. $\mathbf{v} = 2\mathbf{i} + 3\mathbf{j}$, $\mathbf{w} = 5\mathbf{i} + a\mathbf{j}$

57. $\mathbf{v} = 4\mathbf{i} - a\mathbf{j}$, $\mathbf{w} = -2\mathbf{i} + 5\mathbf{j}$

58. $\mathbf{v} = a\mathbf{i} + 5\mathbf{j}$, $\mathbf{w} = a\mathbf{i} - 15\mathbf{j}$

C

59. ***Aviation*** A pilot is flying at an airspeed of 241 mph in a wind blowing 20.4 mph from the east. In what direction must the pilot head in order to fly due north? What is the pilot's speed relative to the ground?

60. ***Aviation*** Answer the questions posed in Problem 59 with the pilot wishing to fly due south.

61. ***Space Science*** The weight of astronauts on the moon is about one-sixth of their weight on earth. This fact has a marked effect on such simple acts as walking, running, and jumping. To study these effects and to train astronauts for working under lunar-gravity conditions, scientists at NASA's Langley Research Center have designed an inclined-plane apparatus to simulate reduced gravity. The apparatus consists of a sling that holds the astronaut in a position perpendicular to the inclined plane (see Figure 10.11). The sling is attached to one end of a long cable that runs parallel to the inclined plane. The other end of the cable is attached to a trolley that runs along an overhead track. This device allows the astronaut to move freely in a plane perpendicular to the inclined plane. Let W be the astronaut's mass and θ be the angle between the inclined plane and the ground. Make a vector diagram showing the tension in the cable and the force exerted by the inclined plane against the feet of the astronaut.

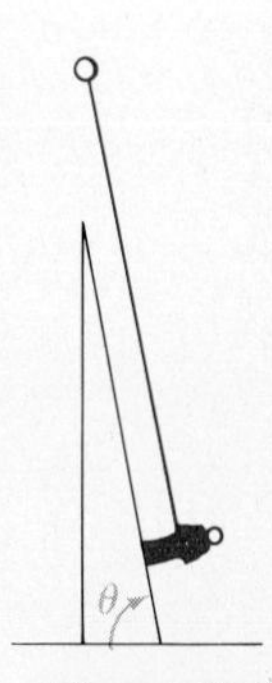

Figure 10.11

62. ***Space Science*** From the point of view of the astronaut in Problem 61, the inclined plane is the ground and the astronaut's simulated mass (that is, the downward force against the inclined plane) is $\mathbf{W}\cos\theta$. What value of θ is required in order to simulate lunar gravity?

10.2 PROPERTIES OF VECTORS

In this section we will look at several properties of vectors that will be useful in more advanced mathematics courses. Suppose you are given a line $ax + by + c = 0$. A vector perpendicular to a line is called a *normal* to the line. We will show that

$$\mathbf{N} = a\mathbf{i} + b\mathbf{j}$$

is normal to

$$L\colon ax + by + c = 0$$

Notice that, once L is given, $\mathbf{N}$ can be found by inspection. For example, if

$$L\colon 3x + 4y + 5 = 0 \text{ then}$$

$$\mathbf{N} = 3\mathbf{i} + 4\mathbf{j}$$

If

$$L\colon 5x - 3y + 7 = 0 \text{ then}$$

$$\mathbf{N} = 5\mathbf{i} - 3\mathbf{j}$$

In general, if $P_1(x_1, y_1)$ and $P_2(x_2, y_2)$ are any two points on L, then P_1P_2 is a representative of the vector determined by the line. That is,

$$P_1P_2 = (x_2 - x_1)\mathbf{i} + (y_2 - y_1)\mathbf{j}$$

We must show that P_1P_2 and $\mathbf{N} = a\mathbf{i} + b\mathbf{j}$ are orthogonal. That is, we must show that

$$P_1P_2 \cdot \mathbf{N} = 0$$

Now,

$$P_1P_2 \cdot \mathbf{N} = a(x_2 - x_1) + b(y_2 - y_1)$$

Since $P_1(x_1, y_1)$ and $P_2(x_2, y_2)$ are on the line, they make the equation true:

$$ax_1 + by_1 + c = 0$$
$$ax_2 + by_2 + c = 0$$

Subtracting,

$$a(x_2 - x_1) + b(y_2 - y_1) = 0$$

Thus,

$$\begin{aligned} P_1P_2 \cdot \mathbf{N} &= a(x_2 - x_1) + b(y_2 - y_1) \\ &= 0 \end{aligned}$$

EXAMPLE 1 Find a vector determined by the line $5x - 3y - 15 = 0$. Also find a normal vector. Show that they are orthogonal.

Solution A normal vector is $5\mathbf{i} - 3\mathbf{j}$.

A vector determined by the line is found by first obtaining two points on the line. If $x = 0$, then $y = -5$; if $y = 0$, then $x = 3$. Thus, two points on the line are $(0, -5)$ and $(3, 0)$. Hence $3\mathbf{i} + 5\mathbf{j}$ is a vector determined by the line.

Finally, $(5\mathbf{i} - 3\mathbf{j}) \cdot (3\mathbf{i} + 5\mathbf{j}) = 15 - 15 = 0$, so they are orthogonal (see Figure 10.12).

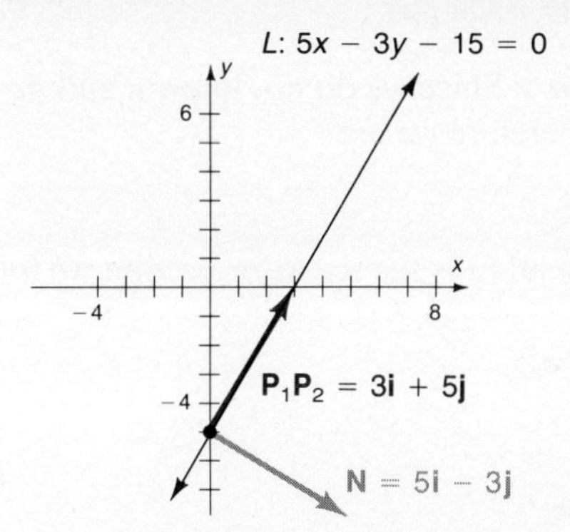

Figure 10.12

■

Let **v** and **w** be two vectors whose representatives have a common base point. If we drop a perpendicular from the head of **v** to the line determined by **w**, we determine a vector called *the vector projection of* **v** *onto* **w**, denoted by **u**, as shown in Figure 10.13.

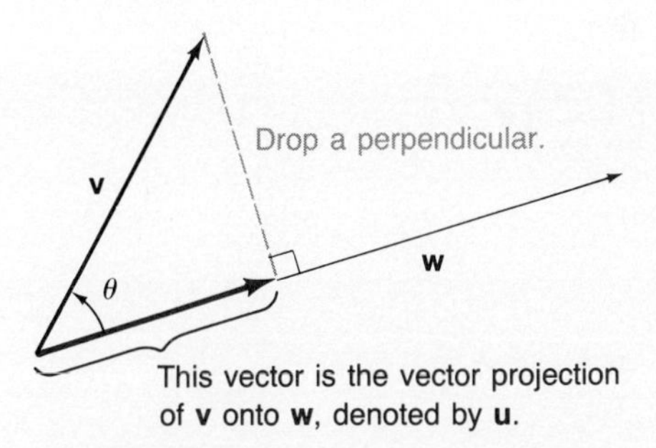

Figure 10.13 Projection of **v** onto **w**

The *scalar projection* is the *length* of the vector projection. Let θ be an acute angle between **v** and **w**. Then

$$\cos\theta = \left|\frac{\mathbf{u}}{\mathbf{v}}\right| \qquad \text{By definition of cosine}$$

Thus,

$$\begin{aligned}|\mathbf{u}| &= |\mathbf{v}|\cos\theta \\ &= |\mathbf{v}|\frac{\mathbf{v}\cdot\mathbf{w}}{|\mathbf{v}|\,|\mathbf{w}|} \\ &= \frac{\mathbf{v}\cdot\mathbf{w}}{|\mathbf{w}|}\end{aligned}$$

If $90^\circ < \theta < 180^\circ$, then $\cos\theta \le 0$ so $|\mathbf{v}|\cos\theta$ is a negative number. Since we want $|\mathbf{u}|$ to be nonnegative (since it is a length), we introduce an absolute value:

$$|\mathbf{u}| = \left|\frac{\mathbf{v}\cdot\mathbf{w}}{|\mathbf{w}|}\right|$$

In order to find the vector projection, we notice

$$\mathbf{u} = s\mathbf{w}$$

for some scalar s. Since we do not know $\mathbf{u}$ and do not know s, this equation is not much help. Therefore we use

$$|\mathbf{u}| = s|\mathbf{w}|$$

since we know $|\mathbf{u}|$ (it is the scalar projection we found above). Thus,

$$\frac{\mathbf{v} \cdot \mathbf{w}}{|\mathbf{w}|} = s|\mathbf{w}|$$

$$s = \frac{\mathbf{v} \cdot \mathbf{w}}{|\mathbf{w}|^2}$$

$$= \frac{\mathbf{v} \cdot \mathbf{w}}{\mathbf{w} \cdot \mathbf{w}}$$

We can now find $\mathbf{u}$:

$$\mathbf{u} = s\mathbf{w}$$

$$= \left(\frac{\mathbf{v} \cdot \mathbf{w}}{\mathbf{w} \cdot \mathbf{w}}\right)\mathbf{w}$$

In summary:

Projections of v onto w

Scalar Projection (a number): $\left|\dfrac{\mathbf{v} \cdot \mathbf{w}}{|\mathbf{w}|}\right|$

Vector Projection (a vector): $\left(\dfrac{\mathbf{v} \cdot \mathbf{w}}{\mathbf{w} \cdot \mathbf{w}}\right)\mathbf{w}$

EXAMPLE 2 Find the scalar and vector projections of $\mathbf{v} = 5\mathbf{i} - 3\mathbf{j}$ onto $\mathbf{w} = 7\mathbf{i} + 4\mathbf{j}$.

Solution Vector projection:

$$\left(\frac{\mathbf{v} \cdot \mathbf{w}}{\mathbf{w} \cdot \mathbf{w}}\right)\mathbf{w} = \frac{35 - 12}{49 + 16}\mathbf{w}$$

$$= \frac{23}{65}(7\mathbf{i} + 4\mathbf{j})$$

$$= \frac{161}{65}\mathbf{i} + \frac{92}{65}\mathbf{j}$$

Scalar projection: We could find the length of the vector projection by calculating

$$\sqrt{\left(\frac{161}{65}\right)^2 + \left(\frac{92}{65}\right)^2}$$

or we can use

$$\left|\frac{\mathbf{v} \cdot \mathbf{w}}{|\mathbf{w}|}\right|$$

Since the latter is easier to calculate, we find

$$\left|\frac{23}{\sqrt{49+16}}\right| = \frac{23}{\sqrt{65}}$$

■

EXAMPLE 3 Find the scalar projection of $\mathbf{v} = 3\mathbf{i} - 2\mathbf{j}$ onto $\mathbf{w} = 2\mathbf{i} + 4\mathbf{j}$.

Solution The scalar projection of $\mathbf{v}$ onto $\mathbf{w}$ is

$$\begin{aligned}\left|\frac{\mathbf{v}\cdot\mathbf{w}}{|\mathbf{w}|}\right| &= \left|\frac{6-8}{\sqrt{4+16}}\right| \\ &= \left|\frac{-2}{2\sqrt{5}}\right| \\ &= \frac{1}{\sqrt{5}}\end{aligned}$$

We can check this result by finding the length of the vector projection of $\mathbf{v}$ onto $\mathbf{w}$:

$$\begin{aligned}\mathbf{u} &= \left(\frac{\mathbf{v}\cdot\mathbf{w}}{\mathbf{w}\cdot\mathbf{w}}\right)\mathbf{w} \\ &= \left(\frac{-2}{4+16}\right)\mathbf{w} \\ &= \frac{-1}{10}(2\mathbf{i}+4\mathbf{j}) \\ &= -\frac{1}{5}\mathbf{i} - \frac{2}{5}\mathbf{j} \\ |\mathbf{u}| &= \sqrt{\left(-\frac{1}{5}\right)^2 + \left(-\frac{2}{5}\right)^2} \\ &= \sqrt{\frac{1+4}{25}} \\ &= \frac{1}{5}\sqrt{5}\end{aligned}$$

■

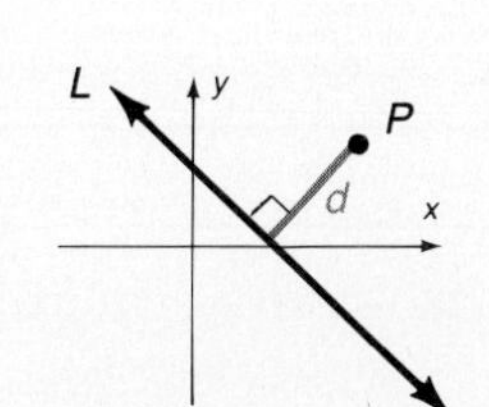

Figure 10.14a Distance from P to L

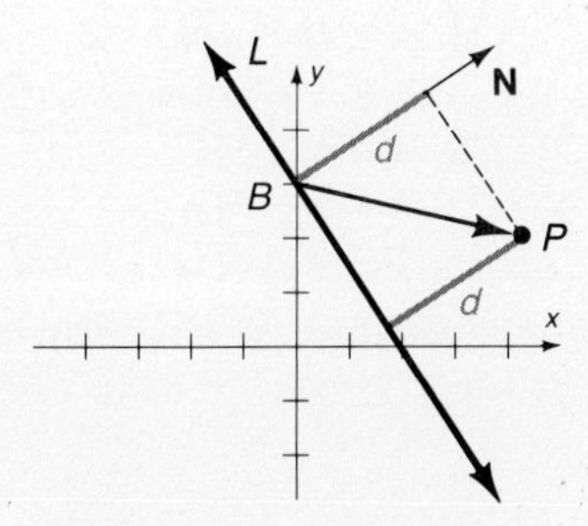

Figure 10.14b Procedure for finding the distance from a point to a line

The final property of vectors we will consider in this section enables us to derive a formula for finding the distance from a point to a line. In Section 9.1 (Problem 65) this formula was derived with a great deal of effort without using vectors. We can now derive the same formula quite easily by using vector ideas.

Let L be any given line and P any given point not on L. By the distance from P to L we mean the perpendicular distance d as shown in Figure 10.14. We wish to find this distance.

If L is a vertical line, then the distance from P to L is easy to find. (Why?) If L is not vertical, then we let B be the y intercept and $\mathbf{N}$ a normal to L. Let the base of $\mathbf{N}$ be drawn at B, as shown in Figure 10.14b.

The distance we seek is seen to be the scalar projection of $\overrightarrow{BP}$ onto **N**. Thus

Vector Formula for the Distance from a Point to a Line

$$d = \left|\frac{\overrightarrow{BP}\cdot \mathbf{N}}{|\mathbf{N}|}\right|$$

EXAMPLE 4 Find the distance from the point $(5, -3)$ to the line $4x + 3y - 15 = 0$.

Solution P is $(5, -3)$ and B is $(0, 5)$. Then

$$\overrightarrow{BP} = 5\mathbf{i} - 8\mathbf{j}$$
$$\mathbf{N} = 4\mathbf{i} + 3\mathbf{j}$$
$$|\mathbf{N}| = \sqrt{4^2 + 3^2} = 5$$
$$\overrightarrow{BP}\cdot\mathbf{N} = 20 - 24 = -4$$

Thus,

$$d = \left|\frac{\overrightarrow{BP}\cdot \mathbf{N}}{|\mathbf{N}|}\right| = \left|\frac{-4}{5}\right| = \frac{4}{5}$$

■

PROBLEM SET 10.2

A

Find a vector normal to each line given in Problems 1–12.

1. $2x - 3y + 4 = 0$
2. $x + y - 1 = 0$
3. $x - y + 3 = 0$
4. $4x + 5y - 3 = 0$
5. $3x - 2y + 1 = 0$
6. $5x + y - 3 = 0$
7. $9x + 7y - 5 = 0$
8. $6x - 3y + 2 = 0$
9. $4x - y - 12 = 0$
10. $y = \frac{2}{3}x - 5$
11. $y = -\frac{1}{2}x - 10$
12. $y = -\frac{5}{8}x + 4$

Find a vector determined by each line given in Problems 13–24.

13. $2x - 3y + 4 = 0$
14. $x + y - 1 = 0$
15. $x - y + 3 = 0$
16. $4x + 5y - 3 = 0$
17. $3x - 2y + 1 = 0$
18. $5x + y - 3 = 0$
19. $9x + 7y - 5 = 0$
20. $6x - 3y + 2 = 0$
21. $4x - y - 12 = 0$
22. $y = \frac{2}{3}x - 5$
23. $y = -\frac{1}{2}x - 10$
24. $y = -\frac{5}{8}x + 4$

In Problems 25–30, find the scalar projection of **v** *onto* **w**.

25. $\mathbf{v} = 3\mathbf{i} + 4\mathbf{j}$, $\mathbf{w} = 5\mathbf{i} + 12\mathbf{j}$
26. $\mathbf{v} = 8\mathbf{i} - 6\mathbf{j}$, $\mathbf{w} = -5\mathbf{i} + 12\mathbf{j}$
27. $\mathbf{v} = 7\mathbf{i} - \sqrt{15}\mathbf{j}$, $\mathbf{w} = 2\sqrt{15}\mathbf{i} + 14\mathbf{j}$
28. $\mathbf{v} = 2\mathbf{i} + \sqrt{5}\mathbf{j}$, $\mathbf{w} = 3\sqrt{5}\mathbf{i} - 3\mathbf{j}$
29. $\mathbf{v} = -2\mathbf{i} + 3\mathbf{j}$, $\mathbf{w} = 6\mathbf{i} + 5\mathbf{j}$
30. $\mathbf{v} = 3\mathbf{i} + 9\mathbf{j}$, $\mathbf{w} = 2\mathbf{i} - 5\mathbf{j}$

In Problems 31–36, find the vector projection of **v** *onto* **w**.

31. $\mathbf{v} = 3\mathbf{i} + 4\mathbf{j}$, $\mathbf{w} = 5\mathbf{i} + 12\mathbf{j}$
32. $\mathbf{v} = 8\mathbf{i} - 6\mathbf{j}$, $\mathbf{w} = -5\mathbf{i} + 12\mathbf{j}$
33. $\mathbf{v} = 7\mathbf{i} - \sqrt{15}\mathbf{j}$, $\mathbf{w} = 2\sqrt{15}\mathbf{i} + 14\mathbf{j}$
34. $\mathbf{v} = 2\mathbf{i} + \sqrt{5}\mathbf{j}$, $\mathbf{w} = 3\sqrt{5}\mathbf{i} - 3\mathbf{j}$
35. $\mathbf{v} = -2\mathbf{i} + 3\mathbf{j}$, $\mathbf{w} = 6\mathbf{i} + 5\mathbf{j}$
36. $\mathbf{v} = 3\mathbf{i} + 9\mathbf{j}$, $\mathbf{w} = 2\mathbf{i} - 5\mathbf{j}$

B

Find the distance from the given point to the given line in Problems 37–50.

37. $3x - 4y + 8 = 0$; $(4, 5)$
38. $5x - 12y + 15 = 0$; $(6, -3)$
39. $3x - 4y + 8 = 0$; $(9, -3)$
40. $5x - 12y + 15 = 0$; $(-2, 6)$
41. $4x + 3y - 5 = 0$; $(-1, -1)$

42. $12x + 5y - 2 = 0; (3, 5)$

43. $4x + 3y - 5 = 0; (6, 1)$

44. $12x + 5y - 2 = 0; (4, -3)$

45. $x - 3y + 15 = 0; (1, -6)$

46. $6x - y - 10 = 0; (-5, 6)$

47. $x - 3y + 15 = 0; (8, 14)$

48. $6x - y - 10 = 0; (8, 10)$

49. $2x - 5y = 0; (4, 5)$

50. $4x + 7y = 0; (5, 10)$

Find the area of the triangle determined by the given points in Problems 51–56.

51. $(1, 2), (4, 5), (-5, 3)$

52. $(-1, 1), (4, 3), (1, -1)$

53. $(0, 0), (5, -3), (-2, -7)$

54. $(5, 6), (-3, 5), (0, 0)$

55. $(3, 0), (0, 8), (-4, 6)$

56. $(0, 6), (-5, 2), (-3, -6)$

C

57. Prove the commutative law $\mathbf{u} \cdot \mathbf{v} = \mathbf{v} \cdot \mathbf{u}$ for any vectors **u** and **v**.

58. Prove the distributive law $\mathbf{u} \cdot (\mathbf{v} + \mathbf{w}) = \mathbf{u} \cdot \mathbf{v} + \mathbf{u} \cdot \mathbf{w}$ for any vectors **u**, **v**, and **w**.

59. Let **u** be the projection of **v** onto **w**. Show that $\mathbf{v} - \mathbf{u}$ is orthogonal to **w**.

60. Let $\mathbf{v} = a\mathbf{i} + b\mathbf{j}$ and $\mathbf{w} = c\mathbf{i} + d\mathbf{j}$ be two vectors. Use the Law of Cosines to show that

$$\cos\theta = \frac{ac + bd}{|\mathbf{v}||\mathbf{w}|}$$

where θ is the angle between the vectors. *Hint:* The Law of Cosines states that $c^2 = a^2 + b^2 - 2ab\cos\gamma$, where a, b, and c are sides of a triangle and γ is the angle opposite side c.

10.3 VECTORS IN THREE DIMENSIONS

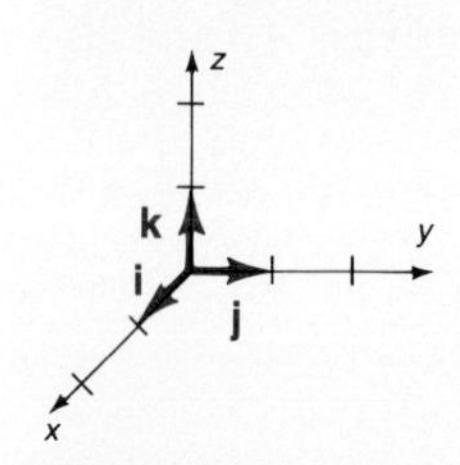

Figure 10.15 A three-dimensional coordinate system with vectors **i**, **j**, and **k**.

A three-dimensional coordinate system can be established by drawing a line, called the **z axis**, so that it passes through the origin of a Cartesian plane and also so that it is perpendicular to that plane (see Figure 10.15). You will study three-dimensional analytic geometry in a more advanced course, and the idea and properties of vectors provide a very useful tool for dealing with three dimensions. In this section we will consider vectors in three dimensions.

Define a unit vector (a vector with length 1) called **k** as the vector in the direction of the positive z axis. The **k** vector is used in conjunction with the **i** and **j** vectors to define a three-dimensional vector as illustrated by Example 1.

EXAMPLE 1 Let O be the point $(0, 0, 0)$ and P be $(5, 4, 3)$. Find the vector $\overrightarrow{OP}$.

Solution

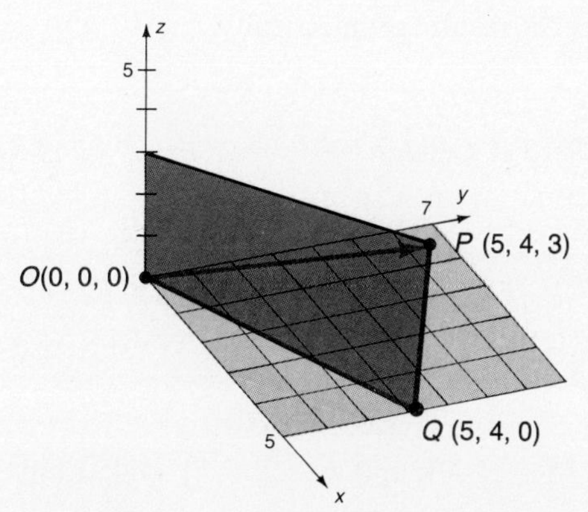

Figure 10.16 Vector $\overrightarrow{OP}$

Let Q be the point $(5, 4, 0)$. Then $\overrightarrow{OQ} = 5\mathbf{i} + 4\mathbf{j}$. Also, we note that $\overrightarrow{OP}$ is the diagonal of a triangle in the plane determined by $\overrightarrow{OP}$ and $\overrightarrow{OQ}$ (see Figure 10.16). Then $\overrightarrow{OP} = (5\mathbf{i} + 4\mathbf{j}) + 3\mathbf{k}$. Thus.

$$\mathbf{v} = 5\mathbf{i} + 4\mathbf{j} + 3\mathbf{k}.$$

Representation of a Vector Determined by Two Points

If A is the point (x_A, y_A, z_A) and B is (x_B, y_B, z_B), then the vector **v** determined by $\overrightarrow{AB}$ is

$$\mathbf{v} = (x_B - x_A)\mathbf{i} + (y_B - y_A)\mathbf{j} + (z_B - z_A)\mathbf{k}$$

Let $a_1 = (x_B - x_A)$, $a_2 = (y_B - y_A)$, and $a_3 = (z_B - z_A)$. Then

$$\mathbf{v} = a_1\mathbf{i} + a_2\mathbf{j} + a_3\mathbf{k}$$

The terminology and processes in three dimensions are identical or similar to those used in two dimensions. For example, you are asked in the problem set to derive the following formula for the magnitude of a vector in three dimensions.

Magnitude of a Vector

The **magnitude** of a vector is the length of the vector. If $\mathbf{v} = a_1\mathbf{i} + a_2\mathbf{j} + a_3\mathbf{k}$, then

$$|\mathbf{v}| = \sqrt{a_1^2 + a_2^2 + a_3^2}$$

EXAMPLE 2 Given $A(4, -2, 3)$ and $B(-1, 3, 5)$, find a vector whose representative is $\overrightarrow{AB}$ and also find its magnitude.

Solution

$$\begin{aligned}\mathbf{v} &= (-1 - 4)\mathbf{i} + [3 - (-2)]\mathbf{j} + (5 - 3)\mathbf{k} \\ &= -5\mathbf{i} + 5\mathbf{j} + 2\mathbf{k} \\ |\mathbf{v}| &= \sqrt{(-5)^2 + (5)^2 + (2)^2} \\ &= \sqrt{25 + 25 + 4} \\ &= \sqrt{54}\end{aligned}$$

■

In two dimensions vectors are defined geometrically and then the algebraic characterization of those operations is considered. In three dimensions, the procedure is exactly reversed. We define the operations algebraically and then interpret the results geometrically.

Addition, Subtraction, and Scalar Multiplication of Vectors

Let $\mathbf{v} = a_1\mathbf{i} + a_2\mathbf{j} + a_3\mathbf{k}$ and $\mathbf{w} = b_1\mathbf{i} + b_2\mathbf{j} + b_3\mathbf{k}$ and let s be any scalar. Then,

ADDITION: $\mathbf{v} + \mathbf{w} = (a_1 + b_1)\mathbf{i} + (a_2 + b_2)\mathbf{j} + (a_3 + b_3)\mathbf{k}$

SUBTRACTION: $\mathbf{v} - \mathbf{w} = (a_1 - b_1)\mathbf{i} + (a_2 - b_2)\mathbf{j} + (a_3 - b_3)\mathbf{k}$

SCALAR MULTIPLICATION: $s\mathbf{v} = sa_1\mathbf{i} + sa_2\mathbf{j} + sa_3\mathbf{k}$

EXAMPLE 3 Let $\mathbf{v} = 3\mathbf{i} + 2\mathbf{j} - \mathbf{k}$ and $\mathbf{w} = 2\mathbf{i} - 5\mathbf{j} + 2\mathbf{k}$. Find $|\mathbf{v}|$, $\mathbf{v} + \mathbf{w}$, $\mathbf{v} - \mathbf{w}$, $3\mathbf{v}$, and $3\mathbf{v} - 2\mathbf{w}$.

Solution

$$\begin{aligned}|\mathbf{v}| &= \sqrt{3^2 + 2^2 + (-1)^2} \\ &= \sqrt{9 + 4 + 1} \\ &= \sqrt{14}\end{aligned}$$

$$\mathbf{v} + \mathbf{w} = (3 + 2)\mathbf{i} + (2 - 5)\mathbf{j} + (-1 + 2)\mathbf{k}$$
$$= 5\mathbf{i} - 3\mathbf{j} + \mathbf{k}$$
$$\mathbf{v} - \mathbf{w} = (3 - 2)\mathbf{i} + (2 + 5)\mathbf{j} + (-1 - 2)\mathbf{k}$$
$$= \mathbf{i} + 7\mathbf{j} - 3\mathbf{k}$$
$$3\mathbf{v} = (3)3\mathbf{i} + (3)2\mathbf{j} + (3)(-1)\mathbf{k}$$
$$= 9\mathbf{i} + 6\mathbf{j} - 3\mathbf{k}$$
$$3\mathbf{v} - 2\mathbf{w} = (9\mathbf{i} + 6\mathbf{j} - 3\mathbf{k}) + (-4\mathbf{i} + 10\mathbf{j} - 4\mathbf{k})$$
$$= 5\mathbf{i} + 16\mathbf{j} - 7\mathbf{k}$$ ■

Geometrically, multiplying by a positive scalar changes the magnitude of a vector but does not change its direction. Multiplication by a negative scalar reverses its direction as well as changing its magnitude.

EXAMPLE 4 Find a vector in the direction of $\mathbf{v} = 3\mathbf{i} - 4\mathbf{j} + 12\mathbf{k}$ with unit length.

Solution

$$|\mathbf{v}| = \sqrt{9 + 16 + 144} = 13$$

Thus, $\frac{1}{13}\mathbf{v} = \frac{3}{13}\mathbf{i} - \frac{4}{13}\mathbf{j} + \frac{12}{13}\mathbf{k}$ is a vector in the same direction (since we multiplied by a scalar) but with unit length. ■

Unit Vector in a Given Direction

If $\mathbf{v}$ is a nonzero vector, a unit vector in the direction of $\mathbf{v}$ is given by

$$\frac{\mathbf{v}}{|\mathbf{v}|}$$

The definition of **scalar product** (also called *dot product* or *inner product*) is the same regardless of the dimension of the vectors $\mathbf{v}$ and $\mathbf{w}$. Remember:

Scalar Product

The scalar product, denoted by $\mathbf{v} \cdot \mathbf{w}$, is defined by

$$\mathbf{v} \cdot \mathbf{w} = |\mathbf{v}||\mathbf{w}| \cos \theta$$

where θ is the angle between $\mathbf{v}$ and $\mathbf{w}$.

The main properties of scalar product are:

Parallel and Orthogonal Vectors

Two vectors are parallel if and only if

$$\mathbf{v} \cdot \mathbf{w} = \pm|\mathbf{v}||\mathbf{w}|$$

Two vectors are orthogonal if and only if

$$\mathbf{v} \cdot \mathbf{w} = 0$$

To calculate the scalar product, we first need to find $\cos \theta$ in three dimensions. Let

$$\mathbf{v} = a_1\mathbf{i} + a_2\mathbf{j} + a_3\mathbf{k} \quad \text{and} \quad \mathbf{w} = b_1\mathbf{i} + b_2\mathbf{j} + b_3\mathbf{k}$$

Using the Law of Cosines, it can be shown that

$$\cos\theta = \frac{a_1b_1 + a_2b_2 + a_3b_3}{|\mathbf{v}||\mathbf{w}|}$$

Thus we see that

Procedure for Finding Scalar Product

$$\mathbf{v}\cdot\mathbf{w} = a_1b_1 + a_2b_2 + a_3b_3$$

For Examples 5–10, let $\mathbf{v} = 3\mathbf{i} + 2\mathbf{j} - \mathbf{k}$ and $\mathbf{w} = 4\mathbf{i} - 3\mathbf{j} + 2\mathbf{k}$.

EXAMPLE 5 Find $\mathbf{v}\cdot\mathbf{w}$.

Solution

$$\begin{aligned}\mathbf{v}\cdot\mathbf{w} &= 3(4) + 2(-3) + (-1)(2)\\ &= 12 - 6 - 2\\ &= \mathbf{4}\end{aligned}$$

■

EXAMPLE 6 Find $|\mathbf{v}|$ and $|\mathbf{w}|$.

Solution

$$\begin{aligned}|\mathbf{v}| &= \sqrt{9+4+1} & \qquad |\mathbf{w}| &= \sqrt{16+9+4}\\ &= \boldsymbol{\sqrt{14}} & &= \boldsymbol{\sqrt{29}}\end{aligned}$$

■

EXAMPLE 7 Find $\cos\theta$, where θ is the angle between $\mathbf{v}$ and $\mathbf{w}$.

Solution

$$\begin{aligned}\cos\theta &= \frac{\mathbf{v}\cdot\mathbf{w}}{|\mathbf{v}||\mathbf{w}|}\\ &= \frac{4}{\sqrt{14}\sqrt{29}} \qquad \text{From Examples 5 and 6}\\ &= \frac{4}{\sqrt{406}}\\ &= \boldsymbol{\frac{2}{203}\sqrt{406}}\end{aligned}$$

■

EXAMPLE 8 Find $\mathbf{v} - s\mathbf{w}$.

Solution

$$\begin{aligned}\mathbf{v} - s\mathbf{w} &= (3\mathbf{i} + 2\mathbf{j} - \mathbf{k}) - (4s\mathbf{i} - 3s\mathbf{j} + 2s\mathbf{k})\\ &= \mathbf{(3 - 4s)i + (2 + 3s)j + (-1 - 2s)k}\end{aligned}$$

■

EXAMPLE 9 Find s so that $\mathbf{v}$ and $\mathbf{v} - s\mathbf{w}$ are orthogonal.

Solution If $\mathbf{v}$ and $\mathbf{v} - s\mathbf{w}$ are orthogonal, then

$$\begin{aligned}\mathbf{v}\cdot(\mathbf{v} - s\mathbf{w}) &= 0\\ 3(3 - 4s) + 2(2 + 3s) + (-1)(-1 - 2s) &= 0\\ 9 - 12s + 4 + 6s + 1 + 2s &= 0\\ -4s &= -14\\ s &= \boldsymbol{\frac{7}{2}}\end{aligned}$$

■

EXAMPLE 10 Find $\mathbf{v} \cdot \mathbf{v}$ and $|\mathbf{v}|^2$

Solution

$$\begin{aligned}\mathbf{v} \cdot \mathbf{v} &= 3(3) + 2(2) + (-1)(-1)\\ &= 9 + 4 + 1\\ &= \mathbf{14}\\ |\mathbf{v}|^2 &= (\sqrt{14})^2\\ &= \mathbf{14}\end{aligned}$$

■

Notice in Example 10 that $\mathbf{v} \cdot \mathbf{v} = |\mathbf{v}|^2$. In general, for any vector $\mathbf{v}$,

$$\mathbf{v} \cdot \mathbf{v} = |\mathbf{v}|^2$$

The formulas for both vector and scalar projections are the same in three dimensions as we found them to be in two dimensions.

Vector and Scalar Projections

VECTOR PROJECTION OF V ONTO W:

$$\left(\frac{\mathbf{v} \cdot \mathbf{w}}{\mathbf{w} \cdot \mathbf{w}}\right)\mathbf{w}$$

SCALAR PROJECTION OF V ONTO W:

$$|\mathbf{v}|\cos\theta = \left|\frac{\mathbf{v} \cdot \mathbf{w}}{|\mathbf{w}|}\right|$$

EXAMPLE 11 Let $\mathbf{v} = 2\mathbf{i} - 6\mathbf{j} + 3\mathbf{k}$ and $\mathbf{w} = \mathbf{i} + 2\mathbf{j} - 2\mathbf{k}$. Find the vector and scalar projections of $\mathbf{v}$ onto $\mathbf{w}$.

Solution The vector projection of $\mathbf{v}$ onto $\mathbf{w}$ is

$$\begin{aligned}\left(\frac{\mathbf{v} \cdot \mathbf{w}}{\mathbf{w} \cdot \mathbf{w}}\right)\mathbf{w} &= \left(\frac{2 - 12 - 6}{1 + 4 + 4}\right)(\mathbf{i} + 2\mathbf{j} - 2\mathbf{k})\\ &= \frac{-16}{9}(\mathbf{i} + 2\mathbf{j} - 2\mathbf{k})\\ &= \frac{-16}{9}\mathbf{i} - \frac{32}{9}\mathbf{j} + \frac{32}{9}\mathbf{k}\end{aligned}$$

For the scalar projection we could find the length of this vector directly (heaven forbid), or use the formula:

$$\begin{aligned}|\mathbf{v}|\cos\theta &= \left|\frac{\mathbf{v} \cdot \mathbf{w}}{|\mathbf{w}|}\right|\\ &= \left|\frac{-16}{\sqrt{1 + 4 + 4}}\right| = \left|\frac{-16}{\sqrt{9}}\right| = \frac{\mathbf{16}}{\mathbf{3}}\end{aligned}$$

■

We have seen that scalar product is an important vector operation which we have defined for vectors in both two and three dimensions. The result of multiplying two vectors using scalar multiplication is a real number, or scalar. There is another operation, called **vector product**, which is defined only for three-dimensional vectors and gives a vector as the result.

Vector Product

If $\mathbf{v} = a_1\mathbf{i} + a_2\mathbf{j} + a_3\mathbf{k}$ and $\mathbf{w} = b_1\mathbf{i} + b_2\mathbf{j} + b_3\mathbf{k}$, the vector product, written $\mathbf{v} \times \mathbf{w}$, is the vector

$$(a_2b_3 - a_3b_2)\mathbf{i} + (a_3b_1 - a_1b_3)\mathbf{j} + (a_1b_2 - a_2b_1)\mathbf{k}$$

These terms can be obtained by using a determinant:

$$\mathbf{v} \times \mathbf{w} = \begin{vmatrix} \mathbf{i} & \mathbf{j} & \mathbf{k} \\ a_1 & a_2 & a_3 \\ b_1 & b_2 & b_3 \end{vmatrix}$$

Sometimes this product is called the *cross product* or *outer product*.

Vector product is *not* commutative, since $\mathbf{v} \times \mathbf{w} \neq \mathbf{w} \times \mathbf{v}$. Using properties of determinants we see that $\mathbf{v} \times \mathbf{w} = -(\mathbf{w} \times \mathbf{v})$: Now,

$$\mathbf{v} \times \mathbf{w} = \begin{vmatrix} \mathbf{i} & \mathbf{j} & \mathbf{k} \\ a_1 & a_2 & a_3 \\ b_1 & b_2 & b_3 \end{vmatrix} = -\begin{vmatrix} \mathbf{i} & \mathbf{j} & \mathbf{k} \\ b_1 & b_2 & b_3 \\ a_1 & a_2 & a_3 \end{vmatrix} = -(\mathbf{w} \times \mathbf{v})$$

The vector $(\mathbf{v} \times \mathbf{w})$ is orthogonal with both the vectors $\mathbf{v}$ and $\mathbf{w}$. To see this we consider the dot product of the vectors $\mathbf{v}$ and $(\mathbf{v} \times \mathbf{w})$. If this product is zero, then the vectors are orthogonal.

$$\begin{aligned} \mathbf{v} \cdot (\mathbf{v} \times \mathbf{w}) &= (a_1\mathbf{i} + a_2\mathbf{j} + a_3\mathbf{k}) \cdot [(a_2b_3 - a_3b_2)\mathbf{i} + (a_3b_1 - a_1b_3)\mathbf{j} \\ &\quad + (a_1b_2 - a_2b_1)\mathbf{k}] \\ &= a_1a_2b_3 - a_1a_3b_2 + a_2a_3b_1 - a_1a_2b_3 + a_1a_3b_2 - a_2a_3b_1 \\ &= 0 \end{aligned}$$

Similarly, we can show that $\mathbf{w} \cdot (\mathbf{v} \times \mathbf{w}) = 0$

We have just proved a geometric property of vectors by using algebra. The vector product of two vectors is a vector orthogonal to the two given vectors. The only way this can occur is in a three-dimensional setting. Any two distinct nonzero three-dimensional vectors that are not parallel (that is, not scalar multiples of one another) can be arranged so that they have a common base and will therefore determine a plane. Then the vector product *must* be orthogonal to this plane as shown in Figure 10.17.

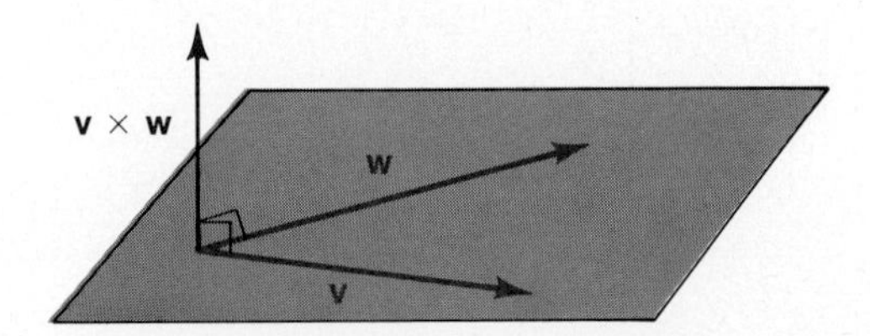

Figure 10.17 Vector product of two vectors

Geometric Property of Vector Product

If $\mathbf{v}$ and $\mathbf{w}$ are not parallel, then they determine a plane and $\mathbf{v} \times \mathbf{w}$ is a *vector orthogonal to this plane. That is, it is orthogonal to both* $\mathbf{v}$ *and* $\mathbf{w}$.

The magnitude, or length, of this vector is denoted by $|\mathbf{v} \times \mathbf{w}|$. Using the formula for magnitude, we can show that

$$|\mathbf{v} \times \mathbf{w}| = |\mathbf{v}|\,|\mathbf{w}| \sin\theta$$

The distributive property holds for vector product:

$$\mathbf{u} \times (\mathbf{v} + \mathbf{w}) = (\mathbf{u} \times \mathbf{v}) + (\mathbf{u} \times \mathbf{w})$$
$$(\mathbf{v} + \mathbf{w}) \times \mathbf{u} = (\mathbf{v} \times \mathbf{u}) + (\mathbf{w} \times \mathbf{u})$$

EXAMPLE 12 Find $\mathbf{i} \times \mathbf{j}$.

Solution

$$\mathbf{i} \times \mathbf{j} = \begin{vmatrix} \mathbf{i} & \mathbf{j} & \mathbf{k} \\ 1 & 0 & 0 \\ 0 & 1 & 0 \end{vmatrix} = \mathbf{k}$$ ■

Similarly:

$$\mathbf{j} \times \mathbf{k} = \mathbf{i} \qquad \mathbf{k} \times \mathbf{i} = \mathbf{j} \qquad \mathbf{i} \times \mathbf{j} = \mathbf{k}$$
$$\mathbf{i} \times \mathbf{i} = \mathbf{j} \times \mathbf{j} = \mathbf{k} \times \mathbf{k} = 0$$

EXAMPLE 13 Find $\mathbf{v} \times \mathbf{w}$, where $\mathbf{v} = 2\mathbf{i} + \mathbf{j} + \mathbf{k}$ and $\mathbf{w} = -\mathbf{i} + 2\mathbf{j} + 3\mathbf{k}$.

Solution

$$\mathbf{v} \times \mathbf{w} = \begin{vmatrix} \mathbf{i} & \mathbf{j} & \mathbf{k} \\ 2 & 1 & 1 \\ -1 & 2 & 3 \end{vmatrix}$$

$$= \begin{vmatrix} 1 & 1 \\ 2 & 3 \end{vmatrix}\mathbf{i} - \begin{vmatrix} 2 & 1 \\ -1 & 3 \end{vmatrix}\mathbf{j} + \begin{vmatrix} 2 & 1 \\ -1 & 2 \end{vmatrix}\mathbf{k}$$

$$= \mathbf{i} - 7\mathbf{j} + 5\mathbf{k}$$ ■

PROBLEM SET 10.3

A

In Problems 1–6, find the vector $\overrightarrow{AB}$ determined by the given points. Also find its magnitude.

1. $A(2, -1, 3), B(4, 5, -3)$ **2.** $A(2, -1, 3), B(4, 5, 3)$

3. $A(4, -3, 1), B(-1, 3, -2)$ **4.** $A(4, 1, 5), B(-3, 0, 2)$

5. $A(7, 1, 0), B(3, -2, 5)$ **6.** $A(2, -1, 2), B(4, -1, -1)$

Let $\mathbf{v} = 3\mathbf{i} - 2\mathbf{j} + \mathbf{k}$ and $\mathbf{w} = 4\mathbf{i} + \mathbf{j} - 3\mathbf{k}$. Find the scalars or vectors requested in Problems 7–9.

7. $|\mathbf{v}|$ **8.** $\mathbf{v} + \mathbf{w}$ **9.** $2\mathbf{v} + 3\mathbf{w}$

Let $\mathbf{v} = 5\mathbf{i} - 3\mathbf{j} + 2\mathbf{k}$ and $\mathbf{w} = -\mathbf{i} + 2\mathbf{j} - 3\mathbf{k}$. Find the scalars or vectors requested in Problems 10–12.

10. $|\mathbf{w}|$ **11.** $\mathbf{v} - \mathbf{w}$ **12.** $3\mathbf{v} - \mathbf{w}$

Let $\mathbf{v} = 5\mathbf{i} + 4\mathbf{k}$ and $\mathbf{w} = \mathbf{j} + 3\mathbf{k}$. Find the vectors requested in Problems 13–15.

13. $\mathbf{v} + \mathbf{w}$ **14.** $\mathbf{v} - \mathbf{w}$ **15.** $5\mathbf{v} - 3\mathbf{w}$

In Problems 16–24, find $\mathbf{v} \cdot \mathbf{w}$ for the given vectors.

16. $\mathbf{v} = \mathbf{i}$, $\mathbf{w} = \mathbf{j}$ **17.** $\mathbf{v} = \mathbf{k}$, $\mathbf{w} = \mathbf{k}$ **18.** $\mathbf{v} = 3\mathbf{i} + 2\mathbf{k}$, $\mathbf{w} = 2\mathbf{i} + \mathbf{j}$

19. $\mathbf{v} = \mathbf{i} - 3\mathbf{j}$, $\mathbf{w} = \mathbf{i} + 5\mathbf{k}$ **20.** $\mathbf{v} = 3\mathbf{i} - 2\mathbf{j} + 4\mathbf{k}$, $\mathbf{w} = \mathbf{i} + 4\mathbf{j} - 7\mathbf{k}$

21. $\mathbf{v} = 5\mathbf{i} - \mathbf{j} + 2\mathbf{k}$, $\mathbf{w} = 2\mathbf{i} + \mathbf{j} - 3\mathbf{k}$ **22.** $\mathbf{v} = 3\mathbf{i} - \mathbf{j} + 2\mathbf{k}$, $\mathbf{w} = 2\mathbf{i} + 3\mathbf{j} - 4\mathbf{k}$

23. $\mathbf{v} = -\mathbf{j} + 4\mathbf{k}$, $\mathbf{w} = 5\mathbf{i} + 6\mathbf{k}$ **24.** $\mathbf{v} = \mathbf{i} - 6\mathbf{j} + 10\mathbf{k}$, $\mathbf{w} = -\mathbf{i} + 5\mathbf{j} - 6\mathbf{k}$

In Problems 25–33, find $\mathbf{v} \times \mathbf{w}$ for the given vectors.

25. $\mathbf{v} = \mathbf{i}$, $\mathbf{w} = \mathbf{j}$ **26.** $\mathbf{v} = \mathbf{k}$, $\mathbf{w} = \mathbf{k}$ **27.** $\mathbf{v} = 3\mathbf{i} + 2\mathbf{k}$, $\mathbf{w} = 2\mathbf{i} + \mathbf{j}$

28. $\mathbf{v} = \mathbf{i} - 3\mathbf{j}$, $\mathbf{w} = \mathbf{i} + 5\mathbf{k}$ **29.** $\mathbf{v} = 3\mathbf{i} - 2\mathbf{j} + 4\mathbf{k}$, $\mathbf{w} = \mathbf{i} + 4\mathbf{j} - 7\mathbf{k}$

30. $\mathbf{v} = 5\mathbf{i} - \mathbf{j} + 2\mathbf{k}$, $\mathbf{w} = 2\mathbf{i} + \mathbf{j} - 3\mathbf{k}$ **31.** $\mathbf{v} = 3\mathbf{i} - \mathbf{j} + 2\mathbf{k}$, $\mathbf{w} = 2\mathbf{i} + 3\mathbf{j} - 4\mathbf{k}$

32. $\mathbf{v} = -\mathbf{j} + 4\mathbf{k}$
$\mathbf{w} = 5\mathbf{i} + 6\mathbf{k}$

33. $\mathbf{v} = \mathbf{i} - 6\mathbf{j} + 10\mathbf{k}$
$\mathbf{w} = -\mathbf{i} + 5\mathbf{j} - 6\mathbf{k}$

B

In Problems 34–39, find a unit vector in the direction of the given vector.

34. $3\mathbf{i} + 4\mathbf{j}$

35. $5\mathbf{i} + 12\mathbf{k}$

36. $\mathbf{i} + \mathbf{j} + \mathbf{k}$

37. $3\mathbf{i} + 12\mathbf{j} - 4\mathbf{k}$

38. $2\mathbf{i} - 2\mathbf{j} + \mathbf{k}$

39. $4\mathbf{i} + 2\mathbf{j} - 3\mathbf{k}$

In Problems 40–43, let $\mathbf{v} = 4\mathbf{i} - \mathbf{j} + \mathbf{k}$ *and* $\mathbf{w} = 2\mathbf{i} + 3\mathbf{j} - \mathbf{k}$. *Find:*

40. $\mathbf{v} \cdot \mathbf{w}$

41. $\cos\theta$, where θ is the angle between $\mathbf{v}$ and $\mathbf{w}$

42. a scalar s such that $\mathbf{v}$ is orthogonal to $\mathbf{v} - s\mathbf{w}$

43. scalars s and t such that $s\mathbf{v} + t\mathbf{w}$ is orthogonal to $\mathbf{w}$

In Problems 44–47, let $\mathbf{v} = 2\mathbf{i} + 3\mathbf{k}$ *and* $\mathbf{w} = 2\mathbf{j} - 3\mathbf{k}$. *Find:*

44. $\mathbf{v} \cdot \mathbf{w}$

45. $\cos\theta$, where θ is the angle between $\mathbf{v}$ and $\mathbf{w}$

46. a scalar s such that $\mathbf{v}$ is orthogonal to $\mathbf{v} - s\mathbf{w}$

47. scalars s and t such that $s\mathbf{v} + t\mathbf{w}$ is orthogonal to $\mathbf{w}$

In Problems 48–51, let $\mathbf{v} = \mathbf{i} - 3\mathbf{j} + 2\mathbf{k}$ *and* $\mathbf{w} = \mathbf{i} + \mathbf{j} + 5\mathbf{k}$. *Find:*

48. $\mathbf{v} \times \mathbf{w}$

49. $\cos\theta$, where θ is the angle between $\mathbf{v}$ and $\mathbf{w}$

50. a scalar s such that $\mathbf{v}$ is orthogonal to $\mathbf{v} - s\mathbf{w}$

51. scalars s and t such that $s\mathbf{v} + t\mathbf{w}$ is orthogonal to $\mathbf{w}$

In Problems 52–55, let $\mathbf{v} = 2\mathbf{i} - 3\mathbf{j} + 6\mathbf{k}$ *and* $\mathbf{w} = 4\mathbf{i} + 3\mathbf{k}$. *Find:*

52. $\mathbf{v} \times \mathbf{w}$

53. $\cos\theta$, where θ is the angle between $\mathbf{v}$ and $\mathbf{w}$

54. a scalar s such that $\mathbf{v}$ is orthogonal to $\mathbf{v} - s\mathbf{w}$

55. scalars s and t such that $s\mathbf{v} + t\mathbf{w}$ is orthogonal to $\mathbf{w}$

C

In Problems 56–61, find the vector and scalar projections of $\mathbf{v}$ *onto* $\mathbf{w}$.

56. $\mathbf{v} = 3\mathbf{i} + 4\mathbf{j} + 12\mathbf{k}$
$\mathbf{w} = 2\mathbf{i} + \mathbf{j} + \mathbf{k}$

57. $\mathbf{v} = 5\mathbf{i} + 2\mathbf{j} + \mathbf{k}$
$\mathbf{w} = -\mathbf{i} + \mathbf{j} + \mathbf{k}$

58. $\mathbf{v} = \mathbf{i} - 2\mathbf{j} + \mathbf{k}$
$\mathbf{w} = \mathbf{i} + 2\mathbf{j} + 5\mathbf{k}$

59. $\mathbf{v} = 2\mathbf{i} + \mathbf{j} - 3\mathbf{k}$
$\mathbf{w} = 5\mathbf{i} + \mathbf{j} - 3\mathbf{k}$

60. $\mathbf{v} = -\mathbf{i} - 2\mathbf{k}$
$\mathbf{w} = 3\mathbf{i} - 4\mathbf{j} + 2\mathbf{k}$

61. $\mathbf{v} = 2\mathbf{i} + 3\mathbf{j} - \mathbf{k}$
$\mathbf{w} = 3\mathbf{i} - 4\mathbf{j} + 2\mathbf{k}$

62. Explain whether each of the following products is a scalar or a vector, or does not exist.
a. $\mathbf{u} \times (\mathbf{v} \cdot \mathbf{w})$ **b.** $\mathbf{u} \cdot (\mathbf{v} \cdot \mathbf{w})$ **c.** $\mathbf{u} \times (\mathbf{v} \times \mathbf{w})$

63. Explain whether each of the following products is a scalar or a vector, or does not exist.
a. $(\mathbf{u} \times \mathbf{v}) \cdot (\mathbf{u} \times \mathbf{w})$ **b.** $(\mathbf{u} \times \mathbf{v}) \times (\mathbf{u} \times \mathbf{w})$
c. $\mathbf{u} \cdot (\mathbf{v} \times \mathbf{w})$

10.4 POLAR COORDINATES*

Up to this point in the book, we have used a rectangular coordinate system. Now we will consider a different system: the **polar coordinate system**. In this system, fix a point O, called the *origin* or **pole**, and represent a point in the plane by an ordered pair $P(r, \theta)$, where θ measures the angle from the positive x axis and r represents the directed distance from the pole to the point P. Both r and θ can be any real number. When plotting points, first measure an angle θ and then measure out a length r along the ray from the pole through P (along **OP**). If θ is positive, the angle is measured in a counterclockwise direction; if θ is negative, it is measured in a clockwise direction. If r is positive, the distance is measured along the ray **OP**; if r is negative, it is measured along a ray extended back through the pole in a direction exactly opposite **OP**.

EXAMPLE 1 Figure 10.18 shows several points. Make sure you understand how each one can be plotted if you are given the ordered pair: $A(4, \frac{\pi}{3})$, $B(3, 3)$, $C(-4, \frac{\pi}{3})$, $D(-3, 3)$, $E(8, -\frac{\pi}{6})$, $F(-8, -\frac{\pi}{6})$, $G(6, \frac{5\pi}{6})$, $H(6, \frac{3\pi}{2}) = I(-6, \frac{\pi}{2}) = J(6, -\frac{\pi}{2})$.

* This section does not require Sections 10.1–10.3.

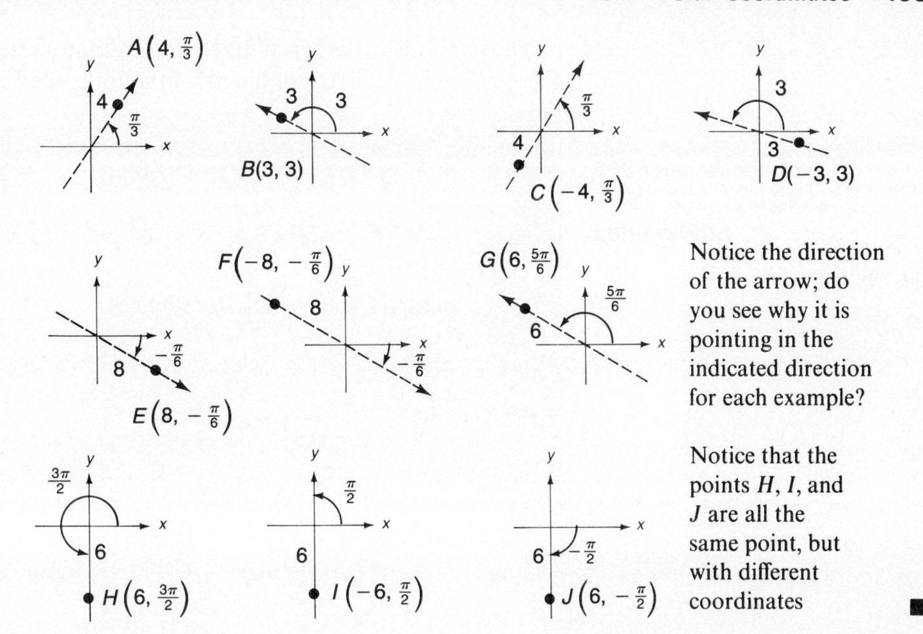

Figure 10.18 Plotting polar-form points

Notice the direction of the arrow; do you see why it is pointing in the indicated direction for each example?

Notice that the points H, I, and J are all the same point, but with different coordinates ■

One thing you will notice from Figure 10.18 is that ordered pairs in polar form are not associated in a one-to-one fashion with points in the plane. Indeed, given any point in the plane, there are infinitely many ordered pairs of polar coordinates associated with that point in polar form. If you are given a point (r, θ) other than the pole in polar coordinates, then $(-r, \theta + \pi)$ also represents the same point. In addition, there are also infinitely many others, all of which have the same first component as one of these, and have second components that are multiples of 2π added to these angles. We call (r, θ) and $(-r, \theta + \pi)$ the **primary representations of the point** if the angles θ and $\theta + \pi$ are between zero and 2π.

Primary Representations of a Point in Polar Form

Every point in polar form has two primary representations:

(r, θ), where $0 \leq \theta < 2\pi$ and $(-r, \pi + \theta)$, where $0 \leq \pi + \theta < 2\pi$

EXAMPLE 2 Give both primary representations for each of the given points:

a. $(3, \frac{\pi}{4})$ has primary representations $(3, \frac{\pi}{4})$ and $(-3, \frac{5\pi}{4})$.

$\frac{\pi}{4} + \pi = \frac{5\pi}{4}$

b. $(5, \frac{5\pi}{4})$ has primary representations $(5, \frac{5\pi}{4})$ and $(-5, \frac{\pi}{4})$.

$\frac{5\pi}{4} + \pi = \frac{9\pi}{4}$, but $(-5, \frac{9\pi}{4})$ is not a primary representation of the point $(5, \frac{5\pi}{4})$ since $\frac{9\pi}{4} > 2\pi$. Use $\frac{\pi}{4}$ since it is coterminal with $\frac{9\pi}{4}$ and satisfies $0 \leq \frac{\pi}{4} < 2\pi$.

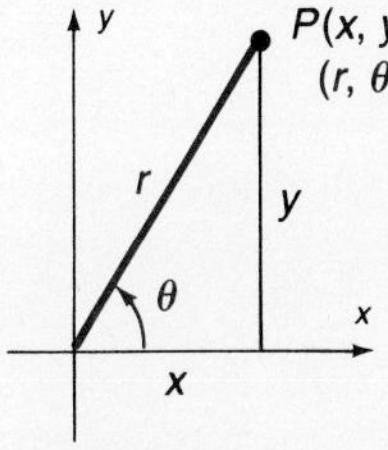

c. $(-6, -\frac{2\pi}{3})$ has primary representations $(-6, \frac{4\pi}{3})$ and $(6, \frac{\pi}{3})$.

d. $(9, 5)$ has primary representations $(9, 5)$ and $(-9, 5 - \pi)$; a point like $(-9, 5 - \pi)$ is usually approximated by writing $(-9, 1.86)$.

Notice that $(-9, 5 + \pi)$ is not a primary representation, since $5 + \pi > 2\pi$.

e. $(9, 7)$ has primary representations $(9, 7 - 2\pi)$ or $(9, 0.72)$ and $(-9, 7 - \pi)$ or $(-9, 3.86)$. ■

Figure 10.19 Relationship between rectangular and polar coordinates

The relationship between the two coordinate systems can easily be found by using the definition of the trigonometric functions (see Figure 10.19).

Relationship between Rectangular and Polar Coordinates

1. To change **from polar to rectangular:**

$$x = r\cos\theta$$
$$y = r\sin\theta$$

2. To change **from rectangular to polar:**

$$r = \sqrt{x^2 + y^2} \qquad \theta' = \tan^{-1}\left|\frac{y}{x}\right|, x \neq 0$$

where θ' is the reference angle for θ. Place θ in the proper quadrant by noting the signs of x and y. If $x = 0$, then $\theta' = \frac{\pi}{2}$.

EXAMPLE 3 Change the polar coordinates $(-3, \frac{5\pi}{4})$ to rectangular coordinates.

Solution

$$x = -3\cos\frac{5\pi}{4} = -3\left(-\frac{\sqrt{2}}{2}\right) = \frac{3\sqrt{2}}{2}$$

$$y = -3\sin\frac{5\pi}{4} = -3\left(-\frac{\sqrt{2}}{2}\right) = \frac{3\sqrt{2}}{2}$$

$$\underbrace{\left(-3, \frac{5\pi}{4}\right)}_{\text{polar form}} = \underbrace{\left(\frac{3\sqrt{2}}{2}, \frac{3\sqrt{2}}{2}\right)}_{\text{rectangular form}}$$

■

EXAMPLE 4 Write both primary representations of the polar-form coordinates for the point $\left(\frac{5\sqrt{3}}{2}, -\frac{5}{2}\right)$

Solution

$$r = \sqrt{\left(\frac{5\sqrt{3}}{2}\right)^2 + \left(-\frac{5}{2}\right)^2} = \sqrt{\frac{75}{4} + \frac{25}{4}} = 5$$

$$\theta' = \tan^{-1}\left|\frac{-\frac{5}{2}}{\frac{5\sqrt{3}}{2}}\right| = \tan^{-1}\left(\frac{1}{\sqrt{3}}\right) = \frac{\pi}{6}$$

$$\theta = \frac{11\pi}{6} \text{ (Quadrant IV)}$$

$$\underbrace{\left(\frac{5\sqrt{3}}{2}, -\frac{5}{2}\right)}_{\text{rectangular form}} = \underbrace{\left(5, \frac{11\pi}{6}\right) = \left(-5, \frac{5\pi}{6}\right)}_{\text{polar form}}$$

$\frac{11\pi}{6} + \pi = \frac{17\pi}{6}$, and $\frac{17\pi}{6}$ is coterminal with $\frac{5\pi}{6}$

■

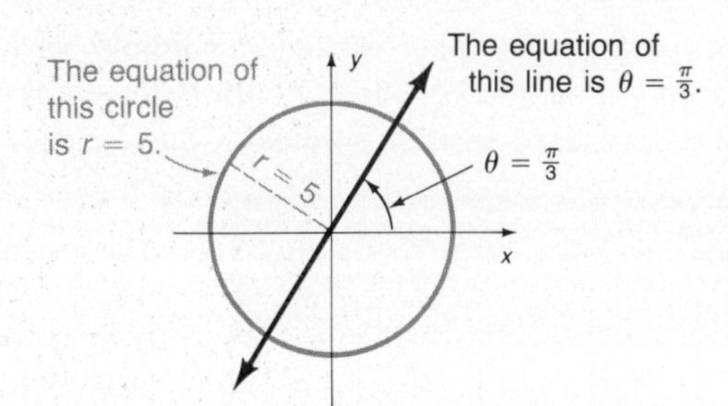

Figure 10.20 Graphs of $r = 5$ and $\theta = \frac{\pi}{3}$

If r and θ are related by an equation, we can speak of the *graph of the equation.* For example, $r = 5$ is the equation of a circle with center at the origin and radius 5. Also, $\theta = \frac{\pi}{3}$ is the equation of the line passing through the pole (as drawn in Figure 10.20).

We now turn our attention to polar-form graphing. The basic procedure is to plot points to determine a curve's general shape and then to make some generalizations that will simplify the graphing of similar curves. Because the representation of points in polar form by ordered pairs of real numbers is not unique, we need the following definition of what it means for a polar-form point to **satisfy** an equation.

A Point Satisfying an Equation

An ordered pair representing a polar-form point (other than the pole) **satisfies an equation** involving a trigonometric function of $n\theta$ (n an integer) if and only if at least one of its primary representations satisfies the given equation.

EXAMPLE 5 Graph $r = 2(1 - \cos\theta)$.

Solution First construct a table of values by choosing values for θ and approximating the corresponding values for r:

θ	0	$\frac{\pi}{6}$	$\frac{\pi}{3}$	$\frac{\pi}{2}$	$\frac{2\pi}{3}$	$\frac{5\pi}{6}$	π	$\frac{7\pi}{6}$	$\frac{4\pi}{3}$	$\frac{3\pi}{2}$	$\frac{5\pi}{3}$	$\frac{11\pi}{6}$
r (approx. value)	0	0.27	1	2	3	3.7	4	3.7	3	2	1	0.27

These points are plotted and then connected as in Figure 10.21.

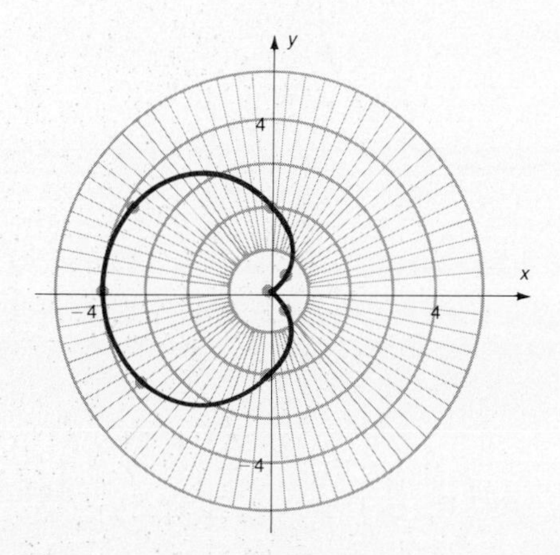

Figure 10.21 Graph of $r = 2(1 - \cos\theta)$

■

The curve in Example 5 is called a **cardioid** because it is heart-shaped. Compare the curve graphed in Example 5 with the general curve $r = a(1 - \cos\theta)$, which is called the **standard-position cardioid**. Consider the following table of values:

θ	0	$\frac{\pi}{2}$	π	$\frac{3\pi}{2}$
r	0	a	$2a$	a

These values for θ should be included whenever you are making a graph in polar coordinates and using the method of plotting points.

These reference points are all that is necessary to plot for future standard-position cardioids, because they will all have the same shape as the one shown in Figure 10.21.

What about cardioids that are not in standard position? In Chapter 2 translations were considered, but the translation of a polar-form curve is rather difficult since points are not labeled in a rectangular fashion. You can, however, easily rotate a polar-form curve.

Rotation of Polar-Form Graphs

The polar graph $r = f(\theta - \alpha)$ is the same as the polar graph of $r = f(\theta)$ that has been rotated through an angle α.

EXAMPLE 6 Graph $r = 3 - 3\cos(\theta - \frac{\pi}{6})$.

Solution Recognize this as a cardioid with $a = 3$ and a rotation of $\frac{\pi}{6}$. Plot the four points shown in Figure 10.22 and draw the cardioid. ■

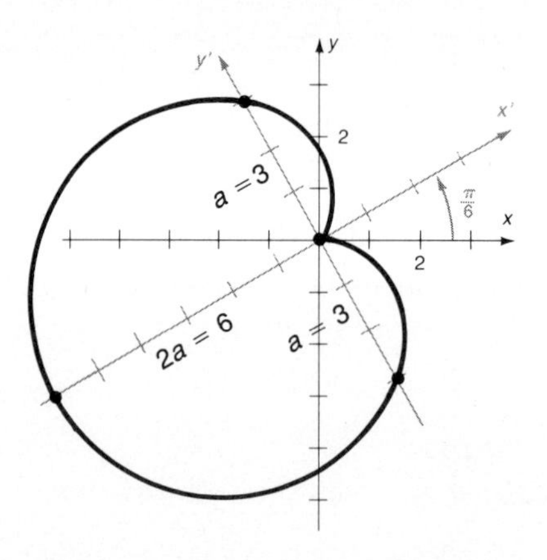

Figure 10.22 Graph of $r = 3 - 3\cos(\theta - \frac{\pi}{6})$

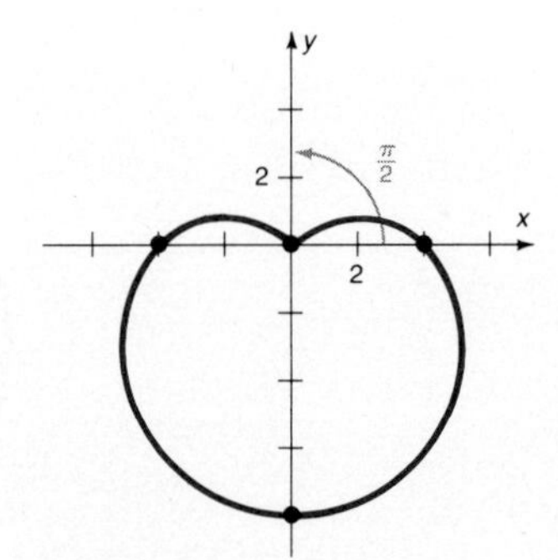

Figure 10.23 Graph of $r = 4(1 - \sin\theta)$

EXAMPLE 7 Graph $r = 4(1 - \sin\theta)$.

Solution Notice that

$$\sin\theta = \cos\left(\frac{\pi}{2} - \theta\right)$$

$$= \cos\left(\theta - \frac{\pi}{2}\right)$$

Thus $r = 4[1 - \cos(\theta - \frac{\pi}{2})]$. This is a cardioid with $a = 4$ that has been rotated $\frac{\pi}{2}$. The graph is shown in Figure 10.23. A curve of the form $r = a(1 - \sin\theta)$ is a cardioid that has been rotated $\frac{\pi}{2}$. ■

The cardioid is only one of the interesting polar-form curves. It is a special case of a curve called a **limacon**, which is developed in the exercises (see Problem 55). The next example illustrates a curve called a **rose curve**.

EXAMPLE 8 Graph $r = 4\cos 2\theta$.

Solution When presented with a polar-form curve that you do not recognize, graph the curve by plotting points. You can use tables, exact values, or a calculator. The following table uses a calculator.

θ	0	$\frac{\pi}{12}$	$\frac{\pi}{6}$	$\frac{\pi}{4}$	$\frac{\pi}{3}$	$\frac{5\pi}{12}$	$\frac{\pi}{2}$	$\frac{7\pi}{12}$	$\frac{2\pi}{3}$	$\frac{3\pi}{4}$	$\frac{5\pi}{6}$	$\frac{11\pi}{12}$
r (approx. value)	4	3.5	2	0	−2	−3.5	−4	−3.5	−2	0	2	3.5

θ	π	$\frac{13\pi}{12}$	$\frac{7\pi}{6}$	$\frac{5\pi}{4}$	$\frac{4\pi}{3}$	$\frac{17\pi}{12}$	$\frac{3\pi}{2}$	$\frac{19\pi}{12}$	$\frac{5\pi}{3}$	$\frac{7\pi}{4}$	$\frac{11\pi}{6}$	$\frac{23\pi}{12}$
r (approx. value)	4	3.5	2	0	−2	−3.5	−4	−3.5	−2	0	2	3.5

The graph is shown in Figure 10.24.

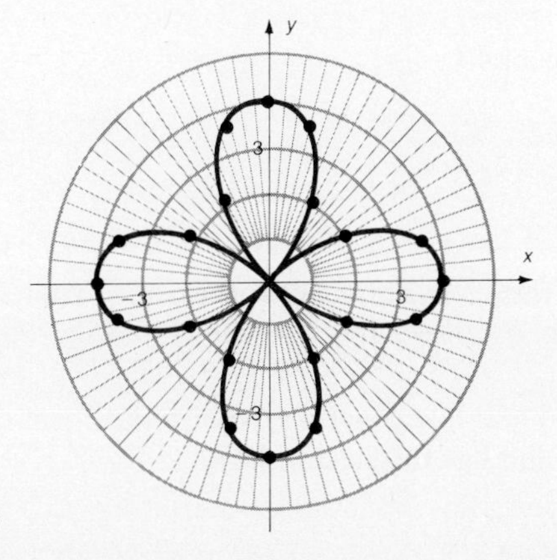

Figure 10.24 Graph of $r = 4\cos 2\theta$ ■

In general,

$$r = a\cos n\theta$$

is a four-leaved rose if $n = 2$ and the length of the leaves is a. If n is an even number, the curve has $2n$ leaves; if n is odd, the number of leaves is n. These leaves are equally spaced on a circle of radius a.

EXAMPLE 9 $r = 4\cos 2(\theta - \frac{\pi}{4})$

Solution This is a rose curve with four leaves of length 4 equally spaced on a circle. However, this curve has been rotated $\frac{\pi}{4}$, as shown in Figure 10.25.

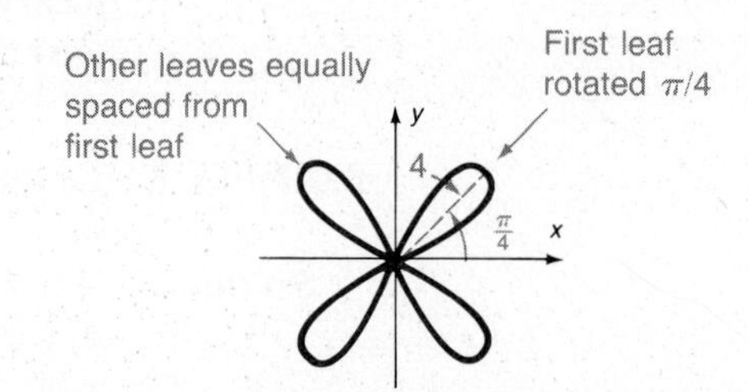

Figure 10.25 Graph of $r = 4\cos 2(\theta - \frac{\pi}{4})$. ■

Notice that Example 9 can be rewritten as a sine curve:

$$\begin{aligned} r &= 4\cos 2\left(\theta - \frac{\pi}{4}\right) \\ &= 4\cos\left(2\theta - \frac{\pi}{2}\right) \\ &= 4\cos\left(\frac{\pi}{2} - 2\theta\right) \\ &= 4\sin 2\theta \end{aligned}$$

These steps can be reversed to graph a rose curve written in terms of a sine function.

EXAMPLE 10 Graph $r = 5\sin 4\theta$.

Solution

$$\begin{aligned} r &= 5\sin 4\theta \\ &= 5\cos\left(\frac{\pi}{2} - 4\theta\right) \\ &= 5\cos\left(4\theta - \frac{\pi}{2}\right) \\ &= 5\cos 4\left(\theta - \frac{\pi}{8}\right) \end{aligned}$$

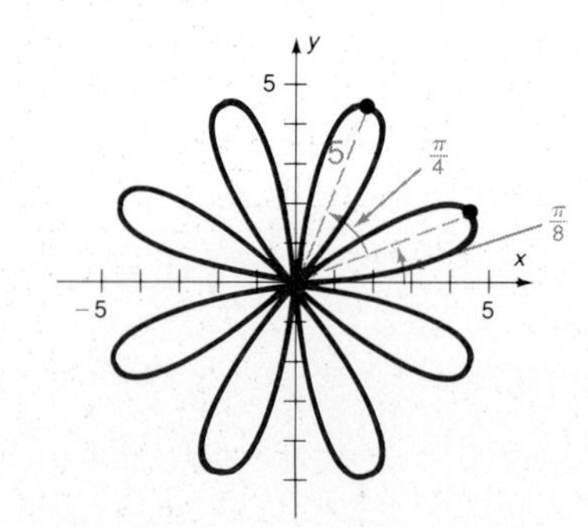

Figure 10.26 Graph of $r = 5\sin 4\theta$

Recognize this as a rose curve rotated $\pi/8$. There are eight leaves of length 5. The leaves are a distance $\frac{\pi}{4}$ apart, as shown in Figure 10.26. ■

The third and last general type of polar-form curve we will consider is called a **lemniscate** and has the general form

$$r^2 = a^2\cos 2\theta$$

EXAMPLE 11 Graph $r^2 = 9\cos 2\theta$.

Solution As before, when graphing a curve for the first time begin by plotting points. For this example be sure to obtain two values for r when solving this quadratic equation. For example, if $\theta = 0$, then $\cos 2\theta = 1$ and $r^2 = 9$, so $r = 3$ or -3.

θ	0	$\frac{\pi}{12}$	$\frac{\pi}{6}$	$\frac{\pi}{4}$	$\frac{\pi}{4}$ to $\frac{3\pi}{4}$	$\frac{5\pi}{6}$	$\frac{11\pi}{12}$	π
r (approx. value)	± 3	± 2.8	± 2.1	0	undefined	± 2.1	± 2.8	± 3

Notice that for $\frac{\pi}{4} < \theta < \frac{3\pi}{4}$ there are no values for r, since $\cos 2\theta$ is negative. For $\pi \le \theta \le 2\pi$, the values repeat the sequence given above, so these points are plotted and then connected, as shown in Figure 10.27.

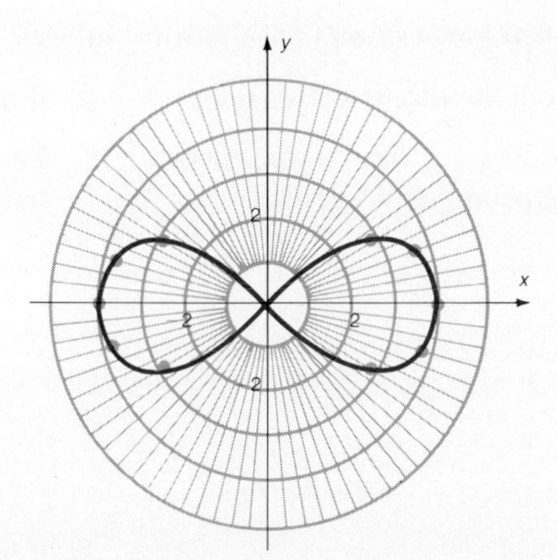

Figure 10.27 Graph of $r^2 = 9\cos 2\theta$

The graphs of $r^2 = a^2\cos 2\theta$ and $r^2 = a^2\sin 2\theta$ are called **lemniscates**. There are always two leaves to a lemniscate and the length of the leaves is a. The sine function can be considered as a rotation of the cosine function.

EXAMPLE 12 Graph $r^2 = 16\sin 2\theta$.

Solution

$$\begin{aligned} r^2 &= 16\sin 2\theta \\ &= 16\cos\left(\frac{\pi}{2} - 2\theta\right) \\ &= 16\cos\left(2\theta - \frac{\pi}{2}\right) \\ &= 16\cos 2\left(\theta - \frac{\pi}{4}\right) \end{aligned}$$

This is a lemniscate whose leaf has length $\sqrt{16} = 4$ and is rotated $\frac{\pi}{4}$ as shown in Figure 10.28.

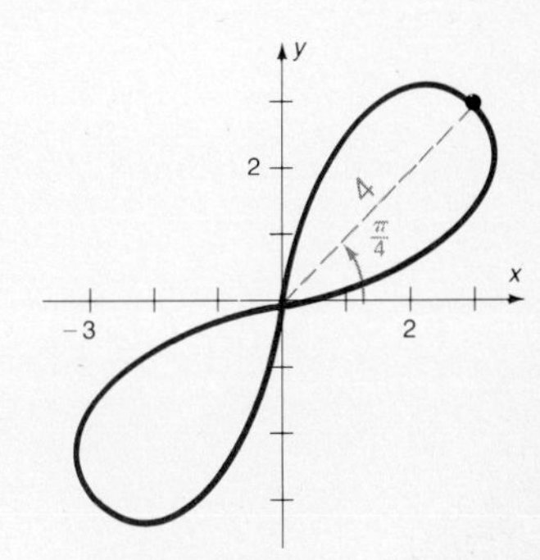

Figure 10.28 Graph of $r^2 = 16\sin 2\theta$

We conclude this section by summarizing the special types of polar-form curves we have developed. There are many others, some of which are presented in the problems, but these three special types are the most common.

Cardioid:

$r = a(1 \pm \cos\theta)$ or $r = a(1 \pm \sin\theta)$

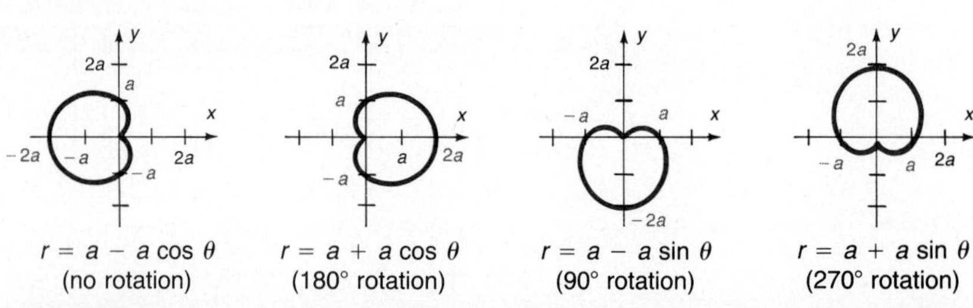

Rose Curve:

$r = a\cos n\theta$ or $r = a\sin n\theta$ (n is a positive integer). Length of each leaf is a.

1. If n is odd, the rose is n-leaved.

   One leaf: If $n = 1$, the rose is a curve with one petal and is circular.

2. If n is even, the rose is $2n$-leaved.

   Two leaves: See the lemniscate described below.

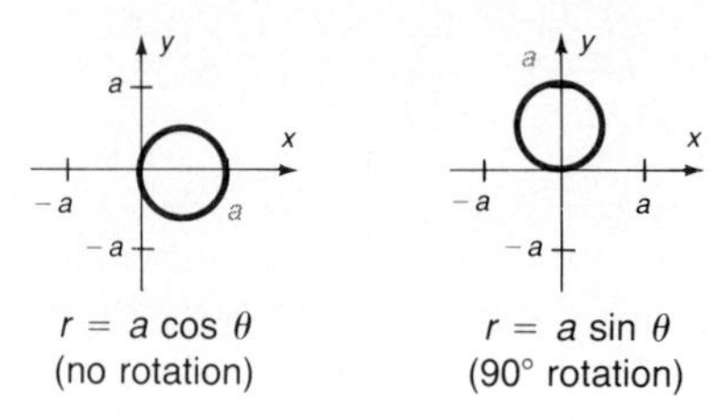

Three leaves: $n = 3$

Four leaves: $n = 2$

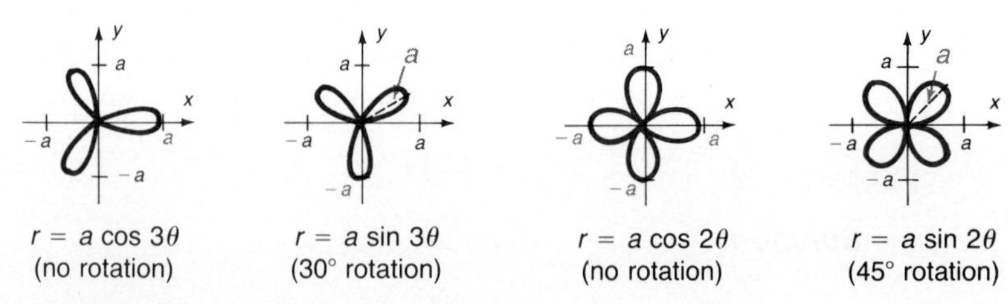

Lemniscate:

$r^2 = a^2\cos 2\theta$ or $r^2 = a^2\sin 2\theta$

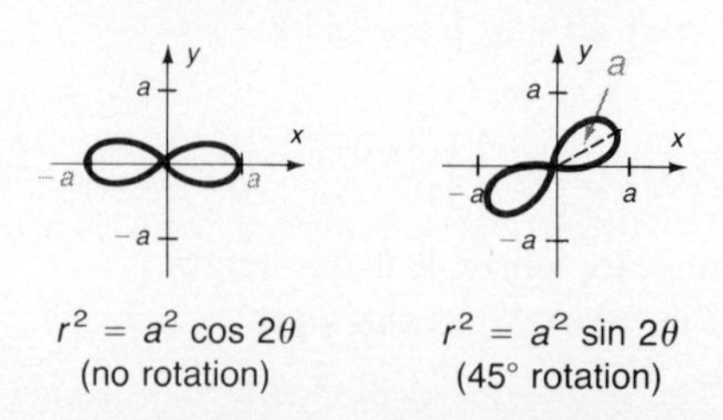

PROBLEM SET 10.4

A

In Problems 1–10, plot each of the given polar-form points. Give both primary representations, and give the rectangular coordinates of the point. In Problems 7–9 approximate values to two decimal places. Otherwise use exact values.

1. $(4, \frac{\pi}{4})$
2. $(6, \frac{\pi}{3})$
3. $(5, \frac{2\pi}{3})$
4. $(3, -\frac{\pi}{6})$
5. $(\frac{3}{2}, -\frac{5\pi}{6})$
6. $(5, -\frac{\pi}{2})$
7. $(-4, 4)$
8. $(4, 10)$
9. $(-4, 5\pi)$
10. $(-3, -\frac{7\pi}{3})$

In Problems 11–20, plot the given rectangular-form points and give both primary representations in polar form. In Problems 17–20 approximate values to the nearest hundredth.

11. $(5, 5)$
12. $(-1, \sqrt{3})$
13. $(2, -2\sqrt{3})$
14. $(-2, -2)$
15. $(3, -3)$
16. $(-6, 6)$
17. $(-\sqrt{3}, 1)$
18. $(4, 3)$
19. $(-12, 5)$
20. $(3, 7)$

Identify each of the curves in Problems 21–35 as a cardioid, a rose curve (state number of leaves), a lemniscate, or none of the above.

21. $r^2 = 9\cos 2\theta$
22. $r = 2\sin 2\theta$
23. $r = 3\sin 3\theta$
24. $r^2 = 2\cos 2\theta$
25. $r = 2 - 2\cos\theta$
26. $r = 3 + 3\sin\theta$
27. $r^2 = \sin 3\theta$
28. $r = 4\sin 30^\circ$
29. $r = 5\cos 60^\circ$
30. $r = 5\sin 8\theta$
31. $r = 3\theta$
32. $r\theta = 3$
33. $\theta = \tan\frac{\pi}{4}$
34. $r = 5(1 - \sin\theta)$
35. $\cos\theta = 1 - r$

B

Sketch each of the curves given in Problems 36–50. If it is not one of the types discussed in the text, graph it by plotting points.

36. $r = 2(1 + \cos\theta)$
37. $r = 3(1 - \sin\theta)$
38. $r = 4(1 + \sin\theta)$
39. $r = 4\cos 2\theta$
40. $r = 5\sin 3\theta$
41. $r = 3\cos 3\theta$
42. $r = 2\cos\theta$
43. $r^2 = 9\cos 2\theta$
44. $r^2 = 16\cos 2\theta$
45. $r^2 = 16\sin 2\theta$
46. $r = 5\sin\frac{\pi}{6}$
47. $r = 9\cos\frac{\pi}{3}$
48. $r = \theta$
49. $r = 3\theta$
50. $r = \tan\theta$

51. Derive the equations for changing from polar coordinates to rectangular coordinates.

52. Derive the equations for changing from rectangular coordinates to polar coordinates.

C

53. What is the distance between $(3, \frac{\pi}{3})$ and $(7, \frac{\pi}{4})$?

54. What is the distance between (r, θ) and (c, α)?

55. The limaçon is a curve of the form $r = b \pm a\cos\theta$ or $r = b \pm a\sin\theta$ where $a > 0$, $b > 0$. (This problem is required for Section 10.5.) There are four types of limaçons.
a. $b/a < 1$ (limaçon with inner loop): graph $r = 2 - 3\cos\theta$ by plotting points.
b. $b/a = 1$ (cardioid): graph $r = 2 - 2\cos\theta$.
c. $1 < b/a < 2$ (limaçon with a dimple): graph $r = 3 - 2\cos\theta$ by plotting points.
d. $b/a \geq 2$ (convex limaçon): graph $r = 3 - \cos\theta$ by plotting points.

56. Graph the following limaçons (see Problem 55):
a. $r = 1 - 2\cos\theta$
b. $r = 2 - \cos\theta$
c. $r = 2 + 3\cos\theta$
d. $r = 2 - 3\sin\theta$
e. $r = 2 + 3\sin\theta$
f. $r = 3 - 2\sin\theta$

57. *Spirals* are interesting mathematical curves. There are three general types of spirals:
a. A spiral of Archimedes has the form $r = a\theta$; graph $r = 2\theta$ $(\theta > 0)$ by plotting points.
b. A hyperbolic spiral has the form $r\theta = a$; graph $r\theta = 2$ $(\theta > 0)$ by plotting points.
c. A logarithmic spiral has the form $r = a^{k\theta}$; graph $r = 2^\theta$ $(\theta > 0)$ by plotting points.

58. Identify and graph the following spirals (see Problem 57). Assume $\theta > 0$.
a. $r = \theta$
b. $r = -\theta$
c. $r\theta = 1$
d. $r\theta = -1$
e. $r = 2^{2\theta}$
f. $r = 3^\theta$

59. The *strophoid* is a curve of the form $r = a\cos 2\theta\sec\theta$; graph this curve where $a = 2$ by plotting points.

60. The *bifolium* has the form $r = a\sin\theta\cos^2\theta$; graph this curve where $a = 1$ by plotting points.

61. The *folium of Descartes* has the form

$$r = \frac{3a\sin\theta\cos\theta}{\sin^3\theta + \cos^3\theta}$$

Graph this curve where $a = 2$ by plotting points.

10.5 INTERSECTION OF POLAR-FORM CURVES*

In order to find the points of intersection of graphs in rectangular form, you need only find the simultaneous solution of the equations that define those graphs. It is not always necessary to draw the graphs. This follows because in a rectangular coordinate system there is a one-to-one correspondence between ordered pairs satisfying an equation and points on its graph.

However, as you saw in the last section, this one-to-one property is lost when you are working with polar coordinates. This means that the simultaneous solution of two equations in polar form may introduce extraneous points of intersection or may even fail to yield all points of intersection. For this reason, our method for finding the intersection of polar-form curves will include sketching the graphs.

EXAMPLE 1 Find the points of intersection of the curves $r = 2\cos\theta$ and $r = 2\sin\theta$.

Solution First consider the simultaneous solution of the system of equations:

$$\begin{cases} r = 2\cos\theta \\ r = 2\sin\theta \end{cases}$$

By substitution,

$$\begin{aligned} 2\sin\theta &= 2\cos\theta \\ \sin\theta &= \cos\theta \\ \theta &= \frac{\pi}{4} + 2n\pi, \frac{5\pi}{4} + 2n\pi \qquad (n \text{ an integer}) \end{aligned}$$

Then find r using the primary representations for θ:

$$\begin{aligned} r &= 2\cos\frac{\pi}{4} & r &= 2\cos\frac{5\pi}{4} \\ &= 2\left(\frac{\sqrt{2}}{2}\right) & &= 2\left(\frac{-\sqrt{2}}{2}\right) \\ &= \sqrt{2} & &= -\sqrt{2} \end{aligned}$$

This gives the points $(\sqrt{2}, \frac{\pi}{4})$ and $(-\sqrt{2}, \frac{5\pi}{4})$. Writing the primary representations for these points, we see that they represent the same point. Thus the simultaneous solution yields one point of intersection. Next consider the graphs of these curves as shown in Figure 10.29.

---

* Optional section

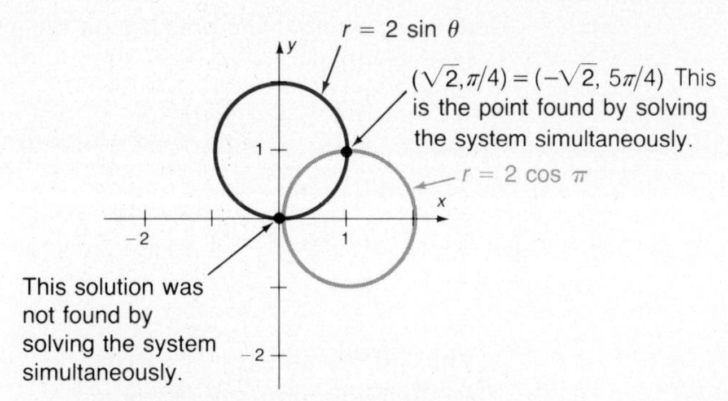

Figure 10.29 Graphs of $r = 2\cos\theta$ and $r = 2\sin\theta$

It looks like $(0, 0)$ is also a point of intersection. Check this point in each of the given equations:

$$r = 2\cos\theta: \quad \text{If } r = 0, \text{ then } \theta = \frac{\pi}{2}$$

$$r = 2\sin\theta: \quad \text{If } r = 0, \text{ then } \theta = \pi$$

At first it does not seem that $(0, 0)$ satisfies the equations since $r = 0$ gives $(0, \frac{\pi}{2})$ and $(0, \pi)$, respectively. Notice that these coordinates are different and do not satisfy the equations simultaneously. But if you plot these coordinates you will see that $(0, 0)$, $(0, \frac{\pi}{2})$, and $(0, \pi)$ are all the same point. Points of intersection for the given curves are $(0, 0)$, $(\sqrt{2}, \frac{\pi}{4})$. ■

The pole is often a solution for a system of equations even though it may not satisfy the equations simultaneously. This is because when $r = 0$, all values of θ will yield the same point—namely the pole. For this reason it is necessary to check separately to see if the pole lies on the given graph.

Graphical Solution of the Intersection of Polar Curves

1. Find the simultaneous solution of the given system of equations.
2. Determine whether the pole lies on the two graphs.
3. Graph the curves and look for other points of intersection.

EXAMPLE 2 Find the points of intersection of the curves $r = 2 + 4\cos\theta$ and $r = 6\cos\theta$.

Solution *Step 1* Solve the equations simultaneously:

$$\begin{cases} r = 2 + 4\cos\theta \\ r = 6\cos\theta \end{cases}$$

By substitution,

$$\begin{aligned} 6\cos\theta &= 2 + 4\cos\theta \\ 2\cos\theta &= 2 \\ \cos\theta &= 1 \\ \theta &= 0 \end{aligned}$$

If $\theta = 0$ then $r = 6$, so an intersection point is $(0, 6)$.

Step 2 Determine whether the pole lies on the graphs.

1. If $r = 0$, then

$$0 = 2 + 4\cos\theta$$

$$\cos\theta = -\frac{1}{2}$$

$$\theta = \frac{2\pi}{3}, \frac{4\pi}{3}$$

Thus $(0, \frac{2\pi}{3})$ and $(0, \frac{4\pi}{3})$ satisfy the first equation and the pole lies on this graph.

2. If $r = 0$, then

$$0 = 6\cos\theta$$

$$\theta = \frac{\pi}{2}, \frac{3\pi}{2}$$

Thus $(0, \frac{\pi}{2})$ and $(0, \frac{3\pi}{2})$ satisfy the second equation so the pole also lies on this graph.

Step 3 Graph the curves and look for other points of intersection. The first curve is a limaçon; the second is a circle as shown in Figure 10.30.

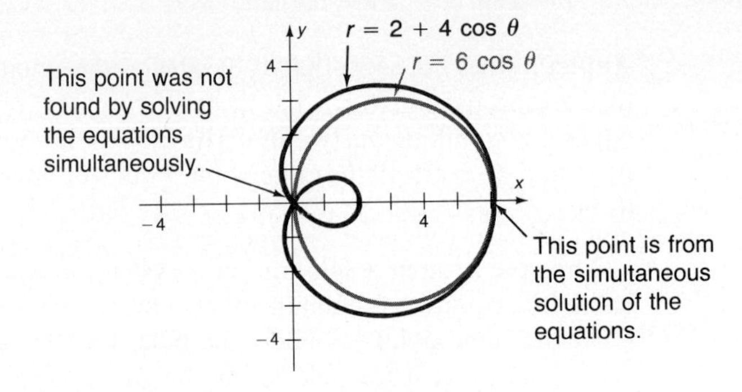

Figure 10.30 Graphs of $r = 2 + 4\cos\theta$ and $r = 6\cos\theta$

Notice that there are no other points of intersection, so the intersection points are $(6, 0)$ and $(0, 0)$. ■

EXAMPLE 3 Find the points of intersection of the curves $r = \frac{3}{2} - \cos\theta$ and $\theta = \frac{2\pi}{3}$.

Solution Solve

$$\begin{cases} r = \dfrac{3}{2} - \cos\theta \\ \theta = \dfrac{2\pi}{3} \end{cases}$$

By substitution,

$$\begin{aligned} r &= \frac{3}{2} - \cos\frac{2\pi}{3} \\ &= \frac{3}{2} - \left(-\frac{1}{2}\right) \\ &= 2 \end{aligned}$$

Solution: $(2, \frac{2\pi}{3})$.

If $r = 0$, the first equation has no solution since

$$0 = \frac{3}{2} - \cos\theta$$

$$\cos\theta = \frac{3}{2}$$

and $\cos\theta$ cannot be larger than 1. Now look at the graph (Figure 10.31).

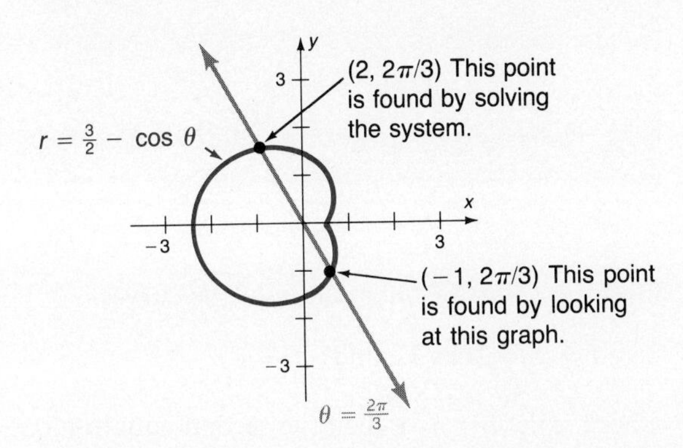

Figure 10.31 Graphs of $r = \frac{3}{2} - \cos\theta$ and $\theta = \frac{2\pi}{3}$

From the graph we see that $(-1, \frac{2\pi}{3})$ looks like a point of intersection. It satisfies the equation $\theta = \frac{2\pi}{3}$, but what about $r = \frac{3}{2} - \cos\theta$?

$$\text{Check}\left(-1, \frac{2\pi}{3}\right): \quad -1 \stackrel{?}{=} \frac{3}{2} - \cos\left(\frac{2\pi}{3}\right)$$

$$= \frac{3}{2} - \left(-\frac{1}{2}\right)$$

$$= 2 \qquad \text{(not satisfied)}$$

But check the other representation:

$$\text{Check}\left(-1, \frac{2\pi}{3}\right) = \left(1, \frac{5\pi}{3}\right): \quad 1 \stackrel{?}{=} \frac{3}{2} - \cos\left(\frac{5\pi}{3}\right)$$

$$= \frac{3}{2} - \left(\frac{1}{2}\right)$$

$$= 1 \quad \text{(satisfied)}$$

You would not have found this other point without checking the graph. ■

Now you might ask if there is a procedure that will yield all the points of intersection without relying on the graph. The answer is yes—if you realize that polar-form equations can have different representations just as we found for polar-form points. Recall that a polar-form point has two primary representations:

$$(r, \theta) \quad \text{and} \quad (-r, \theta + \pi)$$

where the angle (θ or $\theta + \pi$, respectively) is between zero and 2π. For every polar-form equation $r = f(\theta)$, there are equations

$$r = (-1)^n f(\theta + n\pi)$$

for n any integer that yields exactly the same curve.

Analytic Solution for the Intersection of Polar Curves

> 1. Solve each equation of one graph simultaneously with each equation of the other graph. If $r = f(\theta)$ is the given equation, the other equations of the same graph are
>
> $$r = (-1)^n f(\theta + n\pi)$$
>
> 2. Determine if the pole lies on the two graphs.

EXAMPLE 4 Find the points of intersection of the curves

$$r = 1 - 2\cos\theta \quad \text{and} \quad r = 1$$

Solution Check pole: If $r = 0$, then the second equation ($r = 1$) is not satisfied. That is, the graph of the second equation does not pass through the origin. Next solve the equations simultaneously; write out the alternative forms of the equations.

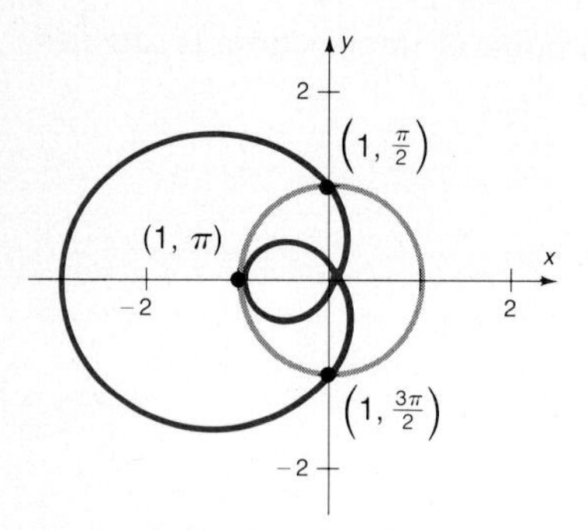

Figure 10.32 Graphs of $r = 1 - 2\cos\theta$ and $r = 1$

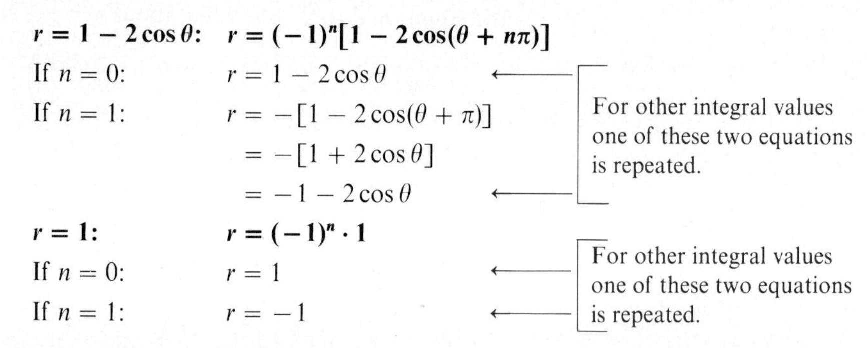

$\mathbf{r = 1 - 2\cos\theta}$: $\quad \mathbf{r = (-1)^n[1 - 2\cos(\theta + n\pi)]}$

If $n = 0$: $\quad r = 1 - 2\cos\theta$

If $n = 1$: $\quad r = -[1 - 2\cos(\theta + \pi)]$
$\quad = -[1 + 2\cos\theta]$
$\quad = -1 - 2\cos\theta$

For other integral values one of these two equations is repeated.

$\mathbf{r = 1}$: $\quad \mathbf{r = (-1)^n \cdot 1}$

If $n = 0$: $\quad r = 1$

If $n = 1$: $\quad r = -1$

For other integral values one of these two equations is repeated.

Solve the systems

$\begin{cases} r = 1 - 2\cos\theta \\ r = 1 \end{cases}$	$\begin{cases} r = 1 - 2\cos\theta \\ r = -1 \end{cases}$	$\begin{cases} r = -1 - 2\cos\theta \\ r = 1 \end{cases}$	$\begin{cases} r = -1 - 2\cos\theta \\ r = -1 \end{cases}$
$1 = 1 - 2\cos\theta$	$-1 = 1 - 2\cos\theta$	$1 = -1 - 2\cos\theta$	$-1 = -1 - 2\cos\theta$
$0 = \cos\theta$	$1 = \cos\theta$	$-1 = \cos\theta$	$0 = \cos\theta$
$\theta = \frac{\pi}{2}, \frac{3\pi}{2}$	$\theta = 0$	$\theta = \pi$	Same as first equation

Points: $(1, \frac{\pi}{2})$, $(1, \frac{3\pi}{2})$, $(1, \pi)$.

The graphs of these curves are shown in Figure 10.32. ■

PROBLEM SET 10.5

A

Sketch each pair of equations in Problems 1–15 on the same coordinate axes.

1. $\begin{cases} r = 4\cos\theta \\ r = 4\sin\theta \end{cases}$
2. $\begin{cases} r = 8\cos\theta \\ r = 8\sin\theta \end{cases}$
3. $\begin{cases} r = 2\cos\theta \\ r = 1 \end{cases}$
4. $\begin{cases} r = 4\cos\theta \\ r = 2 \end{cases}$
5. $\begin{cases} r^2 = 9\cos\theta \\ r = 3 \end{cases}$
6. $\begin{cases} r^2 = 4\sin\theta \\ r = 2 \end{cases}$
7. $\begin{cases} r = 2(1+\cos\theta) \\ r = 2(1-\cos\theta) \end{cases}$
8. $\begin{cases} r = 2(1+\sin\theta) \\ r = 2(1-\sin\theta) \end{cases}$
9. $\begin{cases} r^2 = 4\cos 2\theta \\ r = 2 \end{cases}$
10. $\begin{cases} r^2 = 9\sin 2\theta \\ r = 3 \end{cases}$
11. $\begin{cases} r^2 = 4\sin 2\theta \\ r = 2\sqrt{2}\cos\theta \end{cases}$
12. $\begin{cases} r^2 = 9\cos 2\theta \\ r = 3\sqrt{2}\sin\theta \end{cases}$
13. $\begin{cases} r^2 = \cos 2\theta \\ r^2 = -\cos 2\theta \end{cases}$
14. $\begin{cases} r = 1+\sin\theta \\ r = 1+\cos\theta \end{cases}$
15. $\begin{cases} r = a(1+\sin\theta) \\ r = a(1-\sin\theta) \end{cases}$

Find the points of intersection of the curves given by the equations in Problems 16–30. You need to give only one primary representation for each point of intersection; give the one with a positive r. You graphed each of these systems in Problems 1–15.

16. $\begin{cases} r = 4\cos\theta \\ r = 4\sin\theta \end{cases}$
17. $\begin{cases} r = 8\cos\theta \\ r = 8\sin\theta \end{cases}$
18. $\begin{cases} r = 2\cos\theta \\ r = 1 \end{cases}$
19. $\begin{cases} r = 4\cos\theta \\ r = 2 \end{cases}$
20. $\begin{cases} r^2 = 9\cos\theta \\ r = 3 \end{cases}$
21. $\begin{cases} r^2 = 4\sin\theta \\ r = 2 \end{cases}$
22. $\begin{cases} r = 2(1+\cos\theta) \\ r = 2(1-\cos\theta) \end{cases}$
23. $\begin{cases} r = 2(1+\sin\theta) \\ r = 2(1-\sin\theta) \end{cases}$
24. $\begin{cases} r^2 = 4\cos 2\theta \\ r = 2 \end{cases}$
25. $\begin{cases} r^2 = 9\sin 2\theta \\ r = 3 \end{cases}$
26. $\begin{cases} r^2 = 4\sin 2\theta \\ r = 2\sqrt{2}\cos\theta \end{cases}$
27. $\begin{cases} r^2 = 9\cos 2\theta \\ r = 3\sqrt{2}\sin\theta \end{cases}$
28. $\begin{cases} r^2 = \cos 2\theta \\ r^2 = -\cos 2\theta \end{cases}$
29. $\begin{cases} r = 1+\sin\theta \\ r = 1+\cos\theta \end{cases}$
30. $\begin{cases} r = a(1+\sin\theta) \\ r = a(1-\sin\theta) \end{cases}$

B

Sketch each pair of equations in Problems 31–45 on the same coordinate axes.

31. $\begin{cases} r = 2\theta \\ \theta = \frac{\pi}{6} \end{cases}$
32. $\begin{cases} r = 3\theta \\ \theta = \frac{\pi}{3} \end{cases}$
33. $\begin{cases} r^2 = \cos 2\theta \\ r = \sqrt{2}\sin\theta \end{cases}$
34. $\begin{cases} r = 2(1-\cos\theta) \\ r = 4\sin\theta \end{cases}$
35. $\begin{cases} r = 2(1-\cos\theta) \\ r = 4\sin\theta \end{cases}$
36. $\begin{cases} r = 2(1-\sin\theta) \\ r = 4\cos\theta \end{cases}$
37. $\begin{cases} r = 2\cos\theta + 1 \\ r = \sin\theta \end{cases}$
38. $\begin{cases} r = 2\sin\theta + 1 \\ r = \cos\theta \end{cases}$
39. $\begin{cases} r = \dfrac{5}{3-\cos\theta} \\ r = 2 \end{cases}$
40. $\begin{cases} r = \dfrac{2}{1+\cos\theta} \\ r = 2 \end{cases}$
41. $\begin{cases} r = \dfrac{4}{1-\cos\theta} \\ r = 2\cos\theta \end{cases}$
42. $\begin{cases} r = \dfrac{1}{1+\cos\theta} \\ r = 2(1-\cos\theta) \end{cases}$
43. $\begin{cases} r = a\cos\theta \\ r = a\sec\theta \end{cases}$
44. $\begin{cases} r = a\sin\theta \\ r = a\csc\theta \end{cases}$
45. $\begin{cases} r\sin\theta = 1 \\ r = 4\sin\theta \end{cases}$

Find the points of intersection of the curves given by the equations in Problems 46–60. You need to give only one primary representation for each point of intersection; give the one with a positive r. You graphed each of these systems in Problems 31–45.

46. $\begin{cases} r = 2\theta \\ \theta = \frac{\pi}{6} \end{cases}$
47. $\begin{cases} r = 3\theta \\ \theta = \frac{\pi}{3} \end{cases}$
48. $\begin{cases} r^2 = \cos 2\theta \\ r = \sqrt{2}\sin\theta \end{cases}$
49. $\begin{cases} r = 2(1-\cos\theta) \\ r = 4\sin\theta \end{cases}$
50. $\begin{cases} r = 2(1-\cos\theta) \\ r = 4\sin\theta \end{cases}$
51. $\begin{cases} r = 2(1-\sin\theta) \\ r = 4\cos\theta \end{cases}$
52. $\begin{cases} r = 2\cos\theta + 1 \\ r = \sin\theta \end{cases}$
53. $\begin{cases} r = 2\sin\theta + 1 \\ r = \cos\theta \end{cases}$
54. $\begin{cases} r = \dfrac{5}{3-\cos\theta} \\ r = 2 \end{cases}$
55. $\begin{cases} r = \dfrac{2}{1+\cos\theta} \\ r = 2 \end{cases}$
56. $\begin{cases} r = \dfrac{4}{1-\cos\theta} \\ r = 2\cos\theta \end{cases}$
57. $\begin{cases} r = \dfrac{1}{1+\cos\theta} \\ r = 2(1-\cos\theta) \end{cases}$
58. $\begin{cases} r = a\cos\theta \\ r = a\sec\theta \end{cases}$
59. $\begin{cases} r = a\sin\theta \\ r = a\csc\theta \end{cases}$
60. $\begin{cases} r\sin\theta = 1 \\ r = 4\sin\theta \end{cases}$

10.6 PARAMETRIC EQUATIONS

Up to now, the curves we have dicussed have been represented by a single equation. However, there is another way of representing curves which is often useful. This new representation defines the x and y in (x, y) so that they are *each* functions of some other variable, say t. That is, let

$$x = g(t) \quad \text{and} \quad y = h(t)$$

for functions g and h, where the domain of these functions is some interval I. For example, let

$$g(t) = 1 + 3t \quad \text{and} \quad h(t) = 2t \qquad (\text{for } 0 \le t \le 5)$$

Then, if $t = 1$,

$$x = g(1) = 1 + 3(1) = (4)$$
$$y = h(1) = 2(1) = 2$$

Then the point $(x, y) = (4, 2)$ when $t = 1$. Other values are shown in the following table and plotted in Figure 10.33.

t	0	1	2	3	4	5
x	1	4	7	10	13	16
y	0	2	4	6	8	10

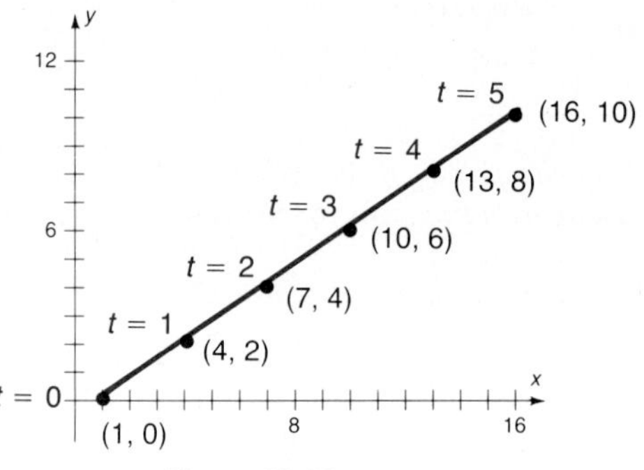

Figure 10.33 Graph of $x = 1 + 3t$, $y = 2t$

The variable t is called a **parameter** and the equations $x = 1 + 3t$ and $y = 2t$ are called **parametric equations** for the line segment shown in Figure 10.33.

Parameter and Parametric Equations

Let t be a number in an interval I. Consider the curve defined by the set of ordered pairs (x, y), where

$$x = f(t) \quad \text{and} \quad y = g(t)$$

for functions f and g defined on I. Then the variable t is called a **parameter** and the equations $x = f(t)$ and $y = g(t)$ are called **parametric equations** for the curve defined by (x, y).

EXAMPLE 1 Plot the curve represented by the parametric equations

$$x = \cos\theta$$
$$y = \sin\theta$$

Solution The parameter is θ and you can generate a table of values by using Table B.3, exact values, or a calculator:

θ	0°	15°	30°	45°	60°	75°	90°	120°	⋯
x	1.00	0.97	0.87	0.71	0.50	0.26	0.00	−0.50	⋯
y	0.00	0.26	0.50	0.71	0.87	0.97	1.00	0.87	⋯

These points are plotted in Figure 10.34. If the plotted points are connected, you can see that the curve is a circle.

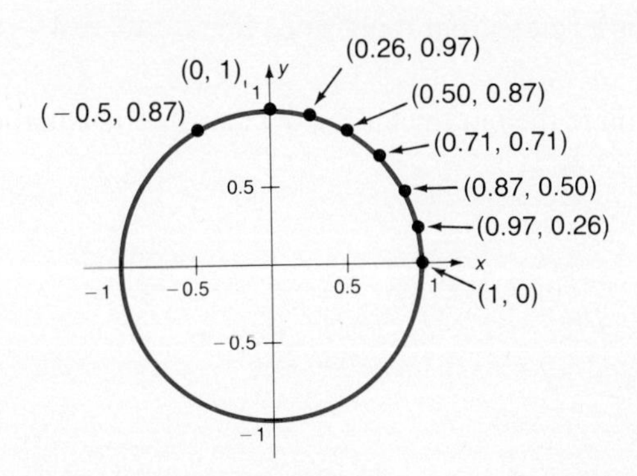

Figure 10.34 Graph of $x = \cos\theta$, $y = \sin\theta$

■

It is possible to recognize the parametric equations in Example 1 as a unit circle if you square both sides of the equation and add:

$$x^2 = \cos^2\theta$$
$$y^2 = \sin^2\theta$$
$$x^2 + y^2 = \cos^2\theta + \sin^2\theta$$

So

$$x^2 + y^2 = 1$$

This process is called **eliminating the parameter**.

EXAMPLE 2 Eliminate the parameter for the parametric equations $x = t + 2$, $y = t^2 + 2t - 1$.

Solution Solve the first equation for t: $t = x - 2$. Substitute into the second equation:

$$\begin{aligned} y &= (x-2)^2 + 2(x-2) - 1 \\ &= x^2 - 4x + 4 + 2x - 4 - 1 \\ &= x^2 - 2x - 1 \end{aligned}$$

You can recognize this as a parabola. It can now be graphed by completing the square or by using the parametric equations and plotting points as illustrated by Example 1. ■

EXAMPLE 3 Eliminate the parameter for the parametric equations

$$x = t^2 - 3t + 1, \; y = -t^2 + 2t + 3.$$

Solution It is not as easy to solve one of these equations for t as it was in Example 2. You can, however, add one equation to the other:

$$x + y = -t + 4 \quad \text{or} \quad t = 4 - x - y$$

This can be substituted into either equation to give

$$\begin{aligned} x &= (4 - x - y)^2 - 3(4 - x - y) + 1 \\ &= 16 - 4x - 4y - 4x + x^2 + xy - 4y + xy + y^2 - 12 + 3x + 3y + 1 \\ &= x^2 + 2xy + y^2 - 5x - 5y + 5 \end{aligned}$$

This is a rotated parabola since $B^2 - 4AC = 4 - 4(1)(1) = 0$. ■

EXAMPLE 4 Eliminate the parameter for the parametric equations $x = 2^t$, $y = 2^{t+1}$

Solution

$$\begin{aligned} \frac{y}{x} &= \frac{2^{t+1}}{2^t} \\ &= 2^{t+1-t} \\ &= 2 \\ y &= 2x \end{aligned}$$

■

Consider the parametric equations given in Example 4 a little more closely. You can plot a curve by using the parametric equations or by eliminating the parameter. Plot the equations in Example 4 both ways:

1. Parametric form:

t	0	1	2	3
x	1	2	4	8
y	2	4	8	16

The values on this table are found by substitution. For example, if $t = 0$ then

$$x = 2^0 \quad \text{and} \quad y = 2^0 + 1$$
$$= 1 \qquad\qquad = 2$$

Can $x = 3$? Solve

$$\begin{aligned} 3 &= 2^t \\ \log 3 &= t \log 2 \\ t &= \frac{\log 3}{\log 2} \\ &\approx 1.5850 \end{aligned}$$

Figure 10.35 Graph of $x = 2^t$, $y = 2^{t+1}$

and then

$$y = 2^{1.5850+1} \approx 6$$

Can $x = 0$? No, since $0 = 2^t$ has no solution.
Can $x = -1$? No, since $2^t > 0$
The graph is shown in Figure 10.35.

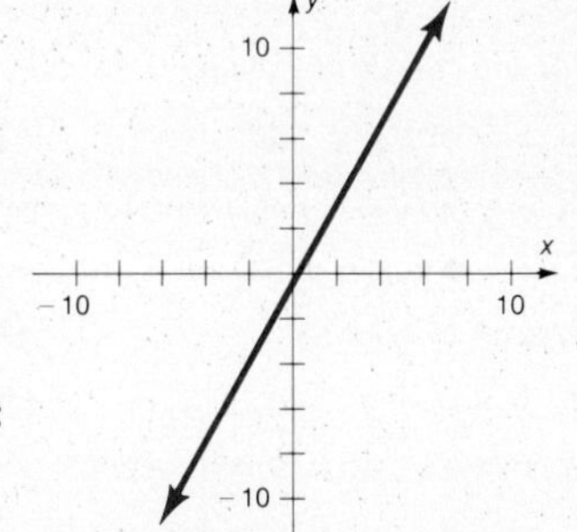

Figure 10.36 Graph of $y = 2x$

2. Eliminate the parameter as shown in Example 4:

$$y = 2x$$

Can $x = 3$? Solve $y = 2(3) = 6$
Can $x = 0$? Solve $y = 2(0) = 0$
Can $x = -1$? Solve $y = 2(-1) = -2$
The graph is shown in Figure 10.36.

Of course, what you notice by comparing Figures 10.35 and 10.36 is that **you must be very careful about the domain for x when eliminating the parameter**. Thus when you eliminate the parameter t and write $y = 2x$ in Example 4, you must include the condition $x > 0$ (from $x = 2^t$, x is nonnegative).

Note also that sometimes it is impossible to eliminate the parameter in any simple way, as in the equations

$$x = 3t^5 - 4t^2 + 7t + 11$$
$$y = 4t^{13} + 12t^5 + 6t^4 + 16$$

You must therefore be able to graph parametric equations using *both* methods since one or the other might not be appropriate for a particular graph.

PROBLEM SET 10.6

A

Plot the curves in Problems 1–20 by plotting points.

1. $x = 4t,\ y = -2t$
2. $x = t + 1,\ y = 2t$
3. $x = t,\ y = 2 + \frac{2}{3}(t - 1)$
4. $x = t,\ y = 3 - \frac{3}{5}(t + 2)$
5. $x = 2t,\ y = t^2 + t + 1$
6. $x = 3t,\ y = t^2 - t + 6$
7. $x = t,\ y = t^2 + 2t + 3$
8. $x = t,\ y = 2t^2 - 5t + 6$
9. $x = 3\cos\theta,\ y = 3\sin\theta$
10. $x = 2\cos\theta,\ y = 2\sin\theta$
11. $x = 4\cos\theta,\ y = 3\sin\theta$
12. $x = 5\cos\theta,\ y = 2\sin\theta$
13. $x = t^2 + 2t + 3,\ y = t^2 + t - 4$
14. $x = t^2 - 2t + 3,\ y = t^2 - t + 4$
15. $x = t^2 + 3t - 4,\ y = 2t^2 + 4t - 1$
16. $x = 2t^2 + t + 6,\ y = t^2 + t + 6$
17. $x = 3^t,\ y = 3^{t+1}$
18. $x = 2^t,\ y = 2^{1-t}$
19. $x = e^t,\ y = e^{t+1}$
20. $x = e^t,\ y = e^{1-t}$

Eliminate the parameter in Problems 21–40 and plot the resulting equations.

21. $x = 4t,\ y = -2t$
22. $x = t + 1,\ y = 2t$
23. $x = t,\ y = 2 + \frac{2}{3}(t - 1)$
24. $x = t,\ y = 3 - \frac{3}{5}(t + 2)$
25. $x = 2t,\ y = t^2 + t + 1$
26. $x = 3t,\ y = t^2 - t + 6$
27. $x = t,\ y = t^2 + 2t + 3$
28. $x = t,\ y = 2t^2 - 5t + 6$
29. $x = 3\cos\theta,\ y = 3\sin\theta$
30. $x = 2\cos\theta,\ y = 2\sin\theta$
31. $x = 4\cos\theta,\ y = 3\sin\theta$
32. $x = 5\cos\theta,\ y = 2\sin\theta$
33. $x = t^2 + 2t + 3,\ y = t^2 + t - 4$
34. $x = t^2 - 2t + 3,\ y = t^2 - t + 4$
35. $x = t^2 + 3t - 4,\ y = 2t^2 + 4t - 1$
36. $x = 2t^2 + t + 6,\ y = t^2 + t + 6$
37. $x = 3^t,\ y = 3^{t+1}$
38. $x = 2^t,\ y = 2^{1-t}$
39. $x = e^t,\ y = e^{t+1}$
40. $x = e^t,\ y = e^{1-t}$

B

Plot the curves in Problems 41–60 by any convenient method.

41. $x = 60t,\ y = 80t - 16t^2$
42. $x = 30t,\ y = 60t - 9t^2$
43. $x = 10\cos t,\ y = 10\sin t$
44. $x = 8\sin t,\ y = 8\cos t$
45. $x = 5\cos\theta,\ y = 3\sin\theta$
46. $x = 4\cos\theta,\ y = 2\sin\theta$
47. $x = t^2,\ y = t^3$
48. $x = t^3 + 1,\ y = t^3 - 1$
49. $x = e^t,\ y = e^{t+2}$
50. $x = e^t,\ y = e^{t-2}$
51. $x = \theta - \sin\theta,\ y = 1 - \cos\theta$
52. $x = \theta + \sin\theta,\ y = 1 - \cos\theta$
53. $x = 4\tan 2t,\ y = 3\sec 2t$
54. $x = 2\tan 2t,\ y = 4\sec 2t$
55. $x = 1 + \cos t,\ y = 3 - \sin t$
56. $x = 2 - \sin t,\ y = -3 + \cos t$
57. $x = 3\cos\theta + \cos 3\theta,\ y = 3\sin\theta - \sin 3\theta$
58. $x = \cos t + t\sin t,\ y = \sin t - t\cos t$
59. $x = \sin t,\ y = \csc t$
60. $x = \cos\theta,\ y = \sec\theta$

C

61. ***Physics*** Suppose a light is attached to the edge of a bike wheel. The path of the light is shown in Figure 10.37. If the radius of the wheel is a, find the equation for the path of the light. Such a curve is called a *cycloid*. *Hint:* Consider Figure 10.38 and find the coordinates of $P(x, y)$. Notice that

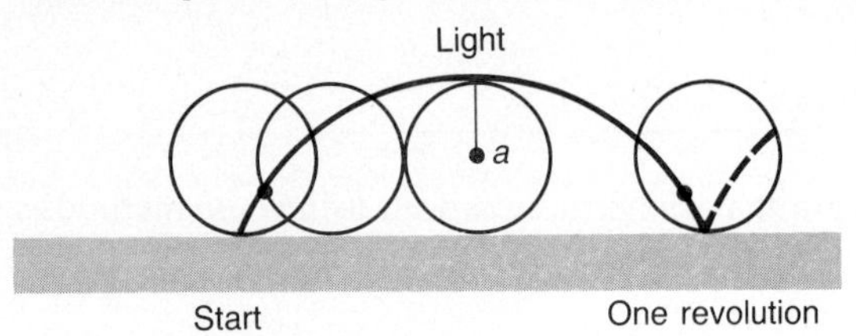

Figure 10.37 Graph of a cycloid

$$x = |OA|$$
$$y = |PA|$$

Find x and y in terms of θ, the amount of rotation in radians.

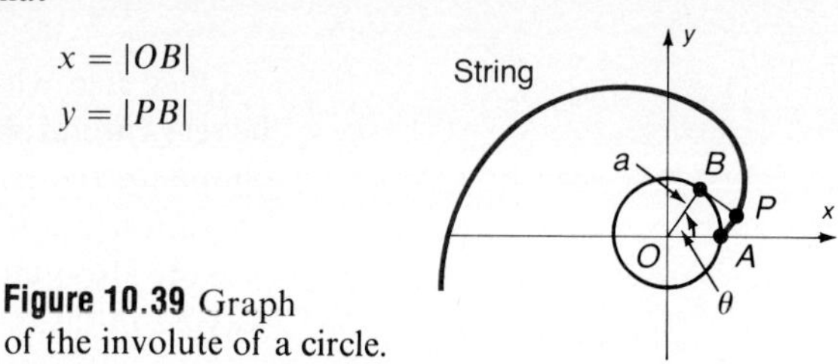

Figure 10.38

62. Suppose a string is wound around a circle of radius a. The string is then unwound in the plane of the circle while it is held tight, as shown in Figure 10.39. Find the equation for this curve, called the *involute of a circle. Hint:* Consider Figure 10.40 and find the coordinates of $P(x, y)$. Notice that

$$x = |OB|$$
$$y = |PB|$$

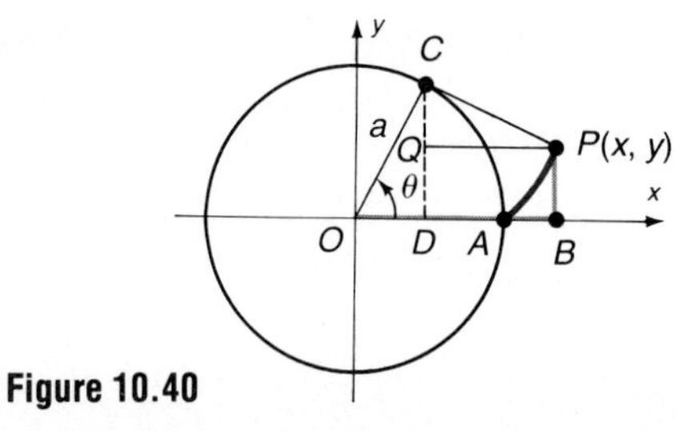

Figure 10.39 Graph of the involute of a circle.

Find x and y in terms of θ, the amount of rotation in radians.

Figure 10.40

CHAPTER 10 SUMMARY

The material of this chapter is reviewed in the following list of objectives. After each objective there are some practice questions. For a sample test, select the first question of each set and check your answers with the answer section. For a sample test without answers, use the second question of each set. Additional practice is given by the other questions in each set. If you are having trouble with a particular type of problem, look back to that section for extra help.

10.1 VECTORS *Objective 1 Find the resultant vector of two given forces.*

1. If $\mathbf{v} = 5\mathbf{i} - 2\mathbf{j}$ and $\mathbf{w} = -7\mathbf{i} + 3\mathbf{j}$, find the resultant vector.
2. If $\mathbf{v} = -3\mathbf{i} + 2\mathbf{j}$ and $\mathbf{w} = 8\mathbf{i} - 3\mathbf{j}$, find the resultant vector.
3. Consider two forces, one with magnitude 8.0 in a N20°E direction and the other with a magnitude of 9.0 in a S50°E direction. What is the resultant vector?

4. An object is hurled from a catapult due east with a velocity of 38 feet per second (fps). If the wind is blowing due south at 15 mph (22 fps), what are the true bearing and the velocity of the object?

Objective 2 Resolve a given vector and find its algebraic representation.

5. Resolve the vector with magnitude 4.5 and $\theta = 51^\circ$ into horizontal and vertical components.
6. Resolve the vector with magnitude 12 and $\theta = 30^\circ$ into horizontal and vertical components.
7. Find the algebraic representation of a vector determined by $\overrightarrow{AB}$ with $A(0,0)$ and $B(5,12)$.
8. Find the algebraic representation of a vector determined by $\overrightarrow{AB}$ with $A(-3,-5)$, $B(-6,-10)$.

Objective 3 Find the magnitude of a vector.

9. $\mathbf{v} = 5\mathbf{i} - 2\mathbf{j}$
10. $\mathbf{w} = -7\mathbf{i} + 3\mathbf{j}$
11. $\mathbf{v} = 6\mathbf{i}$
12. $\mathbf{w} = \cos 42^\circ\mathbf{i} + \sin 42^\circ\mathbf{j}$

Objective 4 Know the definition of scalar product and be able to find the scalar product of two vectors.

13. Define scalar product.
14. Find the scalar product of $\mathbf{v}$ and $\mathbf{w}$ where $\mathbf{v} = 4\mathbf{i} - 2\mathbf{j}$ and $\mathbf{w} = -7\mathbf{i} + 3\mathbf{j}$.
15. Find $(3\mathbf{i} + 5\mathbf{j}) \cdot (-2\mathbf{i} + 3\mathbf{j})$
16. Find $(e\mathbf{i} - e^{-1}\mathbf{j}) \cdot (e^2\mathbf{i} - e\mathbf{j})$

Objective 5 Find the cosine of the angle between two vectors, or use the inverse cosine function to find that angle.

17. Find the cosine of the angle between the vectors $3\mathbf{i} - 2\mathbf{j}$ and $5\mathbf{i} + 12\mathbf{j}$.
18. Find the angle between the vectors $5\mathbf{i} - 2\mathbf{j}$ and $-7\mathbf{i} + 3\mathbf{j}$.
19. Find the angle between the vectors $\cos 15^\circ\mathbf{i} + \sin 15^\circ\mathbf{j}$ and $\cos 20^\circ\mathbf{i} + \sin 20^\circ\mathbf{j}$.
20. Find the cosine of the angle between the vectors $5\mathbf{i}$ and $-3\mathbf{i} - \mathbf{j}$.

10.2 PROPERTIES OF VECTORS

Objective 6 Find a vector normal to a given line and a vector determined by a line. After you have found the vector normal to the given line and the vector determined by the line, check your answer by showing that the dot product of these vectors is 0.

21. $5x - 12y + 3 = 0$
22. $2x + 3y + 12 = 0$
23. $x - 5y + 4 = 0$
24. $x + y = 0$

Objective 7 Find scalar and vector projections. For the given vectors, find the vector and scalar projections of $\mathbf{v}$ *onto* $\mathbf{w}$.

25. $\mathbf{v} = 2\mathbf{i} + \sqrt{5}\mathbf{j}$
    $\mathbf{w} = 3\sqrt{5}\mathbf{i} + 3\mathbf{j}$
26. $\mathbf{v} = 5\mathbf{i} - 12\mathbf{j}$
    $\mathbf{w} = 3\mathbf{i} + 4\mathbf{j}$
27. $\mathbf{v} = 9\mathbf{i} + \mathbf{j}$
    $\mathbf{w} = \mathbf{i} - 9\mathbf{j}$
28. $\mathbf{v} = -2\mathbf{i} - 7\mathbf{j}$
    $\mathbf{w} = 4\mathbf{i} + 3\mathbf{j}$

Objective 8 Find the distance from a given point to a given line.

29. $5x + 12y + 8 = 0$; $(1,5)$
30. $6x - 2y + 5 = 0$; $(-5,-10)$
31. $2x - 3y - 5 = 0$; $(-10,2)$
32. $5x + 3y = 0$; $(8,10)$

Objective 9 Find a vector $\overrightarrow{AB}$ *given the points A and B.*

33. $A(0,0,0)$, $B(4,8,-2)$
34. $A(-2,3,1)$, $B(-8,5,-4)$
35. $A(6,1,2)$, $B(-2,0,5)$
36. $A(0,0,5)$, $B(8,1,0)$

10.3 VECTORS IN THREE DIMENSIONS

Objective 10 *Carry out vector operations, including magnitude, for three-dimensional vectors. Let* $\mathbf{v} = 3\mathbf{i} - 4\mathbf{j} + \mathbf{k}$ *and* $\mathbf{w} = -\mathbf{i} + 3\mathbf{j} - \mathbf{k}$.

37. $\mathbf{v} - \mathbf{w}$
38. $|\mathbf{v}| - |\mathbf{w}|$
39. $2\mathbf{v} + 3\mathbf{w}$
40. $\mathbf{v} - 2\mathbf{w}$

Objective 11 *Find the scalar product for three-dimensional vectors.*

41. $\mathbf{i} \cdot 2\mathbf{j}$
42. $(\mathbf{i} + \mathbf{j} + \mathbf{k}) \cdot (-2\mathbf{i} + 5\mathbf{k})$
43. $(2\mathbf{i} - \mathbf{j} - \mathbf{k}) \cdot (3\mathbf{i} + \mathbf{j} + \mathbf{k})$
44. $(\mathbf{i} + \mathbf{j} - 5\mathbf{k}) \cdot (4\mathbf{j} - \mathbf{k})$

Objective 12 *Find the vector product for three-dimensional vectors.*

45. $\mathbf{i} \times 2\mathbf{j}$
46. $(\mathbf{i} + \mathbf{j} + \mathbf{k}) \times (-2\mathbf{i} + 5\mathbf{k})$
47. $(2\mathbf{i} - \mathbf{j} - \mathbf{k}) \times (3\mathbf{i} + \mathbf{j} + \mathbf{k})$
48. $(\mathbf{i} + \mathbf{j} - 5\mathbf{k}) \times (4\mathbf{j} - \mathbf{k})$

Objective 13 *Find a unit vector in a given direction; find the angle between two three-dimensional vectors; apply vector operations to find scalars; find vector and scalar projections. Let* $\mathbf{v} = \mathbf{i} + \mathbf{j} - \sqrt{7}\mathbf{k}$ *and* $\mathbf{w} = -\mathbf{i} - 2\mathbf{j} + \mathbf{k}$.

49. Find a unit vector in the direction of $\mathbf{v}$.
50. Find the cosine of the angle between $\mathbf{v}$ and $\mathbf{w}$.
51. Find a scalar s such that $\mathbf{v}$ is orthogonal to $s\mathbf{v} + \mathbf{w}$.
52. Find the vector and scalar projections of $\mathbf{v}$ onto $\mathbf{w}$.

10.4 POLAR COORDINATES

Objective 14 *Plot points in polar form and give both primary representations. Use exact values if possible; otherwise approximate to four decimal places.*

53. $(5, \sqrt{75})$
54. $(3, -\frac{2\pi}{3})$
55. $(-2, 2)$
56. $(-5, 9.4248)$

Objective 15 *Change from rectangular to polar form and from polar form to rectangular form. Use exact values if possible; otherwise approximate to four decimal places.*

57. Change the polar-form point $(3, -\frac{2\pi}{3})$ to rectangular form.
58. Change the polar-form point $(5, \sqrt{75})$ to rectangular form.
59. Change the rectangular-form point $(3, -3)$ to polar form.
60. Change the rectangular-form point $(3, \sqrt{3})$ to polar form.

Objective 16 *Sketch polar-form curves.*

61. $r = 2\cos 2\theta$
62. $r = 2 + 2\cos\theta$
63. $r^2 = 25\cos 2\theta$
64. $r = \tan 45°$

Objective 17 *Identify cardioids, rose curves, and lemniscates by looking at the equation.*

65. a. $r^2 = 5\cos 2\theta$ **b.** $r = 5\cos 2\theta$
66. a. $r = 5\theta$ **b.** $r = 5\cos 3\theta$
67. a. $r = 3 + 5\cos\theta$ **b.** $r = 5 + 5\cos\theta$
68. a. $\theta = \frac{\pi}{2}$ **b.** $\frac{\pi}{2}\cos\theta = \frac{\pi}{2} - r$

*10.5 INTERSECTION OF POLAR-FORM CURVES

Objective 18 *Find the intersection of polar-form curves using both the graphing and the analytic methods.*

69. Show the intersection of the curves $r = 4 - 4\sin\theta$ and $4r = 4 - 8\sin\theta$ using the graphing method.
70. Show the intersection of the curves $r = 5\sin\theta$ and $r = 6\cos\theta$ using the graphing method.
71. Find the intersection of the curves in Problem 69 using the analytic method.
72. Find the intersection of the curves in Problem 70 using the analytic method. (Give answer correct to two significant figures.)

---

* Optional section

10.6 PARAMETRIC EQUATIONS

Objective 19 *Sketch a curve represented by parametric equations by plotting points.*

73. $x = 2 + 5t, y = -1 - 3t$

74. $x = 3\cos\theta, y = 5\sin\theta$

75. $x = e^{4t}, y = e^{4t-2}$

76. $x = t^2 + 3t - 1, y = t^2 + 2t + 5$

Objective 20 *Eliminate the parameter, if possible, to sketch a curve defined by parametric equations.*

77. $x = 2 + 5t, y = -1 - 3t$

78. $x = 3\cos\theta, y = 5\sin\theta$

79. $x = e^{4t}, y = e^{4t-2}$

80. $x = t^2 + 3t - 1, y = t^2 + 2t + 5$

APPENDIX A

Significant Digits

When you work with numbers that arise from counting, the numbers are **exact**. If you work with measurements, however, the quantities are necessarily **approximations**. The digits known to be correct in a number obtained by a measurement are called **significant digits**. The digits 1, 2, 3, 4, 5, 6, 7, 8, and 9 are always significant, whereas the digit 0 may or may not be significant, as described by the following two rules:

1. Zeros that come between two other digits are significant, as in 203 or 10.04.
2. If the only function of the zero is to place the decimal point, it is not significant, as in

$$0.\underbrace{000}_{\uparrow}23 \quad \text{or} \quad 23{,}\underbrace{000}_{\uparrow}$$

placeholders   placeholders

If it does more than fix the decimal point, it is significant, as in

$$0.0023\underset{\uparrow}{0} \quad \text{or} \quad 23{,}\underbrace{000.0}_{\uparrow}1$$

This digit is significant.   These are significant since they come between two other digits.

This second rule can, of course, result in certain ambiguities, as in 23,000 (measured to the *exact* unit). To avoid confusion, we use scientific notation in this case:

2.3×10^4 two significant digits
2.3000×10^4 five significant digits

EXAMPLE 1 Two significant digits: 46, 0.00083, 4.0×10^1, 0.050. ■

EXAMPLE 2 Three significant digits: 523, 403, 4.00×10^2, 0.000800. ■

EXAMPLE 3 Four significant digits: 600.1, 4.000×10^1, 0.0002345. ■

When we are doing calculations with approximate numbers (particularly when using a calculator), it is often necessary to round off results.

Procedure for Rounding

To round off numbers:

1. Increase the last retained digit by 1 if the residue is 5 or greater.
2. Retain the last digit unchanged if the residue is less than 5.

Elaborate rules for calculating approximate data can be developed (when it is necessary for some applications, as in chemistry), but there are two simple rules that will work satisfactorily for the material in this book.

Significant-Digit Operations

ADDITION–SUBTRACTION: Add or subtract in the usual fashion, and then round off the result so that the last digit retained is in the column farthest to the right in which both given numbers have significant digits.

MULTIPLICATION–DIVISION: Multiply or divide in the usual fashion, and then round off the result to the smaller number of significant digits found in either of the given numbers.

These rules are particularly important when we are using a calculator, since the results obtained will look much more accurate on the calculator than they actually are. Suppose we want to calculate

$$b = \frac{50}{\tan 35^\circ}$$

1. By division: From Table B.3, $\tan 35^\circ \approx 0.7002$. Thus

$$b \approx \frac{50}{0.7002} = 71.408169090\ldots$$

2. By multiplication:

$$b = \frac{50}{\tan 35^\circ} = 50 \cot 35^\circ$$

$$\approx 50(1.4281) \qquad \text{from Table B.3}$$

$$= 71.405$$

3. By calculator with algebraic logic (set to degrees): [50] [÷] [35] [tan] [=]

   By calculator with RPN logic: [50] [ENTER] [35] [tan] [÷]
   The answer is now displayed: 71.40740034

$$b = \frac{50}{\tan 35^\circ} \approx 71.40740034$$

Notice that all these answers differ. Now use the multiplication–division rule.

50: This number has one or two significant digits; there is no ambiguity if we write 5×10^1 or 5.0×10^1. If the given data include a number with a doubtful degree of accuracy, *in this book we will assume the maximum degree of accuracy*. Thus 50 has two significant digits.

tan 35°: From Table B.3 we find this number to four significant digits; on a calculator you may have 8, 10, or 12 significant digits (depending on the calculator).

The result of this division is correct to two significant digits:

$$b = \frac{50}{\tan 35^\circ} = 71$$

which agrees with all the preceding methods of solution.

In solving triangles in this text, we will assume a certain relationship in the accuracy of the measurement between the sides and the angles.

Significant Digits in Solving Triangles

Accuracy in Sides	Accuracy in Angles
Two significant digits	Nearest degree
Three significant digits	Nearest tenth of a degree
Four significant digits	Nearest hundredth of a degree

This chart means that if the data include one side given with two significant digits and another with three significant digits, the angle would be computed to the nearest degree. If one side is given to four significant digits and an angle to the nearest tenth of a degree, the other sides would be given to three significant digits and the angles computed to the nearest tenth of a degree. In general, results should not be more accurate than the least accurate item of the given data.

If you have access only to a four-function calculator, you can use Table B.3 in conjunction with your calculator. For example, to find b, you first find

$$\tan 35^\circ \approx 0.7002$$

and then calculate

$$b \approx \frac{50}{0.7002}$$ Algebraic logic: [50] [÷] [0.7002] [=]

$$\approx 71.41$$ RPN logic: [50] [ENTER] [0.7002] [÷]

or, to two significant digits, $b = 71$. Of course, one of the great advantages of calculators is that they enable us to work with much greater accuracy without having to use interpolation or tables. See Problems 22–31 of Problem Set 4.3 for some practice problems with significant digits.

APPENDIX B

Tables

TABLE B.1
Powers of e

x	e^x	e^{-x}	x	e^x	e^{-x}	x	e^x	e^{-x}
0.00	1.000	1.000	0.50	1.649	0.607	1.00	2.718	0.368
0.01	1.010	0.990	0.51	1.665	0.600	1.01	2.746	0.364
0.02	1.020	0.980	0.52	1.682	0.595	1.02	2.773	0.361
0.03	1.031	0.970	0.53	1.699	0.589	1.03	2.801	0.357
0.04	1.041	0.961	0.54	1.716	0.583	1.04	2.829	0.353
0.05	1.051	0.951	0.55	1.733	0.577	1.05	2.858	0.350
0.06	1.062	0.942	0.56	1.751	0.571	1.06	2.886	0.346
0.07	1.073	0.932	0.57	1.768	0.566	1.07	2.915	0.343
0.08	1.083	0.923	0.58	1.786	0.560	1.08	2.945	0.340
0.09	1.094	0.914	0.59	1.804	0.554	1.09	2.974	0.336
0.10	1.105	0.905	0.60	1.822	0.549	1.10	3.004	0.333
0.11	1.116	0.896	0.61	1.840	0.543	1.11	3.034	0.330
0.12	1.127	0.887	0.62	1.859	0.538	1.12	3.065	0.326
0.13	1.139	0.878	0.63	1.878	0.533	1.13	3.096	0.323
0.14	1.150	0.869	0.64	1.896	0.527	1.14	3.127	0.320
0.15	1.162	0.861	0.65	1.916	0.522	1.15	3.158	0.317
0.16	1.174	0.852	0.66	1.935	0.517	1.16	3.190	0.313
0.17	1.185	0.844	0.67	1.954	0.512	1.17	3.222	0.310
0.18	1.197	0.835	0.68	1.974	0.507	1.18	3.254	0.307
0.19	1.209	0.827	0.69	1.994	0.502	1.19	3.287	0.304
0.20	1.221	0.819	0.70	2.014	0.497	1.20	3.320	0.301
0.21	1.234	0.811	0.71	2.034	0.492	1.21	3.353	0.298
0.22	1.246	0.803	0.72	2.054	0.487	1.22	3.387	0.295
0.23	1.259	0.795	0.73	2.075	0.482	1.23	3.421	0.292
0.24	1.271	0.787	0.74	2.096	0.477	1.24	3.456	0.289
0.25	1.284	0.779	0.75	2.117	0.472	1.25	3.490	0.287
0.26	1.297	0.771	0.76	2.138	0.468	1.26	3.525	0.284
0.27	1.310	0.763	0.77	2.160	0.463	1.27	3.561	0.281
0.28	1.323	0.756	0.78	2.182	0.458	1.28	3.597	0.278
0.29	1.336	0.748	0.79	2.203	0.454	1.29	3.633	0.275
0.30	1.350	0.741	0.80	2.226	0.449	1.30	3.669	0.273
0.31	1.363	0.733	0.81	2.248	0.445	1.31	3.706	0.270
0.32	1.377	0.726	0.82	2.270	0.440	1.32	3.743	0.267
0.33	1.391	0.719	0.83	2.293	0.436	1.33	3.781	0.264
0.34	1.405	0.712	0.84	2.316	0.432	1.34	3.819	0.262
0.35	1.419	0.705	0.85	2.340	0.427	1.35	3.857	0.259
0.36	1.433	0.698	0.86	2.363	0.423	1.36	3.896	0.257
0.37	1.448	0.691	0.87	2.387	0.419	1.37	3.935	0.254
0.38	1.462	0.684	0.88	2.441	0.415	1.38	3.975	0.252
0.39	1.477	0.677	0.89	2.435	0.411	1.39	4.015	0.249
0.40	1.492	0.670	0.90	2.460	0.407	1.40	4.055	0.247
0.41	1.507	0.664	0.91	2.484	0.403	1.41	4.096	0.244
0.42	1.522	0.657	0.92	2.509	0.399	1.42	4.137	0.242
0.43	1.537	0.651	0.93	2.535	0.395	1.43	4.179	0.239
0.44	1.553	0.644	0.94	2.560	0.391	1.44	4.221	0.237
0.45	1.568	0.638	0.95	2.586	0.387	1.45	4.263	0.235
0.46	1.584	0.631	0.96	2.612	0.383	1.46	4.306	0.232
0.47	1.600	0.625	0.97	2.638	0.379	1.47	4.349	0.230
0.48	1.616	0.619	0.98	2.664	0.375	1.48	4.393	0.228
0.49	1.632	0.613	0.99	2.691	0.372	1.49	4.437	0.225

TABLE B.1 Powers of e (continued)

x	e^x	e^{-x}	x	e^x	e^{-x}	x	e^x	e^{-x}
1.50	4.482	0.223	2.00	7.389	0.135	2.50	12.182	0.082
1.51	4.527	0.221	2.01	7.463	0.134	2.51	12.305	0.081
1.52	4.572	0.219	2.02	7.538	0.133	2.52	12.429	0.080
1.53	4.618	0.217	2.03	7.614	0.131	2.53	12.554	0.080
1.54	4.665	0.214	2.04	7.691	0.130	2.54	12.680	0.079
1.55	4.712	0.212	2.05	7.768	0.129	2.55	12.807	0.078
1.56	4.759	0.210	2.06	7.846	0.127	2.56	12.936	0.077
1.57	4.807	0.208	2.07	7.925	0.126	2.57	13.066	0.077
1.58	4.855	0.206	2.08	8.004	0.125	2.58	13.197	0.076
1.59	4.904	0.204	2.09	8.085	0.124	2.59	13.330	0.075
1.60	4.953	0.202	2.10	8.166	0.122	2.60	13.464	0.074
1.61	5.003	0.200	2.11	8.248	0.121	2.61	13.599	0.074
1.62	5.053	0.198	2.12	8.331	0.120	2.62	13.736	0.073
1.63	5.104	0.196	2.13	8.415	0.119	2.63	13.874	0.072
1.64	5.155	0.194	2.14	8.499	0.118	2.64	14.013	0.071
1.65	5.207	0.192	2.15	8.585	0.116	2.65	14.154	0.071
1.66	5.259	0.190	2.16	8.671	0.115	2.66	14.296	0.070
1.67	5.312	0.188	2.17	8.758	0.114	2.67	14.440	0.069
1.68	5.366	0.186	2.18	8.846	0.113	2.68	14.585	0.069
1.69	5.420	0.185	2.19	8.935	0.112	2.69	14.732	0.068
1.70	5.474	0.183	2.20	9.025	0.111	2.70	14.880	0.067
1.71	5.529	0.181	2.21	9.116	0.110	2.71	15.029	0.067
1.72	5.585	0.179	2.22	9.207	0.109	2.72	15.180	0.066
1.73	5.641	0.177	2.23	9.300	0.108	2.73	15.333	0.065
1.74	5.697	0.176	2.24	9.393	0.106	2.74	15.487	0.065
1.75	5.755	0.174	2.25	9.488	0.105	2.75	15.643	0.064
1.76	5.812	0.172	2.26	9.583	0.104	2.76	15.800	0.063
1.77	5.871	0.170	2.27	9.679	0.103	2.77	15.959	0.063
1.78	5.930	0.169	2.28	9.777	0.102	2.78	16.119	0.062
1.79	5.989	0.167	2.29	9.875	0.101	2.79	16.281	0.061
1.80	6.050	0.165	2.30	9.974	0.100	2.80	16.445	0.061
1.81	6.110	0.164	2.31	10.074	0.099	2.81	16.610	0.060
1.82	6.172	0.162	2.32	10.176	0.098	2.82	16.777	0.060
1.83	6.234	0.160	2.33	10.278	0.097	2.83	16.945	0.059
1.84	6.297	0.159	2.34	10.381	0.096	2.84	17.116	0.058
1.85	6.360	0.157	2.35	10.486	0.095	2.85	17.288	0.058
1.86	6.424	0.156	2.36	10.591	0.094	2.86	17.462	0.057
1.87	6.488	0.154	2.37	10.697	0.093	2.87	17.637	0.057
1.88	6.553	0.153	2.38	10.805	0.093	2.88	17.814	0.056
1.89	6.619	0.151	2.39	10.913	0.092	2.89	17.993	0.056
1.90	6.686	0.150	2.40	11.023	0.091	2.90	18.174	0.055
1.91	6.753	0.148	2.41	11.134	0.090	2.91	18.357	0.054
1.92	6.821	0.147	2.42	11.246	0.089	2.92	18.541	0.054
1.93	6.890	0.145	2.43	11.359	0.088	2.93	18.728	0.053
1.94	6.959	0.144	2.44	11.473	0.087	2.94	18.916	0.053
1.95	7.029	0.142	2.45	11.588	0.086	2.95	19.016	0.052
1.96	7.099	0.141	2.46	11.705	0.085	2.96	19.298	0.052
1.97	7.171	0.139	2.47	11.822	0.085	2.97	19.492	0.051
1.98	7.243	0.138	2.48	11.941	0.084	2.98	19.688	0.051
1.99	7.316	0.137	2.49	12.061	0.083	2.99	19.886	0.050
						3.00	20.086	0.050

TABLE B.2
Common logarithms

N	0	1	2	3	4	5	6	7	8	9
1.0	.0000	.0043	.0086	.0128	.0170	.0212	.0253	.0294	.0334	.0374
1.1	.0414	.0453	.0492	.0531	.0569	.0607	.0645	.0682	.0719	.0755
1.2	.0792	.0828	.0864	.0899	.0934	.0969	.1004	.1038	.1072	.1106
1.3	.1139	.1173	.1206	.1239	.1271	.1303	.1335	.1367	.1399	.1430
1.4	.1461	.1492	.1523	.1553	.1584	.1614	.1644	.1673	.1703	.1732
1.5	.1761	.1790	.1818	.1847	.1875	.1903	.1931	.1959	.1987	.2014
1.6	.2041	.2068	.2095	.2122	.2148	.2175	.2201	.2227	.2253	.2279
1.7	.2304	.2330	.2355	.2380	.2405	.2430	.2455	.2480	.2504	.2529
1.8	.2553	.2577	.2601	.2625	.2648	.2672	.2695	.2718	.2742	.2765
1.9	.2788	.2810	.2833	.2856	.2878	.2900	.2923	.2945	.2967	.2989
2.0	.3010	.3032	.3054	.3075	.3096	.3118	.3139	.3160	.3181	.3201
2.1	.3222	.3243	.3263	.3284	.3304	.3324	.3345	.3365	.3385	.3404
2.2	.3424	.3444	.3464	.3483	.3502	.3522	.3541	.3560	.3579	.3598
2.3	.3617	.3636	.3655	.3674	.3692	.3711	.3729	.3747	.3766	.3784
2.4	.3802	.3820	.3838	.3856	.3874	.3892	.3909	.3927	.3945	.3962
2.5	.3979	.3997	.4014	.4031	.4048	.4065	.4082	.4099	.4116	.4133
2.6	.4150	.4166	.4183	.4200	.4216	.4232	.4249	.4265	.4281	.4298
2.7	.4314	.4330	.4346	.4362	.4378	.4393	.4409	.4425	.4440	.4456
2.8	.4472	.4487	.4502	.4518	.4533	.4548	.4564	.4579	.4594	.4609
2.9	.4624	.4639	.4654	.4669	.4683	.4698	.4713	.4728	.4742	.4757
3.0	.4771	.4786	.4800	.4814	.4829	.4843	.4857	.4871	.4886	.4900
3.1	.4914	.4928	.4942	.4955	.4969	.4983	.4997	.5011	.5024	.5038
3.2	.5051	.5065	.5079	.5092	.5105	.5119	.5132	.5145	.5159	.5172
3.3	.5185	.5198	.5211	.5224	.5237	.5250	.5263	.5276	.5289	.5302
3.4	.5315	.5328	.5340	.5353	.5366	.5378	.5391	.5403	.5416	.5428
3.5	.5441	.5453	.5465	.5478	.5490	.5502	.5514	.5527	.5539	.5551
3.6	.5563	.5575	.5587	.5599	.5611	.5623	.5635	.5647	.5658	.5670
3.7	.5682	.5694	.5705	.5717	.5729	.5740	.5752	.5763	.5775	.5786
3.8	.5798	.5809	.5821	.5832	.5843	.5855	.5866	.5877	.5888	.5899
3.9	.5911	.5922	.5933	.5944	.5955	.5966	.5977	.5988	.5999	.6010
4.0	.6021	.6031	.6042	.6053	.6064	.6075	.6085	.6096	.6107	.6117
4.1	.6128	.6138	.6149	.6160	.6170	.6180	.6191	.6201	.6212	.6222
4.2	.6232	.6243	.6253	.6263	.6274	.6284	.6294	.6304	.6314	.6325
4.3	.6335	.6345	.6355	.6365	.6375	.6385	.6395	.6405	.6415	.6425
4.4	.6435	.6444	.6454	.6464	.6474	.6484	.6493	.6503	.6513	.6522
4.5	.6532	.6542	.6551	.6561	.6571	.6580	.6590	.6599	.6609	.6618
4.6	.6628	.6637	.6646	.6656	.6665	.6675	.6684	.6693	.6702	.6712
4.7	.6721	.6730	.6739	.6749	.6758	.6767	.6776	.6785	.6794	.6803
4.8	.6812	.6821	.6830	.6839	.6848	.6857	.6866	.6875	.6884	.6893
4.9	.6902	.6911	.6920	.6928	.6937	.6946	.6955	.6964	.6972	.6981
5.0	.6990	.6998	.7007	.7016	.7024	.7033	.7042	.7050	.7059	.7067
5.1	.7076	.7084	.7093	.7101	.7110	.7118	.7126	.7135	.7143	.7152
5.2	.7160	.7168	.7177	.7185	.7193	.7202	7210	.7218	.7226	.7235
5.3	.7243	.7251	.7259	.7267	.7275	.7284	.7292	.7300	.7308	.7316
5.4	.7324	.7332	.7340	.7348	.7356	.7364	.7372	.7380	.7388	.7396
N	0	1	2	3	4	5	6	7	8	9

TABLE B.2 Common logarithms (continued)

N	0	1	2	3	4	5	6	7	8	9
5.5	.7404	.7412	.7419	.7427	.7435	.7443	.7451	.7459	.7466	.7474
5.6	.7482	.7490	.7497	.7505	.7513	.7520	.7528	.7536	.7543	.7551
5.7	.7559	.7566	.7574	.7582	.7589	.7597	.7604	.7612	.7619	.7627
5.8	.7634	.7642	.7649	.7657	.7664	.7672	.7679	.7686	.7694	.7701
5.9	.7709	.7716	.7723	.7731	.7738	.7745	.7752	.7760	.7767	.7774
6.0	.7782	.7789	.7796	.7803	.7810	.7818	.7825	.7832	.7839	.7846
6.1	.7853	.7860	.7868	.7875	.7882	.7889	.7896	.7903	.7910	.7917
6.2	.7924	.7931	.7938	.7945	.7952	.7959	.7966	.7973	.7980	.7987
6.3	.7993	.8000	.8007	.8014	.8021	.8028	.8035	.8041	.8048	.8055
6.4	.8062	.8069	.8075	.8082	.8089	.8096	.8102	.8109	.8116	.8122
6.5	.8129	.8136	.8142	.8149	.8156	.8162	.8169	.8176	.8182	.8189
6.6	.8195	.8202	.8209	.8215	.8222	.8228	.8235	.8241	.8248	.8254
6.7	.8261	.8267	.8274	.8280	.8287	.8293	.8299	.8306	.8312	.8319
6.8	.8325	.8331	.8338	.8344	.8351	.8357	.8363	.8370	.8376	.8382
6.9	.8388	.8395	.8401	.8407	.8414	.8420	.8426	.8432	.8439	.8445
7.0	.8451	.8457	.8463	.8470	.8476	.8482	.8488	.8494	.8500	.8506
7.1	.8513	.8519	.8525	.8531	.8537	.8543	.8549	.8555	.8561	.8567
7.2	.8573	.8579	.8585	.8591	.8597	.8603	.8609	.8615	.8621	.8627
7.3	.8633	.8639	.8645	.8651	.8657	.8663	.8669	.8675	.8681	.8686
7.4	.8692	.8698	.8704	.8710	.8716	.8722	.8727	.8733	.8739	.8745
7.5	.8751	.8756	.8762	.8768	.8774	.8779	.8785	.8791	.8797	.8802
7.6	.8808	.8814	.8820	.8825	.8831	.8837	.8842	.8848	.8854	.8859
7.7	.8865	.8871	.8876	.8882	.8887	.8893	.8899	.8904	.8910	.8915
7.8	.8921	.8927	.8932	.8938	.8943	.8949	.8954	.8960	.8965	.8971
7.9	.8976	.8982	.8987	.8993	.8998	.9004	.9009	.9015	.9020	.9025
8.0	.9031	.9036	.9042	.9047	.9053	.9058	.9063	.9069	.9074	.9079
8.1	.9085	.9090	.9096	.9101	.9106	.9112	.9117	.9122	.9128	.9133
8.2	.9138	.9143	.9149	.9154	.9159	.9165	.9170	.9175	.9180	.9186
8.3	.9191	.9196	.9201	.9206	.9212	.9217	.9222	.9227	.9232	.9238
8.4	.9243	.9248	.9253	.9258	.9263	.9269	.9274	.9279	.9284	.9289
8.5	.9294	.9299	.9304	.9309	.9315	.9320	.9325	.9330	.9335	.9340
8.6	.9345	.9350	.9355	.9360	.9365	.9370	.9375	.9380	.9385	.9390
8.7	.9395	.9400	.9405	.9410	.9415	.9420	.9425	.9430	.9435	.9440
8.8	.9445	.9450	.9455	.9460	.9465	.9469	.9474	.9479	.9484	.9489
8.9	.9494	.9499	.9504	.9509	.9513	.9518	.9523	.9528	.9533	.9538
9.0	.9542	.9547	.9552	.9557	.9562	.9566	.9571	.9576	.9581	.9586
9.1	.9590	.9595	.9600	.9605	.9609	.9614	.9619	.9624	.9628	.9633
9.2	.9638	.9643	.9647	.9652	.9657	.9661	.9666	.9671	.9675	.9680
9.3	.9685	.9689	.9694	.9699	.9703	.9708	.9713	.9717	.9722	.9727
9.4	.9731	.9736	.9741	.9745	.9750	.9754	.9759	.9763	.9768	.9773
9.5	.9777	.9782	.9786	.9791	.9795	.9800	.9805	.9809	.9814	.9818
9.6	.9823	.9827	.9832	.9836	.9841	.9845	.9850	.9854	.9859	.9863
9.7	.9868	.9872	.9877	.9881	.9886	.9890	.9894	.9899	.9903	.9908
9.8	.9912	.9917	.9921	.9926	.9930	.9934	.9939	.9943	.9948	.9952
9.9	.9956	.9961	.9965	.9969	.9974	.9978	.9983	.9987	.9991	.9996
N	0	1	2	3	4	5	6	7	8	9

TABLE B.3 Trigonometric functions

Rad	Deg	cos	sin	tan		
.00	**.0**	1.0000	.0000	.0000	**90.0**	1.57
.00	.1	1.0000	.0017	.0017	89.9	1.57
.00	.2	1.0000	.0035	.0035	89.8	1.57
.01	.3	1.0000	.0052	.0052	89.7	1.57
.01	.4	1.0000	.0070	.0070	89.6	1.56
.01	**.5**	1.0000	.0087	.0087	**89.5**	1.56
.01	.6	.9999	.0105	.0105	89.4	1.56
.01	.7	.9999	.0122	.0122	89.3	1.56
.01	.8	.9999	.0140	.0140	89.2	1.56
.02	.9	.9999	.0157	.0157	89.1	1.56
.02	**1.0**	.9998	.0175	.0175	**89.0**	1.55
.02	1.1	.9998	.0192	.0192	88.9	1.55
.02	1.2	.9998	.0209	.0209	88.8	1.55
.02	1.3	.9997	.0227	.0227	88.7	1.55
.02	1.4	.9997	.0244	.0244	88.6	1.55
.03	**1.5**	.9997	.0262	.0262	**88.5**	1.54
.03	1.6	.9996	.0279	.0279	88.4	1.54
.03	1.7	.9996	.0297	.0297	88.3	1.54
.03	1.8	.9995	.0314	.0314	88.2	1.54
.03	1.9	.9995	.0332	.0332	88.1	1.54
.03	**2.0**	.9994	.0349	.0349	**88.0**	1.54
.04	2.1	.9993	.0366	.0367	87.9	1.53
.04	2.2	.9993	.0384	.0384	87.8	1.53
.04	2.3	.9992	.0401	.0402	87.7	1.53
.04	2.4	.9991	.0419	.0419	87.6	1.53
.04	**2.5**	.9990	.0436	.0437	**87.5**	1.53
.05	2.6	.9990	.0454	.0454	87.4	1.53
.05	2.7	.9989	.0471	.0472	87.3	1.52
.05	2.8	.9988	.0488	.0489	87.2	1.52
.05	2.9	.9987	.0506	.0507	87.1	1.52
.05	**3.0**	.9986	.0523	.0524	**87.0**	1.52
.05	3.1	.9985	.0541	.0542	86.9	1.52
.06	3.2	.9984	.0558	.0559	86.8	1.51
.06	3.3	.9983	.0576	.0577	86.7	1.51
.06	3.4	.9982	.0593	.0594	86.6	1.51
.06	**3.5**	.9981	.0610	.0612	**86.5**	1.51
.06	3.6	.9980	.0628	.0629	86.4	1.51
.06	3.7	.9979	.0645	.0647	86.3	1.51
.07	3.8	.9978	.0663	.0664	86.2	1.50
.07	3.9	.9977	.0680	.0682	86.1	1.50
.07	**4.0**	.9976	.0698	.0699	**86.0**	1.50
.07	4.1	.9974	.0715	.0717	85.9	1.50
.07	4.2	.9973	.0732	.0734	85.8	1.50
.08	4.3	.9972	.0750	.0752	85.7	1.50
.08	4.4	.9971	.0767	.0769	85.6	1.49
.08	**4.5**	.9969	.0785	.0787	**85.5**	1.49
.08	4.6	.9968	.0802	.0805	85.4	1.49
.08	4.7	.9966	.0819	.0822	85.3	1.49
.08	4.8	.9965	.0837	.0840	85.2	1.49
.09	4.9	.9963	.0854	.0857	85.1	1.49
.09	**5.0**	.9962	.0872	.0875	**85.0**	1.49
		sin	cos	cot	Deg	Rad

Rad	Deg	cos	sin	tan		
.09	**5.0**	.9962	.0872	.0875	**85.0**	1.48
.09	5.1	.9960	.0889	.0892	84.9	1.48
.09	5.2	.9959	.0906	.0910	84.8	1.48
.09	5.3	.9957	.0924	.0928	84.7	1.48
.09	5.4	.9956	.0941	.0945	84.6	1.48
.10	**5.5**	.9954	.0958	.0963	**84.5**	1.47
.10	5.6	.9952	.0976	.0981	84.4	1.47
.10	5.7	.9951	.0993	.0998	84.3	1.47
.10	5.8	.9949	.1011	.1016	84.2	1.47
.10	5.9	.9947	.1028	.1033	84.1	1.47
.10	**6.0**	.9945	.1045	.1051	**84.0**	1.47
.11	6.1	.9943	.1063	.1069	83.9	1.46
.11	6.2	.9942	.1080	.1086	83.8	1.46
.11	6.3	.9940	.1097	.1104	83.7	1.46
.11	6.4	.9938	.1115	.1122	83.6	1.46
.11	**6.5**	.9936	.1132	.1139	**83.5**	1.46
.12	6.6	.9934	.1149	.1157	83.4	1.46
.12	6.7	.9932	.1167	.1175	83.3	1.45
.12	6.8	.9930	.1184	.1192	83.2	1.45
.12	6.9	.9928	.1201	.1210	83.1	1.45
.12	**7.0**	.9925	.1219	.1228	**83.0**	1.45
.12	7.1	.9923	.1236	.1246	82.9	1.45
.13	7.2	.9921	.1253	.1263	82.8	1.45
.13	7.3	.9919	.1271	.1281	82.7	1.44
.13	7.4	.9917	.1288	.1299	82.6	1.44
.13	**7.5**	.9914	.1305	.1317	**82.5**	1.44
.13	7.6	.9912	.1323	.1334	82.4	1.44
.13	7.7	.9910	.1340	.1352	82.3	1.44
.14	7.8	.9907	.1357	.1370	82.2	1.43
.14	7.9	.9905	.1374	.1388	82.1	1.43
.14	**8.0**	.9903	.1392	.1405	**82.0**	1.43
.14	8.1	.9900	.1409	.1423	81.9	1.43
.14	8.2	.9898	.1426	.1441	81.8	1.43
.14	8.3	.9895	.1444	.1459	81.7	1.43
.15	8.4	.9893	.1461	.1477	81.6	1.42
.15	**8.5**	.9890	.1478	.1495	**81.5**	1.42
.15	8.6	.9888	.1495	.1512	81.4	1.42
.15	8.7	.9885	.1513	.1530	81.3	1.42
.15	8.8	.9882	.1530	.1548	81.2	1.42
.16	8.9	.9880	.1547	.1566	81.1	1.42
.16	**9.0**	.9877	.1564	.1584	**81.0**	1.41
.16	9.1	.9874	.1582	.1602	80.9	1.41
.16	9.2	.9871	.1599	.1620	80.8	1.41
.16	9.3	.9869	.1616	.1638	80.7	1.41
.16	9.4	.9866	.1633	.1655	80.6	1.41
.17	**9.5**	.9863	.1650	.1673	**80.5**	1.40
.17	9.6	.9860	.1668	.1691	80.4	1.40
.17	9.7	.9857	.1685	.1709	80.3	1.40
.17	9.8	.9854	.1702	.1727	80.2	1.40
.17	9.9	.9851	.1719	.1745	80.1	1.40
.17	**10.0**	.9848	.1736	.1763	**80.0**	1.40
		sin	cos	cot	Deg	Rad

TABLE B.3 Trigonometric functions (continued)

Rad	Deg	cos	sin	tan		
.17	**10.0**	.9848	.1736	.1763	**80.0**	1.40
.18	10.1	.9845	.1754	.1781	79.9	1.39
.18	10.2	.9842	.1771	.1799	79.8	1.39
.18	10.3	.9839	.1788	.1817	79.7	1.39
.18	10.4	.9836	.1805	.1835	79.6	1.39
.18	**10.5**	.9833	.1822	.1853	**79.5**	1.39
.19	10.6	.9829	.1840	.1871	79.4	1.39
.19	10.7	.9826	.1857	.1890	79.3	1.38
.19	10.8	.9823	.1874	.1908	79.2	1.38
.19	10.9	.9820	.1891	.1926	79.1	1.38
.19	**11.0**	.9816	.1908	.1944	**79.0**	1.38
.19	11.1	.9813	.1925	.1962	78.9	1.38
.20	11.2	.9810	.1942	.1980	78.8	1.38
.20	11.3	.9806	.1959	.1998	78.7	1.37
.20	11.4	.9803	.1977	.2016	78.6	1.37
.20	**11.5**	.9799	.1994	.2035	**78.5**	1.37
.20	11.6	.9796	.2011	.2053	78.4	1.37
.20	11.7	.9792	.2028	.2071	78.3	1.37
.21	11.8	.9789	.2045	.2089	78.2	1.36
.21	11.9	.9785	.2062	.2107	78.1	1.36
.21	**12.0**	.9871	.2079	.2126	**78.0**	1.36
.21	12.1	.9778	.2096	.2144	77.9	1.36
.21	12.2	.9774	.2113	.2162	77.8	1.36
.21	12.3	.9770	.2130	.2180	77.7	1.36
.22	12.4	.9767	.2147	.2199	77.6	1.35
.22	**12.5**	.9763	.2164	.2217	**77.5**	1.35
.22	12.6	.9759	.2181	.2235	77.4	1.35
.22	12.7	.9755	.2198	.2254	77.3	1.35
.22	12.8	.9751	.2215	.2272	77.2	1.35
.23	12.9	.9748	.2233	.2290	77.1	1.35
.23	**13.0**	.9744	.2250	.2309	**77.0**	1.34
.23	13.1	.9740	.2267	.2327	76.9	1.34
.23	13.2	.9736	.2284	.2345	76.8	1.34
.23	13.3	.9732	.2300	.2364	76.7	1.34
.23	13.4	.9728	.2317	.2382	76.6	1.34
.24	**13.5**	.9724	.2334	.2401	**76.5**	1.34
.24	13.6	.9720	.2351	.2419	76.4	1.33
.24	13.7	.9715	.2368	.2438	76.3	1.33
.24	13.8	.9711	.2385	.2456	76.2	1.33
.24	13.9	.9707	.2402	.2475	76.1	1.33
.24	**14.0**	.9703	.2419	.2493	**76.0**	1.33
.25	14.1	.9699	.2436	.2512	75.9	1.32
.25	14.2	.9694	.2453	.2530	75.8	1.32
.25	14.3	.9690	.2470	.2549	75.7	1.32
.25	14.4	.9686	.2487	.2568	75.6	1.32
.25	**14.5**	.9681	.2504	.2586	**75.5**	1.32
.25	14.6	.9677	.2521	.2605	75.4	1.32
.26	14.7	.9673	.2538	.2623	75.3	1.31
.26	14.8	.9668	.2554	.2642	75.2	1.31
.26	14.9	.9664	.2571	.2661	75.1	1.31
.26	**15.0**	.9659	.2588	.2679	**75.0**	1.31
		sin	cos	cot	Deg	Rad

Rad	Deg	cos	sin	tan		
.26	**15.0**	.9659	.2588	.2679	**75.0**	1.31
.26	15.1	.9655	.2605	.2698	74.9	1.31
.27	15.2	.9650	.2622	.2717	74.8	1.31
.27	15.3	.9646	.2639	.2736	74.7	1.30
.27	15.4	.9641	.2656	.2754	74.6	1.30
.27	**15.5**	.9636	.2672	.2773	**74.5**	1.30
.27	15.6	.9632	.2689	.2792	74.4	1.30
.27	15.7	.9627	.2706	.2811	74.3	1.30
.28	15.8	.9622	.2723	.2830	74.2	1.30
.28	15.9	.9617	.2740	.2849	74.1	1.29
.28	**16.0**	.9613	.2756	.2867	**74.0**	1.29
.28	16.1	.9608	.2773	.2886	73.9	1.29
.28	16.2	.9603	.2790	.2905	73.8	1.29
.28	16.3	.9598	.2807	.2924	73.7	1.29
.29	16.4	.9593	.2823	.2943	73.6	1.28
.29	**16.5**	.9588	.2840	.2962	**73.5**	1.28
.29	16.6	.9583	.2857	.2981	73.4	1.28
.29	16.7	.9578	.2874	.3000	73.3	1.28
.29	16.8	.9573	.2890	.3019	73.2	1.28
.29	16.9	.9568	.2907	.3038	73.1	1.28
.30	**17.0**	.9563	.2924	.3057	**73.0**	1.27
.30	17.1	.9558	.2940	.3076	72.9	1.27
.30	17.2	.9553	.2957	.3096	72.8	1.27
.30	17.3	.9548	.2974	.3115	72.7	1.27
.30	17.4	.9542	.2990	.3134	72.6	1.27
.31	**17.5**	.9537	.3007	.3153	**72.5**	1.27
.31	17.6	.9532	.3024	.3172	72.4	1.26
.31	17.7	.9527	.3040	.3191	72.3	1.26
.31	17.8	.9521	.3057	.3211	72.2	1.26
.31	17.9	.9516	.3074	.3230	72.1	1.26
.31	**18.0**	.9511	.3090	.3249	**72.0**	1.26
.32	18.1	.9505	.3107	.3269	71.9	1.25
.32	18.2	.9500	.3123	.3288	71.8	1.25
.32	18.3	.9494	.3140	.3307	71.7	1.25
.32	18.4	.9489	.3156	.3327	71.6	1.25
.32	**18.5**	.9483	.3173	.3346	**71.5**	1.25
.32	18.6	.9478	.3190	.3365	71.4	1.25
.33	18.7	.9472	.3206	.3385	71.3	1.24
.33	18.8	.9466	.3223	.3404	71.2	1.24
.33	18.9	.9461	.3239	.3424	71.1	1.24
.33	**19.0**	.9455	.3256	.3443	**71.0**	1.24
.33	19.1	.9449	.3272	.3463	70.9	1.24
.34	19.2	.9444	.3289	.3482	70.8	1.24
.34	19.3	.9438	.3305	.3502	70.7	1.23
.34	19.4	.9432	.3322	.3522	70.6	1.23
.34	**19.5**	.9426	.3338	.3541	**70.5**	1.23
.34	19.6	.9421	.3355	.3561	70.4	1.23
.34	19.7	.9415	.3371	.3581	70.3	1.23
.35	19.8	.9409	.3387	.3600	70.2	1.23
.35	19.9	.9403	.3404	.3620	70.1	1.22
.35	**20.0**	.9397	.3420	.3640	**70.0**	1.22
		sin	cos	cot	Deg	Rad

TABLE B.3 Trigonometric functions (continued)

Rad	Deg	cos	sin	tan		
.35	**20.0**	.9397	.3420	.3640	**70.0**	1.22
.35	20.1	.9391	.3437	.3659	69.9	1.22
.35	20.2	.9385	.3453	.3679	69.8	1.22
.35	20.3	.9379	.3469	.3699	69.7	1.22
.36	20.4	.9373	.3486	.3719	69.6	1.21
.36	**20.5**	.9367	.3502	.3739	**69.5**	1.21
.36	20.6	.9361	.3518	.3759	69.4	1.21
.36	20.7	.9354	.3535	.3779	69.3	1.21
.36	20.8	.9348	.3551	.3799	69.2	1.21
.36	20.9	.9342	.3567	.3819	69.1	1.21
.37	**21.0**	.9336	.3584	.3839	**69.0**	1.20
.37	21.1	.9330	.3600	.3859	68.9	1.20
.37	21.2	.9323	.3616	.3879	68.8	1.20
.37	21.3	.9317	.3633	.3899	68.7	1.20
.37	21.4	.9311	.3649	.3919	68.6	1.20
.38	**21.5**	.9304	.3665	.3939	**68.5**	1.20
.38	21.6	.9298	.3681	.3959	68.4	1.19
.38	21.7	.9291	.3697	.3979	68.3	1.19
.38	21.8	.9285	.3714	.4000	68.2	1.19
.38	21.9	.9278	.3730	.4020	68.1	1.19
.38	**22.0**	.9272	.3746	.4040	**68.0**	1.19
.39	22.1	.9265	.3762	.4061	67.9	1.19
.39	22.2	.9259	.3778	.4081	67.8	1.18
.39	22.3	.9252	.3795	.4101	67.7	1.18
.39	22.4	.9245	.3811	.4122	67.6	1.18
.39	**22.5**	.9239	.3827	.4142	**67.5**	1.18
.39	22.6	.9232	.3843	.4163	67.4	1.18
.40	22.7	.9225	.3859	.4183	67.3	1.17
.40	22.8	.9219	.3875	.4204	67.2	1.17
.40	22.9	.9212	.3891	.4224	67.1	1.17
.40	**23.0**	.9205	.3907	.4245	**67.0**	1.17
.40	23.1	.9198	.3923	.4265	66.9	1.17
.40	23.2	.9191	.3939	.4286	66.8	1.17
.41	23.3	.9184	.3955	.4307	66.7	1.16
.41	23.4	.9178	.3971	.4327	66.6	1.16
.41	**23.5**	.9171	.3987	.4348	**66.5**	1.16
.41	23.6	.9164	.4003	.4369	66.4	1.16
.41	23.7	.9157	.4019	.4390	66.3	1.16
.42	23.8	.9150	.4035	.4411	66.2	1.16
.42	23.9	.9143	.4051	.4431	66.1	1.15
.42	**24.0**	.9135	.4067	.4452	**66.0**	1.15
.42	24.1	.9128	.4083	.4473	65.9	1.15
.42	24.2	.9121	.4099	.4494	65.8	1.15
.42	24.3	.9114	.4115	.4515	65.7	1.15
.43	24.4	.9107	.4131	.4536	65.6	1.14
.43	**24.5**	.9100	.4147	.4557	**65.5**	1.14
.43	24.6	.9092	.4163	.4578	65.4	1.14
.43	24.7	.9085	.4179	.4599	65.3	1.14
.43	24.8	.9078	.4195	.4621	65.2	1.14
.43	24.9	.9070	.4210	.4642	65.1	1.14
.44	**25.0**	.9063	.4226	.4663	**65.0**	1.14
		sin	cos	cot	Deg	Rad

Rad	Deg	cos	sin	tan		
.44	**25.0**	.9063	.4226	.4663	**65.0**	1.13
.44	25.1	.9056	.4242	.4684	64.9	1.13
.44	25.2	.9048	.4258	.4706	64.8	1.13
.45	25.3	.9041	.4274	.4727	64.7	1.13
.45	25.4	.9033	.4289	.4748	64.6	1.13
.45	**25.5**	.9026	.4305	.4770	**64.5**	1.13
.45	25.6	.9018	.4321	.4791	64.4	1.12
.45	25.7	.9011	.4337	.4813	64.3	1.12
.45	25.8	.9003	.4352	.4834	64.2	1.12
.45	25.9	.8996	.4368	.4856	64.1	1.12
.45	**26.0**	.8988	.4384	.4877	**64.0**	1.12
.46	26.1	.8980	.4399	.4899	63.9	1.12
.46	26.2	.8973	.4415	.4921	63.8	1.11
.46	26.3	.8965	.4431	.4942	63.7	1.11
.46	26.4	.8957	.4446	.4964	63.6	1.11
.46	**26.5**	.8949	.4462	.4986	**63.5**	1.11
.46	26.6	.8942	.4478	.5008	63.4	1.11
.47	26.7	.8934	.4493	.5029	63.3	1.10
.47	26.8	.8926	.4509	.5051	63.2	1.10
.47	26.9	.8918	.4524	.5073	63.1	1.10
.47	**27.0**	.8910	.4540	.5095	**63.0**	1.10
.47	27.1	.8902	.4555	.5117	62.9	1.10
.47	27.2	.8894	.4571	.5139	62.8	1.10
.48	27.3	.8886	.4586	.5161	62.7	1.09
.48	27.4	.8878	.4602	.5184	62.6	1.09
.48	**27.5**	.8870	.4617	.5206	**62.5**	1.09
.48	27.6	.8862	.4633	.5228	62.4	1.09
.48	27.7	.8854	.4648	.5250	62.3	1.09
.49	27.8	.8846	.4664	.5272	62.2	1.09
.49	27.9	.8838	.4679	.5295	62.1	1.08
.49	**28.0**	.8829	.4695	.5317	**62.0**	1.08
.49	28.1	.8821	.4710	.5340	61.9	1.08
.49	28.2	.8813	.4726	.5362	61.8	1.08
.49	28.3	.8805	.4741	.5384	61.7	1.08
.50	28.4	.8796	.4756	.5407	61.6	1.08
.50	**28.5**	.8788	.4772	.5430	**61.5**	1.07
.50	28.6	.8780	.4787	.5452	61.4	1.07
.50	28.7	.8771	.4802	.5475	61.3	1.07
.50	28.8	.8763	.4818	.5498	61.2	1.07
.50	28.9	.8755	.4833	.5520	61.1	1.07
.51	**29.0**	.8746	.4848	.5543	**61.0**	1.06
.51	29.1	.8738	.4863	.5566	60.9	1.06
.51	29.2	.8729	.4879	.5589	60.8	1.06
.51	29.3	.8721	.4894	.5612	60.7	1.06
.51	29.4	.8712	.4909	.5635	60.6	1.06
.51	**29.5**	.8704	.4924	.5658	**60.5**	1.06
.52	29.6	.8695	.4939	.5681	60.4	1.05
.52	29.7	.8686	.4955	.5704	60.3	1.05
.52	29.8	.8678	.4970	.5727	60.2	1.05
.52	29.9	.8669	.4985	.5750	60.1	1.05
.52	**30.0**	.8660	.5000	.5774	**60.0**	1.05
		sin	cos	cot	Deg	Rad

TABLE B.3 Trigonometric functions (continued)

Rad	Deg	cos	sin	tan		
.52	**30.0**	.8660	.5000	.5774	**60.0**	1.05
.53	30.1	.8652	.5015	.5797	59.9	1.05
.53	30.2	.8643	.5030	.5820	59.8	1.04
.53	30.3	.8634	.5045	.5844	59.7	1.04
.53	30.4	.8625	.5060	.5867	59.6	1.04
.53	**30.5**	.8616	.5075	.5890	**59.5**	1.04
.53	30.6	.8607	.5090	.5914	59.4	1.04
.54	30.7	.8599	.5105	.5938	59.3	1.03
.54	30.8	.8590	.5120	.5961	59.2	1.03
.54	30.9	.8581	.5135	.5985	59.1	1.03
.54	**31.0**	.8572	.5150	.6009	**59.0**	1.03
.54	31.1	.8563	.5165	.6032	58.9	1.03
.54	31.2	.8554	.5180	.6056	58.8	1.03
.55	31.3	.8545	.5195	.6080	58.7	1.02
.55	31.4	.8536	.5210	.6104	58.6	1.02
.55	**31.5**	.8526	.5225	.6128	**58.5**	1.02
.55	31.6	.8517	.5240	.6152	58.4	1.02
.55	31.7	.8508	.5255	.6176	58.3	1.02
.56	31.8	.8499	.5270	.6200	58.2	1.02
.56	31.9	.8490	.5284	.6224	58.1	1.01
.56	**32.0**	.8480	.5299	.6249	**58.0**	1.01
.56	32.1	.8471	.5314	.6273	57.9	1.01
.56	32.2	.8462	.5329	.6297	57.8	1.01
.56	32.3	.8453	.5344	.6322	57.7	1.01
.57	32.4	.8443	.5358	.6346	57.6	1.01
.57	**32.5**	.8434	.5373	.6371	**57.5**	1.00
.57	32.6	.8425	.5388	.6395	57.4	1.00
.57	32.7	.8415	.5402	.6420	57.3	1.00
.57	32.8	.8406	.5417	.6445	57.2	1.00
.57	32.9	.8396	.5432	.6469	57.1	1.00
.58	**33.0**	.8387	.5446	.6494	**57.0**	.99
.58	33.1	.8377	.5461	.6519	56.9	.99
.58	33.2	.8368	.5476	.6544	56.8	.99
.58	33.3	.8358	.5490	.6569	56.7	.99
.58	33.4	.8348	.5505	.6594	56.6	.99
.58	**33.5**	.8339	.5519	.6619	**56.5**	.99
.59	33.6	.8329	.5534	.6644	56.4	.98
.59	33.7	.8320	.5548	.6669	56.3	.98
.59	33.8	.8310	.5563	.6694	56.2	.98
.59	33.9	.8300	.5577	.6720	56.1	.98
.59	**34.0**	.8290	.5592	.6745	**56.0**	.98
.60	34.1	.8281	.5606	.6771	55.9	.98
.60	34.2	.8271	.5621	.6796	55.8	.97
.60	34.3	.8261	.5635	.6822	55.7	.97
.60	34.4	.8251	.5650	.6847	55.6	.97
.60	**34.5**	.8241	.5664	.6873	**55.5**	.97
.60	34.6	.8231	.5678	.6899	55.4	.97
.61	34.7	.8221	.5693	.6924	55.3	.97
.61	34.8	.8211	.5707	.6950	55.2	.96
.61	34.9	.8202	.5721	.6976	55.1	.96
.61	**35.0**	.8192	.5736	.7002	**55.0**	.96
		sin	cos	cot	Deg	Rad

Rad	Deg	cos	sin	tan		
.61	**35.0**	.8192	.5736	.7002	**55.0**	.96
.61	35.1	.8181	.5750	.7028	54.9	.96
.61	35.2	.8171	.5764	.7054	54.8	.96
.62	35.3	.8161	.5779	.7080	54.7	.95
.62	35.4	.8151	.5793	.7107	54.6	.95
.62	**35.5**	.8141	.5807	.7133	**54.5**	.95
.62	35.6	.8131	.5821	.7159	54.4	.95
.62	35.7	.8121	.5835	.7186	54.3	.95
.62	35.8	.8111	.5850	.7212	54.2	.95
.63	35.9	.8100	.5864	.7239	54.1	.94
.63	**36.0**	.8090	.5878	.7265	**54.0**	.94
.63	36.1	.8080	.5892	.7292	53.9	.94
.63	36.2	.8070	.5906	.7319	53.8	.94
.63	36.3	.8059	.5920	.7346	53.7	.94
.64	36.4	.8049	.5934	.7373	53.6	.94
.64	**36.5**	.8039	.5948	.7400	**53.5**	.93
.64	36.6	.8028	.5962	.7427	53.4	.93
.64	36.7	.8018	.5976	.7454	53.3	.93
.64	36.8	.8007	.5990	.7481	53.2	.93
.64	36.9	.7997	.6004	.7508	53.1	.93
.65	**37.0**	.7986	.6018	.7536	**53.0**	.93
.65	37.1	.7976	.6032	.7563	52.9	.92
.65	37.2	.7965	.6046	.7590	52.8	.92
.65	37.3	.7955	.6060	.7618	52.7	.92
.65	37.4	.7944	.6074	.7646	52.6	.92
.65	**37.5**	.7934	.6088	.7673	**52.5**	.92
.66	37.6	.7923	.6101	.7701	52.4	.91
.66	37.7	.7912	.6115	.7729	52.3	.91
.66	37.8	.7902	.6129	.7757	52.2	.91
.66	37.9	7891	.6143	.7785	52.1	.91
.66	**38.0**	.7880	.6157	.7813	**52.0**	.91
.66	38.1	.7869	.6170	.7841	51.9	.91
.67	38.2	.7859	.6184	.7869	51.8	.90
.67	38.3	.7848	.6198	.7898	51.7	.90
.67	38.4	.7837	.6211	.7926	51.6	.90
.67	**38.5**	.7826	.6225	.7954	**51.5**	.90
.67	38.6	.7815	.6239	.7983	51.4	.90
.68	38.7	.7804	.6252	.8012	51.3	.90
.68	38.8	.7793	.6266	.8040	51.2	.89
.68	38.9	.7782	.6280	.8069	51.1	.89
.68	**39.0**	.7771	.6293	.8098	**51.0**	.89
.68	39.1	.7760	.6307	.8127	50.9	.89
.68	39.2	.7749	.6320	.8156	50.8	.89
.69	39.3	.7738	.6334	.8185	50.7	.88
.69	39.4	.7727	.6347	.8214	50.6	.88
.69	**39.5**	.7716	.6361	.8243	**50.5**	.88
.69	39.6	.7705	.6374	.8273	50.4	.88
.69	39.7	.7694	.6388	.8302	50.3	.88
.69	39.8	.7683	.6401	.8332	50.2	.88
.70	39.9	.7672	.6414	.8361	50.1	.87
.70	**40.0**	.7660	.6428	.8391	**50.0**	.87
		sin	cos	cot	Deg	Rad

TABLE B.3 Trigonometric functions (continued)

Rad	Deg	cos	sin	tan		
.70	**40.0**	.7660	.6428	.8391	**50.0**	.87
.70	40.1	.7649	.6441	.8421	49.9	.87
.70	40.2	.7638	.6455	.8451	49.8	.87
.70	40.3	.7627	.6468	.8481	49.7	.87
.71	40.4	.7615	.6481	.8511	49.6	.87
.71	**40.5**	.7604	.6494	.8541	**49.5**	.86
.71	40.6	.7593	.6508	.8571	49.4	.86
.71	40.7	.7581	.6521	.8601	49.3	.86
.71	40.8	.7570	.6534	.8632	49.2	.86
.71	40.9	.7559	.6547	.8662	49.1	.86
.72	**41.0**	.7547	.6561	.8693	**49.0**	.86
.72	41.1	.7536	.6574	.8724	48.9	.85
.72	41.2	.7524	.6587	.8754	48.8	.85
.72	41.3	.7513	.6600	.8785	48.7	.85
.72	41.4	.7501	.6613	.8816	48.6	.85
.72	**41.5**	.7490	.6626	.8847	**48.5**	.85
.73	41.6	.7478	.6639	.8878	48.4	.84
.73	41.7	.7466	.6652	.8910	48.3	.84
.73	41.8	.7455	.6665	.8941	48.2	.84
.73	41.9	.7443	.6678	8972	48.1	.84
.73	**42.0**	.7431	.6691	.9004	**48.0**	.84
.73	42.1	.7420	.6704	.9036	47.9	.84
.74	42.2	.7408	.6717	.9067	47.8	.83
.74	42.3	.7396	.6730	.9099	47.7	.83
.74	42.4	.7385	.6743	.9131	47.6	.83
.74	**42.5**	.7373	.6756	.9163	**47.5**	.83
		sin	cos	cot	Deg	Rad

Rad	Deg	cos	sin	tan		
.74	**42.5**	.7373	.6756	.9163	**47.5**	.83
.74	42.6	.7361	.6769	.9195	47.4	.83
.75	42.7	.7349	.6782	.9228	47.3	.83
.75	42.8	.7337	.6794	.9260	47.2	.82
.75	42.9	.7325	.6807	.9293	47.1	.82
.75	**43.0**	.7314	.6820	.9325	**47.0**	.82
.75	43.1	.7302	.6833	.9358	46.9	.82
.75	43.2	.7290	.6845	.9391	46.8	.82
.76	43.3	.7278	.6858	.9424	46.7	.82
.76	43.4	.7266	.6871	.9457	46.6	.81
.76	**43.5**	.7254	.6884	.9490	**46.5**	.81
.76	43.6	.7242	.6896	.9523	46.4	.81
.76	43.7	.7230	.6909	.9556	46.3	.81
.76	43.8	.7218	.6921	.9590	46.2	.81
.77	43.9	.7206	.6934	.9623	46.1	.80
.77	**44.0**	.7193	.6947	.9657	**46.0**	.80
.77	44.1	.7181	.6959	.9691	45.9	.80
.77	44.2	.7169	.6972	.9725	45.8	.80
.77	44.3	.7157	.6984	.9759	45.7	.80
.77	44.4	.7145	.6997	.9793	45.6	.80
.78	**44.5**	.7133	.7009	.9827	**45.5**	.79
.78	44.6	.7120	.7022	.9861	45.4	.79
.78	44.7	.7108	.7034	.9896	45.3	.79
.78	44.8	.7096	.7046	.9930	45.2	.79
.78	44.9	.7083	.7059	.9965	45.1	.79
.79	**45.0**	.7071	.7071	1.0000	**45.0**	.79
		sin	cos	cot	Deg	Rad

APPENDIX C

Answers

PROBLEM SET 1.1, PAGES 6–7

1. a. $Z, Q,$ and R **b.** Q and R **c.** $N, W, Z, Q,$ and R **d.** Q' and R **e.** Q' and R **3. a.** Q and R **b.** Q and R **c.** Q' and R **d.** Q and R **e.** Q' and R **5. a.** $N, W, Z, Q,$ and R **b.** Q' and R **c.** Q' and R **d.** Q and R **e.** Q and R **7.** 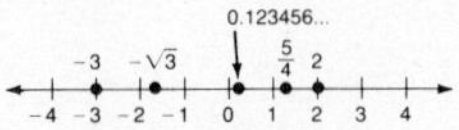 **9.** 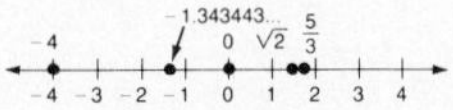 **11.**

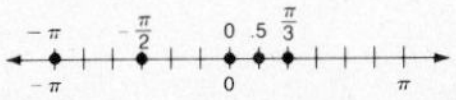

13. a. $<$ **b.** $<$ **15. a.** $<$ **b.** $=$ **17. a.** $=$ **b.** $<$ **19.** reflexive **21.** distributive **23.** closure **25.** transitive **27.** substitution **29.** closure **31.** multiplicative inverse **33.** distributive **35.** identity **37.** no **39.** no **41.** yes **43.** no **45.** Answers vary; $\frac{12}{4} \neq \frac{4}{12}$ **47.** Answers vary; $(12 \div 6) \div 2 \neq 12 \div (6 \div 2)$

49. $-\dfrac{\pi}{3} - 1$ **51.** $\dfrac{3}{\pi + 3}$

53. addition properties: closure, commutative, associative; multiplication properties: closure, commutative, associative, identity; distributive for multiplication over addition.

55. addition properties: closure, commutative, associative, identity, inverse; multiplication properties: closure, commutative, associative, identity; distributive for multiplication over addition

57. not closed for either addition or multiplication; commutative and associative for those sums and products in the set; distributive for multiplication over addition for those sums and products in the set

59. multiplication properties: closure, commutative, associative, identity and inverse

PROBLEM SET 1.2, PAGES 15–16

1. a. $(3, 7)$  **b.** $(-4, -1)$  **c.** $[-2, 6]$  **d.** $(-3, 0]$

3. a. $(-\infty, -3]$  **b.** $[-2, \infty)$ **c.** $(-\infty, 0)$ **d.** $(2, \infty)$ **5. a.**

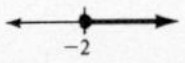

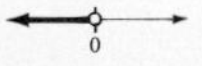

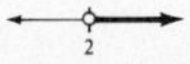

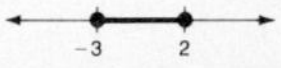

b.  **c.** 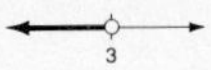 **d.** 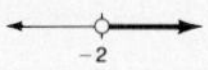 **7. a.** 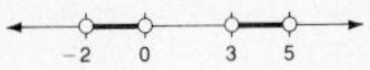 **b.**

c. 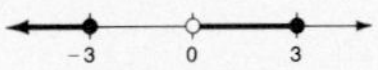 **d.** 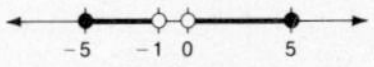 **9. a.** $-4 \le x \le 2$ **b.** $-1 \le x \le 2$ **c.** $0 < x < 8$

d. $-5 < x \le 3$ **11. a.** $x < 2$ **b.** $x > 6$ **c.** $x > -1$ **d.** $x \le 3$ **13. a.** $\pi - 2$ **b.** $5 - \pi$ **c.** $2\pi - 6$

d. $7 - 2\pi$ **15. a.** $\sqrt{2} - 1$ **b.** $2 - \sqrt{2}$ **c.** $1 - \dfrac{\pi}{6}$ **d.** $\dfrac{2\pi}{3} - 1$ **17. a.** $\pi - 3$ **b.** 119 **c.** $3 - \sqrt{5}$

19. $\{5, -5\}$ **21.** $\{\ \}$ or $\varnothing$ **23.** $\{7, -1\}$ **25.** $\{24, -6\}$ **27.** $\{\ \}$ or $\varnothing$ **29.** $\{\frac{2}{5}, -2\}$ **31.** $[7, \infty)$ **33.** $(-\infty, -41]$ **35.** $(-\infty, 2)$ **37.** $(-\frac{8}{5}, 0)$ **39.** $(1, 3]$ **41.** $[-4, 2)$ **43.** $(-\frac{15}{2}, 4]$ **45.** $[-2, 8]$ **47.** $(2.999, 3.001)$ **49.** $\{\ \}$ or $\varnothing$ **51.** $(\frac{1}{3}, \frac{7}{3})$ **53.** $(-\infty, -6) \cup (-1, \infty)$ **55.** You must drive less than 200 miles per day. **57.** $68 < F < 86$

59. values in thousands of dollars **a.** $s < 10$ or $s > 20$ **b.** $|s - 15| > 5$ **61.** $(-\infty, -3) \cup (9, \infty)$ **63.** $[-2, 0]$ **65.** $\{1, -\frac{3}{2}\}$ **67.** Answers vary. **69.** Answers vary.

PROBLEM SET 1.3, PAGE 19

1. $6i$ **3.** $7i$ **5.** $2i\sqrt{5}$ **7.** $8 + 7i$ **9.** $-5i$ **11.** $1 - 6i$ **13.** $3 + 3i$ **15.** 10 **17.** $2 + 5i$ **19.** $7 + i$ **21.** 29 **23.** 34 **25.** 1 **27.** $-i$ **29.** -1 **31.** 1 **33.** $-i$ **35.** -1 **37.** $32 - 24i$ **39.** $-9 + 40i$ **41.** $-198 - 10i$ **43.** $-\frac{3}{2} + \frac{3}{2}i$ **45.** $1 + i$ **47.** $-2i$ **49.** $\frac{1}{5} - \frac{3}{5}i$ **51.** $\frac{1}{5} - \frac{2}{5}i$ **53.** $1 - i$ **55.** $-1 + i$

57. $\frac{-45}{53} + \frac{28}{53}i$ **59.** $0.3131 + 2.2281i$ **61.** $(-1 - \sqrt{3}) + 2i$ **63.** $\left(\frac{5 + 2\sqrt{3}}{13}\right) + \left(\frac{1 + 3\sqrt{3}}{13}\right)i$

65. $(13 + 8\sqrt{3}) + (12 + 4\sqrt{3})i$ **67.** Answers vary.

PROBLEM SET 1.4, PAGES 25–26

1. $\{3, -5\}$ **3.** $\{2, -9\}$ **5.** $\{\frac{4}{5}, -\frac{1}{2}\}$ **7.** $\{4, -\frac{2}{9}\}$ **9.** $\{\frac{4}{3}\}$ **11.** $\{0, 2\}$ **13.** $\{1, -5\}$ **15.** $\{2, -4\}$

17. $\{-3, -4\}$ **19.** $\{5 \pm 3\sqrt{3}\}$ **21.** $\left\{\frac{3 \pm \sqrt{13}}{2}\right\}$ **23.** $\{\frac{2}{3}, -\frac{1}{2}\}$ **25.** $\{1, -6\}$ **27.** $\{5\}$ **29.** $\{\frac{1}{4}, -\frac{2}{3}\}$

31. $\varnothing$ over R; $\{\frac{2}{5} \pm \frac{1}{5}i\}$ over C **33.** $\left\{\pm\frac{\sqrt{5}}{2}\right\}$ **35.** $\left\{0, \frac{7}{3}\right\}$ **37.** $\{2, -\frac{1}{3}\}$ **39.** $\left\{\frac{-5 \pm \sqrt{73}}{8}\right\}$

41. $\left\{\frac{-3 \pm \sqrt{17}}{4}\right\}$ **43.** $\left\{\frac{-3 \pm \sqrt{9 + 16\sqrt{5}}}{8}\right\}$ **45.** $\left\{\frac{2 \pm \sqrt{4 + 3\sqrt{5}}}{3}\right\}$ **47.** $\left\{\frac{-w \pm \sqrt{w^2 - 40}}{4}\right\}$

49. $\left\{\frac{-5 \pm \sqrt{12y - 23}}{6}\right\}$ **51.** $\left\{2, \frac{3t + 2}{4}\right\}$ **53.** $\left\{\frac{-1 \pm \sqrt{8y - 47}}{4}\right\}$ **55.** $\{3 \pm \sqrt{4y - y^2}\}$

57. a. 1454 ft (let $x = 0$) **b.** about 11.1 sec. **59.** It will be in the air for 8 sec. **61.** It will take 7.9 sec.

PROBLEM SET 1.5, PAGES 29–30

1. $(-3, 0)$  **3.** $(-\infty, 2] \cup [6, \infty)$ **5.** $(-7, 8)$

7. $[-2, \frac{1}{2}]$  **9.** $(-\infty, -\frac{2}{3}) \cup (3, \infty)$ 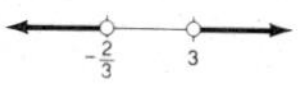 **11.** $(-\infty, -2] \cup [8, \infty)$

13. $(-\infty, \frac{1}{3}) \cup (4, \infty)$ 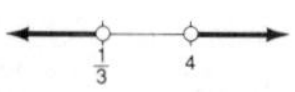 **15.** $(-\infty, -4] \cup [0, 3]$

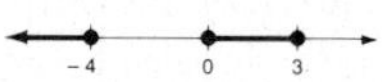

17. $[-3, 2] \cup [4, \infty]$ 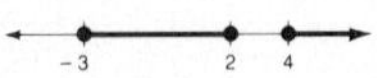 **19.** $(-\infty, -\frac{5}{2}) \cup (-1, \frac{7}{3})$ 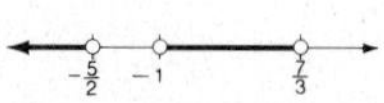 **21.** $(-2, 0)$

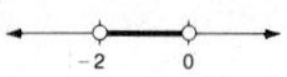

23. $(-\infty, 0) \cup (8, \infty)$ 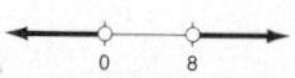 **25.** $(-5, 2]$ 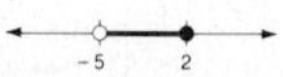 **27.** $(-\infty, 0) \cup (\frac{1}{2}, 5)$

29. $(-\infty, -2) \cup (0, 3)$ 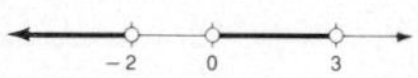 **31.** $(-\infty, -3] \cup [3, \infty)$ **33.** $(-\infty, \infty)$

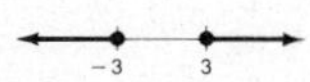

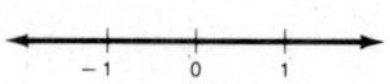

35. $(-\infty, -2) \cup (3, \infty)$ 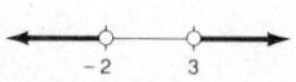 **37.** $[2, 3]$ 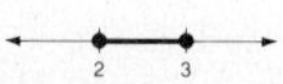 **39.** $[-5, 1]$

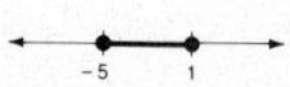

41. $(-1 - \sqrt{3}, 1 + \sqrt{3})$ 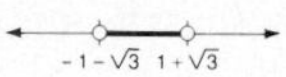 **43.** $(-\infty, \infty)$

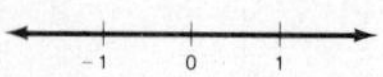

45. $\left(-\infty, \frac{-3 - \sqrt{37}}{2}\right] \cup \left[\frac{-3 + \sqrt{37}}{2}, \infty\right)$ **47.** $[-5, -3) \cup [0, 3) \cup (4, \infty)$

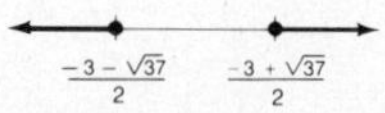

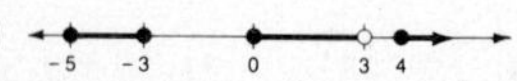

49. $(-3, 2) \cup [12, \infty)$ **51.** $[-13, -\frac{1}{2}) \cup [0, \frac{1}{3})$

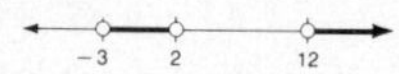

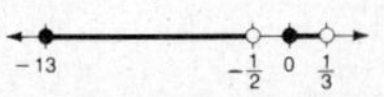

53. $(-\infty, 2) \cup (3, 4]$ 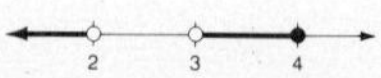 **55.** all numbers except those between -17 and 20 **57.** all numbers except those between -3 and 0, inclusive **59.** The width must be greater than 3 and the length greater than 6.

PROBLEM SET 1.6, PAGES 36–38

1. 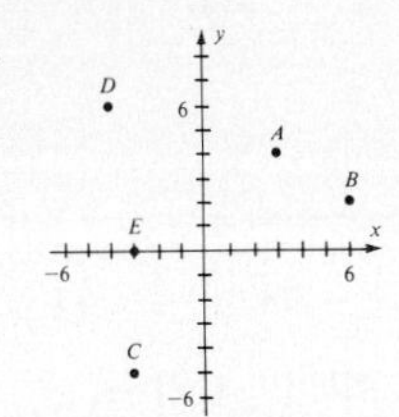
3. 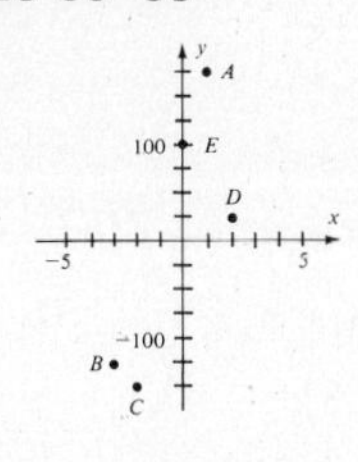
5. 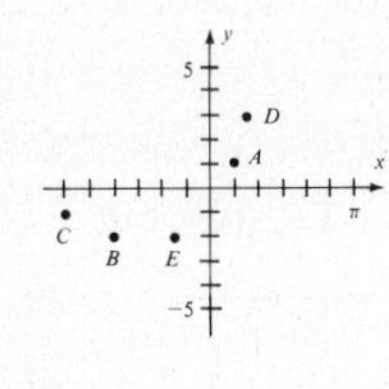
7. Answers vary. 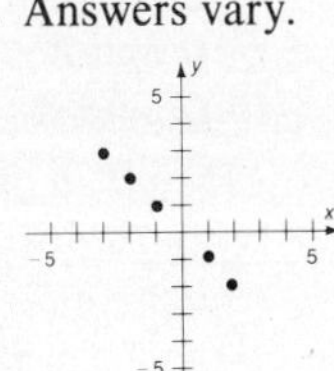
9. Answers vary. 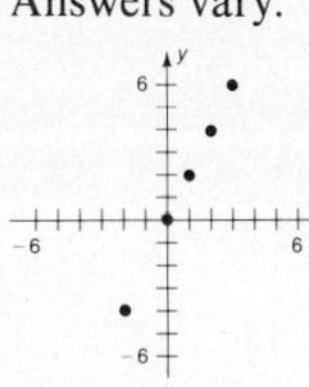

11. Answers vary. **13.** 5 **15.** $2\sqrt{5}$ **17.** $\sqrt{37}$ **19.** $-5x$ **21.** $5x$ **23.** $(7, \frac{13}{2})$ **25.** $(\frac{5}{2}, -1)$

27. $(-\frac{3}{2}, -2)$ **29.** $(-x, \frac{7}{2}x)$ **31.** 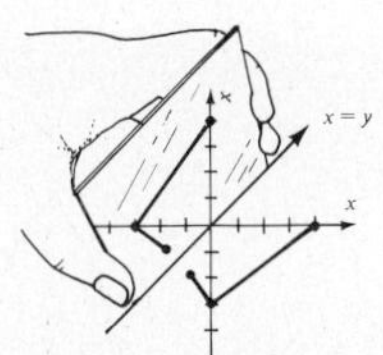
33. 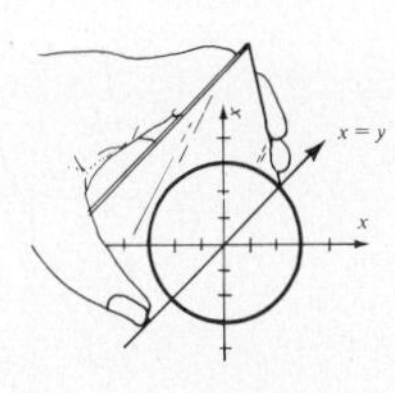
35. 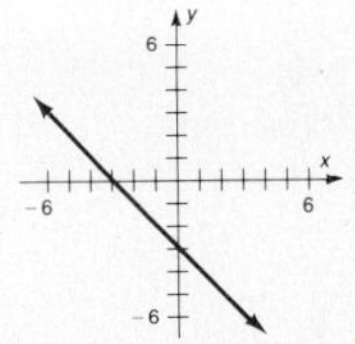

37. 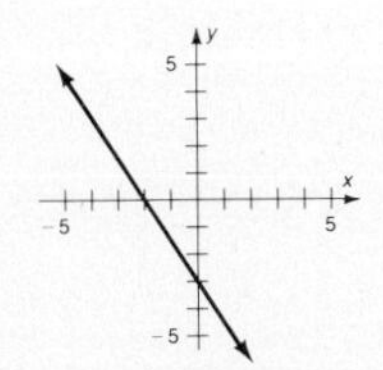
39. 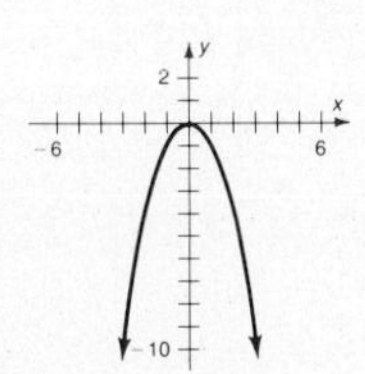
41. 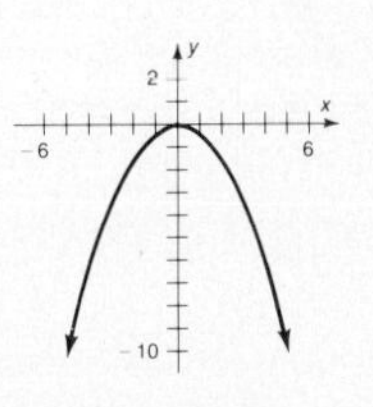
43. 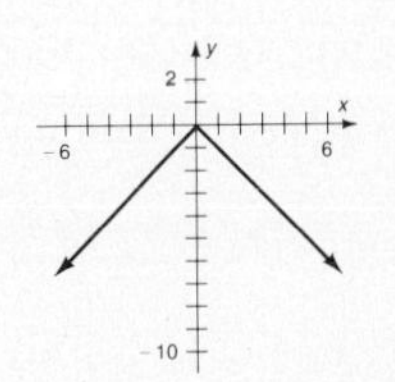
45. 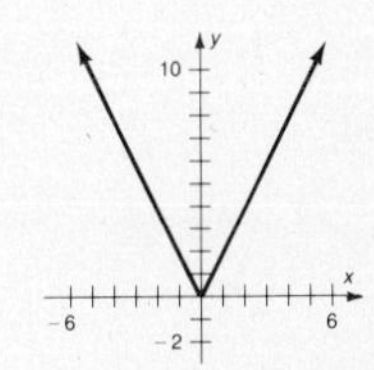

47. 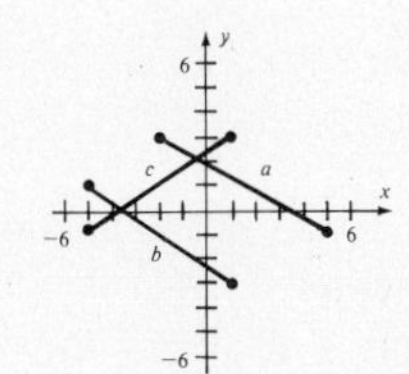
49. 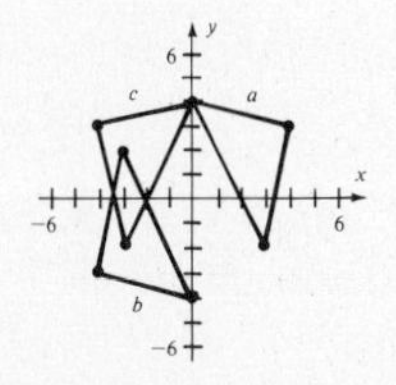
51. 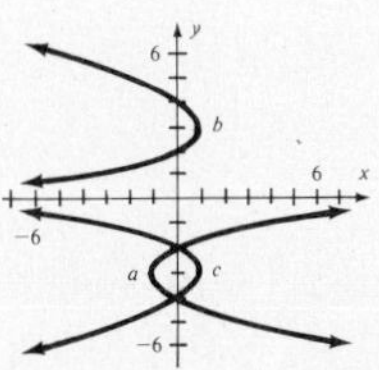
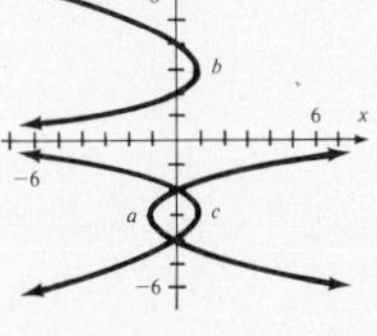

53. $d_1 = \sqrt{40}, d_2 = \sqrt{85}, d_3 = 9$; not a right triangle **55.** $d_1 = 5, d_2 = \sqrt{73}, d_3 = \sqrt{74}$; not a right triangle
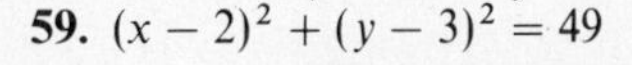
57. $d_1 = \sqrt{52}, d_2 = \sqrt{65}, d_3 = \sqrt{13}$; it is a right triangle. **59.** $(x-2)^2 + (y-3)^2 = 49$ **61.** $(x+4)^2 + (y+1)^2 = 9$
63. $(0, 4 \pm 2\sqrt{15})$ 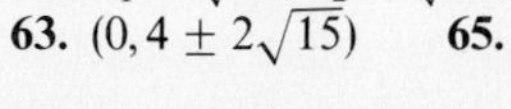
65. 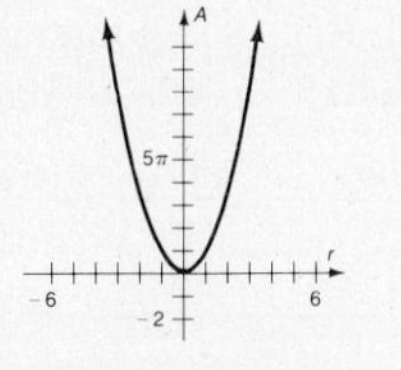
67. 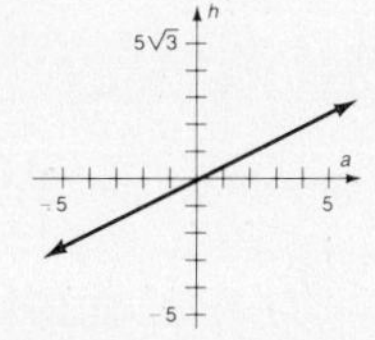
69. Answers vary. **71.** $9x^2 + 25y^2 = 225$

73. $9y^2 - 16x^2 = 144$

CHAPTER 1 SUMMARY, PAGES 38–40

Let N = {counting numbers}, W = {whole numbers}, J = {integers}, Q = {rationals}, Q' = {irrationals} and R = {reals}.

1. $\frac{14}{7}$: N, W, J, Q, and R
$\sqrt{144}$: N, W, J, Q, and R
$6.\overline{2}$: N, J, Q, and R
π: Q' and R
3. $3.\overline{1}$: Q and R
$\frac{5\pi}{6}$: Q' and R
$\frac{22}{7}$: Q and R
$\sqrt{10}$: Q' and R
5. **7.**

9. $=$ **11.** $<$ **13.** $a(b + c)$ **15.** $a(b + c) = 5$ **17.** $(b + c)a$ **19.** $ab + ac$ **21.** $(-4, 2)$
23. $(-3, \infty)$ **25.** **27.** **29.** $-8 \le x < -5$ **31.** $x < 3$ **33.** $(-\infty, 4]$
35. $(-8, -4]$ **37.** $\sqrt{11}$ **39.** $\sqrt{11} - 3$ **41.** 8 **43.** $\pi + 2$ **45.** $\{-8, 8\}$ **47.** $\{-\frac{11}{2}, \frac{5}{2}\}$ **49.** $(-1, 9)$
51. $[-10, 15]$ **53.** i **55.** 29 **57.** $\{-3, 4\}$ **59.** $\{3, -\frac{5}{2}\}$ **61.** $\{-3, 5\}$ **63.** $\{-4, -5\}$ **65.** $\left\{\frac{5 \pm \sqrt{13}}{2}\right\}$
67. $\{-1 \pm \sqrt{6}\}$ **69.** $(-\frac{1}{3}, 1)$ **71.** $(-\infty, \infty)$ **73.** $(-1, 0) \cup (3, \infty)$ **75.** $(-\infty, -1) \cup (0, 2) \cup (2, \infty)$
77–79. **81.** $d = \sqrt{(\gamma - \alpha)^2 + (\delta - \beta)^2}$ **83.** $-5x$ **85.** $(\frac{\alpha+\gamma}{2}, \frac{\beta+\delta}{2})$ **87.** $(3x, \frac{5x}{2})$

89. a set of ordered pairs **91.** satisfies **93.**

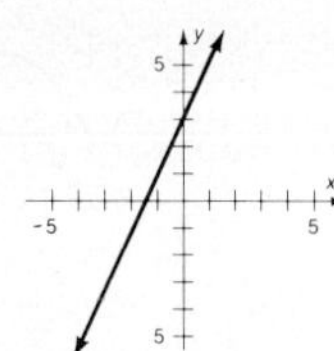

95.

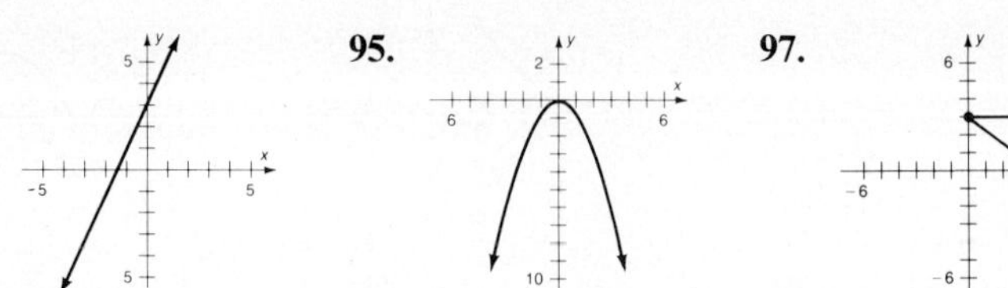

97.

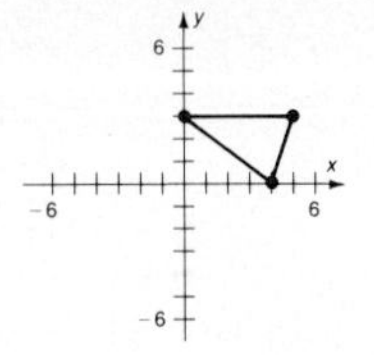

99.

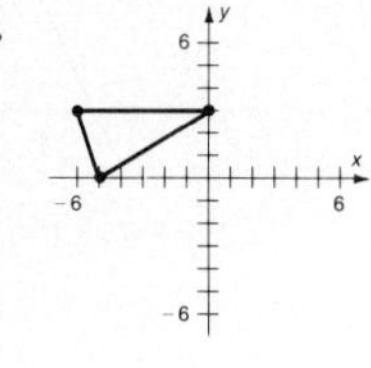

PROBLEM SET 2.1, PAGES 46–48

1. onto, one-to-one function **3.** not a function **5.** into function **7.** function **9.** function **11.** not a function **13.** function **15.** not a function **17.** function **19.** function **21.** not a function **23.** function **25.** function **27.** not a function **29.** function **31.** $R(9, g(9))$; $S(a, g(a))$ **33.** $P(x_0, f(x_0))$; $Q(x_0 + h, f(x_0 + h))$ **35.** $T(x_0, K(x_0))$; $U(x_0 + h, K(x_0 + h))$ **37.** D: $[-5, 6]$; R: $[-6, 3]$; intercepts: $(-4\frac{1}{2}, 0)$, $(4, 0)$ and $(0, 3)$; increasing on $[-5, -3]$; constant on $[-3, 3]$; decreasing on $[3, 6]$ **39.** D: $[-4, 7]$; R: $[-4, 5]$; intercepts: $(0, 5)$, $(3, 0)$, $(6, 0)$; constant on $[-4, 1]$; decreasing on $[-1, 3]$ and $[5, 7]$; increasing on $[3, 5]$ **41.** D: $[-5, \infty)$; R: $[-3, 6) \cup (6, \infty)$; intercept: $(0, -3)$; constant on $[-5, -2]$; decreasing on $[-2, 0]$; increasing on $[0, \infty)$ **43.** D: $[-6, 6]$; R: $[-5, 5]$; intercepts: $(-3, 0)$, $(0, 5)$ and $(3, 0)$; increasing on $[-6, 0]$; decreasing on $[0, 6]$ **45. a.** \$0.92 **b.** \$0.45 **47.** \$1.15 **49. a.** \$0.51 **b.** $e(1984) - e(1944)$
51. a. \$0.02225 **b.** average annual change of price of gasoline from 1944 to 1984 **53. a.** \$.018, $\frac{s(1954) - s(1944)}{10}$
b. \$0.0125; $\frac{s(1964) - s(1944)}{20}$ **c.** \$0.058; $\frac{s(1974) - s(1944)}{30}$ **d.** \$0.02875; $\frac{s(1984) - s(1944)}{40}$ **e.** $\frac{s(1944 + h) - s(1944)}{h}$
55. a. 63,800 **b.** the average annual change in the number of marriages from 1977 to $(1977 + h)$
57. It also has 5 elements. **59.** Answers vary.

PROBLEM SET 2.2, PAGES 54–55

1. a. 1 **b.** 5 **c.** -5 **d.** $2\sqrt{5} + 1$ **e.** $2\pi + 1$ **3. a.** $2w + 1$ **b.** $2w^2 - 1$ **c.** $2t^2 - 1$ **d.** $2v^2 - 1$ **e.** $2m + 1$ **5. a.** $3 + 2\sqrt{2}$ **b.** $5 + 4\sqrt{2}$ **c.** $2t^2 + 12t + 17$ **d.** $2t^2 + 4t + 3$ **e.** $2m^2 - 4m + 1$ **7.** 2 **9.** 2

11. $4t + 2h$ **13. a.** $w^2 - 1$ **b.** $h^2 - 1$ **c.** $w^2 + 2wh + h^2 - 1$ **d.** $w^2 + h^2 - 2$ **15. a.** $x^4 - 1$ **b.** $x - 1$ **c.** $x^2 + 2xh + h^2 - 1$ **d.** $x^2 - 1$ **17.** 2 **19.** $2x + h$ **21.** -2 **23.** $\dfrac{|2x + 2h + 1| - |2x + 1|}{h}$

25. $6x + 2 + 3h$ **27.** $\dfrac{-1}{x(x+h)}$ **29.** $(-\infty, \infty)$ **31.** $(-\infty, -2) \cup (-2, \infty)$ **33.** $[-\frac{1}{2}, \infty)$ **35.** $[-2, 1]$
37. $(-\infty, -2) \cup (-2, 2) \cup (2, \infty)$ **39.** not equal **41.** not equal **43.** not equal **45.** even **47.** neither
49. neither **51.** even **53.** even **55.** For 50 units the average cost is \$13; for 100 units the average cost is \$7. $\dfrac{C(100) - C(50)}{50} = \$1$ **57.** $\dfrac{C(x + h) - C(x)}{1} = -0.40x + 4 - 0.02h$ **59.** Answers vary. It is the average rate of change for an object falling between time x and $x + h$. **61.** Answers vary. **63.** Answers vary.

PROBLEM SET 2.3, PAGES 61–62

The graphs for 1–9 should also be shown.
1. 1 **3.** $\frac{13}{5}$ **5.** $-\frac{7}{3}$ **7.** Slope is undefined. **9.** 0 **11.** $m = -4; b = -1$ **13.** $m = \frac{1}{5}; b = -\frac{6}{5}$

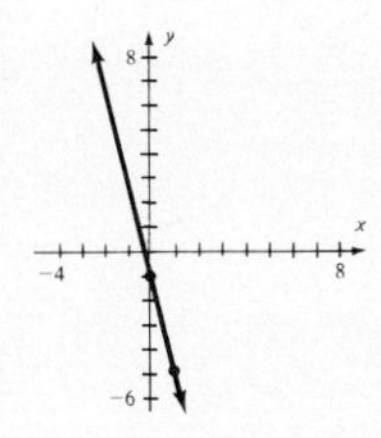

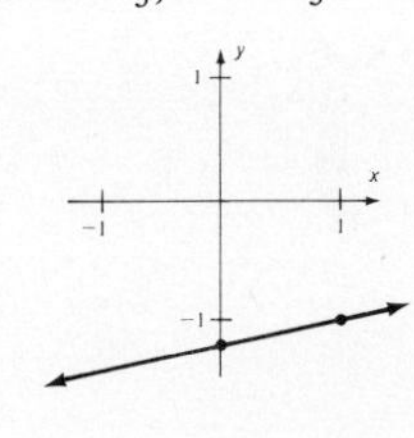

15. $m = 300; b = 0$ **17.** $m = 0; b = -2$ **19.** $m = \frac{1}{3}; b = \frac{2}{3}$ **21.** $m = \frac{2}{5}; b = -240$ **23.**

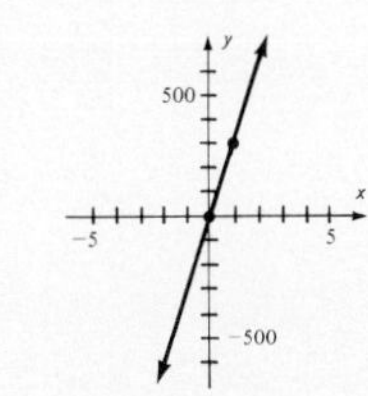

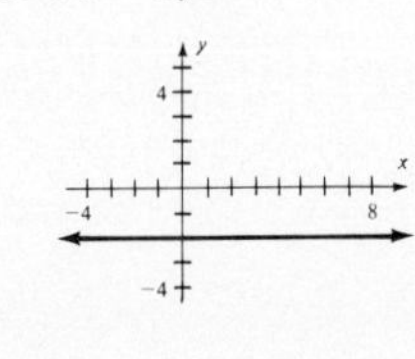

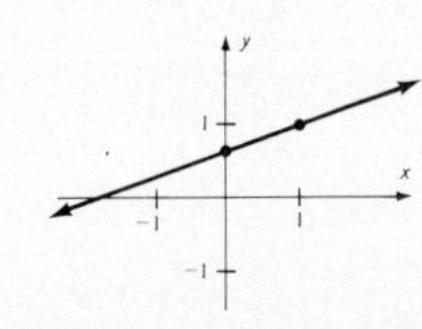

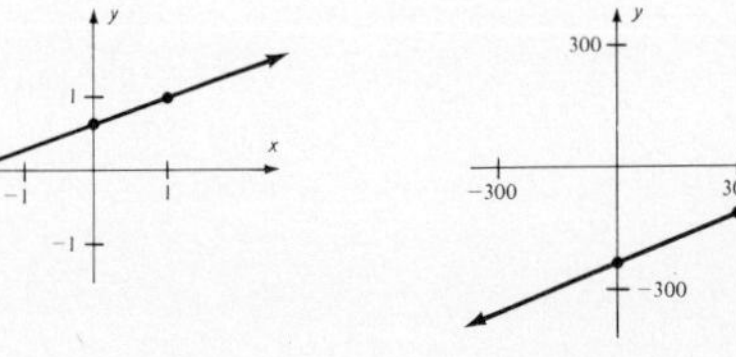

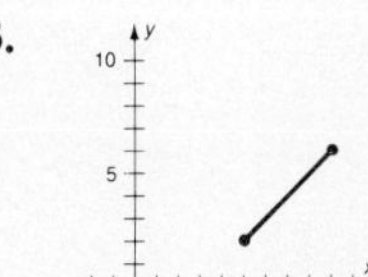

25.

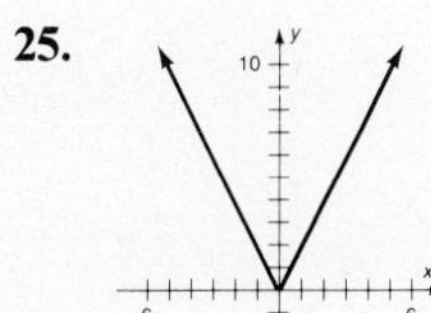

27.

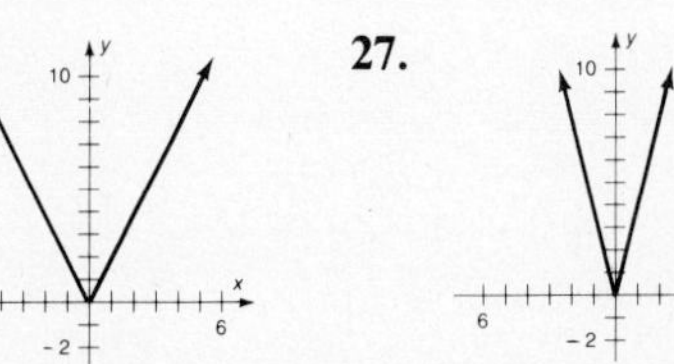

29. $m_{AN} = \frac{2}{3}; m_{AG} = -8; m_{NG} = -\frac{3}{2}$; thus, $m_{AN} \cdot m_{NG} = -1$. Thus AN and NG are perpendicular and it is a right triangle.
31. $m_{RE} = \frac{7}{3}; m_{EC} = -\frac{1}{2}; m_{CT} = \frac{6}{5}$; not a parallelogram **33.** $m_{PA} = 1; m_{AR} = -7; m_{RL} = 1; m_{LP} = -7$; it is a parallelogram.
35. $m_{PR} = 3; m_{AL} = -\frac{1}{3}$; yes **37.** $5x - y + 6 = 0$ **39.** $y = 0$ **41.** $3x - y - 3 = 0$ **43.** $x - 2y + 3 = 0$
45. $x - 2y + 2 = 0$ **47.** $2x - y - 4 = 0$ **49.** $2x + 3y - 16 = 0$ **51.** $2x + y + 4 = 0$
53. a. $A(x_0, f(x_0)); B(x_0 + \Delta x, f(x_0 + \Delta x))$ **b.** $\dfrac{f(x_0 + \Delta x) - f(x_0)}{\Delta x}$ **55. a.** $A(x_0, H(x_0)); B(x_0 + h, H(x_0 + h))$
b. $\dfrac{H(x_0 + h) - H(x_0)}{h}$ **57.** $(1, 10), (2, 20)$; $10x - y = 0$ **59.** $(20, 11.2)$ and $(30, 14.2)$; $3x - 10y + 52 = 0$; if $x = 40$, then $y = 17.2$, so the projected population in 1990 is 17.2 million **61.** $y - 60 = 0$; the cost is \$60. **63–67.** Answers vary.

PROBLEM SET 2.4, PAGES 66–67

1. $(6, 3)$ **3.** $(6, -1)$ **5.** $(0, \sqrt{2})$ **7.** $(3, -4)$ **9.** $(0, 0)$ **11.** $(-3, 0)$ **13.** $y - 3 = (x - 2)^2$ **15.** $y + 1 = x^2$
17. $y + 2 = (x + \sqrt{3})^2$ **19.** $y + 6 = |x|$ **21.** $y + \sqrt{3} = |x - \sqrt{2}|$ **23.** $y' = x'^2$ **25.** $y' = -5x'^2$
27. $y' = -2x'^2$ **29.** $y' = |x'|$ **31.** $y' = -2|x'|$

33
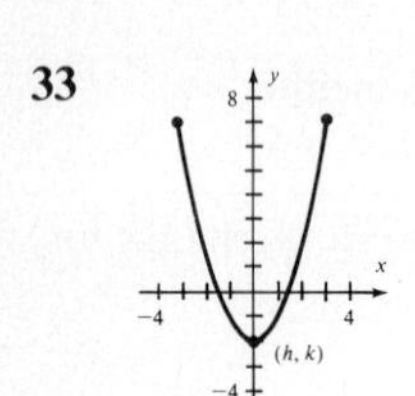

35.
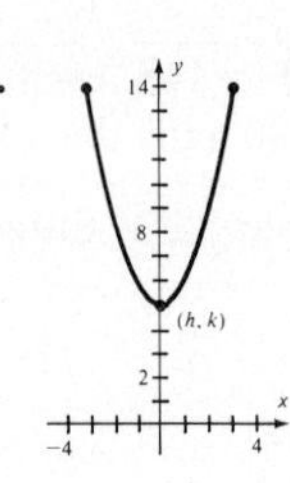

37.
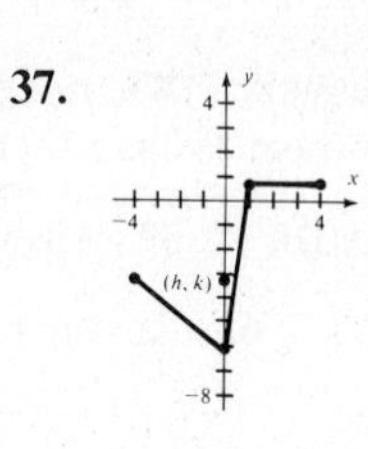

39.
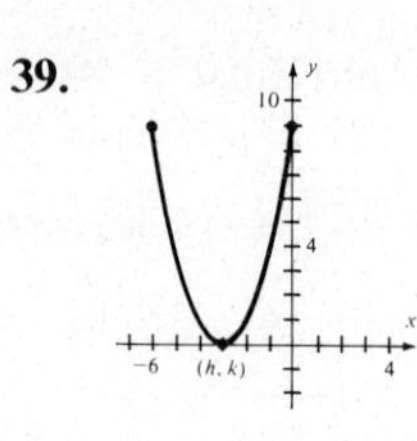

41.
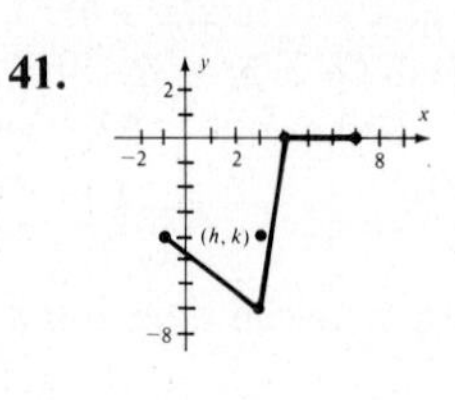

43.
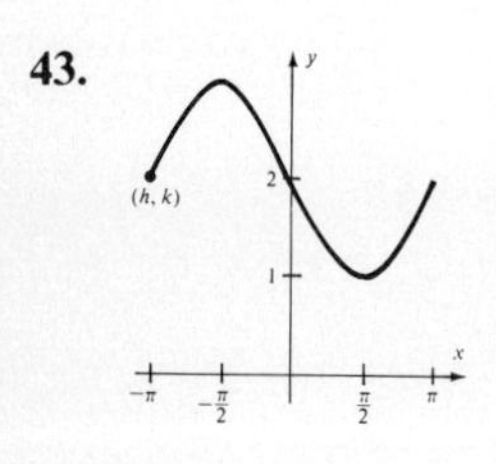

45.
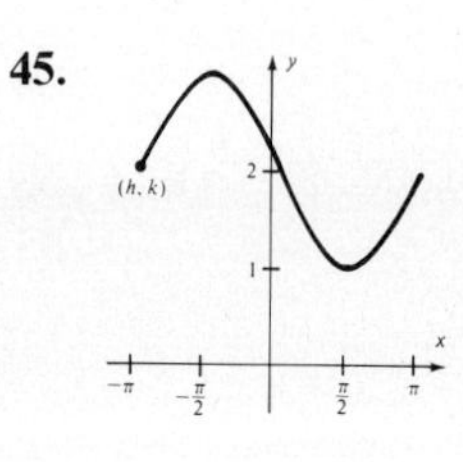

47.
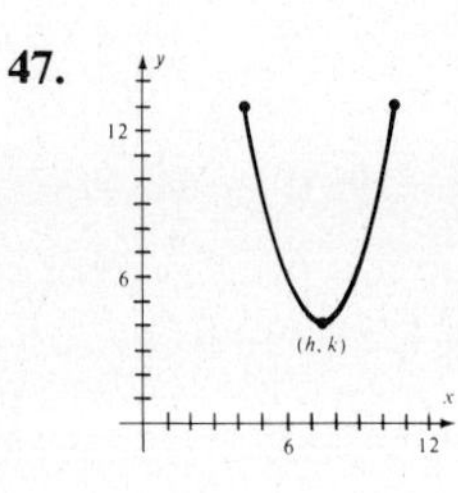

49.
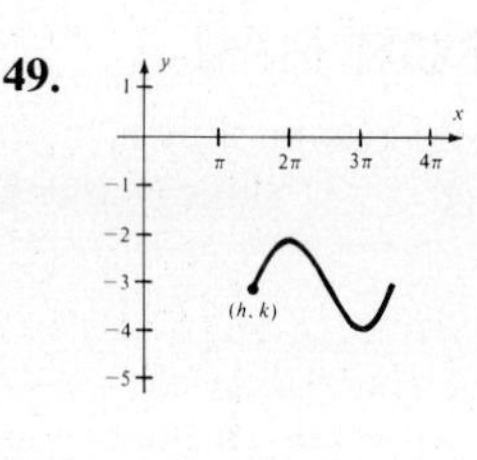

51.
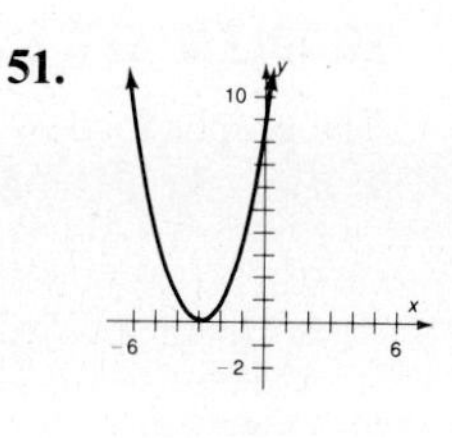

53.
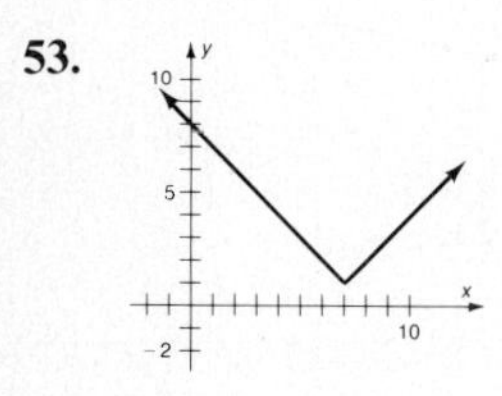

55.
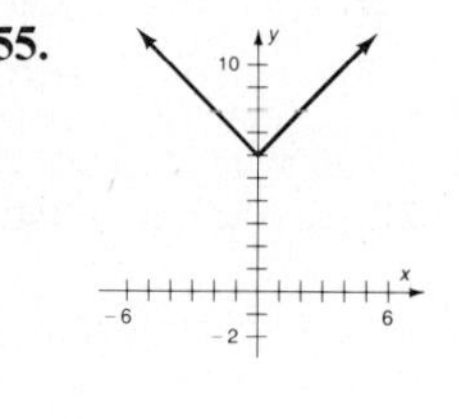

57.
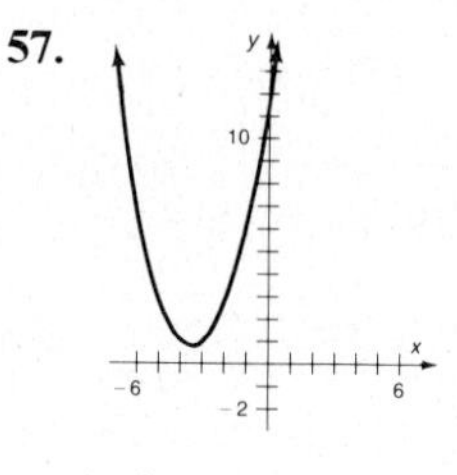

59.
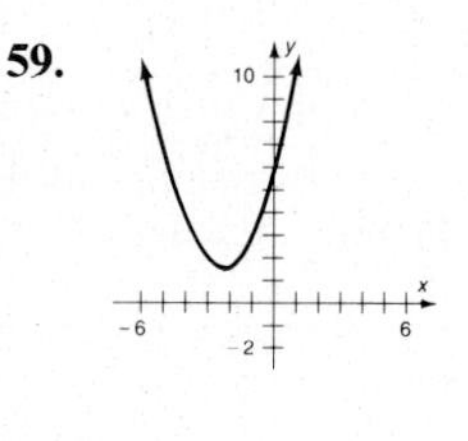

61.
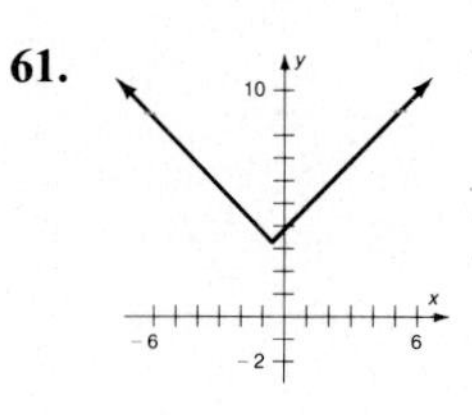

63.
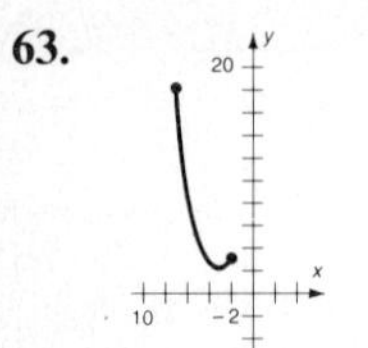

65.
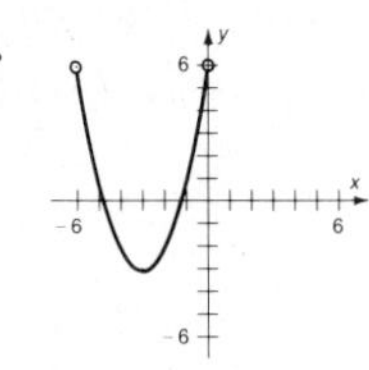

PROBLEM SET 2.5, PAGES 72–73

1.
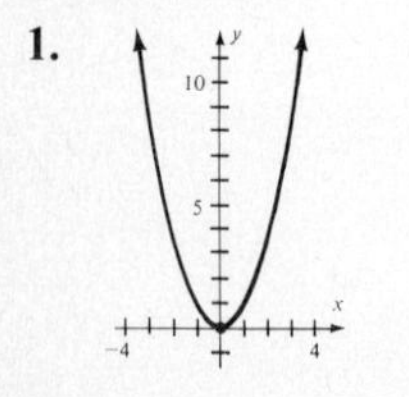

3.
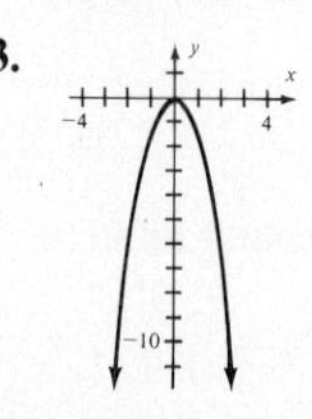

5.
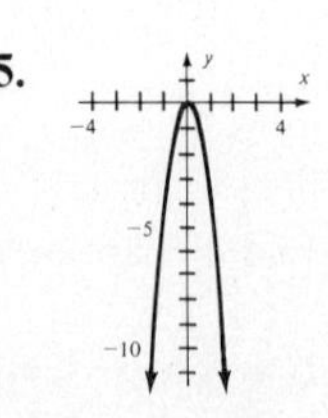

7.
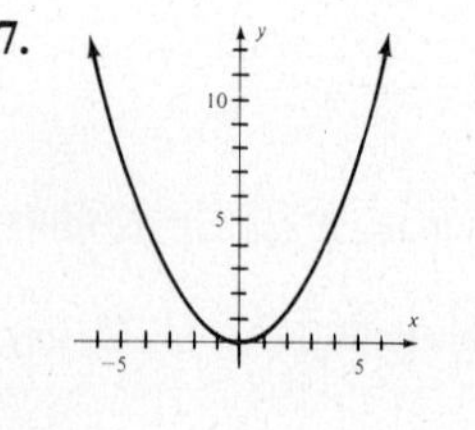

9.
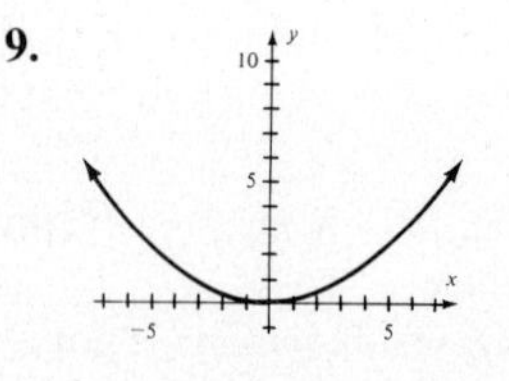

11.
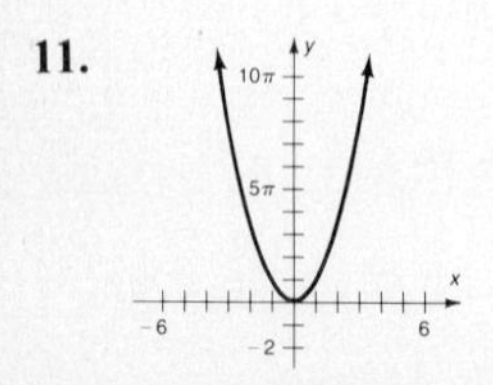

13.
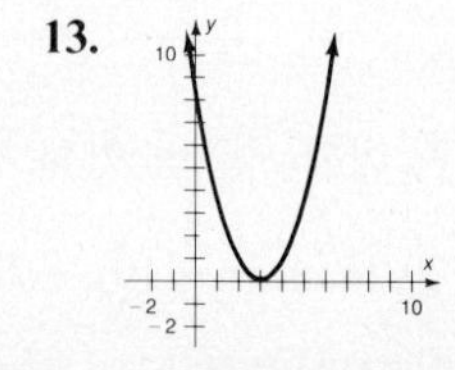

15.
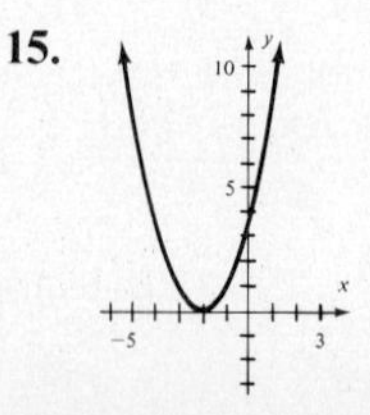

17.
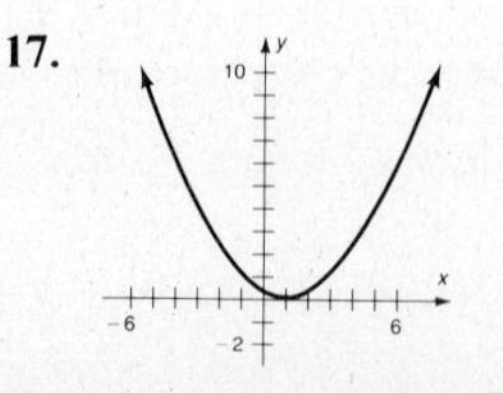

19.
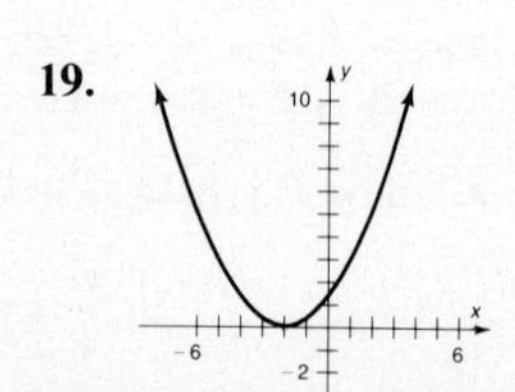

21. 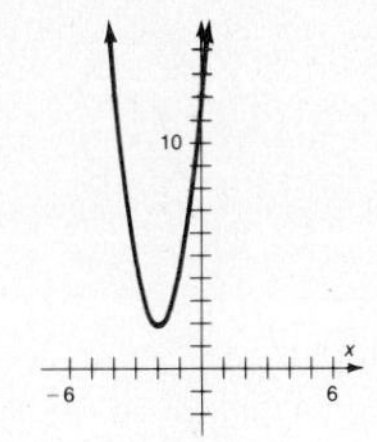

23. 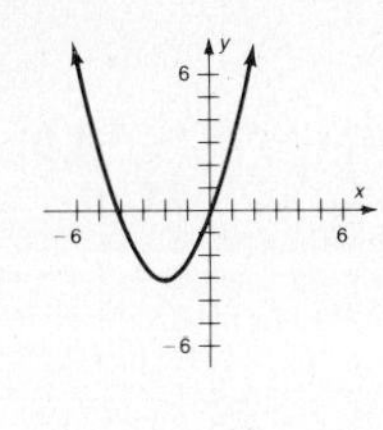

25. 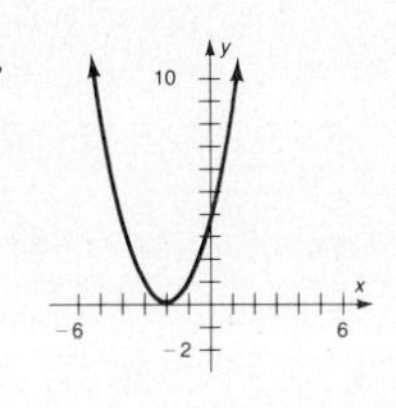

27. 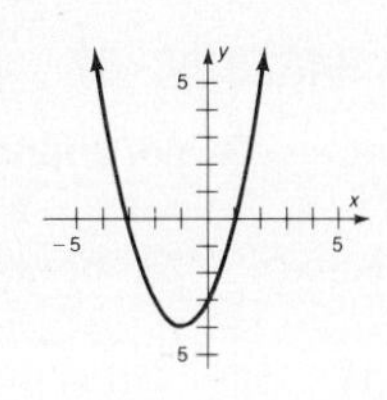

29. 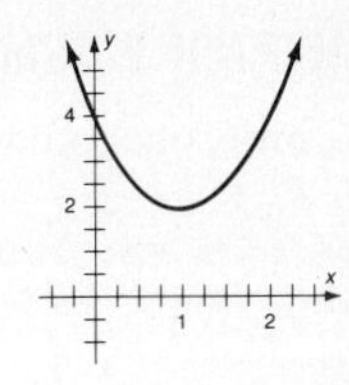

31. 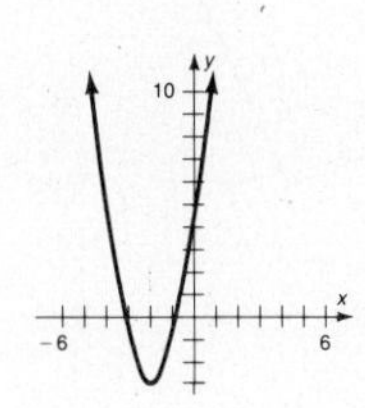

33. 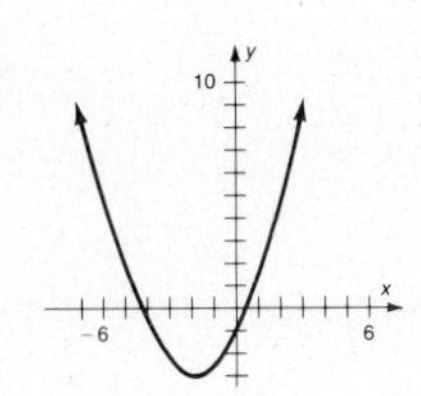

35. 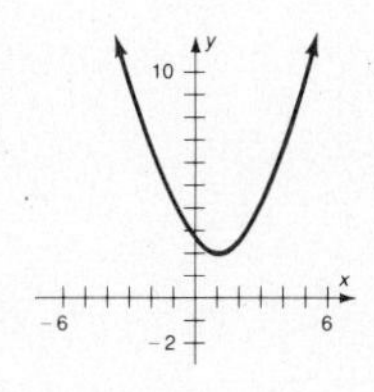

37. 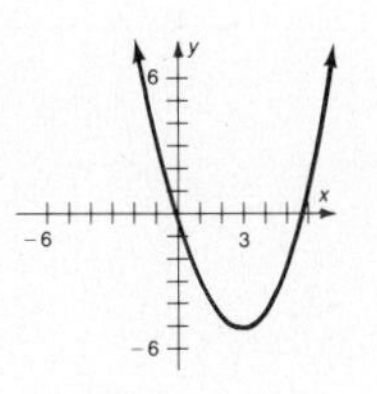

39. 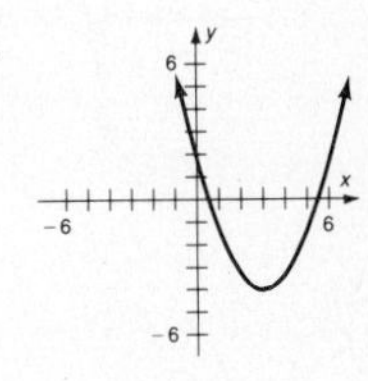

41. 3 **43.** -15 **45.** $\frac{2}{3}$  **47. a.** 375 **b.** \$250,000 loss **c.** \$1,156,250 **49.** \$650
51. a. If $x = 9$, then $y - 18 = -2$, so $y = 16$; height is 16 ft. **b.** If $x = 18$, then $y - 18 = -\frac{2}{81}(18)^2$ so $y = 10$; height is 10 ft.
53. Dimensions are 25′ by 25′; area = 625 ft^2. **55.** The maximum height is about 3,456 ft (to the nearest foot).
55. The maximum height is about 3,456 ft (to the nearest foot).

57. 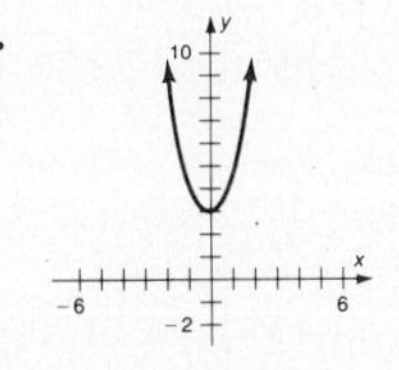

59. 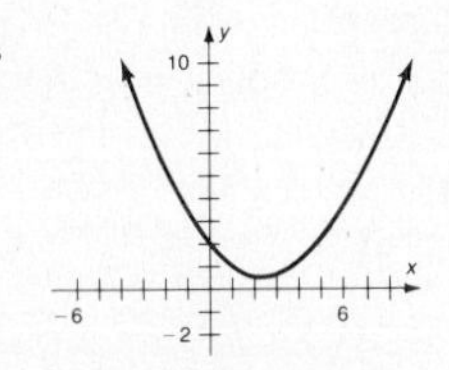

61. 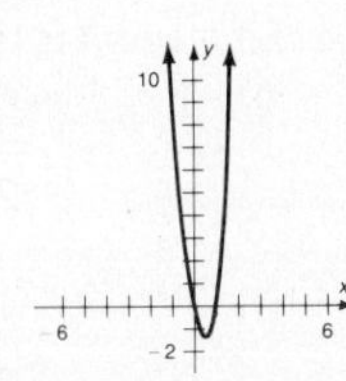

63. a. $y = 2x$ **b.** 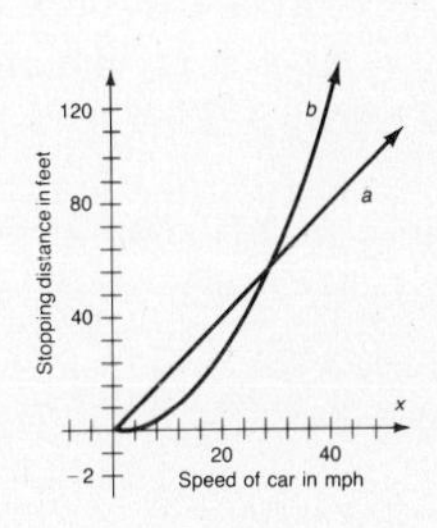

c. Answers vary; at 0 mph and at about 28 mph.

PROBLEM SET 2.6, PAGES 78–80

1. $(f \circ g)(x) = 2x + 9$; $(g \circ f)(x) = 2x + 3$ **3.** $(f \circ g)(x) = x$; $(g \circ f)(x) = x$
5. $(f \circ g)(x) = \dfrac{x^2 - x - 2}{x^2 - x + 1}$; $(g \circ f)(x) = \dfrac{-3x + 6}{(x + 1)^2}$ **7.** $(f \circ g)(x) = \dfrac{1}{x^2 - 2}$; $(g \circ f)(x) = \dfrac{-x^2 + 2x}{(x - 1)^2}$
9. $(f \circ g)(x) = 3$; $(g \circ f)(x) = 45$ **11.** $f \circ g = \{(5, 12), (6, 3)\}$; $g \circ f = \varnothing$ **13.** inverses **15.** not inverses
17. not inverses **19.** $f^{-1} = \{(5, 4), (3, 6), (1, 7), (4, 2)\}$ **21.** $f^{-1} = \{(3, 1), (5, 2), (6, 4), (9, 5)\}$ **23.** $f^{-1}(x) = x - 3$
25. $g^{-1}(x) = \frac{x}{5}$ **27.** The inverse function does not exist because h is not a one-to-one function. **29.** $f^{-1}(x) = x$
31. The inverse function does not exist because f is not a one-to-one function. **33.** $f^{-1}(x) = \dfrac{1 + 2x}{x}$; $x \neq 0$
35. a. $4x^2 - 4x + 1$ **b.** $6x + 3$ **c.** $36x^2 + 36x + 9$ **d.** $36x^2 + 36x + 9$ **37. a.** x **b.** $\frac{1}{2}x^2 - \frac{3}{2}$ **c.** $x^2 + 1$
d. $x^2 + 1$ **39. a.** $6x - 13$ **b.** $2x - 3$ **c.** $6x - 7$ **d.** $6x - 7$ **41. a.** $4x$ **b.** $4x$ **c.** $8x$ **d.** $8x$
43. a. -4 **b.** -2 **c.** 1 **d.** 5 **e.** 8 **45. a.** 6 **b.** 7 **c.** 12 **d.** 4 **e.** 0 **47. a.** 1 **b.** -1
c. $7\frac{1}{2}$ **d.** 14 **e.** 7 **49. a.** D: $[0, 15]$; R: $[-6, 9]$ **b.** D: $[-6, 9]$; R: $[0, 15]$ **51.** inverses **53.** not inverses
55. a. \$300 **b.** \$84 **c.** $(p \circ c)(n) = \dfrac{60n + 240}{n}$ **57. a.** 144π **b.** $(S \circ r)(t) = 36\pi t^2$ **c.** $(0, \frac{8}{3}]$
59. $f^{-1}(x) = -\sqrt{x}$ on $[0, \infty)$ **61.** $f^{-1}(x) = \sqrt{x} - 1$ on $[1, \infty)$ **63.** $f^{-1}(x) = -\frac{1}{2}\sqrt{2x}$ on $[2, 200]$
65. $f^{-1}(x) = \dfrac{1 + x}{2x}$ on $(0, \infty)$ **67.** $f^{-1}(x) = x - 1$ on $[0, \infty)$ **69. a.** 1 **b.** 4 **c.** 9 **d.** k^2

CHAPTER 2 SUMMARY, PAGES 80–82

1. onto, one-to-one function **3.** not a function **5.** $D: [-5, 11]$; $R: [-3, 7]$ **7.** $(-3.4, 0), (9, 0), (0, 2), (3.4, 0)$
9. a. 11 **b.** -11 **11. a.** $3w - 1$ **b.** $5 - w^2 - 2wh - h^2$ **13.** $-2x - h$ **15.** 0 **17.** $(-\infty, 1] \cup [6, \infty)$
19. $(-\infty, \infty)$ **21.** not equal **23.** not equal **25.** neither **27.** even

29. **31.** **33.** $-\frac{5}{4}$ **35.** $\frac{7}{5}$ **37.** **39.**

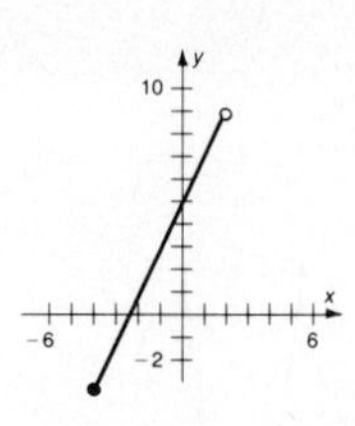

41.

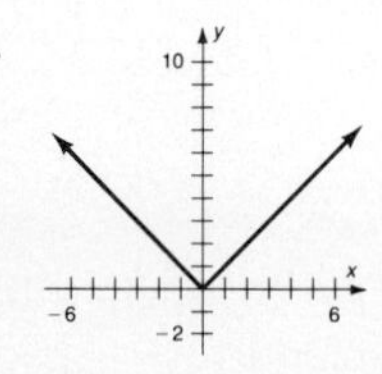

43.

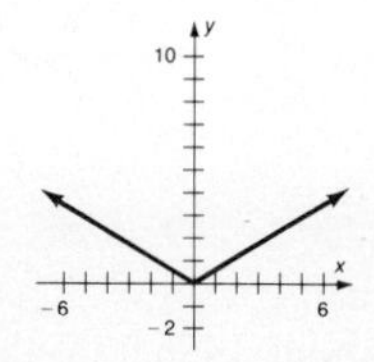

45. $m_{AB} = -\frac{5}{4}$; $m_{BC} = \frac{5}{7}$; $m_{CA} = -\frac{1}{3}$; not a right triangle **47.** $-\frac{2}{3}$

49. $Ax + By + C = 0$; (x, y) is any point on the line and A, B, C are any constants (not all zero).
51. $y = mx + b$; (x, y) is any point on the line, m is the slope and b is the y-intercept. **53.** $2x - 3y - 27 = 0$
55. $5x + 8y + 23 = 0$ **57.** $(-\pi, 6)$ **59.** $(4, 0)$ **61.** $y - 3 = 9(x + \sqrt{2})^2$ **63.** $y = -2(x - \pi)^2$ **65.** $y' = 3x'^2$

67. $y' = 5x'^2$ **69.**

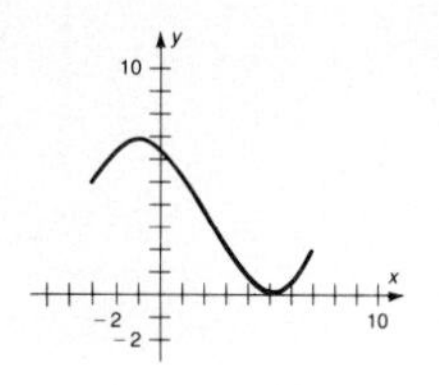

71.

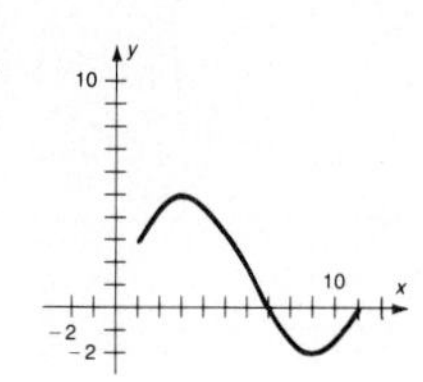

73.

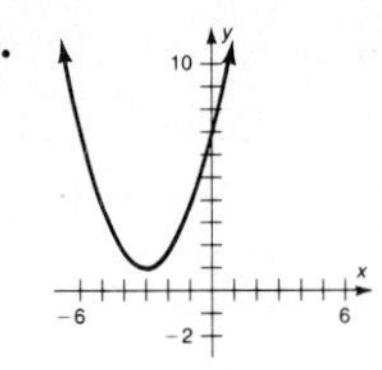

75.

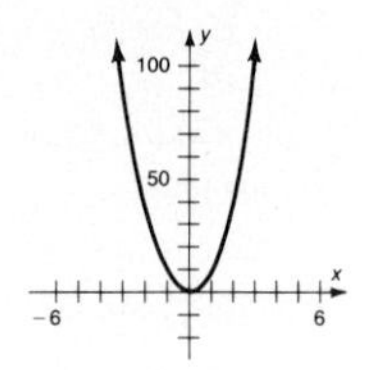

77. $y + 2 = (x + 1)^2$

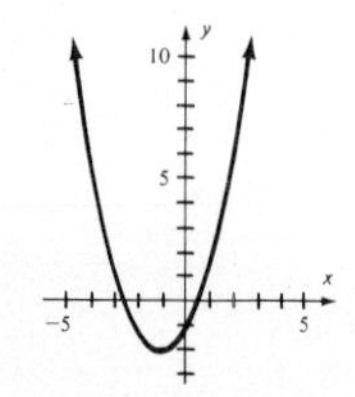

79. $y - 1 = (x - 3)^2$

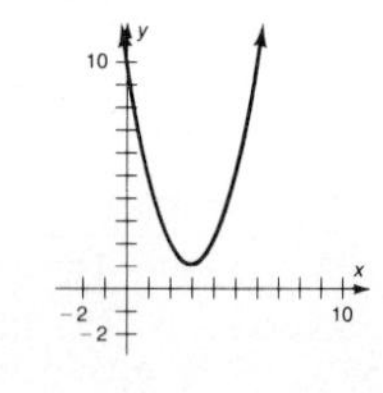

81. maximum value 250 at $x = -6$ **83.** minimum value 850 at $x = 5$

85. $14 - 3x^2$ **87.** 17 **89.** no **91.** no **93.** $f^{-1}(x) = \dfrac{x + 1}{3}$ **95.** $f^{-1}(x) = 2x - 10$

PROBLEM SET 3.1, PAGES 89–90

1. a. 11 **b.** 4 **c.** -10 **3. a.** 26 **b.** 0 **c.** -4 **5. a.** $x^3 - 7x^2 + 6x - 6$ **b.** $15x^3 - 22x^2 + 5x + 2$
7. a. $x^2 + 3x + 2$ **b.** $y^2 + y - 6$ **c.** $x^2 - x - 2$ **d.** $y^2 - y - 6$ **9. a.** $2x^2 - x - 1$ **b.** $2x^2 - 5x + 3$
c. $3x^2 + 4x + 1$ **d.** $3x^2 + 5x + 2$ **11. a.** $a^2 + 4a + 4$ **b.** $b^2 - 4b + 4$ **c.** $x^2 + 8x + 16$ **d.** $y^2 - 6y + 9$
13. a. $m(e + i + y)$ **b.** $(a - b)(a + b)$ **c.** irreducible **d.** $(a - b)(a^2 + ab + b^2)$ **15. a.** $(a + b)^3$ **b.** $(p - q)^3$
c. $(d - c)^3$ **d.** $xy(x + y)$ **17. a.** $(3x + 1)(x - 2)$ **b.** $(3y - 2)(2y - 1)$ **c.** $b(4a - 1)(2a + 3)$ **d.** $2(s - 8)(s + 3)$
19. a. $3x^3 + 8x^2 - 9x + 2$ **b.** $2x^3 + 5x^2 - 8x - 5$ **21. a.** $2x^3 - 3x^2 - 8x - 3$ **b.** $6x^3 + 17x^2 - 4x - 3$
23. a. $x^3 - 3x^2 + 4$ **b.** $x^3 - 3x - 2$ **25.** $(x - y - 1)(x - y + 1)$ **27.** $5(a - 1)(5a + 1)$
29. $-\frac{1}{25}(3x + 10)(7x + 10)$ **31.** $\dfrac{1}{y^8}(x^3 - 13y^4)(x^3 + 13y^4)$ **33.** $(a + b - x - y)(a + b + x + y)$

35. $(2x-3)(x+2)$ **37.** $(6x-1)(x+8)$ **39.** $(6x+1)(x+8)$ **41.** $(4x-3)(x+4)$ **43.** $(9x-2)(x-6)$
45. $(2x-1)(2x+1)(x+2)(x-2)$ **47.** $(x+2)(x+1)(x^2-2x+4)(x^2-x+1)$
49. $\frac{1}{36}(2x+1)(2x-1)(3x-1)(3x+1)$ **51.** $\frac{1}{32}(2x+1)^2(2x-1)(4x^2-2x+1)$

53. **55.** **57.** **59.** **61.**

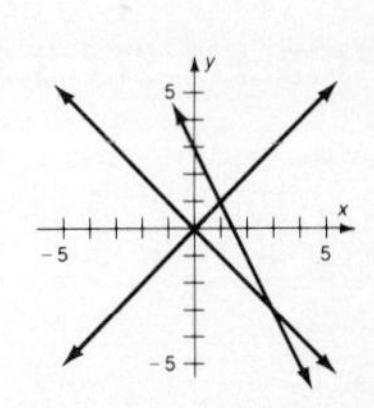

63. $2x^6-23x^5+103x^4-232x^3+306x^2-297x+189$ **65.** $6x^4-x^3-6x^2+25x-12$
67. $x^9-6x^8+15x^7-35x^6+75x^5-96x^4+136x^3-165x^2+75x-125$ **69.** $(x^n-y^n)(x^{2n}+x^ny^n+y^{2n})$
71. $(x^n-y^n)^2$ **73.** $x(x+5)$ **75.** $(x-1)(x+1)(x+2)(x^2-2x+4)$ **77.** $(x+y-a-b)(x+y+a+b)$
79. $(x+y+a+b)(x^2+2xy+y^2-ax-by-ay-bx+a^2+2ab+b^2)$ **81.** $(x+1)(x-4)(x^2-3x-8)$
83. $s(3s+5t)$

PROBLEM SET 3.2, PAGES 94–96

1. a. 3 **b.** -9 **c.** -6 **d.** $3x^2-3x+5$ **3. a.** -2 **b.** -4 **c.** -10 **d.** 16 **e.** 12 **f.** -12
g. $2x^2-8x+6$ **5. a.** -4 **b.** 1 **c.** 3 **d.** 8 **e.** -4 **f.** -20 **g.** -4 **h.** $x^2+3x-4+\dfrac{-4}{x+4}$
7. a. 1 **b.** 1 **c.** 2 **d.** 0 **e.** -5 **f.** -2 **g.** 2 **h.** 0 **i.** x^3+3x^2+3x-2 **9. a.** -2 **b.** 1
c. 0 **d.** 0 **e.** 0 **f.** -8 **g.** 0 **h.** 16 **i.** 16 **j.** -32 **k.** -32 **l.** -64
m. $x^4-2x^3+4x^2-8x+16+\dfrac{-64}{x+2}$ **11.** $3x^2+7x+25$ **13.** x^3-7x^2+8x-8 **15.** $2x^3-6x^2+3x-1$
17. $4x^4+x^3-4x^2-4x-4$ **19.** $3x^2+4x+12$ **21.** x^2+x-6 **23.** x^2+x-12
25. $4x^3+8x^2-7x-7$ **27.** x^3-2x^2+3x-4 **29.** $3x^3-2x^2-5$ **31.** $Q(x)=2x^3-6x^2+2x-2$; $R(x)=0$
33. $Q(x)=x^3-x^2+1$; $R(x)=1-x$ **35.** $Q(x)=3x^2-7x+5$; $R(x)=0$ **37.** $3x^2+2x+1+\dfrac{5}{x-3}$
39. $x^3+x^2+2+\dfrac{3}{x-4}$ **41.** $5x^3-5x^2-5x+3+\dfrac{-11}{x+3}$ **43.** $x^4-5x^3+10x^2-18x+36+\dfrac{-77}{x+2}$
45. $4x^2-5x+7+\dfrac{-8}{x+1}$ **47.** $5x^4-10x^3+20x^2-40x+78+\dfrac{155}{x+2}$ **49.** $x^4+3x^2-7x+\dfrac{-6}{x-3}$
51. x^2+2x-5 **53.** $2x-1$ **55.** $2x^2+19x+93$

57. $x^2+4x-5=(x+5)(x-1)$

1⌋	1	7	5	−23	10
		1	8	13	−10
	1	8	13	−10	0
−5⌋		−5	−15	10	
	1	3	−2	0	

$Q(x)=x^2+3x-2$

59. $K=7$ **61. a.**

$$\begin{array}{r} x^2 - x + 3 \\ 2x+1\,\overline{)\,2x^3 - x^2 + 5x + 3} \\ \underline{2x^3 + x^2} \\ -2x^2 \\ \underline{-2x^2 - x} \\ 6x \\ \underline{6x + 3} \\ 0 \end{array}$$

b.

$-\frac{1}{2}$⌋	2	−1	5	3
		−1	1	−3
	2	−2	6	0

$2x^2-2x+6$

c. The answer to part b is double that of part a. **d.** It will be doubled.

e. Divide by $x + \frac{b}{a}$ synthetically ($a \neq 0$) and then divide the quotient by a and keep the remainder.

PROBLEM SET 3.3, PAGES 100–101

1. a. -3 **b.** -19 **c.** -4 **d.** 842 **e.** -448 **3. a.** -824 **b.** -758 **c.** -812 **d.** -890 **e.** -1022 **5. a.** -10 **b.** -8 **c.** 68 **d.** 2140 **e.** 380 **7. a.** -3 **b.** 8 **c.** 0 **d.** $-\frac{5}{2}$ **e.** 840 **9. a.** 0 **b.** 42 **c.** 0 **d.** 0 **e.** 0 **11. a.** 0 **b.** 0 **c.** 0 **d.** -54 **e.** -10

13. 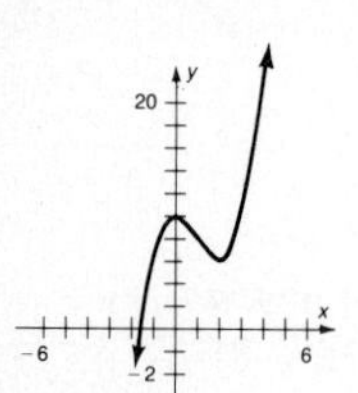

15. 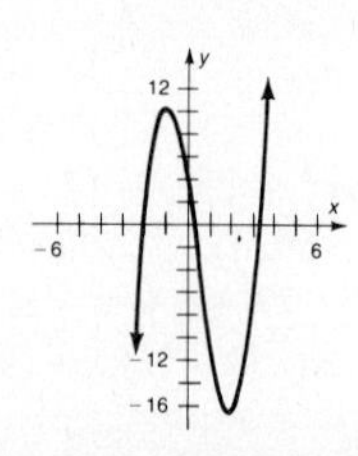

17. 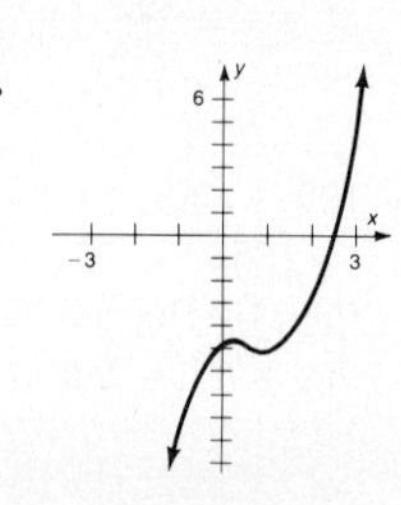

19. 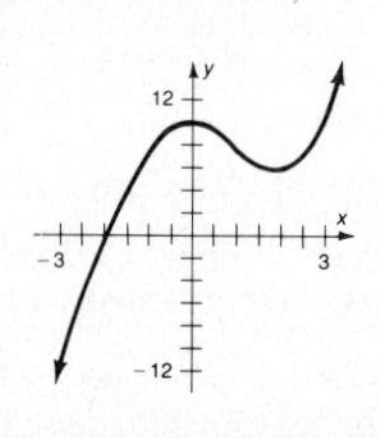

21. 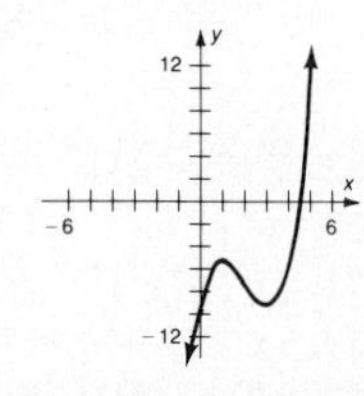

23. 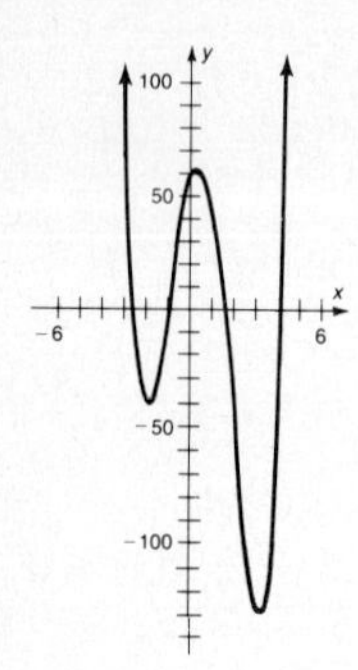

25. 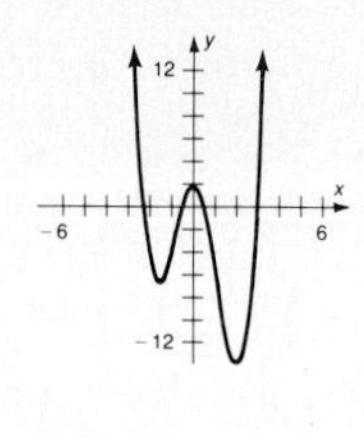

27. 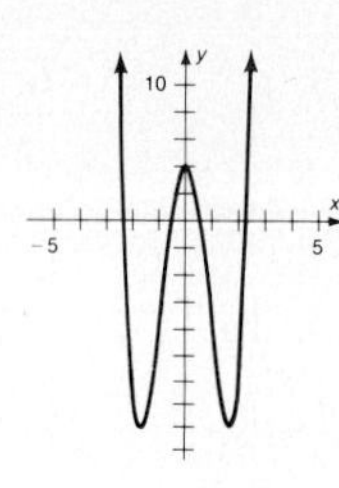

29. 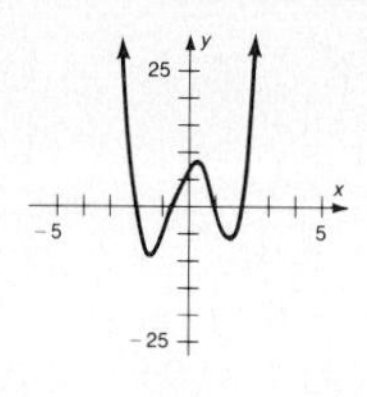

31. 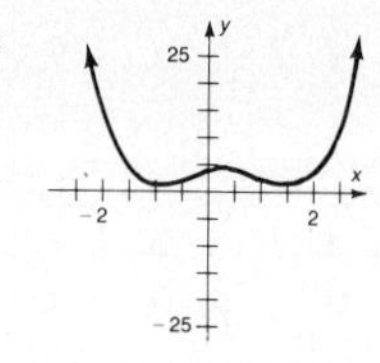

33. 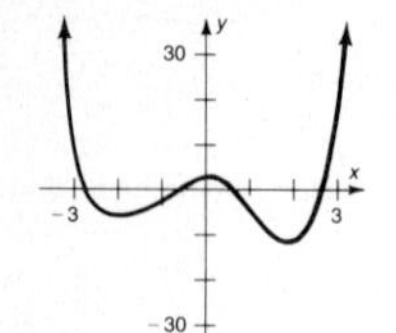

35. 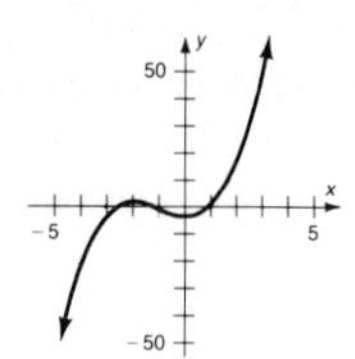

37. 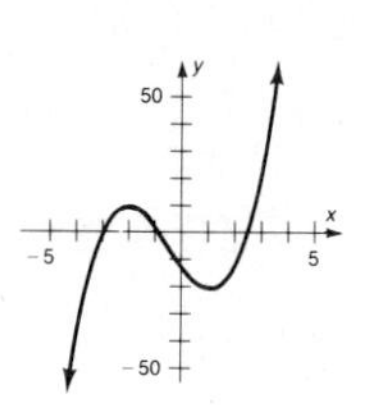

39. 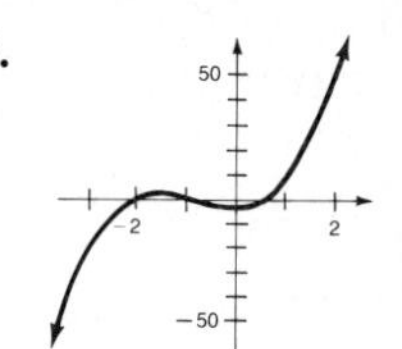

41. 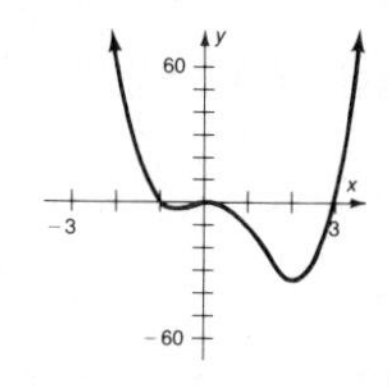

43. 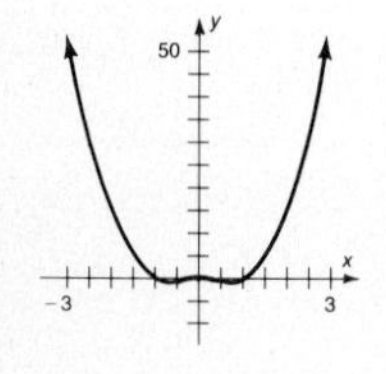

45. 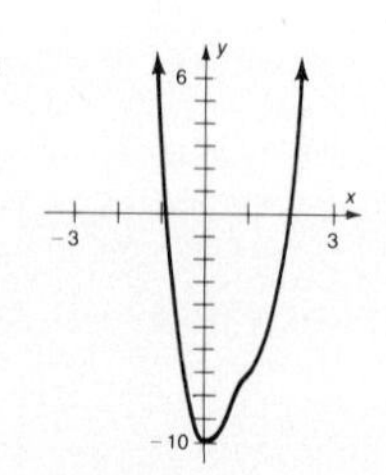

47. 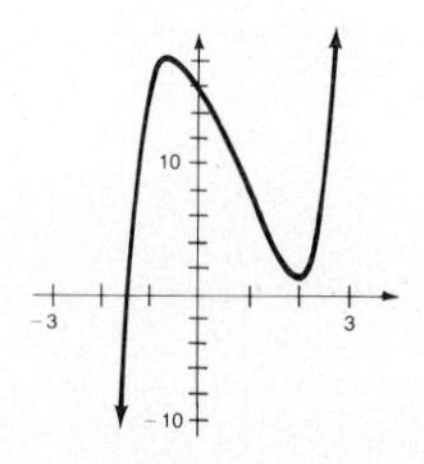

49. 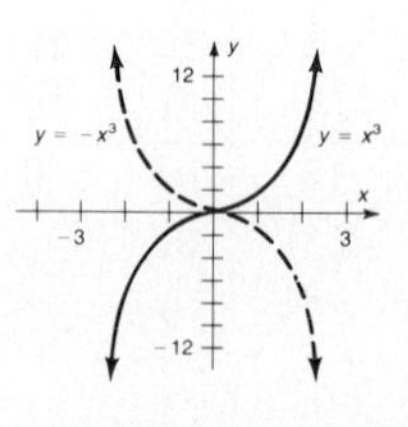

51. 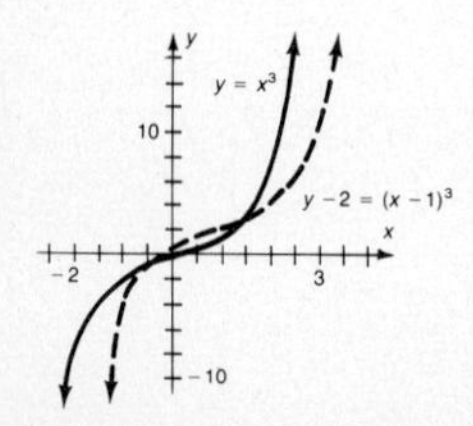

53. 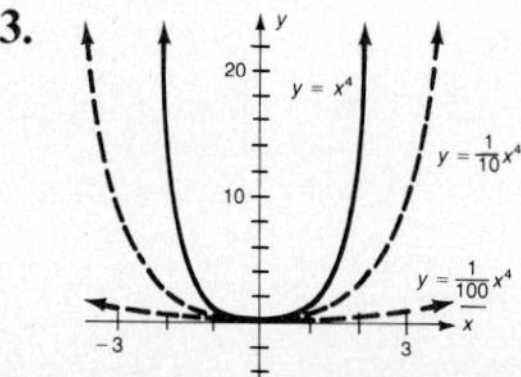

55. Answers vary.

57. 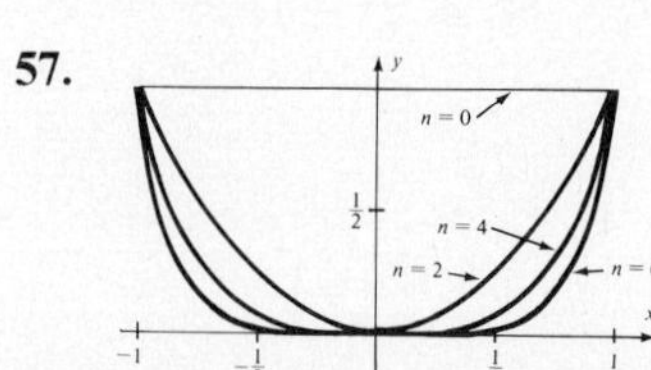

59.

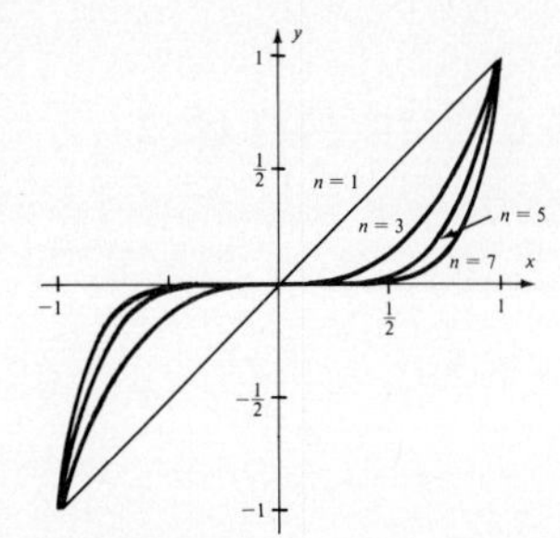

61. Answers vary.

PROBLEM SET 3.4, PAGES 109–110

1. 2, -3 (mult. 2) **3.** 0 (mult. 3), $\frac{3}{2}$ (mult. 2) **5.** -2, 1 (mult. 2), -1 (mult. 2) **7.** -5 (mult. 2), 3 (mult. 2)
9. 0 (mult. 2), 4 (mult. 2) **11.** 0 (mult. 2), 3 (mult. 2), -3 (mult. 2) **13.** 4, 2, or 0 pos; 0 neg **15.** 2 or 0 pos; 1 neg
17. 1 pos; 2 or 0 neg **19.** 1 pos; 2 or 0 neg **21.** 2 or 0 pos; 2 or 0 neg **23.** 2 or 0 pos; 1 neg **25.** $\pm 1, \pm 3, \pm 5, \pm 15$
27. $\pm 1, \pm 2, \pm 3, \pm 4, \pm 6, \pm 12, \pm\frac{1}{2}, \pm\frac{3}{2}$ **29.** $\pm 1, \pm 2, \pm 3, \pm 4, \pm 6, \pm 12$
31. $\pm 1, \pm 3, \pm 5, \pm 15, \pm\frac{1}{2}, \pm\frac{3}{2}, \pm\frac{5}{2}, \pm\frac{15}{2}$ **33.** $\pm 1, \pm 2, \pm 3, \pm 6, \pm 9, \pm 18$ **35.** $\pm 1, \pm\frac{1}{2}, \pm\frac{1}{3}, \pm\frac{1}{6}$
37. $\{1, 2, -2\}$ **39.** $\{2, -3, 3\}$ **41.** $\{2, -2, -3\}$ **43.** $\{-3, -\frac{1}{2}, 5\}$ **45.** $\{1, -1, -6, 3\}$ **47.** $\{-3, -5, -7\}$
49. $\{\frac{1}{2}, -\frac{5}{2}, \frac{7}{2}\}$ **51.** $\{-1, 3, -3, -4\}$ **53.** $\{0, 1, -1, -3\}$ **55.** $\{2, -2\}$ **57.** yes
59. The dimensions of the original cube are 8 cm. **61.** 1.76224... or 1.8 to the nearest tenth
63. The roots, to the nearest tenth, are -0.2, 2.6, 0.4, and -4.8.

PROBLEM SET 3.5, PAGES 114–115

1. $-24 + 4i$ **3.** $24 - 14\sqrt{2}$ **5.** $12 + 16i$ **7.** $-30 + 6i$ **9.** $62 - 38\sqrt{3}$ **11.** 0 **13.** $-48i$ **15.** 0 **17.** 0
19. 5 **21.** 0 **23.** 0 **25.** -3 **27.** 0 **29.** $30 + 6i$ **31.** not a root **33.** yes; $1 + 2i$ **35.** not a root

37. $\{2, -1 \pm \sqrt{3}i\}$ **39.** $\left\{5, \dfrac{-5 \pm 5\sqrt{3}i}{2}\right\}$ **41.** $\{\pm 3, \pm 3i\}$ **43.** $\{\pm 2i, \pm\sqrt{5}i\}$ **45.** $\{\pm 3i, \pm 2i\}$

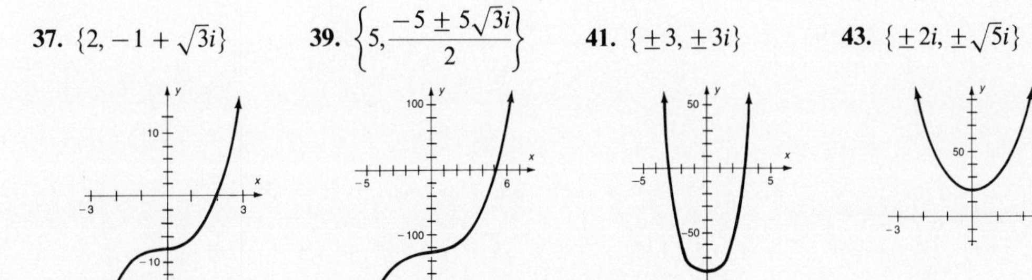

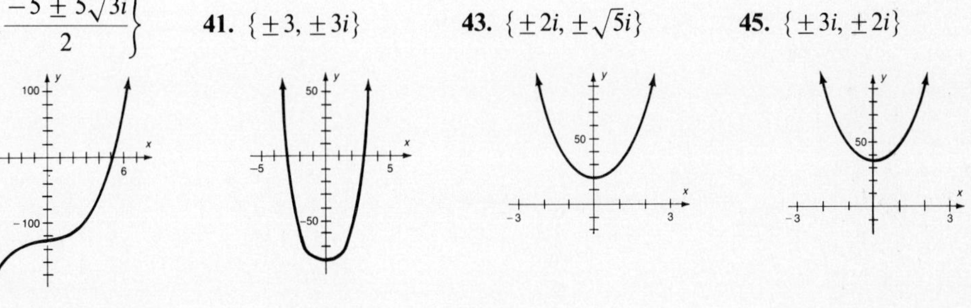

47. $\{3 \pm i, 4 \pm i\}$ **49.** $\left\{2 \pm \sqrt{5}, \dfrac{3 \pm \sqrt{11}i}{2}\right\}$ **51.** $\{-\frac{1}{2}, 1 \pm \sqrt{2}i\}$ **53.** $\{\pm 1, 1 \pm 2i\}$ **55.** $\left\{5, -\dfrac{1}{3}, \dfrac{3 \pm \sqrt{7}}{3}\right\}$

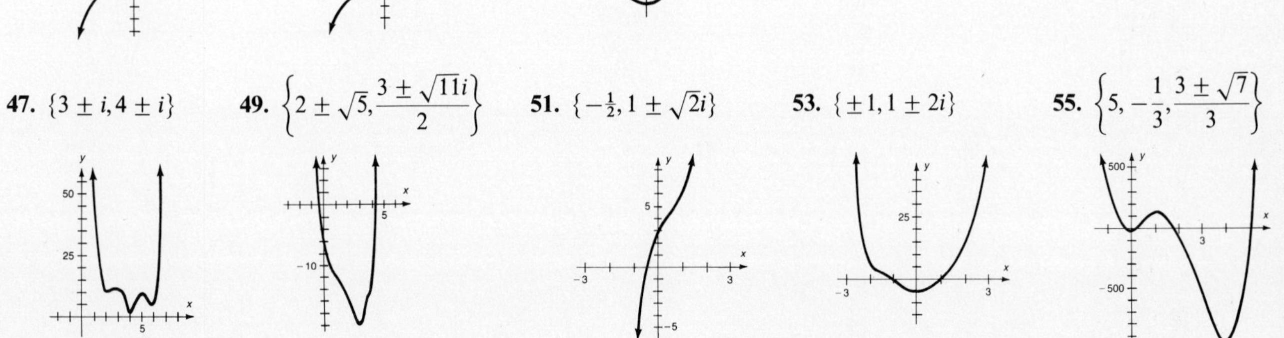

57. $\{\pm 3i, \pm 2i\}$ **59.** $\{2 \pm i, \pm\sqrt{2}\}$ **61.** $\{-1, \frac{3}{2}, \pm\sqrt{5}\}$ **63.** $\{\frac{3}{2}, -3 \pm \sqrt{2}, \pm i\sqrt{2}\}$ **65.** $\left\{\pm\sqrt{2}, \dfrac{-1 \pm \sqrt{3}i}{2}\right\}$

PROBLEM SET 3.6, PAGES 124–125

1. 0 **3.** $-\frac{1}{6}$ **5.** $\frac{1}{2}$ **7.** doesn't exist **9.** 12 **11.** -2 **13.** doesn't exist **15.** 0 **17.** $\frac{1}{4}$ **19.** 2
21. doesn't exist **23.** 5 **25.** doesn't exist **27.** $\frac{4}{3}$ **29.** $\frac{3}{5}$ **31.** $x = 0; y = 0$ **33.** $x = 0; y = 1$
35. $x = 0; y = 2$ **37.** $x = -3; y = 0$ **39.** $x = 4; y = x + 4$ **41.** $x = 1; y = -x - 1$
43. $x = 2; x = -3; y = 0$ **45.** $x = 5; x = -4; y = 0$ **47.** none **49.** none
51. $x = \frac{1}{3}; x = 1, y = 5x - 5$ **53.** none **55.** none **57.** none **59.** $x = 2; x = -4; y = 1$

PROBLEM SET 3.7, PAGES 129–130

1.
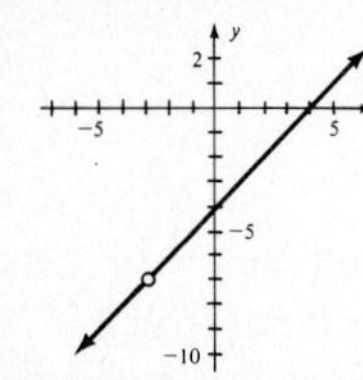

3.
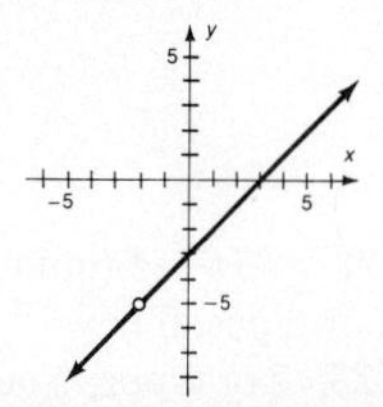

5.
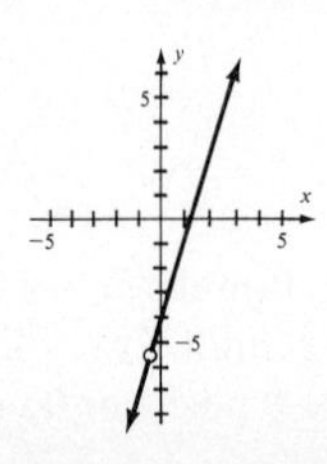

7.
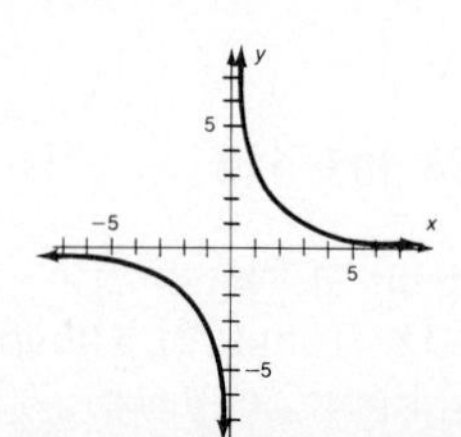

9.
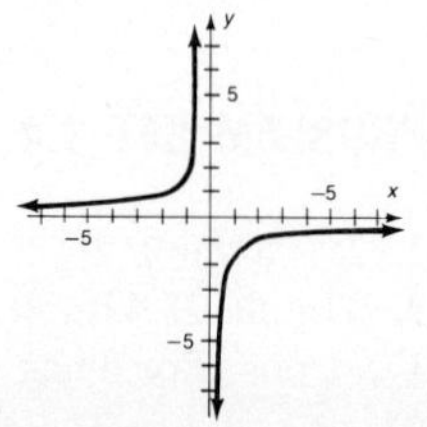

11.
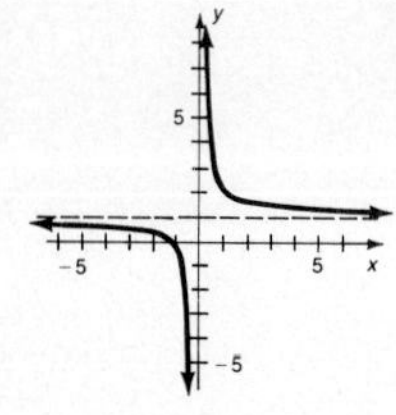

13.
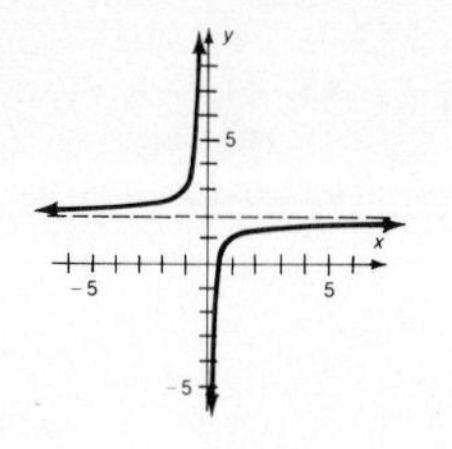

15.
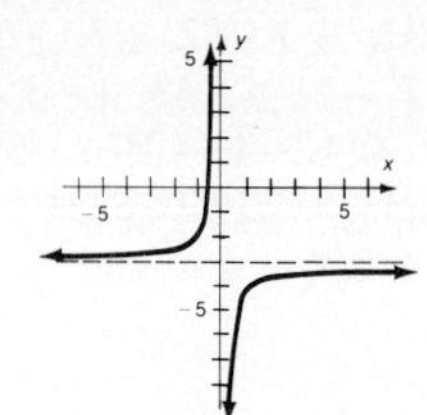

17.
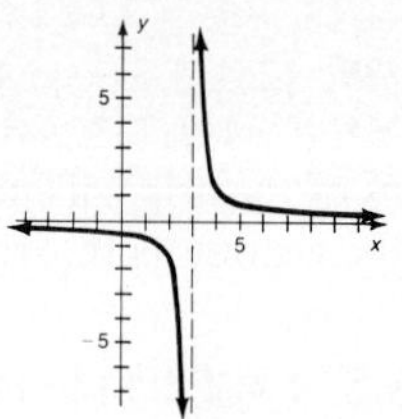

19.
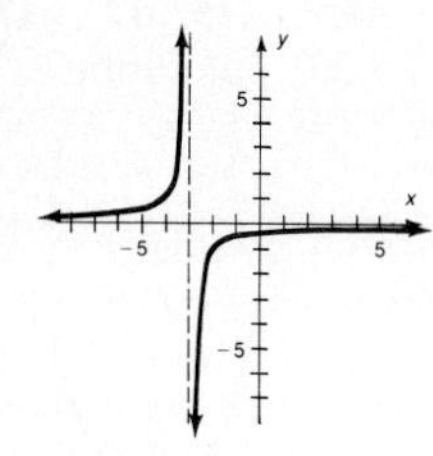

21.
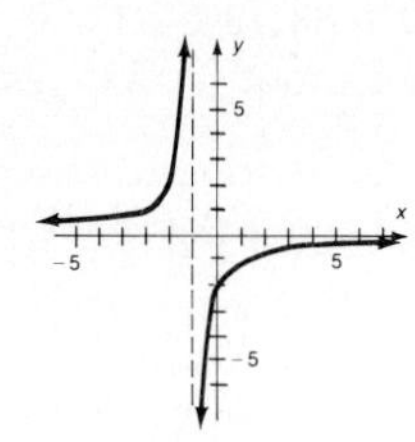

23.
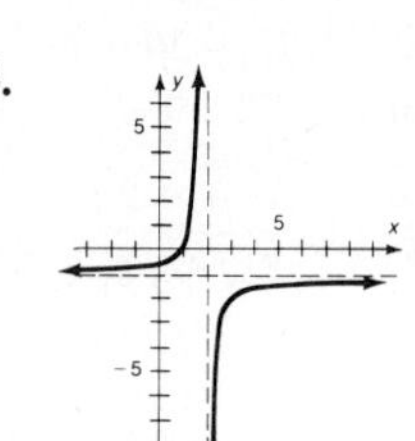

25.
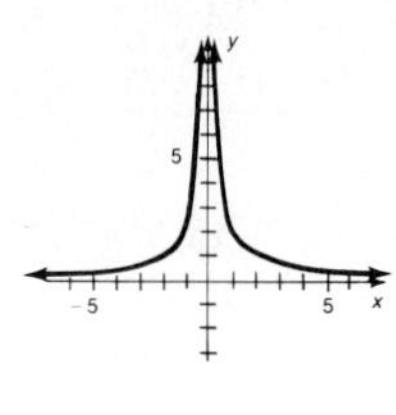

27.
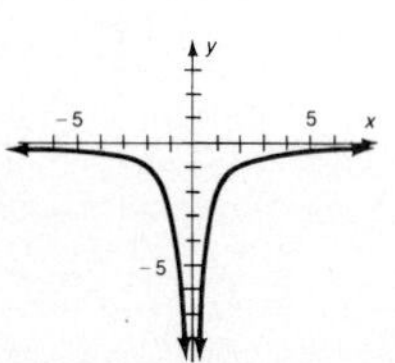

29.
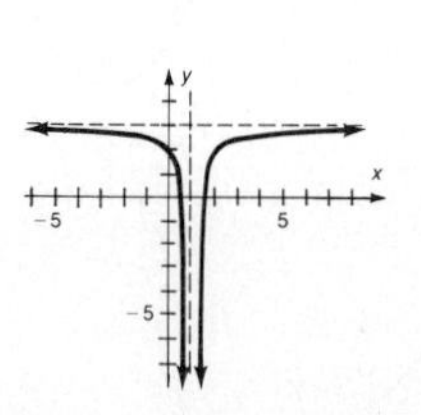

31.
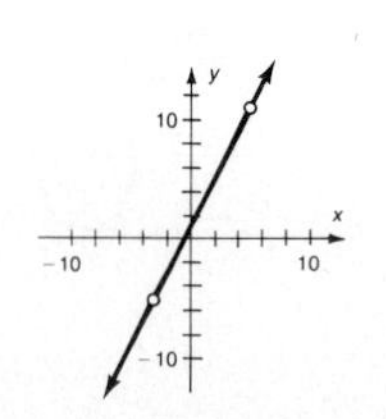

33.
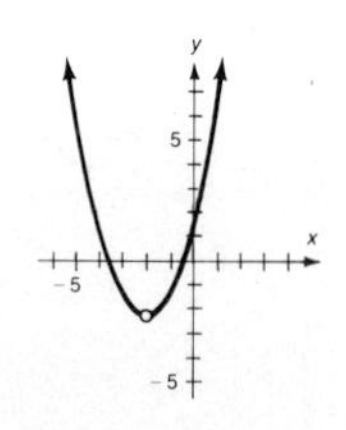

35.
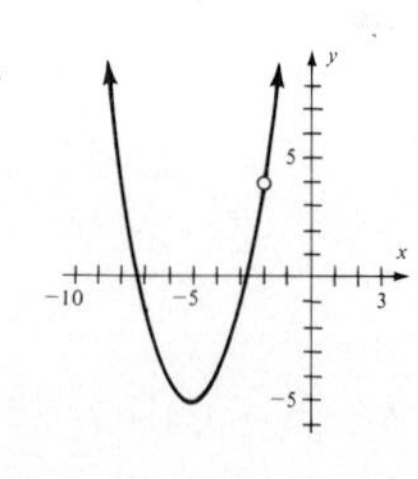

37.
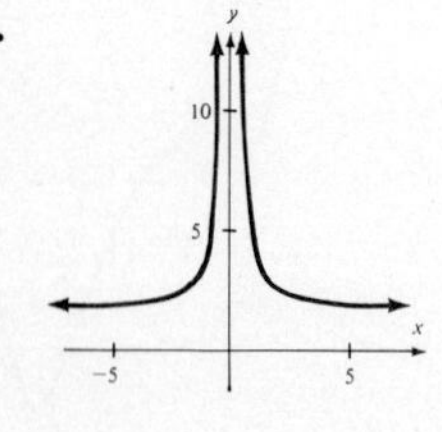

39.
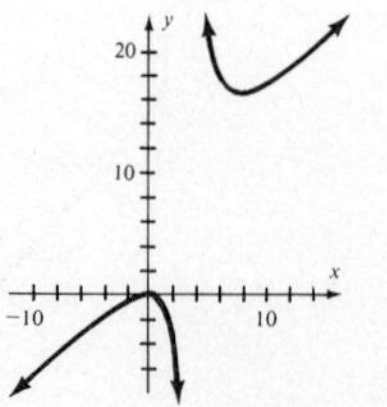

41.
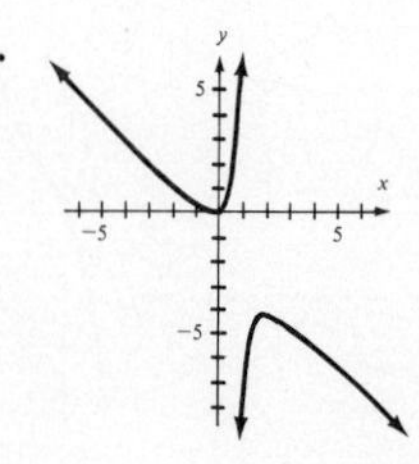

43.
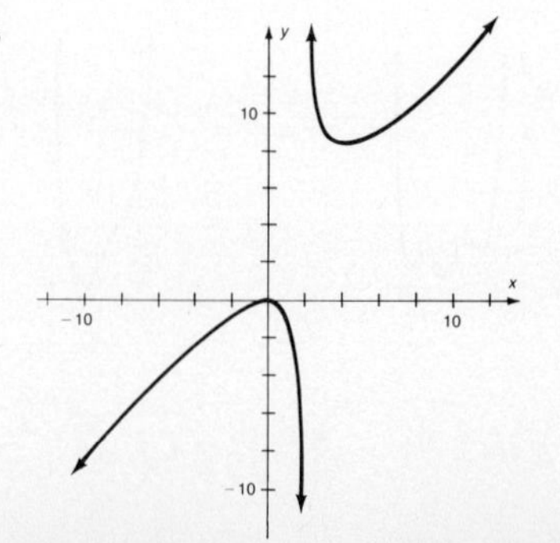

45.

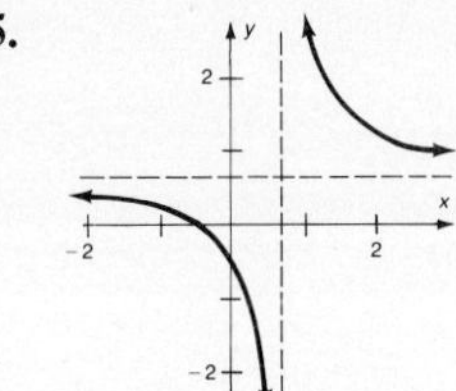

47.

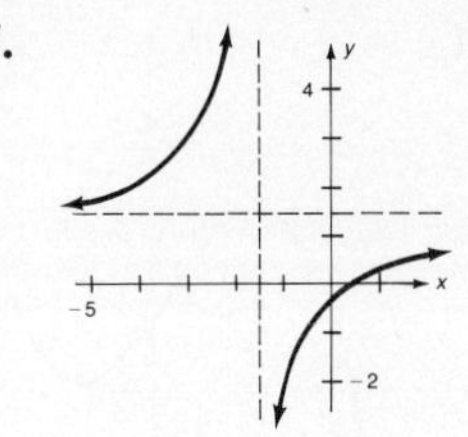

49.

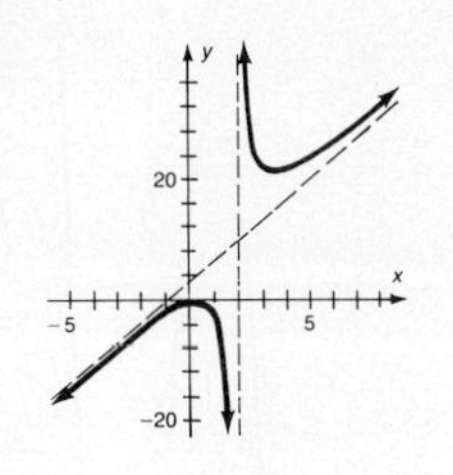

51.

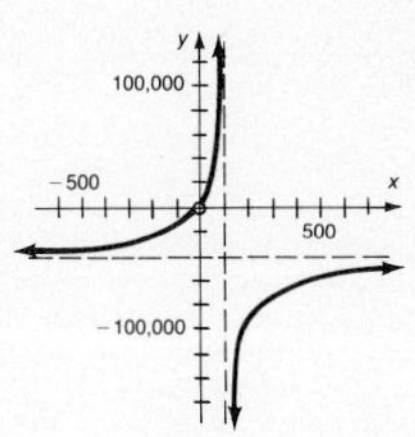

53.

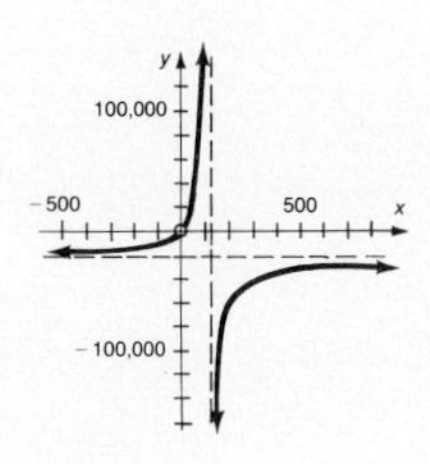

55.

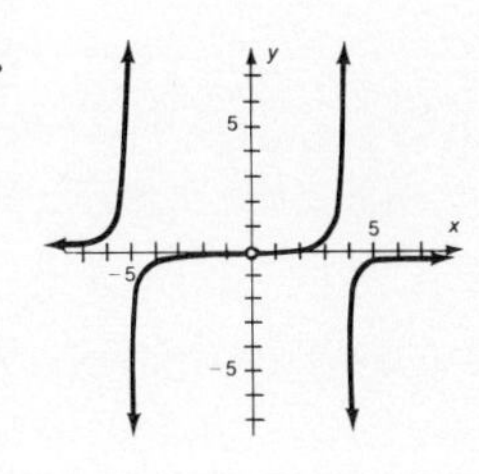

57.

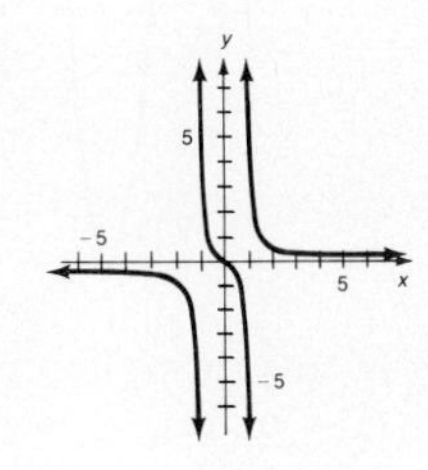

59.

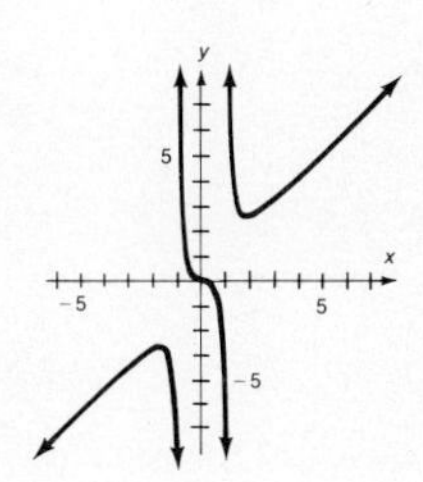

61. Answers vary.

PROBLEM SET 3.8, PAGE 134

1. $(x + 3); (x + 4)$ **3.** $(x - 3); (2x + 1)$ **5.** $(x - 8); (x - 6)$ **7.** $x; (x - 1); (x + 1)$ **9.** $x; (x - 2); (x + 2)$ **11.** $(x - 5); (x - 5)^2$ **13.** $x; (x - 1); (x - 1)^2$ **15.** $(x - 2); (x - 2)^2; (x - 2)^3$ **17.** $(x - 3); (x + 1)$ **19.** $x; (x + 1); (x - 2)$ **21.** $x; (x - 4); (x - 1)$ **23.** $(1 - x); (1 + x); (1 + x^2)$ **25.** $\frac{1}{x} + \frac{2}{x^2} + \frac{5}{x^3}$

27. $\frac{2}{x} - \frac{5}{x^2} + \frac{4}{x^3}$ **29.** $\frac{1}{x + 4} - \frac{1}{x + 5}$ **31.** $\frac{4}{x - 2} + \frac{3}{x - 1}$ **33.** $\frac{2}{x + 2} + \frac{5}{x - 4}$ **35.** $\frac{4}{x + 3} - \frac{2}{x - 2}$

37. $\frac{1}{x - 2} + \frac{3}{x + 2}$ **39.** $\frac{3}{x - 4} - \frac{2}{x - 5}$ **41.** $\frac{3}{x} - \frac{3}{x - 1} + \frac{4}{x + 1}$ **43.** $\frac{2}{x - 2} + \frac{3}{(x - 2)^2}$ **45.** $\frac{1}{x} + \frac{3}{(x + 1)^2}$

47. $\frac{2}{x + 1} + \frac{4}{(x + 1)^2} - \frac{3}{(x + 1)^3}$ **49.** $\frac{5}{6(x + 5)} + \frac{1}{6(x - 1)}$ **51.** $\frac{13}{3(x - 2)} + \frac{8}{3(x + 1)}$ **53.** $\frac{1}{x} + \frac{3}{x + 3} - \frac{4}{x - 2}$

55. $\frac{3}{x - 1} + \frac{2x - 4}{x^2 + 1}$ **57.** $x + 2 + \frac{3}{x - 1} + \frac{1}{(x - 1)^2}$ **59.** $\frac{1}{1 - x} - \frac{3}{1 + x} + \frac{2x + 1}{1 + x^2}$

CHAPTER 3 SUMMARY, PAGES 135–137

1. $P(x) = a_n x^n + a_{n-1}x^{n-1} + a_{n-2}x^{n-2} + \cdots + a_1 x + a_0$, where n is an integer greater than or equal to zero and the coefficients are real numbers. **3.** b^{m+n} **5.** $9x^4 + 15x^3 - 101x^2 - 71x - 12$ **7.** $-3w^3 - 4w^2 + 38w + 13$

9. $\frac{1}{y^2}(2x - 2xy - y^2)(2x + 2xy + y^2)$ **11.** $(2x - 1)(x + 1)(4x^2 + 2x + 1)(x^2 - x + 1)$ **13.**

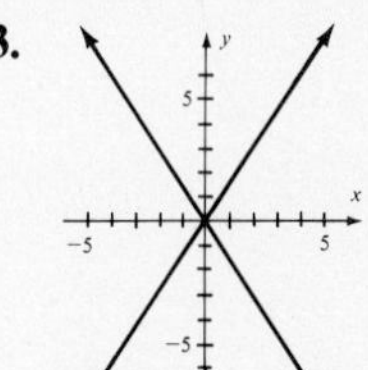

15.

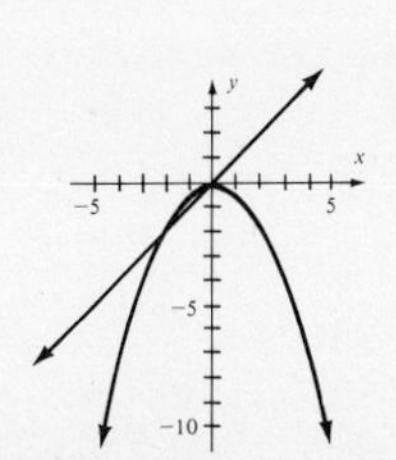

17. $2x + 3$ **19.** $3x^3 - 2x^2 + 3x + \frac{-2}{2x + 1}$ **21.** $x^3 - 2x^2 + 6x - 13$

23. $3x^2 + 8x + 4$ **25.** -42 **27.** 0 **29.**

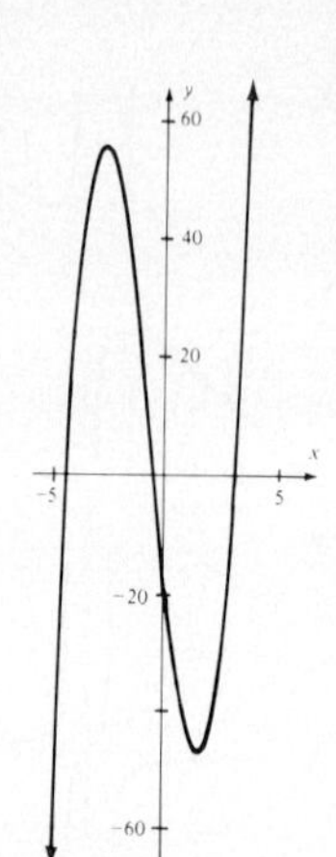

31.

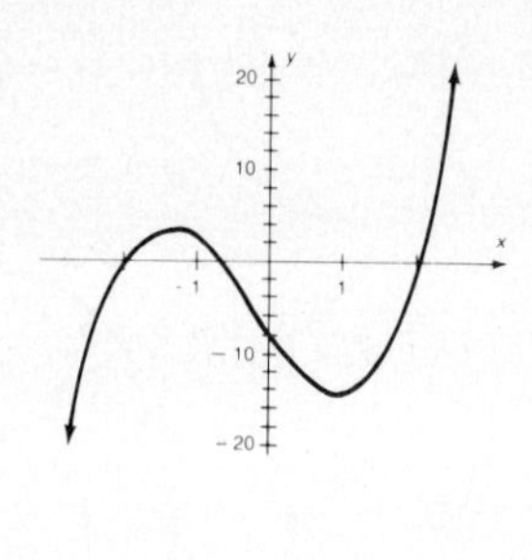

33. 1 (multiplicity 3); -2 (multiplicity 2) **35.** 0 (multiplicity 2); 4 (multiplicity 2); -4 (multiplicity 2)
37. $\pm 1, \pm 2, \pm 3, \pm 4, \pm 6, \pm 12, \pm \frac{1}{3}, \pm \frac{2}{3}, \pm \frac{4}{3}$ **39.** $\pm 1, \pm 2, \pm 4, \pm 8, \pm \frac{1}{3}, \pm \frac{2}{3}, \pm \frac{4}{3}, \pm \frac{8}{3}$
41. 1 pos real root; 2 or 0 neg real roots **43.** 1 pos real root; 2 or 0 neg real roots **45.** $\{-\frac{1}{3}, -4, 3\}$
47. $\{2, -2, -\frac{2}{3}\}$ **49.** 0 **51.** 0 **53.** $\{2, -3, 2 \pm \sqrt{3}\}$ **55.** $\{0, -1, \pm\sqrt{14}\}$ **57.** $\{1 \pm i, \pm\sqrt{3}i\}$
59. $\{1 \pm \sqrt{3}i, \pm\sqrt{5}\}$ **61.** 2 **63.** $-\frac{1}{2}$ **65.** $x = 2$; $y = 2$ **67.** $x = 2$; $y = 2x + 1$

69.

71.

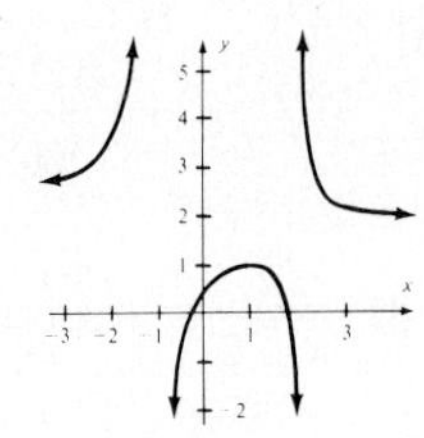

73. $\dfrac{3}{x-1} + \dfrac{2}{x-2} + \dfrac{-1}{(x-2)^2}$ **75.** $2 + \dfrac{3}{x+5} + \dfrac{6}{x-3}$

PROBLEM SET 4.1, PAGES 144–145

1. 5 **3.** -3 **5.** 6 **7.** 16 **9.** -8^3 or -512 **11.** not defined **13.** 8 **15.** 108 **17.** $\frac{1}{100}$
19. $\frac{1}{1000}$ **21.** $x + 1$ **23.** $1 + x$ **25.** $\dfrac{x^2}{y}$ **27.** $x + 2x^{1/2}y^{1/2} + y$ **29.** $x + y$

31.

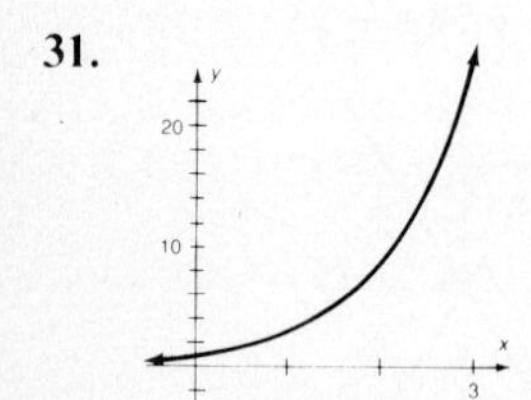

33.

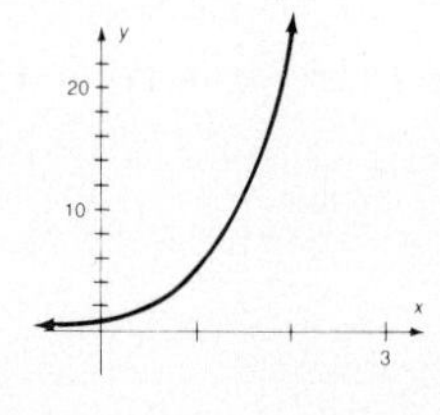

35.

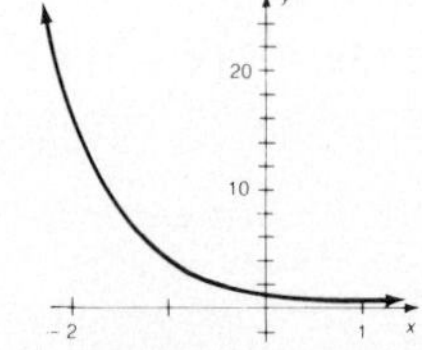

37.

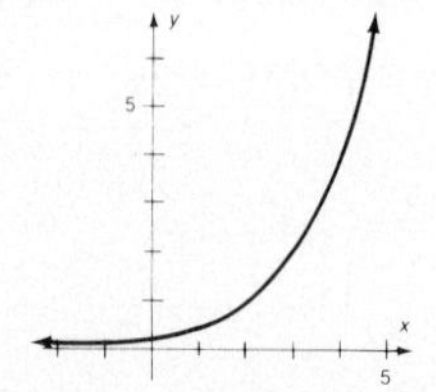

39.

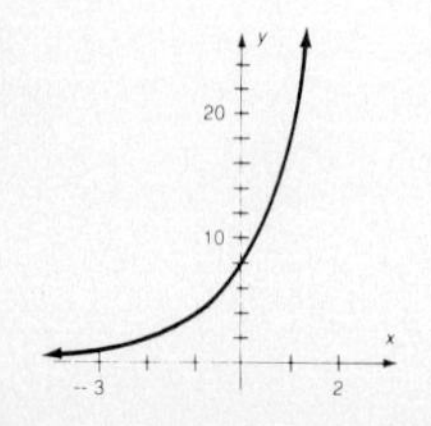

41.

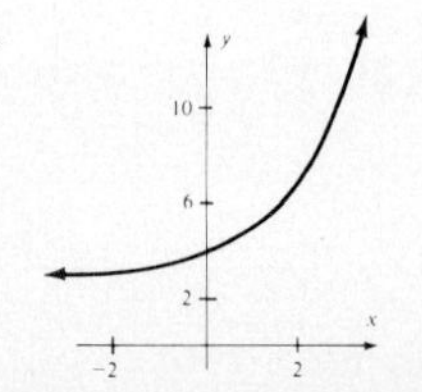

43.

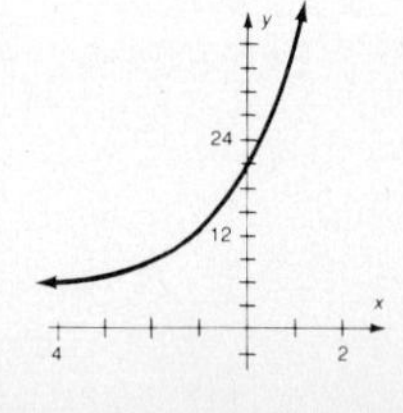

45.

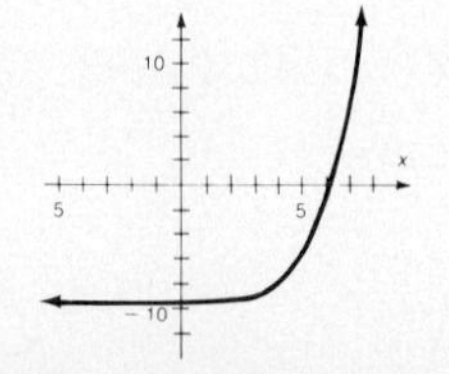

47. 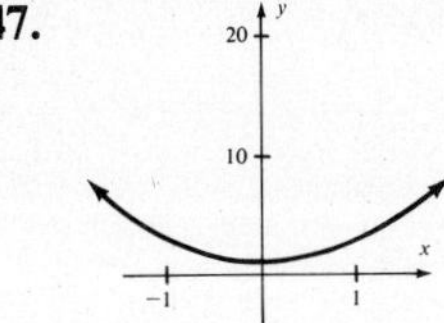

49. 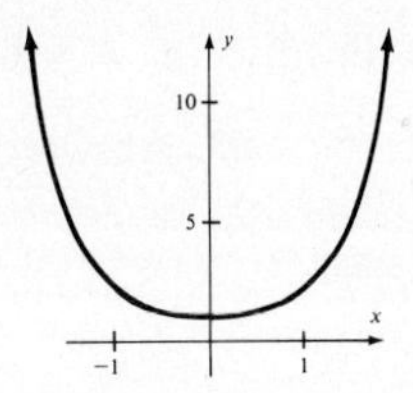

51. 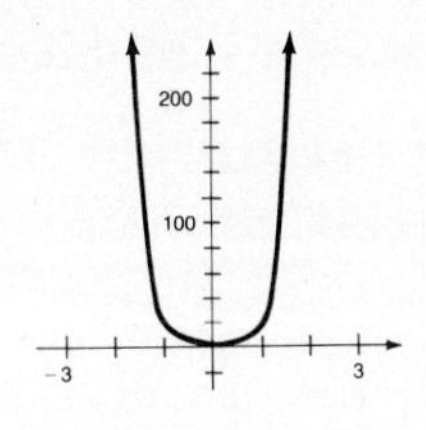

53. 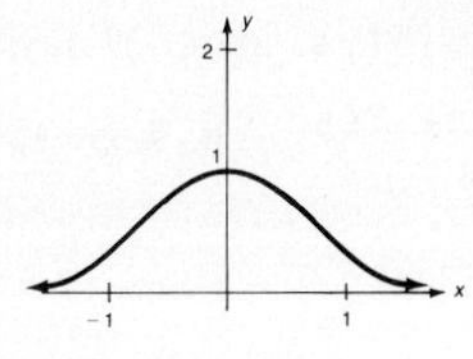

55. $2^{\sqrt{2}} \approx 2.7$ **57.** $10^{\sqrt{2}} \approx 26.0$

59. **a.** If $b = 1$, then $b^x = 1$ is a constant function; algebraic
b. If $b = 0$, then $b^x = 0$ is a constant function; algebraic

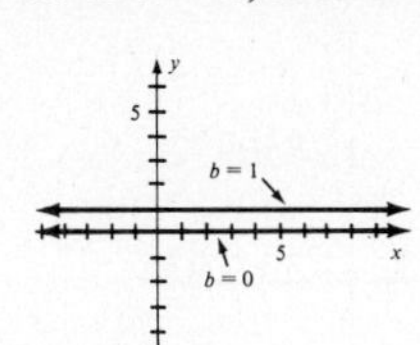

61. 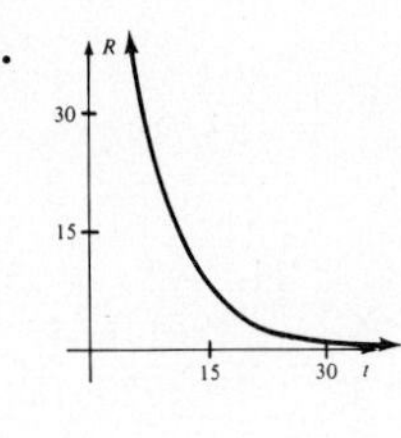

63. 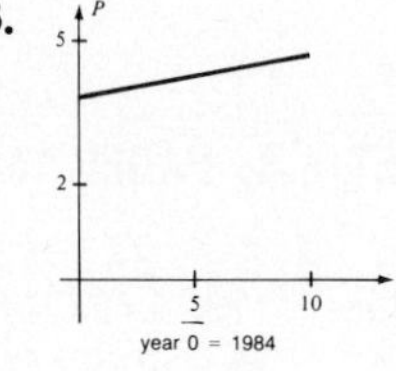

PROBLEM SET 4.2, PAGES 155–156

1. $\log_2 64 = 6$ **3.** $\log 1000 = 3$ **5.** $\log_5 125 = 3$ **7.** $\log_n m = p$ **9.** $\log_{1/3} 9 = -2$ **11.** $\log_9 \frac{1}{3} = -\frac{1}{2}$
13. $10^4 = 10{,}000$ **15.** $10^0 = 1$ **17.** $e^2 = e^2$ **19.** $e^5 = x$ **21.** $2^{-3} = \frac{1}{8}$ **23.** $4^{\frac{1}{2}} = 2$ **25.** $m^p = n$
27. $x^3 = 8$ **29.** 2 **31.** 4 **33.** -1 **35.** 2 **37.** 0.63042788 **39.** 0.92582757 **41.** 4.85491302
43. -0.49349497 or $9.50650503 - 10$ **45.** 0.81977983 **47.** 0.69314718 **49.** 2.56494936 **51.** 1.98787435
53. Answers vary. **55.** \$3,207.14 **57.** \$4,055.20 **59.** \$10,285.33 **61.** **a.** 3,572 **b.** 4,746 **c.** \$116,619
63. **a.** $m \approx 3.89$; about 3.9 or 4.0 on the Richter scale **b.** $m \approx 8.8$; about 8.8 or about 9 on the Richter scale
65. about 320,000 times noisier

PROBLEM SET 4.3, PAGES 161–162

1. 2 **3.** -1 **5.** $2\sqrt{7}$ **7.** e^2 **9.** e^4 **11.** 9.3 **13.** $5\sqrt{5}$ **15.** $\sqrt{70}$ **17.** $\frac{7}{2}$ **19.** $\varnothing$ or no values of x
21. 10^5 or 100,000

In Problems 23–31, the first number shown is what is seen on the calculator display, the second number is the answer shown to the correct number of significant digits.

23. $57{,}334 \approx 57{,}300$ **25.** $4914 \approx 4910$ **27.** $53{,}045 \approx 53{,}000$ **29.** $93{,}877.1 \approx 94{,}000$ **31.** $0.9405645491 \approx 0.941$

33. 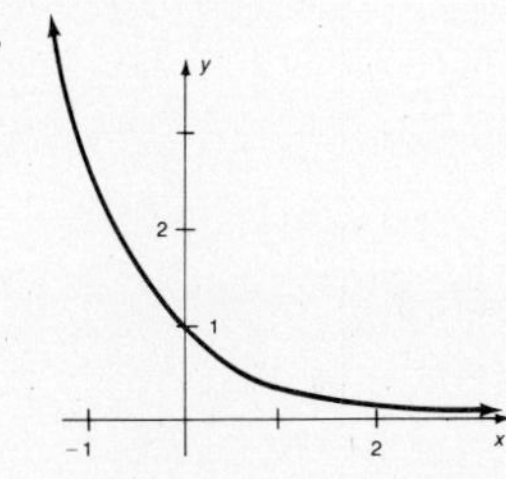

35. 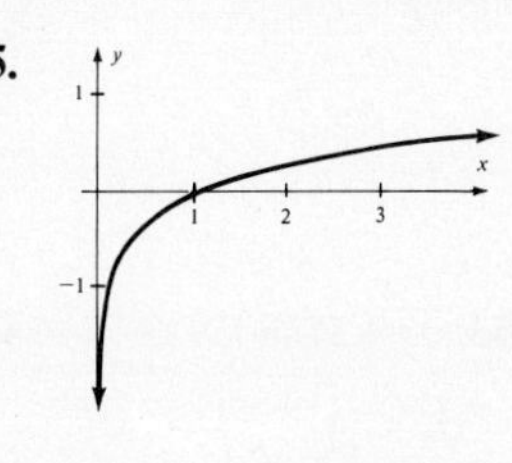

37. 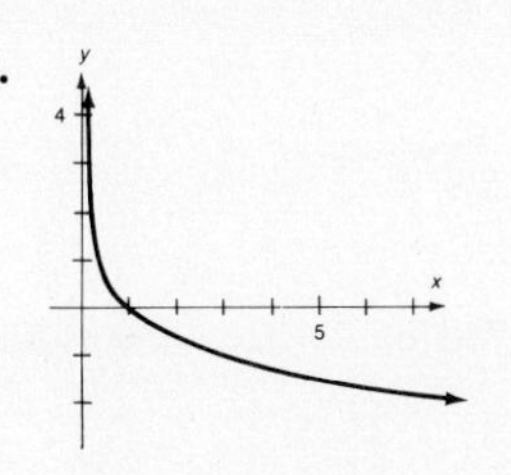

39. 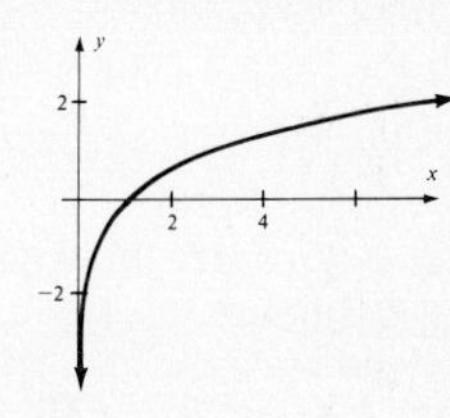

41. 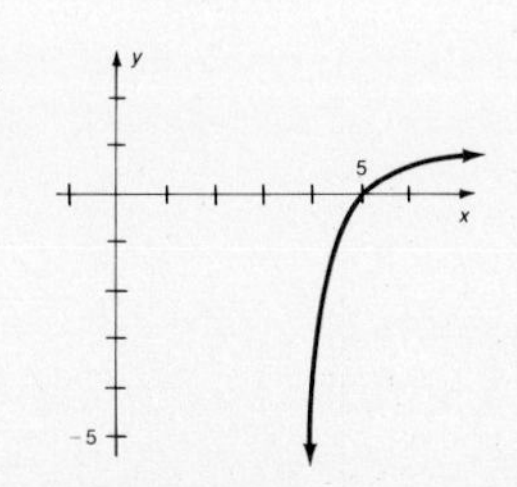

43. 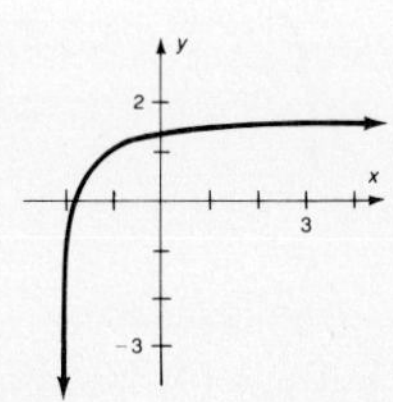
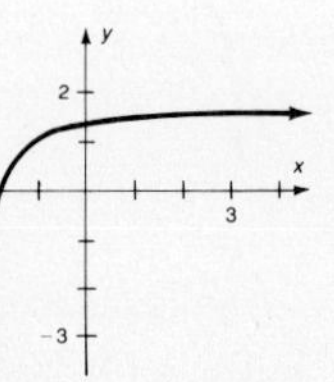

45. 2, 6 **47.** 25 **49.** 3, 6 **51.** $\dfrac{2}{e^4}$ **53.** \$1,131.47

55. \$1,530.69 **57. a.** about 29 days **b.** No; if $N = 80$, then t is not defined. **c.** $N = 80(1 - e^{-t/62.5})$

59. a. $E = 10^{1.5m+11.8}$ **b.** $3.55 \cdot 10^{24}$ ergs **61.**

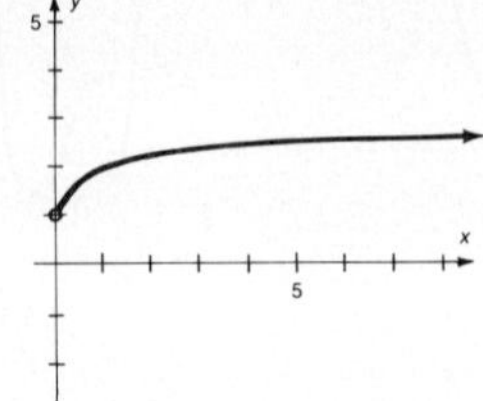

63. Answers vary.

PROBLEM SET 4.4, PAGES 168–169

1. 7 **3.** $\frac{5}{3}$ **5.** $\frac{2}{3}$ **7.** $-\frac{4}{3}$ **9.** 3 **11.** -2 **13.** $\frac{1}{4}$ **15.** $-\frac{2}{15}$ **17.** 2.1132828 **19.** -1.2850972
21. 2.011465868 **23.** 1.62324929 **25.** -2.48811664 **27.** -0.66958623 **29.** 0.78746047 **31.** 2.12907205
33. 1.15129255 **35.** -0.57564627 **37.** -0.049306144 **39.** 2.7429396 **41.** 1.4306766 **43.** -2.321928
45. -1.2920297 **47.** 11.89566 periods; about 6 years **49.** 31 quarters; about $7\frac{3}{4}$ years
51. $k \approx 0.023104906$ **53.** $t \approx 10.813529$ **55.** $t \approx 11{,}400$ **57.** about 17,400 years ago

59. It is about $\frac{1}{2}$ mile above sea level. **61.** $t = -250 \ln\left(\frac{P}{50}\right)$ **63. a.** $\frac{x^5}{5!} + \frac{x^6}{6!}$ **b.** $\frac{x^{r-1}}{(r-1)!}$

c. $$e = 1 + 1 + \frac{1}{2} + \frac{1}{2 \cdot 3} + \frac{1}{2 \cdot 3 \cdot 4} + \frac{1}{5!} + \frac{1}{6!} + \cdots$$
$$= 2 + 0.5 + 0.1666\ldots + 0.041666\ldots + 0.008333\ldots + 0.0013888\ldots + 0.0001984127\ldots + \cdots$$
$$\approx 2.718$$

d. $$e^{0.5} = 1 + 0.5 + \frac{(0.5)^2}{2} + \frac{(0.5)^3}{2 \cdot 3} + \frac{(0.5)^4}{2 \cdot 3 \cdot 4} + \frac{(0.5)^5}{5!} + \cdots$$
$$\approx 1.649$$

CHAPTER 4 SUMMARY, PAGES 169–171

1. 25 **3.** $\sqrt{3}$ **5.** **7.**

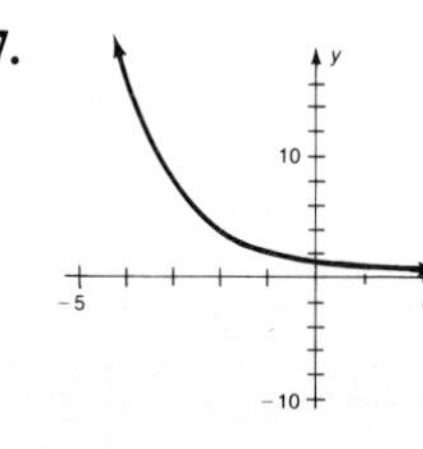

9. $\log \sqrt{10} = 0.5$ **11.** $\log_9 729 = 3$

13. $10^0 = 1$ **15.** $2^6 = 64$
17. 0.4771212547 **19.** -2.677780705 **21.** 1.098612289 **23.** -4.342805922 **25.** 3 **27.** 5

29. $\log_b A + \log_b B$ **31.** $p \log_b A$ **33.** **35.** **37.** 2 **39.** 243

41. $831{,}450 \approx 831{,}000$ **43.** 47,045,881 **45.** 0.096910013 **47.** ± 0.5888866497 **49.** 1.304007668
51. -2.768345799 **53.** 3 **55.** 1.771243749 **57.** 9.67957033 **59.** 4.191806549 **61.** $k \approx 0.026$
63. \$131,894.24 **65.** \$36,018.87

EXTENDED APPLICATION—EXPONENTIAL POPULATION GROWTH, PAGES 174–175

1. About 0.9% **3.** About 1.8% **5.** August 1988 **7.** No; it is closer to 12 than it is to 13 (actually 12.4 years). **9. a.** about 1.3% decline **b.** about 494,000 **11–15.** Answers vary.

PROBLEM SET 5.1, PAGES 184–185

1. a. $\frac{\pi}{6}$ **b.** $\frac{\pi}{2}$ **c.** $\frac{3\pi}{2}$ **d.** $\frac{\pi}{4}$ **3. a.** 180° **b.** 45° **c.** 60° **d.** 360°

5. a. **b.**

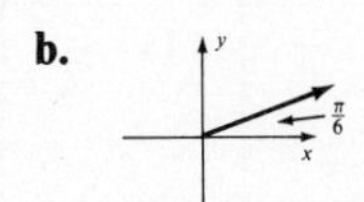

c. **d.**

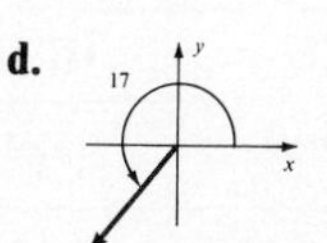

7. a. **b.**

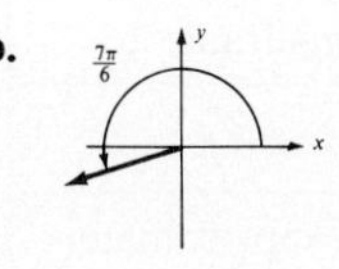

c. **d.**

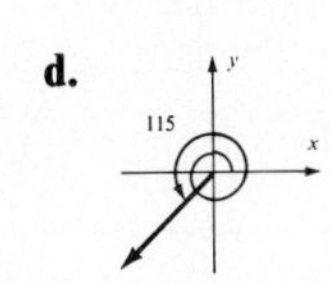

9. a. **b.**

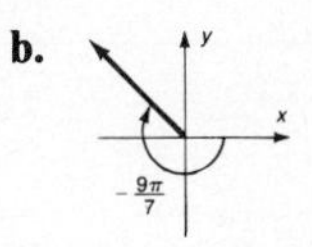

c. **d.**

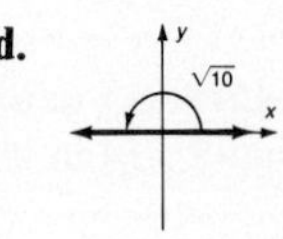

11. a. 40° **b.** 180° **c.** 30° **d.** 330° **13. a.** 240° **b.** 140° **c.** 180° **d.** 280° **15. a.** $\frac{7\pi}{4}$ **b.** $\frac{\pi}{4}$ **c.** $\frac{5\pi}{3}$ **d.** $2\pi - 2$ **17. a.** $9 - 2\pi \approx 2.7168$ **b.** $2\pi - 5 \approx 1.2832$ **c.** $\sqrt{50} - 2\pi \approx 0.7879$ **d.** $2\pi - 6 \approx 0.2832$ **19. a.** 0.5236 **b.** 5.4978 **c.** 5.4867 **d.** 3.1415 **21. a.** 60° **b.** 60° **c.** 60° **d.** 45° **23. a.** $\frac{\pi}{12}$ **b.** $\frac{\pi}{3}$ **c.** $\frac{\pi}{6}$ **d.** $\frac{\pi}{4}$ **25. a.** $\frac{\pi}{4}$ **b.** 0 **c.** $\frac{\pi}{6}$ **d.** $\frac{\pi}{3}$ **27. a.** 0.7879 **b.** 0.4250 **c.** 0.5236 **d.** 0.7854 **29.** 18.00° **31.** 300.00° **33.** 30.00° **35.** −171.89° **37.** −143.24° **39.** 29.22° **41.** $\frac{\pi}{9}$ **43.** $\frac{-11\pi}{9}$ **45.** $\frac{17\pi}{36}$ **47.** 5.48 **49.** 6.11 **51.** 17.19 **53.** 14.04 cm **55.** 4.19 m **57.** 4.89 ft **59.** 14.07 cm **61.** 87.27 cm **63.** about 440 km **65.** The diameter of the moon is about 5200 km.

PROBLEM SET 5.2, PAGES 191–192

1. 0.6 **3.** −0.6 **5.** −0.4 **7.** 2.9 **9.** 0.4 **11.** 0.34202014 **13.** 2.9238044 **15.** 0.36397023 **17.** −0.9961947 **19.** 0.20527052 **21.** 0.84147098 **23.** 0.88699492 **25.** 0.41929364 **27.** −0.49996641 **29.** + **31.** − **33.** − **35.** − **37.** − **39.** − **41.** II, III **43.** III **45.** III

47. $\cos\theta = -\frac{3}{5}$ $\sec\theta = -\frac{5}{3}$ $\sin\theta = \frac{4}{5}$ $\csc\theta = \frac{5}{4}$ $\tan\theta = -\frac{4}{3}$ $\cot\theta = -\frac{3}{4}$

49. $\cos\theta = \frac{5}{13}$ $\sec\theta = \frac{13}{5}$ $\sin\theta = \frac{12}{13}$ $\csc\theta = \frac{13}{12}$ $\tan\theta = \frac{12}{5}$ $\cot\theta = \frac{5}{12}$

51.

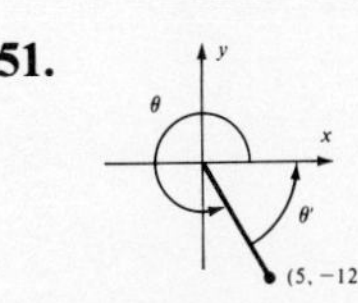

$\cos\theta = \frac{5}{13}$ $\sec\theta = \frac{13}{5}$ $\sin\theta = -\frac{12}{13}$ $\csc\theta = -\frac{13}{12}$ $\tan\theta = -\frac{12}{5}$ $\cot\theta = -\frac{5}{12}$

53. $\cos\theta = -\frac{6}{37}\sqrt{37}$ $\sec\theta = -\frac{1}{6}\sqrt{37}$ $\sin\theta = \frac{1}{37}\sqrt{37}$ $\csc\theta = \sqrt{37}$ $\tan\theta = -\frac{1}{6}$ $\cot\theta = -6$

55. $|x|\sqrt{2}$ **57.** $-x\sqrt{2}$ **59.** $3x$ **61.** $\frac{1}{2}$ **63.** −4 **65.**

$$\sin 1 = 1 - \frac{1^3}{3!} + \frac{1^5}{5!} - \frac{1^7}{7!} + \cdots$$
$$\approx 1 - 0.166667 + 0.008333 - 0.0001984 + \cdots$$
$$\approx 0.8415$$

PROBLEM SET 5.3, PAGES 197–198

1. a. 1 **b.** 1 **c.** $\frac{\sqrt{3}}{2}$ **d.** $\frac{\sqrt{3}}{2}$ **3. a.** 0 **b.** 1 **c.** 0 **d.** $\frac{\sqrt{2}}{2}$ **5. a.** $\sqrt{2}$ **b.** undefined **c.** 2 **d.** $\sqrt{3}$ **7. a.** 1 **b.** −1 **c.** −1 **d.** 0 **9. a.** $\frac{1}{2}$ **b.** $\frac{1}{2}$ **c.** $\frac{\sqrt{2}}{2}$ **d.** 1 **11. a.** 0.5650 **b.** 0.5850 **13. a.** 0.6428 **b.** 0.1763 **15. a.** −0.3640 **b.** −0.1736 **17. a.** 0.7337 **b.** −0.4791 **19. a.** 0.9320 **b.** 0.7961 **21. a.** 1.5574 **b.** 0.0709 **23. a.** −0.7470 **b.** 0.1411 **25. a.** 0.9004 **b.** 0.3555 **27.** Answers vary.

29. $\cos\frac{5\pi}{4} = \frac{-\sqrt{2}}{2}$ **31.** $\sin\frac{-\pi}{4} = \frac{-\sqrt{2}}{2}$ **33.** $\cos 210° = \frac{-\sqrt{3}}{2}$ **35.** 1 **37.** -1 **39.** $\sqrt{2}$ **41.** $\frac{1}{2}$ **43.** 1 **45.** $\frac{1}{2}$
47. 1 **49.** $\frac{\sqrt{2}}{2}$ **51.** $\frac{2}{3}\sqrt{3}$ **53.** $\frac{\sqrt{3}}{3}$ **55.** $\frac{\sqrt{3}}{2}$ **57.** $-\sqrt{3}$ **59.** $\frac{1}{2}$ **61–63.** Answers vary.

PROBLEM SET 5.4, PAGES 206–207

1.

x = angle	$\frac{2\pi}{3}$	$\frac{3\pi}{4}$	$\frac{5\pi}{6}$	$\frac{7\pi}{6}$	$\frac{5\pi}{4}$	$\frac{4\pi}{3}$	$\frac{7\pi}{4}$	$\frac{11\pi}{6}$
Quadrant	II; −	II; −	II; −	III; −	III; −	III; −	IV; +	IV; +
$y = \cos x$	$-\frac{1}{2}$	$\frac{-\sqrt{2}}{2}$	$\frac{-\sqrt{3}}{2}$	$\frac{-\sqrt{3}}{2}$	$\frac{-\sqrt{2}}{2}$	$-\frac{1}{2}$	$\frac{\sqrt{2}}{2}$	$\frac{\sqrt{3}}{2}$
y (approximate)	−0.5	−0.71	−0.87	−0.87	−0.71	−0.5	0.71	0.87

See Figure 5.17 in text for graph.

3. See Figure 5.14 in text for graph.

5.

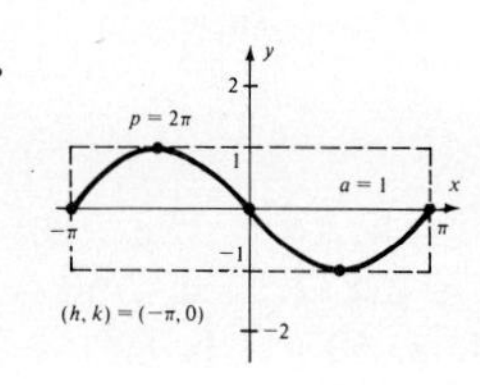

7.

9.

11.

13.

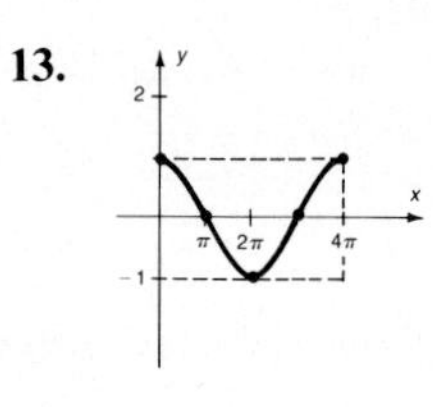

15.

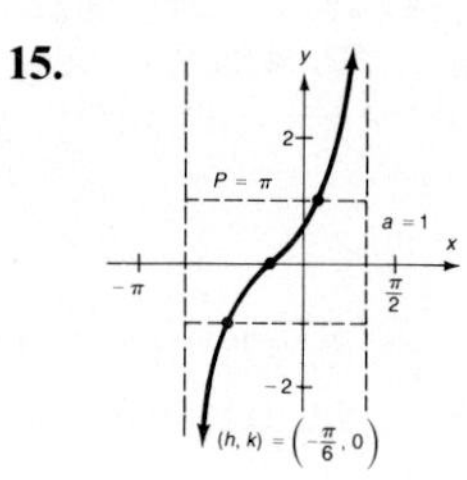

17.

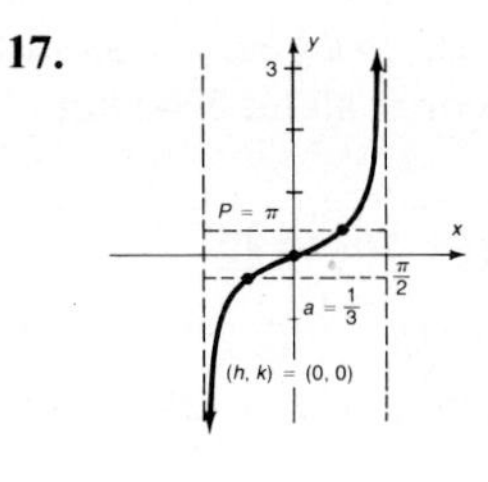

19.

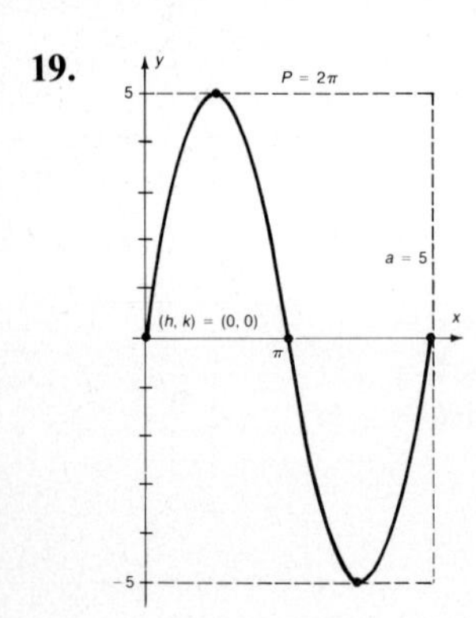

21.

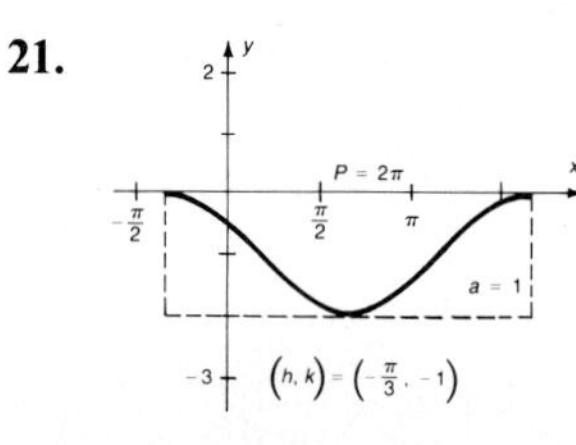

23.

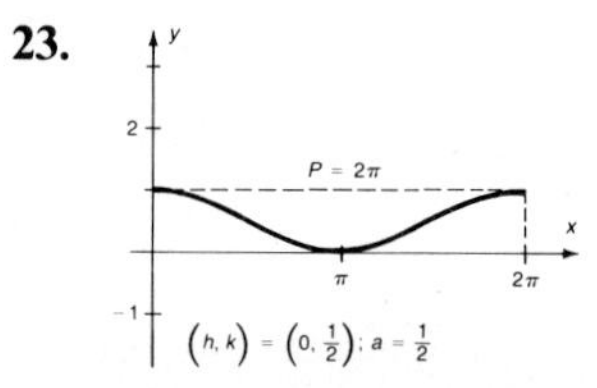

25.

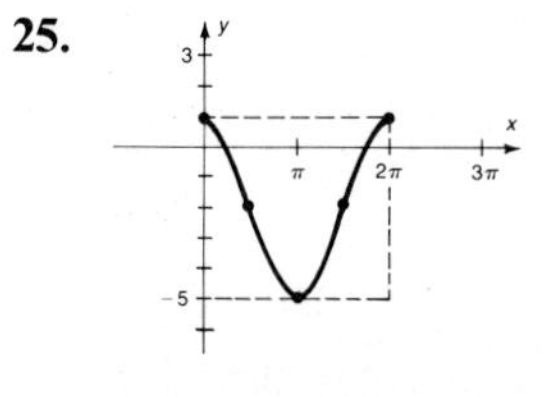

27.

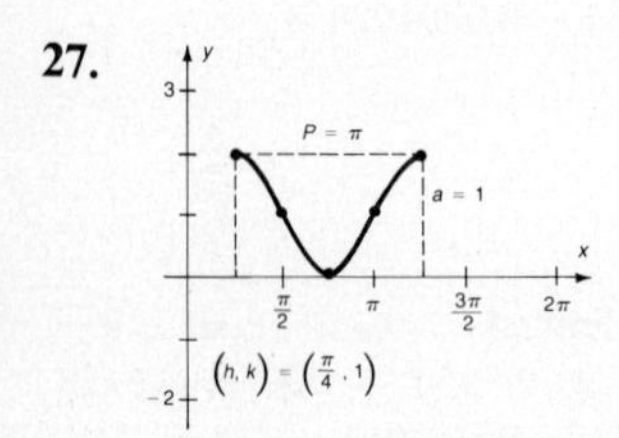

29.

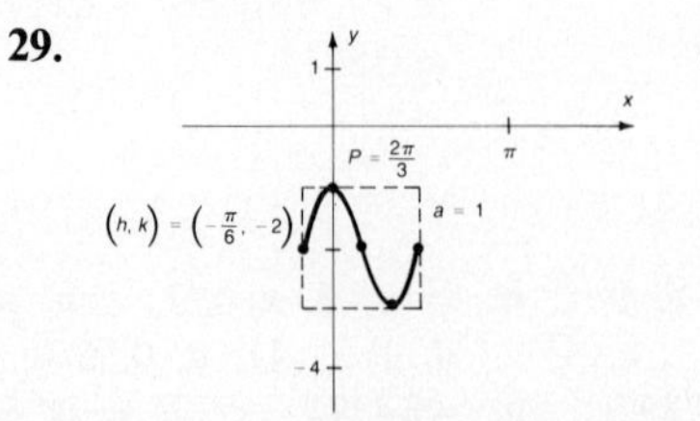

31.

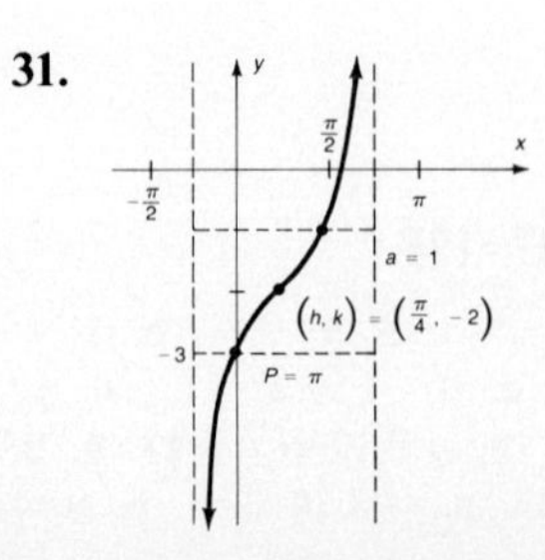

33.

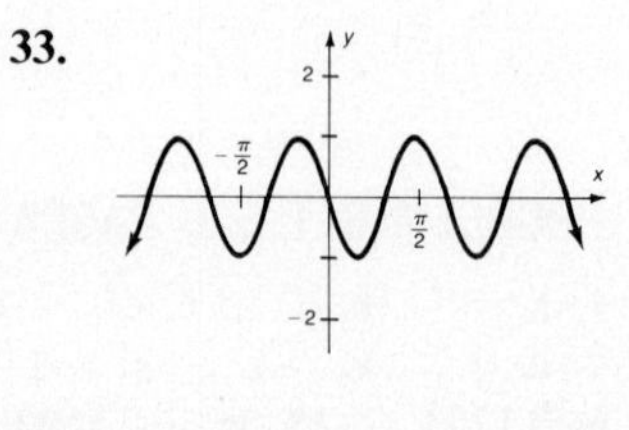

35. 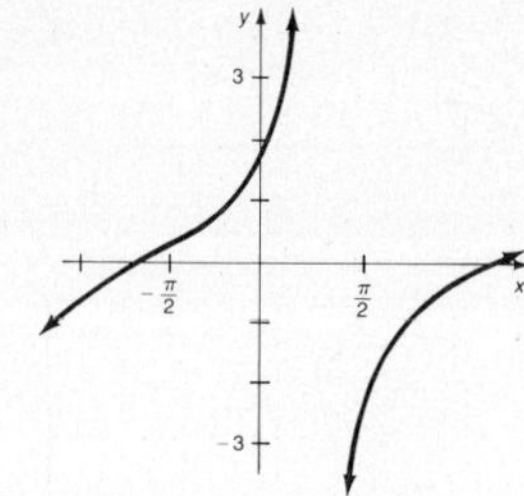

37. 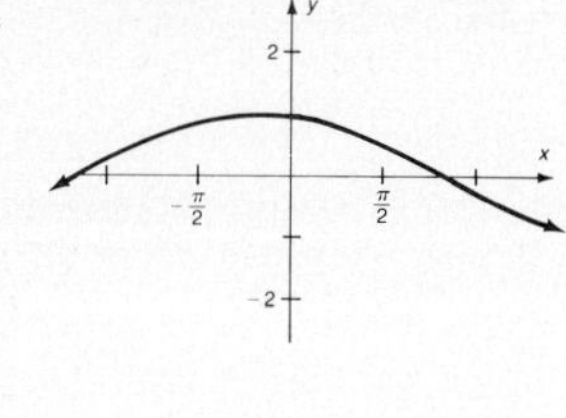

39. 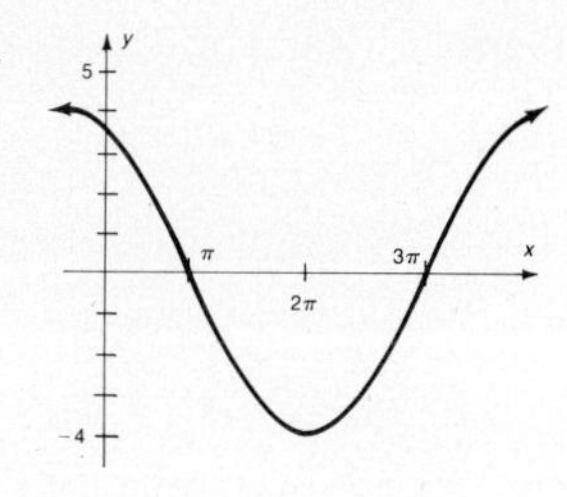

41. 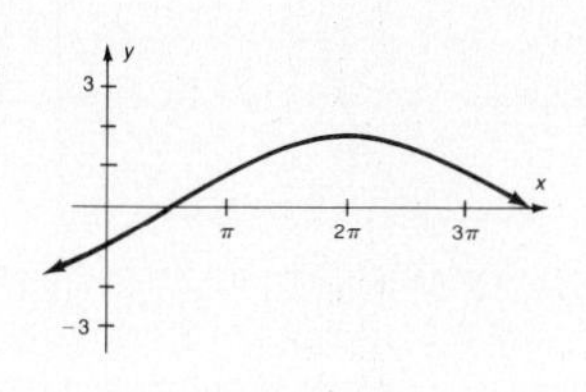

43. 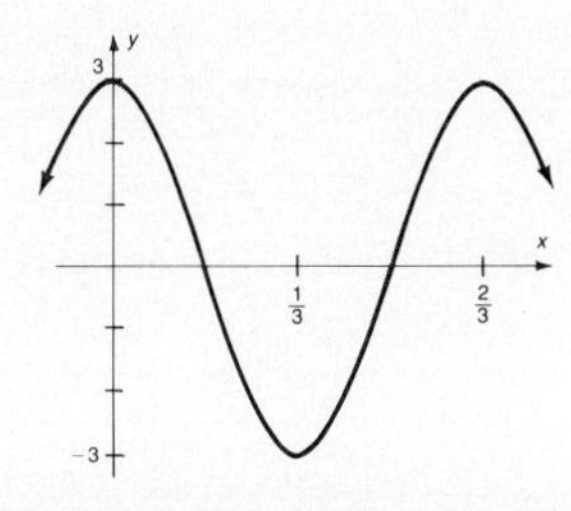

45. 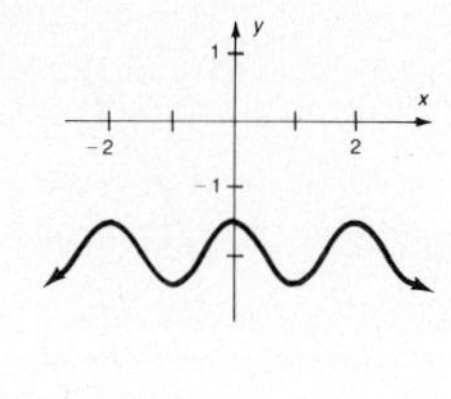

47. 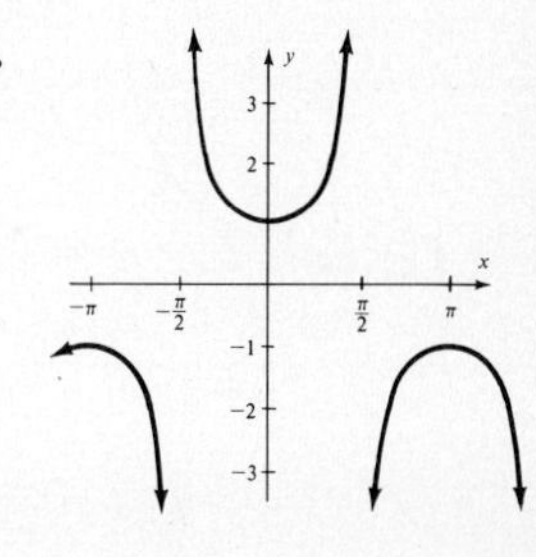

49. 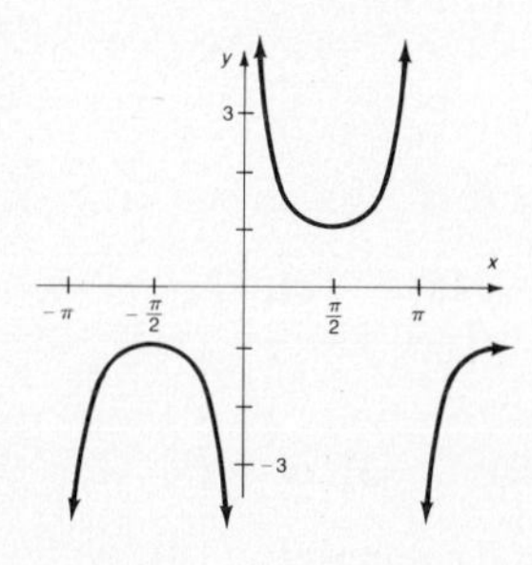

51. 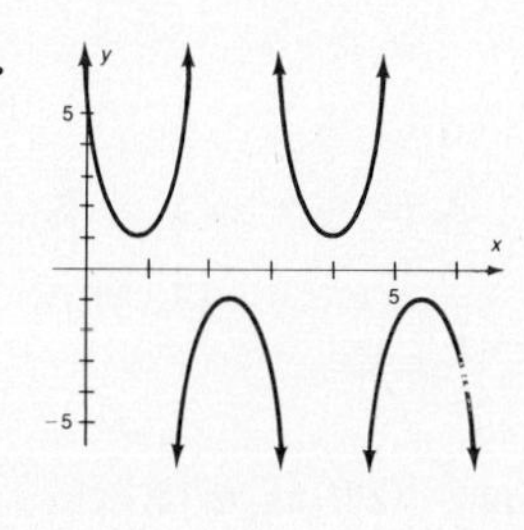

53. 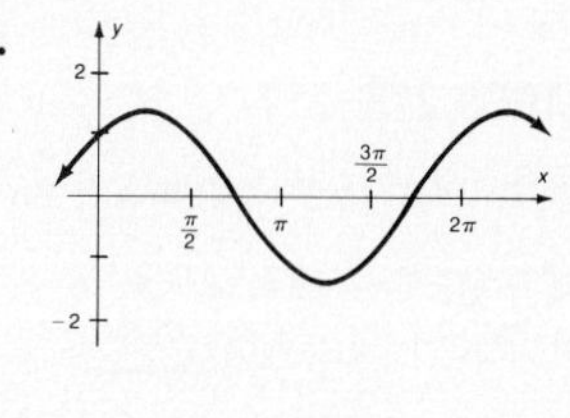

55. 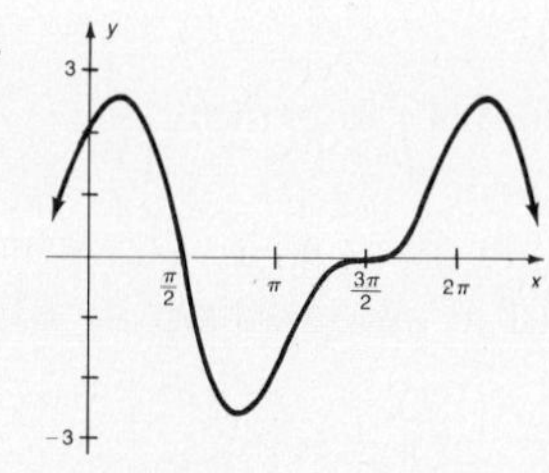

57. 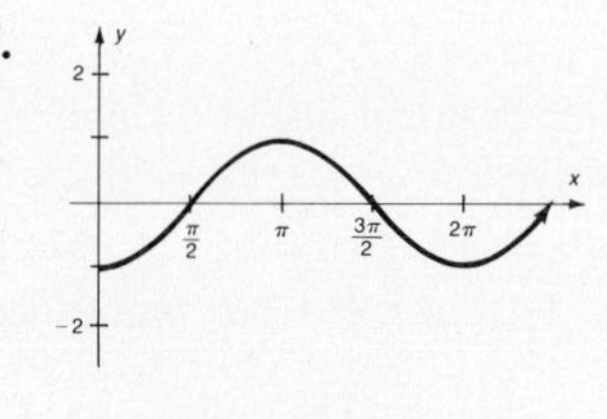

59. 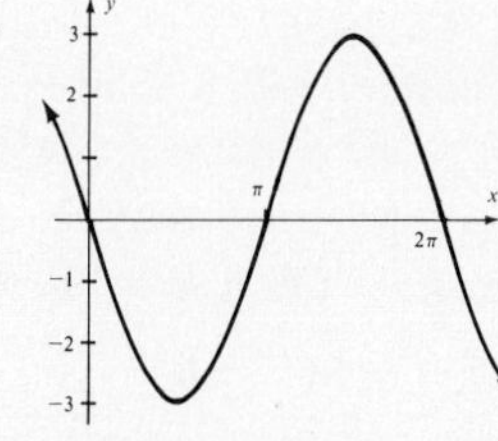

61. 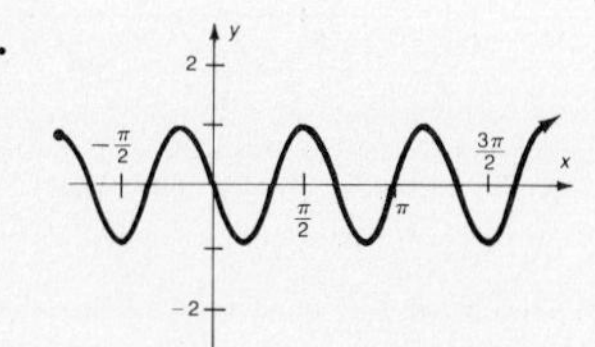

63. 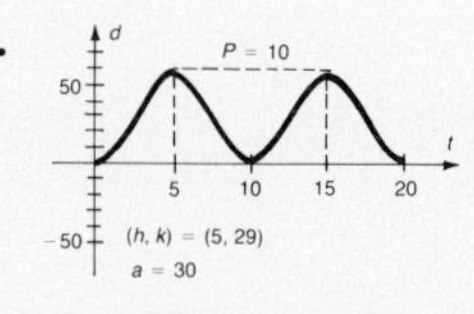

65. 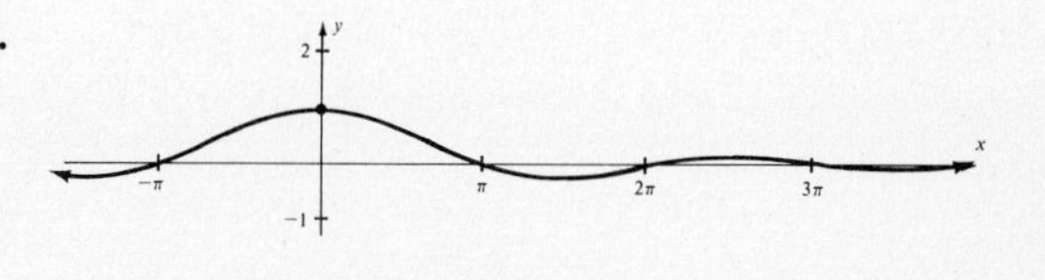

PROBLEM SET 5.5, PAGE 214

1. a. 0° or 0 **b.** 30° or $\frac{\pi}{6}$ **c.** 30° or $\frac{\pi}{6}$ **d.** 0° or 0 **3. a.** 60° or $\frac{\pi}{3}$ **b.** 45° or $\frac{\pi}{4}$ **c.** 45° or $\frac{\pi}{4}$ **d.** 45° or $\frac{\pi}{4}$ **5. a.** 150° or $\frac{5\pi}{6}$ **b.** $-45°$ or $-\frac{\pi}{4}$ **c.** $-45°$ or $-\frac{\pi}{4}$ **d.** 120° or $\frac{2\pi}{3}$ **7. a.** 0° or 0 **b.** 60° or $\frac{\pi}{3}$ **c.** 60° or $\frac{\pi}{3}$ **d.** 60° or $\frac{\pi}{3}$ **9.** 0.58 **11.** 1.65 **13.** 2.81 **15.** 1.32 **17.** 0.98 **19.** 1.34 **21.** .34 **23.** 2.90 **25.** 1.77 **27.** 69° **29.** $-28°$ **31.** 46° **33.** 85° **35.** 54° **37.** 141° **39.** 135° **41.** 153.43° 43. $-71.57°$ **45.** $\frac{\pi}{6}$

47. $\frac{\pi}{15}$ **49.** $\frac{2\pi}{15}$ **51.** 0.4163 **53.** 1.28 **55.** 0.2836 **57.** -0.64 **59.**

61.

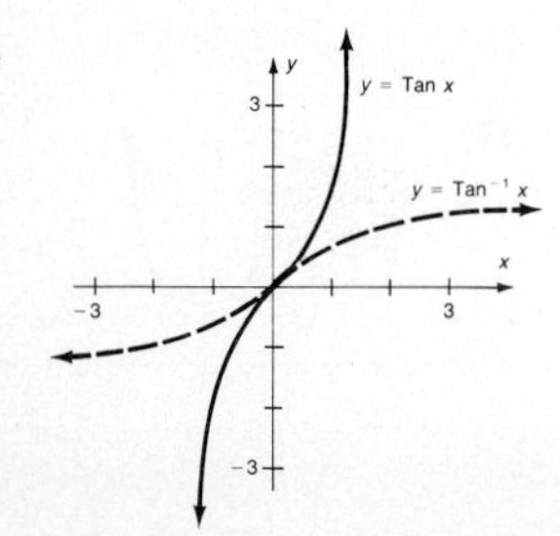

63.

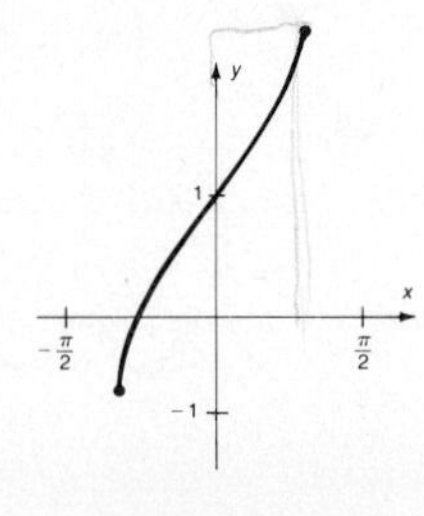

65.

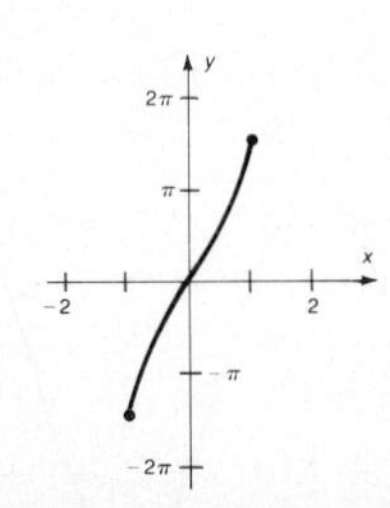

67.

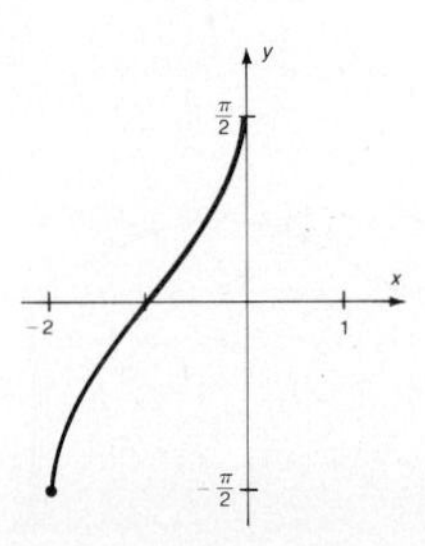

CHAPTER 5 SUMMARY, PAGES 215–217

1. a. lambda **b.** theta **c.** phi **d.** alpha **e.** beta **3.** $x^2 + y^2 = 1$ **5.** 145° **7.** $\frac{7\pi}{6}$

9. **11.** **13.** **15.** **17.** 270° **19.** $-315°$ **21.** $\frac{5\pi}{3}$

23. $\frac{3\pi}{10}$ or 0.9425 **25.** $s = r\theta$; radius; angle measured in radians **27.** $5\pi \approx 15.7$ cm **29.** 60° **31.** $4 - \pi$

33. an angle in standard position **35.** $\sin\theta = b$; $\csc\theta = \frac{1}{b}$; $b \neq 0$ **37.** all **39.** tangent and cotangent **41.** $\frac{1}{3}$

43. secant **45.** 1.0896 **47.** 1.4587

49. θ is an angle in standard position with a point $P(x, y)$ on the terminal side of θ a nonzero distance of r from the origin.

51. $\sin\theta = \frac{y}{r}$; $\csc\theta = \frac{r}{y}$, $y \neq 0$ **53.** $\cos\theta = \frac{5}{13}$; $\sin\theta = -\frac{12}{13}$; $\tan\theta = -\frac{12}{5}$; $\sec\theta = \frac{13}{5}$; $\csc\theta = -\frac{13}{12}$; $\cot\theta = -\frac{5}{12}$

55. $\cos\theta = -\frac{5}{\sqrt{29}}$; $\sin\theta = \frac{2}{\sqrt{29}}$; $\tan\theta = -\frac{2}{5}$; $\sec\theta = -\frac{\sqrt{29}}{5}$; $\csc\theta = \frac{\sqrt{29}}{5}$; $\cot\theta = -\frac{5}{2}$ **57. a.** 1 **b.** $\frac{\sqrt{3}}{2}$ **c.** $\frac{\sqrt{2}}{2}$ **d.** $\frac{1}{2}$ **e.** 0 **f.** -1 **g.** 0 **59. a.** 0 **b.** $\frac{\sqrt{3}}{3}$ **c.** 1 **d.** $\sqrt{3}$ **e.** undefined **f.** 0 **g.** undefined **61.** $\frac{1}{2}$

63. -1 **65.** 1.4587 **67.** 1.0896 **69.** **71.**

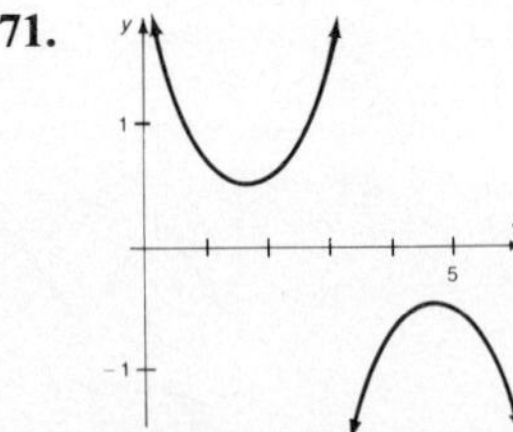

73.

75.

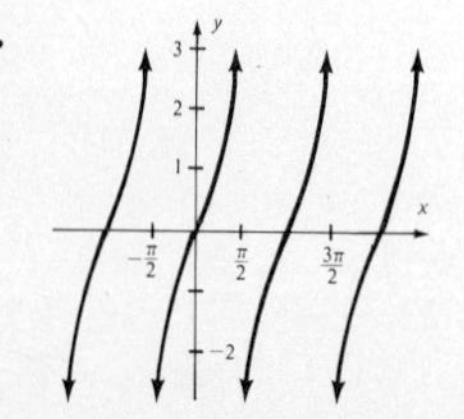

77.

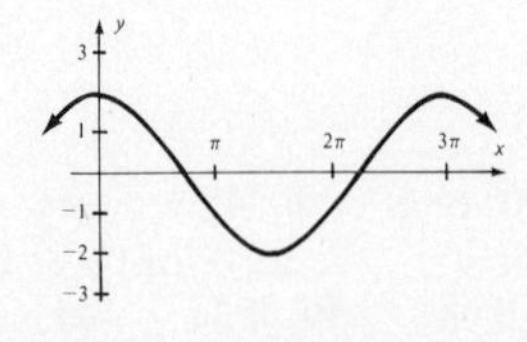

79.

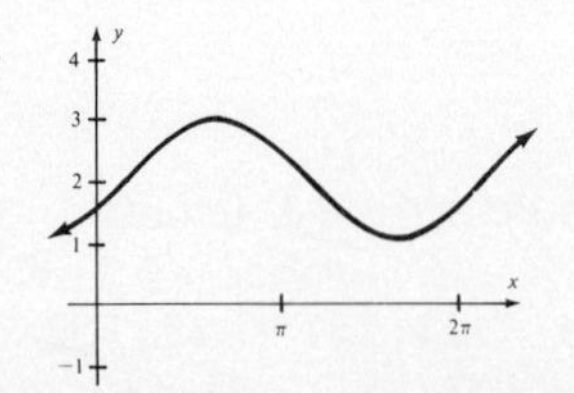

81. $-\frac{\pi}{2} < y < \frac{\pi}{2}$

83. $0 < y < \pi$ **85.** $\frac{\pi}{6}$ or $30°$ **87.** $\frac{\pi}{6}$ or $30°$ **89.** 0.3194 **91.** 1.2741 **93.**

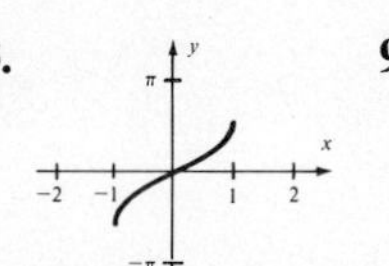

95.

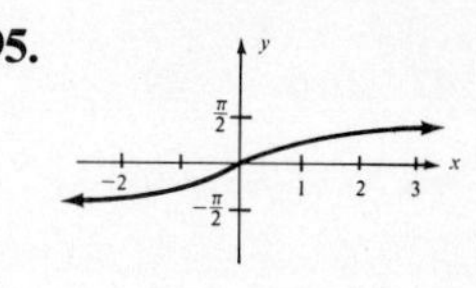

97. 1 **99.** −0.4292

PROBLEM SET 6.1, PAGES 223–224

1. $\frac{\pi}{6}$ **3.** $-\frac{\pi}{6}$ **5.** $\begin{cases} \frac{\pi}{6} + 2n\pi \\ \frac{5\pi}{6} + 2n\pi \end{cases}$ **7.** $\frac{\pi}{3}$ **9.** $-60°$ **11.** $\begin{cases} \frac{\pi}{3} + 2n\pi \\ \frac{5\pi}{3} + 2n\pi \end{cases}$ **13.** $\frac{\pi}{6}, \frac{5\pi}{6}, \frac{7\pi}{6}, \frac{11\pi}{6}$ **15.** $\frac{\pi}{3}, \frac{2\pi}{3}, \frac{4\pi}{3}, \frac{5\pi}{3}$
17. $\frac{2\pi}{3}, \frac{5\pi}{6}, \frac{5\pi}{3}, \frac{11\pi}{6}$ **19.** $\frac{\pi}{12}, \frac{5\pi}{12}, \frac{3\pi}{4}, \frac{13\pi}{12}, \frac{17\pi}{12}, \frac{7\pi}{4}$ **21.** $\frac{5\pi}{12}, \frac{7\pi}{12}, \frac{17\pi}{12}, \frac{19\pi}{12}$ **23.** $0, \pi$ **25.** $\frac{\pi}{2}, \frac{3\pi}{2}$
27. $\frac{\pi}{6}, \frac{\pi}{3}, \frac{5\pi}{6}, \frac{5\pi}{3}$ **29.** $0, \pi, \frac{\pi}{4}, \frac{5\pi}{4}$ **31.** $\frac{\pi}{4}, \frac{3\pi}{4}, \frac{5\pi}{4}, \frac{7\pi}{4}$ **33.** $0, \pi, \frac{\pi}{3}, \frac{5\pi}{3}$ **35.** 2.2370, 4.0461
37. 0.3649, 1.2059, 3.5065, 4.3475 **39.** 0.6662, 2.4754 **41.** 0.8213, 2.3203 **43.** 0, 1.57, 2.09, 3.14, 4.19, 4.71
45. 0.41, 1.16, 3.55, 4.30 **47.** 0.32, 1.89, 3.46, 5.03 **49.** 0.00, 2.09, 4.19 **51.** 0.00, 0.52, 2.62, 3.14
53. 0.00, 1.05, 1.22, 1.92, 2.09, 3.14, 3.32, 4.01, 4.19, 5.24, 5.41, 6.11 **55.** 0.21, 1.05, 1.47, 2.30, 2.72, 3.56, 3.98, 4.82, 5.24, 6.07
57. $\begin{cases} \frac{5\pi}{6} + 2n\pi \\ \frac{7\pi}{6} + 2n\pi \end{cases}$ **59.** $\begin{cases} 0.4014 + 2n\pi \\ 2.7402 + 2n\pi \end{cases}$ **61.** $0.9423 + n\pi$ **63.** 0.185

PROBLEM SET 6.2, PAGES 227–228

1. See inside back cover. **3.** III, IV **5.** I, IV **7.** II **9.** I **11.** $\cot(A + B)$ **13.** $\tan(\frac{\pi}{15})$
15. $\cos(\frac{\pi}{8})$ **17.** $\cos 127°$ **19.** 1 **21.** 1 **23.** −1

25.
$$\begin{aligned} \frac{1}{\sin\theta} &= \frac{1}{y/r} && \text{By definition of the trig functions} \\ &= \frac{r}{y} && \text{Dividing fractions} \\ &= \csc\theta && \text{By definition of the trig functions} \end{aligned}$$

27.
$$\begin{aligned} x^2 + y^2 &= r^2 && \text{By the Pythagorean Theorem} \\ 1 + \frac{y^2}{x^2} &= \frac{r^2}{x^2} && \text{Dividing both sides by } x^2, x \neq 0 \\ 1 + \left(\frac{y}{x}\right)^2 &= \left(\frac{r}{x}\right)^2 && \text{Properties of exponents} \\ 1 + \tan^2\theta &= \sec^2\theta && \text{By definition of the trig functions} \end{aligned}$$

29.
$$\begin{aligned} \cos\theta &= \cos\theta \\ \sin\theta &= \pm\sqrt{1 - \cos^2\theta} \\ \tan\theta &= \frac{\pm\sqrt{1 - \cos^2\theta}}{\cos\theta} \\ \sec\theta &= \frac{1}{\cos\theta} \\ \csc\theta &= \frac{1}{\pm\sqrt{1 - \cos^2\theta}} \\ \cot\theta &= \frac{\cos\theta}{\pm\sqrt{1 - \cos^2\theta}} \end{aligned}$$

31.
$$\begin{aligned} \cot\theta &= \cot\theta \\ \tan\theta &= \frac{1}{\cot\theta} \\ \csc\theta &= \pm\sqrt{1 + \cot^2\theta} \\ \sin\theta &= \frac{1}{\pm\sqrt{1 + \cot^2\theta}} \\ \cos\theta &= \frac{\cot\theta}{\pm\sqrt{1 + \cot^2\theta}} \\ \sec\theta &= \frac{\pm\sqrt{1 + \cot^2\theta}}{\cot\theta} \end{aligned}$$

33.
$$\begin{aligned} \csc\theta &= \csc\theta \\ \sin\theta &= \frac{1}{\csc\theta} \\ \cot\theta &= \pm\sqrt{\csc^2\theta - 1} \\ \tan\theta &= \frac{1}{\pm\sqrt{\csc^2\theta - 1}} \\ \cos\theta &= \frac{\pm\sqrt{\csc^2\theta - 1}}{\csc\theta} \\ \sec\theta &= \frac{\csc\theta}{\pm\sqrt{\csc^2\theta - 1}} \end{aligned}$$

35. $\cos\theta = \frac{3}{5}$; $\csc\theta < 0$

Quad IV; use Problem 29

$\sin\theta = -\sqrt{1 - \frac{9}{25}} = -\frac{4}{5}$

$\tan\theta = -\frac{4}{3}$ $\cot\theta = -\frac{3}{4}$

$\sec\theta = \frac{5}{3}$ $\csc\theta = -\frac{5}{4}$

	$\cos\theta$	$\sin\theta$	$\tan\theta$	$\sec\theta$	$\csc\theta$	$\cot\theta$
37.	$\frac{5}{13}$	$\frac{12}{13}$	$\frac{12}{5}$	$\frac{13}{5}$	$\frac{13}{12}$	$\frac{5}{12}$
39.	$\frac{-12}{13}$	$\frac{-5}{13}$	$\frac{5}{12}$	$\frac{-13}{12}$	$\frac{-13}{5}$	$\frac{12}{5}$
41.	$\frac{-\sqrt{5}}{3}$	$\frac{2}{3}$	$\frac{-2\sqrt{5}}{5}$	$\frac{-3\sqrt{5}}{5}$	$\frac{3}{2}$	$\frac{-\sqrt{5}}{2}$
43.	$\frac{5\sqrt{34}}{34}$	$\frac{3\sqrt{34}}{34}$	$\frac{3}{5}$	$\frac{\sqrt{34}}{5}$	$\frac{\sqrt{34}}{3}$	$\frac{5}{3}$
45.	$\frac{-\sqrt{10}}{10}$	$\frac{-3\sqrt{10}}{10}$	3	$-\sqrt{10}$	$\frac{-\sqrt{10}}{3}$	$\frac{1}{3}$

47. $\sin\theta$ **49.** $\dfrac{2}{1-\cos^2\theta}$ or $\dfrac{2}{\sin^2\theta}$ **51.** $\csc\theta$ **53.** $\dfrac{\sin^2\theta-\cos^2\theta}{\sin\theta\cos\theta}$ **55.** $\dfrac{\sin^2\theta+\cos\theta}{\sin\theta}$ **57.** 1 **59.** $\dfrac{1+\sin^2\theta}{\cos^2\theta}$
61. $\cos^2\theta-\sin\theta$

PROBLEM SET 6.3, PAGES 232–233

Proofs of trigonometric identities vary.

PROBLEM SET 6.4, PAGES 240–241

1. $\dfrac{\sqrt{3}\cos\theta-\sin\theta}{2}$ **3.** $\dfrac{1+\tan\theta}{1-\tan\theta}$ **5.** $\dfrac{\sqrt{2}}{2}(\cos\theta+\sin\theta)$ **7.** $\cos^2\theta-\sin^2\theta$ **9.** $\dfrac{2\tan\theta}{1-\tan^2\theta}$ **11.** $\cos 52^\circ$
13. $\cos\frac{\pi}{3}$ **15.** $-\tan\frac{\pi}{6}$ **17.** $-\tan 49^\circ$ **19.** $-\sin 31^\circ$ **21.** $-\tan 24^\circ$ **23.** 0.8746 **25.** 0.6561 **27.** 0.6745

	angle θ	$\cos\theta$	$\sin\theta$	$\tan\theta$
29.	-15°	$\frac{\sqrt{6}+\sqrt{2}}{4}$	$\frac{\sqrt{2}-\sqrt{6}}{4}$	$-2+\sqrt{3}$
31.	75°	$\frac{\sqrt{6}-\sqrt{2}}{4}$	$\frac{\sqrt{2}+\sqrt{6}}{4}$	$2+\sqrt{3}$
33.	105°	$\frac{\sqrt{2}-\sqrt{6}}{4}$	$\frac{\sqrt{2}+\sqrt{6}}{4}$	$-2-\sqrt{3}$

35. a. $\cos(\theta-\frac{2\pi}{3})$ **b.** $-\sin(\theta-\frac{2\pi}{3})$ **c.** $-\tan(\theta-\frac{2\pi}{3})$

37.
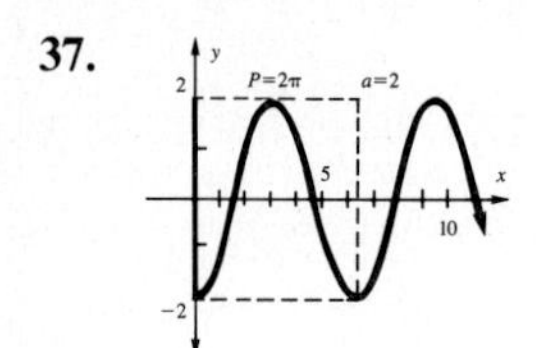

39.
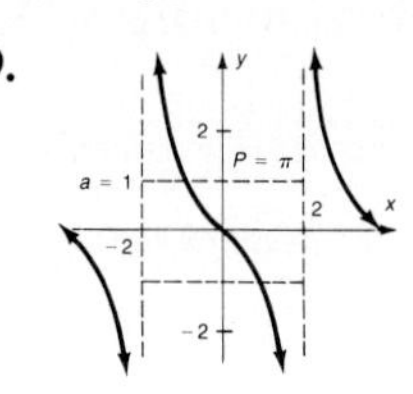

41.
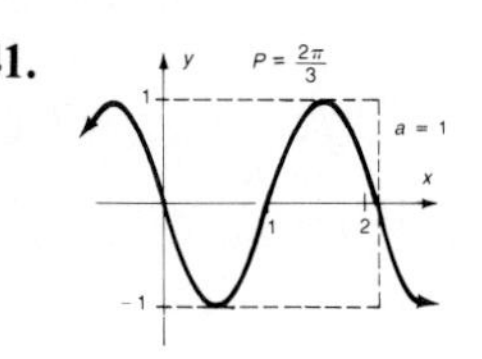

43.
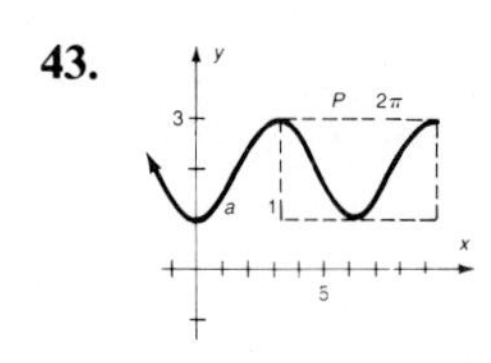

45.
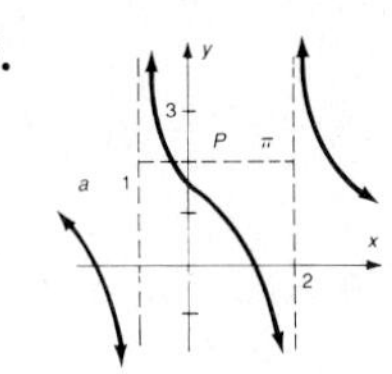

47–67. Proofs of identities vary.

PROBLEM SET 6.5, PAGES 248–249

1. $\frac{\sqrt{2}}{2}$ **3.** $\frac{1}{2}$ **5.** -1 **7.** $\frac{1}{2}\sqrt{2-\sqrt{2}}$ **9.** $\sqrt{2}-1$

	$\cos 2\theta$	$\sin 2\theta$	$\tan 2\theta$
11.	$\frac{119}{169}$	$\frac{-120}{169}$	$\frac{-120}{119}$
13.	$\frac{7}{25}$	$\frac{-24}{25}$	$\frac{-24}{7}$
15.	$\frac{-119}{169}$	$\frac{120}{169}$	$-\frac{120}{119}$

	$\cos\frac{1}{2}\theta$	$\sin\frac{1}{2}\theta$	$\tan\frac{1}{2}\theta$
17.	$\frac{\sqrt{26}}{26}$	$\frac{5\sqrt{26}}{26}$	5
19.	$\frac{\sqrt{10}}{10}$	$\frac{3\sqrt{10}}{10}$	3
21.	$\frac{-2\sqrt{13}}{13}$	$\frac{3\sqrt{13}}{13}$	$-\frac{3}{2}$

23. $\cos 28^\circ+\cos 64^\circ$ **25.** $\sin 77^\circ+\sin 29^\circ$ **27.** $\frac{1}{2}\cos 17^\circ+\frac{1}{2}\cos 123^\circ$ **29.** $\frac{1}{2}\sin 315^\circ+\frac{1}{2}\sin 85^\circ$
31. $\frac{1}{2}\cos 3\theta-\frac{1}{2}\cos 7\theta$ **33.** $\frac{1}{2}\sin 5\theta-\frac{1}{2}\sin\theta$ **35.** $2\sin 8^\circ\cos 14^\circ$ **37.** $2\cos 51.5^\circ\cos 26.5^\circ$ **39.** $2\sin 62.5^\circ\sin 37.5^\circ$
41. $-2\sin 4x\sin x$ **43.** $2\cos\frac{5\theta}{2}\cos\frac{\theta}{2}$ **45.** $-2\sin\frac{3z}{2}\cos\frac{15z}{2}$

	$\cos\theta$	$\sin\theta$	$\tan\theta$
47.	$\frac{\sqrt{2}}{2}$	$\frac{\sqrt{2}}{2}$	1
49.	$\frac{1}{2}$	$\frac{\sqrt{3}}{2}$	$\sqrt{3}$
51.	$\frac{3\sqrt{10}}{10}$	$\frac{\sqrt{10}}{10}$	$\frac{1}{3}$

53–65. Proofs of identities vary.

PROBLEM SET 6.6, PAGES 255–256

The complex numbers in Problems 1–27 should also be plotted.

1. a. $\sqrt{10}$ **b.** $5\sqrt{2}$ **c.** $\sqrt{13}$ **d.** $\sqrt{13}$ **3. a.** $\sqrt{29}$ **b.** $\sqrt{41}$ **c.** $\sqrt{17}$ **d.** $\sqrt{2}$ **5.** $\sqrt{2}$ cis 315°
7. 2 cis 30° **9.** 2 cis 240° **11.** 5 cis 0° **13.** 6 cis 345° **15.** 2 cis 320° **17.** $\frac{3}{2} + \frac{3\sqrt{3}}{2}i$ **19.** $-\frac{5}{2} - \frac{5\sqrt{3}}{2}i$
21. $-5i$ **23.** -2 **25.** $7.3084 + 3.2539i$ **27.** $-8.8633 - 1.5628i$ **29.** 15 cis 140° **31.** $\frac{5}{2}$ cis 267°
33. 3 cis 130° **35.** 81 cis 240° **37.** 64 cis 180° or -64 **39.** 256 cis 120° or $-128 + 128\sqrt{3}i$
41. 2 cis 80°, 2 cis 200°, 2 cis 320° **43.** 2 cis 40°, 2 cis 112°, 2 cis 184°, 2 cis 256°, 2 cis 328°
45. 2 cis 32°, 2 cis 104°, 2 cis 176°, 2 cis 248°, 2 cis 320° **47.** 3 cis 0°, 3 cis 120°, 3 cis 240°
49. cis 22.5°, cis 112.5°, cis 202.5°, cis 292.5° **51.** $\sqrt[8]{2}$ cis 56.25°, $\sqrt[8]{2}$ cis 146.25°, $\sqrt[8]{2}$ cis 236.25°, $\sqrt[8]{2}$ cis 326.25°
53. 2 cis 15°, 2 cis 75°, 2 cis 135°, 2 cis 195°, 2 cis 255°, 2 cis 315°
55. $2^{\frac{1}{18}}$ cis 15°, $2^{\frac{1}{18}}$ cis 55°, $2^{\frac{1}{18}}$ cis 95°, $2^{\frac{1}{18}}$ cis 135°, $2^{\frac{1}{18}}$ cis 175°, $2^{\frac{1}{18}}$ cis 215°, $2^{\frac{1}{18}}$ cis 255°, $2^{\frac{1}{18}}$ cis 295°, $2^{\frac{1}{18}}$ cis 335°
Note: Problems 59–63 should also be illustrated graphically.
59. $1, i, -1, -i$ **61.** $-0.6840 + 1.8794i$, $-1.2856 - 1.5321i$, $1.9696 - 0.3473i$
63. $1.9696 + 0.3473i$, $-0.3473 + 1.9696i$, $-1.9696 - 0.3473i$, $0.3473 - 1.9696i$
65. 8 cis 0°, 8 cis 72°, 8 cis 144°, 8 cis 216°, 8 cis 288° **67.** cis 72°, cis 144°, cis 216°, cis 288° **69–71.** Answers vary.

PROBLEM SET 6.7, PAGES 261–263

	α	β	γ	a	b	c
1.	30°	60°	90°	80	140	160
3.	45°	45°	90°	9.0	9.0	13
5.	37°	53°	90°	11	15	19
7.	71°	19°	90°	69	24	73
9.	54°	36°	90°	18	13	22
11.	40°	50°	90°	24	29	38
13.	76°	14°	90°	29	7.2	30
15.	45°	45°	90°	49	49	69
17.	77°	13°	90°	390	90	400
19.	50°	40°	90°	98	82	130
21.	6°	84°	90°	3.8	36	36
23.	56.00°	34.00°	90.00°	3484	2350	4202
25.	27.66°	62.34°	90.00°	1625	3100	3500
27.	42°	48°	90°	320	350	470
29.	17.54°	72.46°	90.00°	1296	4100	4300

31. The building is 23 m tall. **33.** The ship is 200 m away. **35.** The car is 340 ft away.
37. The distance is 350 ft. **39.** It is 13 ft above the ground. **41.** The angle of elevation is about 56°.

43. The height is 1251 ft. **45.** 263 m **47.** The distance across the river is 170 m. **49.** The tower is 222.0 ft high. **51.** The height of the center is 14.7 ft. **53.** The distance from the earth to Venus is 63.4 million miles (or 6.34×10^7 mi). **55.** The radius of the inscribed circle is 633.9 ft. **57.** Answers vary. **59.** The shadow will be about 12 ft. **61.** Answers vary.

PROBLEM SET 6.8, PAGES 272–274

1. $\alpha = 54^\circ$ $a = 7.0$; $\beta = 113^\circ$ $b = 8.0$; $\gamma = 13^\circ$ $c = 2.0$

3. $\alpha = 82^\circ$ $a = 17$; $\beta = 54^\circ$ $b = 14$; $\gamma = 44^\circ$ $c = 12$

5. no triangle formed

7. $\alpha = 80^\circ$ $a = 5.0$; $\beta = 52^\circ$ $b = 4.0$; $\gamma = 48^\circ$ $c = 3.8$

9. ambiguous case
$\alpha = 47.0^\circ$ $a = 10.2$; $\beta = 57.8^\circ$ $b = 11.8$; $\gamma = 75.2^\circ$ $c = 13.5$
$\alpha = 47.0^\circ$ $a = 10.2$; $\beta' = 122.2^\circ$ $b = 11.8$; $\gamma' = 10.8^\circ$ $c' = 2.61$

11. ambiguous case
$\alpha = 56^\circ$ $a = 4.5$; $\beta = 67^\circ$ $b = 5.0$; $\gamma = 57^\circ$ $c = 4.6$
$\alpha = 56^\circ$ $a = 4.5$; $\beta' = 113^\circ$ $b = 5.0$; $\gamma' = 11^\circ$ $c' = 1.0$

13. $\alpha = 103^\circ$ $a = 78$; $\beta = 67^\circ$ $b = 74$; $\gamma = 10^\circ$ $c = 14$

15. $\alpha = 65^\circ$ $a = 38$; $\beta = 78^\circ$ $b = 41$; $\gamma = 37^\circ$ $c = 25$

17. $\alpha = 110.5^\circ$ $a = 38.2$; $\beta = 21.3^\circ$ $b = 14.8$; $\gamma = 48.2^\circ$ $c = 30.4$

19. $\alpha = 14^\circ$ $a = 26$; $\beta = 42^\circ$ $b = 71$; $\gamma = 123^\circ$ $c = 88$

21. $\alpha = 60.2^\circ$ $a = 14.2$; $\beta = 84.8^\circ$ $b = 16.3$; $\gamma = 35.0^\circ$ $c = 9.39$

23. $\alpha = 147^\circ$ $a = 49.5$; $\beta = 15.0^\circ$ $b = 23.5$; $\gamma = 18.0^\circ$ $c = 28.1$

25. ambiguous case
$\alpha = 90.8^\circ$ $a = 98.2$; $\beta = 57.1^\circ$ $b = 82.5$; $\gamma = 32.1^\circ$ $c = 52.2$
$\alpha' = 25.0^\circ$ $a' = 41.5$; $\beta' = 122.9^\circ$ $b = 82.5$; $\gamma = 32.1^\circ$ $c = 52.2$

27. ambiguous case
$\alpha = 23.1^\circ$ $a = 21.3$; $\beta = 64.5^\circ$ $b = 49.0$; $\gamma = 92.4^\circ$ $c = 54.2$
$\alpha = 23.1^\circ$ $a = 21.3$; $\beta' = 115.5^\circ$ $b = 49.0$; $\gamma' = 41.4^\circ$ $c' = 35.9$

29–33. Answers vary. **35.** The boats are 56 km apart.

37. Ambiguous case; it would be there at about 12:06 P.M. and at 12:21 P.M. **39.** 34° and 146° **41.** 415 km **43.** The height of the tree is 58 ft. **45.** He is 5.39 miles from the target. **47.** The distance from the first city is 1.926 mi (10,170 ft) and from the second city is 0.5363 mi (2832 ft). The altitude is 1850 ft. **49.** The angle of elevation is 19.0° **51.** The jetty is about 5030 ft long. **53.** The cost is \$253. **55.** Answers vary. **57.** 66.4 sq units **59.** 1140 sq units **61–65.** Answers vary.

CHAPTER 6 SUMMARY, PAGES 274–278

1. $\frac{\pi}{12}, \frac{7\pi}{12}, \frac{13\pi}{12}, \frac{19\pi}{12}$ **3.** $\frac{\pi}{12}, \frac{\pi}{4}, \frac{3\pi}{4}, \frac{11\pi}{12}, \frac{17\pi}{12}, \frac{19\pi}{12}$ **5.** $\frac{\pi}{3}, \frac{2\pi}{3}, \frac{4\pi}{3}, \frac{5\pi}{3}$ **7.** No solution

9. $\sec\theta = \dfrac{1}{\cos\theta}$; $\csc\theta = \dfrac{1}{\sin\theta}$; $\cot\theta = \dfrac{1}{\tan\theta}$ **11.** $\cos^2\theta + \sin^2\theta = 1$; $\tan^2\theta + 1 = \sec^2\theta$; $1 + \cot^2\theta = \csc^2\theta$

13. Quad II: $\csc\delta = \frac{5}{3}$; $\cos\delta = -\frac{4}{5}$; $\sec\delta = -\frac{5}{4}$; $\tan\delta = -\frac{3}{4}$; $\cot\delta = -\frac{4}{3}$

15. Quad III; $\csc\omega = -\frac{5}{3}$; $\cos\omega = -\frac{4}{5}$; $\sec\omega = -\frac{5}{4}$; $\tan\omega = \frac{3}{4}$; $\cot\omega = \frac{4}{3}$ **17.** $\dfrac{\sin^2\theta + \cos\theta}{\sin\theta\cos\theta}$ **19.** $\dfrac{2\sin^2\theta + \cos^2\theta}{\cos^2\theta}$

21–27. Answers vary. **29.** $\cos 52^\circ$ **31.** $\sin 1.115$ **33.** $\cos(\theta - \frac{\pi}{6})$ **35.**

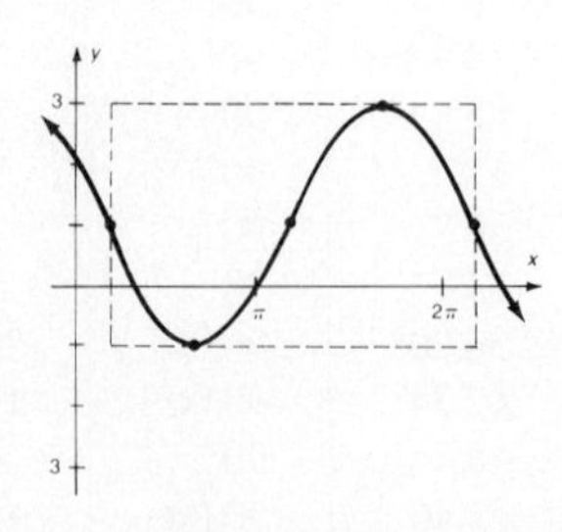

37. $\cos(\theta - 30^\circ) = \frac{1}{2}(\sqrt{3}\cos\theta + \sin\theta)$ **39.** Write as $\tan(23^\circ - 85^\circ) = \tan(-62^\circ) \approx -1.881$.

41. Write as $\tan 2(\frac{\pi}{6}) = \tan\frac{\pi}{3} = \sqrt{3}$. **43.** $-\frac{24}{25}$ **45.** Write as $\cos 120^\circ = -\frac{1}{2}$. **47.** $\frac{\sqrt{10}}{10}$ **49.** $\frac{1}{2}\sin 4\theta + \frac{1}{2}\sin 2\theta$ **51.** $2\sin\frac{h}{2}\cos(\frac{2x+h}{2})$ **53.** $7\sqrt{2}\,\text{cis}\,315^\circ$ **55.** $7\,\text{cis}\,330^\circ$ **57.** $2\sqrt{2} - 2\sqrt{2}i$ **59.** $-5i$ **61.** $-128 - 128\sqrt{3}i$ **63.** $\text{cis}\,154^\circ$

65. If n is any positive integer, then the nth roots of $r \operatorname{cis} \theta$ are given by $\sqrt[n]{r} \operatorname{cis}\left(\frac{\theta + 2\pi k}{n}\right)$ for $k = 0, \ldots, n-1$.

67. $-2.5556 + 0.6848i$; $2.5556 - 0.6848i$; also plot points which are $\sqrt{7} \operatorname{cis} 165°$ and $\sqrt{7} \operatorname{cis} 345°$.

69. If θ is an acute angle in a right triangle, then $\cos\theta = \frac{\text{adj}}{\text{hyp}}$; $\sin\theta = \frac{\text{opp}}{\text{hyp}}$; $\tan\theta = \frac{\text{opp}}{\text{adj}}$

71. $a = 475, b = 678, c = 828, \alpha = 35.0°, \beta = 55.0°, \gamma = 90.0°$ **73.** See page 263 **75.** See page 264

77. $\cos\alpha = \frac{b^2 + c^2 - a^2}{2bc}$; $\alpha = \cos^{-1}\left(\frac{b^2 + c^2 - a^2}{2bc}\right)$ **79.** $\alpha + \beta + \gamma = 180°$; $\alpha = 180° - \beta - \gamma$ **81.** $a = 34, b = 52, c = 61,$ $\alpha = 34°, \beta = 58°, \gamma = 88°$ **83.** Ambiguous case: $a = 51, a' = 13, b = 34, c = 21, \alpha = 137°, \alpha' = 11°, \beta = 27°, \beta' = 153°, \gamma = 16°$ 16°

85. 430 ft; 48.9° **87.** 278 ft

EXTENDED APPLICATION—SOLAR POWER, PAGES 282–283

1. 16 h 50 m or 4:50 P.M. **3.** 19 h 1 m or 7:01 P.M. **5.** 6 h 18 m or 6:18 A.M. **9–11.** Answers vary.

7.

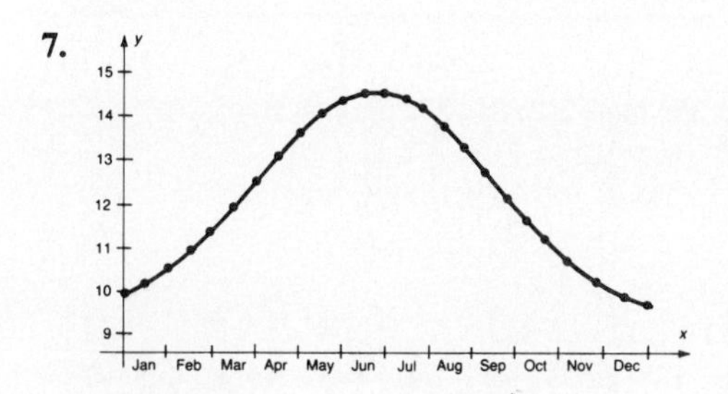

PROBLEM SET 7.1, PAGES 291–292

1.

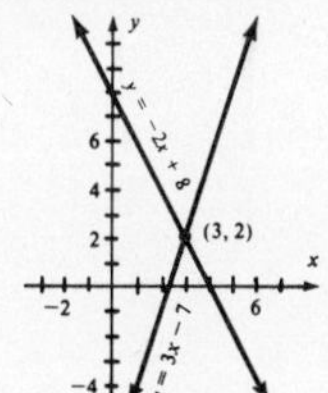

3.

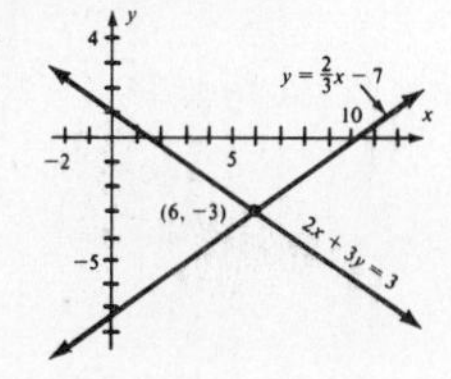

5.

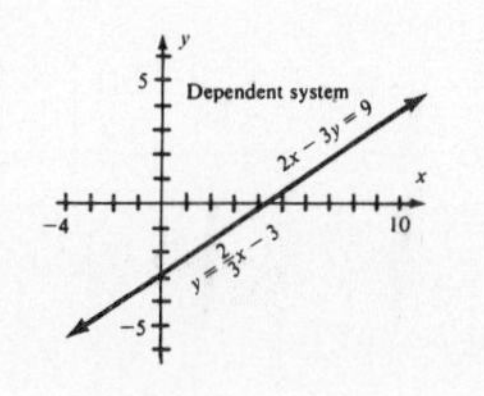

7.

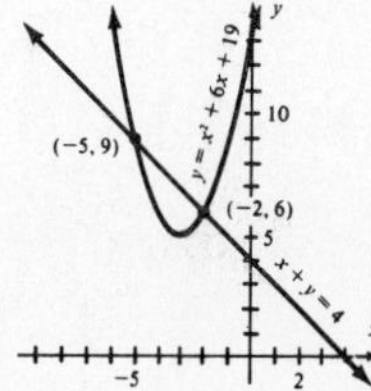

9.

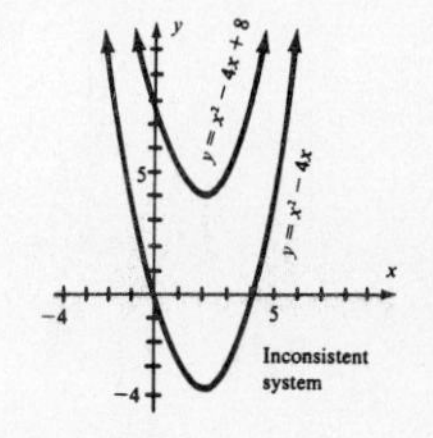

The answers for variables and constants in Problems 11–27 may vary, but the solution to the problems will not vary.

11. The variables are $s = x_1$ and $t = x_2$; the constants are
$a_{11} = 1$ $a_{12} = 1$ $b_1 = 1$
$a_{21} = 3$ $a_{22} = 1$ $b_2 = -5$
The solution is $(s, t) = (-3, 4)$.

13. The variables are $a = x_1$ and $v = x_2$; the constants are
$a_{11} = 3$ $a_{22} = -5$ $b_1 = -10$
$a_{21} = 3$ $a_{22} = -5$ $b_2 = 10$
The system is inconsistent.

15. The variables are $t_1 = x_1$ and $t_2 = x_2$; the constants are
$a_{11} = 3$ $a_{12} = 5$ $b_1 = 1541$
$a_{21} = 2$ $a_{22} = -1$ $b_2 = -160$
The solution is $(t_1, t_2) = (57, 274)$.

17. The variables are $\gamma = x_1$ and $\sigma = x_2$; the constants are
$a_{11} = 1$ $a_{12} = -3$ $b_1 = -4$
$a_{21} = 5$ $a_{22} = -4$ $b_2 = -9$
The solution is $(\gamma, \sigma) = (-1, 1)$.

19. The variables are $c = x_1$ and $d = x_2$; the constants are
$a_{11} = 1$ $a_{12} = 1$ $b_1 = 2$
$a_{21} = 2$ $a_{22} = -1$ $b_2 = 1$
The solution is $(c, d) = (1, 1)$.

21. The variables are $q_1 = x_1$ and $q_2 = x_2$; the constants are
$a_{11} = 3$ $a_{12} = -4$ $b_1 = 3$
$a_{21} = 5$ $a_{22} = 3$ $b_2 = 5$
The solution is $(q_1, q_2) = (1, 0)$.

23. The variables are $x = x_1$ and $y = x_2$; the constants are
$a_{11} = 7$ $a_{12} = 1$ $b_1 = 5$
$a_{21} = 14$ $a_{22} = -2$ $b_2 = -2$
The solution is $(\frac{2}{7}, 3)$.

25. The variables are $\alpha = x_1$ and $\beta = x_2$; the constants are
$a_{11} = 1$ $a_{12} = 1$ $b_1 = 12$
$a_{21} = 1$ $a_{22} = -2$ $b_2 = -4$
The solution is $(\alpha, \beta) = (\frac{20}{3}, \frac{16}{3})$.

27. The variables are $\theta = x_1$ and $\phi = x_2$; the constants are
$a_{11} = 2$ $a_{12} = 5$ $b_1 = 7$
$a_{21} = 3$ $a_{22} = 4$ $b_2 = 0$
The solution is $(\theta, \phi) = (-4, 3)$.

29. dependent system **31.** $(3, 15)$ **33.** $(6, -1)$ **35.** $(2\gamma - \delta, \gamma - \delta)$ **37.** $\left(\frac{a+b}{2}, \frac{a-b}{2}\right)$ **39.** $\left(\frac{c+d}{5}, \frac{2d-3c}{5}\right)$
41. $(-2, 2), (-5, -4)$ **43.** $(1, 2), (1, -2), (-1, 2), (-1, -2)$ **45.** $(\sqrt{13}, 1), (-\sqrt{13}, 1), (5, 2), (-5, 2)$
47. Answers vary. **49.** Subtract to obtain given identity. **51.** The numbers are 1 and -3 or 3 and -1.
53. The sides are 5 and 12 ft. **55.** $\left(\frac{\pi}{3}, \frac{7\pi}{6}\right), \left(\frac{\pi}{3}, \frac{11\pi}{6}\right), \left(\frac{5\pi}{3}, \frac{7\pi}{6}\right), \left(\frac{5\pi}{3}, \frac{11\pi}{6}\right)$ **57.** inconsistent system
59. $(2\sqrt{2}, 0), (-2\sqrt{2}, 0)$ **61.** $\left(\frac{17}{21}, -\frac{20}{9}\right)$ **63.** $\left(\frac{-9+\sqrt{113}}{2}, \frac{-15+\sqrt{113}}{8}\right), \left(\frac{-9-\sqrt{113}}{2}, \frac{-15-\sqrt{113}}{8}\right)$
65. $(h, k) = (1, 8)$ **67.** $\left(\frac{a+\sqrt{a^2-4}}{2}, \frac{a-\sqrt{a^2-4}}{2}\right), \left(\frac{a-\sqrt{a^2-4}}{2}, \frac{a+\sqrt{a^2-4}}{2}\right)$

PROBLEM SET 7.2, PAGES 295–296

1. 10 **3.** -10 **5.** 36 **7.** 10 **9.** -28 **11.** -1 **13.** 1 **15.** $\cot^2\theta$ **17.** $(s, t) = (5, -3)$ **19.** $\left(\frac{2}{13}, \frac{3}{13}\right)$
21. $(q_1, q_2) = \left(\frac{25}{29}, -\frac{3}{29}\right)$ **23.** $\left(\frac{21}{19}, -\frac{24}{19}\right)$ **25.** inconsistent system **27.** dependent system **29.** $(3, -3)$
31. $(31, 19)$ **33.** $(-54, -31)$ **35.** $\left(\frac{5a+3b}{13}, \frac{a-2b}{13}\right)$ **37.** $\left(\frac{a}{a^2-b^2}, \frac{-b}{a^2-b^2}\right)$ **39.** $\left(\frac{sf-td}{cf-ed}, \frac{ct-es}{cf-ed}\right)$
41. $\left(\frac{\alpha d-b\beta}{ad-bc}, \frac{a\beta-c\alpha}{ad-bc}\right)$ **43.** $(800, 200)$ **45.** $(1800, 200)$ **47.** $(63, 84)$ **49.** $\left(-\frac{1}{5}, \frac{3}{10}\right)$ **51.** $\left(\frac{11}{15}, \frac{1}{5}\right)$
53–55. Answers vary. **57.** $(2, 1), (-2, 1), (2, -1), (-2, -1)$ **59.** $(3, 2)$

PROBLEM SET 7.3, PAGES 302–304

1. -75 **3.** -20 **5.** -44 **7.** 3 **9.** -3 **11.** 38 **13.** -4 **15.** 21 **17.** -260 **19.** -28 **21.** 10
23. -5 **25.** 150 **27.** 0 **29.** 24 **31.** $(5, 4, -3)$ **33.** $(3, -1, 2)$ **35.** $(4, -3, 2)$ **37.** $(-4, 0, 3)$
39. $\left(5, \frac{15}{2}, \frac{3}{2}\right)$ **41.** $x + 4y + 11 = 0$ **43.** $8x - 3y + 7 = 0$ **45.** 84 **47.** $(w, x, y, z) = (3, 1, -2, 1)$
49. $(s, t, u, x) = (-3, 1, -2, -1)$ **51–59.** Answers vary.

PROBLEM SET 7.4, PAGES 312–313

1. $\begin{bmatrix} 2 & 4 & 2 \\ 6 & -2 & 4 \\ 2 & 2 & 5 \end{bmatrix}$ **3.** $\begin{bmatrix} 16 & 19 & 10 \\ 29 & 16 & 15 \\ 35 & 9 & 31 \end{bmatrix}$ **5.** $\begin{bmatrix} 13 & 25 & 20 \\ 25 & 31 & 23 \\ 42 & 24 & 33 \end{bmatrix}$ **7.** $\begin{bmatrix} 20 & 21 & 34 \\ 29 & 16 & 15 \\ 7 & 48 & 5 \end{bmatrix}$ **9.** $\begin{bmatrix} 34 & 117 & 44 \\ 45 & 143 & 97 \\ 109 & 100 & 151 \end{bmatrix}$

11. $\begin{bmatrix} 2 & 8 \\ 16 & 8 \\ -7 & 7 \\ 7 & 7 \end{bmatrix}$ **13.** $\begin{bmatrix} 14 & 14 \\ -7 & 7 \end{bmatrix}$ **15.** $\begin{bmatrix} 14 & 4 & 8 & 30 \\ -7 & -1 & -9 & 3 \end{bmatrix}$ **17.** not conformable **19.** $\begin{bmatrix} 1 & 0 & 0 & 0 \\ 0 & 1 & 0 & 0 \\ 0 & 0 & 1 & 0 \\ 0 & 0 & 0 & 1 \end{bmatrix}$

21. $\begin{bmatrix} 2 & 7 \\ 1 & 4 \end{bmatrix}$ **23.** $\begin{bmatrix} 0 & \frac{1}{2} \\ \frac{1}{3} & -\frac{1}{6} \end{bmatrix}$ **25.** $\begin{bmatrix} 3 & -17 & -20 \\ 3 & -18 & -20 \\ -1 & 6 & 7 \end{bmatrix}$ **27.** $\begin{bmatrix} 1 & 0 & -1 & 0 \\ 0 & \frac{1}{2} & 0 & 0 \\ -2 & 0 & 2 & 1 \\ 0 & 0 & 1 & 0 \end{bmatrix}$ **29.** $\begin{bmatrix} \frac{3}{5} & 0 & -\frac{2}{5} & \frac{1}{5} \\ \frac{1}{5} & 0 & \frac{1}{5} & -\frac{1}{10} \\ 0 & 1 & 0 & 0 \\ -\frac{6}{5} & 0 & \frac{4}{5} & \frac{1}{10} \end{bmatrix}$

31. $(-4, 7)$ **33.** $(25, 14)$ **35.** $(50, 29)$ **37.** $(-1, 4)$ **39.** $(-5, 2)$ **41.** $(-3, -2)$ **43.** $(4, -2)$ **45.** $(12, -5)$
47. $(0, 4)$ **49.** $(-2, 4, 3)$ **51.** $(3, -5, 2)$ **53.** $(-5, 18, 5)$ **55.** $(5, 4, -1)$ **57.** $(1, 2, 3)$ **59.** $(-3, 2, 7)$
61. $(w, x, y, z) = (3, -2, 1, -4)$ **63–65.** Answers vary. **67.**

$$\begin{aligned} MN &= M \\ M^{-1}(MN) &= M^{-1}M \\ (M^{-1}M)N &= M^{-1}M \\ IN &= I \\ N &= I \end{aligned}$$

PROBLEM SET 7.5, PAGES 317–319

Answers for Problems 1–27 may vary.

1. Interchange Rows 1 and 3. **3.** Multiply first row by $\frac{1}{2}$. **5.** Add (-1) times second row to first row.

$\left[\begin{array}{ccc|c} 1 & 3 & -4 & 9 \\ 0 & 2 & 4 & 5 \\ 3 & 1 & 2 & 1 \end{array}\right]$ $\left[\begin{array}{ccc|c} 1 & 2 & 5 & -6 \\ 6 & 3 & 4 & 6 \\ 10 & -1 & 0 & 1 \end{array}\right]$ $\left[\begin{array}{ccc|c} 1 & 5 & -12 & 2 \\ 4 & 1 & 9 & 2 \\ 7 & 6 & 1 & 3 \end{array}\right]$

7. $\left[\begin{array}{ccc|c} 1 & 2 & -3 & 0 \\ 0 & 3 & 1 & 4 \\ 0 & 1 & 7 & 6 \end{array}\right]$ **9.** $\left[\begin{array}{ccc|c} 1 & 2 & 4 & 1 \\ 0 & 9 & 8 & 4 \\ 0 & 13 & 17 & 7 \end{array}\right]$ **11.** $\left[\begin{array}{cccc|c} 1 & 4 & -1 & 3 & 3 \\ 0 & 16 & 3 & 13 & 9 \\ 0 & -19 & 14 & -14 & 17 \\ 0 & -4 & 3 & -3 & -3 \end{array}\right]$

13. Multiply Row 2 by $\frac{1}{2}$. $\left[\begin{array}{ccc|c} 1 & 3 & 5 & 2 \\ 0 & 1 & 3 & -4 \\ 0 & 3 & 4 & 1 \end{array}\right]$

15. Add (-1) times third row to second row. $\left[\begin{array}{ccc|c} 1 & 4 & -1 & 6 \\ 0 & 1 & -5 & -2 \\ 0 & 4 & 6 & 5 \end{array}\right]$

17. Multiply Row 2 by $\frac{1}{5}$. $\left[\begin{array}{ccc|c} 1 & 3 & -2 & 4 \\ 0 & 1 & \frac{1}{5} & \frac{3}{5} \\ 0 & 7 & 9 & 2 \end{array}\right]$

19. $\left[\begin{array}{ccc|c} 1 & 5 & -3 & 2 \\ 0 & 1 & 4 & 5 \\ 0 & 0 & -8 & -13 \end{array}\right]$ **21.** $\left[\begin{array}{cccc|c} 1 & 7 & 6 & 6 & 2 \\ 0 & 1 & 9 & 2 & 1 \\ 0 & 0 & 42 & 9 & 5 \\ 0 & 0 & -37 & 0 & -2 \end{array}\right]$

23. Multiply Row 3 by $\frac{1}{8}$. $\left[\begin{array}{ccc|c} 1 & -3 & 4 & -5 \\ 0 & 1 & 3 & 6 \\ 0 & 0 & 1 & \frac{3}{2} \end{array}\right]$

25. $\left[\begin{array}{ccc|c} 1 & 3 & 0 & 9 \\ 0 & 1 & 0 & -2 \\ 0 & 0 & 1 & 4 \end{array}\right]$ **27.** $\left[\begin{array}{cccc|c} 1 & 0 & 0 & 0 & -2 \\ 0 & 1 & 0 & 0 & 1 \\ 0 & 0 & 1 & 0 & 0 \\ 0 & 0 & 0 & 1 & -6 \end{array}\right]$ **29.** $\left[\begin{array}{ccc|c} 1 & 0 & 0 & 35 \\ 0 & 1 & 0 & -4 \\ 0 & 0 & 1 & -21 \end{array}\right]$ $(x, y, z) = (35, 4, -21)$

31. $(3, 2)$ **33.** $(1, \frac{3}{2})$ **35.** $(3, 1)$ **37.** $(1, 2, 3)$ **39.** $(3, -1, 2)$ **41.** $(5, 6, 1)$ **43.** $(1, 1, 1)$ **45.** $(6, 2, -9)$
47. $(\frac{9}{7}, \frac{1}{7}, -\frac{4}{7})$ **49.** dependent system **51.** $(1, 0, 0)$ **53.** $(w, x, y, z) = (1, 2, -3, -2)$ **55.** $y = x^2 + 4x + 5$
57. $y = x^2 - 10x + 20$ **59.** Mix 7 bars of first alloy and 3 bars of second alloy.
61. 3 units of Type I, 7 units of Type II, and 5 units of Type III

PROBLEM SET 7.6, PAGES 323–324

1.

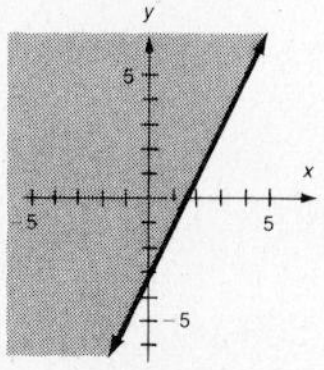

3.
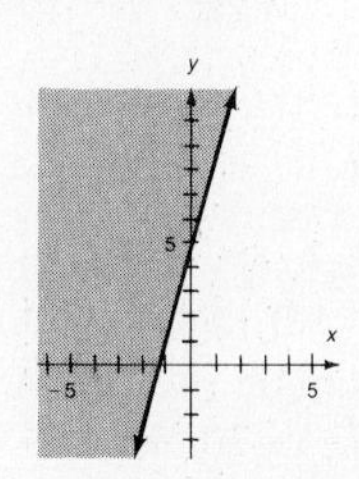

5.
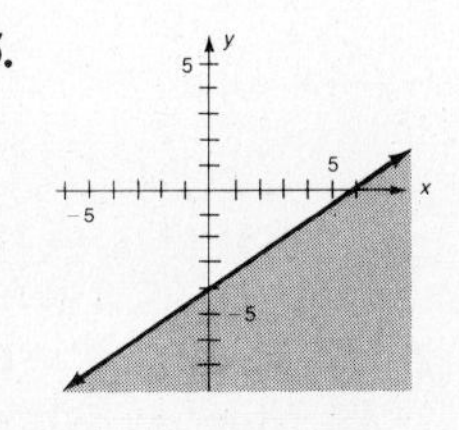

7.
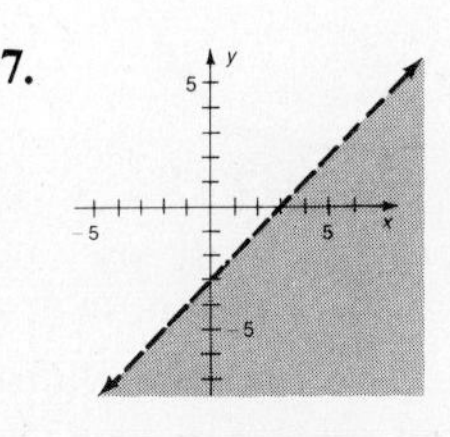

9.
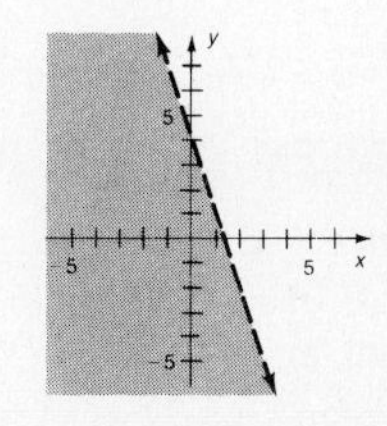

11.
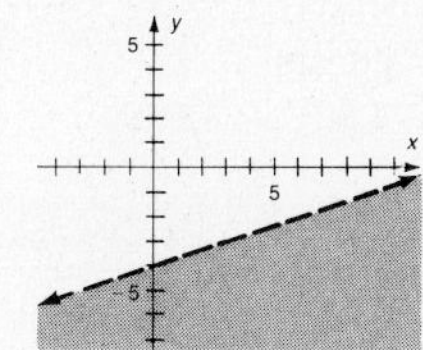

13.
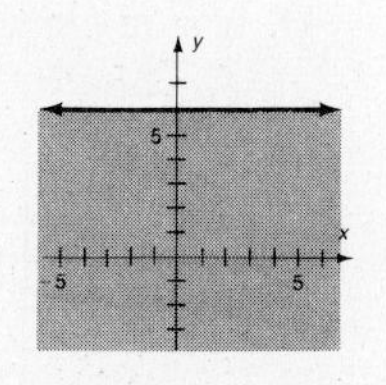

15.
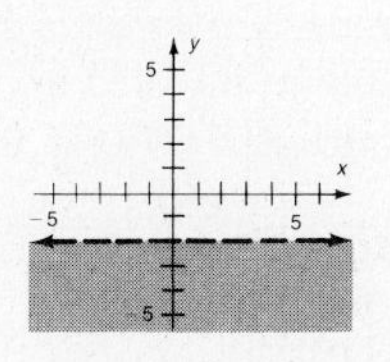

17.
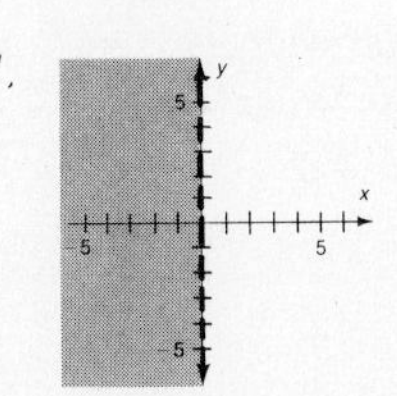

19.
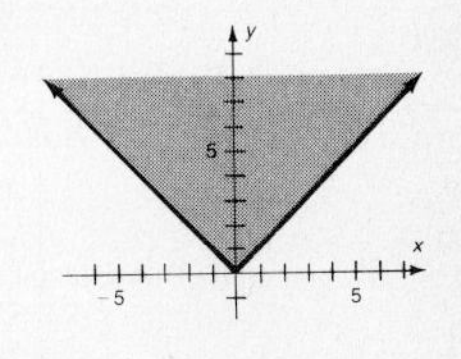

21.
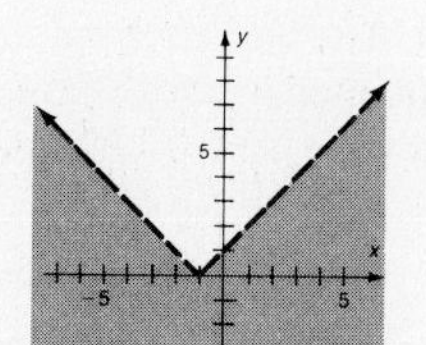

23.
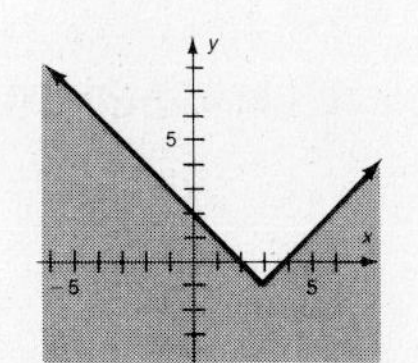

25.
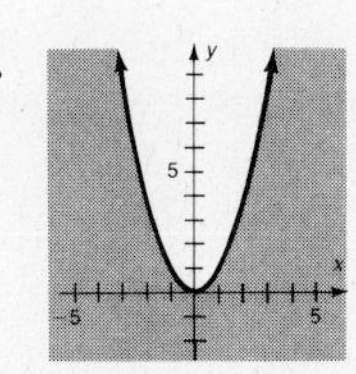

27.
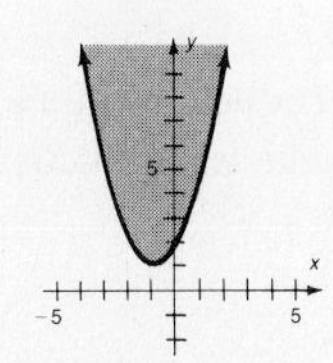

29.
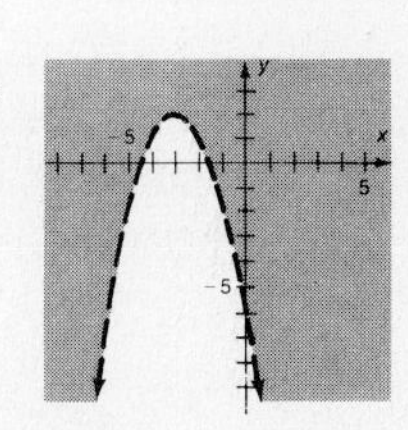

31.

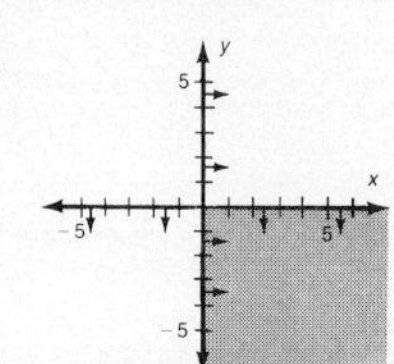

33.

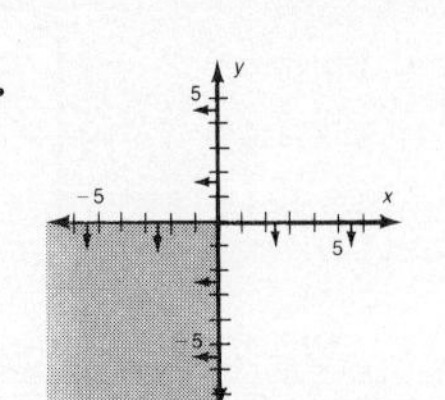

35.

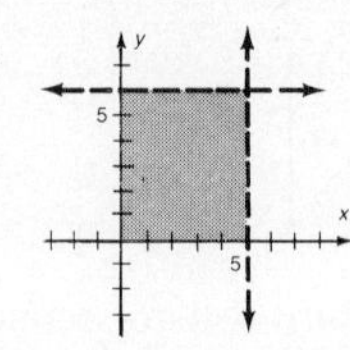

37.

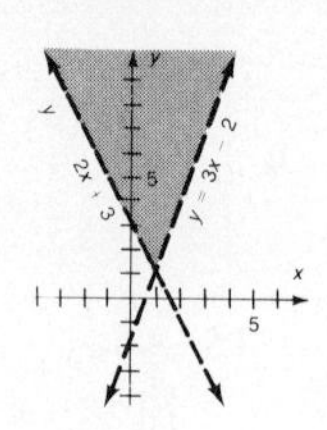

39.

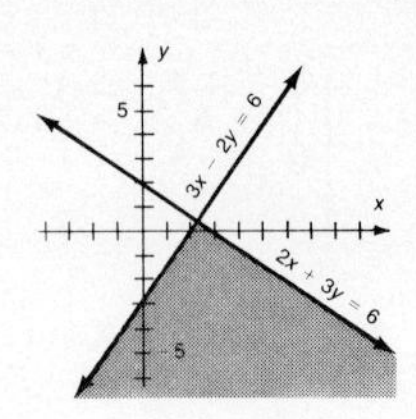

41.

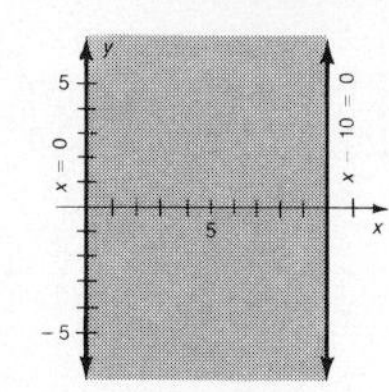

43.

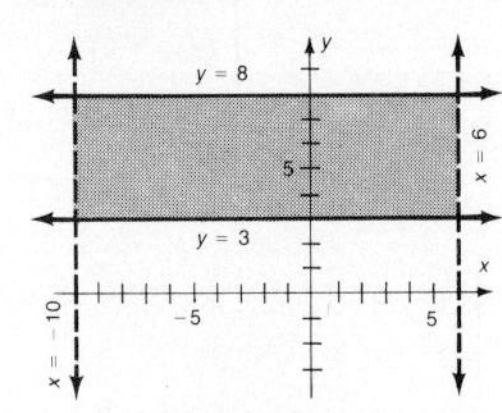

45.

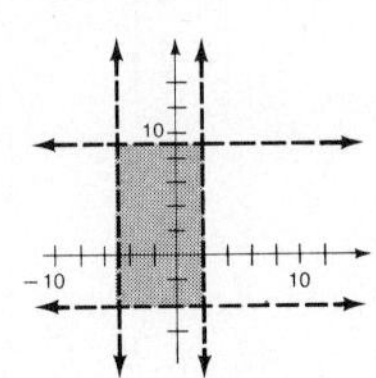

47.

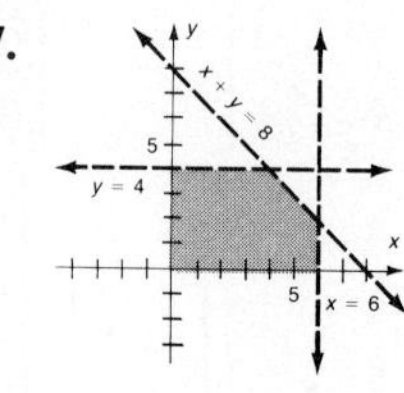

49.

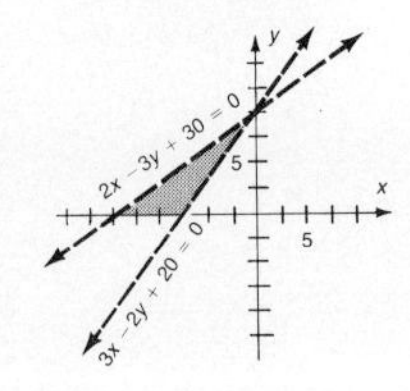

51.

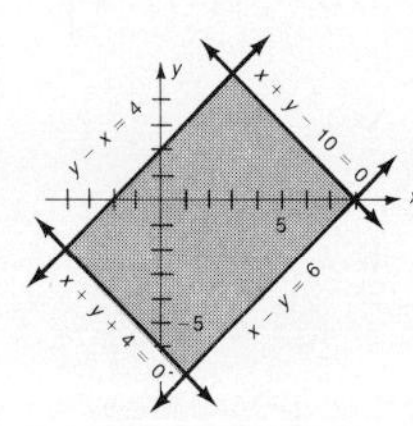

53.

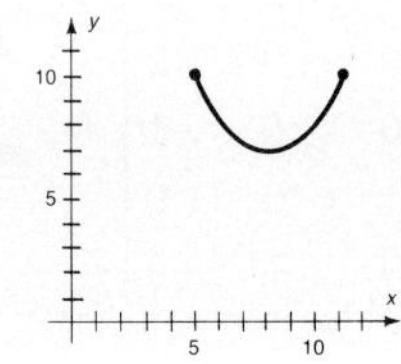

55.

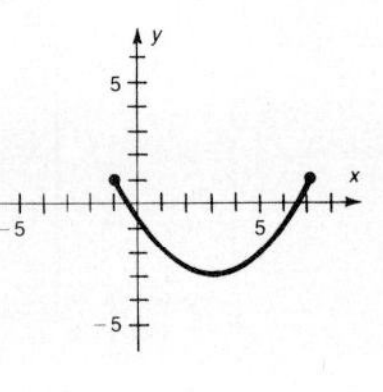

57.

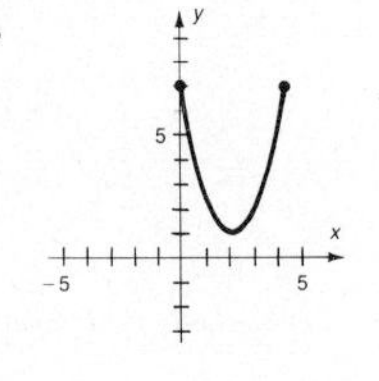

59.

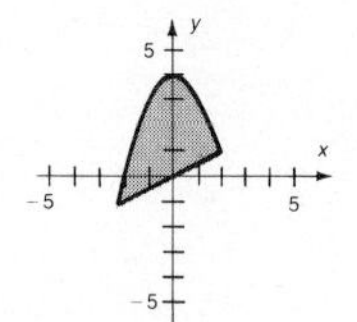

61.

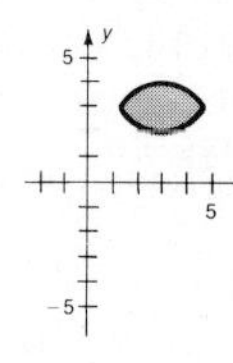

63.

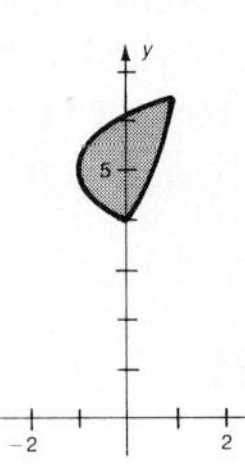

PROBLEM SET 7.7, PAGES 332–335

1. no **3.** no **5.** no **7.** no **9.** yes **11.** yes **13.** $(0,0), (0,\frac{9}{2}), (5,2), (6,0)$ **15.** $(0,0), (0,4), (2,3), (4,0)$
17. $(0,0), (0,4), (4,4), (6,2), (6,0)$ **19.** $(0,0), (0,\frac{8}{5}), (\frac{8}{3},0), (\frac{24}{13},\frac{16}{13})$ **21.** $(50,0), (\frac{200}{7},\frac{60}{7}), (8,24), (0,40)$
23. $(3,2), (5,5), (7,5), (\frac{10}{3},\frac{4}{3})$ **25.** Max $W = 190$ at $(5,2)$. **27.** Max $T = 600$ at $(6,0)$. **29.** Max $P = 500$ at $(2,3)$.
31. Min $C = 72$ at $(0,6)$. **33.** Max $F = 12$ at $(6,0)$. **35.** Min $I = 120$ at $(0,6)$. **37.** Max $P = 701.5$ at $(\frac{25}{2},9)$.
39. Max $P = 45$ at $(6,3)$. **41.** Min $X = \frac{62}{3}$ at $(\frac{10}{3},\frac{4}{3})$. **43.** Min $K = 560$ at $(4,0)$.

45. x = number of regular widgets
y = number of deluxe widgets
Maximize $P = 25x + 30y$

$$\text{subject to: } \begin{cases} x \geq 0 \\ y \geq 0 \\ 3x + 2y \leq 8 \\ 2x + 4y \leq 8 \end{cases}$$

47. x = amount invested in stocks
y = amount invested in bonds
Maximize $R = 0.12x + 0.08y$

$$\text{subject to: } \begin{cases} x \geq 0 \\ y \geq 0 \\ x \leq 8 \\ y \geq 2 \\ x + y \leq 10 \\ x \leq 3y \end{cases}$$

49. x = number of commercial guests
y = number of other guests
Maximize profit $P = 4.50x + 3.50y$

$$\text{subject to: } \begin{cases} x \geq 0 \\ y \geq 0 \\ x + y \leq 200 \\ 0.4x + 0.2y \leq 0.50 \end{cases}$$

51. x = number of Alpha products produced per day
y = number of Beta products produced per day
Maximize profit $P = 5x + 8y$

$$\text{subject to: } \begin{cases} 0 \leq x \leq 700 \\ 0 \leq y \\ x + 3y \leq 1200 \\ x + 2y \leq 1000 \end{cases}$$

53. Max profit $P = 14{,}300$ with all 100 acres planted in corn.

55. Max profit = \$100 with 4 regular widgets and 0 deluxe widgets produced. **57.** Minimum cost = \$5.92 obtained with 8 oz of food *A* and 24 oz of food *B*. **59.** Maximum return = \$900,000 with \$2,500,000 invested in stocks and \$7,500,000 invested in bonds.

CHAPTER 7 SUMMARY, PAGES 335–338

1.  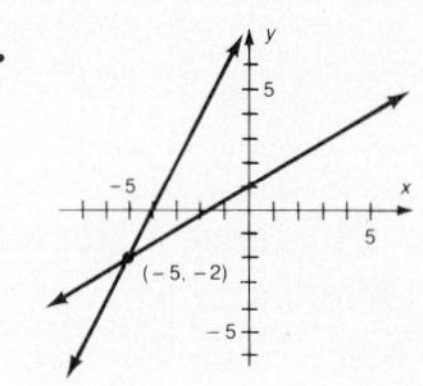

3. 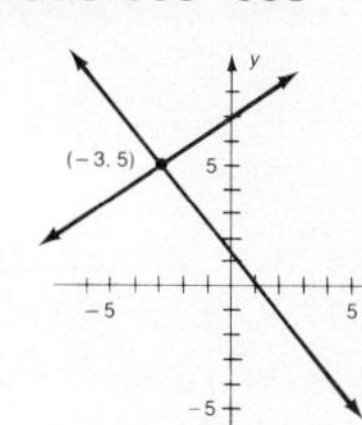

5. $(-6, 2)$ **7.** $(-16, 21)$ **9.** $(-3, 4)$ **11.** $(5, -9)$ **13.** $(\frac{7}{16}, -\frac{11}{16})$

15. dependent system **17.** 29 **19.** -57 **21.** $(4, 0, -3)$ **23.** $(-1, -3, 5)$ **25.** $\begin{bmatrix} 5 & -1 & 1 \\ -3 & 8 & 7 \\ 3 & -6 & 6 \end{bmatrix}$

27. $\begin{bmatrix} 10 & -14 & -1 \\ 3 & -7 & 23 \\ 7 & -23 & -4 \end{bmatrix}$ **29.** $\begin{bmatrix} 7 & 2 \\ 3 & 1 \end{bmatrix}$ **31.** does not exist **33.** $(1, 7, 0)$ **35.** $(8, -2, 3)$ **37.** $(7, -5, 1)$

39. $(1, 2, -3)$ **41.** 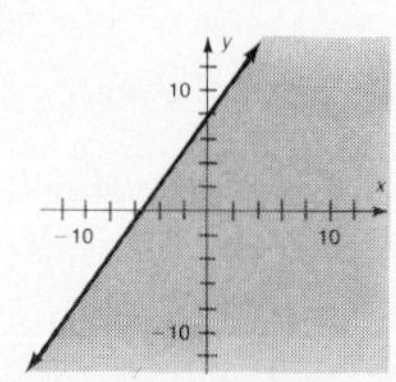

43. 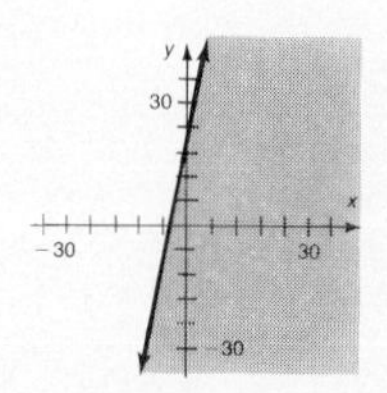

45. 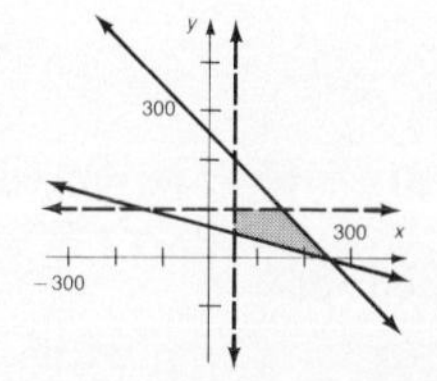

47. 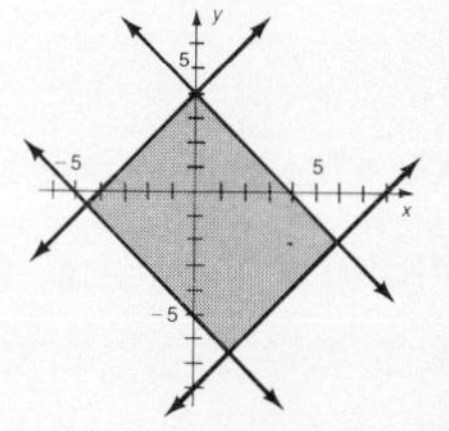

49. Maximum $P = 52.5$ at $(7, \frac{5}{2})$. **51.** Maximum $M = 1170$ at $(170, 80)$. **53.** 20 g of *I* and 40 g of II
55. Minimum cost $C = \$2.52$ with 12 g of food *A* and 14 g of food *B*.

PROBLEM SET 8.1, PAGES 344–345

Proofs for Problems 1–10 vary.

1. true **3.** true **5.** true **7.** true **9.** true

11. $P(k)$: $5 + 9 + 12 + \cdots + (4k + 1) = k(2k + 3)$
$P(k + 1)$: $5 + 9 + 13 + \cdots + (4k + 5) = (k + 1)(2k + 5)$

13. $P(k)$: $2^2 + 4^2 + 6^2 + \cdots + (2k)^2 = \dfrac{2k(k + 1)(2k + 1)}{3}$
$P(k + 1)$: $2^2 + 4^2 + 6^2 + \cdots + (2k + 2)^2 = \dfrac{(2k + 2)(k + 2)(2k + 3)}{3}$

15. $P(k)$: $\cos(\theta + k\pi) = (-1)^k \cos\theta$
$P(k + 1)$: $\cos(\theta + k\pi + \pi) = (-1)^{k+1} \cos\theta$

17. $P(k)$: $k^2 + k$ is even
$P(k + 1)$: $(k + 1)^2 + (k + 1)$

19. $P(k)$: $(\frac{2}{3})^{k+1} < (\frac{2}{3})^k$
$P(k + 1)$: $(\frac{2}{3})^{k+2} < (\frac{2}{3})^{k+1}$

21–53. Proofs vary.

55. Conjecture: $1^3 + 2^3 + \cdots + n^3 = \left[\dfrac{n(n + 1)}{2}\right]^2$ or $\dfrac{n^2(n + 1)^2}{4}$. Proofs vary.

57. Conjecture: $2 + 6 + 18 + 54 + \cdots + (2 \cdot 3^{n-1}) = 3^n - 1$. Proofs vary. **59–61.** Proofs vary.

PROBLEM SET 8.2, PAGES 351–352

1. 22 **3.** 2 **5.** 72 **7.** 132 **9.** 220 **11.** 1140 **13.** 8 **15.** 28 **17.** 56 **19.** 70 **21.** 1326
23. 1,000 **25.** $(a + b)^6 = a^6 + 6a^5b + 15a^4b^2 + 20a^3b^3 + 15a^2b^4 + 6ab^5 + b^6$ **27.** $(x + 3)^3 = x^3 + 9x^2 + 27x + 27$

29. $(x+y)^5 = x^5 + 5x^4y + 10x^3y^2 + 10x^2y^3 + 5xy^4 + y^5$ **31.** $(x+2)^5 = x^5 + 10x^4 + 40x^3 + 80x^2 + 80x + 32$
33. $(x+y)^4 = x^4 + 4x^3y + 6x^2y^2 + 4xy^3 + y^4$ **35.** $(\frac{1}{2}x + y^3)^3 = \frac{1}{8}x^3 + \frac{3}{4}x^2y^3 + \frac{3}{2}xy^6 + y^9$
37. $(x^{\frac{1}{2}} + y^{\frac{1}{2}})^4 = x^2 + 4x^{\frac{3}{2}}y^{\frac{1}{2}} + 6xy + 4x^{\frac{1}{2}}y^{\frac{3}{2}} + y^2$ **39.** $1 - 8x + 28x^2 - 56x^3 + 70x^4 - 56x^5 + 28x^6 - 8x^7 + x^8$
41. $a^{10} + 10a^9b + 45a^8b^2 + 120a^7b^3 + \cdots$ **43.** $a^{14} + 14a^{13}b + 91a^{12}b^2 + 364a^{11}b^3 + \cdots$
45. $x^{16} + 32x^{15}y + 480x^{14}y^2 + 4480x^{13}y^3 + \cdots$ **47.** $x^{12} - 24x^{11}y + 264x^{10}y^2 - 1760x^9y^3 + \cdots$
49. $1 - 0.24 + 0.0264 - 0.00176 + \cdots$ **51.** $\binom{8}{1} = 8$; term is $\binom{8}{1}(a^2)^2(-2b)^1 = -16a^4b$; coefficient is -16.
53. 4032 **55.** 160 **57–63.** Proofs vary.

PROBLEM SET 8.3, PAGES 357–358

1. a. arithmetic **b.** $d = 3$ **c.** 17 **3. a.** geometric **b.** $r = 2$ **c.** 96 **5. a.** neither **b.** Difference increases by one between each term. **c.** 85 **7. a.** geometric **b.** $r = q$ **c.** pq^5 **9. a.** neither **b.** Difference increases by two between each term. **c.** 36 **11. a.** neither **b.** Number of fives between twos increases by one. **c.** 2
13. a. neither **b.** Systematic listing of fractions (excluding those previously listed); that is, $\frac{2}{6}, \frac{3}{6}, \frac{4}{6}$ are previously listed in another form. **c.** $\frac{5}{6}$ **15. a.** neither **b.** perfect cubes **c.** 216 **17.** 1, 5, 9 **19.** 10, 5, $\frac{5}{2}$
21. a, ar, ar^2 **23.** $-1, 1, -1$ **25.** $2, \frac{3}{2}, \frac{4}{3}$ **27.** 2, 2, 2 **29.** 20 **31.** 49 **33.** 12 **35.** 11 **37.** 80
39. 10 **41.** 57 **43.** $\frac{5}{2^8}$ or $5 \cdot 2^{-8}$ or 0.01953125 **45.** 5^4 or 625 **47.** 2, 6, 18, 54, 162 **49.** 1, 1, 2, 3, 5
51. $\sum_{k=1}^{7} \frac{1}{2^k}$ **53.** $\sum_{k=1}^{5} 6^{k-1}$ **55.** arithmetic **57.** geometric **59.** geometric
61. a. $a_1b_1 + a_2b_2 + a_3b_3 + \cdots + a_rb_r$ **b.** $\sum_{j=1}^{r} ka_j = ka_1 + ka_2 + ka_3 + \cdots + ka_r = k(a_1 + a_2 + a_3 + \cdots + a_r) = k\sum_{j=1}^{r} a_j$ **c.** $\sum_{j=1}^{r} k = \underbrace{k + k + k + k + \cdots + k}_{r \text{ factors}} = kr$
63. 47 (add preceding two terms) **65.** 7 (arranged alphabetically)

PROBLEM SET 8.4, PAGES 363–364

1. 5, 9, 13, 17 **3.** 85, 88, 91, 94 **5.** 100, 95, 90, 85 **7.** $-\frac{5}{2}, -2, -\frac{3}{2}, -1$ **9.** $2\sqrt{3}, 3\sqrt{3}, 4\sqrt{3}, 5\sqrt{3}$
11. $x, x+y, x+2y, x+3y$ **13.** $a_1 = 5; d = 3$ **15.** $a_1 = 6; d = 5$ **17.** $a_1 = -8; d = 7$ **19.** $a_1 = x; d = x$
21. $a_1 = x - 5b; d = 2b$ **23.** $a_n = 5$ **25.** $a_n = 24 + 11n$ **27.** $a_n = -3 + 2n$ **29.** $a_n = x + n\sqrt{3}$ **31.** 101
33. 25 **35.** $-10{,}600$ **37.** -2 **39.** 2 **41.** -140 **43.** 4 **45.** 12 **47.** 11,500 **49.** 2030 **51.** n^2
53. a, b, c, and d are harmonic. **55. a.** 12 **b.** $\frac{19}{2}$ **c.** $\frac{5}{12}$ **d.** -6 **e.** $\frac{1}{15}$ **57.** Option D
59. Option A: \$43,440.
Option B: \$44,160; \$720 more than A
Option C: \$44,520; \$1080 more than A and \$360 more than B
Option D: \$44,760; \$1320 more than A, \$600 more than B, and \$240 more than C.

PROBLEM SET 8.5, PAGES 369–370

1. 5, 15, 45 **3.** 1, -2, 4 **5.** $-15, -3, -\frac{3}{5}$ **7.** $8, 8x, 8x^2$ **9.** x, xy, xy^2 **11.** $g_1 = 7; r = 2$ **13.** $g_1 = 100; r = \frac{1}{2}$ **15.** $g_1 = xyz; r = \frac{1}{z}$ **17.** $g_n = 7 \cdot 2^{n-1}$ **19.** $g_n = 200(\frac{1}{2})^n$ or $200 \cdot 2^{-n}$ or $100(\frac{1}{2})^{n-1}$ **21.** $g_n = xyz^{2-n}$
23. 2000 **25.** $-\frac{40}{3}$ **27.** $-\frac{1296}{5}$ **29.** 10^{-7} or 0.0000001 **31.** 2 **33.** $\frac{93}{25}$ **35.** $\frac{1}{3}$ **37.** $\frac{1111}{10{,}000}$ **39.** $\frac{85}{256}$
41. $\frac{5}{9}$ **43.** $\frac{3}{11}$ **45.** $\frac{5}{11}$ **47.** $\frac{218}{999}$ **49.** $\frac{27}{11}$ **51.** $\frac{22{,}309}{9900}$ **53.** 9 mailings **55.** 50 ft
57. She must pay \$25 in taxes, and they must pay her \$125. **59.** $4 + 2\sqrt{2}$ **61.** $\frac{4 + 3\sqrt{2}}{2}$ **63.** $3 - 3\sqrt{2}, 3, 3 + 3\sqrt{2}$
65. The area of the shaded portion is $\frac{1}{4}$ the area of the original square.

CHAPTER 8 SUMMARY, PAGES 370–371

1. If a given proposition $P(n)$ is true for $P(1)$ and if the truth of $P(k)$ implies the truth of $P(k+1)$, then $P(n)$ is true for all positive integers. **3.** $P(k)$: $4 + 8 + 12 + \cdots + 4k = 2k(k+1)$
$P(k+1)$: $4 + 8 + 12 + \cdots + 4(k+1) = 2(k+1)(k+2)$

5. $a^5 + 5a^4b + 10a^3b^2 + 10a^2b^3 + 5ab^4 + a^5$ **7.** $32x^5 + 80x^4y + 80x^3y^2 + 40x^2y^3 + 10xy^4 + y^5$ **9.** 2,598,960

11. 70 **13.** $(a+b)^n = \sum_{k=0}^{n} \binom{n}{k} a^{n-k}b^k$ **15.** $\binom{15}{r}x^{15-r}(-y)^r$ **17.** arithmetic; $a_n = -9 + 10n$

19. neither; 11111, 111111; number of ones increases by 1. **21.** 40 **23.** $3^{10} - 1$ or 59,048 **25.** $A_{10} = 460$

27. $G_{10} = 81 - 3^{-6} \approx 80.99862826$ **29.** $d = 2$; $A_{10} = 110$ **31.** $g_{10} = 2560$; $G_5 = 155$ **33.** 2000 **35.** $\frac{24}{11}$

PROBLEM SET 9.1, PAGES 379–381

1.

3.

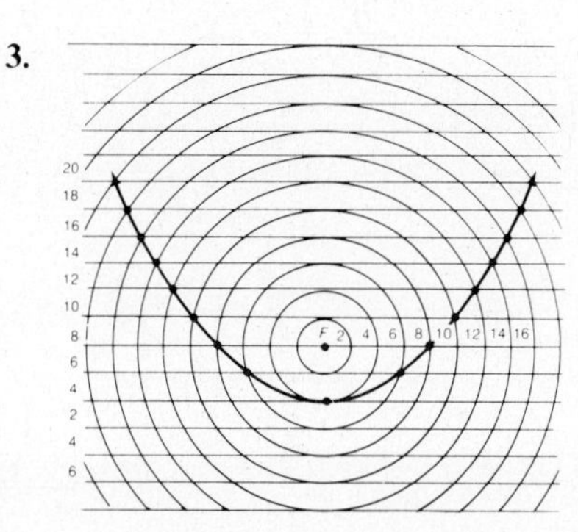

5.

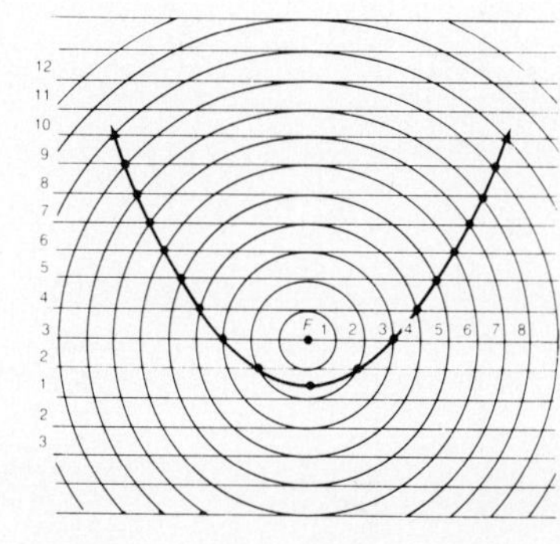

7.

9.

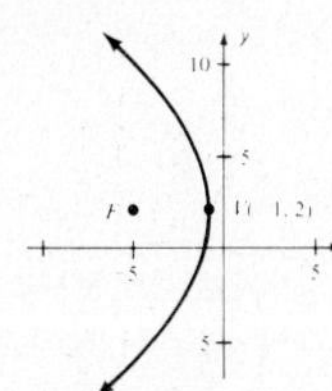

11.

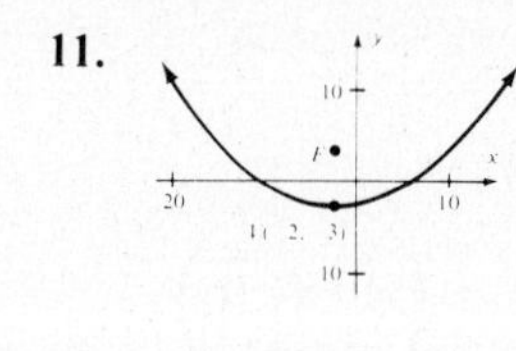

13.

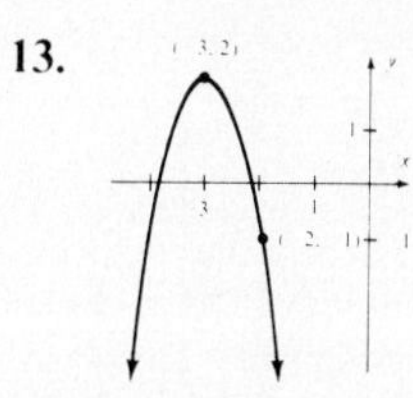

15.

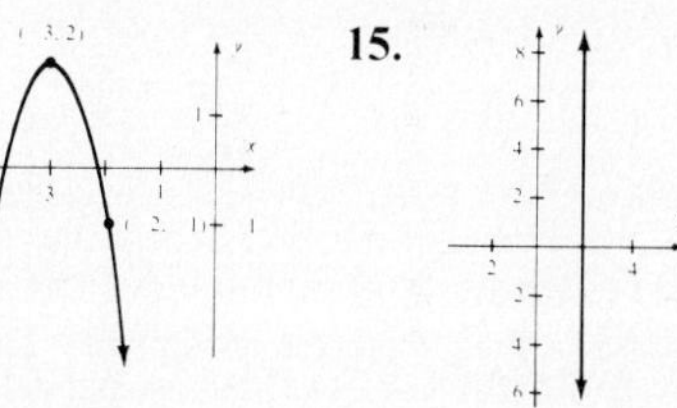

17.

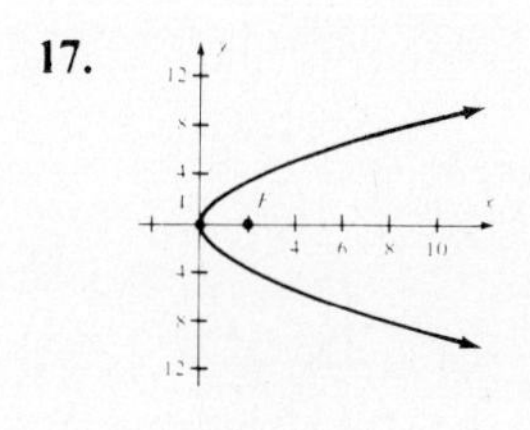

19.

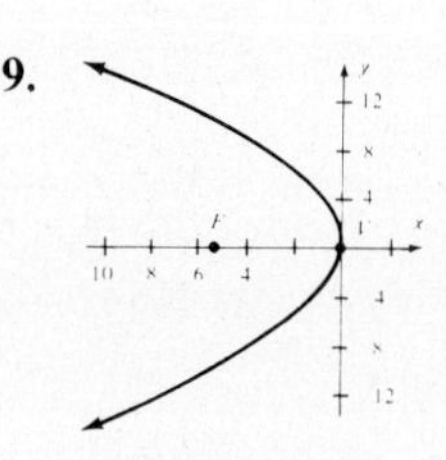

21.

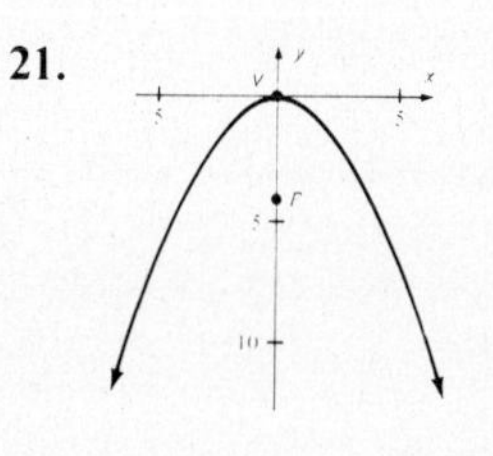

23.

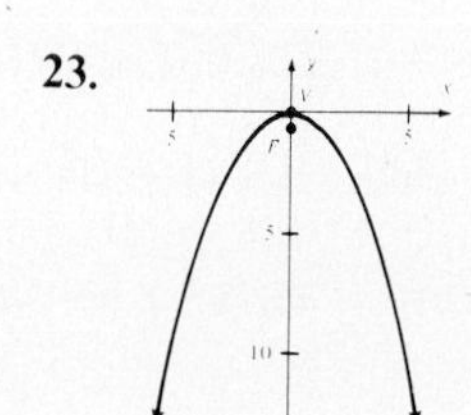

25.

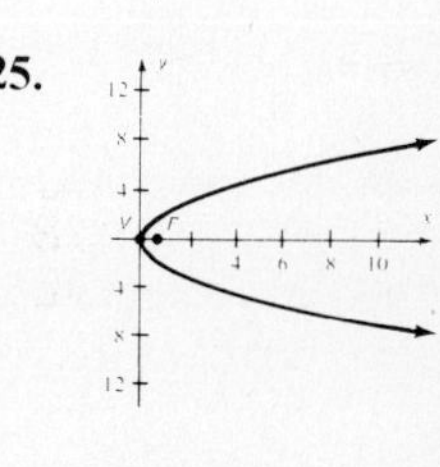

27.

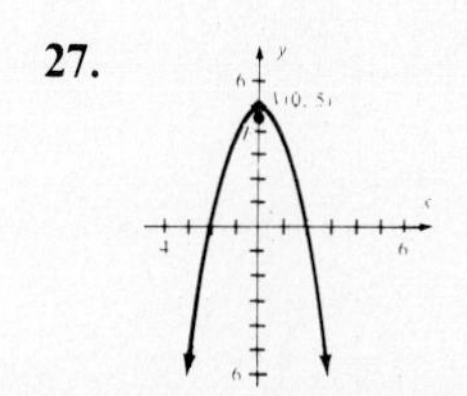

29.

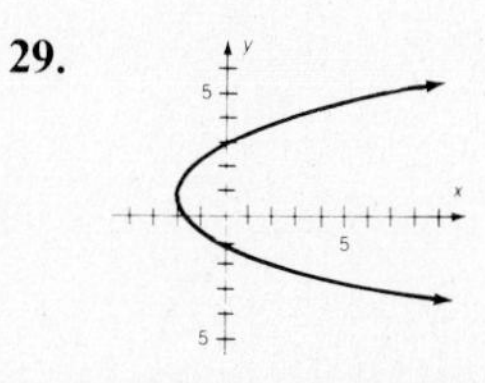

31.

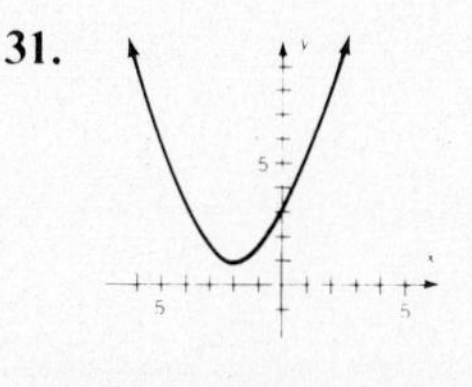

33.

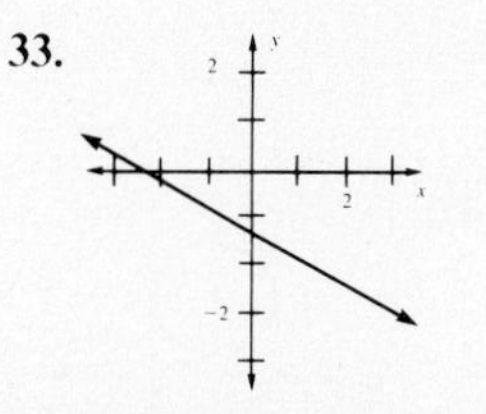

35.

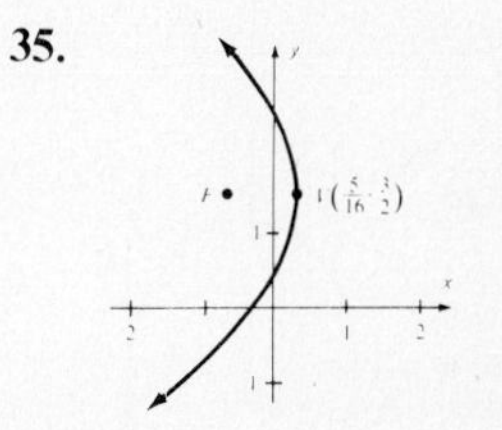

37.

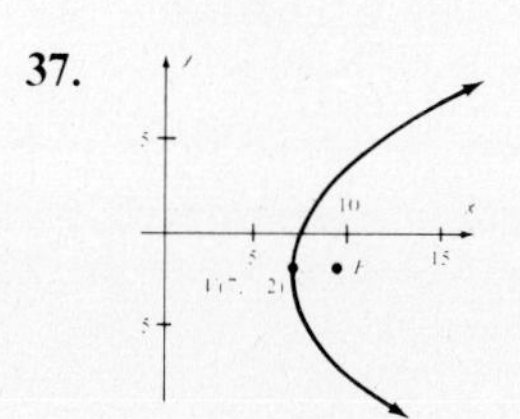

39.

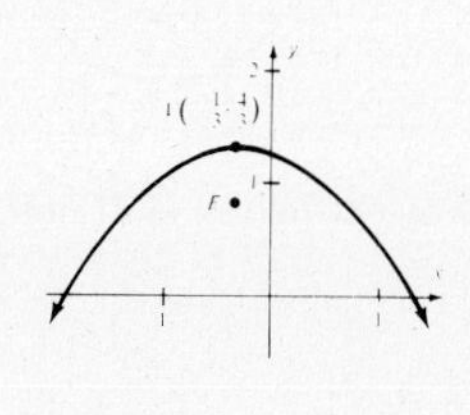

41. $y^2 = 10(x - \frac{5}{2})$ **43.** $(y-2)^2 = -16(x+1)$

45. $(x+2)^2 = 24(y+3)$ **47.** $(x+3)^2 = -\frac{1}{3}(y-2)$ **49.** $x = 2$

51. $(x-100)^2 = -200(y-50)$. Also may add: $0 \leq x \leq 200$ **53.** 2.25 m from the vertex on the axis of the parabola

55. $(-3, 4)$ and $(1, 12)$ **57.** $(2, 3)$ and $(3, 1)$ **59.** 21.0 seconds at age 40 **61–63.** Answers vary. **65. a.** $m = -\frac{A}{B}$
b. $m' = \frac{B}{A}$ **c.** $Bx - Ay - Bx_0 + Ay_0 = 0$ **d.** $x = \dfrac{B^2x_0 - ABy_0 - AC}{A^2 + B^2}$, $y = \dfrac{-ABx_0 + A^2y_0 - BC}{A^2 + B^2}$
e. $d = \dfrac{|Ax_0 + By_0 + C|}{\sqrt{A^2 + B^2}}$ **67.** $25x^2 + 120xy + 144y^2 - 1110x + 1730y + 4209 = 0$

PROBLEM SET 9.2, PAGES 389–391

1.

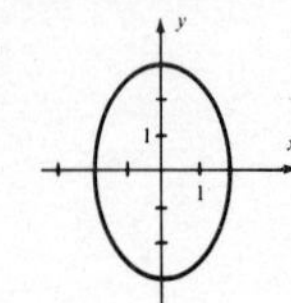

3.
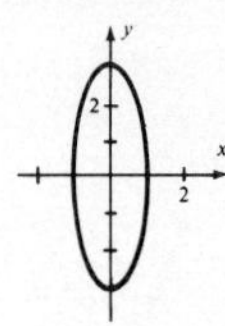

5.
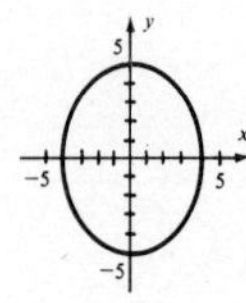

7.
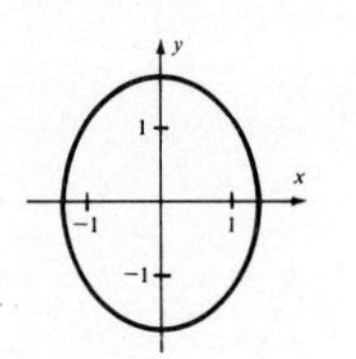

9.
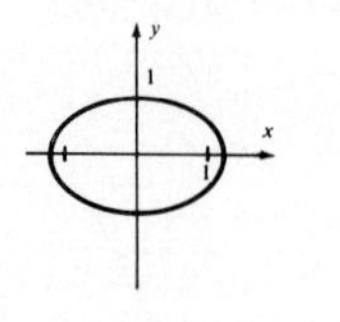

11.
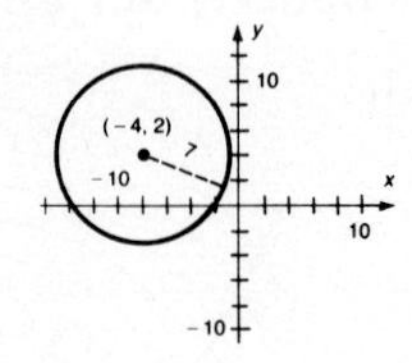

13.
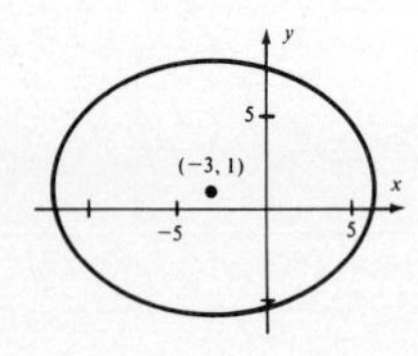

15.
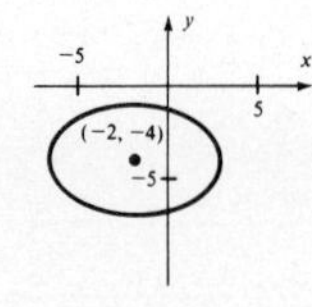

17.
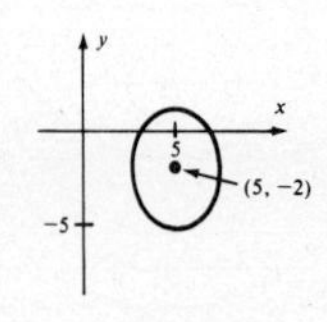

19.
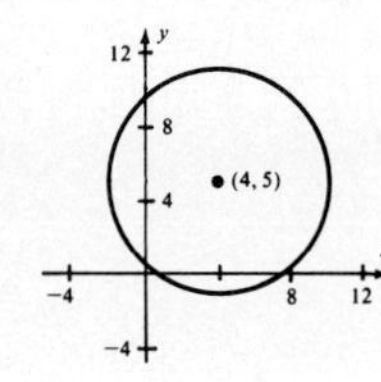

21.
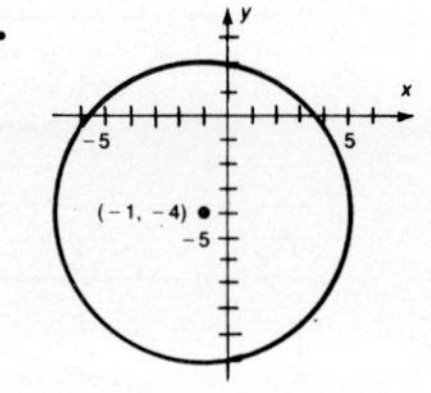

23.
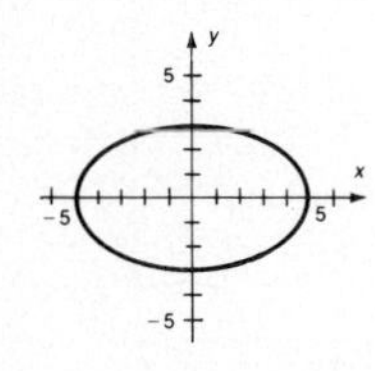

25.
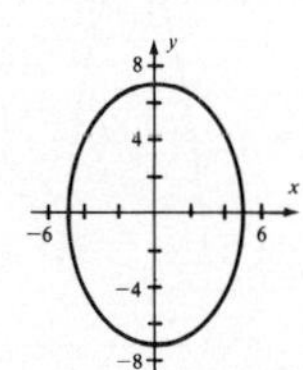

27.
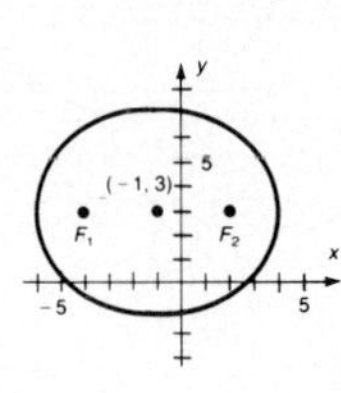

29.
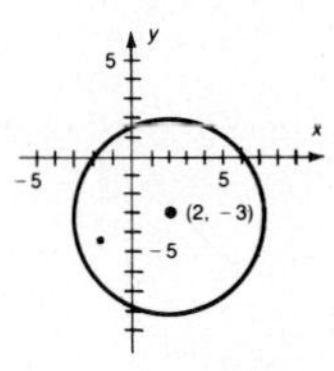

31. $(x - 4)^2 + (y - 5)^2 = 36$ **33.** $(x + 1)^2 + (y + 4)^2 = 36$ **35.** $\dfrac{x^2}{9} + \dfrac{y^2}{25} = 1$ **37.** $\dfrac{x^2}{24} + \dfrac{y^2}{49} = 1$

39. $\dfrac{(x + 1)^2}{25} + \dfrac{(y - 3)^2}{21} = 1$ **41.** $x^2 + y^2 - 4x + 6y - 12 = 0$ or $(x - 2)^2 + (y + 3)^2 = 25$ **43.**

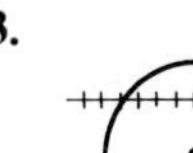

45.
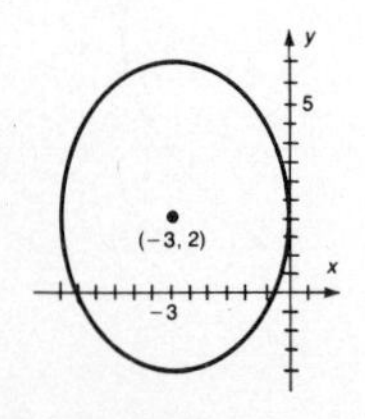

47.
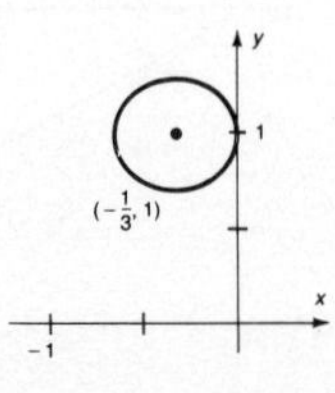

49.
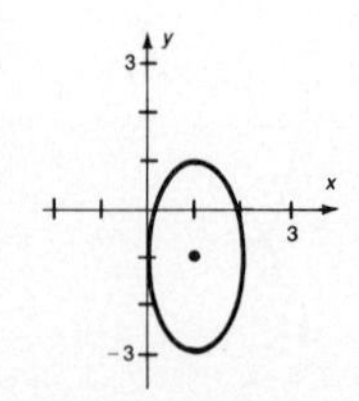

51.
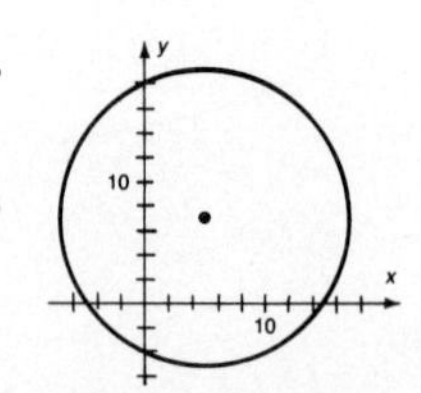

53.
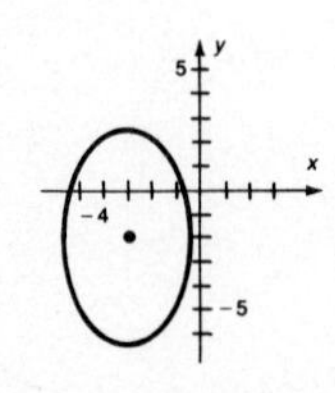

55. Answers vary. **57.** The distance at aphelion is 94,500,000 miles and at perihelion is 91,500,000 miles. **59.** 0.053

61. **63.** Answers vary.

PROBLEM SET 9.3, PAGES 399–401

1.

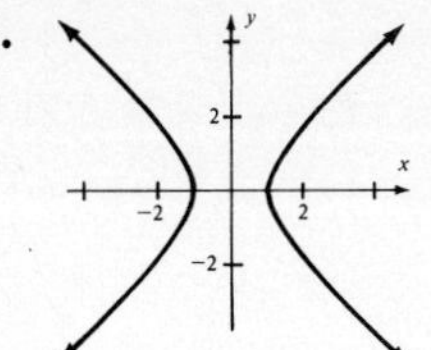

3.

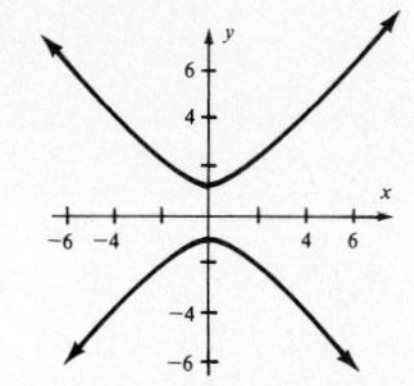

5.

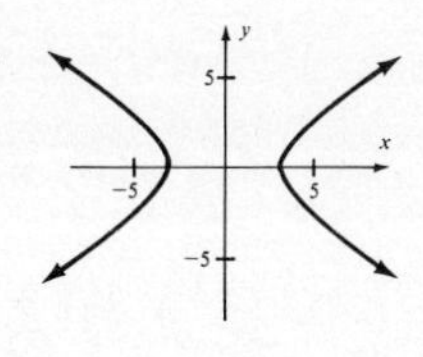

7.

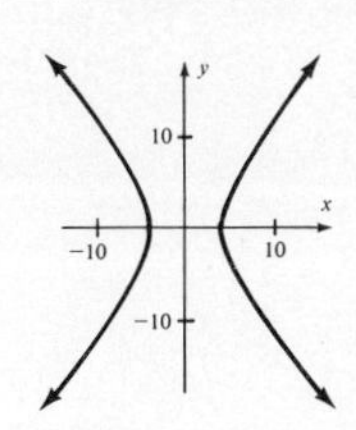

9.

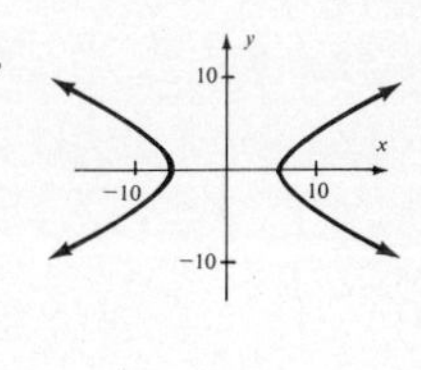

11.

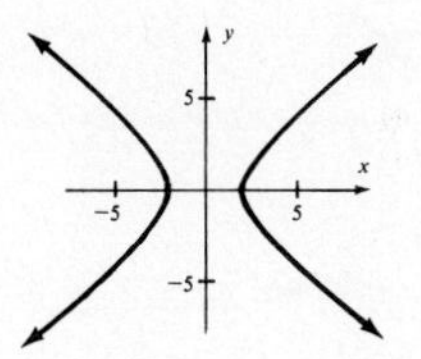

13.

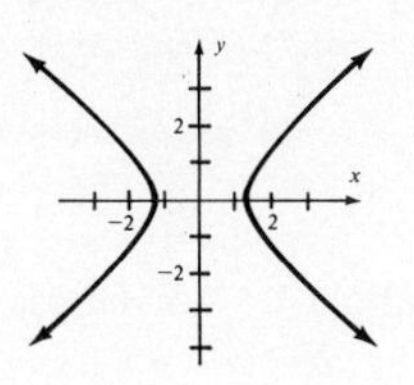

15.

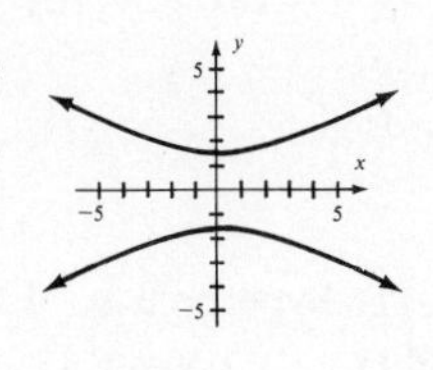

17.

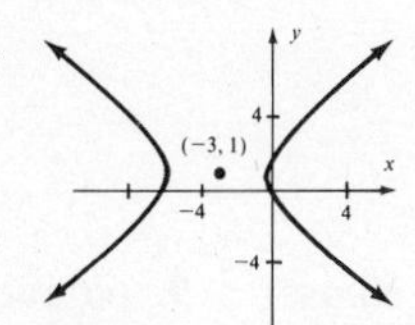

19.

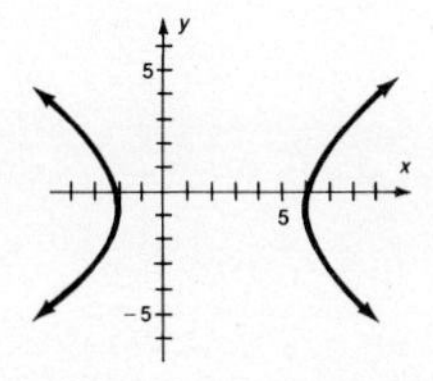

21.

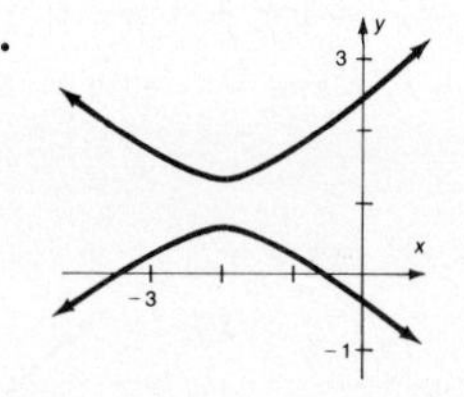

23. line

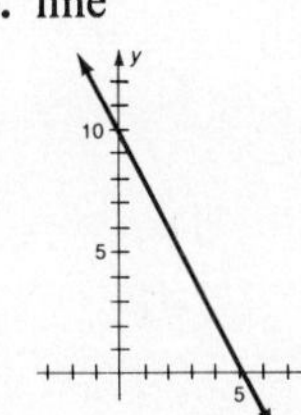

25. parabola

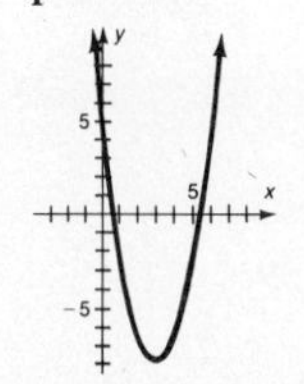

27. line

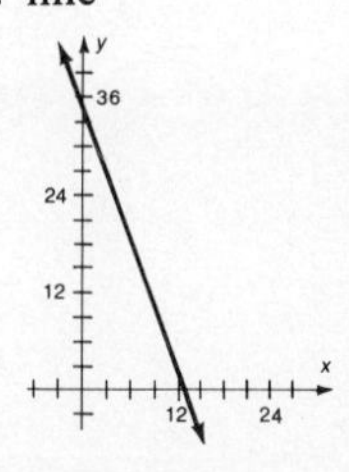

29. parabola

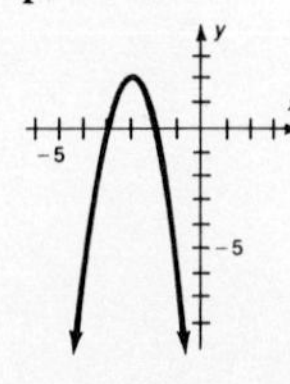

31. ellipse

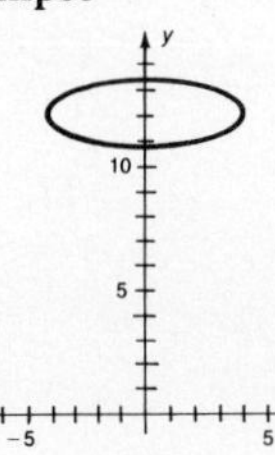

33. circle

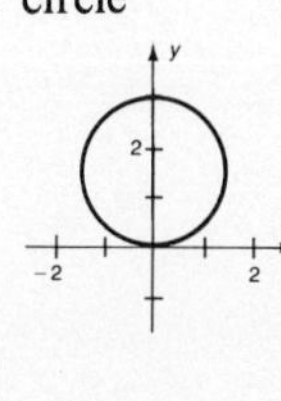

35. $\dfrac{x^2}{25} - \dfrac{y^2}{11} = 1$

37. $\dfrac{(y-6)^2}{4} - \dfrac{(x-4)^2}{5} = 1$

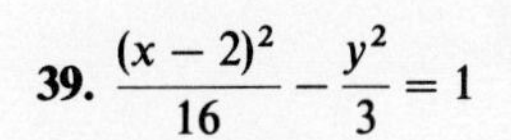

39. $\dfrac{(x-2)^2}{16} - \dfrac{y^2}{3} = 1$

41.

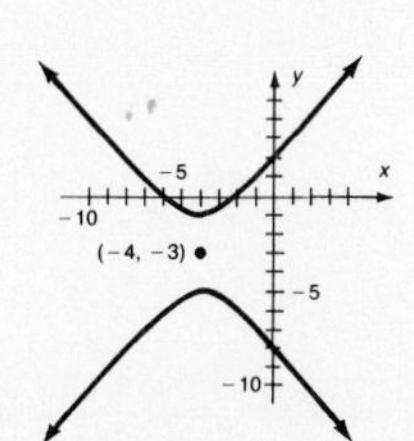

43.

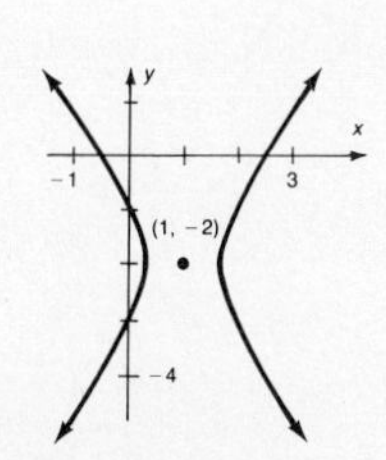

45.

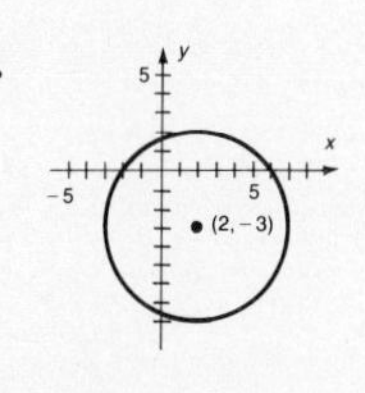

47.

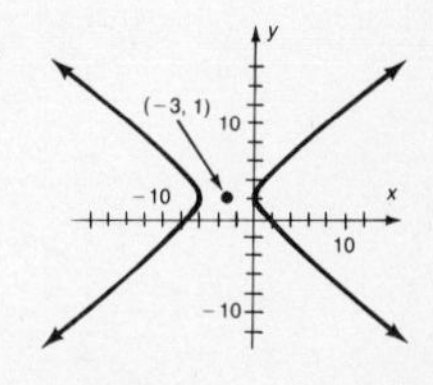

49. parabola

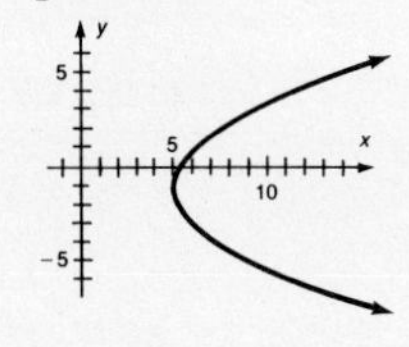

51. parabola

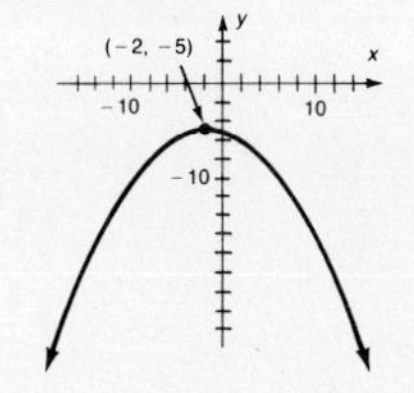

53. hyperbola

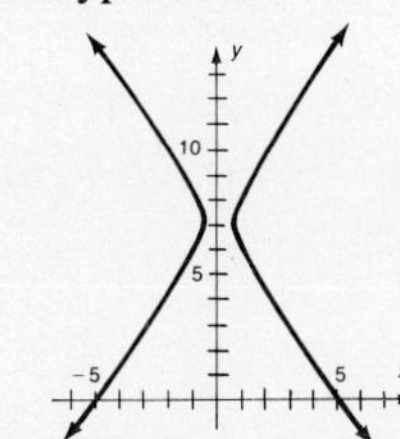

55. circle

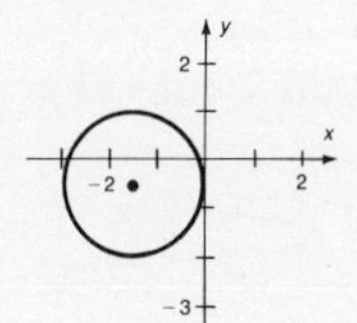

57. ellipse

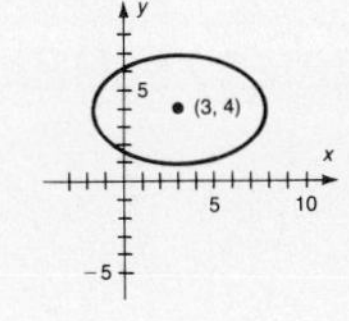

59–65. Answers vary.

PROBLEM SET 9.4, PAGES 407–408

1. $xy = 6$

$$\left[\tfrac{1}{\sqrt{2}}(x' - y')\right]\left[\tfrac{1}{\sqrt{2}}(x' + y')\right] = 6$$
$$\tfrac{1}{2}(x' - y')(x' + y') = 6$$
$$x^2 - y^2 = 12$$

3.
$$x^2 - 4xy + 4y^2 + 5\sqrt{5}y - 10 = 0$$
$$\tfrac{1}{5}(2x' - y')^2 - 4(\tfrac{1}{5})(2x' - y')(x' + 2y') + 4(\tfrac{1}{5})(x' + 2y')^2 + 5\sqrt{5}(\tfrac{1}{\sqrt{5}})(x' + 2y') - 10 = 0$$
$$4x'^2 - 4x'y' + y'^2 - 8x'^2 - 12x'y' + 8y'^2 + 4x'^2 + 16x'y' + 16y'^2 + 25x' + 50y' - 50 = 0$$
$$25y'^2 + 50y' = -25x' + 50$$
$$y'^2 + 2y' = -x' + 2$$
$$y'^2 + 2y' + 1 = -x' + 3$$
$$(y' + 1)^2 = -(x' - 3)$$

5. hyperbola **7.** ellipse **9.** parabola **11.** hyperbola **13.** parabola **15.** ellipse **17.** ellipse **19.** ellipse **21.** ellipse **23.** $\theta = 45°$; $x = \frac{1}{\sqrt{2}}(x' - y')$; $y = \frac{1}{\sqrt{2}}(x' + y')$ **25.** $\theta = 45°$; $x = \frac{1}{\sqrt{2}}(x' - y')$; $\frac{1}{\sqrt{2}}(x' + y')$ **27.** $\theta \approx 63.4°$; $x = \frac{1}{\sqrt{5}}(x' - 2y')$; $y = \frac{1}{\sqrt{5}}(2x' + y')$ **29.** $\theta = 30°$; $x = \frac{1}{2}(\sqrt{3}x' - y')$; $y = \frac{1}{2}(x' + \sqrt{3}y')$ **31.** $x = \frac{1}{2}(\sqrt{3}x' - y')$; $y = \frac{1}{2}(x' + \sqrt{3}y')$ **33.** $\theta = 60°$; $x = \frac{1}{2}(x' - \sqrt{3}y')$; $y = \frac{1}{2}(\sqrt{3}x' + y')$ **35.** $\theta \approx 71.6°$; $x = \frac{1}{\sqrt{10}}(x' - 3y')$; $y = \frac{1}{\sqrt{10}}(3x' + y')$

37. 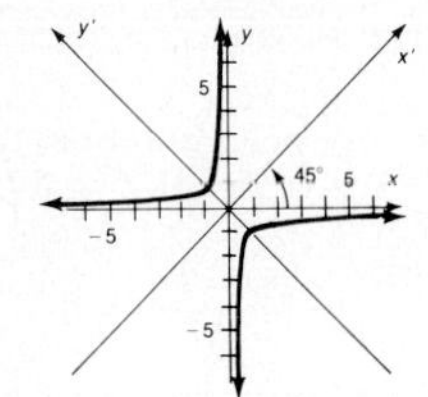

39. 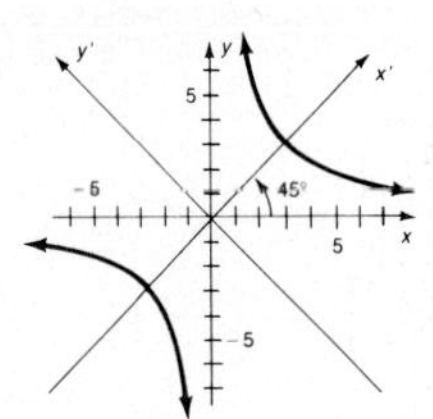

41. 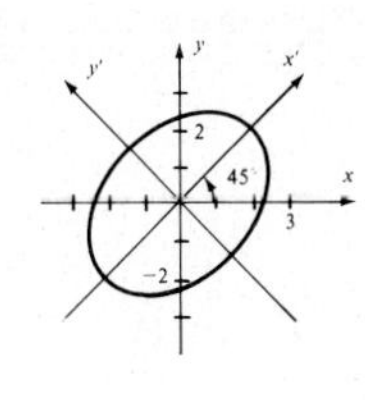

43. 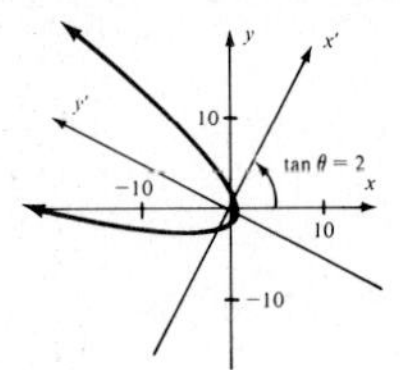

45. 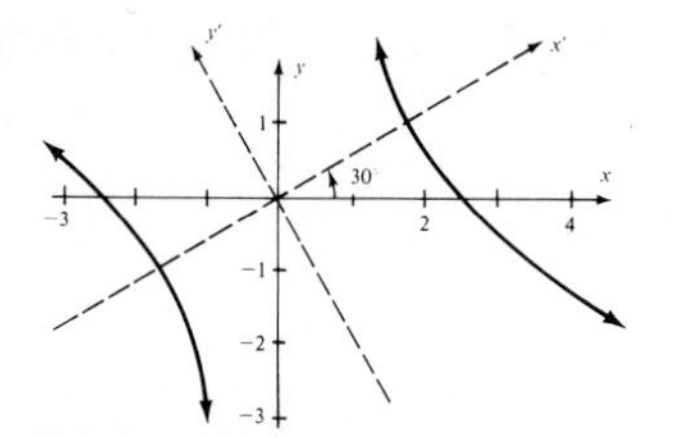

47. 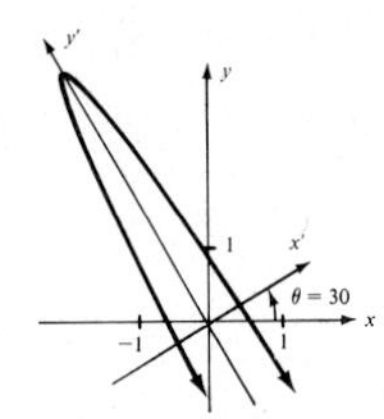

49. 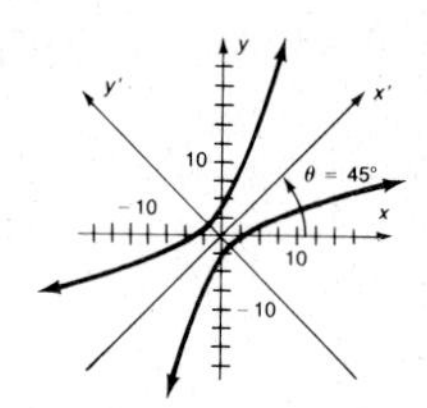

51. 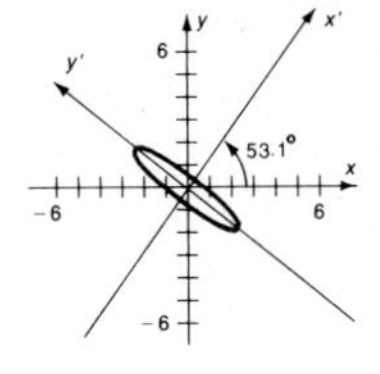

53. 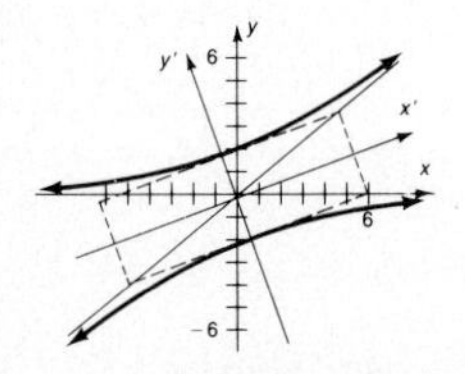

55. 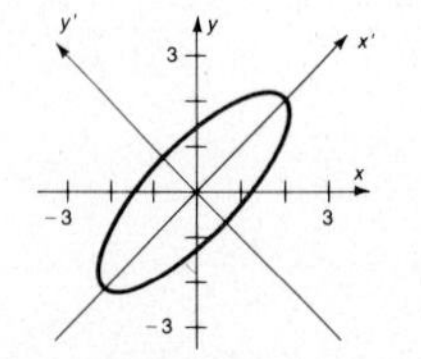

57. 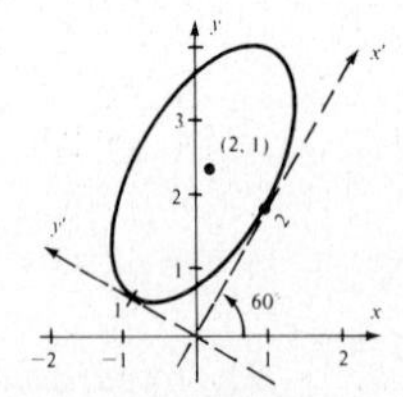

59. 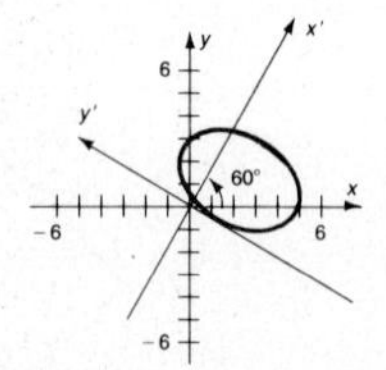

CHAPTER 9 SUMMARY, PAGES 408–409

1. 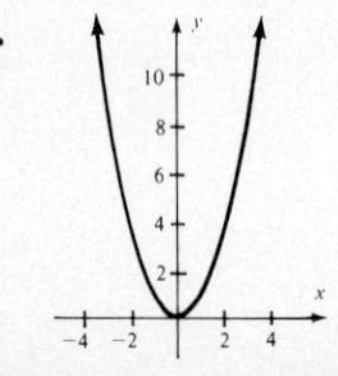

3. 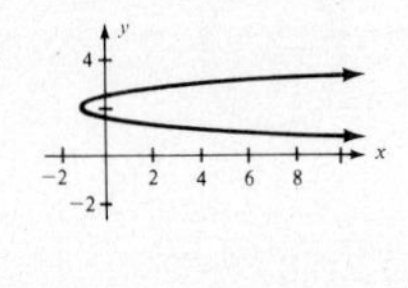

5. $(y - 3)^2 = 20(x - 6)$ **7.** $(x + 3)^2 = -24(y - 5)$

9. 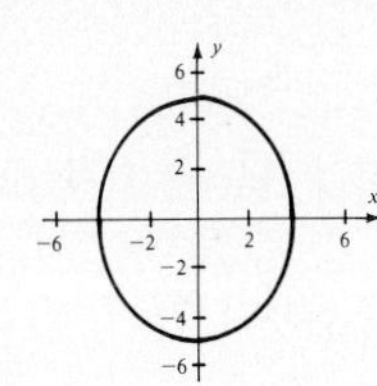

11. 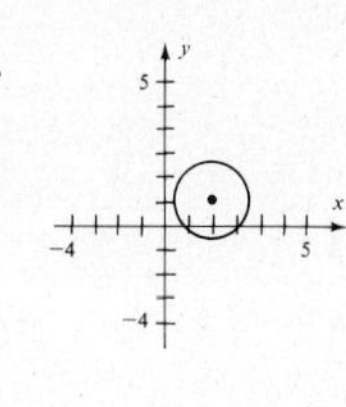

13. $\dfrac{(x-4)^2}{4} + \dfrac{(y-1)^2}{3} = 1$

15. $(x+1)^2 + (y+2)^2 = 64$

17. 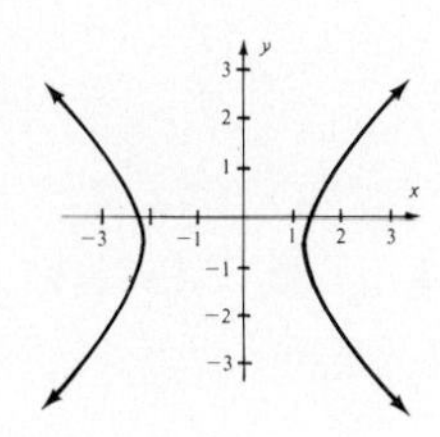

19. 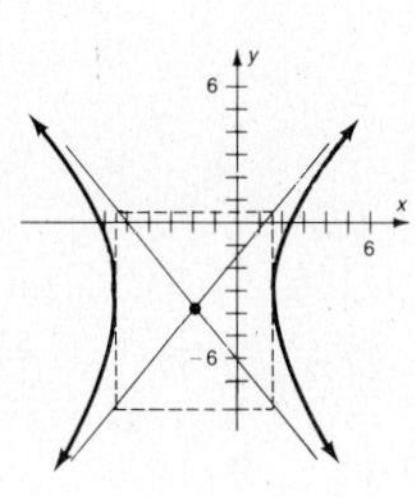

21. $\dfrac{(x+5)^2}{1} - \dfrac{(y-4)^2}{3} = 1$

23. $\dfrac{y^2}{9} - \dfrac{x^2}{16} = 1$

25. $\dfrac{(x-h)^2}{a^2} + \dfrac{(y-h)^2}{b^2} = 1$

27. $(y-k)^2 = 4c(x-h)^2$

29. parabola

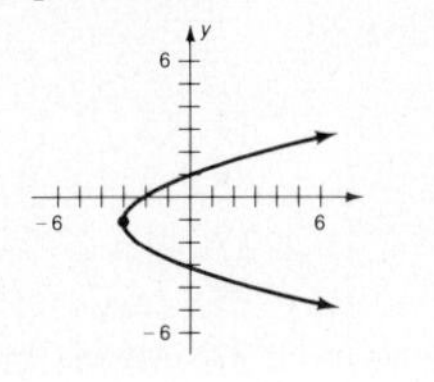

31. ellipse

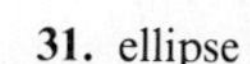

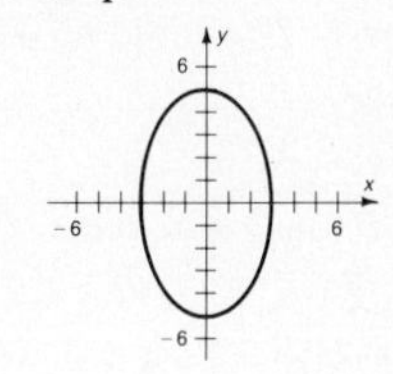

33. 45° **35.** 26.6° **37.** hyperbola **39.** parabola

EXTENDED APPLICATION—PLANETARY ORBITS, PAGE 412

1. The greatest distance is $a + c \approx 1.5 \cdot 10^8$, and the least distance is $a - c \approx 1.3 \cdot 10^8$.
3. The greatest distance is $3.7 \cdot 10^9$, and the least distance is $3.6 \cdot 10^9$. **5.** 5500 miles
7. $1.2x^2 + 1.3y^2 = 1.6 \cdot 10^{15}$ **9.** $8.802x^2 + 8.805y^2 = 7.751 \cdot 10^{16}$

PROBLEM SET 10.1, PAGES 420–421

Answers to Problems 1–11 should also include a vector diagram.
1. $\mathbf{v} = 6\mathbf{i} + 6\sqrt{3}\mathbf{j}$ **3.** $\mathbf{v} = \mathbf{i} + \mathbf{j}$ **5.** $\mathbf{v} = 7\cos 23°\mathbf{i} + 7\sin 23°\mathbf{j} \approx 6.4435\mathbf{i} + 2.7351\mathbf{j}$
7. $\mathbf{v} = 4\cos 112°\mathbf{i} + 4\sin 112°\mathbf{j} \approx -1.4984\mathbf{i} + 3.7087\mathbf{j}$ **9.** $\mathbf{v} = -2\mathbf{i} + 2\mathbf{j}$ **11.** $\mathbf{v} = -6\mathbf{i} - 5\mathbf{j}$ **13.** $\mathbf{v} = 8\mathbf{i} - 10\mathbf{j}$
15. $\mathbf{v} = -7\mathbf{i} - \mathbf{j}$ **17.** $\mathbf{v} = -9\mathbf{i} - 6\mathbf{j}$ **19.** 5 **21.** $\sqrt{85} \approx 9.2195$ **23.** $2\sqrt{2} \approx 2.8284$ **25.** $\sqrt{10} \approx 3.1623$
27. $\sqrt{41} \approx 6.4031$ **29.** not orthogonal **31.** orthogonal **33.** orthogonal
35. The plane is traveling at 244 mph with a heading of 260°.

	$\mathbf{v} \cdot \mathbf{w}$	$\lvert\mathbf{v}\rvert$	$\lvert\mathbf{w}\rvert$	$\cos\theta$
39.	-112	10	13	$\frac{-56}{65}$
41.	0	8	16	0
43.	-39	$3\sqrt{10}$	$\sqrt{29}$	$\frac{-13\sqrt{290}}{290}$
45.	1	1	1	1
47.	7	$\sqrt{26}$	$\sqrt{13}$	$\frac{7\sqrt{2}}{26}$
49.	1	$\sqrt{2}$	1	$\frac{\sqrt{2}}{2}$

51. 75° **53.** 45° **55.** 90° **57.** $-\frac{8}{5}$
59. The pilot must fly in the direction of 4.9°; the speed relative to the ground is 240 mph.
61. The weight of the astronaut is resolved into two components, one parallel to the inclined plane with length y and the other perpendicular to it with length x. The weight of the astronaut is $|\mathbf{x}|$.

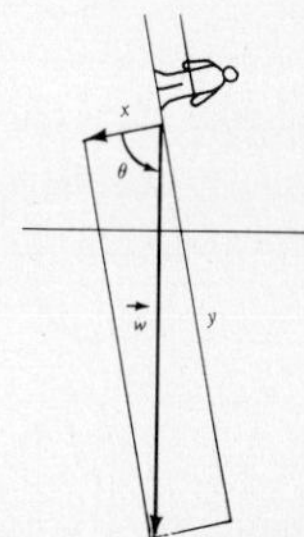

PROBLEM SET 10.2, PAGES 426–427

Answers to Problems 1–24 may vary.

1. $2\mathbf{i} - 3\mathbf{j}$ **3.** $\mathbf{i} - \mathbf{j}$ **5.** $3\mathbf{i} - 2\mathbf{j}$ **7.** $9\mathbf{i} + 7\mathbf{j}$ **9.** $4\mathbf{i} - \mathbf{j}$ **11.** $\mathbf{i} + 2\mathbf{j}$ **13.** $3\mathbf{i} + 2\mathbf{j}$ **15.** $\mathbf{i} + \mathbf{j}$ **17.** $2\mathbf{i} + 3\mathbf{j}$
19. $7\mathbf{i} - 9\mathbf{j}$ **21.** $\mathbf{i} + 4\mathbf{j}$ **23.** $2\mathbf{i} - \mathbf{j}$ **25.** $\frac{63}{13}$ **27.** 0 **29.** $\frac{3}{61}\sqrt{61}$ **31.** $\frac{315}{169}\mathbf{i} + \frac{756}{169}\mathbf{j}$ **33.** 0
35. $\frac{18}{61}\mathbf{i} + \frac{15}{61}\mathbf{j}$ **37.** $d = 0$ **39.** $d = \frac{47}{5}$ **41.** $d = \frac{12}{5}$ **43.** $d = \frac{22}{5}$ **45.** $d = \frac{17}{5}\sqrt{10}$ **47.** $d = \frac{19}{10}\sqrt{10}$
49. $d = \frac{17}{29}\sqrt{29}$ **51.** $\frac{21}{2}$ **53.** $\frac{41}{2}$ **55.** 19 **57–59.** Answers vary.

PROBLEM SET 10.3, PAGES 433–434

1. $\mathbf{v} = 2\mathbf{i} + 6\mathbf{j} - 6\mathbf{k}$; $|\mathbf{v}| = 2\sqrt{19}$ **3.** $\mathbf{v} = -5\mathbf{i} + 6\mathbf{j} - 3\mathbf{k}$; $|\mathbf{v}| = \sqrt{70}$ **5.** $\mathbf{v} = -4\mathbf{i} - 3\mathbf{j} + 5\mathbf{k}$; $|\mathbf{v}| = 5\sqrt{2}$ **7.** $\sqrt{14}$
9. $18\mathbf{i} - \mathbf{j} - 7\mathbf{k}$ **11.** $6\mathbf{i} - 5\mathbf{j} + 5\mathbf{k}$ **13.** $5\mathbf{i} + \mathbf{j} + 7\mathbf{k}$ **15.** $25\mathbf{i} - 3\mathbf{j} + 11\mathbf{k}$ **17.** 1 **19.** 1 **21.** 3 **23.** 24
25. $\mathbf{k}$ **27.** $-2\mathbf{i} + 4\mathbf{j} + 3\mathbf{k}$ **29.** $-2\mathbf{i} + 25\mathbf{j} + 14\mathbf{k}$ **31.** $-2\mathbf{i} + 16\mathbf{j} + 11\mathbf{k}$ **33.** $-14\mathbf{i} - 4\mathbf{j} - \mathbf{k}$ **35.** $\frac{5}{13}\mathbf{i} + \frac{12}{13}\mathbf{k}$
37. $\frac{3}{13}\mathbf{i} + \frac{12}{13}\mathbf{j} - \frac{4}{13}\mathbf{k}$ **39.** $\frac{4}{29}\sqrt{29}\mathbf{i} + \frac{2}{29}\sqrt{29}\mathbf{j} - \frac{3}{29}\sqrt{29}\mathbf{k}$ **41.** $\frac{2}{21}\sqrt{7}$ **43.** $s = -\frac{7}{2}t$ **45.** $-\frac{9}{13}$
47. $s = \frac{13}{9}t$ **49.** $\frac{4}{63}\sqrt{42}$ **51.** $s = -\frac{27}{8}t$ **53.** $\frac{26}{35}$ **55.** $s = -\frac{25}{26}t$ **57.** $\frac{2}{3}\mathbf{i} - \frac{2}{3}\mathbf{j} - \frac{2}{3}\mathbf{k}$; $\frac{2}{3}\sqrt{3}$
59. $\frac{20}{7}\mathbf{i} + \frac{4}{7}\mathbf{j} - \frac{12}{7}\mathbf{k}$; $\frac{4}{7}\sqrt{35}$ **61.** $-\frac{24}{29}\mathbf{i} + \frac{32}{29}\mathbf{j} - \frac{16}{29}\mathbf{k}$; $\frac{8}{29}\sqrt{29}$ **63. a.** scalar **b.** vector **c.** scalar

PROBLEM SET 10.4, PAGE 443

Points in Problems 1–20 should also be plotted.

	Polar	Rectangular
1.	$(4, \frac{\pi}{4}) = (-4, \frac{5\pi}{4})$	$(2\sqrt{2}, 2\sqrt{2})$
3.	$(5, \frac{2\pi}{3}) = (-5, \frac{5\pi}{3})$	$(-\frac{5}{2}, \frac{5\sqrt{3}}{2})$
5.	$(\frac{3}{2}, \frac{7\pi}{6}) = (-\frac{3}{2}, \frac{\pi}{6})$	$(\frac{-3\sqrt{3}}{4}, -\frac{3}{4})$
7.	$(-4, 4) = (4, 0.86)$	$(2.61, 3.03)$
9.	$(-4, \pi) = (4, 0)$	$(4, 0)$

11. $(5\sqrt{2}, \frac{\pi}{4}) = (-5\sqrt{2}, \frac{5\pi}{4})$ **13.** $(4, \frac{5\pi}{3}) = (-4, \frac{2\pi}{3})$
15. $(3\sqrt{2}, \frac{7\pi}{4}) = (-3\sqrt{2}, \frac{3\pi}{4})$ **17.** $(2, \frac{5\pi}{6}) = (-2, \frac{11\pi}{6})$
19. $(13, 2.75) = (-13, 5.89)$ **21.** lemniscate **23.** 3-leaved rose
25. cardioid **27.** none **29.** nonc (circle) **31.** none
33. none (line) **35.** cardioid

37.

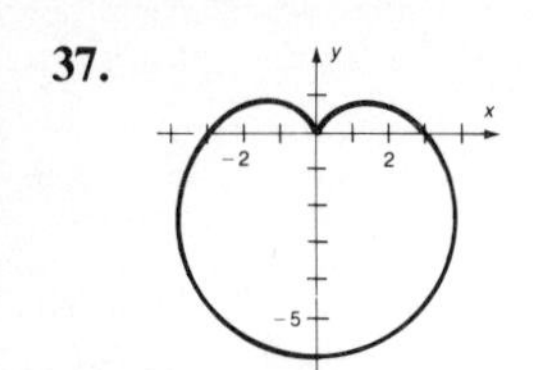

39.

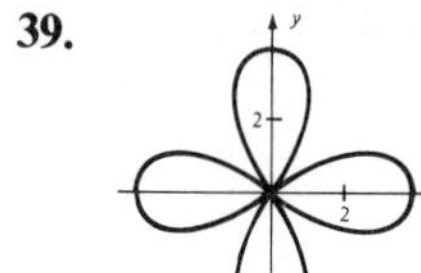

41.

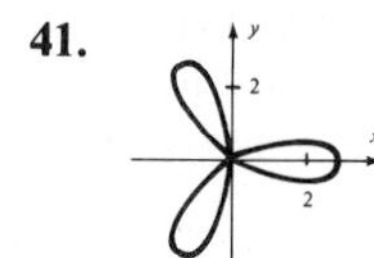

43.

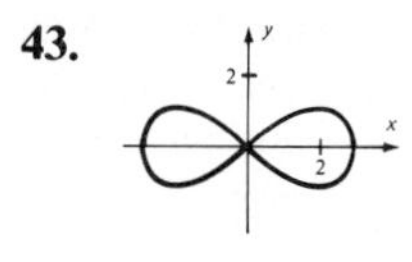

45.

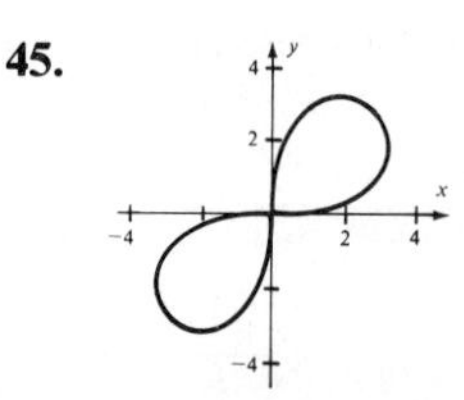

47.

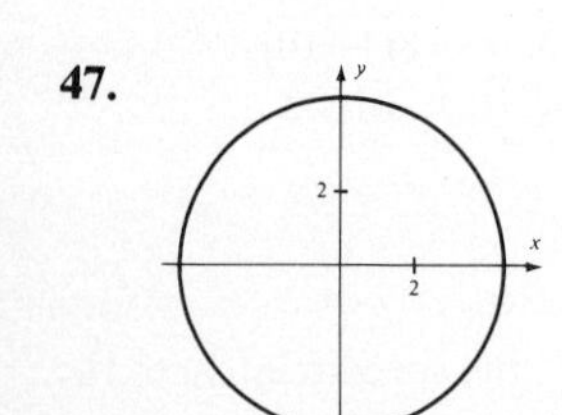

49.

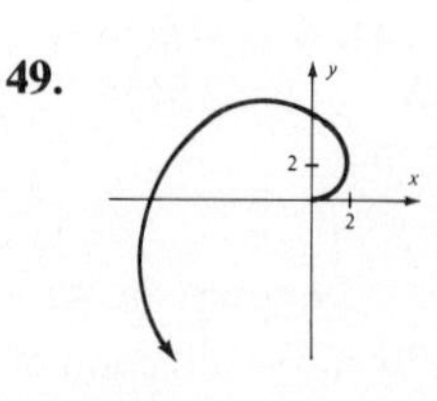

51. Answers vary. **53.** $d \approx 4.1751$ **55. a.**

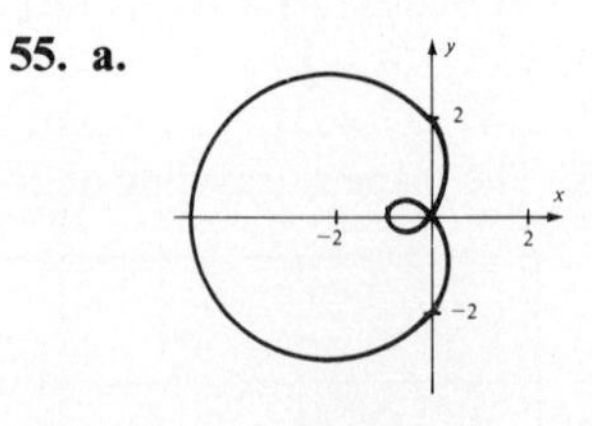

b.

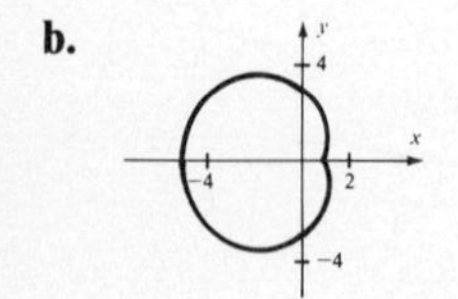

c.

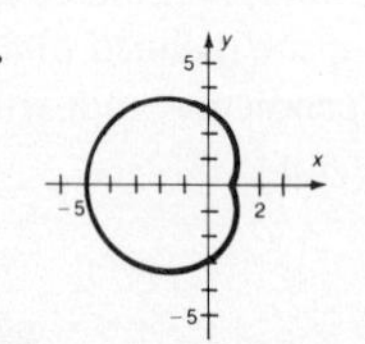

d.

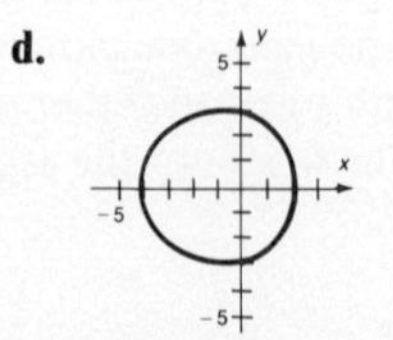

57. a.

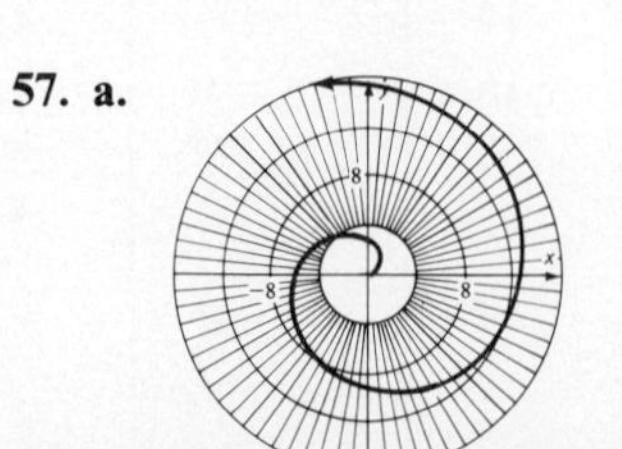

b.

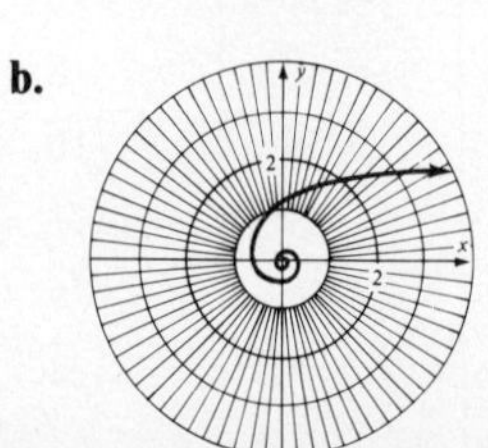

c.

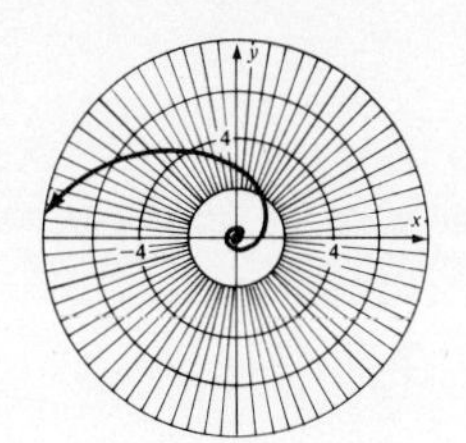

59.

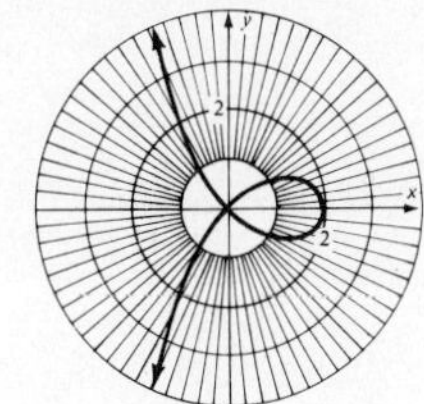

61.

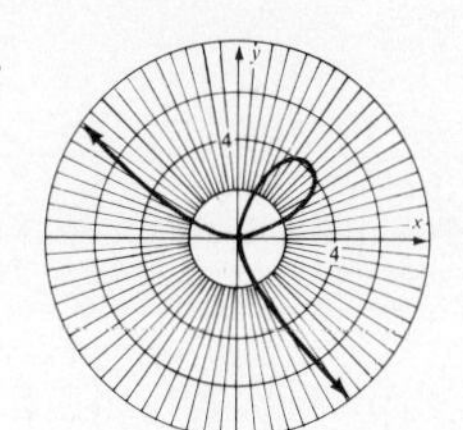

PROBLEM SET 10.5, PAGE 449

1.

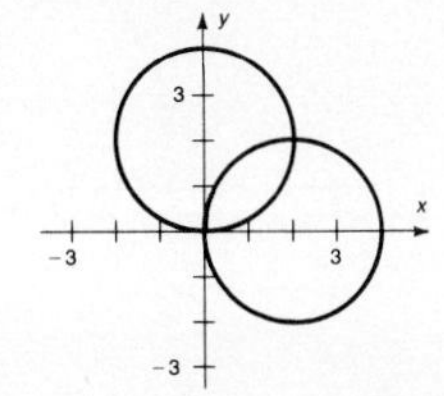

3.

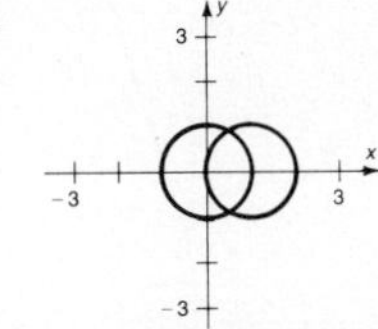

5.

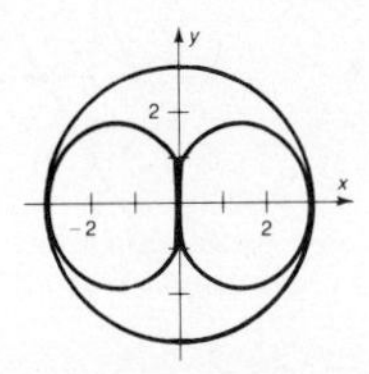

7.

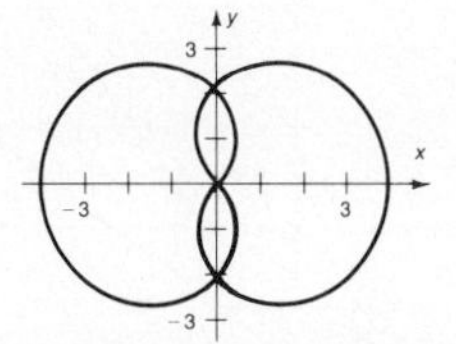

9.

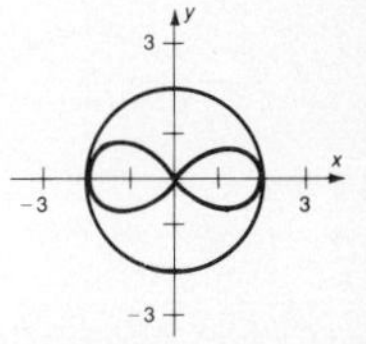

11.

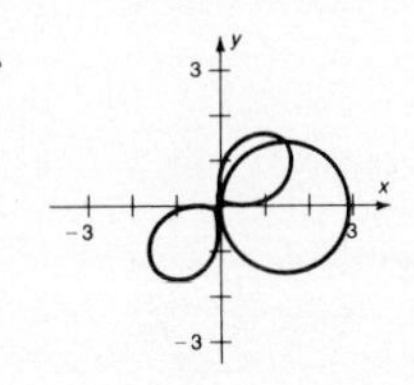

13.

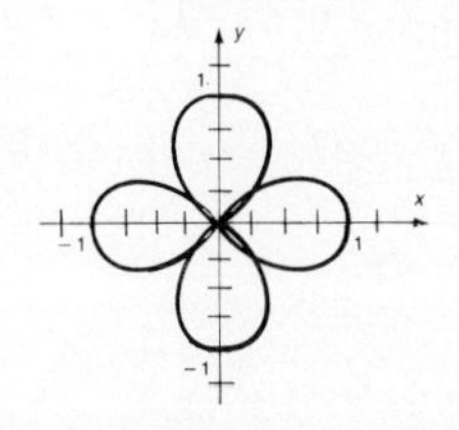

15.

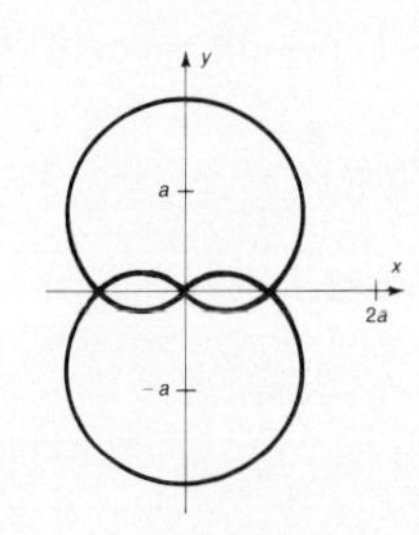

17. $(0,0), (4\sqrt{2}, \frac{\pi}{4})$ **19.** $(2, \frac{\pi}{3}), (2, \frac{5\pi}{3})$

21. $(2, \frac{\pi}{2}), (2, \frac{3\pi}{2})$ **23.** $(2,0), (2,\pi), (0,0)$ **25.** $(3, \frac{\pi}{4}), (3, \frac{5\pi}{4})$ **27.** $(0,0), (\frac{3}{2}\sqrt{2}, \frac{\pi}{6}), (\frac{3}{2}\sqrt{2}, \frac{5\pi}{6})$ **29.** $(0,0), (\frac{2+\sqrt{2}}{2}, \frac{\pi}{4}), (\frac{2-\sqrt{2}}{2}, \frac{5\pi}{4})$

31.

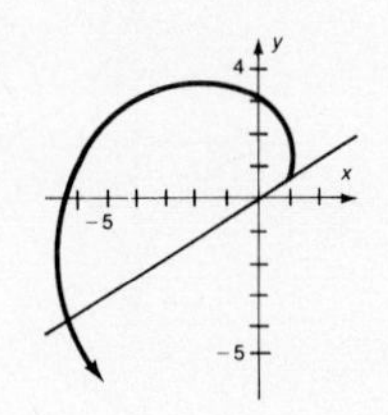

33.

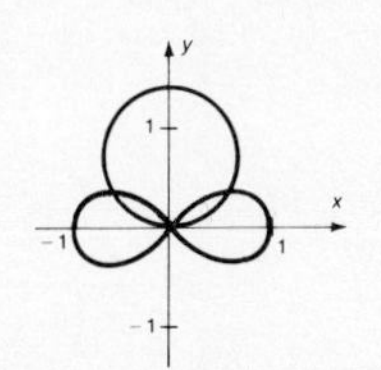

35.

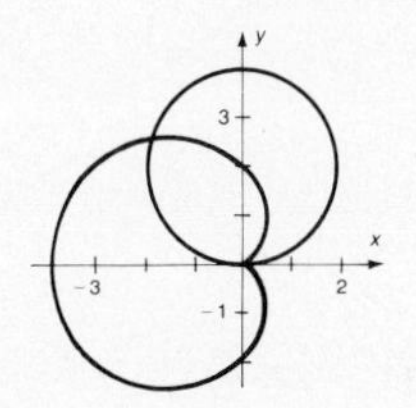

37.

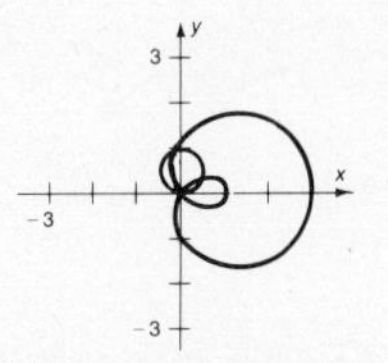

39.

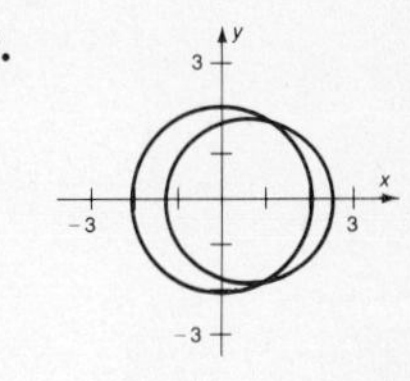

41.

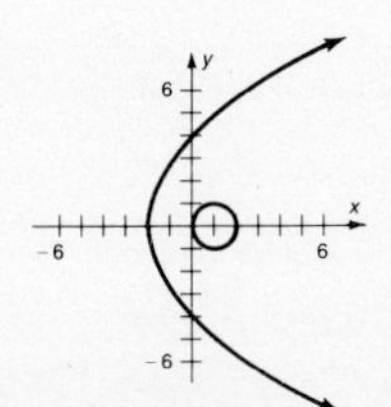

43.

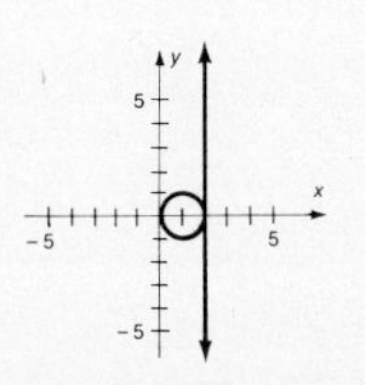

45.

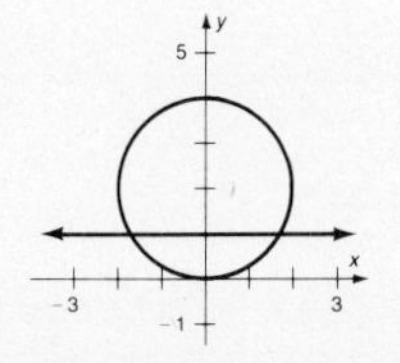

47. $(0,0), (\pi, \frac{\pi}{3}), (4\pi, \frac{4\pi}{3})$

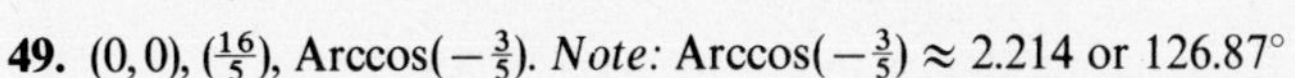

49. $(0,0), (\frac{16}{5})$, Arccos$(-\frac{3}{5})$. *Note:* Arccos$(-\frac{3}{5}) \approx 2.214$ or $126.87°$
51. $(0,0)$, $(3.2, 2\pi -$ Arccos $0.8)$. *Note:* $2\pi -$ Arccos $0.8 \approx 5.640$ or $323.13°$
53. $(0,0), (1,0)$, $(0.6$, Arcsin $0.8)$. *Note:* Arcsin $0.8 \approx 0.927$ or $53.1°$ **55.** $(2, \frac{\pi}{2}), (2, \frac{3\pi}{2})$
57. $(2+\sqrt{2}, \frac{5\pi}{4}), (2+\sqrt{2}, \frac{3\pi}{4})(2-\sqrt{2}, \frac{\pi}{4}), (2-\sqrt{2}, \frac{7\pi}{4})$ **59.** $(a, \frac{\pi}{2})$

PROBLEM SET 10.6, PAGES 453–454

The graphs for Problems 1–20 are the same as for Problems 21–40.

1.

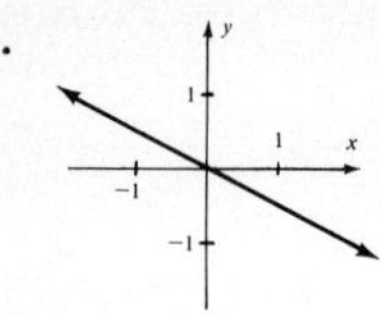

3.

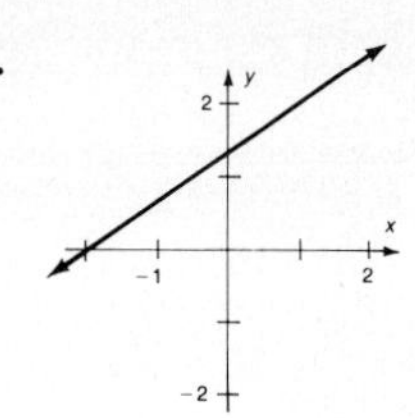

5.

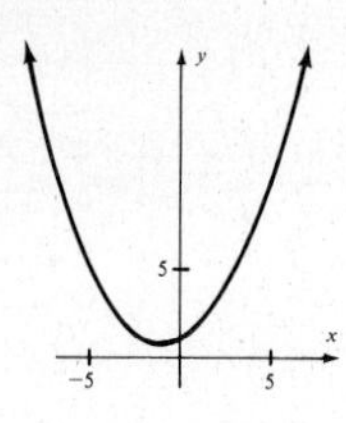

7.

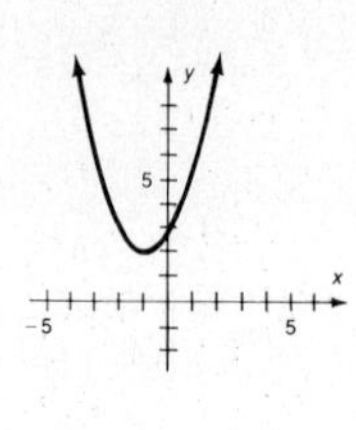

9.

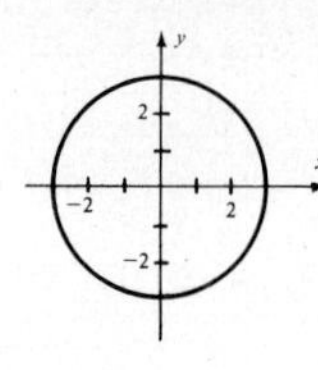

11.

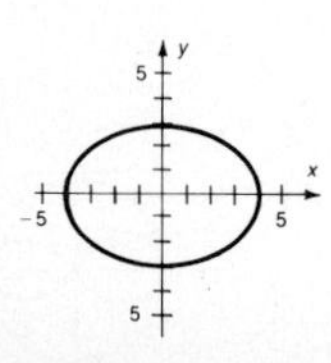

13.

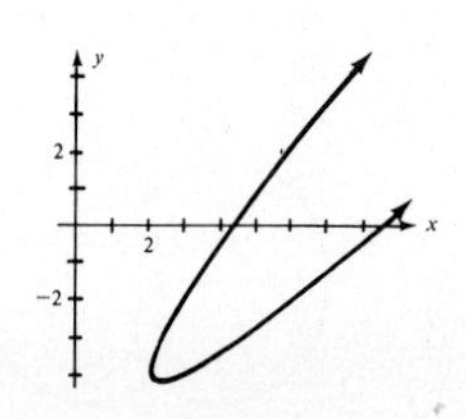

15.

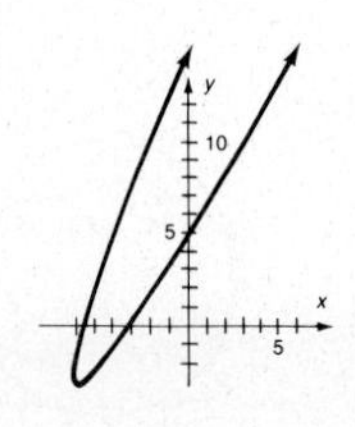

17.

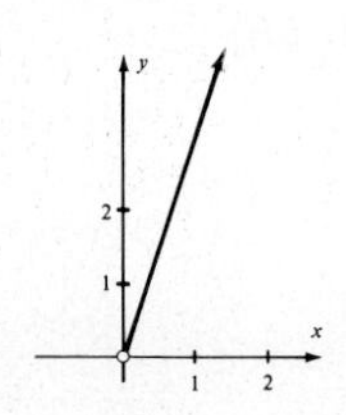

19.

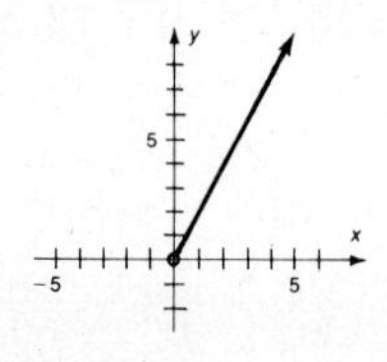

21. $y = -\frac{1}{2}x$ **23.** $y = \frac{2}{3}x + \frac{4}{3}$ **25.** $(x+1)^2 = 4(y - \frac{3}{4})$ **27.** $y = x^2 + 2x + 3$ **29.** $x^2 + y^2 = 9$

31. $\frac{x^2}{16} + \frac{y^2}{9} = 1$ **33.** $x^2 - 2xy + y^2 - 13x + 12y + 38 = 0$; $\cot 2\theta = \frac{1-1}{-2} = 0$; $\theta = 45°$

35. $4x^2 - 4xy + y^2 + 36x - 20y + 75 = 0$; $\cot 2\theta = \frac{4-1}{-4} = -\frac{3}{4}$; $\theta \approx 63.4°$ **37.** $y = 3x, x > 0$

39. $y = ex, x > 0$ **41.** **43.**

45.

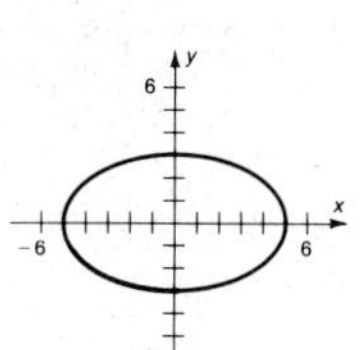

47.

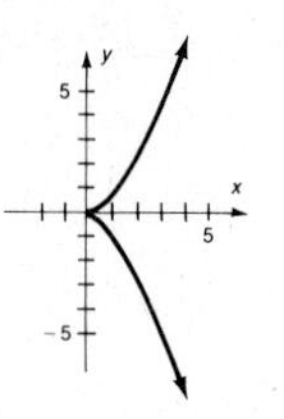

49.

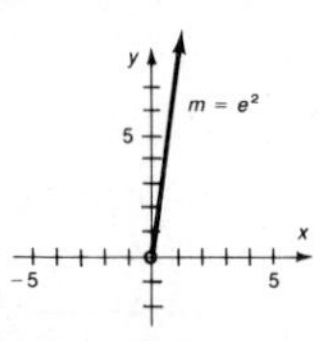

51.

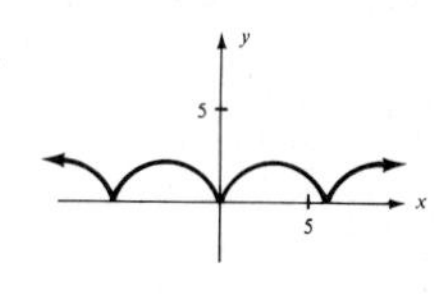

53.

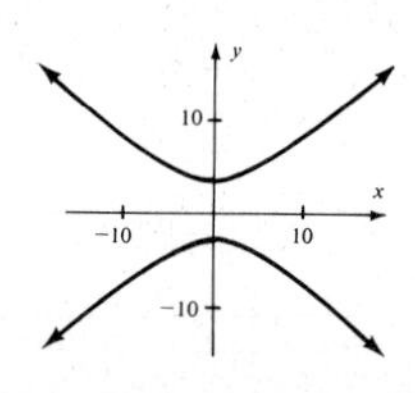

55.

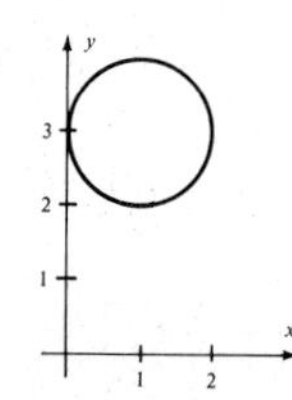

57.

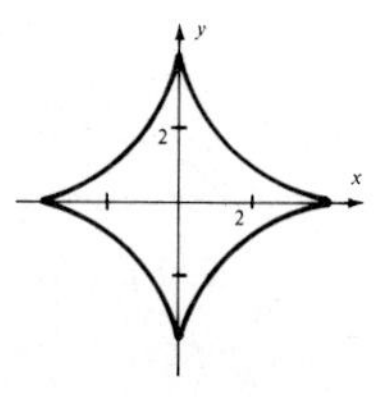

59.

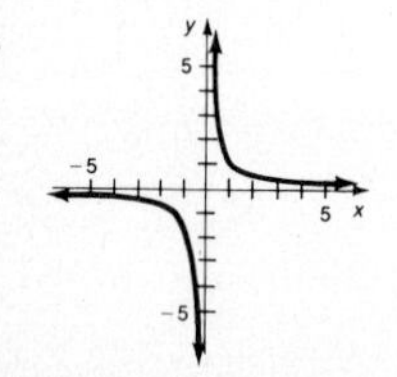

61. $x = a(\theta - \sin\theta)$, $y = a(1 - \cos\theta)$

CHAPTER 10 SUMMARY, PAGES 454–457

1. $-2\mathbf{i} + \mathbf{j}$ **3.** The magnitude is 9.8 and the direction is N80° E. **5.** $\mathbf{v}_x \approx 2.8$; $\mathbf{v}_y \approx 3.5$ **7.** $5\mathbf{i} + 12\mathbf{j}$
9. $\sqrt{29}$ **11.** 6 **13.** Let $\mathbf{v} = a\mathbf{i} + b\mathbf{j}$ and $\mathbf{w} = c\mathbf{i} + d\mathbf{j}$. Then the scalar product is $\mathbf{v} \cdot \mathbf{w} = ac + bd$ **15.** 9
17. $-\frac{9}{13}\sqrt{13}$ **19.** 5° **21.** $\mathbf{N} = 5\mathbf{i} - 12\mathbf{j}$; $\mathbf{v} = -\frac{3}{5}\mathbf{i} - \frac{1}{4}\mathbf{j}$; $\mathbf{N} \cdot \mathbf{v} = -3 + 3 = 0$ **23.** $\mathbf{N} = \mathbf{i} - 5\mathbf{j}$; $\mathbf{v} = -4\mathbf{i} - \frac{4}{5}\mathbf{j}$; $\mathbf{N} \cdot \mathbf{v} = -4 + 4 = 0$ **25.** $\frac{5}{2}\mathbf{i} + \frac{\sqrt{5}}{2}\mathbf{j}$; $\frac{\sqrt{30}}{2}$

27. $\mathbf{0}$; 0 **29.** $\frac{73}{13}$ **31.** $\frac{24}{13}\sqrt{13}$ **33.** $4\mathbf{i} + 8\mathbf{j} - 2\mathbf{k}$ **35.** $-8\mathbf{i} - \mathbf{j} + 3\mathbf{k}$ **37.** $4\mathbf{i} - 7\mathbf{j} + 2\mathbf{k}$ **39.** $3\mathbf{i} + \mathbf{j} - \mathbf{k}$
41. 0 **43.** 4 **45.** $2\mathbf{k}$ **47.** $-5\mathbf{j} + 5\mathbf{k}$ **49.** $\frac{1}{3}\mathbf{i} + \frac{1}{3}\mathbf{j} - \frac{\sqrt{7}}{3}\mathbf{k}$ **51.** $\frac{3+\sqrt{7}}{9}$
Points in Problems 53–56 should also be plotted.
53. $(5, 2.3771) = (-5, 5.5187)$ **55.** $(-2, 2) = (2, 5.1416)$ **57.** $(-1.50000, -2.5981)$ **59.** $(3\sqrt{2}, \frac{7\pi}{4})$

61.

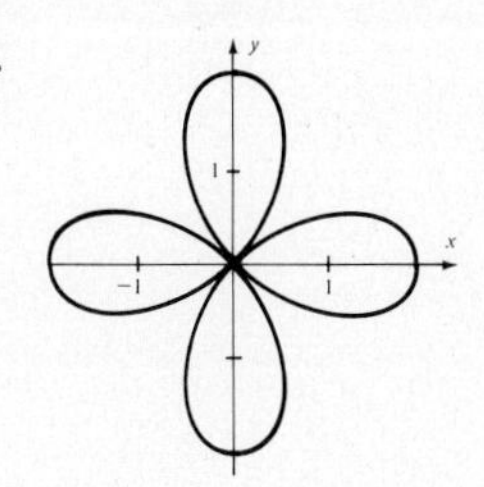

63.

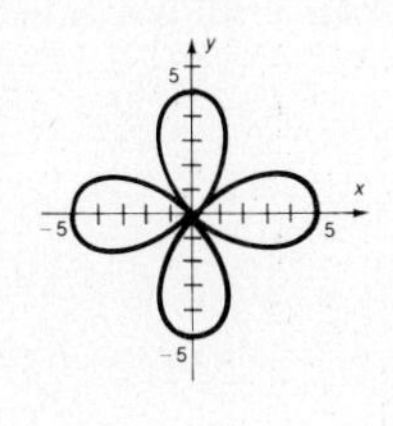

65. a. lemniscate **b.** 4-leaved rose **67. a.** none (lemniscon) **b.** cardioid **69.** $(0, 0), (4, 0), (4, \pi)$ **71.** $(4, \pi), (0, 0), (4, 0)$

73. **75.**

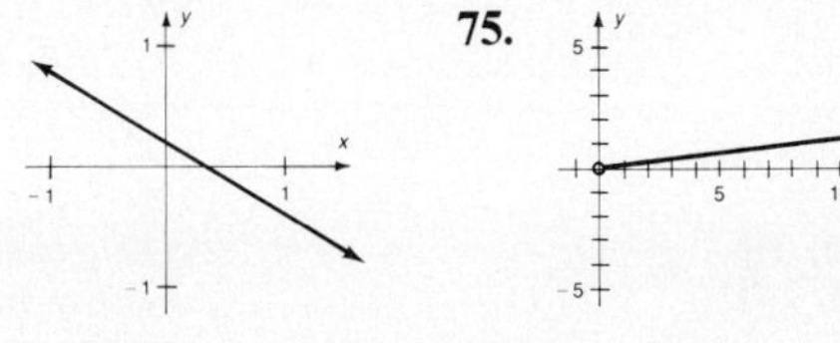

77. $3x + 5y - 1 = 0$ (See Problem 73 for graph.) **79.** $y = e^{-2}x$, $x > 0$ (See Problem 75 for graph.)

Index

A

B

C

D

E

F

G

H

I

O

P

T

U

V

W

X

Y

Z

Applications Index

Calculus

The purpose of this text is to prepare you for calculus. This list gives examples which are not typical of precalculus textbooks, but which provide practice of concepts which will be particularly useful in your study of calculus.

Chemistry

Consumer

Earth Science

Economics

Engineering

Extended Applications: Each extended application is introduced by a news clipping and examines a topic in more than the usual depth.

General Interest

Medicine

Navigation

Physics

Police Science

Psychology

Social Science

Space Science

Sports

Surveying

Conic Sections

[illegible] is the parabola with vertex (0, 0) opening upward if $a > 0$, downward if $a < 0$.

[illegible] is the parabola with vertex (0, 0) opening to the right if $a > 0$ and to the left if $a < 0$.

[illegible] is the ellipse with center (0, 0), x-intercepts $\pm a$, y-intercepts $\pm b$ and foci $(\pm c, 0)$ with constant sum $2a$; $c^2 = a^2 - b^2$.

[illegible] is the ellipse with center (0, 0), x-intercepts $\pm b$, y-intercepts $\pm a$ and foci $(0, \pm c)$ with constant sum $2a$; $c^2 = a^2 - b^2$.

[illegible] is the hyperbola with center (0, 0), x-intercepts $\pm a$, foci $(\pm c, 0)$ and constant difference $2a$; $c^2 = a^2 + b^2$.

[illegible] is the hyperbola with center (0, 0), y-intercepts $\pm a$, foci $(0, \pm c)$ and constant difference $2a$; $c^2 = a^2 + b^2$.

[illegible] are the asymptotes in both cases.

These standard conic sections can be translated or rotated.

Translation substitution:

$x = x' + h$

$y = y' + k$

Rotation substitution:

$x = x' \cos\theta - y' \sin\theta$

$y = x' \sin\theta + y' \cos\theta$

where $\cot 2\theta = [illegible]$

[illegible]

Polynomials

[illegible]

Remainder Theorem: If $P(x)$ is divided by $x - c$, then the remainder is equal to $P(c)$.

Intermediate Value Theorem for Polynomial Functions: If P is a polynomial function and $P(a) \neq P(b)$ for $a < b$, then P takes on every value between $P(a)$ and $P(b)$ on the interval $[a, b]$.

Factor Theorem: If c is a root of the polynomial equation $P(x) = 0$, then $x - c$ is a factor of $P(x)$. Also, if $x - c$ is a factor of $P(x)$, then c is a root of the polynomial equation $P(x) = 0$.

Root Limitation Theorem: A polynomial function of degree n has at most n distinct roots.

Location Theorem: If P is a polynomial function and $P(a)$ and $P(b)$ are opposite in sign, then there is at least one real root on the interval $[a, b]$.

Rational Root Theorem: If $P(x)$ has integer coefficients and has a rational root p/q (where p/q is reduced), then p is a factor of a_0 and q is a factor of a_n.

Upper and Lower Bound Theorem: If $a > 0$ and all the numbers in the last row have the same sign in the synthetic division of $P(x)$ by $x - a$, then a is an upper bound for the roots of $P(x) = 0$. If $b < 0$ and the numbers in the last row alternate in sign in the synthetic division of $P(x)$ by $x - b$, then b is a lower bound for the roots of $P(x) = 0$.

Descartes' Rule of Signs: Let $P(x)$ be written in descending powers of x.

1. The number of positive real zeros is equal to the number of sign changes or is equal to that number decreased by an even integer.
2. The number of negative real zeros is equal to the number of sign changes in $P(-x)$ or is equal to that number decreased by an even integer.

Fundamental Theorem of Algebra: If $P(x)$ is a polynomial of degree ≥ 1 with complex coefficients, then $P(x) = 0$ has at least one complex root.

Number of Roots Theorem: If $P(x)$ is a polynomial of degree $n \geq 1$ with complex coefficients, then $P(x) = 0$ has exactly n roots (if roots are counted according to their multiplicity).

Trigonometry

Trigonometric Functions

Let θ be an angle in standard position with a point $P(x,y)$ on the terminal side a distance of r from the origin ($r \neq 0$). Then the trigonometric functions are defined by

$$\cos\theta = \frac{x}{r} \qquad \sec\theta = \frac{r}{x}\ (x \neq 0)$$

$$\sin\theta = \frac{y}{r} \qquad \csc\theta = \frac{r}{y}\ (y \neq 0)$$

$$\tan\theta = \frac{y}{x}\ (x \neq 0) \qquad \cot\theta = \frac{x}{y}\ (y \neq 0)$$

Inverse Trigonometric Functions

Inverse Function	*Domain*	*Range*
$y = \text{Arccos } x$	$-1 \leq x \leq 1$	$0 \leq y \leq \pi$
$y = \text{Arcsin } x$	$-1 \leq x \leq 1$	$-\frac{\pi}{2} \leq y \leq \frac{\pi}{2}$
$y = \text{Arctan } x$	all reals	$-\frac{\pi}{2} < y < \frac{\pi}{2}$
$y = \text{Arccot } x$	all reals	$0 < y < \pi$

Exact Values

function \ angle θ	0	$\frac{\pi}{6}$	$\frac{\pi}{4}$	$\frac{\pi}{3}$	$\frac{\pi}{2}$	π	$\frac{3\pi}{2}$
$\cos\theta$	1	$\frac{\sqrt{3}}{2}$	$\frac{\sqrt{2}}{2}$	$\frac{1}{2}$	0	-1	0
$\sin\theta$	0	$\frac{1}{2}$	$\frac{\sqrt{2}}{2}$	$\frac{\sqrt{3}}{2}$	1	0	-1
$\tan\theta$	0	$\frac{\sqrt{3}}{3}$	1	$\sqrt{3}$	undef.	0	undef.
$\sec\theta$	1	$\frac{2}{\sqrt{3}}$	$\frac{2}{\sqrt{2}}$	2	undef.	-1	undef.
$\csc\theta$	undef.	2	$\frac{2}{\sqrt{2}}$	$\frac{2}{\sqrt{3}}$	1	undef.	-1
$\cot\theta$	undef.	$\frac{3}{\sqrt{3}}$	1	$\frac{\sqrt{3}}{3}$	0	undef.	0

Solving Triangles

Given	Conditions on Given Information	Law to Use for Solution
1. SSS	a. The sum of the lengths of the two smaller sides is less than or equal to the length of the larger side.	No solution
	b. The sum of the lengths of the two smaller sides is greater than the length of the larger side.	Law of cosines
2. SAS	a. The angle is greater than or equal to 180°.	No solution
	b. The angle is less than 180°.	Law of cosines
3. ASA or AAS	a. The sum of the angles is greater than or equal to 180°.	No solution
	b. The sum of the angles is less than 180°.	Law of sines
4. SSA	Let θ be the given angle with adjacent (adj) and opposite (opp) sides given; the height h is found by $h =$ (adj) $\sin\theta$.	
	a. $\theta > 90°$	
	i. opp $\leq$ adj	No solution
	ii. opp $>$ adj	Law of sines
	b. $\theta < 90°$	
	i. opp $< h <$ adj	No solution
	ii. opp $= h <$ adj	Right-triangle solution
	iii. $h <$ opp $<$ adj	*Ambiguous case:* Use law of sines to find two solutions.
	iv. opp $\geq$ adj	Law of sines
5. AAA		No solution

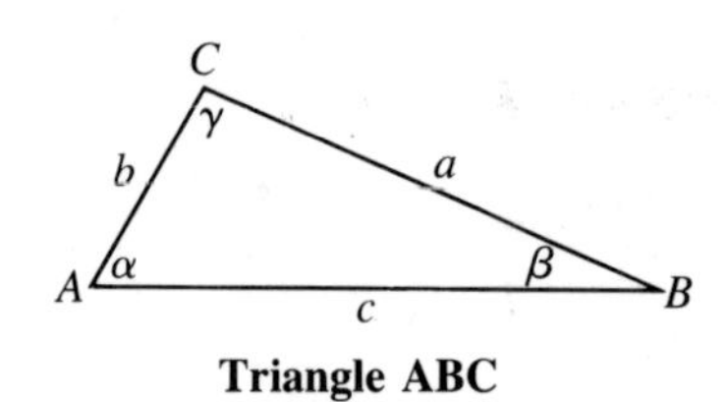

Triangle ABC

Pythagorean Theorem

In a right triangle, $c^2 = a^2 + b^2$

Law of Cosines

$$a^2 = b^2 + c^2 - 2bc\cos\alpha$$
$$b^2 = a^2 + c^2 - 2ac\cos\beta$$
$$c^2 = a^2 + b^2 - 2ab\cos\gamma$$

$$\cos\alpha = \frac{b^2 + c^2 - a^2}{2bc}$$

$$\cos\beta = \frac{a^2 + c^2 - b^2}{2ac}$$

$$\cos\gamma = \frac{a^2 + b^2 - c^2}{2ab}$$

Law of Sines

$$\frac{\sin\alpha}{a} = \frac{\sin\beta}{b} = \frac{\sin\gamma}{c}$$